ADVANCES IN
X-RAY ANALYSIS

Volume 35B

A Continuation Order Plan is available for this series. A continuation order will bring delivery of each new volume immediately upon publication. Volumes are billed only upon actual shipment. For further information please contact the publisher.

ADVANCES IN X-RAY ANALYSIS

Volume 35B

Edited by

Charles S. Barrett
University of Denver
Denver, Colorado

John V. Gilfrich
Sachs/Freeman Associates
Washington, D.C.

Ting C. Huang
IBM Almaden Research Center
San Jose, California

Ron Jenkins
JCPDS–ICDD
Swarthmore, Pennsylvania

Gregory J. McCarthy
North Dakota State University
Fargo, North Dakota

Paul K. Predecki
University of Denver
Denver, Colorado

Richard Ryon
Lawrence Livermore National Laboratory
Livermore, California

Deane K. Smith
Pennsylvania State University
University Park, Pennsylvania

Sponsored by
University of Denver Department of Engineering
and
JCPDS – International Centre for Diffraction Data
Springer Science+Business Media, LLC

The Library of Congress cataloged the first volume of this title as follows:

Conference on Application of X-ray Analysis.
Proceedings 6th-   1957-   [Denver]

v. illus. 24-28 cm. annual.
No proceedings published for the first 5 conferences.
Vols. for 1958-   called also: Advances in X-ray analysis, v. 2-
Proceedings for 1957 issued by the conference under an earlier name: Conference
on Industrial Applications of X-ray Analysis. Other slight variations in name of
conference.
Vol. for 1957 published by the University of Denver, Denver Research Institute,
Metallurgy Division.
Vols. for 1958-   distributed by Plenum Press, New York.
Conferences sponsored by University of Denver, Denver Research Institute.
1. X-rays — Industrial applications — Congresses. I. Denver University. Denver
Research Institute   II. Title: Advances in X-ray analysis.
TA406.5.C6   58-35928

ISBN 978-1-4613-6532-7   ISBN 978-1-4615-3460-0 (eBook)
DOI 10.1007/978-1-4615-3460-0

ISBN 0-306-44249-3

Proceedings of combined First Pacific–International Congress
on X-Ray Analytical Methods (PICXAM) and Fortieth Annual Conference
on Applications of X-Ray Analysis, held August 7–16, 1991,
in Hilo and Honolulu, Hawaii

MATHEMATICAL CORRECTION PROCEDURES IN XRF -

THE LONG AND THE SHORT

Gerald R. Lachance

Nepean, Ontario, Canada

The processes of absorption and enhancement in x-ray fluorescence spectrometry preclude the direct conversion of measured intensities to concentrations unless the specimens being analyzed are practically identical to the reference materials that were used as standards for calibration. While well known techniques of spectrometry such as the use of internal standards or techniques specific to x-ray fluorescence such as Compton scatter may be used in some instances, the analysis of alloys for example is not amenable to these techniques. The goal of mathematical correction procedures is to compensate adequately for matrix effects and therefore to extend the concentration range within which analysis may be accomplished. The evolution of the concepts and expressions that have been proposed for this purpose spans some 50 years, is somewhat fascinating and examined in outline after a brief nostalgic look at where it all began.

Having set out to "examine spectra of a few elements in greater detail," Moseley[1] describes how he "... used as targets a number of substances mounted on a truck inside an exhausted tube ..." with provision for "... each target to be brought in turn into the line of fire." Of the elements chosen as forming a continuous series (Z 20-30) "... calcium alone gave any trouble. In this case ... the layer of lime which covered the surface of the metal gave off such a quantity of gas that the x-ray could only be excited for a second or two at-a-time." Prophetically, Moseley concluded that "the prevalence of lines due to impurities suggest that this may prove a powerful method of chemical analysis."

In a second paper[2], the instrumentation is described: "... the aluminium trolley that carries the targets can be drawn to and fro by means of silk fishing-lines wound on brass bobbins. The slit (which defines the x-ray beam) should be fixed exactly opposite the focus-spot of the cathode stream, though a slight error can be remedied by deflecting the cathode rays with a magnet." Consider some of Moseley's comments: "... the examination of keltium would be of exceptional interest, as no place ..., ... lines due to impurities were frequently present but caused little trouble except in the rare-earth group ... The x-ray spectrum of praesodymia showed that it consisted roughly 50% La, 35% Ce and 15% Pr ..."

Some twenty years later, von Hevesy[3] determined lead in a number of alloys. The procedure may be summarized as follows:

- one sample of zinc with 0.087% Pb used as standard
- equal exposures made with other samples
- Pb concentration calculated by direct ratio

with the comment, "Results for the alloys of unknown lead content ranged from 0.064 to 0.008 percent and, although no analysis by another method was carried out, these values were in good agreement with other evidence as to the composition of the samples." Thus was the stage set for x-ray fluorescence to emerge as a technique for quantitative elemental determination and the derivation of mathematical expressions for converting intensities to concentration. For easier comparison, a common set of symbols (see Appendix) is used rather than those in the original expressions.

One of the first theoretically derived expressions relating emitted intensity to specimen composition is due to von Hamos.[4] For the restricted analytical context: binary specimen irradiated by a monochromatic incident source and secondary fluorescence emission (enhancement) is not present, the emitted intensity is given by

$$I_i = I_{(i)} C_i K_i \tag{1}$$

with the variable K_i defined as a function of C_i, μ_i, μ_j and instrument geometry only. Current practice for influence coefficient models tends towards the calculation of the RECIPROCAL of K_i as a function of all the elements EXCEPT C_i.

Some nine years later, Sherman[5] derived the following expression for a monochromatic incident source wavelength λ

$$I_i = \frac{\mu_i' C_i g_i}{C_i \mu_i^* + C_j \mu_j^* + ...} \tag{2}$$

where

g_i is a proportionality constant

$\mu_i' = \mu_i (\lambda) \, csc \, \psi'$

$\mu_i^* = \mu_i (\lambda) \, csc \, \psi' + \mu_i (\lambda_i) \, csc \, \psi''$

$\mu_j^* = \mu_j (\lambda) \, csc \, \psi' + \mu_j (\lambda_i) \, csc \, \psi''$

thereby extending Eq (1) to multi-element systems. The terms for absorption of the incident and emergent x-radiations were combined for each element individually, the asterisk indicating that the path lengths of each having been taken into consideration. Sherman[6] and Shiraiwa and Fujino[7] further developed the theory of emitted intensity as a function of specimen composition and fundamental parameters for a polychromatic incident beam but their treatment is beyond the scope of this presentation. It incorporates an expression for enhancement previously reported, without proof, by Gillam and Heal.[8] However it is interesting to note that in a discussion of the problem of extending monochromatic to polychromatic formulations, Sherman concluded "Instead of measuring intensity... as counts per second, it is advantageous to represent the intensity by the reciprocal of the counting rate, i.e., time required for the counter to reach a predetermined fixed count." This led to the expression

$$\frac{t_i}{t_{(i)}} = 1 + (\frac{C_j}{C_i})\frac{\mu_j^*}{\mu_i^*} + (\frac{C_k}{C_i})\frac{\mu_k^*}{\mu_i^*} + ... \tag{3}$$

It is noted that $t_i / t_{(i)} = I_{(i)} / I_i$, the reciprocal of the generally current practice of expressing relative intensity as

$$R_i = I_i / I_{(i)} \tag{4}$$

Beattie and Brissey[9] retained Sherman's definition of relative intensity and reformulated Eq (3) yielding

$$- (\frac{I_{(i)}}{I_i} - 1)\, C_i + (\frac{\mu_j^*}{\mu_i^*})\, C_j + ... \ = 0 \tag{5}$$

Faced with the prospect of working with a polychromatic incident source and lacking the practical means of dealing with integrals over the wavelength range of interest, Beattie and Brissey opted to calculate the values of the influence coefficients from experimental intensity data using the expression

$$K_{ij} = \frac{C_i}{C_j} (\frac{I_{(i)}}{I_i} - 1) \tag{6}$$

which are then introduced in Eq (5)

$$- (\frac{I_{(i)}}{I_i} - 1)C_i + K_{ij}\, C_j + ... \ = 0 \tag{7}$$

For the determination of tantalum in niobium oxides, Campbell and Carl[10] proposed the following expression to correct for the matrix effect due to the presence of impurities

$$C_i = C_{i,expr} [\, 1 + r_{ij}\, C_j + r_{ik}\, C_k + ... \,] \tag{8}$$

where $C_{i,expr}$ is the value determined experimentally from a calibration of tantalum oxide in niobium oxide and C_j is given by

$$C_j = \frac{I_j\ (for\ specimen)}{I_j\ (for\ 1\%\ C_j)}\ , \quad (or\ C_j\ determined\ chemically).$$

It was observed that the experimental data agreed qualitatively with those expected from calculations of the linear absorption coefficients for the various matrices. Mitchell[11] working with the same oxide system also generated influence coefficient values experimentally (termed absorption-enhancement indices) and examined their pattern for a given analyte by observing plots of indices versus matrix element atomic number.

A quite different concept was proposed by Lucas-Tooth and Price[12] for compensating mathematically for matrix effects. Here the individual corrections are a function of intensities of the matrix elements rather than their concentrations. In a subsequent work, Lucas-Tooth and Pyne[13] proposed a model in which apparent concentrations are substituted for the intensities of the matrix elements. In either case the influence coefficients are obtained by regression of experimental data. The coefficients are empirical, i.e., have little or no physical meaning, and it is generally acknowledged that this approach is only applicable over very limited concentration ranges.

In the mid 60's Lachance and Traill[14] proposed a correction model that circumvents the over-definition problem of the Beattie-Brissey model Eq (5), namely

$$C_i = R_i [1 + \alpha_{ij} C_j + \alpha_{ik} + ...] \tag{9}$$

with the stipulation that the values of the influence coefficients may be calculated provided the assumption can be made that a single wavelength may represent (i.e., be equivalent to) a polychromatic incident beam. This was an oversimplification that was subsequently abandoned in favor of the concept that α coefficients may be related to 'fundamental influence coefficients,' thus retaining a close link to the basic principles of x-ray physics.

From 1968 on, the emergence of the fundamental parameters approach proposed by Criss and Birks[15] provided a method for the correction of matrix effects based on physical principles of x-ray fluorescence emission and led to the re-evaluation of older influence coefficient models. Criss and Birks compared two mathematical approaches for corrected matrix effects: the empirical regression method which had been in use for many years, and a new method that accounts for matrix effects by means of measured spectral distributions of x-ray tube generated incident sources, mass attenuation coefficients, fluorescence yields, etc. The model used for the regression method was that of Beattie and Brissey [Eq (5)] but simplified, updated and optimized for computer evaluation. The coefficients are determined using experimental data by solving sets of simultaneous equations of the type

$$C_i / R_i = r_{ii} C_i + r_{ij} C_j + r_{ik} C_k + ... \tag{10}$$

For a three component system, nine influence coefficients are required and unknowns are analyzed by rewriting the above equation in the following form and solving for C_i, C_j, ...

$$(R_i r_{ii} - 1)C_i + R_i r_{ij} C_j + R_i r_{ik} C_k = 0$$

$$R_j r_{ji} C_i + (R_j r_{jj} - i) C_j + R_j r_{jk} C_k = 0$$

etc., with the stipulation that $C_i + C_j + C_k = 1.0$

It was observed that influence coefficients selected in this way do not represent the effect of one element on another but are the best set of numbers to describe the intensities measured.

The fundamental parameters approach for the correction of matrix effects proposed by Criss and Birks is based on the concept of using either measured or calculated spectral distributions of x-ray tubes and replacing integrals by summations over a discrete number of effective wavelength intervals. This led to the expression for primary fluorescence emission.

$$P_i = g_i\, C_i \sum_\lambda \frac{\mu_i(\lambda)\, I(\lambda)\, \Delta\lambda}{\mu_s(\lambda)\, csc\,\psi' + \mu_s(\lambda_i)\, csc\,\psi''} \tag{11}$$

where

$I(\lambda)$ is the incident intensity for the interval $\Delta\lambda$

$$\mu_s = \sum_i C_i\, \mu_i\,; \quad \Sigma\, C_i = 1.0$$

and to the expression (abridged) that includes both primary and secondary fluorescence emission, i.e., enhancement present.

$$P_i + S_i = P_i \left(1 + \sum_j e_{ij}\, C_j\right) \tag{12}$$

An iteration procedure is used to determine the weight fractions in unknowns. The measured relative intensities are scaled to 1.0 and used as the first estimate of composition. A set of theoretical relative intensities are calculated and compared to the measured values. The differences are used to make a better estimate of composition. The process is repeated until for some set of assumed concentrations, the calculated relative intensities agree with the measured values.

Subsequently it was shown, Lachance,[16] that the expressions for $P_i + S_i$ and $P_{(i)}$ could be combined leading to the formalism

$$P_i + S_i = P_{(i)}\, C_i - \sum_j X_{j,a}\, C_j + \sum_j X_{j,e}\, C_j \tag{13}$$

where $X_{j,a}$ and $X_{j,e}$ are the sums of the prorated absorption and enhancement monochromatic terms, respectively. Thus, the primary and secondary fluorescence emitted intensity is equal to product $P_{(i)}\, C_i$ minus the sum of each individual absorption effect plus the sum of each individual enhancement effect. Solving the $P_{(i)}\, C_i$ leads to

$$P_{(i)}\, C_i = P_i + S_i + \sum_j X_{j,a}\, C_j - \sum_j X_{j,e}\, C_j \tag{14}$$

which can be interpreted as 'the intensity corrected for matrix effect is equal to the emitted intensity plus a correction term that compensates for the decrease in intensity due to absorption minus a correction term that compensates for the increased intensity due to secondary enhancement.' Defining $X_{j,n} = X_{j,a} - X_{j,e}$ and substituting in Eq (14) gives

$$P_{(i)}\, C_i = P_i + S_i + \sum_j X_{j,n}\, C_j \tag{15}$$

which can also be expressed as

$$C_i = \frac{P_i + S_i}{P_{(i)}} \left[1 + \sum_j \frac{X_{j,n}}{P_i + S_i}\, C_j\right] \tag{16}$$

or

$$C_i = R_i [1 + \sum_j n_{ij} C_j] \tag{17}$$

i.e., Criss and Birks' fundamental parameters equation expressed in influence coefficient formalism. Two concepts that have been proposed in the domain of fundamental influence coefficients are examined next, namely the option to choose an element other than the analyte for elimination in the correction for matrix effect term, and the option to calculate influence coefficients specific to absorption and enhancement.

Given that the experimental determination of influence coefficients requires a large number of standards and that these are not always available, de Jongh[17,18] proposed a model wherein an element other than the analyte is arbitrarily chosen for elimination in contrast to Eq (17) which is based on the elimination of the analyte. The process is described as follows:

C$_i$ / R$_i$ is calculated using a fundamental parameters expression for an average composition.
C$_i$ / R$_i$ is calculated with respect to the average composition due to 0.1% change of element j, leading to the computation of delta-coefficients. The elimination of a selected components yields beta-coefficient values. Betas are converted to 'ALPHAS' so that they may be used in expressions similar to Eq (17), i.e., .

$$C_i = R_i [1 + \sum_j n_{ijn} C_j] \tag{18}$$

where n is retained as the symbol for the influence coefficients to indicate that the coefficients are in the domain of fundamental influence coefficients. The subscript ijn refers to the influence of an exchange of j and n on the value of C$_i$ / R$_i$ of the analyte i. It is noted that j ≠ n in the summation term Eq (18) whereas j ≠ i in the summation term Eq (17).

The fact that Eq (14) may be broken down to

$$P_{(i)} C_i = P_i + \sum_j X_{j,a} C_j \tag{19a}$$

and

$$P_i + S_i = P_i + \sum_j X_{j,e} C_j \tag{19b}$$

and transformed to

$$C_i = R_i' (1 + \sum_j a_{ij,P} C_j) \tag{20a}$$

and

$$R_i = R_i' (1 + \sum_j e_{ij,P} C_j) \tag{20b}$$

where

$$R_i' = \frac{P_i}{P_{(i)}}; \quad a_{ij,P} = \frac{X_{j,a}}{P_i}; \quad e_{ij,P} = \frac{X_{j,e}}{P_i}$$

thus yielding fundamental influence coefficients for absorption and enhancement, respectively. Ratioing Eqs (20a) and (20b) results in the cancellation of R_i' while cross-multiplication and solving for C_i yields the model proposed by Broll and Tertian[19]

$$C_i = R_i \left[1 + \sum_j \{a_{ij,P} - e_{ij,P} \frac{C_i}{R_i}\} C_j\right] \tag{21}$$

The algorithm proposed by Rousseau[20] may also be obtained by ratioing Eqs (20a) and (20b) and solving for C_i

$$C_i = R_i \left[\frac{1 + \sum_j a_{ij,P} C_j}{1 + \sum_j e_{ij,P} C_j}\right] \tag{22}$$

Although fundamental influence coefficient algorithms have common origins, readers are referred to the original publications for details as to their mode of applications.

The computation of fundamental influence coefficients implies that the composition of the specimen is known or can be very closely estimated. An alternate approach has gradually evolved based on the concept that the value of α_{ij} influence coefficients can be 'customized' as a function of composition for each specimen during the iteration process. In one case, Lachance,[21] the algorithm is identical to Eq (17) except that the symbol α is retained to indicate that approximations are involved, i.e.,

$$C_i = R_i \left[1 + \sum_j \alpha_{ij} C_j\right] \tag{23}$$

where

$$\alpha_{ij} = \alpha_{j1} + \frac{\alpha_{j2} C_M}{1 + C_i \alpha_{j3}} + \sum_k \alpha_{ijk} C_k \tag{24}$$

$$C_M = C_j + C_k + ...$$

The coefficients α_{j1}, α_{j2} and α_{j3} coefficients are functions of binary n_{ij} coefficients while the cross-coefficient α_{ijk} is an average value to compensate for 'third-element effects.' Tertian,[22] on the other hand, proposed the model

$$C_i = \frac{R_i}{1 + \epsilon_i} \left[1 + \sum_j \{\alpha + \frac{\beta C_i}{1 + \gamma (1 - C_i)}\} C_j\right] \tag{25}$$

where the coefficients α, β and γ are also functions of binary n_{ij} coefficients and the term $1 + \epsilon_1$ compensates for 'third-element effects.'

Eq (17) formalism may also be retained for the Rasberry and Heinrich[23] model in which case the symbol r_{ij} is retained to indicate that the coefficients are obtained by regression of experimental date and are empirical in nature. In this case

$$r_{ij} = A_{ij} + \frac{B_{ij}}{1 + C_i} \tag{26}$$

where coefficients A_{ij} are used when the significant effect is absorption; in such cases the corresponding B_{ij} coefficients are zero. The B_{ij} coefficients are used when the predominant effect is enhancement, in which case the A_{ij} coefficients are zero.

Due to the fact that it is customary in the analysis of steels not to determine the main constituent Fe, Ito, et. al.[24] proposed the following model

$$C_i = (a + bI_i + cI_i^2) [1 + \sum_j r_{ij} C_j] \tag{27}$$

termed the Japanese Industrial Standard method. Although their accuracies were found to be similar, a number of advantages were noted when the method was compared to the α-correction model, namely: fewer iteration steps are required; the error due to dead time is corrected to some extent, and it offers greater convenience.

APPENDIX

Symbols

I_i, I_j	net fluorescence intensity of element i, element j
C_i, C_j	weight fraction of element i, element j
$I_{(i)}$	net fluorescence intensity of pure element i
t_i	time taken to accumulate a fixed number of counts for element i
$t_{(i)}$	time taken to accumulate an identical number of counts for pure element i
P_i	intensity due to primary fluorescence emission
S_i	intensity due to secondary fluorescence emission
μ	mass attenuation coefficient
λ	wavelength
λ_i	wavelength characteristic line element i
ψ'	angle of incidence
ψ''	angle of emergence
a_{ij}	absorption influence coefficient element j on analyte i
e_{ij}	enhancement influence coefficient element j on analyte i
n_{ij}	net influence coefficient element j on analyte i
α_{ij}	influence coefficient element j on analyte i
r_{ij}	empirical influence coefficient element j on analyte i

REFERENCES

1. Moseley, H.G.J., Philos Mag. 26, 1024 (1913).
2. Moseley, H.G.J., Philos Mag. 27, 703 (1914).
3. von Hevesy, G., Chemical Analysis by X-Rays and its Applications, McGraw-Hill Book Co., New York (1932).
4. von Hamos, L., Arkiv. Math. Astron. Fys. 31a, 25 (1945).
5. Sherman, J., ASTM Special Tech. Publ. No. 157 (1954).
6. Sherman, J., Spectrochim. Acta. 7, 283 (1955).
7. Shiraiwa, T., and N. Fujino, Japan J. of Appl. Phys. 5, 886 (1966).
8. Gillam, E., and H.T. Heal, Brit. Jour. Appl. Physics 3, 353 (1952).

9.   Beattie, H.J., and R.M. Brissey, Anal. Chem. 26, 981 (1954).
10.   Campbell, W.J., and H.F. Carl, Norelco Reporter, v.III (1956).
11.   Mitchell, B.J., Anal. Chem. 33, 917 (1961).
12.   Lucas-Tooth, H.J., and B.J. Price, Metallurgia 64, 149 (1961).
13.   Lucas-Tooth, H.J., and C. Pyne, Adv. in X-Ray Anal., v.7, 523
          Plenum Press, New York (1964).
14.   Lachance, G.R., and R.J. Traill, Can. Spectrosc. 11, 43 (1966).
15.   Criss, J.W., and L.S. Birks, Anal. Chem. 40, 1080 (1968).
16.   Lachance, G.R., Adv. in X-Ray Anal. v.31, 471, Plenum Press, New York
          (1988).
17.   de Jongh, W.K., X-Ray Spectrom. 2, 151 (1973).
18.   de Jongh, W.K., X-Ray Spectrom. 8, 52 (1979).
19.   Broll, N., and R. Tertian, X-Ray Spectrom. 12, 1 (1983).
20.   Rousseau, R.M., X-Ray Spectrom. 13, 115 (1984).
21.   Lachance, G.R., paper presented at the Intern'l Conf. on Industrial Inorganic
          Elemental Analysis, Metz, France, June 1981.
22.   Tertian, R., X-Ray Spectrom. 17, 89 (1988).
23.   Rasberry, S.D., and K.F.G. Heinrich, Anal. Chem. 46, 81 (1974).
24.   Ito, M., et al., X-Ray Spectrom. 10, 3 (1981).

NEW DEVELOPMENTS IN FP-BASED SOFTWARE FOR BOTH

BULK AND THIN-FILM XRF ANALYSIS

Liangyuan Feng, Brian J. Cross,
and Richard Wong

Kevex Instruments
355 Shoreway Rd.
San Carlos, CA 94070

INTRODUCTION

In the last decade, the Fundamental Parameter (FP) method
has been increasingly used in quantitative XRF analysis as an
effective means for interelement effect corrections. Many authors have contributed to its continuous development both in
theory and practice [1 - 26]. Among various implementations, the
FP-alpha method has been used widely in bulk-sample analysis
because of its unique advantage of speed [1 - 6]. In recent
years, however, with the advent of faster computers, a rigorous
full FP approach has become feasible and practical. This is especially true in the field of multilayer thin-film analysis,
where the FP-alpha approach has not yet been applied.

Although several thin-film analysis FP-software packages
have already been developed [7 - 19], there remain a number of
subjects which require further work to satisfy the needs of a
wide variety of applications. For example, in X-Ray Micro-
Fluorescence (XRMF) technology, multilayer thin-film analysis
has been an important application. However, systematic errors
in intensity measurements often lead to inaccuracy of the
results. These may occur because of local variations of the
sample surface or changes in geometry or tube flux, or when
the beam spot is larger than the sample area. A reference monitor can be used to compensate the tube flux change. However, it
is necessary to explore other ways to correct systematic errors
derived from the other sources mentioned above.

We have recently developed a comprehensive full FP package
(TFFP) for both bulk and multilayer thin-film analysis, which
handles up to 6 layers and up to 5 excitation conditions, and
performs rigorous interlayer enhancement corrections. In the

course of this development, we have also formulated a new iteration scheme which works well for both bulk and thin-film
analysis, even when a systematic intensity error exists. This
paper focuses on the discussion of the new iteration scheme.

THE OLD ITERATION SCHEME

Conventionally, during the iteration in FP-based programs,
concentrations and thicknesses are refined by comparing the
measured and the calculated (theoretical) intensities on an
absolute basis. To correct the concentration estimates, the
following equation (described here and elsewhere in computer
programming notation) is often used:

$$C(i) = C(i) * A(i,meas)/A(i,calc) \tag{1}$$

where $C(i)$ is the weight fraction of analyte i, $A(i,meas)$ is the
theoretical equivalent of the measured intensity of analyte i
obtained using calibration coefficients, and $A(i,calc)$ is the
theoretical intensity of i calculated based on the current concentration and thickness estimates. Then, the non-normalized
sum of the weight fractions of all elements in a certain layer
is used for thickness (T) correction:

$$T = T * \sum_i C(i) \tag{2}$$

It is obvious that a systematic error in the measured intensities will inevitably cause a bias in both the calculated concentrations and thicknesses.

THE NEW ITERATION SCHEME

Assuming the measured intensities of all the analytes change
by a factor of k, one may expect a change by the same factor in
both the calculated concentrations and thicknesses if the previous
scheme is used. However, if equation (1) is modified by replacing
the absolute intensity terms with normalized ones, the factor k
cancels out during the concentration refinement:

$$C(i) = C(i) * \frac{k*A(i,meas)/k*\sum_n A(n,meas)}{A(i,calc)/\sum_n A(n,calc)}$$

$$= C(i) * A(i,meas)/A(i,calc) *$$

$$\sum_n A(n,calc)/\sum_n A(n,meas) \tag{3}$$

where $\sum_n A(n,meas)$ is the sum of the measured intensities of all
analytes in a certain layer, and $\sum_n A(n,calc)$ is the sum of the
calculated intensities of the same analytes.

Table 1. Thickness Variations with Intensities

Samp.	Layer	Calculated Thickness (Å)			
		LAMA3	A	B	B/A
T1	1 (FeNi)	2123	2204	1830	0.8303
	2 (Cu)	2470	2462	2022	0.8213
	3 (Cr)	1635	1703	1380	0.8103
T2	1 (Cr)	1698	1734	2085	1.2024
	2 (Cu)	2462	2452	3002	1.2243
	3 (FeNi)	2055	2186	2673	1.2228

Layer 1 is the top layer;
LAMA3 - results from [12];
A - original intensity data [12] used;
B - original intensity data decreased (T1)
    or increased (T2) by 20%;
B/A - ratio of thickness in column B to
    that in column A.

As clear from the above discussion, the systematic error can be removed using the proposed scheme if it (the constant factor k) exists on all analytes of the sample. For bulk sample analysis, the new scheme can be used directly and has proven very effective (Table 2). For multilayer thin-film analysis, however, since there is one more degree of freedom (thickness), one more condition is required in order to achieve the results free of the systematic error. Tests reveal that, although the calculated thicknesses change with the measured intensities, the fractional change is constant for the different layers (Table 1). This implies that it should be possible to use a thickness adjustment factor to correct for the thickness bias due to the systematic error. If one of the thin layers has a known thickness, then this factor is known. If not, other techniques have to be used in order to obtain this factor. One approach to the solution of this problem is to use the intensity signal from the substrate [17 - 19]. This requires an extra calibration step to measure the signal from the free (or unobstructed) substrate. In the new scheme, we employed a similar technique, but utilized a modified algorithm which is more flexible for various applications and requires an adjustment made only once outside the iteration loop.

The new algorithm generates a scaling factor in the final stage of calculation, which is used to scale the measured free substrate signal. The thickness adjustment factor is then calculated by comparing the scaled free substrate signal with the signal from the sample. As the free substrate signal is measured at calibration time, any systematic discrepancy in their intensity measurements should be compensated by this factor. Since the thickness adjustment is made only once, it has only a minor effect on the speed. This procedure can be described using the following expressions:

$$r = I(s,theo)/I°(s,theo) \tag{4}$$

Table 2. Bulk Sample Test Results
(Sample C1154 - NIST High Temperature Alloy)

Component	Certified (%)	Original intensities used		Original intensities decreased by 20%	
		Old Scheme (%)	New Scheme (%)	Old Scheme (%)	New Scheme (%)
Cr	19.06	19.11	19.06	15.95	19.16
Mn	1.42	1.45	1.44	1.21	1.45
Fe	64.42	63.98	63.81	53.02	63.68
Ni	12.92	12.93	12.89	10.75	12.91
Cu	.40	.40	.40	.34	.40
Mo	.07	.05	.05	.04	.05

Note: 1. Original intensity data were cited from [3].
      2. Standards used: C1151, C1153, C1285, C1288, and C1289.

and

$$Tfac = r * I°(s,meas)/I(s,meas) \qquad (5)$$

where r is the scaling factor, Tfac is the thickness adjustment
factor, I(s,theo) and I(s,meas) denote the theoretically calculated and the measured intensities of the substrate signal from
the sample, and I°(s,theo) and I°(s,meas) denote those from the
free substrate, respectively. If the substrate is not analysed
and its composition is unknown, then it is treated as a pure bulk
layer of the substrate element. This approximation accounts adequately for the absorptions of the substrate signal by the overlying layers and should be valid if there are no significant
enhancement effects between the substrate and the upper layers.

DISCUSSION

Extensive tests were made on various bulk and multilayer
thin-film samples. Tables 2 and 3 show some of the results. They
demonstrate how the new scheme eliminates systematic errors in
both FP-based bulk and multilayer thin-film analyses.

It should be noted that, although the new scheme works well
when there are no systematic errors observed, the calculated
thickness adjustment factor is usually different from unity,
and the final results always show minor differences from those
obtained with the old scheme. This is because there are random
errors which also contribute to the calculated factor. On the
other hand, for thin-film analysis the new scheme shows essentially no difference from the old scheme in the calculated
concentrations, because the old scheme also forces the sum of the
calculated concentrations to approach 100% during the simultaneous concentration and thickness refinement. In this respect,
it differs from bulk sample analysis.

Table 3. Multilayer Thin Film Test Results
(Sample T2 - Cr/Cu/Fe10Ni90)

1) Original intensity data used :

Tfac = 1.023419

Layer	Thickness (Å)				Concentration (%)				
	Given	LAMA3	TFFP		Given		LAMA3	TFFP	
			Thk1	Thk2				old sch.	new sch.
1	2000	1698	1734	1775	Cu	100	100.00	100.00	100.00
2	2000	2462	2452	2509	Cr	100	100.00	100.00	100.00
3	2000	2055	2146	2196	Fe	10	10.28	9.81	9.81
					Ni	90	89.72	90.19	90.19

Note: 1. Original intensity data were cited from [12].
      2. Thk1 is the calculated thickness without substrate sig-
         nal adjustment and Thk2 is the thickness calculated with
         substrate signal (Ni-Ka) adjustment (Thk2 = Tfac*Thk1).

2) Original intensities increased by 20% :

 a. Substrate analysed -

Tfac = 0.8280846

Layer	Thickness (Å)				Concentration (%)				
	Given	LAMA3	TFFP		Given		LAMA3	TFFP	
			Thk1	Thk2				old sch.	new sch.
1	2000	1698	2075	1718	Cu	100	100.00	100.00	100.00
2	2000	2462	3002	2486	Cr	100	100.00	100.00	100.00
3	2000	2055	2673	2213	Fe	10	10.28	9.81	9.81
					Ni	90	89.72	90.19	90.19

 b. Substrate used as a backing -

Tfac = 0.8330693

Layer	Thickness (Å)				Concentration (%)				
	Given	Lama3	TFFP		Given		Lama3	TFFP	
			Thk1	Thk2				old sch.	new sch.
1	2000	1698	2134	1777	Cu	100	100.00	100.00	100.00
2	2000	2462	3010	2507	Cr	100	100.00	100.00	100.00
3	assumed pure bulk Ni layer								

CONCLUSIONS

When used in FP-based software, the proposed new iteration scheme corrects for systematic intensity errors as much as 20%, if one of the following conditions exists:

1. Single-layer bulk samples are analysed.

2. One of the thin layers has a known thickness.

3. A free substrate signal is used in either of the following cases:

    a) the substrate is the bottom layer of a multilayer thin-film system, whether it is thin or bulk, analysed or fixed;
    b) the substrate, whether thin or bulk, is a backing of the sample and is not analysed.

Case 3.b requires no knowledge about the substrate composition and thickness, and works well when there are no significant enhancement effects between the substrate and the overlying layers.

REFERENCES

1. W. K. de Jongh, X-ray Spectr., 2:151 (1973).
2. G. R. Lachance and F. Claisse, Adv. X-ray Anal., 23:87 (1980).
3. G. Tao, P. A. Pella and R. M. Rousseau, "NBS Technical Note 1213", (1985).
4. R. M. Rousseau, X-Ray Spectr., 15:207 (1986).
5. R. M. Rousseau, X-Ray Spectr., 16:103 (1987).
6. L. Feng, P. A. Pella and B. J. Cross, Adv. X-ray Anal., 33:509 (1990).
7. D. Laguitton and W. Parrish, Anal. Chem., 49(8) (1977).
8. D. Laguitton and M. Mantler, Adv. X-ray Anal., 20:515 (1977).
9. T. C. Huang, X-Ray Spectr., 10(1) (1981).
10. M. Mantler, Adv. X-ray Anal., 27:433 (1984).
11. M. Mantler, Analytical Chemica Acta, 188:25 (1986).
12. T. C. Huang and W. Parrish, Adv. X-ray Anal., 29:395 (1986).
13. J. E. Willis, X-Ray Spectr., 18:143 (1989).
14. D. K. G. de Boer and P.N. Brouwer, Adv. X-ray Anal., 33:237 (1990).
15. D. K. G. de Boer, X-Ray Spectr., 19:145 (1990).
16. Y. Kataoka and T. Arai, Adv. X-ray Anal., 33:213 (1990).
17. J. R. Maldonado and D. Maydan, The Bell System Technical Journal, October, 1851 (1979).
18. R. Linder, G. Kladnik and J. Augenstine, X-Ray Imaging, Vol.II, Proc. Soc. Photo-Opt. Instrum. Eng., 691:28 (1986).
19. B. J. Cross and D. C. Wherry, Thin Solid Films, 166:263 (1988).
20. J. Sherman, Spectrochimica Acta, 7:283 (1955).
21. J. W. Criss and L. S. Birks, Anal. Chem., 40(7) (1968).
22. T. Shiraiwa and N. Fujino, Adv. X-ray Anal, 12:446 (1969).
23. G. Pollai, M. Mantler and H. Ebel, Spectrochimica Acta, 26(B):747 (1971)

24. J. W. Criss, L. S. Birks and J. V. Gilfrich, Anal. Chem., 50(33) (1978).
25. J. W. Criss, Adv. X-Ray Anal., 23:93 (1980).
26. P. A. Pella, L. Feng and J. A. Small, X-Ray Spectr., 14:125 (1985).

SOFTWARE PACKAGES FOR THE AUTOMATIC ASSESSMENT OF XRF DATA

FOR QUALITATIVE AND SEMI-QUANTITATIVE ANALYSIS

P L Warren, A E Smith, J D v Aalten, N Hodkinson

Wilton Materials Research Centre, ICI Advanced Materials
Middlesbrough, Cleveland, England

Qualitative inorganic analysis is required for the identification of
unknowns, the classification of type, and sometimes to decide what subsequent
quantitative analysis is needed. The traditional way of performing
qualitative XRF analysis on unknown materials is by subjecting the sample to
a full spectral scan. This takes time and an experienced operator to
interpret the spectra. Classifying the elements detected as major, minor or
trace can also be person dependent. Round robin tests have confirmed this by
showing considerable variation in results between laboratories.

To counter these problems, ICI have over the years developed in-house
programs (SQICI) for (1) automatic assessment of qualitative XRF data, and
(2) calculating the rough concentrations of the elements detected. Our
experience covers both the full scan approach - which attempts to identify
all peaks generated from a spectral scan, and the peak hopping routine -
which uses a selected analyte line to represent each chemical element (Z >
9). In each case the algorithm has to be carefully considered to take
account of possible line overlaps from tube lines and other elements present,
differentiation of random noise from small peaks, and instrument blanks at
low concentrations.

Recently, commercial software packages have become available that promise a
significant step forward over previous methods. While the instrument
manufacturers have adopted a similar choice of algorithm for the qualitative
part of the procedure, they have enhanced the quantification stage by
employing mathematical techniques such as fundamental parameters.   We have
used the SQICI software as a bench-mark for comparing the results from the
manufacturers programs. In particular we are looking for certain features
which will give good all-round results and provide a user friendly interface
for both novice and experienced operators. An ideal program, we believe,
would possess the following attributes -:

* reliable algorithm - no false positives or negatives
* improved accuracy over previous qual and semi-quant methods
* wide range of sample types - solid, liquids, powders
* samples with organic or inorganic matrices
* improved lower limits of detection

Advances in X-Ray Analysis, Vol. 35
Edited by C.S. Barrett *et al.*, Plenum Press, New York, 1992

711

    * realistic "less than" values for elements not detected
    * success with difficult elements ie near tube lines
    * extend to lower Z elements
    * operate in either interactive or automatic mode
    * easy calibration with user or commercial standards
    * provide clear and versatile reporting facilities

A number of XRF users in different ICI laboratories have been considering the merits of such systems, particularly when applied to the analysis of plastic compounds with quite complex formulations. Three packages have been installed in various parts of the Company, initially on a trial basis, to gain experience with commercial packages -:

    1   SSQ origin Siemens   test instrument = SRS-303

Features   Optimised Spectral scans
                Detector Overflow Correction
                Simple 2 point Calibration
                Automatic background Subtraction
                Peak Identification and assignment
                REPORTS - all elements to 0.001%

    2   PSA origin Philips   test instrument = PW1400

Features   User defined scan conditions
                User defined standards
                Peak Identification and assignment
                Definable peak significance levels
                Quantification on best free line
                REPORTS   - elements found
                            - elements not quantified
                            - elements possibly present

    3   Uniquant   origin Omega Data Systems.
                              test instrument = PW1404
Features   67 element lines (peaks)
                8 selected background wavelengths
                Choice of background models
                Line overlap corrections
                Exacting calibration procedure
                Intrinsic element sensitivities
                REPORTS - concentrations + standard error

Results

All three packages were well received by the participating laboratories. They were found to be easy to operate and in general gave reliable results for the analysis of a wide range of different types of materials. While it was expected that their main application area would be the detection of major and minor components, they proved to give surprisingly good results for trace components (down to 10 ppm in favourable cases) especially in light matrix materials such as polymers and environmental samples. Most times the algorithm for identifying lines correctly was found to be reliable. However here the systems varied in their approach. The Philips system which looks for the best "free line" produced fewer false positives than the comparative Siemens program which relies on overlap corrections and quantifies on a line pre-determined for each element. The Uniquant uses a peak hopping approach, only measuring at one pre-selected line for each element, plus 8 backgrounds. However, coupled with a stringent line overlap correction routine, the method proved reasonably dependable. Only when the sample contained high concentrations of certain elements that effected the shape of the background curve, did the algorithm show some weaknesses. Then a few false positives in the wavelength region of the rare earth L lines were evident.

The Siemens SSQ was simplest and quickest to calibrate using two set up standards provided. But the Philips PSA allowed the user to include any number of their own standards. Uniquant requires the user to run each of 50 mainly one element standards through the complete measurement cycle - a slow and painstaking procedure, but only needed once. A gold plated reference disc allowed the user to correct for instrument drift quite simply by an internal ratio to the Au La and Au Ma lines. All three systems allow the user to enter values for the non-determined elements (eg C,H,O,N) as fixed percentages or as the balance. Overall the accuracy was better than our in-house comparisons, with figures often within the range 10-50 % of the true concentrations. None of the report formats were entirely satisfactory, nor calculated the lower limit of detection from the data available. The Uniquant approach was best, reporting to two significant figures and expressing a value for the possible error.

Conclusions

Computerisation of the x-ray scans offers the necessary speed and consistancy that is required for what is becoming an increasingly important part of XRF analysis. Fundamental parameter programs are a reliable means of correcting for inter-element effects. Where analysts lack the necessary reference standards for full quantitative analysis, the reliability of the semi-quant mathematical routines can often give acceptable results which would be impossible otherwise.

In-house computer programs have been found to be expensive and slow to develop. Ours has also suffered from limited performance and applicability. The three commercial systems all give good results. However we still believe we have features that the instrument manufacturers have been slow to pick up. This includes the more thorough checking for line identification and the reporting of realistic "less than" figures, rather than very low % levels. The commercial systems now available are still relatively new and we expect that they will improve still further as present developments come to fruition and ideas that are fed back from current users become incorporated in future packages.

A FAST ALGORITHM FOR FUNDAMENTAL PARAMETER CALCULATIONS

Ch. Pöhn and H. Ebel

Institut für Angewandte und Technische Physik
Technische Universität Wien, A 1040 Wien, Austria

ABSTRACT

A fundamental parameter algorithm for quantitative XFA with characteristic $K\alpha$- and $L\alpha$-radiations of the specimens is presented. This allows an application to programmable pocket calculators or a fast evaluation of measured countrates for the unknown composition of the specimen by means of PCs. For the practical utilization we need the incidence angle and the take-off angle of the x-rays in the instrument, the high voltage V in kV which is supplied to the x-ray tube (E_0) and reference measurements on either pure elements or a single reference specimen. The quality of the analytical results depends on the concentration differences of the unknown specimen and the reference specimen.

THEORY

In Sherman's equation[1] for primary excited characteristic x-rays of the specimen

$$n_{i,prim} = c_i \omega_{ij} p_{ijk} \kappa_{ijk} M_{ij} \cdot \int_{E_{ij}}^{E_o} \frac{\tau_{E,i} x_E}{\sum_m c_m \left[\frac{\tau_{E,m}}{\cos\alpha} + \frac{\tau_{ijk,m}}{\cos\beta} \right]} dE$$

we have expressed the spectral distribution x_E of incident x-rays by Kramers's distribution function[2]

$$x_E = const_w I Z_{anode} \left(\frac{E_o}{E} - 1 \right)$$

and the energy dependence of the photoabsorption coefficient $\tau_{E,m}$ of the element m for x-rays of energy E by

$$\tau_{E,m} = T_m E^{-3}$$

Thus, Sherman's equation becomes

$$n_{i,prim} = Const_i \cdot \int_{E_{ij}}^{E_0} \frac{E_0 - E}{E + const_i \cdot E^4} \cdot dE$$

with $Const_i = c_i \omega_{ij} P_{ijk} \kappa_{ijk} M_{ij} T_i const_w IZ_{anode} cos\alpha \frac{1}{\sum_m c_m T_m(E)}$

and $const_i = \frac{cos\alpha}{cos\beta} \cdot \frac{\sum_m c_m T_m(E_{ijk})}{\sum_m c_m T_m(E)} \cdot \frac{1}{E_{ijk}^3}$.

Due to absorption edges in the integration interval E_{ij} to E_0, the integration procedure has to be performed from the energy $E_1 = E_{ij}$ to the energy E_2 of the following absorption edge, from there to the next absorption edge energy E_3, and so on. The last integration interval ranges from E_q to $E_{q+1} = E_0$. Thus, we divided our integration interval into q subintervals. It is worth noting that T_m of the element m which corresponds to one of the absorption edges, changes its numerical value at the edge. Consequently, we describe T_m by $T_m(E)$ and $Const_i$ and $const_i$ are different for adjacent intervals. For an arbitrary integration interval from E_a to E_b ($E_b > E_a$) the result of the integration of the above given expression for $n_{i,prim}$ is described by

$$\Delta n_{i,prim}(E_a, E_b) = Const_i \cdot \sum_{h=1}^{5} \left[A_h(E_b) - A_h(E_a) \right]$$

The coefficients $A_1(E)$ to $A_5(E)$ are given by

$$A_1(E) = + E_0 \cdot lnE$$

$$A_2(E) = - \frac{\sqrt{3}}{3 \cdot \sqrt[3]{const_i}} \ atn \left(\frac{-1 + 2E \cdot \sqrt[3]{const_i}}{\sqrt{3}} \right)$$

$$A_3(E) = - \frac{1}{3 \cdot \sqrt[3]{const_i}} \ ln \left(1 + E \cdot \sqrt[3]{const_i} \right)$$

$$A_4(E) = + \frac{1}{6 \cdot \sqrt[3]{const_i}} \ ln \left(1 - E \cdot \sqrt[3]{const_i} + E^2 \cdot \sqrt[3]{const_i^2} \right)$$

$$A_5(E) = - \frac{E_o}{3} \ln\left(1 + E^3 \text{const}_i \right)$$

The total primary excited characteristic signal $n_{i,prim}$ is the sum of q contributions $\Delta n_{i,prim}(E_a, E_b)$. We used the following symbols: c_i is the concentration of element i in weight fractions, the index j defines the ionized level of the atom, the index k gives the level from where the empty state in j is filled up, α is the angle of incidence and β is the angle of take-off of x-rays, relative to the normal of the specimen plane, E_o is the maximum energy of the white spectrum. The remaining quantities will be discussed in the next section. Photon energies are given in keV.

FUNDAMENTAL PARAMETERS

We have calculated the T_m-values of the photoabsorption coefficients by least squares fits of McMaster's[3] $\tau_{E,m}$.

$$T_m = \frac{\displaystyle\int_{E_a}^{E_b} \tau_{E,m} E^{-3} dE}{\displaystyle\int_{E_a}^{E_b} E^{-6} dE}$$

These fits compare absolute values of τ. Thus, the best agreement between the tabulated values and our values is achieved at the photon energies with high absorption coefficients. This fact is profitable to an application of our approximation to quantitative XFA. A further advantage of our concept is a polynomial description of T_m depending on the atomic number Z_m. Table 1 gives the polynomial coefficients and Fig.1 depicts our $\tau(E)$-response in comparison with the response according to McMaster[3].

$$T_m = \sum_{a=0}^{3} t_a Z_m^a$$

A further contribution to data reduction is the polynomial approximation of the edge energies by

$$E_{ij} = \sum_{m=0}^{3} B_{mj} Z_i^m$$

Table 1. Polynomial coefficients t_0 to t_3 of the universal description of the photoabsorption coefficient τ (cm^2/g) of the chemical elements $11 \le Z \le 83$ for x-radiation of photon energy $1 \le E \le 50$

Range	t_0	t_1	t_2	t_3
K	+3.987E+4	-3.613E+3	+5.072E+1	+9.742E+0
L1	-2.087E+4	+2.934E+3	-1.285E+2	+2.457E+0
L2	+2.983E+4	-2.134E+2	-5.834E+1	+1.717E+0
L3	+3.470E+4	-1.057E+3	-2.110E+1	+1.056E+0
M1	+4.309E+3	+1.397E+2	-1.756E+1	+0.355E+0
M2	+9.383E+4	-4.023E+3	+4.780E+1	-0.156E-1
M3	+1.114E+5	-4.803E+3	+6.096E+1	-0.113E+0
M4	-1.603E+3	+6.143E+1	-6.249E+0	+0.153E+0
M5	+3.043E+3	-1.219E+2	-2.117E+0	+0.935E-1
N	-1.999E+4	+9.405E+2	-1.529E+1	+0.927E-1

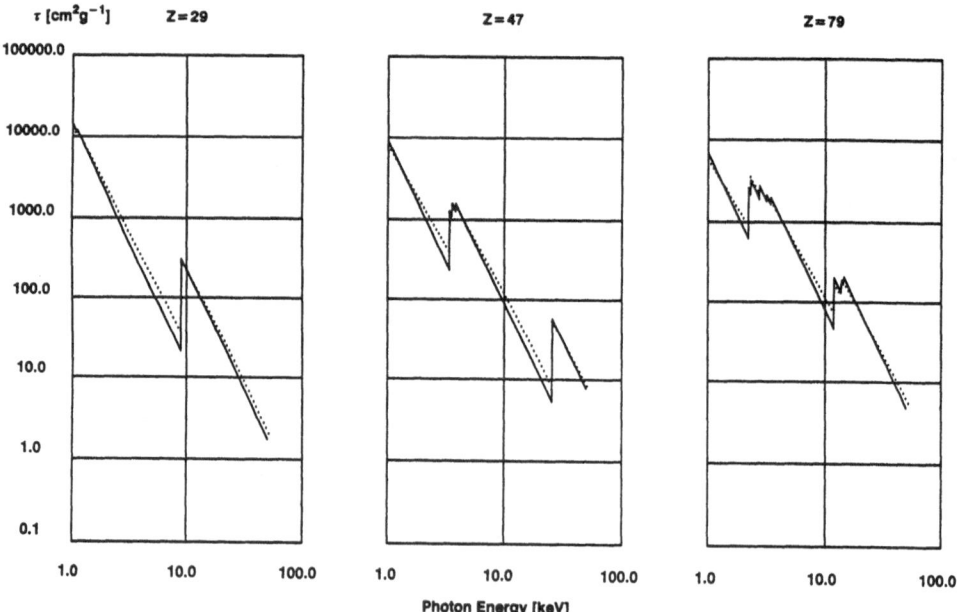

Figure 1. Comparison of the E^{-3}-dependence of photoabsorption coefficients τ with McMaster's response. The full curves represent the approximation of $\tau(E)$ and the broken curves McMaster's data.

Table 2. Polynomial fit coefficients for the universal description of the absorption edge energies of the chemical elements

edge j	B_{0j}	B_{1j}	B_{2j}	B_{3j}	z_i
K	-1.304E-1	-2.633E-3	9.718E-3	4.144E-5	11 - 63
L1	-4.506E-1	1.566E-2	7.599E-4	1.792E-5	28 - 83
L2	-6.018E-1	1.964E-2	5.935E-4	1.843E-5	30 - 83
L3	3.390E-1	-4.931E-2	2.336E-3	1.836E-6	30 - 83
M1	-8.645	3.977E-1	-5.963E-3	3.624E-5	52 - 83
M2	-7.499	3.459E-1	-5.250E-3	3.263E-5	55 - 83
M3	-6.280	2.831E-1	-4.117E-3	2.505E-5	55 - 83
M4	-4.778	2.184E-1	-3.303E-3	2.115E-5	60 - 83
M5	-2.421	1.172E-1	-1.845E-3	1.397E-5	61 - 83

This is valid for the chemical elements from $z_i=11$ to $z_i=83$. Table 2 gives the numerical values of B_{mj}.

The following polynomial presentation[4] of the energy E_{ijk} of characteristic radiations allows to calculate the energy of the $K\alpha$-radiation and the energy of the $L\alpha$-radiation.

$$E_{iK\alpha} = -6.654E-2 - 8.609E-3 \cdot z_i + 9.63E-3 \cdot z_i^2 + 7.268E-6 \cdot z_i^3 + 1.131E-7 \cdot z_i^4 \quad \ldots \quad K\alpha\text{-radiation} \ (\ 11 \leq z_i \leq 50 \ )$$

$$E_{iL\alpha} = +4.776E-2 - 1.706E-2 \cdot z_i + 1.523E-3 \cdot z_i^2 + 4.414E-6 \cdot z_i^3 - 1.739E-8 \cdot z_i^4 \quad \ldots \quad L\alpha\text{-radiation} \ (\ 30 \leq z_i \leq 83 \ )$$

After the quantification of the photoabsorption coefficients, the edge energies and the characteristic energies, there remain the fundamental parameters: fluorescence yield ω_{ij}, transition probability p_{ijk}, detection efficiency κ_{ijk} of the energy dispersive system, absorption edge jump factor M_{ij} and the product $const_w IZ_{anode}$ of the tube spectrum. These fundamental parameters are obtained with measurements either performed on pure elements or on one single specimen of known composition. The pure element i provides a measured signal N_{ijk} which is described by $\Delta n_{i,prim}(E_{ijk}, E_o)$. This can be used for the elimination of the unknown instrumental quantities and fundamental parameters of the element i in $Const_i$.

$$Const_i = c_i \cdot \frac{T_i}{\sum_m c_m T_m(E)} \cdot \frac{N_{ijk}}{\sum_{h=1}^{5} \left[A_h(E_o) - A_h(E_j) \right]}$$

The measured signal $n_{ijk}(ref)$ of a reference specimen gives

$$Const_i = \frac{c_i/c_i(ref)}{\sum_m c_m T_m(E)} \cdot \frac{n_{ijk}(ref)}{\sum_{y=1}^{q} \frac{1}{\sum_m c_m(ref)T_{my}} \sum_{h=1}^{5}\left[A_{hy}(E_y)-A_{hy}(E_{y+1})\right]}$$

Both possibilities do not call any longer for the knowledge of the remaining fundamental parameters. Additionally, this referencing to a specimen of known composition helps to reduce the errors which are introduced by the numerical approximations of the fundamental parameters, the neglection of a possible excitation by characteristic radiation from the x-ray tube and the neglection of secondary and teriary excitation in the specimen.

QUANTITATIVE ANALYSIS[5]

The specimen consists of r elements. Thus, r characteristic countrates n_{ijk} have to be measured. A first assumption of the composition of the specimen is given by the concentrations $c_i(1)$.

$$c_i(1) = n_{ijk}/\sum_m n_{mjk}.$$

The concentrations $c_i(1)$ are used to calculate $n_{i,prim}(1)$. The ratio $r_i(1) = n_{ijk}/n_{i,prim}(1)$ allows to determine $c_i(2)$ by

$$c_i(2) = r_i(1)c_i(1)/\sum_m r_m(1)c_m(1)$$

The iteration is continued until the difference $c_i(p)-c_i(p-1)$ of the concentration of two consecutive steps becomes smaller than a selected value.

REFERENCES

1   J.Sherman, Spectrochimica Acta 7, 283 (1955)

2   H.A.Kramers, Phil.Mag. 46, 836 (1923)

3   W.H.McMaster, N.K.del Grande, J.H.Mallett and J.H.Hubbell
    Compilation of X-Ray Cross-Sections, UCRL-50174, Sect.II,
    Rev.1. Lawrence Radiation Laboratory, University of
    California, Livermore, CA. (1969)

4   Ch.Pöhn, J.Wernisch and W.Hanke,
    X-Ray Spectrometry 14 (1985) 120

5   J.Criss and L.S.Birks, Anal.Chem. 40 (1968) 1080

AN ALGORITHM FOR THE DESCRIPTION OF WHITE AND CHARACTERISTIC TUBE SPECTRA ($11 \leq Z \leq 83$, $10\text{keV} \leq E_o \leq 50\text{keV}$)

H. Ebel, H. Wiederschwinger and J. Wernisch

Institut für Angewandte und Technische Physik
Technische Universität Wien, A 1040 Wien, Austria

P. A. Pella

National Institute of Standards and Technology
Gaithersburg, MD 20899, USA

INTRODUCTION

Kramers[1] described the cross section of electron interaction with target atoms of atomic number Z by

$$\sigma_{K,E} = \text{const}_w Z \left(\frac{E_o}{E} - 1 \right) \tag{1}$$

where E_o is the kinetic energy of impinging electrons, and E the energy of x-ray photons of the continuum. Smith et al[2] modified this equation, introducing an exponent x, so that

$$\sigma_{S,E} = \text{const}_w Z \left(\frac{E_o}{E} - 1 \right)^x \tag{2}$$

We applied the cross-section $\sigma_{S,E}$ to the evaluation of experimental results[3]. The evaluation of the measured spectral responses of the x-ray signals n_E was performed by

$$n_E = \sigma_{S,E} f(d_{eff}) R_E D_E \Omega \cdot i \cdot \Delta E \cdot \Delta t. \tag{3}$$

where $f(d_{eff})$ describes the absorption of x-rays of energy E in the target, R_E accounts for backscattering of electrons, D_E quantifies the efficiency of x-ray detection within the solid angle Ω. Further specific parameters are the electron current i, the energy range ΔE of a selected portion of the continuum at energy E, and the time Δt of data accumulation.

Advances in X-Ray Analysis, Vol. 35
Edited by C.S. Barrett *et al.*, Plenum Press, New York, 1992

All experiments were carried out in a scanning electron microscope using the specimen as "anode material". X-rays were
detected using an energy-dispersive detector. We investigated
the elements C (graphite), Al, Ti, Cu, Rh, Sb, Tb, W, Au and
Pb. The acceleration voltage V was varied from 10 kV to 30
kV, in steps of 5 kV. From these results, a dependence of x
on the acceleration voltage V and on the atomic number Z of
the target material was found. This paper describes how this
relationship is quantified.

WHITE RADIATION

For a well established description of $x(E_o,Z)$, we extended
the range of E_o up to 50 keV and we added the elements Mg,
Si, Sc, Cr, Ge, Y, Zr, Nb, Ag, Sn, Ce, Pr and Yb. Due to the
poor reproducibility of the experimental results for C, we
decided to omit this element. Thus, a total of 22 elements
was measured at nine different E_o-values in the range from 10
keV to 50 keV. However, considering our previous experinces
(see ref.3) the following changes were made as well:

i. Originally, we used relative errors for the least
squares fits of measured spectral responses. We found that
the agreement of the fitted responses with measured values in
the energy range from 5 keV to 15 keV was improved by minimizing the absolute errors, e.g.,

$$\sum_E (n_{E,meas} - n_E)^2 = min \qquad\qquad (4)$$

ii. In our evaluation of white spectra we no longer consider the influence of backscattering. The original equations
given by Kramers[1] and by Kulenkampff[4] contain the anode
current and thus, backscattering is included in their cross-
section $\sigma_{K,E}$. We now express our results in terms of the cup
current of the electron probe micro analyzer, which is identical to the anode current of an x-ray tube.

iii. We described in ref.3 the absorption in the target by
a single effective depth d_{eff} for the total energy range of
white radiation. We have now applied a unique evaluation of
the target absorption factor $f(\chi)$ using a $\phi(\rho z)$-algorithm
from Love and Scott[6-8]. The photoelectric cross sections τ_E
of McMaster[5] were used for x-ray absorption. E_o, E and ΔE
are given in keV.

From the total number of measured spectral responses we
obtain $const_w$, and x (Fig.1), where

$$const_w = 1.36\cdot10^9 \ sr^{-1}mA^{-1}keV^{-1}s^{-1}$$

$$x = 1.0314 - 0.0032\cdot Z + 0.0047\cdot E_o$$

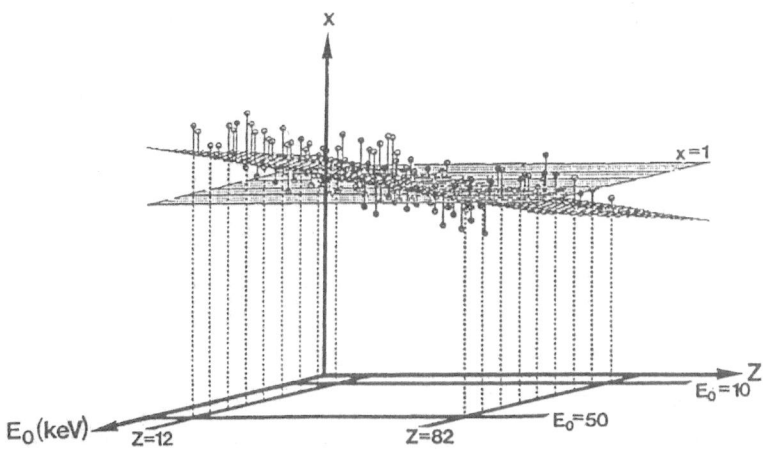

Fig.1 Circles represent the results of the experiments. The dark plane depicts the above given equation for x and the grey plane with x=1 the cross section according to Kramers.

Love and Scott assume a constant depth distribution $\Phi(\rho z)$ of the production rate of photons from the surface to a maximum depth ρz_m.

$$\rho z_m = \frac{A}{Z} \cdot (0.0787 \cdot 10^{-5} J^{1/2} E_o^{3/2} + 0.735 \cdot 10^{-6} E_o^2) \quad (5)$$

A is the atomic mass and the ionization potential $J=0.0135 \cdot Z$ has been given by Bloch[9]. The mean range of penetration $\rho \bar{z}_E$ is given by

$$\rho \bar{z}_E = \rho z_m \cdot \frac{0.49269 - 1.0987\eta + 0.78557\eta^2}{0.70256 - 1.09865\eta + 1.0046\eta^2 + \ln U_o} \cdot \ln U_o \quad (6)$$

with the overvoltage ratio $U_o = E_o/E$, and $\eta = E_o^m \cdot e^c$ according to Hunger and Küchler[10]. The exponent m is equal to $0.1382 - 0.9211 Z^{-1/2}$, and the factor e^c is given by

$$e^c = 0.1904 - 0.2236 \cdot \ln Z + 0.1292 \cdot (\ln Z)^2 - 0.0149 \cdot (\ln Z)^3 \quad (7)$$

The absorption $f(\chi_E)$ of x-rays of energy E in the target of a given element becomes

$$f(\chi_E) = \frac{1 - \exp(-2\chi_E \rho \bar{z}_E \cdot \sin\alpha)}{2\chi_E \rho \bar{z}_E \sin\alpha} \quad (8)$$

where $\chi_E = \tau_E/\sin\varepsilon$, ε is the take-off angle of x-rays and α is the incidence angle of electrons relative to the target surface.

CHARACTERISTIC RADIATION

The following theoretical description of measured characteristic count rates is used.

$$n_{jk} = const \cdot \frac{1}{S_j} \cdot R\omega_j P_{jk} \cdot f(x_{jk}) \cdot D_{jk} \cdot i \cdot \Omega \cdot \Delta t \qquad (9)$$

The index j defines the ionized atomic level of the chemical
element under investigation and k is the origin of the electron filling up the j-vacancy. Thus CuKα1 is described by j=K
and k=L3. $1/S_j$ is the intensity factor, R is the backscattering factor, ω_j is the fluorescence yield, p_{jk} is the transition probability and $f(x_{jk})$ describes the absorption of
characteristic photons of the energy E_{jk} in the target. D_{jk}
is the detector efficiency. The intensity factor $1/S_j$ is

$$\frac{1}{S_j} = \frac{z_j b_j}{z}(U_o \ln U_o + 1 - U_o)\left[1 + 16.5\left(\frac{J}{E_j}\right)^{1/2} \cdot \frac{U_o^{1/2} \ln U_o + 2(1 - U_o^{1/2})}{U_o \ln U_o + 1 - U_o}\right]$$

$$(10)$$

where z_j is the number of electrons in the ionized shell
($z_K=2$, $z_L=8$). According to Mott and Massey[11] the factor b_j
is either $b_K=0.35$, or $b_L=0.25$. The overvoltage ratio is defined by E_o/E_j. E_j is the energy corresponding to the j-
absorption edge of the element. The values of E_j are taken
from McMaster's tables[5]. J is again $0.0135 \cdot Z$. The effect of
electron backscattering is taken into account by the following equation[12] for R.

$$R = 1 + \sum_{j=1}^{5} \sum_{i=1}^{j} a_{i,j-i+1} \cdot \left(\frac{1}{U_o} - 1\right)^i \cdot z^{j-i+1} \qquad (11)$$

where the constants $a_{i,j-i+1}$ are as follows:

$a_{1,1} = 0.5580848699 \cdot 10^{-2}$ $a_{1,2} = 0.2709177328 \cdot 10^{-3}$

$a_{2,1} = 0.3401533559 \cdot 10^{-1}$ $a_{1,3} = -0.5531081141 \cdot 10^{-5}$

$a_{2,2} = -0.1601761397 \cdot 10^{-3}$ $a_{3,1} = 0.9916651666 \cdot 10^{-1}$

$a_{1,4} = 0.5955796251 \cdot 10^{-7}$ $a_{2,3} = 0.2473523226 \cdot 10^{-5}$

$a_{3,2} = -0.4615018255 \cdot 10^{-3}$ $a_{4,1} = 0.1030099792$

$a_{1,5} = -0.3210316856 \cdot 10^{-9}$ $a_{2,4} = -0.3020861042 \cdot 10^{-7}$

$a_{3,3} = -0.4332933627 \cdot 10^{-6}$ $a_{4,2} = -0.3113053618 \cdot 10^{-3}$

$a_{5,1} = 0.3630169747 \cdot 10^{-1}$

Fluorescence yields ω_j have been described by Pöhn et al[13] by least-squares fits of data from the literature.

For $\quad$ j=K, $12 \leq$ Z ≤ 42: $\quad \omega_K = \quad 8.502 \cdot 10^{-2} - 1.458 \cdot 10^{-2} \cdot Z +$
$+ 5.677 \cdot 10^{-4} \cdot Z^2 + 3.174 \cdot 10^{-5} \cdot Z^3 - 6.559 \cdot 10^{-7} \cdot Z^4 \qquad (12)$

and for j=L3, $38 \leq$ Z ≤ 78: $\quad \omega_{L3} = -1.983 + 1.432 \cdot 10^{-1} \cdot Z -$
$- 3.842 \cdot 10^{-3} \cdot Z^2 + + 4.578 \cdot 10^{-5} \cdot Z^3 - 1.959 \cdot 10^{-7} \cdot Z^4 \qquad (13)$

We calculated the transition probabilities p_{jk} from the data of Johnson and White[14]. For $K\alpha$-lines p_{Kk} is the line intensity of $K\alpha 1$ plus $K\alpha 2$ divided by the sum of K-line intensities, and for $L\alpha$-lines $p_{L3,k}$ is obtained from

$$p_{L3,k} = \frac{I(L\alpha 1) + I(L\alpha 2)}{I(L\alpha 1) + I(L\alpha 2) + I(L\beta 2) + I(Ll)} \qquad (14)$$

Finally, there remains the absorption correction $f(\chi_{jk})$. The indices j and k define the energy E_{jk} of the characteristic $K\alpha$-radiation, or of the characteristic $L\alpha$-radiation of the target material. We employed the numerical values from the tables of Johnson and White[14] and represent the characteristic energies by the weighted mean values of $K\alpha 1$ and $K\alpha 2$, or $L\alpha 1$ and $L\alpha 2$, respectively.

The evaluation of the measured results gave identical values for $const_{K\alpha}$ and $const_{L\alpha}$ for $K\alpha$- and $L\alpha$-radiation, and is

$$\boxed{const = 6 \cdot 10^{13} \ sr^{-1} mA^{-1} s^{-1}}$$

For application of our results to the description of white and characteristic x-ray spectra of x-ray tubes we need to know the absorption A of white radiation, and A_{jk} of characteristic radiation by the window.

We compared the results of our investigations with the description of the white and the characteristic spectra by Pella et al[15] and found excellent agreement.

CONCLUSION

We describe the white spectrum of an x-ray tube by the number n_E of photons in the energy range E to $E + \Delta E$

$$\boxed{n_E = const_w \cdot Z \cdot \left(\frac{E_o}{E} - 1 \right)^x \cdot f(\chi_E) \cdot \Omega \cdot A \cdot i \cdot \Delta E \cdot \Delta t}$$

and the characteristic spectrum (Kα or Lα) by the number n_{jk} of characteristic photons

$$n_{jk} = const \cdot \frac{1}{S_j} \cdot R \cdot \omega_j \cdot p_{jk} \cdot f(\chi_{jk}) \cdot A_{jk} \cdot i \cdot \Omega \cdot \Delta t$$

ACKNOWLEDGEMENT: The research work has been supported by "Fonds zu Förderung der wissenschaftlichen Forschung in Österreich, Projekt P 7012-PHY":

REFERENCES

1   H.A.Kramers, Phil.Mag. 46, 836 (1923)
2   D.G.W.Smith, C.M.Gold and D.A.Tomlinson, X-Ray Spectrometry 4, 149 (1975)
3   H.Ebel, M.F.Ebel, J.Wernisch, Ch.Pöhn and H.Wiederschwinger, X-Ray Spectrometry 18, 89 (1989)
4   H.Kulenkampff, Ann.Phys. 69, 548 (1923)
5   W.H.McMaster, N.K.del Grande, J.H.Mallett and J.H.Hubbell, Compilation of X-Ray Cross-Sections, UCRL-50174, Sect.II, Rev.1. Lawrence Radiation Laboratory, University of California, Livermore, CA. (1969)
6   G.Love and V.D.Scott, J.Phys.D 13, 995 (1980)
7   G.Love, M.G.Cox and V.D.Scott, J.Phys.D 11, 7 (1978)
8   G.Love and V.D.Scott, Scanning 4, 111 (1981)
9   F.Bloch, Z.Phys. 81, 363 (1933)
10 H.J.Hunger and L.Küchler, Phys.Status Solidi A56, K45 (1979)
11 N.F.Mott and H.S.W.Massey, The Theory of Atomic Collisions, Oxford University Press, London (1949)
12 H.J.August, R.Razka and J.Wernisch, Scanning 10, 107 (1988)
13 Ch.Pöhn, J.Wernisch and W.Hanke, X-Ray Spectrometry 14, 120 (1985)
14 G.G.Johnson Jr. and E.W.White, X-Ray Emission and keV Tables for Nondiffractive Analysis, ASTM Data series DS 46, Philadelphia (1970)
15 P.A.Pella, L.Y.Feng and J.A.Small, X-Ray Spectrometry 14, 125 (1985)

NCSXRF: A GENERAL GEOMETRY MONTE CARLO SIMULATION CODE

FOR EDXRF ANALYSIS

T. He, R. P. Gardner, and K. Verghese

Center for Engineering Applications of Radioisotopes
Box 7909, North Carolina State University
Raleigh, North Carolina 27695-7909

INTRODUCTION

EDXRF analysis is conveniently split into two parts: (1) the determination of X-ray intensities and (2) the determination of elemental amounts from X-ray intensities. For the first, most EDXRF analysis has been done by some method of integrating the essentially Gaussian distribution of observed full energy pulse heights. This might be done, for example, by least-square fitting of Gaussian distributions superimposed on a straight line or a quadratic background. Recently more elaborate shapes of the energy peaks also have been considered (Kennedy, 1990). After the X-ray intensities have been determined, interelement effects between the analyte element and other elements must be corrected for in order to obtain the elemental amounts from X-ray intensities. This correction can be done either by an empirical correction procedure as in the influence coefficient method which requires measurements on a number of standard samples to determine the required coefficients, or by theoretical calculation as in the fundamental parameters method which does not require standard samples.

An alternate technique that is used to determine X-ray intensities is the library least-squares method. This method is based on assuming that the X-ray intensities from elemental library spectra can be summed linearly at all pulse height intervals (or channels) to give the spectral intensity of a sample that contains a mixture of elements. This approach has the advantage that all of the spectrum information is used.

However, there are several restrictions in the application of the library least-square method for EDXRF analysis. First of all, the basic principle of the library least-squares approach assumes that a linear summation of library spectra will yield the unknown mixture sample spectrum. But, because of interelement effects, EDXRF analysis is a nonlinear problem and both the library spectrum intensities and their shapes are affected by the sample composition. The library least-squares method can be applied to the EDXRF nonlinear analysis problem only when a linear approximation is valid. This will be the case if the library spectra can be obtained for samples with elemental amounts close to those in the unknown sample of interest. However, experimental determination of library spectra is very time consuming, expensive, and often impossible. In this paper, we describe a Monte Carlo code, NCSXRF, which has been written

Advances in X-Ray Analysis, Vol. 35
Edited by C.S. Barrett *et al.*, Plenum Press, New York, 1992

to simulate the complete spectral response of EDXRF systems for samples of any known composition. Application of NCSXRF in calculating library spectra for the solution of the inverse problem (namely, analysis of a sample of unknown composition) is also discussed.

MONTE CARLO LIBRARY LEAST-SQUARES ANALYSIS APPROACH

A new approach, called the Monte Carlo library least-square (MCLLS) analysis principle, has been proposed and investigated (Verghese, Mickael, He, and Gardner, 1988). In this approach, Monte Carlo simulation is used to generate the elemental library spectra, thus eliminating the need for an elaborate and expensive experimental program to measure them. This means, most importantly, that standard samples may not be needed to do EDXRF analysis.

The MCLLS approach consists of the following steps:

1. By Monte Carlo simulation, generate the complete pulse height spectrum for a sample of assumed composition.

2. Within the Monte Carlo computer code, keep track of the individual spectral response for each element within the sample to provide library spectral responses.

3. Use the library least-squares (linear) analysis method to obtain the elemental amounts in any unknown sample or samples for which the complete spectral response has been taken.

4. If the elemental amounts calculated for the unknown sample are not close enough to those assumed for the sample used in the Monte Carlo simulation so that a linear relationship exists, then another Monte Carlo simulation must be performed for an assumed composition closer to the unknown sample or samples. Chi-square values can be used as a indicator that the correct composition has been found.

MONTE CARLO SIMULATION OF EDXRF

The most important requirement of the MCLLS approach for EDXRF is the ability to accurately simulate the EDXRF analyzer system. Monte Carlo simulation accuracy is dependent on the accuracy of the chosen physical model of X-ray transport which includes the description of the physical interaction mechanisms and the fundamental cross section data, and on the correct and efficient use of variance reduction techniques in the Monte Carlo simulation.

Several studies (Gardner and Hawthorne,1975; Hawthorne and Gardner,1975; Doster and Gardner,1982a,1982b; Yacout, Gardner and Verghese,1987) on the use of Monte Carlo simulation to predict EDXRF response have been reported. While the earlier of these studies concentrated on the simulation of characteristic X-ray intensities, the recent studies treat the simulation of the entire observed pulse-height spectrum. The rationale for this latter development is that there is additional useful information in the entire pulse-height spectrum that should be made available to the analyst. These simulation results have shown good agreement with experimental results.

In the present work, an improved EDXRF Monte Carlo simulation code has been developed. The improvements are as follows:

1. A general geometry subroutine named HERMETOR (Prettyman, Gardner and Verghese, 1990) has been added so that very complex geometries like those of secondary fluorescer analyzers can be simulated.

2. Additional variance reduction techniques such as correlated sampling (Gardner, Mickael, and Verghese, 1989) have been added to make the code more efficient.

3. The MCLLS principle has been included and implemented. This requires the categorizing of the scored X-ray weights in such a way that the required library spectrum for each individual element and the elemental X rays scattering from materials surrounding the sample as another library spectrum are obtained.

4. A more exact and extended energy range (1 to 60 keV) Si(Li) detector response function (He, Gardner, and Verghese, 1991) to convert Monte Carlo simulation results to pulse height spectra has been incorporated in the code. All the individual K X rays are now used.

5. The code has been written using standard FORTRAN 77 so that it can be run on VAX/VMS, DEC station, and CRAY Y-MP computers.

MONTE CARLO SIMULATION METHODOLOGY IN NCSXRF

Each simulation history is initiated by sampling the X-ray source photon parameters. The source photon is then tracked through the various surrounding material as well as the sample. At each interaction site the source photon is split and three interactions are forced to occur: (1) the photoelectric interaction is forced to occur for all possible elements (when the source photon energy is larger than the element absorption edge energy); (2) the source photon is forced to backscatter to the detector from all elements [Both Rayleigh and Compton scattering are forced to occur. After a proper weight adjustment, the scattered source photon is scored by the statistical estimation technique (Mickael, Gardner and Verghese, 1988)]; and (3) the source photon is forced to scatter inside the system and the scattering element and the type of scattering are sampled according to the X-ray cross section at the corresponding source photon energy. The scattered source photon energy and scattering angle are sampled according to the chosen scattering type. Then the photon is tracked to the next interaction site. At the new interaction site all three interactions discussed previously are forced to occur again. The history is terminated if the next interaction site is either outside the system boundary or inside the detector. These steps are continued until the photon is killed by either Russian Roulette or a minimum preset cutoff energy.

The primary characteristic X-rays are scored first by the statistical estimation technique and this score is categorized as due to the unscattered primary characteristic X ray of their corresponding elements after proper weight corrections. Then, each primary characteristic X ray is treated like a source photon to produce secondary characteristic X rays and each secondary characteristic X ray is treated in a similar way as the primary characteristic X ray to produce tertiary characteristic X rays, etc. The number of sequential X rays to be tracked is an option in the code. The NCSXRF code can track up to six sequential X rays.

The simulation results are categorized into five groups: (1) the total characteristic X-ray intensity of each element; (2) the primary through sixth sequential characteristic X-ray intensity of each element; (3) the source photon backscattering intensity from each element; (4) the scattered characteristic X-ray intensity of each element; and, finally, (5) the total X-ray spectrum and library spectrum of each element. To obtain the elemental spectra, each photon weight scored in the detector is first categorized according to its originating zone. The spectra are classified into two groups: one is the set of sample elemental spectra, and the other is an extraneous zone spectrum which is the sum of X-ray spectra of all elements from all zones except the sample zone. Then sample element spectra are categorized into elemental library spectra. After the intensities of X rays incident on the detector are predicted by Monte Carlo simulation, the Si(Li) detector response function is used to convolve this into the complete pulse-height spectra.

The complete composition and density correlated sampling (Gardner, Mickael, and Verghese, 1989) approach has been added to the NCSXRF code. It allows one to simulate a "reference" sample in the usual way while simultaneously simulating a number of "comparison" samples which have different compositions and densities in the various zones to that of the reference sample. The relative standard deviations of the differences of the correlated samples from the reference sample are typically two orders of magnitude less than the normal relative standard deviations while computation time is only increased very slightly. This feature allows accurate simulation of small differences in sample composition and density in very short computation times. This is very useful, for instance, in determining sample compositions by a modification of the MCLLS approach when the sample thickness (and/or density) is unknown (He, Gardner, and Verghese, 1991) and also in generating elemental library spectra for a whole suite of compositions that are close to that of the reference sample for use in the normal MCLLS iterations. This feature of the NCSXRF code has not been experimentally verified as yet.

The Rayleigh scattering cumulative distribution function is obtained by multiplying the atomic form factor by the Thomson cross section. The Compton scattering cumulative distribution function is given by multiplying the Klein-Nishina scattering cross section by the incoherent scattering function to correct for the shell binding effect of the electrons. Both cumulative functions are numerically integrated from scattering angle $\theta = 0$ to π by Gaussian quadrature in one degree increments over the energy range from 1 to 150 keV. Values obtained from integration are then fit to two-variable cubic spline functions (Yacout, Verghese, and Gardner, 1986) to reduce the code memory requirements and increase the data search speed. Considering the electron motion and the associated Doppler effect, the Compton scattering photon energy E' is calculated by:

$$(1/E') = 1/E_c + (q/m_0 c)\sqrt{1/E_c^2 + 1/E_i^2 - 2\mu/(E_c E')} \qquad (1)$$

where μ is $\cos\theta$, θ is scattering angle, m_0 is the electron rest mass, c is the speed of light, q is a parameter that is related to the Compton profile data of the atom, E_i is the incident photon energy, and E_c is the scattered photon energy calculated by the Compton formula:

$$E_c(\theta) = E_i/[1 + E_i(1 - \mu)/m_0 c^2] \qquad (2)$$

All of the Compton profile data are from theoretical calculations (Biggs, Mendelsohn and Mann, 1975) assuming free atoms and using the impulse approximation and Hartree-Fock wavefunction. Atomic form factor and incoherent scattering function data are taken from Hubbell, Veigele, Briggs, Brown, Cromer, and Howerton, (1975). X-ray photoelectric, Rayleigh, and Compton scattering cross section data are taken from Veigele (1973). Atomic form factor, incoherent scattering function, and Compton profile data are fitted to cubic spline functions (Yacout, Gardner, and Verghese, 1986).

RESULTS

1. Experimental Verification of NCSXRF

The NCSXRF simulation results have been benchmarked against experimental data collected on an annular ring radioisotope source excitation system. The system as shown in a previous paper (Gardner, and Hawthorne, 1975) is typical of radioisotope source excited energy dispersive X-ray analyzers. The excitation source is either ^{109}Cd or ^{55}Fe depending on the elements to be measured. One pure titanium sample, two known alloy samples from NIST, and one aluminium alloy sample of known composition from ALCOA are used as test samples to validate the Monte Carlo results.

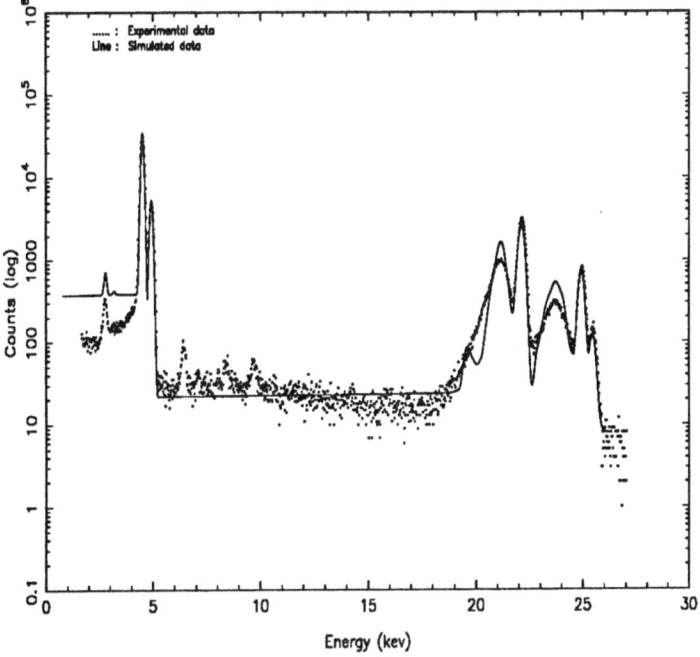

Fig. 1 Titanium X ray Spectra

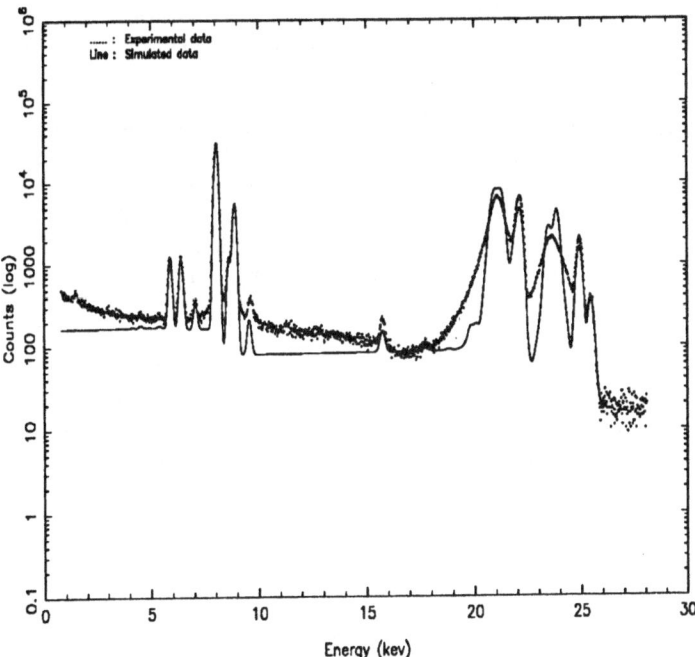

Fig. 2 2024 Aluminium Alloy X ray spectra

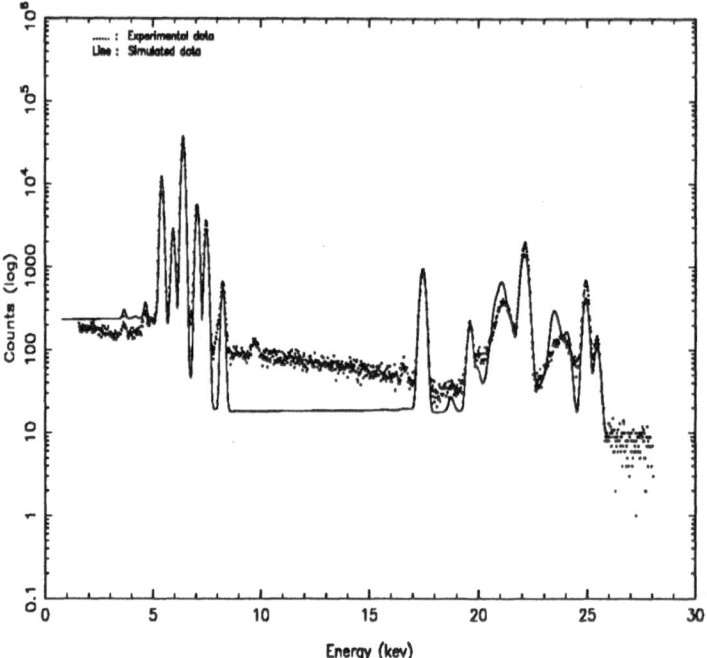

Fig. 3 304 Stainless Steel X Ray Spectra

To verify the NCSXRF code, the X-ray spectrum of one of the standard samples mentioned above is measured first, then the simulated X-ray spectrum is obtained by running the Monte Carlo simulation code based on the known sample elemental composition. The measured X-ray pulse height spectrum (dots) and simulated X-ray pulse height spectrum (line) of the standard samples are shown in Figs. 1 to 4.

In all four figures, the peaks below 10 keV energy are photoelectric peaks due to the characteristic X rays of elements in the samples, the peaks above 20 keV are ^{109}Cd excitation source backscatter peaks. In Figs. 3 and 4 a high background between the characteristic X-ray peaks and the source backscattered peaks due to pulse pile-up can be observed in the measured spectrum, which is not observed (as it should not be) in the simulation spectrum. Pulse pile-up could be corrected for by the model derived and demonstrated by Gardner and Wielopolski (1977a and 1977b).

For all of the standard samples the simulated spectra show very good agreement with measured spectra in the fluorescent X-ray region. But the source photon backscatter region shows some disagreement. The simulated Compton backscatter peaks have a higher amplitude and smaller peak width than in the measured spectra. One of the possible reasons for this disagreement is the atomic Compton profile data. The data which we used originated from theoretical calculations. There are several approximations in such a calculation, one of which is the assumption of a free atom which evidently is not true for the atoms in alloy samples. There are studies (Williams, 1977) which show that the experimental Compton profile value for a pure nickel sample is significantly different from the value which is obtained by theoretical calculation assuming free nickel atoms. Comparison showed that the theoretically calculated Compton profile peak amplitude is about twice as high as the measured value and that the theoretical peak is narrower than the measured peak. These observations coincide with NCSXRF results.

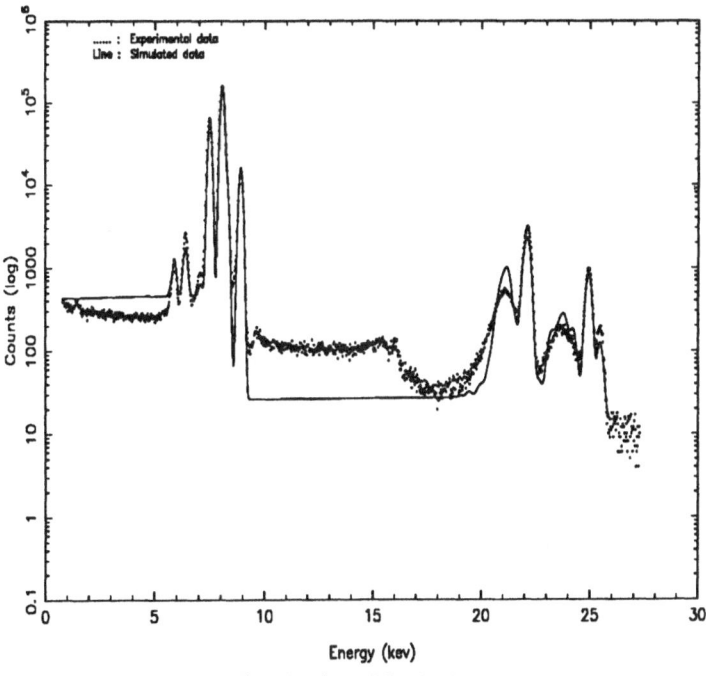

Fig. 4 Cu–Ni Alloy X Ray Spectra

The NCSXRF code can be used to investigate experimental systems to help identify unknown amounts of elements from materials around the sample. The authors had been using an [55]Fe radioisotope source to investigate aluminum alloy sample thickness and composition. The argon in the air not only showed two characteristic X-ray peaks(K_α and K_β) in the XRF spectrum but also enhanced the characteristic X-ray intensities from some of the lighter elements in the samples (Mg, Al, and Si) while absorbing the X rays from higher atomic number elements in the samples such as Ti. The NCSXRF code was used to investigate the argon effect and correctly predicted the XRF spectrum of the aluminum alloy samples.

2. Monte Carlo Library Least-Squares (MCLLS) Analysis

Two zinc elemental library spectra from Cu-Zn alloys (dots represent the spectrum from a 20 % Cu - 80 % Zn alloy and the line represents the one from the 80 % Cu - 20 % Zn alloy.) obtained by Monte Carlo simulation are shown in Fig. 5.

Both the library spectrum shapes (K_α peak to K_β peak ratio) and the intensities are different for these two spectra. Table 1 shows the library least-squares results for two searches. One used Cu and Zn library spectra from the 20 % Cu alloy to fit the simulated 20% Cu alloy X-ray spectrum to obtain Cu and Zn amounts, while the other used the Cu and Zn library spectra from the 80 % Cu alloy to obtain Cu and Zn from the 20 % Cu alloy X-ray spectrum.

It is obvious that the fitted result using library spectra from the 80 % Cu alloy is very inaccurate. The library least-squares method requires that the assumed elemental compositions from which element library spectra are calculated are from samples close in composition to the true element compositions in the samples. It is time consuming and expensive to obtain these library spectra from experimental measurements. The NCSXRF code can be used for obtaining such

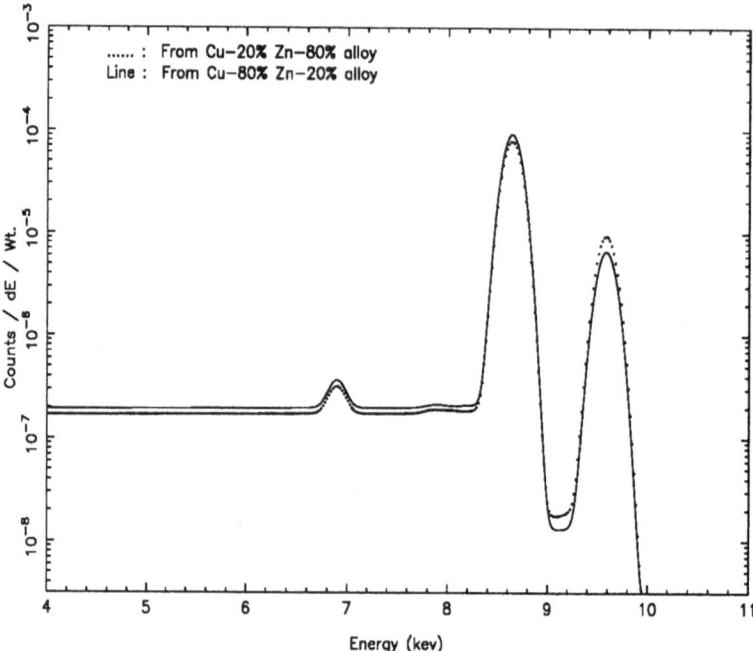

Fig. 5 Zn Element Library Spectra

elemental library spectra directly. The authors have used the MCLLS method to analyze many other sample compositions (Verghese, Mickael, He, and Gardner, 1988) with excellent results. More than one iteration through the Monte Carlo loop is rarely necessary if the initial guessed composition is fairly close to the actual. Further MCLLS development using experimental spectra from standard test samples is underway.

3. Secondary Fluorescer X-Ray Machines

The NCSXRF code has been used for simulating X-ray tube-secondary fluorescer systems to investigate the machine design parameters. The code is very flexible in user specification of the system geometry and machine parameters. Comparison with experimental data on secondary fluorescer systems is not yet available.

DISCUSSION AND CONCLUSIONS

A general geometry Monte Carlo simulation code named NCSXRF has been developed. The code has been tested by comparing the simulated XRF spectrum with the measured XRF spectrum for several standard samples. Good agreement was found in the characteristic X-ray

Table 1. MCLLS Results for Cu-Zn Alloys

| Element | Actual | Compositions (%) | |
		Fitted with 20 % Cu Alloy Library Spectrum	Fitted with 80 % Cu Alloy Library Spectrum
Zn	80.00	80.25	70.12
Cu	20.00	20.03	18.85

part of the spectrum. The disagreement in the source photon Compton backscatter peak region most likely is the result of using inaccurate Compton profile data. A set of more accurate Compton profile data must be obtained and used to improve the quality of the simulated spectra.

The Monte Carlo Library Least-Squares (MCLLS) approach has been implemented in the NCSXRF code. It appears that it is quite accurate and rarely requires more that one Monte Carlo iteration for convergence. The complete composition and density correlation feature in the NCSXRF code should prove very useful in generating a range of library spectra at different compositions for use with the MCLLS approach.

ACKNOWLEDGEMENTS

The authors gratefully acknowledge the financial support provided by the Amoco Production Company, Mobil Research and Development Corporation, ARCO Oil & Gas Company, EXXON Production Research Company, Atlas Wireline Services, and Conoco, Inc. under an Associates Program supporting Nuclear Oil Well Logging research. Computing time on a Cray Y-MP was provided for part of this work by the N.C. Supercomputing Center.

REFERENCES

- Biggs, F., L. B. Mendelsohn, and J. B. Mann, 1975, "Hartree-Fock Compton Profiles for the Elements", Atomic Data and Nuclear Data Tables, 16, 201.

- Doster, J. M. and Gardner, R. P., 1982a, "The Complete Spectral Response for EDXRF Systems - Calculation by Monte Carlo and Analysis Applications. 1. Homogeneous Samples", X-Ray Spectrometry, A11(4), 173-180.

- Ibid., 1982b, "2. Heterogeneous Samples", 181-186.

- Gardner, R. P. and Hawthorne, A. R., 1975, "Monte Carlo Simulation of the X-Ray Fluorescence Excited by Discrete Energy Photons in Homogeneous Samples Including Tertiary Inter-Element Effects", X-Ray Spectrometry, 4, 138-148.

- Gardner, R.P. and L. Wielopolski, 1977a, "A Generalized Method for Correcting Pulse-Height Spectra for the Peak Pile-Up Effect Due to Double Sum Pulses. Part I. Predicting Spectral Distortion for Arbitrary Pulse Shapes", Nuclear Instruments and Methods, 140, 289-296.

- Ibid., 1977b, "Part II. The Inverse Calculation for Obtaining True from Observed Spectra", 297-303.

- Gardner, R.P., M.W. Mickael, and K. Verghese, 1989, "Complete Composition and Density Correlated Sampling in the Specific Purpose Monte Carlo Codes McPNL and McDNL for Simulating Pulsed Neutron and Neutron Porosity Logging Tools", Nuclear Geophysics, Vol. 3, No. 3, pp. 157-165.

- Hawthorne, A. R. and R.P. Gardner, 1975, "Monte Carlo Simulation of X-Ray Fluorescence from Homogeneous Multielement Samples Excited by Continuous and Discrete Energy Photons from X-Ray Tubes", Analytical Chemistry, 47(13), 2220-2225.

- He, T., R.P. Gardner, and K. Verghese, 1990, "An Improved Si(Li) Detector Response Function", Nuclear Instruments and Methods in Physics Research A299, 354-366.

- He, T., C. L. Dobbs, K. Verghese and R. P. Gardner, 1991, "Investigation of Energy-Dispersive X-Ray Fluorescence Analysis for On-Line Aluminum Thickness/Composition Measurement", Transactions of the American Nuclear Society, Vol. 63, 147-148.

- Hubbell, J.H., W.J. Veigele, E.A. Briggs, R.T. Brown, D.T. Cromer, and R.J. Howerton, 1975, "Atomic Form Factors, Incoherent Scattering Functions, and Photon Scattering Cross Sections", J. Phys. Chem. Ref. Data, 4, 471.

- Kennedy, G., 1990, "Comparison of Photopeak Integration Methods", Nuclear Instruments and Methods in Physics Research A299, 342-349.

- Mickael, M., R.P. Gardner, and K. Verghese, 1988, "An Improved Method for Calculating the Expected Value of Particle Scattering to Finite Detectors in Monte Carlo Simulation", Nuclear Science and Engineering, 99, 251-266.

- Prettyman, T. H., R. P. Gardner, and K. Verghese, 1990, "MCPT: A Monte Carlo Code for Simulation of Photon Transport in Tomographic Scanners", Nuclear Instruments and Methods in Physics Research A299, 516-523.

- Scofield, J.H., 1975, ATOMIC INNER-SHELL PROCESSES, Academic Press Inc., 265-288.

- Veigele, W.J., 1973, Atomic Data Tables, 5, 51-111.

- Verghese, K., M. Mickael, T. He, and R.P. Gardner, "A New Analysis Principle for EDXRF: the Monte Carlo - Library Least-Squares Principle", Advances in X-Ray Analysis, Vol. 31, pp. 461-469.

- Yacout, A. M., R.P. Gardner, and K. Verghese, 1984, "Cubic Spline Techniques for Fitting X-Ray Cross Sections", Nuclear Instruments and Methods in Physics Research, 220, 461-472.

- Yacout, A. M., R.P. Gardner, and K. Verghese, 1986, "Cubic Spline Representation of the Two-Variable Cumulative Distribution Functions for Coherent and Incoherent X-Ray Scattering", X-Ray Spectrometry, 15, 259-265.

- Yacout, A. M., R.P. Gardner, and K. Verghese, 1987, "Monte Carlo Simulation of the X-Ray Fluorescence Spectra from Multielement Homogeneous and Heterogeneous Samples", Advances in X-Ray Analysis, 30, 121-132.

- Williams, B., 1977, COMPTON SCATTERING, McGraw-Hill, New York

UNIFICATION OF "STANDARD BACKGROUND" TECHNIQUE USING
SCATTERED RADIATION IN X-RAY FLUORESCENCE ANALYSIS (XRF)

V. I. Smolniakov

Neutron Research Department, Leningrad Nuclear
Physics Institute (LNPI), Academy of Sciences USSR
188350 Gatchina, Leningrad region, U.S.S.R.

ABSTRACT

Some x-ray fluorescence - concentration relationships in the framework of XRF were researched. Fundamental calculation approaches for primary fluorescence and incoherent scattering were realized for evaluation of matrix influence. A new binary approach was produced for the cases considered, and its unification was related to the empirical and regression types of the "standard background" technique, widely used in the analytical practice of XRF. It is confirmed that application of the calculations by fundamental parameters (FP) in combination with the empirical approach allows the reduction of the set of standards (to as few as one) in the analysis procedure with wide variations in matrices and concentrations, without loss of accuracy.

INTRODUCTION

In XRF there are several general procedures which have significant differences: internal standard, comparison with external standard, regression coefficients, and calculations by FP. It is possible to say that new developments in XRF are directed toward reducing the specimen preparation problems and the set of standards even up to standardless calculations by FP. On the other hand, after the characteristic x-ray fluorescent effect was discovered, and after applying it in analytical practice up to the current time all methods in XRF have one basic characteristic - the value of the x-ray radiation intensity (P) depends upon the quantity (C) of the fluorescent element (i). This fact can be written in the following manner:

$$P_i = f(K_i) \cdot C_i , \qquad (1)$$

where K is proportionality coefficient. In this case to produce an infinite library of standards, any type of XRF can be used to determine the relative concentration of the element of interest in an unknown sample by comparing with an adequate standard from this library, i.e., we can use the following kind of algorithm (symbol "*" relates to comparison standard):

$$C_i = [f(K_i)^* / f(K_i)] \cdot [P_i / P_i^*] \cdot C_i^* , \qquad \text{if } f(K_i)^* = f(K_i) . \qquad (2)$$

It is possible to confirm that efforts of all investigators working in quantitative XRF are concentrated in the unification of calculation algorithms, which would work with as many as possible types of matrices, wide range of concentration of elements of interest and would minimize physical and technical restrictions for the analysis procedure. It is possible to express this as:

$$f(K_k)^* \cdot (\eta_k)^* \cdot C_k^* = f(K_k)_n \cdot (\eta_k)_n \cdot (C_k)_n, \qquad n \rightarrow \infty, \ k=i,j.\ldots \qquad (3)$$

where η is measured analytical parameter.

In this paper the author attempts to propose this kind of a solution.

PROBLEMS

The above-mentioned analytical procedures are empirical except for the last one, i.e., they are elaborated, in general, experimentally under instrumental conditions and the range of concrete problems and technologies. Empirical approaches are applied successfully for relatively simple, one type matrices and small inter-element influence, i.e., for limited ranges of element variations in unknowns compared with standards. Also, in this case, it is necessary to have and to support from time to time a remeasured library of standards and artificial specimens. Often it leads to difficulties in the analytical procedure that can be impossible in some cases.

Theoretical calculations for primary, secondary and tertiary fluorescence under conditions of strong influence of matrix and inter-element effects were generalized by Sherman[1] and put into the base of FP method by Criss and Birks.[2] It allows successful complex analysis of multi-component compound unknowns and the reduction of the set of standards right up to standardless semi-quantitative analysis.[3,4] However there are some essential difficulties in that process: the exit of iterative procedure on 100% determination of unknowns when there can be a situation with unmeasured light elements, i.e., for elements below Na; the restriction of the software to 20-25 analyzed elements, when the sample consists of 30-50 elements of interest, i.e., in rocks and ores; considerable increase in computer time for multi-component analysis, even for high-speed computers, which is not acceptable for routine analytical practice. All of these facts lead to reducing the accuracy in quantitative analysis and it may prevent full success in solving the problems of XRF. Also there are some other difficulties in FP method by Criss: the insufficient accuracy in FP data and knowledge of the real instrumental conditions, failure to consider the enhancement effect by scattering, etc.

Some investigators have combined the advantages of empirical and calculation technique, have elaborated and successfully applied semi-empirical analysis method in XRF.[5]

In 1953 Andermann and Kemp[6] proposed to take into consideration the absorbing properties of the matrix by means of a comparison of the analytic line intensities for elements of interest with nearby background. It is the method of "standard background." Later the investigators used the dependence between background and intensity of scattering; it was proposed to use one (generally incoherent scattering) as the standard for comparison.[7,8] The theoretical basis for the "standard background" method is considered next. Let us consider the Sherman approach for primary fluorescence (infinitely bulk homogeneous specimen, plane parallel primary and secondary beams):

$$P_i = q \cdot E_i \cdot C_i \cdot I_0 \cdot [\tau_{i,0}/(\mu_{s,0} + A \cdot \mu_{s,i})], \qquad (4)$$

where $A=\sin\phi/\sin\psi$ is the geometrical factor; ϕ, ψ are angles for primary and fluorescent radiations respectively; $q=A\cdot(d\Omega/4\pi)$ is the collimation factor; I is the intensity of primary radiation (0); τ, μ are mass absorption coefficients; and E is the fluorescent constant. The analogous approach for scattering is:

$$P_{sc}=q\cdot I_0\cdot[(d_m\sigma_0/d\Omega)/(\mu_{s,0}+A\cdot\mu_{s,sc})],\qquad(5)$$

where $d\sigma/d\Omega$ is the mass differential cross-section of primary scattering. As it follows from Reference 9, in the case of small concentration of analyzed elements in rock matrix, when absorption matrix properties depend on absorption buffer properties and inter-element influence is absent, the ratio P_i/P_{sc} can be as

$$P_i/P_{sc}=const\cdot C_i/(d_m\sigma_0/d\Omega).\qquad(6)$$

When the rock matrix is changed, variations in the cross-sections of incoherent component of scattering are considerably less than those of the coherent component. That is why analysts apply the "standard background" method with incoherent scattering in XRF successfully. Also so-called standardization for scattering is very similar to the method of "internal standard" that is available for auto-control of real instrumental conditions.

In Reference 10 the application of the method "standard background" with incoherent scattering is extended for the analysis of the samples with a high concentration of analyzed elements and with conditions of high absorption by high-concentration elements. It is considered by the next algorithm:

$$C_i=[P_i/(P_{inc}-F)]/[K_i+\sum(A_{ik}\cdot P_{ik})],\quad k=i,j,\dots n,.\qquad(7)$$

where A's are coefficients accounting for absorption by high-concentration analyzed and matrix elements. Also there is the example of successful application of algorithm (7) for analyzing concentrations for Cu in Norilsk's Cu-Ni oils and for concentrates (concentration ranges for Cu are from 0.213% to 70.60% and for Ni from 0.53% to 77.34%). In Reference 11 the algorithm (7) is transformed:

$$C_i=[P_i/(P_{inc}-F)]\cdot[(K_i)^{n-1}/(\prod(K_i+A_{ik}\cdot P_{ik}))],\quad k=i,j,\dots n,\qquad(8)$$

From the common analysis of algorithms (7) and (8) we can conclude that they are both regressive and do not have fundamental character, and that the applications of the constant coefficients lead to matrix restrictions in the range of acceptable errors for the analyzed element concentrations. .

The problem of the current paper is the investigation of the conformity to natural laws of the expression $C=f(P/P-inc)$, and getting the fundamental unification for calculation algorithms in accordance with method "standard background" with scattering.

RESULTS AND DISCUSSIONS

A binary physical model was considered: analyzed element - buffer. The calculations were fulfilled on the basis of (4) and (5). As the analyzed element we considered Cu in different matrices when exciting energy was AgKα with A=1. Values of FP data were from References 12-15. Three extreme cases are given: Cu in C (light matrix), Cu in Pb (heavy matrix, in energy range, the Pb L-absorption edge lies between incoherent scattering and Cu

K-edge absorption), Cu in Fe (strong absorber of Cu characteristic radiation). You can observe typical conformities to natural laws for variations in different analytical parameters under a change in concentrations of Cu in various matrices and the logic of unification of "standard background" technique using scattered radiation.

Let us write the common expression for the concentration of the analyzed element

$$C_i = f(K_{s,i}) \cdot (P_i/P_{inc})$$

(9)

where K is the proportionality coefficient for the model "analyzed element-buffer." Let us investigate the function f(K) (later on we shall denote primary fluorescence and incoherent scattering relatively to those for the pure element, i.e., R=P/Ppur.el.). In order to do this, reducing C in evident form, let us rewrite expression (9) as:

$$(R_i/C_i) = \eta_{R_i, C_i} = 1/f(K_{s,i})$$

(10)

Figure 1(a) shows the functions likely (10) for our three cases. The facts that functions develop along the straight line when concentrations are changed from 0.0 to 1.0, and that deviations during "development" from the straight line are submitted in accordance with some functional character - are common for all of them.

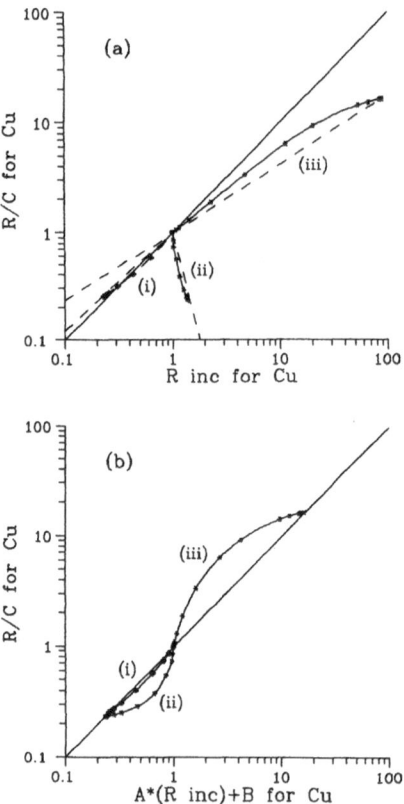

Figure 1.   Illustration of relationships (10) and (11), for (a) and (b) respectively, for three cases: (i) Cu-Pb, (ii) Cu-Fe, (iii) Cu-C. All data values are given in relative units.

Table 1. Numerical values for coefficients A, B, C, D and H for our three
considering cases: Cu in C, Cu in Pb and Cu in Fe (in related
unites)

N	Type model	Coefficients				
		A	B	C	D	H
i	Cu in Pb	0.9684	0.03157	-2.93856	3.69376	0.12041
ii	Cu in Fe	-1.9601	2.9601	-3.25767	16.3580	-24.7034
iii	Cu in C	0.1666	0.8334	0.93955	0.0016777	0.0083926

To reduce all straight lines to one line which is passed from "true zero" up to final value for $C = 1.0$ let us transform R-inc into $(A \cdot R\text{-inc} + B)$ for the i-th element and get the next unified function:

$$\left.\begin{array}{l} \eta_{P_i,C_i} = [1/f(K_{s,i})] \cdot (A \cdot R^i_{inc} + B) \\ f(K_{s,i}) = (D \cdot R^i_{inc} + H)/(1 - C \cdot R_i) \end{array}\right\} \tag{11}$$

These cases are illustrated in figure 1(b). Constants A, B, C, D and H are fundamental for all types of the binary model and geometric conditions for equipment. Values of these constants are listed in table 1.

We point out that all calculations were fulfilled for primary fluorescence and therefore in real analysis it is necessary to consider secondary and tertiary fluorescence following the proposed "standard background" method. It is preferred to apply the technique by Tertian,[5] but it is possible to use any. The analysis procedure is as follows:

(a)   Corresponding to approach (11) theoretical calculations are executed for all available compositions of samples (unknowns).
(b)   By means of simple binary specimens the equipment is calibrated in accordance with theoretical calculations.
(c)   Calculations of concentrations for all analyzed elements are fulfilled beginning from fluorescence, which is nearest to exciting radiation, and then step by step for all others.   For all elements which have enhancement effects, the contribution of secondary and tertiary fluorescence to measured intensity is calculated and the intensity of primarily fluorescence is picked out.
(d)   After the first step in accordance with results obtained, binary model and known and unknown parts of matrix are calculated.
(e)   The process (a)-(d) is repeated iteratively until the set of values differ by less than some arbitrary amount.

SUMMARY

The proposing unification of "standard background" technique with using scattered radiation has the following advantages in analytical practice of XRF compared with others:

(a)   On the basis of real physical processes unification of the "standard background" technique allows the solution with given accuracy, of the straight problem of calculating the intensity for any concentrations of analyzed elements and the reverse problem of the determination of the proportions of analyzed elements for the concentration range from 0.0 up to 1.0 and for any type of matrices.

(b)   Semi-empirical approach to unification "standard background" technique, i.e., combination of the use of a calculation method with FP and comparison with preliminary measured simple binary specimens of known composition, permits the reduction of many disadvantages in well-known and widely applied techniques in XRF.

(c)   Realization of unified "standard background" technique may be fulfilled as a version of the application of cleanly calculated standardless analysis and as a version with the use of only one standard, which consists of all analyzed elements in the unknowns.

(d)   Unlike well-known FP method by Criss, the proposed one allows successful quantitative analysis of unknowns without depending on 100% composition - this fact leads to increasing its importance and regions of application.

(e)   Application of scattered radiation, as a conditional internal standard for comparison, leads to increasing the stability of the analytical procedure with automatic control for the variations in changing factors (i.e., automatic control for real instrumental conditions).

REFERENCES

1. J. Sherman, Spectrochim. Acta 7:283 (1955).
2. J.W. Criss and L.S. Birks, Anal. Chem. 40:1080 (1968).
3. Software EDXRF ver. 1.32 by Tracor X-ray, Inc., USA (1990).
4. V.I. Smolniakov, paper presented for publ. in Journ. Plant Laboratory, Moscow.
5. R. Tertian, X-Ray Spectrom. 17:1989 (1988).
6. G. Anderman, J.V. Kemp, Anal. Chem. 30:1306 (1958).
7. N.F. Losev, "Kolichestvenny rentgenospectralny fluorescentny analiz", Atomizdat, Moscow (1973).
8. P.I. Plotnikov and G.A. Pshenichny, "Fluorescentny rentgeno-spectralny analiz", Nauka, Moscow (1973)
9. A.V. Bahtiarov, "Rentgenospectralny fluorescentny analiz v geologii i geohimii", Nedra, Leningrad (1985).
10. A.V. Bahtiarov, Equipment and Methods of X-Ray Anal. 21;3 (1978).
11. N.A. Verman and D.N. Stroganov, paper presented for publ. in Journ. Ore Concentrations, Leningrad.
12. D.K.G. de Boer, Spectrochim. Acta 448:1171 (1989).
13. E. Browne and R.B.Firestone, "Tables of Radioactive Isotopes", Wiley-Interscience, New York (1986).
14. E.B. Saloman and J.H. Hubbell, "X-ray Attenuation Coefficients", NBSIR 86-3431, Gaithesburg (1986).
15. M.A. Blohin and I.G.Sheveytzer, "Rentgenospectralny spravochnik", Nauka, Moscow (1982).

DECOMPOSITION SPECTROMETRIC DATA OF ENERGY

DISPERSIVE X-RAY FLUORESCENCE ANALYSIS (EDXRF)

V. I. Smolniakov and I. A. Koltun

Neutron Research Department, Leningrad Nuclear
Physics Institute (LNPI), Academy of Sciences USSR
188350 Gatchina, Leningrad region, U.S.S.R.

ABSTRACT

It is well-known that in EDXRF, using high-resolution semiconductor detectors, evaluation of x-ray fluorescence radiation line intensities from multiplex spectrometric information represents definite difficulties, especially in automation of measurement-calculation procedures.

A common spectrum decomposition problem is to get the following parameters: a number of spectral lines and their centroids, intensities from measured experimental data and their errors.

We have developed special software for solving this problem using personal computers and high-level programming language C. It uses profiles of real-form lines of pure chemical elements produced by semiconductor detector spectrometers and these techniques: digital filters with parameters for suppression of background, multiplex structure analysis, and stable linear least-squares fit to get peak intensities. Also it established special criteria for reliability of the results.

We compared our investigation with software "EDXRF" (ver.1.32) and spectrum decomposition with Gaussian peaks.

INTRODUCTION

All investigators and practitioners, who work in energy-dispersive x-ray spectrometry, attach great importance to the quality of spectrum processing. The main successes in deconvolution of x-ray spectra obtained with solid-state detectors belong to a very simple but powerful technique "least-squares fit with digital filtering."[1]

The authors have been processing spectrometric data in accordance with such a technique by means of software "EDXRF"[2] successfully. But there are some problems in its exploitation. For example, sometimes we get absurd results: i.e., negative peak areas when the peak of corresponding element line is obviously present. Also criteria for the quality of spectrum decomposition is absent and it is problematical to trust the results of processing spectrometric data by "EDXRF." That is why the authors decided to elaborate their own software for spectrum deconvolution based on the technique of least-squares fit with digital filtering in order to improve quality and interpretation of the results.

PURPOSES AND PROBLEMS

When authors elaborated the software the purposes were:

(a)   To investigate the main properties of the technique of least-squares fit with digital filtering, such as the quality of suppression of the continuum, deconvolution of the overlapped characteristic peaks under different spectrometric conditions (level of linear background, peaks on abrupt nonlinear continuum, essentially overlapped peaks with different ratio of composite singlets).

(b)   To elaborate objective criteria for the reliability and correctness of the results of spectrum deconvolution.

(c)   To get criteria for the reliability of results based on the physical peculiarities of composition of the real spectrum in different matrices.

As a result, we wished to create flexible software with step-by-step operational control for application with different purposes - both scientific and analytic practice.

The problem of spectrum decomposition in EDXRF is next. Let us suppose that the analyzed spectrum is the linear sum of measured reference spectra in correct proportions, a slowly-varying smooth continuum component and statistical noise. We can express this assumption in the following manner:

$$Y_j = \sum_{i=1}^{m} A_i \cdot P_{ij} + b_j + e_j, \quad j = \overline{1, N} \tag{1}$$

where

Y_j – analyzed spectrum in channel j,

P_{ij} – i-th reference spectrum in channel j,

b_j – smooth continuum component in channel j,

e_j – statistical fluctuation in channel j,

A_i – part (coefficient) of i-th reference spectrum.

The problem is to estimate for the analyzed spectrum the set of proportionality coefficients and their errors $\{A_i, \Delta A_i, i=1, N\}$ that provide the best fit to the analyzed spectrum in accordance with the given reference spectra. The reference spectra may be a set of measured pure element spectra or any set of curves which provide an accurate model of shape and amplitude of the particular x-ray lines of interest.

USING THE METHOD OF LEAST-SQUARES FIT WITH DIGITAL FILTERING FOR SOLVING THE PROBLEM

The method of least-squares fit with digital filtering, and its peculiarities for processing spectrometric data were described in several papers.[3,4,5] Our efforts are connected with investigation of the shape and the parameters of a digital filter (as shown diagrammatically in Figure 1) relative to the real-form spectra of secondary characteristic fluorescence radiation of pure elements and their mixtures in accordance with the above-mentioned purpose (a).

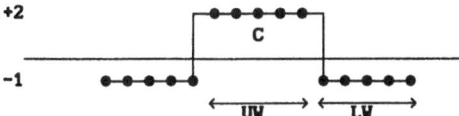

Figure 1. Example of simple digital
filter used in our software
(UW,LW,C - its parameters)

Let us fix attention on the main steps of algorithm. Firstly both analyzed spectrum and reference spectra that are considered in deconvolution are processed by means of a digital filter (symmetrical, with zero area). We can apply the digital filter to both sides of equation (1):

$$\text{FILTR}(Y_j) = \sum_{i=1}^{m} A_i \cdot \text{FILTR}(P_{ij}) + \text{FILTR}(b_j) + \text{FILTR}(e_j), \quad j=\overline{1,N}$$

Taking into consideration the property of a digital filter to suppress the smooth background component, we get a system of linear algebraic equations with correlated data. Let us write it in matrix form:

$$Y^F = P^F \cdot A + e^F \tag{2}$$

The covariance matrix of the left side of equation (2) is next:

$$C = \{C_{ij}\} = \text{COV}(Y_i^F, Y_j^F) = \sum_{k=1}^{2 \cdot LW+UW-(j-1)} H_k \cdot F_{k+j-1} \cdot Y_{i+k-LW-UW/2}, \quad i \le j \tag{3}$$

where $\{H_i, \ i=1,2 \cdot \overline{LW+UW}\}$ - filter coefficients

The solution of the problem (2) with the least-squares method is the following:

$$\hat{A} = \text{Arg min } (Y^F - P^F A)^T C^{-1} (Y^F - P^F A) \tag{4}$$

Using decomposition by Holessky for matrix $C = (D^T)D$, where D is the upper triangular matrix, problem (4) is reduced to a standard model least-squares:

$$\hat{A} = \text{Arg min } (\tilde{Y}^F - \tilde{P}^F A)^T (\tilde{Y}^F - \tilde{P}^F A)$$
$$\text{where } \tilde{Y}^F = (D^T)^{-1} Y^F, \quad \tilde{P}^T = (D^T)^{-1} P^F,$$

where $\tilde{Y}^F$ has noncorrelative components with the identical dispersions. Covariance matrix for the estimated solution is calculated under the condition that $\tilde{P}^F$ has complete rank as follows:

$$\hat{CA} = ((\tilde{Y}^F)^T \tilde{Y}^F)^{-1} \frac{\| \tilde{P}^F \hat{A} - \tilde{Y}^F \|}{N - m} \tag{5}$$

PECULIARITIES OF NEW SOFTWARE AND RESULTS OF PROCESSING DATA

To organize the above-mentioned processes (2)-(5), the authors elaborated special software with a pattern using algorithms taken from reference 6, including different methods of transformation of the data matrix - singular decomposition, transformations by Householder, and also by solving least-squares problem with linear restrictions-inequalities. It permits the use of one of the given methods depending on processing time and accuracy of result. We point out the possibility of taking additional a priori information about the analyzed spectrum during processing data, which leads to increasing stability of the least-squares problem, and therefore the accuracy and reliability of the results. The general idea about our software is represented in the flow-chart in Figure 2. The software is written in computer language MS C ver. 5.1.

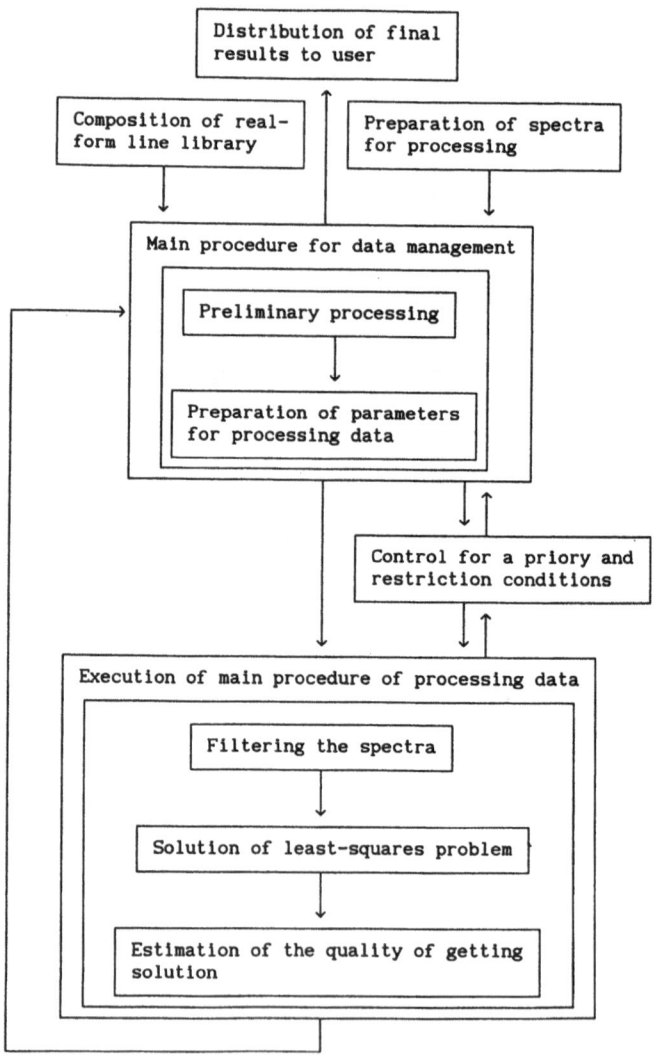

Figure 2. Flow-chart of new software for processing spectrometric data in EDXRF.

Table 1.   Comparison of test data processed by different software - new one, "EDXRF"
           and "AXIL" (all data are given in normalized parts of pure element profiles)

Models		Results of processing data		
type	test data	new software	EDXRF	AXIL
Fe	1.0/1.0	1.0000 / 1.0000	0.9991 / 0.97125	0.97610 / 1.0041
/	1.0/0.1	0.9991 / 0.10000	0.97525/0.08225	0.98610/0.1108410
Co	1.0/0.001	1.0000/0.00098713	0.9991 / 0.000925	0.9991/0.0211525
Rb	1.0/1.0	1.0000 / 1.0000	1.0000 / 1.0000	0.9741 / 1.0241
/	1.0/0.1	0.9991 / 0.09991	0.9982 / 0.08225	0.98110/0.123625
Y	1.0/0.001	0.9991/0.001001	0.9991/0.0017725	0.98210/0.24025
Ba	1.0/1.0/	1.0000 / 0.9991 /	0.99010/ 0.9982 /	
/	1.0/1.0/	1.0000 / 1.0000 /	0.99110/ 0.72525/	——
La	1.0	1.0000	0.86810	
/	1.0/1.0/	1.0000 / 1.0001 /	0.9952 / 1.0021 /	
Ce	1.0/0.1/	1.0000 / 0.10001/	0.9921 / 0.0 /	——
/ Pr	0.1	0.10002	0.44625	
/	1.0/1.0/	0.9991 / 1.0001 /	0.9951 / 1.0011 /	
Nd	1.0/0.001	0.9991/0.0012025/	0.9971 / 0.0 /	——
	0.001	0.0010319	0.01025	

As a criteria for reliability and correctness of solution it is possible not only to get covariance matrix (5) and estimation for χ^2, but to see the curve of weighted residuals in order to control trends, including statistical criteria. The last possibility is very important in processing real data because it is possible to reason about availability and location of unreferenced peaks and to distinguish such a situation from others when incorrect energy calibration takes place. The supplementary collections and sublibraries of real-form line fluorescence radiation of analyzed elements are the other criteria for reliability of results. With their help the choice of reference spectra for decomposition and values of parameters in digital filtering are controlled. All such sublibraries are composed in a special manner taking into consideration variations in peak shape models of real-form lines depending on differential absorption, type of matrices, and so on, right up to a physical model of the analysis algorithm for concentration calculations.

The results of testing the new software as compared with software "EDXRF"[2] and "AXIL"[7] can be seen in Table 1. As test data we used model spectra, which were obtained from reference spectra of pure elements by means of multiplying by different constants and summing up. We considered three widespread cases from practice of EDXRF, when there are unresolved lines with small and large intensities. In the first case they are FeK_β and CoK_α, in the second case - RbK_β and YK_α, in the third case - K_α-lines of Ba, La, Ce and K_β-lines of Nd and Pr. As was mentioned above, the same method, namely least-squares fit with digital filtering and using real-form lines, was used in "EDXRF." The method of spectrum decomposition with Gaussian peaks was used in "AXIL."

In Table 1 you can see that results by the new software are the best compared with the same for "EDXRF" and "AXIL," and correspond to the high requirements of experiments in EDXRF. The case in test 0.001 Pr, 0.001 Nd and our corresponding results - 0.0012025 Pr, 0.0010319 Nd - it is necessary to analyze separately under conditions when peak is less than background. As the result we can confirm that the above-mentioned purposes are fulfilled.

SUMMARY

1. As the results of these investigations the authors elaborated on the base of the method of least-squares fit with digital filtering mathematical algorithm and software for processing spectrometric data from experiments in EDXRF.

2. Quality and reliability of results correspond to the high requirements of EDXRF-spectrometry.

3. In the framework of the new software there are different possibilities for active choice of parameters for processing data, flexible control, objective estimation of current procedures and final results. This software can be used in different spheres - in routine analysis, scientific researches and in practice of experiments with high accuracy.

REFERENCES

1. F.H. Schamber, Proc. 8th Nat. Conf. on Electron Probe Analysis 85 (1973).
2. Software EDXRF ver.1.32 by Tracor X-ray, Inc., USA (1990).
3. F.H. Schamber, "X-Ray Fluorescence Analysis of Environmental Samples", T.G.Dzubay, Ed., Ann. Arbor Sci., p.241 (1977).
4. J.J. McCarthy and F.H. Schamber, Proc. Workshop on Energy Dispersive X-Ray Spectrometry, p. 273 Gaithersburg (1979).
5. J.C. Russ, Proc. Workshop on Energy Dispersive X-Ray Spectrometry, p.297 Gaithersburg (1979).
6. C.L. Lawson, R.J. Hanson, "Solving Least Squares Problems", Prentice-Hall, Inc., Englewood Cliffs, N.J., 1974.
7. AXIL X-ray Analysis Software ver.3.0 by Dept. Chem., University of Antwerp, Belgium (1990).

X-RAY FLUORESCENCE ANALYSIS OF NONHOMOGENEOUS MATERIALS

BY $\Delta\mu$-CORRECTION METHOD

V. I. Karmanov and V. V. Zagorodny

E.O. Paton Electric Welding Institute of the Ukraine
Academy of Sciences, Kiev, Ukraine

ABSTRACT

The fundamental parameters method (FPM) enables one to determine with high accuracy the chemical composition of homogeneous samples, having only one reference sample. However, the reference sample composition should be similar to that of the samples analyzed.

The x-ray fluorescence analysis of multicomponent heterogeneous materials (ores, minerals, their mixtures, welding electrode coating mixtures, fluxes, etc.) is made by the $\Delta\mu$-correction method based on the combined use of the fundamental and empirical correlations maintaining all the advantages of the FPM. Sample composition is calculated on the basis of the element intensities measured in the sample and in the reference specimen and is corrected for the disturbing effect of excitation conditions and heterogeneity as well as the calculated values of one of the fundamental parameters (μ_i). At the preliminary stage of calibration, the coefficients are determined using regression and the absolute fundamental expression for the element fluorescence intensity.

INTRODUCTION

A great advantage of the fundamental parameters method (FPM) is that it requires a minimal number of reference samples. In principle, one multi-component sample is sufficient.[1,2,3]

However, the calculation of absorption and excitation corrections using fundamental correlations for x-ray fluorescence intensity, excited by primary polychromatic radiation, requires in the general form a considerable amount of calculation using a high performance computer. This hinders considerably the application of the FPM for x-ray fluorescence analysis of multicomponent samples, made by the x-ray multichannel spectrometers, connected with a computer.

Unfortunately, the fundamental correlation, relating the intensity of fluorescence excited by the primary polychromatic radiation from the x-ray tube with the elemental composition of the sample is not an exact equation.

By improving the physical model and making the numerical values of parameters more precise, the matching of calculated and experimentally measured intensities can be increased; however, absolute matching can hardly ever be attained.

In reference 4 it is shown that for homogeneous binary samples Fe_2O_3-ZnO the matching can be not less than 3-4%, but depends considerably on the similarity of the reference specimen and the sample analyzed. When analyzing the samples of complex composition, the use of a constant value of the effective wavelength λ_{eff}, as was suggested[3], is also not always acceptable because of the ambiguity of the λ_{eff} value. The calculations of λ^{Fe}_{eff} for binary samples of Fe_2O_3-ZnO, made by the authors, are given in Fig. 1.

$\Delta\mu$-CORRECTION METHOD

The x-ray fluorescence analysis of multicomponent heterogeneous materials (ores, minerals, their mixtures, welding electrode coating mixtures, fluxes, etc.) is made by the $\Delta\mu$-correction method, based on the combined use of the fundamental and empirical correlations while maintaining all the advantages of FPM. Sample composition is calculated on the basis of the element intensities measured in the sample and in the reference specimen and is corrected for the disturbing excitation conditions and heterogeneity as well as the calculated values of one of the fundamental parameters, e.g. the value of $\Delta\mu_i$-μ_{sample}-μ_{ref}. At the preliminary stage the calibration coefficients are determined using regression and the absolute fundamental expression for the element fluorescence intensity.

The mismatching of the calculated and measured intensities is removed by finding the corresponding functions or the transition coefficient using several heterogeneous samples of known composition.

The correlation for calculating the value of C_x the concentration of the element being determined is written in the form:

$$C_x = (I_x/I_{ref})^{meas} \ C_{ref} \ f_1 \ f_2 \ f_3 \tag{1}$$

where: (I_x/I_{ref}) is the ratio of intensities measured in the sample analyzed and in the reference specimen.

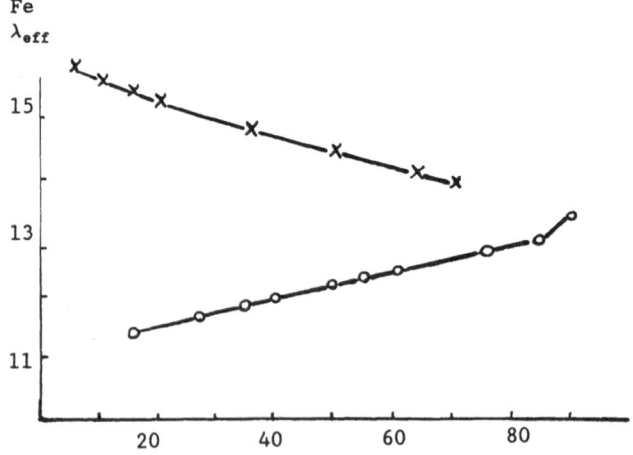

Fig. 1.   Dependence of λ_{eff} for iron on Fe_2O_3 content in binary samples of Fe_2O_3 + ZnO

C_{ref} is the concentration of the element being determined in the reference specimen; f_1, f_2 are functions taking into consideration the dependence of absorption and excitation corrections on such values as $\Delta\mu$, C_B, I_B, etc.

f_3 is the function or the coefficient relating the intensities calculated and measured in the heterogeneous sample.

The calculation algorithm is as follows:

The experimentally measured intensities, corrected for instrument errors for all the elements determined, are transformed into the values of relative theoretical intensities by the regression dependences of the form:

$$I_i^{theor} = a_{io} + a_{i1}I_i^{exp} \tag{2}$$

where: I_i^{theor} and I_i^{exp} are the theoretical and experimental intensities of the element i being determined.

The concentration of a sample element in the zeroth approximation is determined by the equation of the form:

$$C_{io} = b_{io} + b_{i1}I_i^{theor} \tag{3}$$

The concentrations of the elements being determined are corrected to the preset sum of the concentrations of sample elements, i.e. the part of elements not being determined is calculated. Then the effective wavelengths of the primary spectrum for the given elements is calculated by the equation of the form:

$$\lambda_i^{eff} = d_{io} + d_{i1}\Delta F_i \tag{4}$$

where:

$$\Delta F_i = \frac{1}{\mu(\lambda_+^n, C)/\sin\phi + \mu(\lambda_{k\alpha}^i, C)/\sin\psi}$$

$$- \frac{1}{\mu(\lambda_-^n, C)/\sin\phi + \mu(\lambda_{k\alpha}^i, C)/\sin\psi}$$

Here, $\mu(\lambda_+^n, C)$ and $\mu(\lambda_-^n, C)$ are the sample mass coefficients for the radiations with wavelengths near the edge of n elements absorption, whose absorption edge has the wavelength less than the wavelength of the absorption edge of the determined element, i:

$$\lambda_+^n = \lambda_k^n + 0.001, \qquad \lambda_-^n = \lambda_k^n - 0.001 \;;$$

$\mu(\lambda_{k\alpha}^C)$ are the sample mass absorption coefficients for the radiation of the K_α-line of the determined element i.

The difference between the total mass absorption coefficients of the primary and the secondary radiations for the sample and for the reference specimen is calculated as follows:

$$\Delta\mu_i = (\mu(\lambda_{k\alpha}^i, C)/\sin\psi + \mu(\lambda_{eff}^i, C)/\sin\phi)$$
$$- (\mu(\lambda_{k\alpha}^i, C^{ref})/\sin\psi + \mu(\lambda_{eff}^{i,o}, C^{ref})/\sin\phi) \tag{5}$$

The corrections to absorption are calculated by the preliminary calculated dependences:

$$f^i_1 = l_{io} + l_{i1}\Delta\mu_i \tag{6}$$

The correction to the selective excitation is calculated by the dependences of the form:

$$f^i_2 = f^i_o + \sum_k^i f^i_k C_k \tag{7}$$

or

$$f^i_2 = f^i_o + \sum_k^i f^i_k I_k / I^{ref} \tag{8}$$

where the summation is made using all the selectively exciting elements k.

The concentration of sample elements for the next iteration step is determined as follows:

$$C^{n+1}_i = C^n_i \; f^i_1 \; f^i_2, \tag{9}$$

where C^n_i and C^{n+1}_i are the concentration of element i at the preceding and the subsequent steps of iteration. The criterion for terminating the iteration when solving the system of n coupling equations is the inequality:

$$|C^{n+1}_i - C^n_i| \le \epsilon_i \tag{10}$$

where ϵ_i is preset individually for each element.

RESULTS AND DISCUSSIONS

The version of FPM considered above is used in the methodology and the software for the system of analytical control, and for monitoring the technological process of the production of welding consumables (electrodes, flux-cored wires, fluxes). The most important stage of control is the determination of the chemical composition of raw materials and ready powder mixtures.

The x-ray fluorescence analysis of materials, including the elements that selectively absorb the primary and fluorescent radiation of the element studied is very complicated. Among them are ferromanganese, silicomanganese, ferrotitanium, ferrovanadium and other ferroalloys.

Thus, in ferromanganese that contains, depending on the grade, 65-90% manganese, 5-20% iron, 0.3-2% silicon, 0.1-0.8% phosphorus, the selective absorption of the primary radiation is the reason for high values of mass absorption coefficients of the primary radiation (μ_1) for iron and manganese as compared with the absorption of fluorescent radiation (μ_i). The calculations showed that the absorption effects are prevailing and the corrections to the absorption are changed by 10-12%. Since the effective wavelengths of elements λ_{eff} depend on the sample composition, the value of λ_{eff}, corrected by the relation (4), was used in calculations. The comparison of the maximum errors of element concentration calculations with the preset errors admissible in Standards and Specification are given in Table 1.

One of the most complicated problems for x-ray fluorescence analysis is flux mixtures of welding consumables. They are the heterogeneous mechanical mixtures of ores, minerals, ferroalloys and metal powders, dispersed and mixed according to strictly determined proportions, that can be violated because of the technical faults of the equipment performance.

Table 1. Maximum Errors (ΔC, abs. %) of Calculation of Element Concentration in Manganese Materials as Compared with the Admissible Errors (d₃, abs. %) of the Results of Chemical Analysis

Element	ΔC	d_3
ferromanganese		
Mn	0.15	0.5
P	0.001	0.01 - 0.02
S	0.005	0.02 - 0.10
silicomanganese		
Mn	0.15	0.5
P	0.004	0.01 - 0.03
S	0.05	0.2 - 0.3

The flux mixture composition for different types of welding consummables is not the same. Thus, the flux mixture for basic electrodes is a mechanical mixture of powders of marble, fluorspar, rutile, silicate materials and ferroalloys (ferromanganese, ferrosilicon and ferrotitanium). The elements controlled are: iron (3-10%), titanium (1-5%), manganese (3-7%), calcium (25-32%), silicon (5-10%), and fluorine (5-9%). For these compositions the values of λ_{eff} were calculated for iron, manganese, titanium, calcium and silicon (Table 2). The analysis of the calculated absolute values of corrections f_1 and f_2 (Table 2) shows that for all the elements in the compositions studied the correction to excitation f_2 is close to unity and can be neglected without lowering the accuracy of x-ray fluorescence analysis considerably. Absorption correction f_1 for iron and manganese differs from unity by ± 1-2%, and for silicon by ± 5%. It can be concluded from this that in the multicomponent system studied the content of iron and manganese can be calculated by a simple method of external standards. The application of FPM is justified only for calculation of the amounts of titanium, calcium and silicon in flux mixtures, taking into account only absorption corrections.

Absorption corrections for the mentioned elements were correlated with $\Delta\mu$ values and for their calculation the following relations were selected:

Table 2. Values of λ_{eff} Correction to Absorption K_1 and Excitation K_2 for X-Ray Fluorescence Analysis of Welding Electrode Flux Mixtures

Elements	Fe	Mn	Ti	Ca	Si
Effective Wavelength λ_{eff} (A°)	1.37-1.38	1.42-1.45	1.72-1.68	2.25-2.28	3.47-3.49
Absorption Correction	0.98-1.00	0.99-1.01	0.97-1.07	0.96-1.03	0.95-1.05
Excitation Correction	-	-	0.97-1.01	0.99-1.01	0.98-1.01

Table 3. Comparison of X-Ray Fluorescence Analysis Results
          (I) for Welding Consumable Reference Flux Mixtures Using FPM with
          (II) Calculation Results

Content of Element, %	Fe		Mn		Ti		Ca		Si	
Specimen Index	I	II	I	II	I	II	I	II	I	II
36	6.65	6.68	4.84	4.86	2.38	2.42	29.1	29.5	5.5	5.3
37	8.45	8.28	3.34	3.24	2.86	2.36	28.4	28.2	6.97	6.83
38	8.3	8.8	3.8	4.0	2.64	2.84	28.8	28.4	6.6	6.8
43	7.14	7.4	5.6	5.7	2.6	2.5	28.3	27.9	7.0	7.2
44	8.5	8.2	2.4	2.2	3.0	3.1	28.4	28.7	6.9	7.0

$$K_{Ti} = 0.9991 + 0.0028 \, \Delta\mu_{Ti} \qquad R = 0.98 \qquad (11)$$

$$K_{Ca} = 1.0014 + 0.0035 \, \Delta\mu_{Ca} \qquad R = 0.967 \qquad (12)$$

$$K_{Si} = 0.9999 + 0.0008 \, \Delta\mu_{Si} \qquad R = 0.99 \qquad (13)$$

where R is the correlation coefficient.

These relations, as well as the determined function f_3 were used for the calculation, by the measured intensities of iron, manganese, titanium, calcium and silicon contents in the multicomponent samples of welding electrode flux mixtures controlled. The comparison of the x-ray fluorescence analysis results for the reference flux mixtures by FPM in the variant described above with the results of elemental composition calculation, made on the basis of chemical analysis results for the raw materials used is given in Table 3.

CONCLUSIONS

The advantages of the $\Delta\mu$-correction method are as follows:
- the limitations of FPM and other semi-empirical methods, requiring the similarity of the compositions of the sample and the reference specimen are removed;
- the concentration calculation is made with the use of simple relations that help to solve easily the problems of convergence of the iteration process with the use of a personal computer;
- the preparation and use of the method requires a minimal number of reference samples (in principle, one multicomponent sample is sufficient).

REFERENCES

1. Criss, I. W., and L. S. Birks, Anal. Chem. 40:1080 (1968).
2. Karmanov, V. I., I. K. Pokhodnya and A. E. Marchenko, Industr. Lab. 2:167 (1972).
3. Afonin, V. P. and T. N. Gunicheva, "X-Ray Fluorescent Analysis of Ores and Minerals," Nauka, Novosibirsk (1977).
4. Karmanov, V. I., Industr. Lab. 11:25 (1986).

THEORETICAL CALCULATION OF BACKGROUND IN X-RAY SPECTROMETRY

FOR THE DETERMINATION OF SOME HEAVY TRACE ELEMENTS

Jean-Jacques Gruffat

Département de Géologie - Ecole des Mines
158 Cours Fauriel
42023 SAINT-ETIENNE CEDEX 2 - FRANCE

ABSTRACT

The Kulenkampff-Kramers formula giving the spectral distribution of the continuum as a function of wavelength allows a correct calculation of background under the peak. It is only necessary to measure two backgrounds, one on each side of the peak. The true background under the peak is given by multiplying them by adequate coefficients and adding them up. This method has been applied to the determination of low amounts of Ce, La, Ba and Cs in geological samples.

INTRODUCTION

The matrix correction method by means of a Compton scattered characteristic x-ray line[1] for the determination of trace elements is only valid for elements whose characteristic x-ray is shorter in wavelength than the K absorption jump of the heaviest element, generally iron for geological samples. In this case, the theoretical limitation prohibits the use of the Lα ray for determining elements lighter than thulium and of Lβ ray for elements lighter than dysprosium. Owing to a 100 KV generator, it is easy to obtain an intense K spectrum for these elements, but an exact determination of the background under the peak is now difficult because in this range the background is neither flat nor linear. We propose using the Kulenkampff-Kramers formula for determining the true background under peak.

THEORETICAL CALCULATION OF BACKGROUND UNDER PEAK

It is assumed that the emission spectrum shape of the x-ray tube is not distorted by sample absorption and Compton scattering.

When the intensity is expressed in photons or counts as is done in x-ray spectrometry, the Kulenkampff-Kramers formula[2] giving the true background under a peak at a λ wavelength is:

$$N_\lambda = K_1 \frac{1}{\lambda} \left(\frac{1}{\lambda_0} - \frac{1}{\lambda} \right) \quad \text{with} \quad \lambda_0(\text{Å}) = \frac{12.398}{V(KV)} \quad \text{(Duane-Hunt limit)}$$

If the peak position is given by the Θ angle, the formula becomes:

$$N_\Theta = K_2 \frac{1}{\sin \Theta} \left(\frac{1}{\sin \Theta_0} - \frac{1}{\sin \Theta} \right)$$

For two measured backgrounds at $\Theta + \alpha$ and $\Theta - \beta$ angles, we have :

$$N_{\Theta+\alpha} = K_2 \frac{1}{\sin(\Theta+\alpha)} \left[\frac{1}{\sin\Theta_0} - \frac{1}{\sin(\Theta+\alpha)} \right], \quad N_{\Theta-\beta} = K_2 \frac{1}{\sin(\Theta-\beta)} \left[\frac{1}{\sin\Theta_0} - \frac{1}{\sin(\Theta-\beta)} \right]$$

Multiplying $N_{\Theta+\alpha}$ by $\frac{\beta}{\alpha+\beta} \frac{\sin^2(\Theta+\alpha)}{\sin^2\Theta}$ and $N_{\Theta-\beta}$ by $\frac{\alpha}{\alpha+\beta} \frac{\sin^2(\Theta-\beta)}{\sin^2\Theta}$

and adding up those two results gives :

$$N_C = \frac{K_2}{\sin\Theta} \left[\frac{\beta\sin(\Theta+\alpha)+\alpha\sin(\Theta-\beta)}{(\alpha+\beta)\sin\Theta\sin\Theta_0} - \frac{1}{\sin\Theta} \right]$$

After an expansion of N_C using the Taylor's formula we obtain:

$$N_C = \frac{K_2}{\sin\Theta} \left[\frac{1}{\sin\Theta_0} - \frac{1}{\sin\Theta} \right]$$

It is the expression of background at θ angle given by the Kulenkampff-Kramers' formula. This result is exact at all orders if α is equal to β, at the second order if α is different from β.

So, it is possible to obtain a good value for the background under the peak in a spectral range where the continuum against angle is not linear.

APPLICATION: DETERMINATION OF SMALL AMOUNTS OF CE, LA, BA, CS IN GEOLOGICAL SAMPLES

Measurements were made on a Philips PW1404 spectrometer with a Rh target tube operated at 80 KV, 35 mA and a LiF220 crystal. Intensities were measured on CeKα ($2\theta = 14,47°$), LaKα ($2\theta = 15,02°$), BaKα ($2\theta = 15,60°$) and CsKα ($2\theta = 16,22°$). Backgrounds were measured for these four elements at $2\theta-2\beta = 11,9°$, and $2\theta+2\alpha = 16,9°$.

Two calibrations have been done, the first one using our theoretical calculation of background, the second one using a linear background. For cerium (10 standards used, range 4.7 to 260 ppm), the RMS are respectively 4.5 ppm and 21.6 ppm; for lanthanium (12 standards used, range 3 to 1350 ppm), the RMS are 4.1 ppm and 14.9 ppm, for barium (10 standards used, range 1.5 to 1880 ppm), the RMS are 11.1 ppm and 20.0 ppm, and for caesium (10 standards used, range 0.6 to 640 ppm) the RMS are 2.7 ppm and 8.7 ppm.

Our theoretical method for calculating background under peak gives always better results for calibration than the one using a linear background.

REFERENCES

1. R. Vié Le Sage et al. Utilisation du Rayonnement Primaire Diffusé par l'Echantillon pour une Détermination Rapide et Précise des Eléments Traces dans les Roches. X-Ray Spectrom. 8, 121 (1979).

2. R. Tertian, F. Claisse, Principles of Quantitative X-Ray Fluorescence Analyses. Heyden, London (1982).

SYSTEMATIC COMPUTATION OF SCATTERING CORRECTIONS WITH THE CODE SHAPE

J.E. Fernández[*] and V.G. Molinari

Laboratorio di Ingegneria Nucleare di Montecuccolino
University of Bologna
Via dei Colli 16, 40136 Bologna, Italy

ABSTRACT

The recently developed code SHAPE allows the build-up of EDXRF spectra for collimated and monochromatic excitation. In this work the code is applied to investigate the importance of the scattering corrections to the more intense XRF characteristic lines from low and medium Z elements. Results are given graphically for pure elements from Z=11 (Na) to Z=42 (Mo). Heavier element lines from a multi-component geological sample, where light matrix amplifies the influence of scattering, are studied.

INTRODUCTION

Rayleigh or Compton scattering in the target can modify the XRF intensities in an appreciably percentage becoming a source of uncertainty for the computational methods of quantitative analysis. These contributions have been recently studied with great detail with recourse to photon transport theory.[1,2] Analytical relationships[2] are now available for the contributed intensities due to two-collision events, one scattering plus one photoelectric effect, which constitute the most important part of such corrections. Three of them are discrete: photoelectric-Rayleigh, Rayleigh-photoelectric, and Compton-photoelectric [denoted as (P,R), (R,P) and (C,P), respectively], while the photoelectric-Compton [(P,C)] is continuous. This spectrum expands on the wavelength-interval $[\lambda_i, \lambda_i + 2\lambda_c]$ becoming a source of asymmetry for the line shape.[2] Since it is low intensity and the energy width of the interval (for low energies) is short, the spectrum cannot be easily resolved from the main line and, therefore, we can consider it as wavelength-integrated at the same energy of the line.

---

[*] Member of CONICET, Buenos Aires, Argentina. On leave of absence from the Faculty of Mathematics, Astronomy and Physics, University of Córdoba, Argentina.

Advances in X-Ray Analysis, Vol. 35
Edited by C.S. Barrett *et al.*, Plenum Press, New York, 1992

The recently developed computer code SHAPE[3] builds-up the emission
spectrum resulting from excitation with a collimated beam of monochromatic
X-Ray photons. It is based on a more complete set of detailed calculations
obtained with transport theory,[4] from which the preceding contributions
are a part. In this work, we use the code to compute the scattering corrections to the more intense characteristic line of low and medium Z pure
elements in the range Z=11 to 42, i.e., those having $K\alpha_1$ lines which exceed 1 keV. A couple of examples on a multi-component target show that the
total scattering contribution from a low-Z matrix exceeds the secondary XRF
enhancement to the lines of the heavier element and represents an important
part of the overall emission.

RESULTS AND DISCUSSION

All the computations were performed for the same excitation-detection
geometry ($\vartheta_0|\vartheta|\varphi_0|\varphi = 45°|135°|0°|0°$ in the notation of references [1-4])
which corresponds to the well known and widely used geometry where the
source and detected beams form a 90° angle, and both are 45° with respect
to the sample surface. Excitation energy was scanned in the fixed set of
energy values (1, 3, 5, 7, 9, 11, 13, 15, 17, 20, 25, 30, 40, 50, 70, 90,
120, and 150 keV) lying between the K absorption-edge energy of the element
and 150 keV. The $K\alpha_1$ line was chosen for the computations because its higher intensity. $K\alpha$ was used in place of $K\alpha_1$ for unresolved doublets. The
emission energy is that of such line.

Figure 1 shows the plots of the scattering corrections as a function
of excitation energy for targets of pure elements from Z=11 to 42. Scattering contributions are plotted in units of the first-order photoelectric
intensity of the line that they modify (primary XRF-intensity) in order to
quantify their relative importance to change the intensity of the line. The
total scattering contribution (the sum of the four single ones) displayed
in the same units gives the overall importance of the effect of scattering
on XRF-intensity within the second-order of interaction. Since multiple
scattering contributions with only the photoelectric effect (i.e., secondary or tertiary XRF) are not possible in these lines (the more energetic in
a pure element matrix), the scattering correction is the only modification
to the line intensity in these cases. It should be noted that these line
intensities are frequently considered as clean primary XRF-intensities in
standard less quantitative analysis.

(P,R) and (C,P) chains dominate at low and high energies, respectively, for most of the elements considered. However, the lower Z elements (Na
to Si) show a greater (R,P) contribution that prevails at intermediate
energies over the (P,R) one. The (P,C) chain is the less intense for this
range of elements. The entity of the total correction is strongly energy
dependent for the lower Z elements, but becomes more uniform for increasing
Z. Indeed, the range of variation changes from [0.01%@2 keV-60%@150 keV]
for Na to [2%@20 keV-10%@150 keV] for Mo. In all cases, the one percent
correction is early exceeded.

Figure 2 shows, as an example, the total scattering correction on the
$K\alpha_1$ lines of two different elements of a multi-component geological matrix
($Si O_5 Al_2$) as a function of excitation energy. The correction for the Al
$K\alpha_1$ line is compared with the enhancement due to secondary XRF-intensity
produced by Si. It is shown how the scattering contribution grows monotoni-

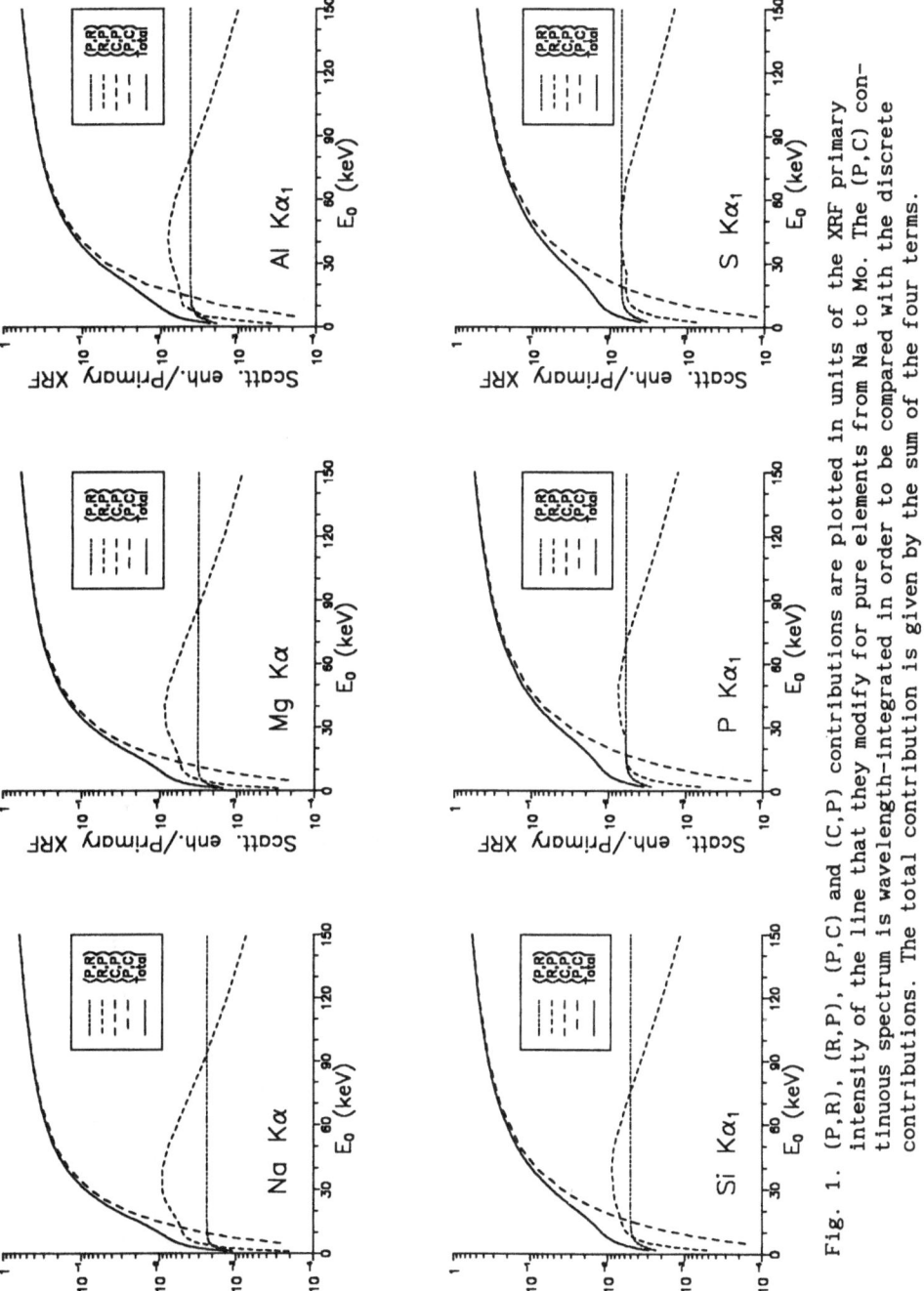

Fig. 1. (P,R), (R,P), (P,C) and (C,P) contributions are plotted in units of the XRF primary intensity of the line that they modify for pure elements from Na to Mo. The (P,C) continuous spectrum is wavelength-integrated in order to be compared with the discrete contributions. The total contribution is given by the sum of the four terms.

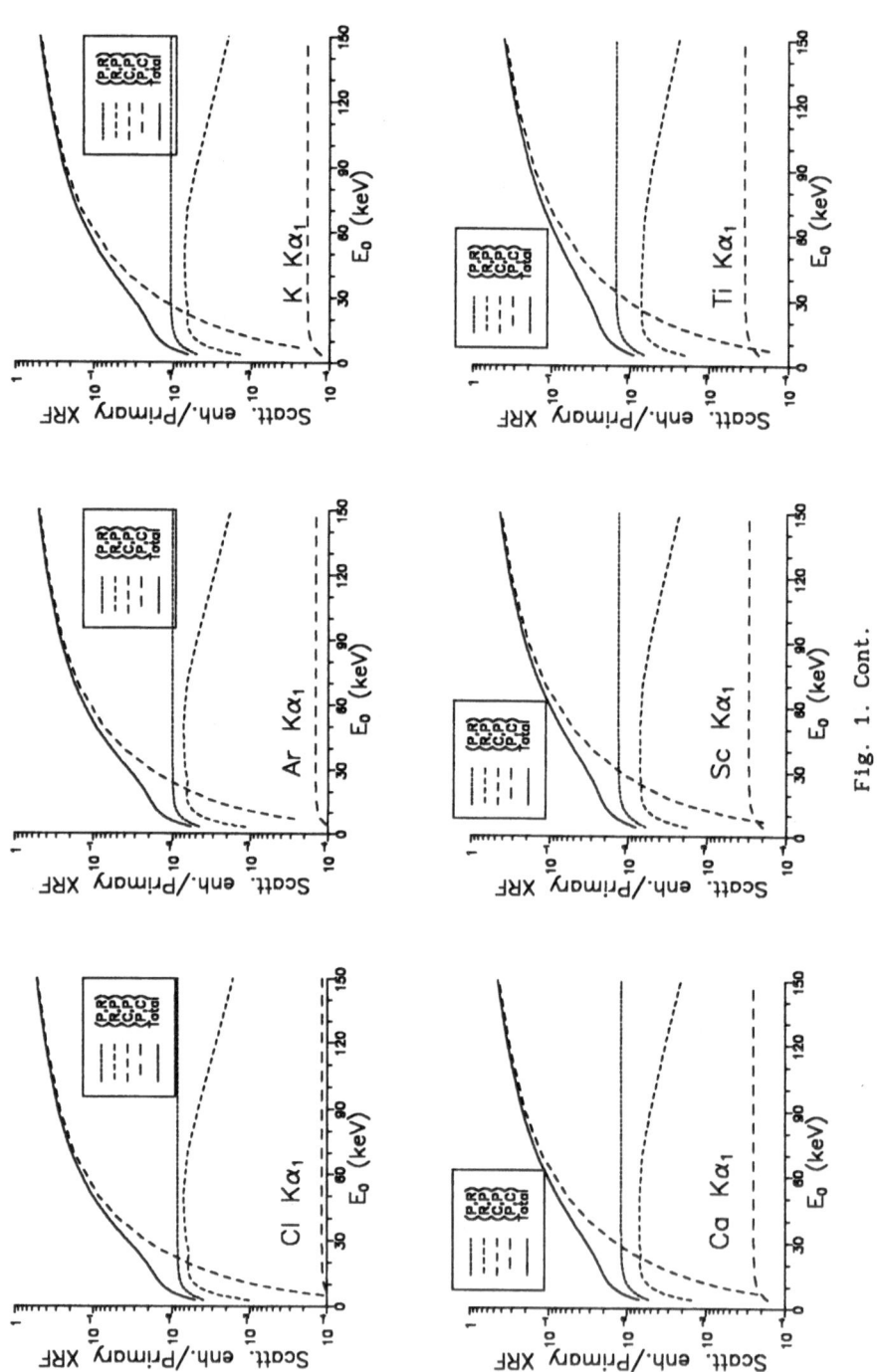

Fig. 1. Cont.

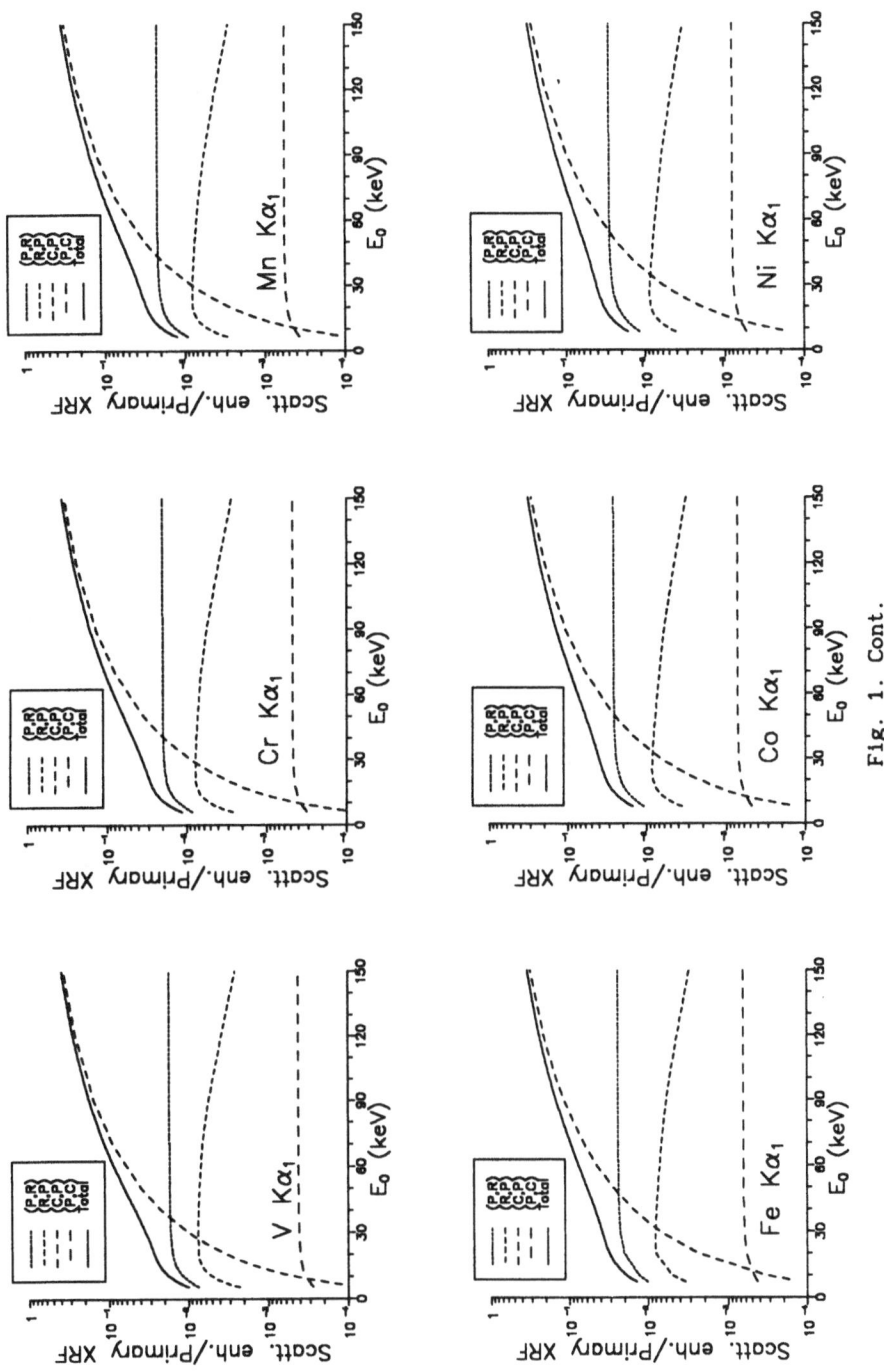

Fig. 1. Cont.

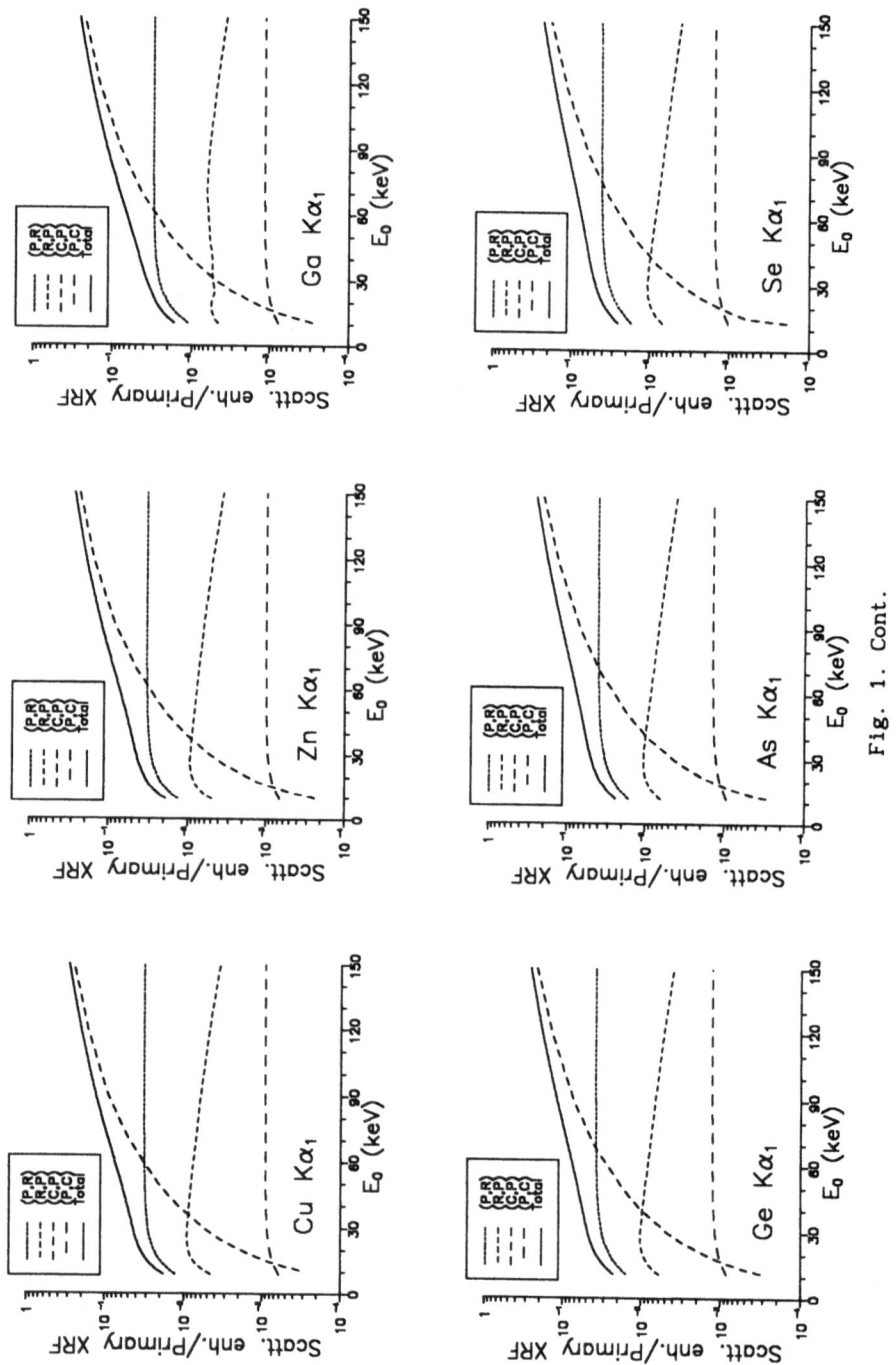

Fig. 1. Cont.

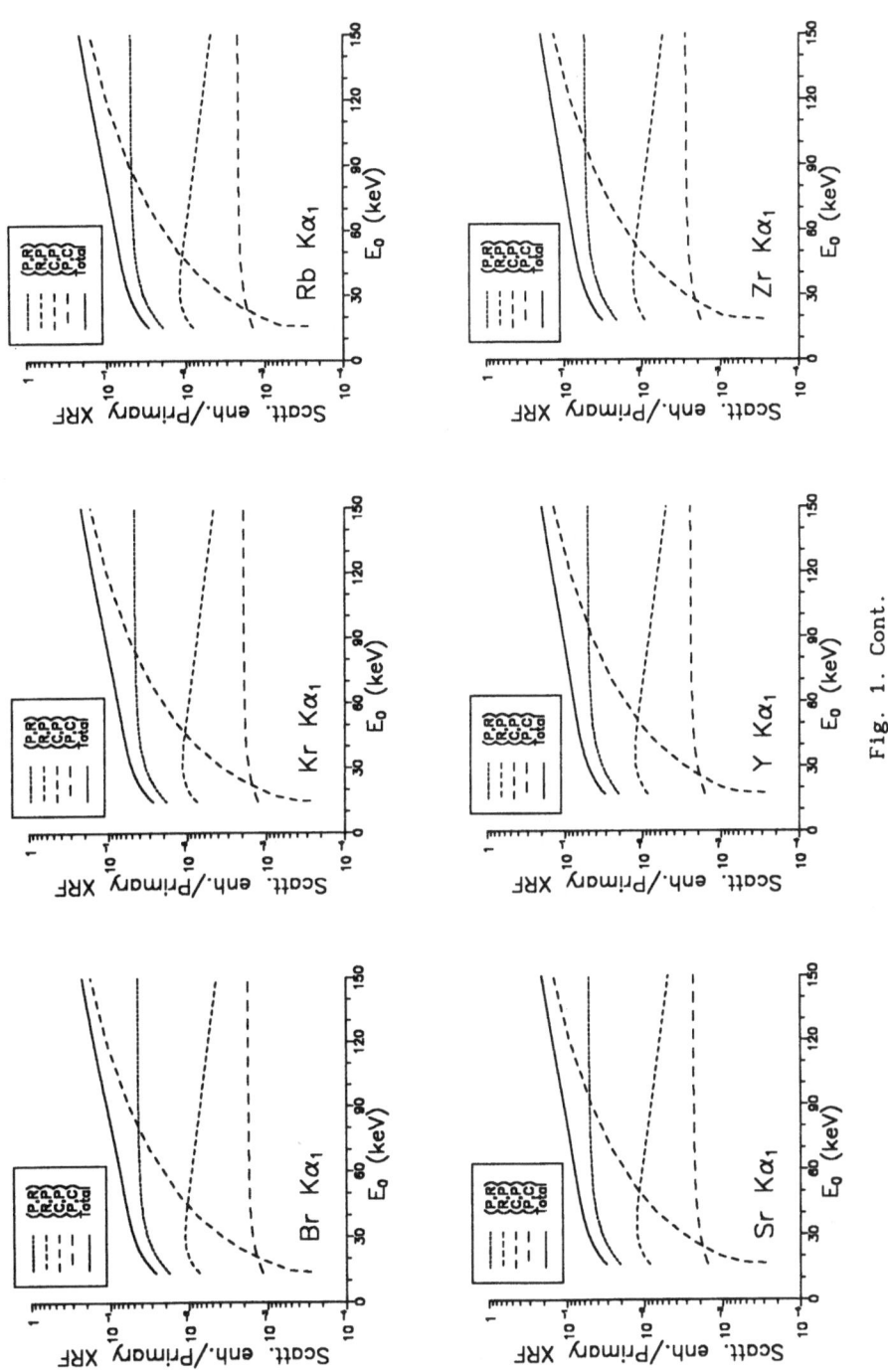

Fig. 1. Cont.

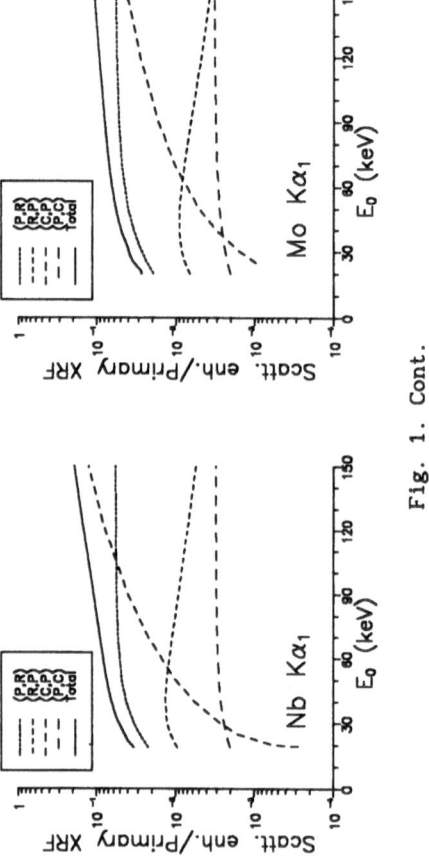

Fig. 1. Cont.

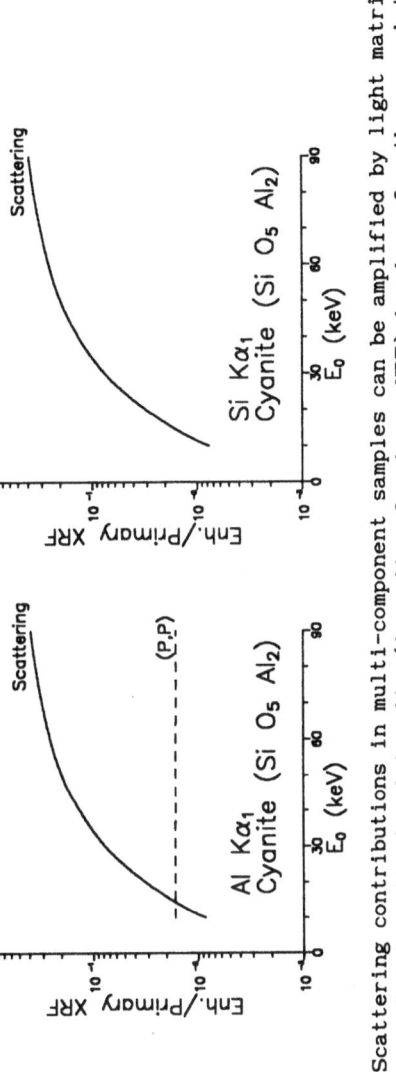

Fig. 2. Scattering contributions in multi-component samples can be amplified by light matrices. Total scattering intensity (in units of primary-XRF) is shown for the more intense lines of Al and Si in cyanite (Si O_5 Al_2). The scattering contribution to the Al $K\alpha_1$ line exceeds the secondary XRF-intensity in Al due to Si. In both cases the modification is significantly high.

cally with the excitation energy and how it exceeds the double photoelectric contribution for $E_0 >$ 13 keV. The scattering correction to the Si $K\alpha_1$
line is the only modification of this line because no secondary XRF-intensity is produced. The correction on both lines is of similar importance and
roughly changes in the interval [2%@10 keV-40%@90 keV]. It is easily seen
that a computation of intensities for a Rh tube excitation ($K\alpha_1$@20 keV)
ignoring this correction will fail in near 6% without considering further
modifications due to continuum excitation below the operation voltage.

CONCLUSIONS

     Scattering corrections to XRF intensity are important in pure as in
multi-component targets. They cannot be safely ignored in computational
methods of quantitative analysis, and should be computed for the given
excitation-detection geometry and source energy. A graphic table is
provided, for widely used geometry, energy range, and pure elements, as a
rapid evaluation means that indicates the entity of the correction in every
case. Multi-component targets present additional complexities that can
deeper the scattering influence. A particular study is advisable in such
cases. A more exhaustive table should complete the present information to
all the elements in the periodic table, and give a considerably greater
number of multi-component examples of study.

REFERENCES

1. J.E. Fernández, V.G. Molinari and M. Sumini, Corrections for the effect
   of scattering on XRF intensity, in: "Advances in X-Ray Analysis," Vol
   33, C.S. Barrett et al. eds., Plenum Press, New York (1990).
2. J.E. Fernández, Rayleigh and Compton scattering contributions to the XRF
   intensity, X-Ray Spectrom. (1991), to be published.
3. J.E. Fernández and M. Sumini, SHAPE: a computer simulation of energy
   dispersive X-Ray spectra, X-Ray Spectrom. (1991), to be published.
4. J.E. Fernández and V.G. Molinari, X-Ray photon spectroscopy calculations, in: "Advances in Nuclear Science and Technology," Vol 22, M.
   Becker and J. Lewins, eds., Plenum Press, New York (1991).

RECENT DEVELOPMENTS IN SURFACE AND THIN FILM ANALYSIS USING

LOW - ENERGY ELECTRON INDUCED X - RAY SPECTROMETRY (LEEIXS)

M.J.Romand, F.Gaillard, M.Charbonnier
Department of Applied Chemistry and Chemical Engineering
(CNRS, URA 417), Université Claude Bernard - Lyon I
69622 Villeurbanne Cedex, FRANCE

ABSTRACT

A review is given of the main gas-discharge sources and auxiliary equipments which were used in soft and ultra-soft X-ray emission spectrometry. Special attention is paid to the basic principles and instrumentation of low-energy electron-induced X-ray spectrometry (LEEIXS) whose excitation source is an electron beam generated in a glow-discharge system. Capabilities of LEEIXS in surface and thin film analysis and characterization are illustrated by examples dealing with control and optimization of surface treatment and thin film deposition processes. Sensitivity of the technique down to the submonolayer range and influence of backscattering phenomena are shown.

INTRODUCTION

Low-pressure glow-discharge plasmas play an ever-increasing role both in material processing and material characterization. In the first case, considerable attention has been given in recent years to plasma processes and their applications in sputtering, plasma-assisted etching, activated reactive evaporation, ion plating, plasma-assisted chemical vapor deposition and plasma polymerization [1]. In the second case, major advancements in the analytical chemistry of liquids, solids and solid surfaces have been made, the glow discharge whatever may be its mode of generation (dc, rf or pulsed), constituting quite a versatile source for atoms, molecules, photons and ions [2]. Among these techniques we can cite inductively coupled plasma-atomic emission spectrometry (ICP-AES) [3], or mass spectrometry (ICP-MS) [4], glow-discharge mass spectrometry (GDMS) [5], sputtered neutral mass spectrometry (SNMS) [6] and glow discharge optical spectrometry (GDOS) [7].

In this context, the purpose of the present paper is (i) to review the basic principles and instrumentation of low-energy electron induced X-ray spectrometry (LEEIXS) [8,9], an X-ray emission technique using an electron beam generated in a glow discharge system and (ii) to present some of its new applications in the field of thin film and solid surface analysis and characterization.

INSTRUMENTAL AND HISTORICAL

Firstly, let us recall that it was the use of gas discharge (or cold cathode) tubes that led to the discovery of X-rays by Roentgen in 1895; however the unsteady nature of operations (due to pressure changes) in such devices led to their being abandoned in favour of the currently more conventional Coolidge (thermoionic or hot cathode) tubes.

Although it can be assumed that the reader is familar with the basic concepts underlying electrical discharges in gases at reduced pressure, it may be useful to describe with some detail the instrumentation used in this work and to briefly recall a few facts concerning the operation of a GD source. The experimental device is shown in figure 1. The discharge takes place inside an insulating (silica) tube, 1.5 cm in diameter and 2 cm in length. The cathode is an aluminum cylinder powered by a dc negative voltage supply. The anode is a grounded nickel electrode with a hole in its center but the nature of the anode material is not particularly critical. Typical operating conditions are in the range 0.5 - 5 kV with a discharge current between 0.1 and 0.5 mA. In the experiments described both previously and in this paper, the windowless tube operates in the residual air of the spectrometer in the pressure range 2-15 Pa but other gases such as neon, argon etc... could be used to feed the source enclosure. The stability of the discharge is obtained by adjusting the pressure by means of an electronically-driven valve.This valve, whatever may be the chosen applied voltage, reacts by admitting air at a rate that keeps the pre-set current in the discharge constant. The voltage applied across the electrodes causes breakdown of the residual air and formation of a light emitting plasma. The two main regions which can be distinguished inside the tube are the cathode dark space and the negative glow. The cathode dark space is a narrow sheath that surrounds the cathode and across which virtually all of the discharge voltage is dropped. The negative glow is a region of bright emission filling the major part of the discharge volume particularly when reduced-size sources are used (this is the case with the device which has been developed in our laboratory). The negative glow is an essentially field-free reactive zone in which electrons, ions and metastable (excited) species are present.

The source described here is of the "front face secondary emission" - type [10]. Electron emission from the cathode is produced following bombardment of this electrode surface both by ions generated in the negative glow and by fast neutrals created by resonant charge transfer in the cathode sheath. It should be noted here that :

(i) the insulating tube surrounding the cathode is designed to confine the electron emission solely to the cathode front face. These "secondary" electrons are then accelerated

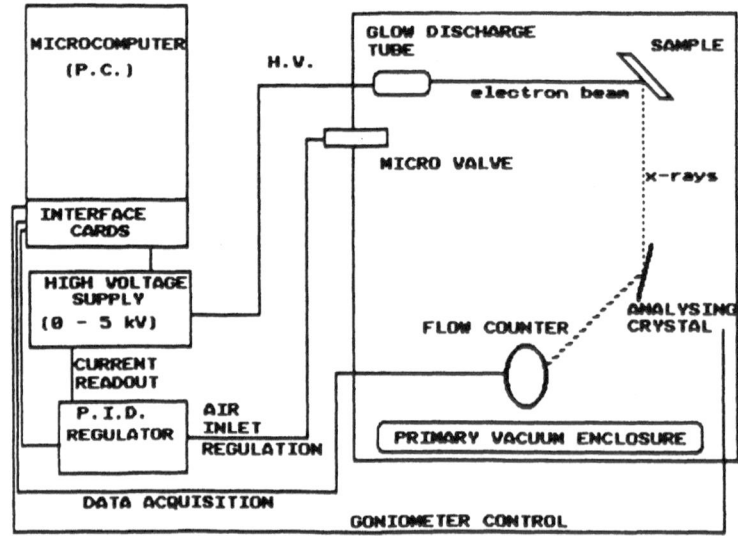

Fig. 1. Schematic diagram of the glow-discharge system

along the electric field lines through the cathode dark space to form a relatively well-collimated electron beam [11], whose energy corresponds to the potential difference between anode and cathode. The hole in the anode allows a part of the electron beam to escape from the discharge tube. The diameter of this beam at the sample surface is less than 1 cm^2, depending upon the size of the hole in the anode.

(ii) A part of the secondary electrons so-accelerated collide with gas species in the negative glow region and then create more ions which sustain the discharge process.

(iii) The bright emission in the negative glow region arises from excitation of atoms or molecules by lower energy electrons that have been slowed down by inelastic collisions.

(iv) The bombardment of the cathode by high energy ions and neutrals, in addition to causing the ejection of secondary electrons, is responsible for the ejection of surface atoms by momentum transfer processes (sputtering processes). Under these circumstances the best materials to be used as cathodes have to possess both a high secondary electron emission coefficient and a low sputtering yield under ion bombardment. Indeed, this last condition is required in order to prolong the cathode lifetime and to minimize the deposition of sputtered species on the insulating parts of the gas discharge tube. Aluminum is a material suited for such purposes and this is the reason why the cathode used in LEEIXS is commonly made of this metal. However it should also be added that the interesting properties of this cathode material as an emitter of secondary electrons result from the thin native oxide layer which always covers the metallic substrate in the working conditions (discharge in air at a pressure of a few Pa and at low current density), the secondary emission coefficient of the aluminum oxide being nearly 10 times as great as that of the pure metal [10].

(v) The source described here operates in the so-called abnormal glow discharge mode, the current - voltage characteristics of the glow discharge showing, for a given pressure, an increasing slope as the applied voltage is raised [11].

(vi) It functions with such electron beam discharge currents and applied voltages that power through the tube is normally lower than a few watts. This small power consumption means that the target of the tube need not be water-cooled. It is also worthy of note here that such devices could operate at higher power but current densities have to be kept at reduced values in order to avoid the cathode material being sputtered at a huge rate and above all to avoid sample damage during analysis.

(vii) The gas-discharge tube described here can be applied to probe non-conducting samples. The charging effects commonly encountered in most surface analysis techniques operating in high and ultra-high vacuum conditions are not present with this source-type working in a primary vacuum.

In addition the LEEIXS spectrometer is a wavelength dispersive instrument equipped with (i) flat analyzing crystals for dispersing soft and ultra-soft radiations (λ = 0.3 - 20 nm) emitted by samples under investigation and (ii) a conventional flow-proportional (P-10) counter, whose window is made of a very thin (0.5 μm) aluminized polypropylene film.

Historically only a few gas tubes have already been used as radiation and/ or electron sources for X-ray spectrometry but none of them was really suited to carry out "true" surface analysis as LEEIXS does.Wyckoff and Davidson [12,13] were the first to incorporate gas-filled tubes in commercial wavelength dispersive X-ray spectrometers. In their sources, electrons extracted from the cathode were used to bombard a grounded anode (Al or Cu). X-rays emitted and electrons scattered by this electrode left the tube through an open window and excited the specimen to be analyzed which acted consequently as a secondary target. Using low Z anodes these authors enhanced the fluorescence excitation. Commonly, investigations were made with a tube power of 50 watts (5 mA at 10 kV). Results concerning both K spectra of light elements and L and M spectra of higher Z elements were given [12-15]. Further Solomon and Baun [16] and Vanhatalo et al. [17] built similar gas-filled X-ray tubes (in their principle even if not in their form and size) . These devices included an aluminum anode and worked in argon at 4.5 kV and 3.2 mA for the former and at 7 kV and 4 mA for the latter. In both cases, intensity stability was achieved by an automatic pressure controller which regulated the vacuum by the anode current. It should also be noted that the device developed by Vanhatalo et al. was used successfully as a "linear" X-ray source in a grating spectrometer operating up to the ultra-soft X-ray range (the Al $L_{2,3}$ band spectrum was taken). In the same way, McGee and Saha [18] built a plasma-controlled X-ray tube using a low pressure atmosphere of helium, a copper anode and a cage- like cathode constructed from a nickel wire-mesh. For the production of Al K_α lines, a thin sheet of aluminum was soldered to the copper anode. Other X-ray lines available for excitation were claimed to be obtainable by attaching other materials of appropriate Z number with good thermal contact . Such a tube operated at 8.61 kV and 5 mA.

Contrary to the above systems, Sahores et al. [19,20] ,Witmer and VanMeijl [21], Mill and Belcher [22] developed sources which worked at 10-15 kV / 2 mA, 5-6 kV / 4-5 mA, and 10 kV / 8mA, respectively, and which privileged the electronic excitation at the expense of the fluorescence one. In all these cases, anodes were made of high Z elements (W or Au) in order to enhance electron backscattering, and tubes were constructed in such a way that it was possible to mount them in commercial X-ray fluorescence spectrometers. In this context a cold cathode tube denoted "Elent 10" and its generator were made available in France from the "Compagnie Générale de Radiologie" (CGR) [20]. In addition, this same Company marketed a portable wavelength-dispersive spectrometer equipped with a gas tube of special shape working at lower power (e.g. 15 W at 5 kV). In this device, electrons were extracted from the gas-tube through a grid-type anode. The spectrometer itself was designed for field work and allowed to introduce 15 pellet samples in the vacuum chamber and to successively place one sample after the other in front of the electron tube [20].

All the studies cited above have unequivocally brought into light the interest of gas-filled tubes in order to extend the analytical capabilities of conventional X-ray spectrometry, in the long wavelength range, especially for the analysis of light elements. In this same range, Romand et al. [23-27] were the first to state that the use of such equipment could be an alternative and convenient way for studying solid surfaces and very thin films in the range of thickness up to about 0.1 µm. First experiments which essentially dealt with quantitative

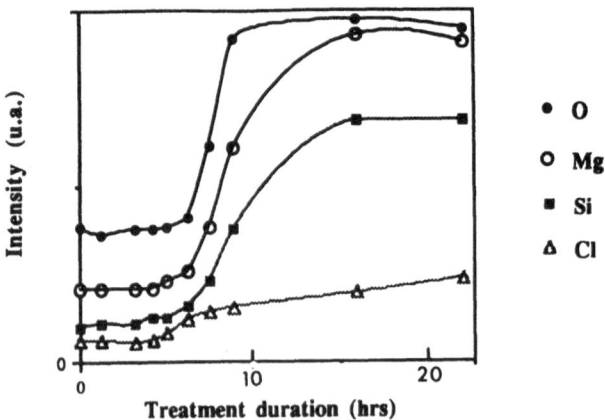

Fig. 2. ClK_α MgK_α, SiK_α and OK_α intensity variations of a stainless steel treated in a MgCl$_2$ medium, versus treatment duration.

analysis of very thin oxide films grown on various metallic substrates, were carried out with an "Elent 20" - model tube from CGR (gas-source similar in its principle to that inserted in the portable spectrometer cited above) and later on with a laboratory-made system. The latter (described at the beginning of the instrumental chapter) has two significant advantages. The first results from the use of an open anode; indeed under these conditions the electron beam extracted from the cathode and accelerated in the interelectrode space reaches the sample surface with the nominal energy E_0 while with a plain anode electrons are backscattered towards the sample and thus have their energy distributed between E_0 and zero. Consequently in the latter case both the estimation of the depth to which incident electrons can penetrate into the sample and that from which characteristic X-rays can be generated are not reasonably accessible, which renders accurate quantitative analyses difficult. The second advantage results from the use of very low current intensities (≤ 0.5 mA) and densities (≤ 0.5 mA. cm^{-2}) largely lower than those employed with all the sources previously mentioned. Let us recall here that electron bombardment of a specimen may also lead to a variety of damage including dissociation, desorption, oxidation, reduction, carbonization, diffusion etc. , depending on the nature of the material and residual gas and the pressure in the spectrometer and that such phenomena have to be minimized by using as low current densities as possible.

RESULTS AND DISCUSSION

The chemical and compositional characterization of surfaces, interfaces and thin films has considerable importance in surface and thin film technology. In this section, we will firstly present a few examples which illustrate the capabilities of LEEIXS in order to control and/or optimize surface treatment and thin film deposition processes. In some cases correlation will also be made between analytical information provided by LEEIXS and the properties of the surface and thin films studied.

The first example is relative to AISI-304 L stainless steels which were subjected to a passivation treatment in a boiling (142°C) concentrated and aerated MgCl$_2$ solution. Figure 2 shows how OK$_\alpha$, MgK$_\alpha$, SiK$_\alpha$ and ClK$_\alpha$ X-ray intensities vary with treatment duration. In each case, incident electron energy is taken high enough above the corresponding ionization threshold energy to probe the whole thickness of the surface layer. Results illustrate that , for treatment duration higher than 6 hours, and for an appropriate conditioning of the alloy surface, a change of the oxidation process kinetics and of the nature of the surface layer occur. An additional study of the fine structure of SiKβ - K β' emission band prove the formation of a magnesium silicate like surface film [28].

In the same way the next example also concerns AISI-304 L stainless steels which were subjected to a galvanostatic anodization in a hot and concentrated sulfuric-chromic acid medium [29]. Figure 3 shows (i) the intensity variation of the OK$_\alpha$ band and of the SK$_\alpha$ line from the oxides formed by this electrochemical treatment and (ii) the CrL$_{\alpha,\beta}$/FeL$_{\alpha,\beta}$ intensity ratio variation , as a function of the treatment duration. It is quite clear from figure 3 that : (i) the largest oxide thicknesses (90 nm or so) are obtained for treatment durations equal to or greater than 15-20 min, (ii) sulfur impurities are incorporated into the oxides (as SO$_4$ $^{2-}$ anions) to an amount roughly proportional to the film thickness, and (iii) the sample surfaces become markedly more and more chromium enriched (as Cr $^{3+}$) when the oxide film thickness increases up to its maximum value. In each case , incident electron energies were taken in order to probe the whole thickness of the surface oxide layer. These results enable the factors that influence the growth of such films to be studied, thus allowing optimal experimental conditions to be established. In the present case, the improvement of the adhesion properties of these electrochemically prepared films has been associated with the surface chromium oxide enrichment [29].

The next example deals with polyaniline films grown on metallic substrates by electropolymerization of aniline in an aqueous solution containing sodium sulfate as the

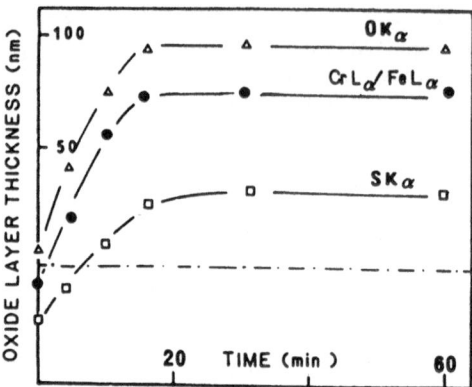

Fig. 3. Anodization of a stainless steel in a sulfuric-chromic bath. Effects
of treatment duration on oxide film thickness, sulfur impurity
incorporation and Cr / Fe surface concentration.

supporting electrolyte. These conducting polymer films were here synthesized by electrolysis
at constant voltage [30]. Such processes are explored in our laboratory as a possible route for
depositing organic primers to which another polymer (adhesive, paint...) can subsequently
be adhered. Figure 4 shows both (i) the intensity variation of the CK_α band and of the SK_α
line and (ii) the current variation through the electrochemical cell, as a function of the
electropolymerization time. As it can be seen there once again exists a remarkable correlation
between the film characteristics (thickness and SO_4^{2-} anion incorporation) and the change of
the electrochemical parameter which permits to follow the film growth [31,32]. It should also be
said concerning these LEEIXS experiments that the current density of the incident electron
beam has been chosen low enough here ($\simeq 0.1$ mA. cm^{-2}) to avoid the degradation of the
organic materials under investigation.

In addition figure 5 shows CK_α, OK_α and FK_α spectra from a polytetrafluoroethylene
(PTFE) sample before and after its surface has been subjected to a sodium-naphtalenide
treatment for various times [11]. This chemical treatment is also designed to enhance adhesion
properties of polymer material of low-energy surfaces. Spectra shown here illustrate the fact
that the reductive treatment is responsible for a defluorination of the polymer surface which
then consists mainly of carbon and oxygen. Let us note here that the presence of oxygen

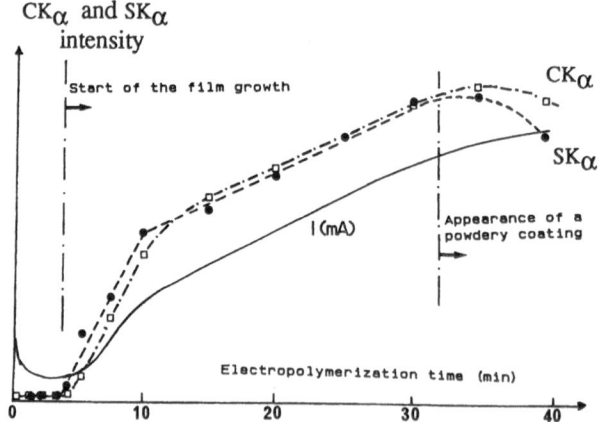

Fig. 4. Electropolymerization of aniline on a Ti6Al4V alloy surface.
Changes in CK_α and SK_α intensities, and in the current through
the electrochemical cell, as a function of treatment duration.

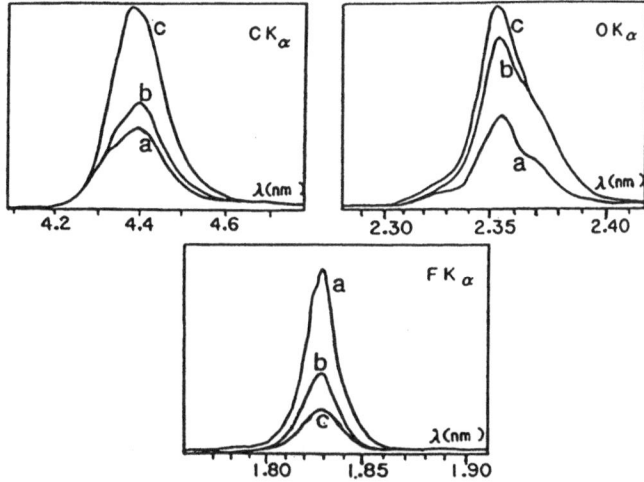

Fig. 5. CK_α, OK_α and FK_α spectra from a PTFE sample : (a) such "as received",
(b) and (c) after a Na-naphtalenide treatment for 30s and 2 min, respectively.

should result from the fact that the treatment is carried out in ambient atmosphere and therefore that chemical groups such as C=O, COH and CO_2H would be produced after exposure of the reactive treated surface to air. As the corresponding polymers have a high chemical and thermal stability, the LEEIXS experiments have been conducted with a higher current density (0.2 mA.cm^{-2}) than in the previous case.

The next example is concerned with polymer films produced this time by a glow-discharge, onto stainless steel substrates. This process, denoted plasma polymerization, permits depositing thin primer or protective films from a vapour phase at a reduced pressure. As such films possess chemical and physical properties that are claimed to be superior to those of conventionally polymerized materials we also explored this route in our laboratory in order to deposit a primer to which another polymer can subsequently be adhered. Films studied here are deposited from a CH_4 plasma in a diode reactor using a 13.56 MHz rf-generator. Figure 6 shows the variation of the CK_α band intensity from the polymer films as

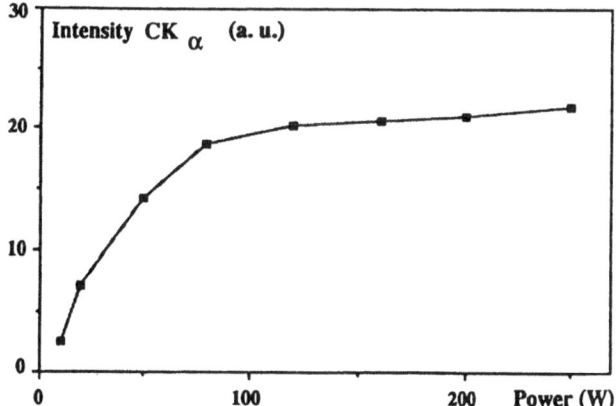

Fig. 6. Thin films of carbon deposited by plasma process on stainless steel.
Changes of CK_α intensity as a function of the power supplied
by the generator.

a function of the power delivered by the generator, other experimental parameters (pressure and flow of methane in the reactor) being maintained constant [33]. Here again (CK_α intensity being directly proportional to film thickness in the thickness range investigated) LEEIXS is well able to control the film deposition process.   Obviously quite similar conclusions can be drawn for inorganic films prepared from plasma [34], gaseous cementation [35], or sol-gel [36] processes. By way of example, figure 7 illustrates the result of a quantitative analysis of a ZrO_2 ceramic layer deposited by a sol-gel process onto a stainless steel substrate. The "k-ratio" method used consists of measuring the ratio of the ZrL_α line intensity of the thin film to that of a thick or bulk oxide of similar composition (here a monolithic gel sample) as a function of the incident electron beam energy E_0. As shown in a recent paper [9] from OK_α intensity measurements of thin oxide films grown on various metallic substrates, this ratio is unity at low electron energies since electrons penetrate identically into the oxide of the thin film or thick standard, and that this ratio decreases at higher energies as some incident electrons enter the substrate of the thin oxide while they continue to progress into the oxide of the thick sample. From the plot in figure 7 it is then possible to determine the energy E_0 associated with the change in slope for which the film thickness is the effective X-ray excitation depth R_e. The corresponding oxide thickness may then be evaluated using one of the range equations giving R_e (Feldman's equation for instance). In the present example a mass thickness of 14.5 μg. cm^{-2} of zirconia is determined, which corresponds to a film thickness of 29 nm, the density of the film material being taken at 5. Quite obviously, a similar result can be obtained by considering the OK_α band intensity instead of the ZrL_α line intensity but a residual oxide layer is present at the substrate surface before the sol-gel film deposition.

Concerning again inorganic films, it should be noted here that an original application of LEEIXS has been developed by Hecq et al. for the in-situ control of the growth of thin films obtained by a sputter deposition process in a plasma environment [37-39]. The corresponding experiments were carried out by associating an X-ray spectrometer with a sputtering chamber. In such systems electrons emitted under ion impact of the target to be sputtered are accelerated across the cathode sheath and a part of the fast electrons so-obtained bombards the substrate on which sputtered species are deposited. By following the intensity variation of a characteristic X-ray line or band from the film or from the substrate as a function of deposition time, it is possible to monitor the deposition process. By way of example this method has been used to control the deposition of a thin film of SiO_2 which results from the sputtering of a Si target in a mixture of Ar and O_2 [37].

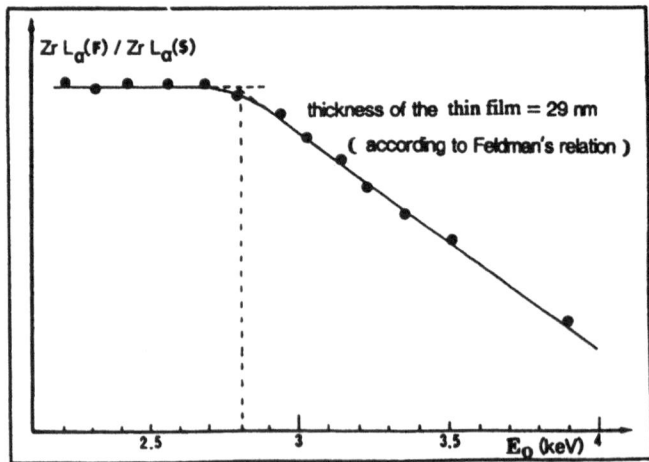

Fig. 7. ZrL_α intensity ratio changes obtained from a thin and a "thick" ZrO_2 films on stainless steel, as a function of the incident electron beam energy E_0 .

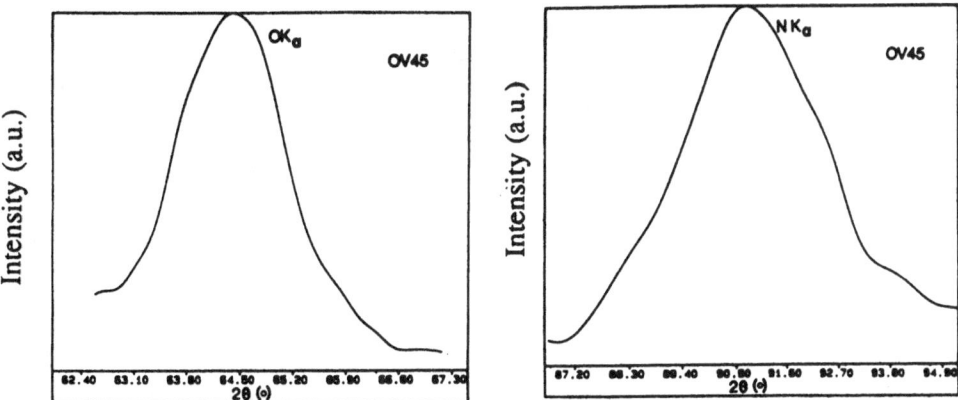

Fig. 8. OK$_\alpha$ and NK$_\alpha$ sectra of "as received" carbon fibers.

All the examples described above concern analysis and characterization of surface-treated materials for which LEEIXS is particularly well suited, samples under investigation being flat and of sufficiently large area. In addition, it should be noted that LEEIXS is also quite able to provide a lot of information about sample surfaces which do not necessarily respond to such criteria and even though quantitative data are much more difficult to obtain. For instance fiber materials such as those used to make composites, belong to this class. Let us recall that here too, surface composition before and after surface treatments has to be characterized in order to improve adhesion between fibers and matrice (polymer in the most common case). In this context, carbon fibers can be investigated by LEEIXS after fiber strands have been joined and stretched across the reference surface of a hollow sample holder. Figure 8 shows OK$_\alpha$ and NK$_\alpha$ spectra of an "as received" specimen. Clearly the signal intensity of oxygen is very low (~ 100 counts . s^{-1}) and is characteristic of a slight oxidation of the material surface. The source of nitrogen is not certain but its presence could be associated with the manufacturing process of the fibers and not at all with the result of a surface treatment. This hypothesis is based on the comparison of the OK$_\alpha$ and NK$_\alpha$ signal intensities (same order of magnitude) and on the fact that the detection sensitivity for NK$_\alpha$ is

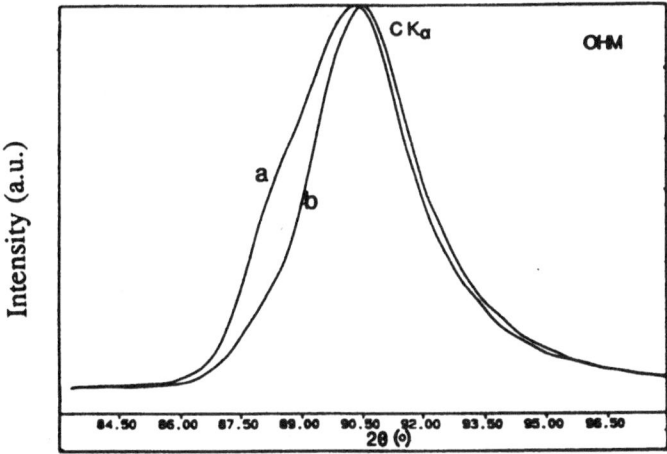

Fig. 9. CK$_\alpha$ bands of "as received" carbon fibers orientated (a) parallel,
(b) perpendicular to the electron beam direction.

much lower than that for OK_α . In addition, it is worthy of note, in spite of what has already been said about the applicability of LEEIXS in quantitative analysis of such samples, that it is possible to get information about surface treatment effects by determining the change in line or band intensity ratios. In the present case a primer-based surface treatment changes the O / N ratio by a factor 2.5. Concerning again the carbon fibers it should also be added that the CK_α band fine structure appears to be dependent on the fiber orientation with respect to the direction of the incident electron beam. As shown in figure 9 the FWHM of the CK_α band is significantly larger when the electron beam direction is parallel to the fiber orientation. No sufficiently satisfying explanation (experimental artifact or physical phenomenon) about this observation can presently be given.

In dealing with all the examples described above, we have demonstrated the specific capabilities of LEEIXS for analyzing various types of thin and very thin films, the information depth being in each case not much more than some tens of nanometers. As the sensitivity of LEEIXS as well as that of electron probe microanalysis (EPMA) is limited ultimately by the presence of the bremsstrahlung (continuum background) on which the characteristic signal is superimposed we have, sooner or later, to deal with signal-to-noise problems and with the definition of the minimum detectable film thickness (or mass thickness) t_{min}. In order to define t_{min}, we note t the thickness of a standard thin film, as well as P and B the average numbers of X-ray counts measured for the net peak of the element of interest and for the background at the peak position, respectively. By considering t_{min} as the film thickness which gives a signal intensity equal to three times the standard deviation of the background (3 σ criterion) we obtain the following simple equation

$$t_{min} = t \ 3\sqrt{B} . P^{-1}$$

Let us recall here that the establishment of this equation assumes that there exists a linear relationship between film thickness and X-ray intensity in the thickness range of the selected standard.

Under these conditions, t_{min} can be improved (decreased) when the $P / \sqrt{B}$ ratio is maximized. As both P and B are dependent on the energy E_0 of the incident electron beam the variation of this ratio with E_0 or with the overvoltage (or reduced energy) $U = E_0 / E_c$ has to be studied (E_c = threshold energy of the inner level ionized)* .

The results presented here were obtained by considering a standard constituted of a very thin ("natural") oxide film on a stainless steel substrate. This film is about 4 nm in thickness which, assuming a density of 5.21 for the Cr_2O_3 / Fe_2O_3 oxide material, corresponds to a mass thickness of about 2 μg. cm^{-2} of oxide. Figure 10 (a) shows the variation of P (OK_α intensity for a counting time of 20 s) as a function of U ($E_c = E_K = 0.531$ keV for oxygen).In such a very thin film, the OK_α absorption within the film itself is obviously negligible. As the influence of oxygen traces incorporated into the substrate can also be considered as negligible the curve shown in figure (10 a) should be very closely

---

* $P = f(\sigma) = \varphi(U)$
  σ is the cross-section for ionization of the inner shell electron of binding energy E_c by an incident electron of energy E_0. σ can be expressed in the form [50] :
  $$\sigma (E_0, E_c) \ E^2_c = 6.51 \ 10^{-14} \ Z^* \ b \ ln \ (cU) / U$$
  where Z^* is here the number of electrons in the inner shell (or subshell) considered, b and c parameters depending on the shell.
  * $B = \Psi(U)$
  The intensity of the continuum background at energy E can be described with an equation due to Kramers [51] :
  $$B(E) = kiZ (E_0 - E) / E$$
  As at the line position [52] $E \sim 0.9 \ E_c$, B can be expressed as a function of U
  $$B(U) \sim k'iZ (U - 0.9)$$
  where i is the electron beam current and Z the atomic munber of the target (essentially the substrate in the present experiments).

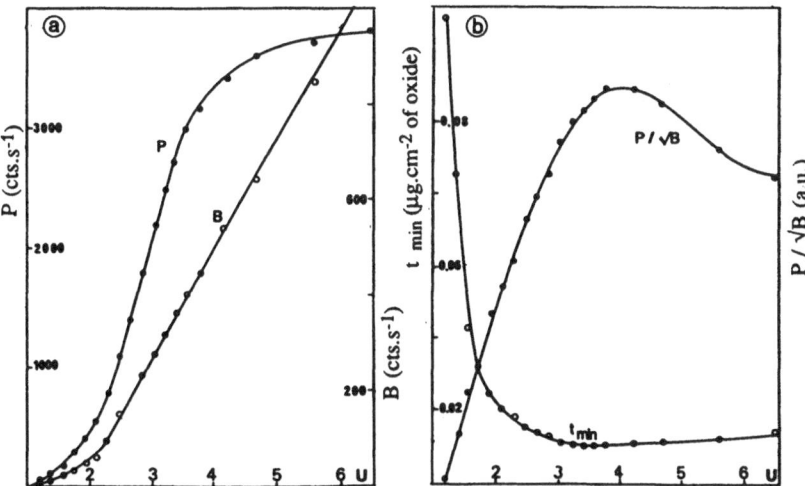

Fig. 10. "Natural" oxide on a stainless steel. Variation of net peak (P),
background B, P / √ B and detection limit t min, as a function
of reduced energy U.

related to the variation of the ionization cross-section σ with energy [40,41,50]. However in this
example two main differences with the theoretical curve can be evidenced. The former
concerns the foot of the experimental curve (just above U = 1) for which the observed slope is
lower than that expected. This is due to the increasing of the number of oxygen atoms
participating in the X-ray signal as the electron beam energy increases and therefore as the
electron beam progresses into the oxide film . By contrast, when the film is wholly crossed
by the incident electron beam, the slope of the curve fits well enough with that of a σ vs. U
theoretical curve. The latter difference concerns the top of the experimental curve. The
maximum is not reached for U around 3 as expected from the theoretical considerations, the
measured intensities continuing to increase beyond this value. Essentially this behaviour can
be explained by the additional ionizations of the oxygen atoms due to electrons backscattered
within the substrate and which return and cross the oxide layer in the upward direction with
an energy $E > E_K$. As will be seen further, backscattering can indeed play an important role
in LEEIXS, its extent depending on the nature of the substrate.

In the same manner figure 10 (a) represents the variation of the background intensity B
(counting time = 20 s) as a function of U. As expected the intensity of B increases with U. B
was here measured in the conventional way i.e. by linear interpolation between the
backgrounds to the left and right of the peak position. It should also be pointed out that we
have been careful to perform these experiments (B and P intensity measurements) while
keeping the sample current constant. This was made possible by measuring, for each value of
E_0, the corresponding current with a conducting specimen electrically isolated from the
sample holder and adjusting the discharge current as necessary. This operation has to take
into account the fact that the number of electrons escaping from the discharge tube and thus
reaching the sample varies with the voltage applied between anode and cathode, especially
when very low voltages are used.

In addition figure 10 (b) shows how P / √ B and t min vary as a function of reduced
energy U. From these two curves it can clearly be observed that there exists an optimum
overvoltage for which the detection limit in terms of film thickness is minimum. This
optimum voltage is significantly lower than the value of U corresponding to the unattained
maximum of P. In the present example the minimum mass detectable corresponds to about 8.
10^{-3} μg. cm^{-2} of oxide i.e. to an equivalent thickness of about 1.6 10^{-2} nm or to a surface
coverage lower than a tenth of monolayer. Theoretically, oxide layer thicknesses on metallic

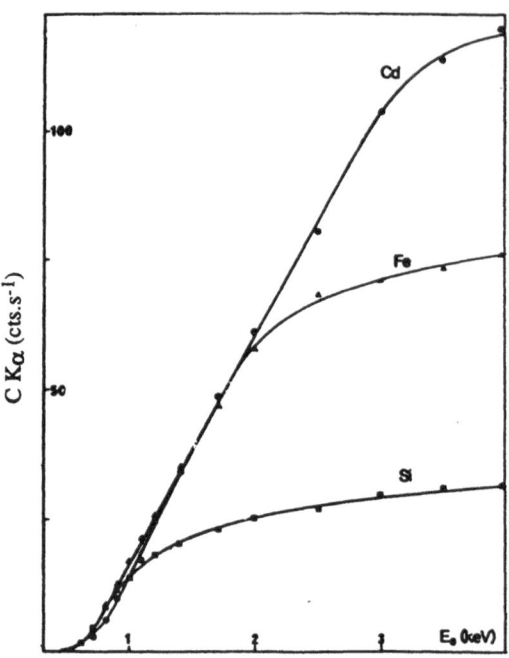

Fig. 11. Ultra-thin films of carbon deposited by plasma process on Si, Fe and Cd substrates. CK_α intensity variations as a function of the incident electron beam energy E_0.

substrates can therefore be detected (via OK_α intensity measurements) down to the submonolayer range. In fact, determinations of oxygen at these levels cannot be encountered in LEEIXS. Indeed these experiments normally operate in air, at about 10^{-1} torr and in these conditions, metallic surfaces in particular are always covered by a residual oxide layer and oxidized species. Nevertheless the conclusion drawn about the values of minimum detectable film thicknesses are quite applicable to the trace analysis of elements (for instance F, Na...) which are not naturally present (e. g. carbon or oxygen) at the sample surfaces. In some favorable cases (high fluorescence yield, high reflecting power of the dispersing device....), minimum detectable film thickness in the range 10^{-2} - 10^{-3} monolayer can even be expected.

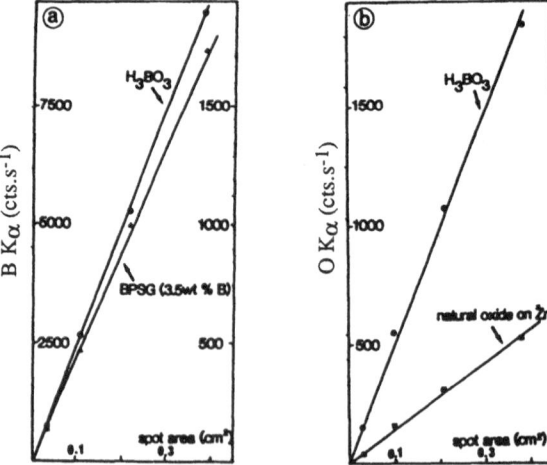

Fig. 12. Changes in BK_α and OK_α intensities emitted by different materials, as a function of the electron beam area on the sample surface.

As was mentioned earlier, backscattering phenomena could possibly play a significant role in the total X-ray emission of the element of interest. As the backscattering coefficient increases with increasing atomic number, very thin film analysis can be influenced by the nature of substrates. The corresponding effects are demonstrated in figure 11 where the "yield" of CK_α from thin carbon films is shown as a function of electron beam energy for three different substrates (Si, Fe, Cd). These substrates were firstly "mirror" polished (except Si), then rinsed , ion-etched in order to eliminate residual carbon contamination, and lastly subjected to a plasma treatment in a CH_4 vapor phase (cf. example associated with figure 6) Thin carbon films were deposited simultaneously in the reactor and it is assumed that a quasi-equivalent thickness was obtained on each substrate. Intensity measurements associated with the beginning of the curves presented in figure 11 tend to prove that this is probably the case. On the contrary, the corresponding curves diverge considerably as the electron beam energy increases. Such differences in CK_α intensity measurements at the same energy (above 2 keV) can then clearly be attributed to the enhancement of the CK_α emission from the films by electrons backscattered from the substrates although a slight contribution of fluorescence excitation by SiK, FeL and CdM radiations cannot be wholly excluded. This result justifies in part why in figure 10 (a)   the maximum of the curve I (OK_α) vs. U is not reached for U around 3 and why errors can be introduced when we try to determine the "true" variation of the ionization cross-section of an inner level as a function of E_0 or U from a thin film deposited onto any substrate.

Lastly it is interesting to make some ·comments about an additional possibility of LEEIXS. Using the electron beam which escapes from the gas-discharge tube and decreasing the size of the anode hole, an unfocused electron beam of reduced diameter bombards the surface sample. Insofar as the decreasing of the material volume providing the information is consistent with obtaining an exploitable signal we could conceive the development of a simple instrument permitting the characterization of localized areas. In this context figure 12 shows the linear decrease of the x-ray signal intensity when the diameter of the electron beam at the sample surface decreases. Figure 12 (a) concerns BK_α detection in H_3BO_3 and in a phosphosilicate glass containing 3.5 % B in weight, respectively. In a similar way, figure 12 (b) concerns OK_α detection in H_3BO_3 and from a naturally surface-oxidized zinc substrate, respectively. Both these experiments were performed with 3 keV electrons at a current intensity of 0.2 mA. BK_α and OK_α radiations were dispersed with a multilayer device (OV 120) and with a TIAP crystal, respectively. In each case, it can be pointed out that signal intensities of boron or oxygen remain at a sufficiently high level when the size of the electron probe decreases. As a result it becomes quite possible to develop an electron beam excited X-ray analyzer that permits the rapid characterization of localized areas with an electron spot of about 1 mm in diameter and therefore to build a mapping instrument (macroprobe - type) using a sample holder, mobile in both the x and y directions. By way of example , such an analyzer could be used to control the local composition and thickness of thin films obtained by the various processes mentioned in this paper and more particularly those obtained par plasma processes (see also paper by Charbonnier et al. [34] concerning chromium nitrides Cr_xN_y deposited by reactive magnetron sputtering).

CONCLUSION

The above examples show that considerable information about chemical composition and thickness of very thin films can be obtained using low-energy electron-induced X-ray spectrometry (LEEIXS) and that this technique is a well-suited tool for controlling and optimizing thin film deposition as well as surface treatment processes. This potential, even though it has not yet been exploited to a great extent, derives from a number of specific advantages that we can summarize as follows :

   * surface analysis in non-UHV conditions

   * possibility of using low-energy electron beams at low current densities in order to avoid damage to some materials (polymers in particular)

* ability to tune the incident electron beam energy in order to probe selected depths (in the range of some tens of nanometers for which other non-destructive analysis methods are often little or badly-suited)

* capability of probing insulating samples

* high sensitivity for light element analysis

* quantifiability [9]

* possibility of obtaining chemical state information [42]

* relative simplicity of the instrumental device.

Finally, it must be said that at a time where an impetus has been given throughout the world towards research aiming at studying surfaces and very thin films and improving light element detection sensitivity by X-ray emission methods, LEEIXS is a technique which is highly-attuned for significant progress in this direction. Apart from researches carried out by EPMA and which are not mentioned here, all the others [43-49] are conducted, as with LEEIXS, using wavelength-dispersive devices (analyzing crystals or gratings) and laboratory-made spectrometers.

REFERENCES

1. J. A. Thornton, in : " Deposition technologies for films coatings .Developments and Applications," R.F. Bunshah et al. eds., Noyes Publications, Park Ridge, USA (1982).
2. W. W. Harrison, in : Proc."XXVIIth Colloquium Spectroscopicum Internationale," Bergen, Norway (1991) p. C-PL- 3.
3. P. W. J. M. Boumans, in : " Inductively coupled plasma emission spectroscopy," Vol 1 and 2, Wiley, New York (1987).
4. A. R. Date, A. L. Gray, in : "Applications of inductively coupled plasma mass spectroscopy," Blackie, Glasgow (1983).
5. W. W. Harrison, K. R. Hess, R. K. Marcus, and F. L. King, Anal. Chem., 58 : 341 (1986).
6. H. Oechsner, in : "Topics in Current Physics. Thin film and Depth Profile Analysis," Vol 37, H.Oechsner ed., Springer-Verlag, Berlin (1984).
7. R. Berneron, Spectrochim. Acta, B 33 : 443 (1978).
8. M. Romand, R. Bador, M. Charbonnier and F. Gaillard, X-ray Spectrometry, 16:7-16 (1987).
9. M. J. Romand, F. Gaillard, M. Charbonnier, and D. S. Urch, Adv. in X-ray Analysis, 34 (1991).
10. J. J. Rocca, J. D. Mayer, M. R. Farrell, and G. L. Collins, J.Appl.Phys., 56: 790 (1984).
11. M. Romand, F. Gaillard, and M. Charbonnier, in: " Plasma Surface Engineering," E.Broszeit et al. eds., Vol.2, DGM Informationgesellschaft -Verlag, Oberursel (1988) p.759-766.
12. R. W. G. Wyckoff, and F. D. Davidson, Nature, 205:969 (1965); Rev. Sci. Instrum., 34: 572 (1963).
13. F. D. Davidson, and R. W. G. Wyckoff, Norelco Rept., 14 :3 (1967).
14. R. W. G. Wyckoff, and F. D. Davidson, J.Appl.Phys , 36: 1883 (1965).
15. F. D. Davidson, and R. W. G. Wyckoff, Adv. in X-ray Analysis , 9: 344 (1966).
16. J. S. Solomon, and W. L. Baun, Rev. Sci.Instrum., 40: 1458 (1969).
17. J. Vanhatalo, L. Kaihola, and E. Suoninen, J.Phys.E : Sci.Instrum., 9: 1156 (1976).
18. J. F. McGee, and T. Saha, Adv. in X-ray Analysis, 22: 241 (1976).
19. E. Larribau, B. Grubis, and J. Sahores, French Patent, 69-31.060 (1969).

20. J. J. Sahores, E. Larribau, and J. Mihura, Adv. in X-ray Analysis , 16: 27 (1973).
21. A. W. Witmer, and E. W. J. M. Van Meijl, Spectrochim. Acta., 34 B: 415 (1979).
22. J. C. Mills, and C. B. Belcher, X-ray Spectrom., 7:138 (1978).
23. M. Romand, G. Bouyssoux, and R. Bador, in: Proc. " XIXth Colloquium Spectroscopicum Internationale," Philadephia (1976), Extended Abstract 203.
24. M. Romand, R. Bador, A. Roche, and G. Bouyssoux, J.Microsc.Spectrosc. Electron., 2 : 627 (1977).
25. R. Bador, A. Roche, G. Bouyssoux, and M. Romand, Spectrochim. Acta., 33 B: 437 (1978).
26. A. Roche , R. Bador, F. Buiguez, G. Bouyssoux, M. Charbonnier, and M. Romand, in: Proc."XXIst Colloquium Spectroscopicum Internationale," Cambridge (1979), Extended Abstract p.214.
27. R. Bador, M. Romand, M. Charbonnier, and A. Roche, Adv. in X-ray Analysis, 24 : 351 (1981).
28. F. Gaillard, M. Romand, A. Roche, M. Charbonnier, R. Bador, and A. Desestret , in: "Passivity of Metals and Semi-conductors," M.Froment ed., Elsevier. Sci. Publi., Amsterdam (1983) p. 181.
29. M. Romand, F. Gaillard, M. Charbonnier, and A. Roche, J.Adhesion , 23:1 (1987).
30. G. Bouyssoux, F. Gaillard, and A. Roche, in : Proc. " International Symposium on Trends and New Applications in thin Films," Strasbourg (1987) p. 369-373.
31. G. Bouyssoux, F. Gaillard, and M. Romand, in: Proc. "XXVIIth Colloquium Spectroscopicum Internationale," Bergen, Norway (1991) p. C-PO - 24.
32. G. Bouyssoux et al., to be submitted.
33. L. Deshayes, M. Charbonnier, and M. Romand, Le vide. les Couches minces, suppl. n° 257 (1991).
34. M. Charbonnier, M. Romand, and A. Roche, These Proceedings.
35. R. Hillel, M. P. Berthet, A. Roche, and J. Bouix, J.Materials Sci., 25:3191 (1990).
36. C. Chino, M. Charbonnier, A. M. de Becdelièvre, C. Guizard, M. Pauthe, and J. F. Quinson, in: Proc. " Eurogel ' 91," Saarbrücken, Germany (1991), Under press.
37. M. Hecq, and J. Leleux, J.Vac.Sci.Techn., A5: 1760 (1987).
38. M. Hecq, and J. Leleux, Anal. chem., 59:440 (1987).
39. M. Hecq, Spectrochim. Acta, 43B : 1 (1988).
40. M. Charbonnier, M. Romand, and F. Gaillard, Analusis, 16 : 17 (1988).
41 A. K. Gyani, P. McClusty, D. S. Urch, M. Charbonnier, F. Gaillard, and M. Romand, Adv. in X-ray Analysis, 33 : 247 (1990).
42. F. Gaillard, and M. Romand, Surf. Interface Anal., 16 : 429-434 (1990).
43. R. D. Carson, C. P. Frank, S. Schnatterly, and F. Zutavern, Rev. Sci. Instrum., 55 : 1973 (1984).
44. A. Szasz, and J. Kojnok, Appl. Surf. Sci., 24 : 34 (1985).
45. G. Andermann, F. Fujiwara, T. C. Huang, J. K. Howard, and N. Staud, Adv. in X-ray Analysis, 32 : 261 (1989).
46. G. Andermann, Appl. Surf. Sci., 31 : 1 (1988).
47. T. Scimeca, and G. Andermann, Surf. Interface Anal., 10 : 321-326 (1987).
48. J. E. Rubensson, N. Wassdahl, G. Gray, J. Rindstedt, R. Nyholm, S. Cramm, N. Martensson, and J. Nordgren, Phys. Rev. Lett. , 60 : 1759 (1988).
49. Y. Claesson, N. Wassdahl, M. Georgson, B. Gray, C. G. Ribbing, and J. Nordgren, Vacuum, 41 : 1275 (1990).
50. C. J. Powell , Rev. Modern. Phys. , 48 : 33-47 (1976).
51. H. A. Kramers, Phil. Mag. , 46 : 836 (1923).
52. T. S. Rao-Sahib, and D. B. Wittry, J. Appl. Phys. , 45 : 5060 (1974).

DEPTH PROFILING BY MEANS OF X-RAY FLUORESCENCE ANALYSIS

H.Ebel, R.Svagera and S.Rezai Afshar

Institut für Angewandte und Technische Physik
Technische Universität Wien, A 1040 Wien, Austria

INTRODUCTION

Sherman[1] described the excitation of characteristic radiations by primary x-rays and by secondary excitation. The derivation has been made assuming a homogeneous sample. Criss and Birks[2] inverted the problem from the calculation of fluorescent countrates to the quantitative XFA by means of fundamental parameters. Theoretical[3] and instrumental[4-6] developments enabled a reduction of the sample area and led to small area XFA and imaging XFA sytems. Depth profiling by means of XFA is a further development. We continue the original concept[7] of variable take-off angle technique for the determination of film thicknesses without reference samples and apply the variation of the incidence angle to depth profiling.

THEORY

A differential contribution $d^2 n_{ijk}$ to measured countrates n_{ijk} from primary excited characteristic x-rays can be obtained from equ.1. It is derived from photoabsorption of a fraction $x_E dE$ of the incident x-ray spectrum in the volume of a thickness dt in a depth t. The composition depends on the depth t. Additionally, there exists a depth dependence of the density ρ. The sample consists of z chemical elements. Fig.1 depicts the directions of the incident x-rays and of the take-off of fluorescent radiations with respect to the sample.

$$d^2n_{ijk} = qx_E dE \frac{\rho c_i \tau_{E,i} dt}{\cos\alpha} M_{ij}\omega_{ij}P_{ijk} \frac{\Omega}{4\pi} \cdot \kappa_{ijk}$$

$$\cdot e^{-\sum_{m=1}^{z}\left(\left[\frac{\tau_{E,m}}{\cos\alpha} + \frac{\tau_{ijk,m}}{\cos\beta}\right]\cdot\int_{x=0}^{t}\rho c_m dx\right)} \quad (1)$$

q is the cross section of the parallel beam of incident radiation, $\tau_{E,i}$ and $\tau_{E,m}$ are the photoabsorption coefficients of x-rays of energy E in the elements i and m, and $\tau_{ijk,m}$ describes the photoabsorption of characteristic radiation of energy E_{ijk} in the element m. The indices i, j and k define the chemical element, the ionized atomic level and the atomic level from where the empty state in j is filled up. The fraction of photoabsorption in the atomic level j of the element i is quantified by M_{ij}, the fluorescence yield by ω_{ij}, the transition probability by P_{ijk} and the efficiency of detection of fluorescent radiation in the solid angle Ω by κ_{ijk}.

Usually, the products ρc remains constant for homogeneous specimens and integration leads to Sherman's equation for primary excited characteristic x-radiation n_{ijk}

$$n_{ijk} = const_{ijk} \cdot \frac{c_i}{\cos\alpha} \int_{E_{ij}}^{E_o} \frac{M_{ij}\tau_{E,i}x_E}{\sum_{m=1}^{z} c_m\left[\frac{\tau_{E,m}}{\cos\alpha} + \frac{\tau_{ijk,m}}{\cos\beta}\right]} dE \quad (2)$$

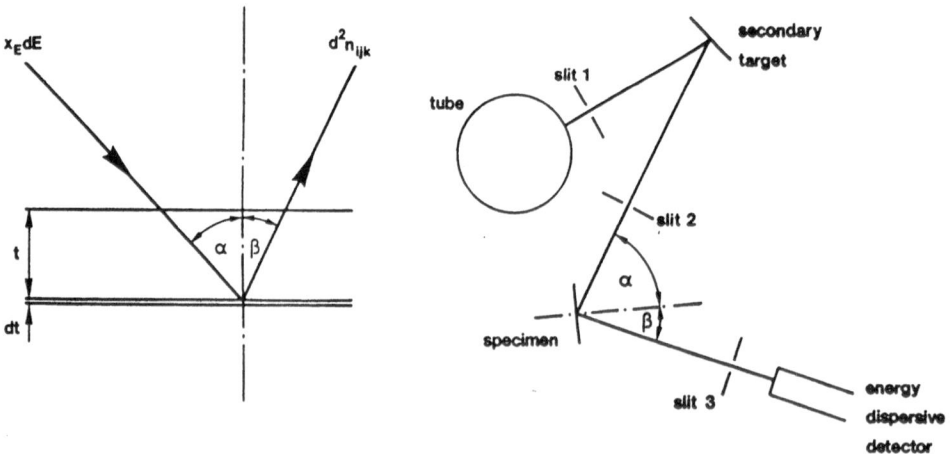

Fig.1   Geometry of the incident x-rays, the sample and the take-off direction of fluorescent x-rays.

with

$$\text{const}_{ijk} = q\omega_{ij}p_{ijk} \cdot \frac{\Omega}{4\pi} \cdot \kappa_{ijk} \cdot \qquad (3)$$

E_{ij} is the energy according to the j-absorption edge of the element i and E_o is the highest energy of the continuous spectrum of the x-ray tube.

But, our attitude is focused on inhomogeneous specimens. Therefore, we have to deal with the depth distribution of the product ρc. In order to simplify the integration of equ.1, we continue by introducing three assumptions into our concept:

   i. The density remains constant for the investigated depth range

  ii. The concentrations are described by model functions.

A further essential goal of our considerations is the influence of the photon energy E on the available results of depth profiling. Thus, we deal with the problem by the usual application of monochromatic excitation. This can be realized either by secondary targets (see Fig.1) or by a tunable monochromator. An additional simplification of the problem of depth profiling arises from the reduction of the excitation mechanisms to primary excitation. This allows to describe the problem just by equ.1. From these considerations the third assumption follows:

iii. The photon energy E(i) for the excitation of the characteristic radiation ijk has to be located in the energy range between the j-absorption edge of the element i and the subsequent absorption edge of another specimen element.

The three assumptions are applied to Zn-layers on hot dip galvanized steel. This is a reduction to the investigation of a binary system. However, the conclusions can be applied to a general problem of depth profiling.

a.RECTANGULAR DEPTH PROFILE (see Fig.11)

The depth profile of Zn-concentration is described by $c_{Zn}=1$ for $0 \le t \le D$ and by $c_{Zn}=0$ for $t>D$. The monochromatic excitation of FeKα is performed by NiK-radiation and of ZnKα by GeK-radiation, respectively. In equ.1 we replace the spectral portion $x_E dE$ of the incident tube spectrum by the intensities $I_{s(1)}$ of $K\alpha_1$, $I_{s(2)}$ of $K\alpha_2$ and $I_{s(3)}$ of $K\beta_{1,3}$ of the characteristic K-spectrum of the secondary target s and by an integration of equ.1 we obtain the expressions for $n_{FeK\alpha}$ (equ.4) from the substrate and $n_{ZnK\alpha}$ (equ.5) from the layer.

$$n_{FeK\alpha} = \frac{\overline{const}_{FeK\alpha}}{\cos\alpha} \sum_{R=1}^{3} I_{Ni(R)} \tau_{Ni(R),Fe} \cdot \frac{e^{-\left(\frac{\tau_{Ni(R),Zn}}{\cos\alpha} + \frac{\tau_{FeK\alpha,Zn}}{\cos\beta}\right)\rho D}}{\frac{\tau_{Ni(R),Fe}}{\cos\alpha} + \frac{\tau_{FeK\alpha,Fe}}{\cos\beta}} \tag{4}$$

$$n_{ZnK\alpha} = \frac{\overline{const}_{ZnK\alpha}}{\cos\alpha} \sum_{R=1}^{3} I_{Ge(r)} \tau_{Ge(R),Zn} \cdot \frac{1 - e^{-\left(\frac{\tau_{Ge(R),Zn}}{\cos\alpha} + \frac{\tau_{ZnK\alpha,Zn}}{\cos\beta}\right)\rho D}}{\frac{\tau_{Ge(R),Zn}}{\cos\alpha} + \frac{\tau_{ZnK\alpha,Zn}}{\cos\beta}} \tag{5}$$

Fig.2 shows the spectral responses for a specimen with a Zn coating of thickness D=20μm, measured under an incidence angle α=45° and a take-off angle β=45°. The selectivity of the different excitations is evident from the ratios of $n_{FeK\alpha}/n_{ZnK\alpha}$ of 4:1 for an excitation by NiK-radiation, and of 1:40 for an excitation by GeK radiation.

$\overline{const}_{ijk} = const_{ijk} M_{ij}$ in equs.4 and 5 is eliminated by forming the ratio

$$r_{ijk}(\alpha) = \frac{n_{iij}(\alpha)}{n_{ijk}(\alpha_{ref})} \tag{6}$$

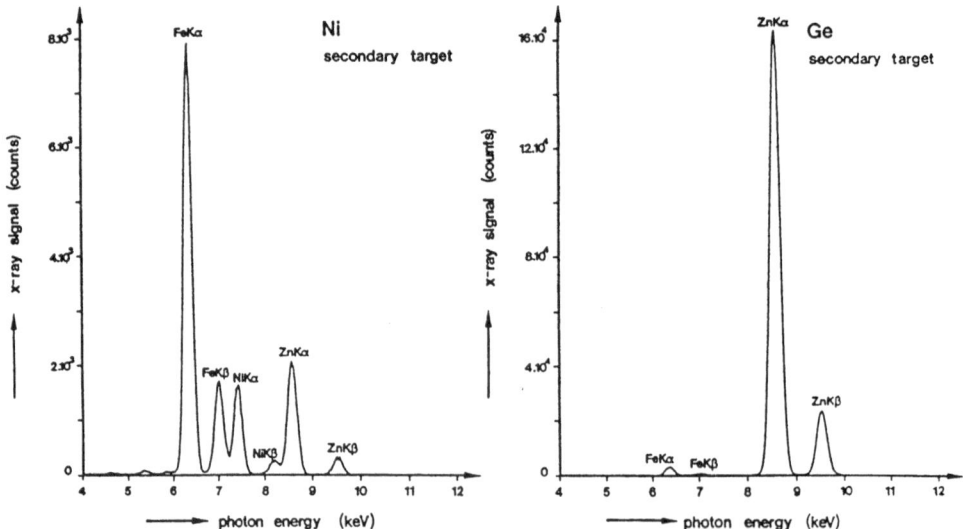

Fig.2 Comparison of the characteristic signals from hot dip galvanized steel with coating thickness of 20μm, excited by NiK and GeK radiation.

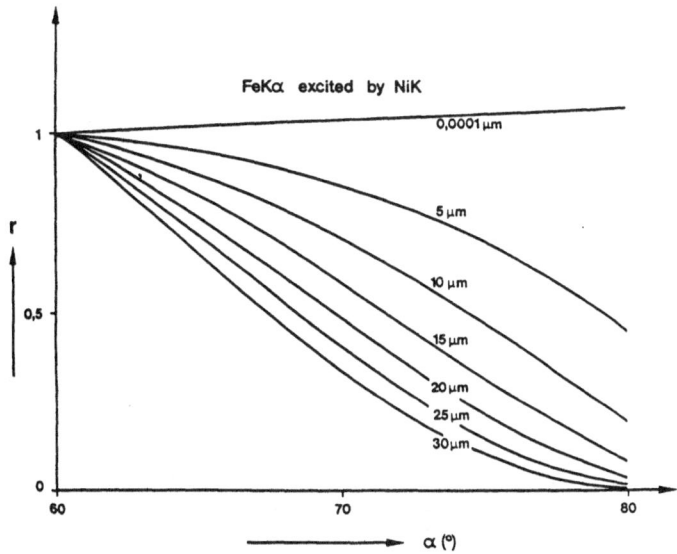

Fig.3 $r_{FeK\alpha}(\alpha)$ with thickness D as parameter ($\alpha_{ref}=60°$, $\alpha+\beta=90°$).

of the countrates of the element i measured under an angle of incidence α and under an arbitrariliy chosen reference angle of incidence α_{ref}. A similar possibility is offered by the ratio of the countrates $n_i(\beta)$ and $n_i(\beta_{ref})$, measured under different take-off angles β.

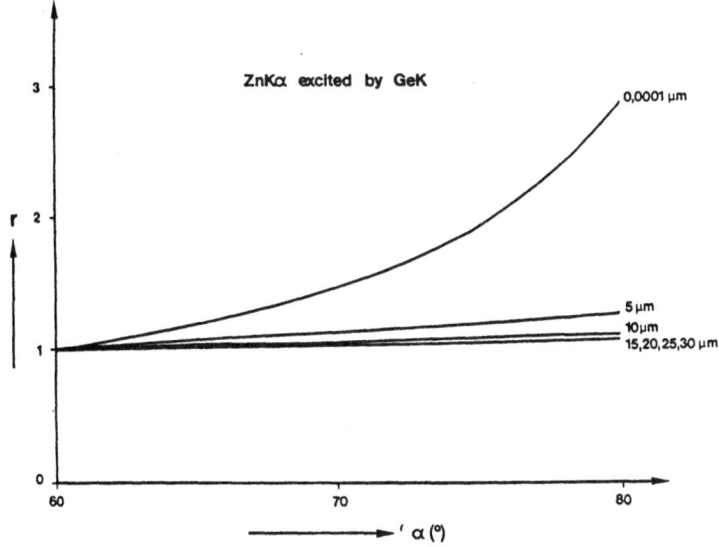

Fig.4 $r_{ZnK\alpha}(\alpha)$ with thickness D as parameter ($\alpha_{ref}=80°$, $\alpha+\beta=90°$).

$$r_{ijk}(\beta) = \frac{n_{ijk}(\beta)}{n_{ijk}(\beta_{ref})} \qquad\qquad (7)$$

The density ρ is $\rho=\rho_{Zn}=7.14g/cm^3$. The relative intensities $I_{s(1)}, I_{s(2)}$ and $I_{s(3)}$ were taken from the tables of Johnson and White[8] and the photoabsorption coefficients from McMaster's tables[9]. Thus, for a thickness determination without reference layers we only need the countrates of one of the two elements, measured under two angles α or β.

Calculated responses of $r_{FeK\alpha}(\alpha)$ (see equ.6) with the thickness D as parameter are given in Fig.3. A reference angle $\alpha_{ref}=60°$ has been chosen and the geometry between the directions of incidence and take-off is fixed to $\alpha+\beta=90°$.

The responses of Fig.3 give an evidence for a good resolution of Zn thicknesses in the range of D>5μm. The $r_{ZnK\alpha}(\alpha)$ responses of Fig.4 show a much poorer resolution at thicknesses D>5μm. But, we expect that thicknesses of D<5μm can be distinguished much better by $r_{ZnK\alpha}$. Thus, we changed the range of incidence angles to $80°\leq\alpha\leq87.5°$ and calculated the responses $r_{FeK\alpha}(\alpha)$ and $r_{ZnK\alpha}(\alpha)$ with $\alpha_{ref}=80°$ of Figs.5 and 6.

From the curves of Figs.3 to 6 we conclude, that this method of XFA performed on steels with Zn coating covers a thickness range from 50nm to 50μm. All results are gained without reference samples. Furthermore it is necessary, to adapt the experiment to the actual problem by a proper choice of the fluorescent radiation (FeKα, ZnKα) and of the range of incidence angles.

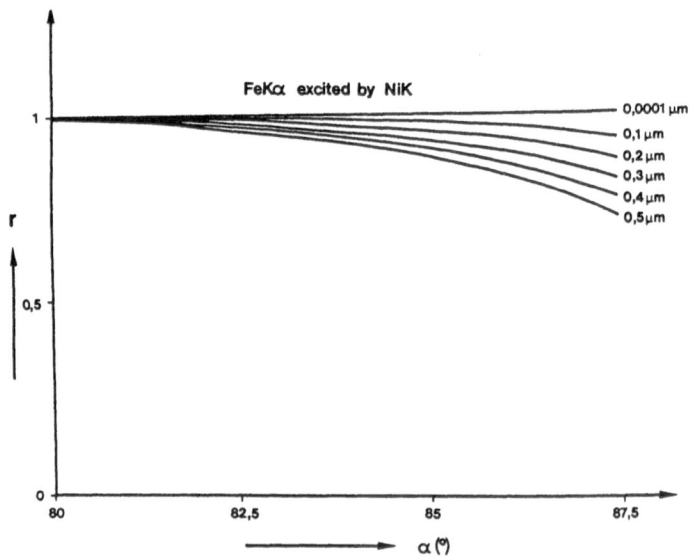

Fig.5 $r_{FeK\alpha}(\alpha)$ with thickness D as parameter ($\alpha_{ref}=80°$, $\alpha+\beta=90°$).

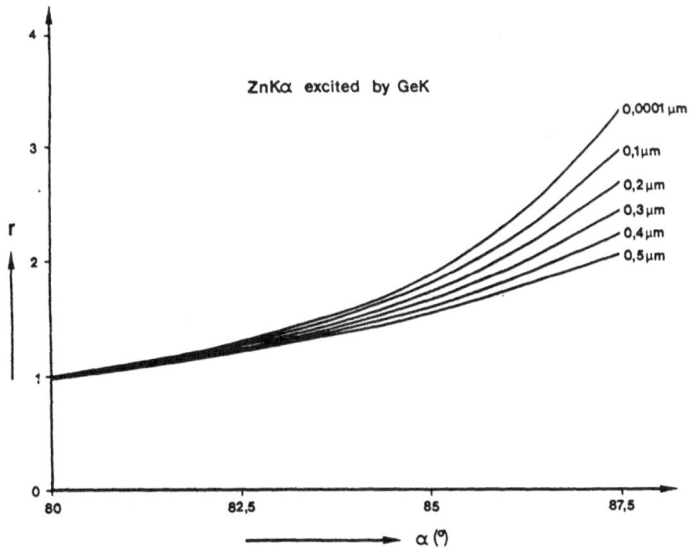

Fig.6 $r_{ZnK\alpha}(\alpha)$ with thickness D as parameter ($\alpha_{ref}=80°$, $\alpha+\beta=90°$).

As already mentioned, there also exists the possibility of monochromatic excitation by means of a crystal monochromator. A specific problem of this method of thickness determination arises from the strength of the measured signals. Only a small portion of the photon flux from secondary targets can be used for the excitation. This is due to the restriction of the beam divergence to well defined angles of incidence. A change to a monochromator is only reasonable for sufficient

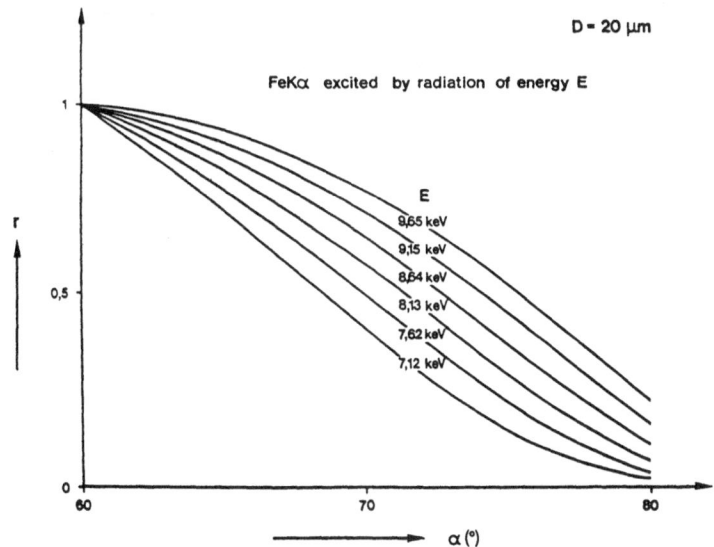

Fig.7 $r_{FeK\alpha}(\alpha)$ with photon energy as parameter ($\alpha_{ref}=60°$, $\alpha+\beta=90°$, $D=20\mu m$).

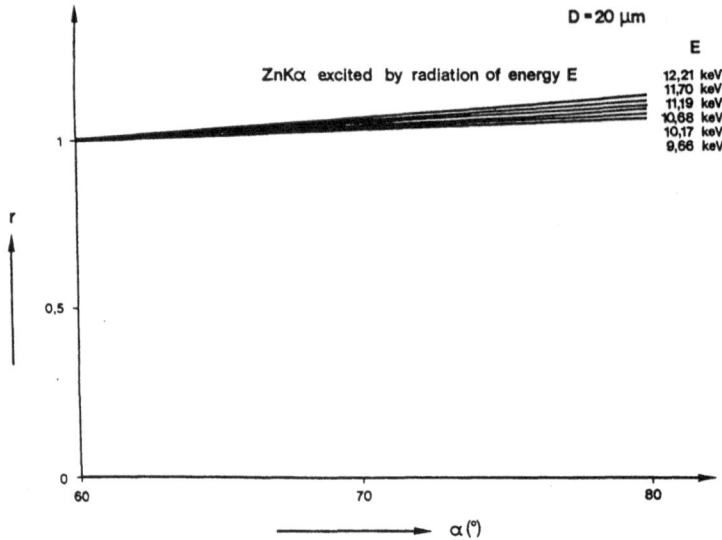

Fig.8 $r_{ZnK\alpha}(\alpha)$ with photon energy as parameter ($\alpha_{ref}=60°$, $\alpha+\beta=90°$, $D=20\mu m$).

photon fluxes from the x-ray tube. An increase of the mono-chromator flux by an insertion of wider slits into the path of x-rays causes an increase of the divergence but also a broadening of the bandwidth of the monochromatic radiation. Consequently, we have to recognize both influences on the responses $r_{ijk}(\alpha)$. For our calculations of the influence of

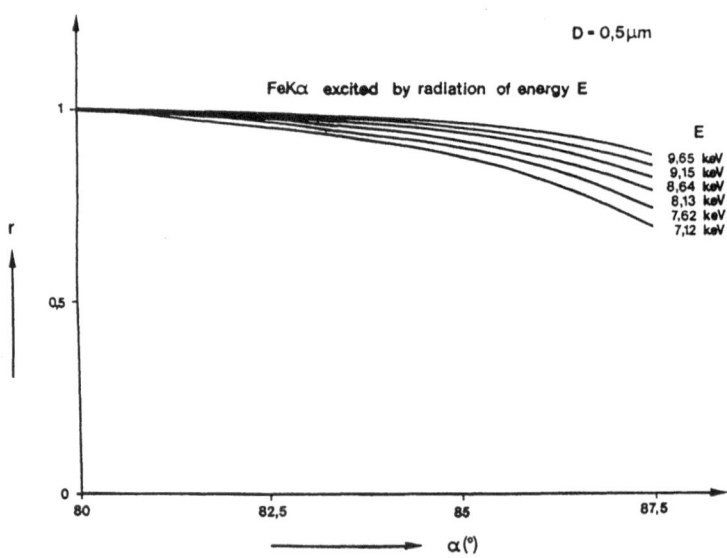

Fig.9 $r_{FeK\alpha}(\alpha)$ with photon energy as parameter ($\alpha_{ref}=80°$, $\alpha+\beta=90°$, $D=0.5\mu m$).

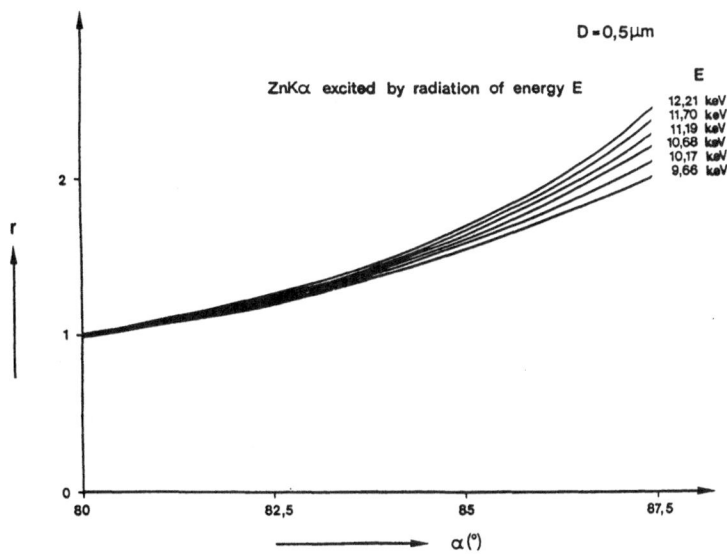

Fig.10 $r_{ZnK\alpha}(\alpha)$ with photon energy as parameter ($\alpha_{ref}=80°$, $\alpha+\beta=90°$, D=0.5μm).

photon energy on $r_{FeK\alpha}(\alpha)$ we represent the energy interval between the FeK and the ZnK absorption edge (7.112keV, 9.659keV) by six equidistant photon energies, separated by 0.51keV. For $r_{ZnK\alpha}(\alpha)$ we employed photon energies with an identical energy separation, starting with the lowest photon energy close to the ZnK absorption edge. Figs.7 to 10 depict the results of the calculations.

The results of Figs.7 to 10 demonstrate the necessity of a precise knowledge of the photon energies of the incident x-radiation.

If there exist experimental results from more than **two** different incidence angles α_u or take-off angles β_u **then** the determination of the thickness D has to be **performed by** a least sqares fit. From a total number of Z+1 **measurements we** obtain Z ratios r_{ijk} and define the error **quantity v for** the least squares fit by

$$v = r - r_{meas} \tag{8}$$

and the standard deviation σ of measured and calculated countrate ratios by

$$\sigma = \sqrt{\frac{1}{Z-1} \sum_{u=1}^{Z} v_u^2} \tag{9}$$

A systematic variation of D allows to plot $\sigma(D)$. The D-value corresponding to the minimum σ_{min} of σ gives the solution. An excellent countrate statistics is reflected by a small numerical value of σ_{min}. But, the numerical value of σ_{min} is not only defined by the statistical significance. Another essential influence is the agreement between the true depth profile and the chosen model function. Hot dip galvanized steel has an interface of intermetallic Fe-Zn compounds between steel an Zn coating. The thickness and the concentration profile in the interface depends on the galvanizing procedure.

b. TRAPEZOIDAL DEPTH PROFILE   (see Fig.11)

A possible description of the true profile is a trapezoidal depth response of the Zn concentration. For the trapezoidal model function the solution of equ.1 becomes

$$n_{ijk} = \frac{\overline{const}_{ijk}}{\cos\alpha}. \tag{10}$$

$$\cdot\sum_{R=1}^{3} I_{s(R)}{}^{\tau}s(R) \cdot \sum_{d=1}^{3} \int_{l(d)}^{u(d)} \rho c_{i} e^{-\sum_{m=1}^{2}\left[\left[\frac{{}^{\tau}s(R),m}{\cos\alpha}+\frac{{}^{\tau}ijk,m}{\cos\beta}\right]\cdot\int_{0}^{t}\rho c_{m}dx\right]} dt$$

First depth interval d=1 $0 \le t \le D_1$ $c_{Zn}=1$
Second depth interval d=2 $D_1 < t \le D_2$ $c_{Zn}=(D_2-t)/(D_2-D_1)$
Third depth interval d=3 $t > D_2$ $c_{Zn}=0$

The lower and the upper limits l(d) and u(d) of the integration intervals depend on the depth interval. The values are $l(1)=0$, $u(1)=l(2)=D_1$, $u(2)=l(3)=D_2$ and $u(3)=\infty$. It should be mentioned that the integration of ρc_m in the exponent has to be performed for d=1 from 0 to t, for d=2 from 0 to D_1 plus

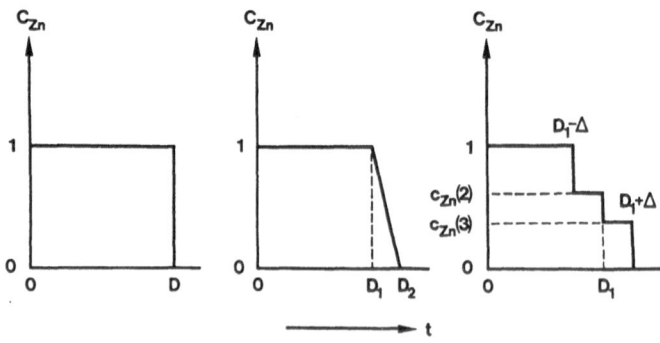

Fig.11 Three model functions of $c_{Zn}(t)$.

the integral from D_1 to t and for d=3 from 0 to D_1 plus the integral from D_1 to D_2 plus the integral from D_2 to t.

The ratios $r_{ijk}(\alpha)$ or $r_{ijk}(\beta)$ are formed in the same way as discussed for the rectangular depth profile. These ratios depend on the unknowns D_1 and D_2. The numerical values D_1 and D_2 are evaluated from the experimental results by the determination of σ_{min}.

Now it becomes possible to compare the numerical values of σ_{min} from the rectangular depth profile and from the trapezoidal depth profile. If σ_{min} of the trapezoidal profile is smaller than σ_{min} of the rectangular profile than the trapezoidal concentration profile gives the better description of the true depth concentration response.

A more detailed information on the concentration response in the interface region can be gained by an intersection of the depth range close to the interface (see Fig.11). In the intervals 1, 2, 3 and 4 we use $c_{Zn}=1$, $c_{Zn}(2)$, $c_{Zn}(3)$ and $c_{Zn}=0$, respectively. Thus, we make use of the results of the preceeding determination of D_1 and D_2.

After the integration of equ.1 we obtain $r_{FeK\alpha}(\alpha)$ and $r_{ZnK\alpha}(\alpha)$ and from the minimum σ_{min} of the standard deviation σ the unknown values of $c_{Zn}(2)$ and $c_{Zn}(3)$. If σ_{min} of the multilayer structure of Fig.11 is smaller than the value from the trapezoidal profile, than the multilayer model comes closer to the true profile. This procedure can be continued by the assumption of more than two layers in the interface region. The limitations are given by countrate statistics, the accuracy of the angular settings, by errors due to beam divergences, the bandwidth of the monochromator and the neglection of the excitation by tube radiation after scattering on the secondary target. A further limitation for an extended application of this procedure comes from lateral inhomogeneities of the specimen.

ABSTRACT

A modified instrumentation allows an application of XFA to a quantification of depth profiles of sample composition. The concept of variable x-ray geometry has been demonstrated for the example of hot dip galvanized steel and a variation of the incidence angle of x-rays. But, the concept can be applied to a variation of the take-off angle too. It is also possible to extend the considerations from a binary system to a multielement system. We investigated the quantification of an interface. The principle of this method allows its application to multilayer structures and different kinds of concentration gradients.

REFERENCES

1   J.Sherman, Spectrochim.Acta 7, 283 (1985)

2   J.W.Criss and L.S.Birks, Anal.Chem. 40, 1080 (1968)

3   N.Gurker, Adv.in X-Ray Anal. 23, 263 (1980)

4   D.Wherry and B.Cross, Analyst 12, 8 (1986)

5   D.R.Boehme, Adv.in X-Ray Anal. 30, 39 (1987)

6   D.A.Carpenter, X-Ray Spectrom. 18, 253 (1989)

7   H.Ebel, Z.Metallk. 56, 802 (1965)

8   G.G.Johnson Jr. and E.W.White, X-Ray Emission and keV Tables for Nondiffractive Analysis, ASTM Data series DS 46, Philadelphia (1970)

9   W.H.McMaster, N.K.del Grande, J.H.Mallett and J.H.Hubbell, Compilation of X-Ray Cross-Sections, UCRL-50174, Sect.II, Rev.1. Lawrence Radiation Laboratory, University of California, Livermore, CA (1969)

GRAZING INCIDENCE X-RAY FLUORESCENCE ANALYSIS

USING SYNCHROTRON RADIATION

Atsuo Iida

Photon Factory
National Laboratory for High Energy Physics
O-ho, Tsukuba-shi, Ibaraki, 305 Japan

ABSTRACT

The X-ray fluorescence analysis of a trace element under a grazing incidence condition has been developed using synchrotron radiation. The interference effect plays an important role for determining the depth distribution of the elemental concentration. The elemental distribution above, on or below the material surface has been studied. The glancing angle dependence of the X-ray fluorescence signal around the critical angle strongly reflects the elemental distribution, and can be used to determine the position of the element of interest.

INTRODUCTION

The development of a characterization technique that is sensitive to the chemical and physical properties of the material surface is vital because of recent advances in both science and technology. X-ray analysis, including x-ray diffraction/scattering and X-ray fluorescence (XRF) has high accuracy and a non-destructive nature. Thin-film analysis has been one of the most successful application fields of X-ray analysis in which the film thickness, composition and crystal structure are determined,[1,2] though the sensitivity to the surface is rather low compared with techniques using electrons and ions.

Recent progress in experimental techniques involving X-ray analysis, especially the use of the grazing-incidence condition, have enabled X-ray analysis to be capable of characterizing the surface of a material. Grazing incidence X-ray diffraction/scattering studies can reveal the surface crystal structure.[3] Total reflection XAFS (X-ray absorption fine structure) is used to analyze the local atomic arrangement and electronic structure near the surface.[4] X-ray reflectometry has long been used to characterize the layered surface structure.[5,6,7] An X-ray standing wave (XSW) method has recently been applied to the determination of a surface atomic arrangement.[8] All of these methods have become powerful and effective techniques using synchrotron radiation (SR), due to the high brilliance.

The detection of an XRF signal under the grazing incidence condition is used in the total-reflection XAFS and XSW. For analytical applications, XRF analysis under the grazing incidence condition has been used for elemental analysis in the dilute solution[9] and the surface contamination analysis of semiconductor wafers.[10] Another application of grazing-incidence XRF (GIF) analysis is to determine the concentration profile of elements of interest along the depth.[11,12,13,14]

Since SR has a narrow angular divergence, a high photon flux density and linear polarization, XRF analysis has extremely high sensitivity for trace elements.[15] GIF using SR is, therefore, useful for both high angular-resolution experiments as well as trace-element analysis. In this paper analyses of the concentration profile of trace elements using SR is reviewed. X-ray intensity modulation above a surface or in a thin film is interpreted in terms of interference between the refracted (incident) and reflected X-rays. Emphasis is put on a qualitative understanding of the oscillation structure observed in the reflectivity as well as the XRF intensity angular dependences.

EXPERIMENTAL AND ANALYTICAL PROCEDURE

Although the experimental procedure using SR has been described in detail elsewhere,[14] a brief summary is given here. Synchrotron X-rays were monochromatized with a sagittal focusing double-crystal monochromator with a horizontal rotation axis. The sample was mounted on a high-precision goniometer with a translation stage for alignment. The incident and the reflected X-rays were monitored with ionization chambers and the XRF signal was detected with a Si(Li) detector.

In order to calculate the XRF intensity under the grazing incidence condition, macroscopic optical theory[16,17,18] has been used. For instance, the reflectivity for layered material (Fig.1) is given by a recursion formula, as follows:

$$R_{n-1,n} = a_{n-1}^4 \, (R_{n,n+1} + F_{n-1,n}) \, / \, (R_{n,n+1}F_{n-1,n} + 1), \qquad (1)$$

where

$$R_{n,n+1} = a_n^2(E_n^r \, / \, E_n)$$

and

$$F_{n-1,n} = (f_{n-1} - f_n) \, / \, (f_{n-1} + f_n).$$

Here, R and F are the reflection coefficient and the Fresnel coefficient, respectively; a_n (the phase factor) and f_n are given by

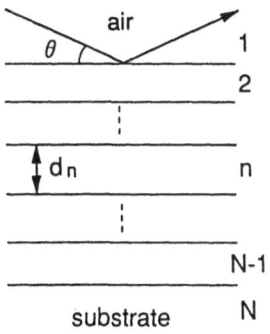

Fig.1 Schematic representation of a multilayer thin film used for the calculation. θ is the glancing angle and d_n the thickness of the n-th layer.

$$a_n = \exp(-i\pi f_n d_n/\lambda)$$

and

$$f_n = (\theta^2 - 2\delta - 2i\beta)^{1/2},$$

where d_n and $n = 1 - \delta - i\beta$ are the thickness and the refractive index of the n-th layer, λ is the X-ray wavelength and θ is the glancing angle. The critical angle (θ_c) is defined as $\theta_c = \sqrt{2\delta}$. The XRF intensity is proportional to the X-ray intensity, which is also calculated according to the same formalism.

XRF INTENSITY FROM HOMOGENEOUS MATERIAL

We first consider the XRF intensity from trace elements which are near the surface of a homogeneous material. In this case, the XRF intensity was calculated from eq.(1) with N=2. This type of material has been studied both extensively and intensively by GIF. There are two types of samples: 1) the element of interest is in or on the sample, and 2) it is above the surface.

Elements in/on the surface

When the element of interest is in or on the sample, the angular dependence of the XRF intensity reflects the elemental concentration along the depth. Under the grazing incidence condition, the X-ray intensity in the material decreases along the depth (evanescent wave). The penetration depth varies from a few nm to a few μm as a function of θ. The XRF intensity from the element is given by[14]

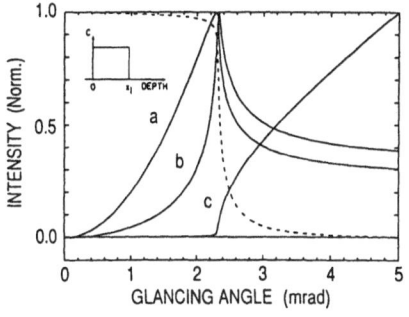

Fig.2 Normalized calculated X-ray fluorescence intensities for various concentration depths. The concentrational profile along the depth is a step function, as shown in an inset. $X_j = 5$ Å, 200 Å and ∞ for curves a, b and c respectively. The reflectivity is also indicated by the broken line. A Si substrate with an incident X-ray energy of 13.4 keV is assumed.

$$I(\theta) \propto M(\theta) \int F(t) \ \exp(-\mu(\theta) t) \ dt, \tag{2}$$

where M(θ) is the surface intensity of X-rays, F(t) the concentrational distribution along the depth (t) and $\mu(\theta)$ the effective absorption coefficient (reciprocal of the penetration depth). The typical angular dependences of the calculated XRF intensities are shown in Fig.2. For simplicity, a step function is assumed for the concentration profile. The angular dependence of the XRF intensity is sensitive to the concentrational profile; the depth analyzed with this technique is between a few nm and a few μm. The application of such analysis to various types of semiconductor materials[13,14] and dissolved polymers[11,12] has been reported.

A typical example of an analysis in which the element of interest is on the sample is contamination analysis of a semiconductor (Si) wafer[10] and trace-element analysis of a solution.[9] Fig.3 shows a spectrum obtained from a contaminated Si wafer. Due to the very low scattered X-ray background, a highly sensitive analysis down to 10^{10} atoms cm^{-2} has been achieved. For a contamination analysis, the trace element is just on the surface, and the XRF intensity is similar to profile a in Fig.2. In solution analysis, however, the sample thickness cannot be neglected, since the XRF intensity is the integral of the X-ray intensity over the thickness. The inhomogeneity of the sample thickness is a source of error for quantitative analysis.[19] When the sample thickness is more than a few thousand Å and is homogeneous, the XRF intensity is doubled below the critical angle, because the sample is excited by the incident and reflected X-rays and the interference effect is averaged.

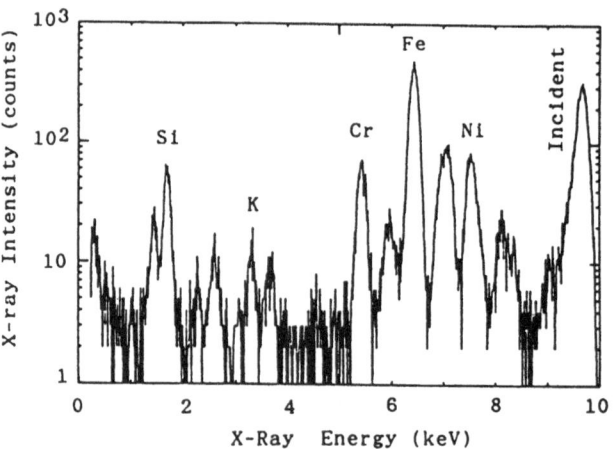

Fig.3 X-ray fluorescence spectrum from a contaminated Si wafer under the total-reflection condition. The excitation X-ray energy was 9.6 keV and counting time was 500 sec. 0.5 mrad glancing angle.

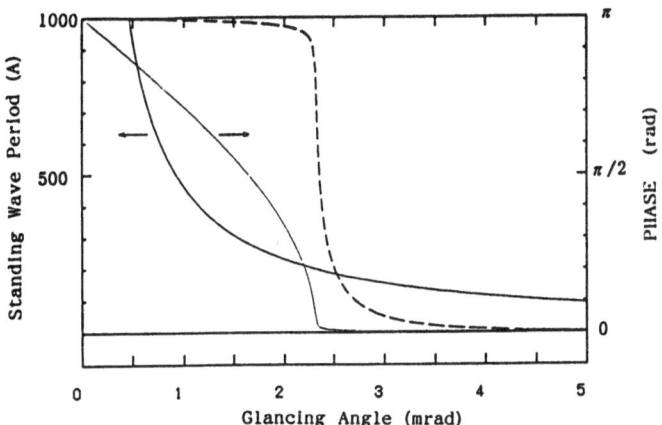

Fig.4 The wavelength of a standing wave above the surface and a phase change at the surface. The calculation condition is the same as that in Fig.2.

Element above the surface

The surface intensity (M(θ)) in eq.(2) is the result of X-ray interference between the incident (E) and the reflected (E^r) X-rays. The X-ray intensity above the surface is approximately given by

$$I = | E + E^r |^2$$
$$= E^2 \{ 1 + b^2 + 2 b \cos(2\pi kr - \gamma) \}, \tag{3}$$

where k is the scattering vector, r is the distance from the surface and $E^r/E = b \exp(i\gamma)$. The wavelength of the standing wave L is given by $L = |k|^{-1} = \lambda / 2 \sin\theta$. L, together with γ, are shown in Fig. 4 as a function of the glancing angle. γ varies from π to 0 as θ changes from 0 to θc, so the standing wave is destructive (node) at the surface where $\theta = 0$, but is constructive (anti-node) for $\theta >= \theta_c$. Fig.5 illustrates the X-ray intensities above and below the surface for several glancing angles. The exponential decay of the X-ray intensity in the substrate implies an evanescent wave.

From Fig.5, it can be seen that the X-ray intensity above the surface is strongly modulated due to an interference effect. Fig.6 shows the calculated XRF intensity from an element located above the surface as a function of the glancing angle. From this figure, one can determine the position of the element (surface XSW). Recent studies concerning a Langmuir-Blodgett (L-B) film[20,21,22] are based on such an XSW above the surface. The advantages of surface XSW, compared to XSW using a crystal, are that it does not require any periodic structure, and that the period of the standing wave is more than a few nm.

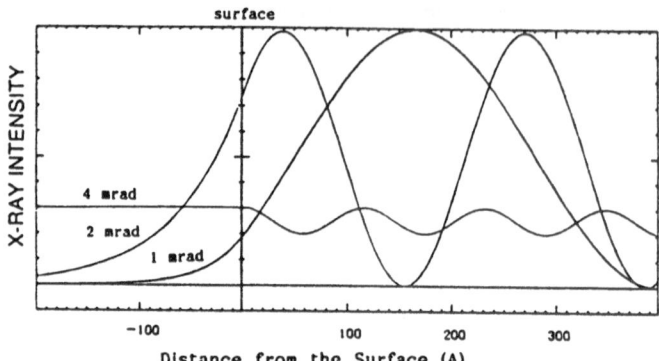

Fig.5 The X-ray intensity below and above the surface. The positive distance is for above the surface. The calculation condition is the same as that in Fig.2.

The angular dependences of the XRF intensity of solution analysis and surface contamination analysis are special cases of an XSW method. In the former, the XRF intensity is integrated over the sample thickness; in the latter, the XRF signal from the element of interest reflects the X-ray intensity at the surface. Furthermore, in the next section we show that the analysis of a L-B film is a special case of a layered structure in which the density of the film is very low.

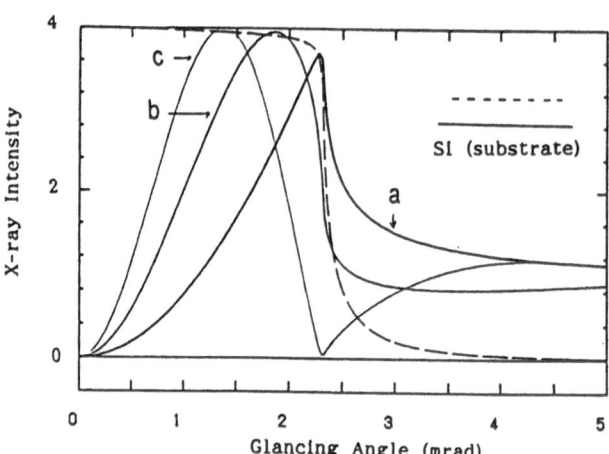

Fig.6 X-ray intensity above the surface as a function of the glancing angle. Curves a, b and c correspond to the X-ray intensity at 0 Å, 50 Å and 100 Å above the surface respectively. The broken line shows the reflectivity.

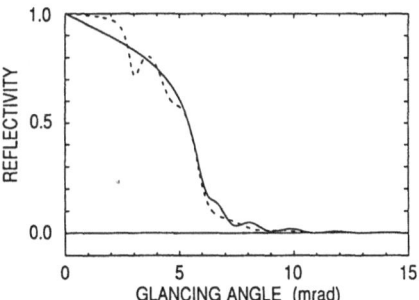

Fig.7 Calculated angular dependences of the reflectivity for a single layer film. A W film/Si substrate (solid line) and a Si film/W substrate (broken line). The film thickness is 200 Å. The incident X-ray energy is 13.4 keV. The critical angles for Si and W are 2.32 and 5.78 mrad, respectively.

XRF INTENSITY FROM A LAYERED STRUCTURE

A thin film

The angular dependence of the reflectivity and the XRF intensity from a layered structure shows the oscillation structure which is usecd for determining the layer thickness, refractive index and the surface roughness.[6,17,18,23] To analyze the oscillation structure, we first consider the simplest case, i.e. a single thin layer on a substrate. We had better consider two cases: (1) $\delta_f > \delta_s$ ($\theta_c^f > \theta_c^s$, high-Z element film on a lower Z substrate) and (2) $\delta_f < \delta_s$ ($\theta_c^f < \theta_c^s$), where f and s stand for the film and substrate, respectively.

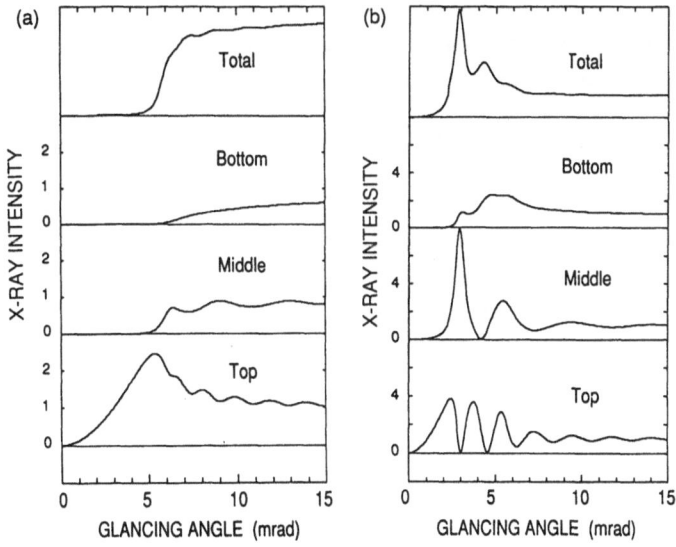

Fig.8 Calculated X-ray fluorescence intensities from elements localized at various depths (top, middle and bottom) in the thin film and from a homogeneous distribution (total). (a)W/Si[s] and (b)Si/W[s].

The calculated reflectivity curves are shown in Fig.7 for a W film on a Si substrate (W/Si[s]) and a Si film on a W substrate (Si/W[s]). The reflectivity curve shows an oscillation structure which is the result of interference between the refracted and reflected X-rays in the film. For W/Si[s], the incident X-rays penetrate into the bottom interface for $\theta > \theta_c{}^W$; however, since there are no appreciable reflected X-rays, so the interference effect is rather weak. In contrast, for Si/W[s], since the reflected X-rays in the film are strong for $\theta_c{}^{Si} < \theta < \theta_c{}^W$, the standing-wave field is significantly enhanced over this angular range.

Figures 8(a) and (b)   show the calculated XRF intensities from an element localized at various depths in the film. As shown in eq.(3), the X-ray standing wave in the film is well described by L and γ.   The wavelength of the standing wave (L) is given by $|k|^{-1} = \lambda/2|f_n|$. The phase change ($\gamma$) at the bottom interface increases from 0 to π (π to 0) as θ increases from 0 to θ_c, and is π (0) for $\theta > \theta_c$ for $\delta_f > \delta_s$ ($\delta_f < \delta_s$). With these parameters, the curves in Fig.8 can be qualitatively understood.

For $\delta_f > \delta_s$ (Fig.8(a)), the XRF intensity from the top surface shows an oscillation structure which reflects the successive change in the standing-wave field; that from the bottom interface however, shows no significant oscillation, since the standing wave has its node at the bottom interface ($\gamma = \pi$). When the elemental distribution is homogeneous in the film, the resultant XRF intensity (total) has a weak oscillation structure due to contributions from various depths.

In Fig.8(b), the oscillation structure   is strongly enhanced for $\theta_c{}^{Si} < \theta < \theta_c{}^W$ and the XRF intensity curve from each depth has its own characteristic profile. For $\theta > \theta_c{}^W$, the oscillation structure is relatively weak, similar to that in Fig.8(a), except that the standing wave has its anti-node at the bottom interface. It can also be seen from both Figs.8(a) and (b) that the XRF intensity sharply increases at higher θ when the element of interest is located deeper within the material.

Fig.9 shows the experimental result for thin films. In Fig.9(a), the oscillation structure of a Cr K XRF signal is weak, and is similar to the calculated results given in Fig.8(a), except that the effect of the roughness is neglected in the calculation. Although the sample shown in Fig.9(b) conprises of 3 layers, for an X-ray analysis, only the top 2 layers are important, since the Au layer is so thick that there is no appreciable reflected wave from this layer. A strong enhancement in the Cr K XRF signal is observed for $\theta_c{}^{Cr} < \theta < \theta_c{}^{Au}$. It was also experimentally confirmed that the number of maxima in the reflectivity and the XRF intensity was dependent on the thickness of the Cr layer, but was independent of λ, except for the anomalous dispersion region (near the absorption edge), as is expected from theory.

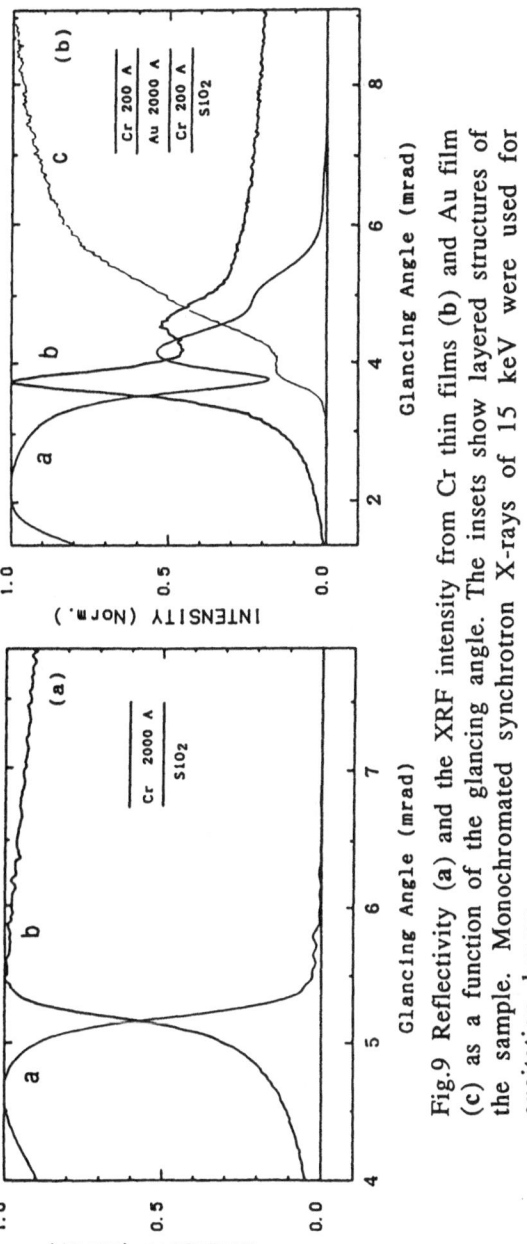

Fig.9 Reflectivity (a) and the XRF intensity from Cr thin films (b) and Au film (c) as a function of the glancing angle. The insets show layered structures of the sample. Monochromated synchrotron X-rays of 15 keV were used for excitation beams.

In summary, the XRF intensity angular dependence for an element in a film can be qualitatively interpreted. The glancing angle at which the XRF intensity increases reflects the depth distribution of the element of interest. The oscillation structure also gives information concerning the position of the element.

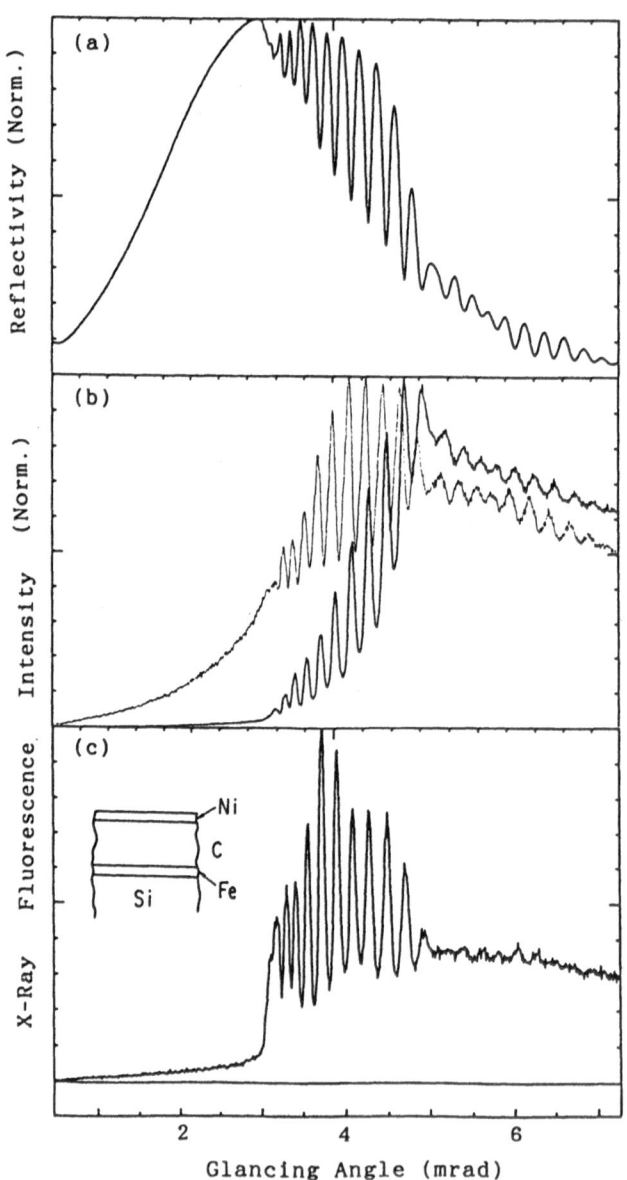

Fig.10 Reflectivity (a) and Fe, Ni (b) and Zn (c) K fluorescence intensity as a function of the glancing angle. Ni(34 A)/C(843 A)/Fe(111 A)/Si(substrate). The excitation energy was 10.5 keV.

When the concentration of the element of interest is high, i.e. the major constituent, the reflectivity curve can be effectively used for an analysis of the layered structure, since the concentrational distribution affects the δ variation. The XRF method used in combination with SR is, however, unique for minor and trace-element analyses near the surface. An example involving the analysis of a minor element is shown in Fig.10. A sample consists of 3 layers of Ni/C/Fe and the interference effect is significant, since Ni and Fe thin layers act as the high-reflectivity layer. The glancing angle at which the oscillation begins corresponds to the critical angle of C. The Zn signal obtained is the contamination during the evaporation process. Because the Zn signal abruptly increases at $\theta > \theta_c C$, it was determined that the greater part of Zn was incorporated in the C layer. A comparison with the calculation showed that the Zn contamination occurred during the C evaporation process and that the Zn concentration in the C layer was a few %.

Since the sensitivity of SRXRF is less than 10^{12} atoms/cm^2, trace element analysis in a layered structure seems to be both useful and effective.

Multilayer structure

As the number of layer increases, although the analysis becomes more complicated, the basic idea of a standing wave still holds. One of the interesting applications is the well-known XSW method for a multilayer material which has a periodic structure. It was confirmed that the dynamical diffraction phenomenon was observed in the Pt-C multilayer[24] and the Langmuir-Blodgett film[25] and that a standing wave was excited in the layered structure. The position of the impurity atoms in the synthetic multilayer (W/Si)[26] and the transition metal element in the L-B film[27] were determined.

SUMMARY

In this review, it was shown how to extract information about the concentrational distribution of a trace element above, on or below the surface of a material with reference to the interference effect. Though most of the applications up to now have been concerned with major or minor element analysis, the SR XRF method will become a unique tool for trace element analysis. Another interesting application is chemical state analysis of a surface or thin film using grazing-incidence condition.[28] The combination of SR and GIF will have many potential applications in fundamental research as well as in practical analysis and provide us with a fruitful analytical method.

ACKNOWLEDGMENT

The author thanks Dr. K.Sakurai for his permission to reproduce figure 9 as well as valuable discussion. Thanks are also due to T.Noma and S.Kojima for their help with the experiments. I wish to thank Prof. Y.Gohshi for his continuous encouragement and the PF staff for their help during the study.

REFERENCES

1)T.C.Huang and W.Parrish, Adv. in X-ray Anal., **22**:43 (1979)

2)D.Laguitton and W.Parrish, Anal. Chem., **49**:1152 (1977)

3)I.K.Robinson, in "Handbook on Synchrotron Radiation", D.Moncton and G.S.Brown, eds. (North-Holland, Amsterdam,1991)

4)G.N. Greaves, Adv. in X-Ray Analysis, **34**:13 (1991)

5)H.Kiessig, Ann. Phys., **5**:715 (1931)

6)P.Boher, P.Houdy and C.Schiller: J. Appl. Phys., **68**:6133 (1990)

7)B.Lengeler, these proceedings

8)M.J.Bedzyk, Nucl. Instrum. and Methods, **A266**:679 (1988)

9)P.Wobrauschek, P.Kregsamer, C.Streli and H.Aiginger, Adv. In X-Ray Anal., **34**:1 (1991)

10)K.Nishihagi, N.Yamashita, N.Fujino, K.Taniguchi and S.Ikeda, Adv. in X-Ray Anal., **34**:81 (1991)

11)J.M.Bloch, M.Sansone, F.Rondelez, D.G.Peifer, P.Pincus, M.W.Kim and P.M.Eisenberger, Phys. Rev. Lett., **54**:1039 (1985)

12)W.B.Yun abd J.M.Bloch, J. Appl. Phys., **68**:1421 (1990)

13)M.Brunnel, Acta Cryst. **A42**:304 (1986)

14)A.Iida, Adv. in X-ray Anal., **34**:23 (1991)

15)A.Iida and Y.Gohshi, in "Handbook on synchrotron radiation", vol.4, p307 (North-Holland, Amsterdam, 1991).

16)L.G. Parratt, Phys. Rev., **95**:359 (1954)

17)A.Krol, C.Sher and Y.H.Kao, Phys. Rev. **B38**:8579 (1988)

18)D.K.G. de Boer, Phys. Rev., **B44**:498 (1991)

19)A.Iida, A.Yoshinaga, K.Sakurai and Y.Gohshi, Anal. Chem. , **58**:394 (1986)

20)M.J.Bedzyk, D.H.Bilderback, G.M.Bommarito, M.Caffrey and J.S.Schildkraut, Science **241**:1788 (1988)

21)M.J.Bedzyk, G.M.Bommarito and J.S.Schildkraut, Phys. Rev. Lett., **62**:1376 (1989)

22)M.J.Bedzyk, G.M.Bommarito, M.Caffrey and T.L.Penner, Science **248**:52 (1990)

23)K.Sakurai and A.Iida, These proceedings

24)T.W.Barbee, Jr. and W.K.Warburton, Materials Letters, **17** (1984)

25)A.Iida, T.Matsushita and T.Ishikawa, Jpn. J. Appl. Phys., **24**:L675 (1985)

26)T.Matsushita, A.Iida, T.Ishikawa, T.Nakagiri and K.Sakai, Nucl. Instr. and Methods **A246**:751(1986)

27)H.D.Abruna, G.M.Bommarito, D.Acevedo, Science, **250**:69 (1990)

28)K.Sakurai and A.Iida, Adv. in X-Ray Anal.,vol.33, p205

GRAZING INCIDENCE X-RAY SPECTROSCOPY

FOR THIN LAYER ANALYSIS

Hideki Hashimoto, Hiroshi Nishioji and Hideo Saisho

Toray Research Center Inc.
Sonoyama 1-Chome, Otsu, Shiga 520, Japan

ABSTRACT

Reflection and fluorescence intensity profile curves for thin films were measured under the grazing incidence conditions using synchrotron radiation. A titanium layer and a carbon / titanium bilayer sputtered on a silicon wafer were subjected to heat treatment. The analysis of the reflection and fluorescence profile curves shows that the sample without the heat treatment has another high-density layer on the surface or interface, and that the heat treatment results in the removal of the high-density layer and the formation of a thick homogeneous layer.

INTRODUCTION

The reflection curve for a stratified material under the grazing incidence conditions has a shape characterizing its layered structure. Modulation in the reflection curve is much affected by the thickness of the layers and the interfacial roughness[1-3]. The same phenomenon as the reflectivity is also observed in the fluorescence intensity profile curve of elements constituting the layered-structure sample. On the basis of the analysis of these curves measured as a function of the the glancing angle, we can determine the following characteristic parameters of layered thin films:

(1) The thickness and density of layers.
(2) The concentration profile of composition elements.
(3) The roughness of surface or interface.
(4) The depth profile of impurity elements.

The present paper describes the characterization of layered thin films by the analysis of both reflection and fluorescence intensity profile curves.

EXPERIMENTAL

Samples

   Two kinds of samples were prepared by a sputtering method : one was a
Ti film 50nm thick on a Si wafer, and another was a C / Ti bilayer film
with 300nm of C and 30nm of Ti on a Si wafer. The samples prepared were
cut into pieces of 2cm square. A part of these sample pieces were
subjected to heat treatment under an argon atmosphere at 500, 750 and
1000K for 2 hr.

Measurements

   All measurements were performed using synchrotron radiation at the
Photon Factory on the beam line 4A. The energy of incident X-rays was
10keV. Incident and reflected X-ray intensities were measured with
ionization chambers ; fluorescence X-ray intensities were detected with a
Si(Li) detector. The beam size of incident X-rays was 0.1mm in height and
2mm in width.

RESULTS AND DISCUSSION

Data Analysis

   Using some models for a single- or multi-layer structure, Vidal and
Vincent's matrix method for theoretical calculation[4-5] is applied to find
an optimal model with a theoretical curve in closer agreement with
experiment. The parameters used in these calculations are the complex
refractive index $n = 1- \delta -i \beta$, the thickness and the interfacial roughness
for the respective layers. The following relationships give δ and β:

$$\delta = ne^2 \lambda^2 / 2\pi mc^2 \qquad\qquad (1)$$

$$\beta = \mu \lambda / 4\pi \qquad\qquad (2)$$

where n is the total number of dispersive electrons per unit volume of
the material and μ is the linear absorption coefficient. Among these
parameters, n and μ can be obtained from the elemental composition and
density. Consequently, analyses of the reflection and fluorescence
profile curves enable us to determine the important characteristic data
on the layered thin films.

Titanium films

   Figure 1(a) shows the experimental reflection curves, where periods of
modulation become shorter as the heat treatment temperature is raised.
This means that the layer thickness is increasing as the heating
proceeds. Since the total Ti amount is not changed, the cause should be
sought in the change of metallic titanium into titanium silicide due to
the heating, resulting in the increase of layer thickness. Figure 1(b)
shows the fluorescence intensity profiles of Ti Kα. Except for the
as-deposited sample, the profiles remain in the same shape. The maximum
fluorescence intensity is at a similar level for all four samples.

   If the layer of the as-deposited sample was of metallic titanium, its
thickness was estimated to be around 57nm on the basis of the modulation
periods of the reflection curve. However, since the amplitude was not

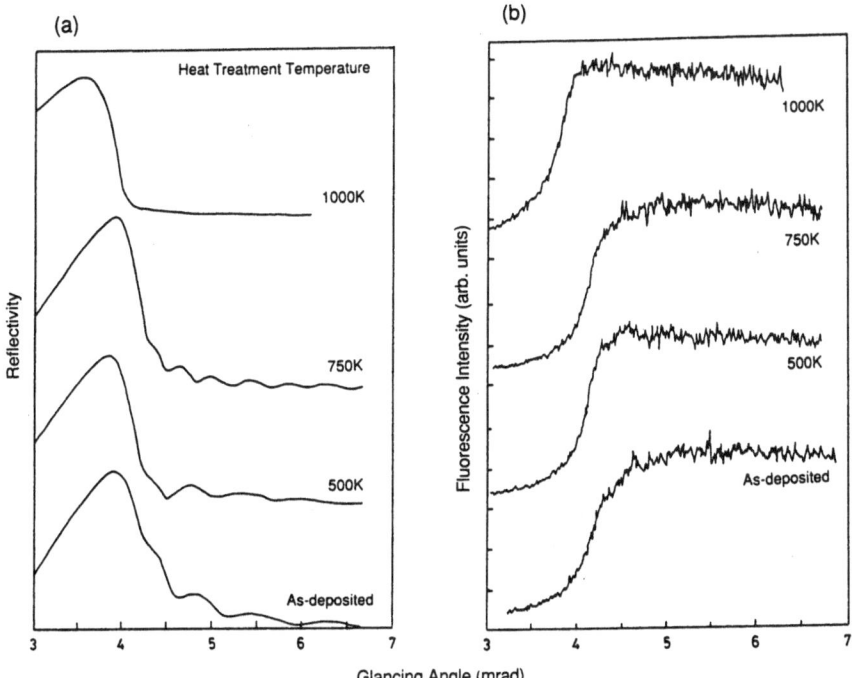

Fig.1.   Experimental reflection curves (a) and Ti Kα fluorescence
          intensity profiles (b) for Ti on Si wafer samples
          (as-deposited and heat treatment).
          Heat treatment temperatures: 500, 750 and 1000K.

fully fit, the experimental results could not be well explained if the
proposed model was of metallic titanium alone. Hence, a bilayer model was
considered, where another layer was present between the Ti film and Si
substrate. The assumption of the presence of 8nm titanium silicide layer
with the   density of $4.13g/cm^3$ beneath 51nm titanium layer led to good
agreement   between   the   calculated   and   experimental   reflection   curves.
However,   the   fluorescence   profile   failed   to   fit.   A   large   discrepancy   was
particularly found around the critical angle. No fit calculation could be
found out even if the composition or density of the silicide layer was
modified or even when the number of silicide layers was increased. Hence,
another layer must be assumed on the Ti layer. This layer should have a
larger critical angle than Ti. Titanium oxide was found to be appropriate
for it. The values of δ and β of the oxide layer and the thickness of
the respective layers were modified. As shown in Fig.2, the final results
indicate that the composition consists of 42% titanium and 58% oxygen and
the density is $4.67g/cm^3$. The composition is similar to that of Ti_2O_3
(density = $4.6g/cm^3$, Ti / O =40 / 60). The total Ti amount calculated from
the above results is $24\mu g/cm^2$. This value is in good agreement with that
determined by chemical   analysis.   This   confirms   the   validity   of   the
proposed structure model.

     The analysis of the reflection and fluorescence   curve for the heat
treatment at 500K indicates that the Ti layer reacts with the silicon
substrate   to   form   the   homogeneous   silicide   layer   64nm   thick.   The
fluorescence intensity profile suggests that such a high-density surface

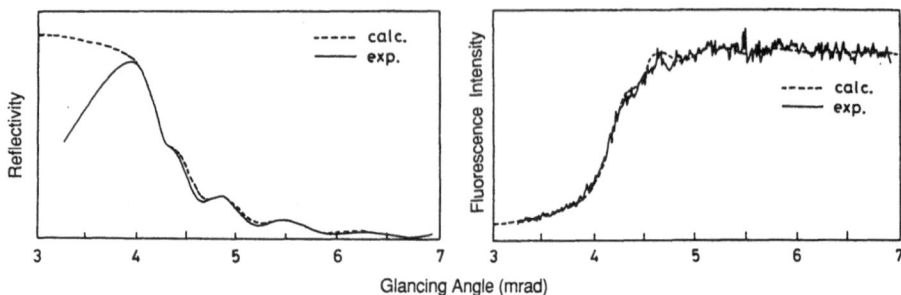

Fig.2.   Comparison of the experimental results for the as-deposited
          sample with the calculations based on the model of
          $Ti_{0.42}O_{0.58}$(11nm) / Ti (43nm) / $TiSi_2$(6nm) / silicon substrate.
          $\delta = 9.16 \times 10^{-6}$, $\beta = 3.55 \times 10^{-7}$ for the oxide layer;
          $\delta = 8.56 \times 10^{-6}$, $\beta = 4.84 \times 10^{-7}$ for the titanium;
          $\delta = 8.08 \times 10^{-6}$, $\beta = 2.67 \times 10^{-7}$ for the silicide;
          $\delta = 4.75 \times 10^{-6}$, $\beta = 7.42 \times 10^{-8}$ for the silicon substrate.

layer as found in the as-deposited sample does not exist or, even if any,
it is so thin that no influence is caused. It is estimated from the
calculation that the silicide layer consists of 76% Ti and 24% Si, and
its density is $4.41g/cm^3$.   The sample treated at 750K also forms the
homogeneous silicide layer of 96nm thickness. It fits well to the
experimental results, when it has the composition of 46% Ti and 54% Si,
and its density is $4.31g/cm^3$.

Bilayer films of carbon / titanium

     Figure 3 shows the experimental results for the as-deposited sample.
The presence of two critical angles is found in the reflection curve. The
smaller angle side corresponds to the carbon layer ; the larger angle
side corresponds to the titanium layer. The fluorescence profile was
measured for Fe as well as for Ti because of detection of Fe regarded as
an impurity in the carbon layer.

     Caluclation was made on the assumption that a bilayer film of C / Ti on
the Si substrate was present. The experimental results can be well

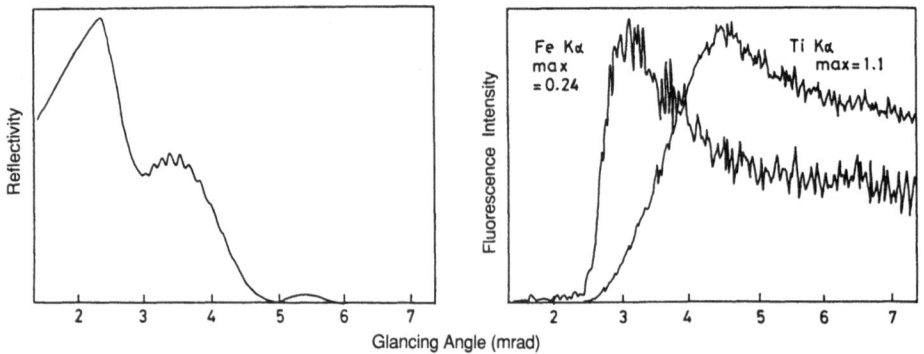

Fig.3.   Experimental results for the carbon and titanium sputtered
          sample without heat treatment.

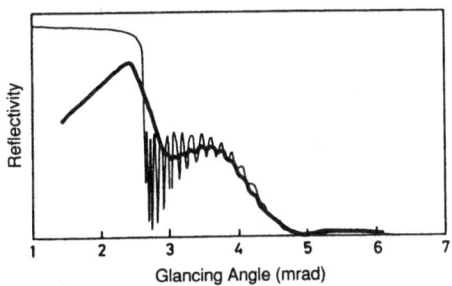

Fig.4.   Comparison of the experimental
reflection curve (bold line)
with the calculation ; the
sample is assumed to be
carbon (280nm) / $Ti_{0.65}Si_{0.35}$
(23nm) / silicon substrate.
$\delta = 3.45 \times 10^{-6}$, $\beta = 2.57 \times 10^{-8}$
for the carbon layer and
$\delta = 8.46 \times 10^{-6}$, $\beta = 3.57 \times 10^{-7}$
for the silicide layer.

reproduced in case the parameters in Fig.4 and the surface roughness of
5nm are applied. The values of δ and β at the carbon layer cannot be
obtained simply by changing the carbon density. The influence of impurity
Fe should be taken into consideration. The carbon layer contains 1.7% Fe
and the density is $1.7g/cm^3$. This density is rather smaller than the
graphite density of $2.26g/cm^3$. The titanium layer is made of titanium
silicide, consisting of 65% Ti and 35% Si, and the density is $4.35g/cm^3$.
When the experimental and calculated results are compared in more detail
as to the reflection curve, the decreasing tendency for the reflectivity
around the critical angle differs. This difference suggests that the
presence of another layer on the surface or C / Ti interface. Such a
layer sould have a higher density, i.e., a larger δ-value than the layer
below.

Figure 5 shows that the reflectivity of the heated sample at 500K
decreases around the critical angle more abruptly than that of the
as-deposited sample. This indicates that the high-density layer either on
the surface or on the interface is almost removed and the heated sample
can be taken as a bilayer structure. This fact can also be confirmed at
the fluorescence profile of Fe and Ti. In the case of the heat-treated
sample at 1000K, Fe and Ti have the same distribution pattern, as shown

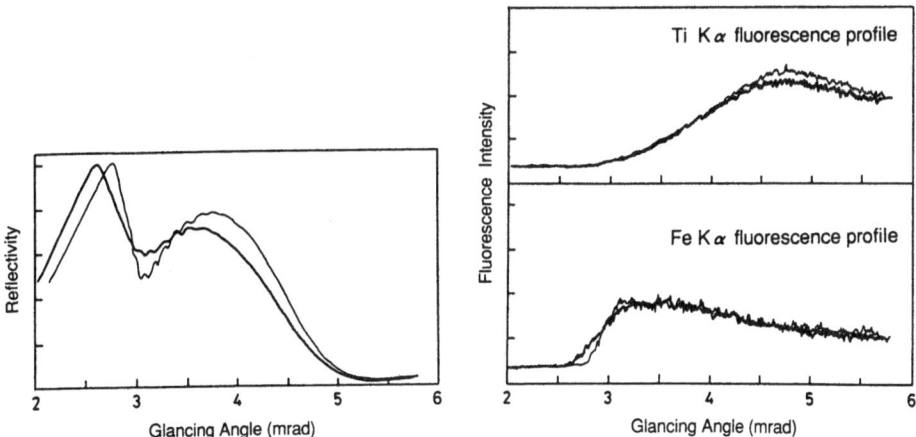

Fig.5.   Experimental results for the as-deposited sample (bold line)
and 500K heat treatment sample.

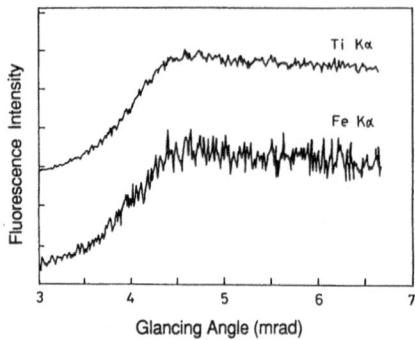

Fig.6.   Comparison of the Ti K α fluorescence profile
with the Fe K α fluorescence profile for the 1000K
heat treatment C / Ti / Si substrate sample.

in Fig.6. It is assumed that C atoms make the same behavior with Fe atoms
and form a single layer by mixture with Ti and Si atoms. The detailed
analysis in this connection will be a task to be investigated in future.

CONCLUSION

Measurement and analysis of the reflectivity and fluorescence under
the grazing incidence conditions using synchrotron radiation provide us
much useful information regarding the composition, thickness and density
of a thin film nondestructively. The present study of the thin films with
Ti and with C / Ti has yielded the following conclusions:

(1) The as-deposited sample has another high-density layer on the
    surface or interface.
(2) The high-density layer can be removed by the heat treatment at 500K.
(3) The heat treatment at high temperatures above 750K results in the
    formation of a single-layer thick film consisting of composition
    elements including the Si substrate.

ACKNOWLEDGMENTS

The authors express their sincere thanks to Dr. A. Iida and Dr. T.
Matsushita of the National Laboratory for High Energy Physics for their
invaluable support and guidance.

REFERENCES

1. L. G. Parratt, Phys. Rev.,95:359(1954)
2. A. Iida and H. Hashimoto, Photon Factory Activity Report, 5:138(1987)
3. H. Hashimoto, H. Nishioji, H. Saisho and A. Iida, Adv. in X-ray Chem.
   Anal. Jpn.,20:143(1989)
4. B. Vidal and P. Vincent, Appl. Opt., 23:1794(1984)
5. A. Król, C. J. Sher and Y. H. Kao, Phys. Rev., B38:8579(1988)

LAYER THICKNESS DETERMINATION OF THIN FILMS BY GRAZING

INCIDENCE X-RAY EXPERIMENTS USING INTERFERENCE EFFECT

Kenji Sakurai and Atsuo Iida[*]

National Research Institute for Metals: 1-2-1 Sengen, Tsukuba
Ibaraki 305, Japan

[*]Photon Factory, National Laboratory for High Energy Physics
1-1 Oho, Tsukuba, Ibaraki 305, Japan

ABSTRACT

A novel method using Fourier transform algorithm is proposed to determine each layer thickness of multi-layered thin films from interference oscillation observed in X-ray specular reflection. The peak position in Fourier space gives each layer thickness of the film. The principle of the present technique as well as its applications are described.

INTRODUCTION

Grazing incidence X-ray techniques recently have been extensively used for the near-surface study of materials and the characterization of thin films.[1-9] In those experiments, as is often the case in practical analysis of multi-layered thin films, a complicated oscillating structure is observed in the angular/energy dependence of X-ray reflectivity and the signal proportional to the amplitude of the electric field (for example, fluorescent X-rays, electron yield). This is due to interference caused by multiple reflections of X-rays at each interface,[1,10,11] and therefore, such oscillation includes the information on layer thickness and on interface roughness.

In the earlier works,[12-14] the period of the reflectivity oscillation, which directly relate to layer thickness, were analyzed by direct reading of the position of maxima and minima. This method is effective for a single film layer, but has inherent limitations in the analysis of the complicated curve for increased layer numbers. On the other hand, recently, least-squares curve fitting procedures has been often employed using theoretical models.[4,15] Layer thicknesses as well as surface/interface roughness are determined automatically. However, it is difficult to obtain a best fit for the entire reflectivity curve, when neither every layer thickness nor every interface roughness is known.

In the present paper, a new technique for determining each layer thickness of multi-layered thin films is reported. A frequency analysis of reflectivity oscillation based on Fourier transform is promising.[16] This works well when any interface roughness is not given.

FORMULATION OF OSCILLATING PART OF REFLECTIVITY

The reflectivity from a multi-layered thin film is usually calculated using a recursive equation.[1,17] For the simple three-layered model, i.e., a single film layer (2nd layer) besides air (1st layer) and the substrate (3rd layer), reflectivity R is written as follows, when absorption is negligible :

$$R = R_{1,2}{}^2 = (R_{2,3} + F_{1,2})^2 / (R_{2,3} F_{1,2} + 1)^2$$
$$= (\exp(-i\gamma_2) F_{2,3} + F_{1,2})^2 / (\exp(-i\gamma_2) F_{2,3} F_{1,2} + 1)^2$$

where

$$\gamma_j = 4\pi \, d_j \, \{(\theta^2 - \theta c_j{}^2)^{1/2} / \lambda\}$$

$R_{j-1,j}$ and $F_{j-1,j}$ are the reflection coefficient and the Fresnel coefficient at the interface between (j-1)th and jth layers, respectively; θ is the glancing angle; d_j and θc_j are the critical angle and the layer thickness of jth layer, respectively; λ is the wavelength of the X-rays.

When R is small, this equation is simply rewritten:

$$R = (F_{1,2}{}^2 + F_{2,3}{}^2 + 2 F_{1,2} F_{2,3} \cos\gamma_2) / \{(1 - F_{1,2}{}^2) (1 - F_{2,3}{}^2)\} \qquad (1)$$

This equation indicates that the oscillation part of reflectivity is expressed as a simple cosine function. Furthermore, it is important that γ_2 in $\cos\gamma_2$ is the product of d_2 and $(\theta^2-\theta c_2{}^2)^{1/2}/\lambda$. That is, the cosine oscillating part of the data, when plotted as a function of $(\theta^2-\theta c_2{}^2)^{1/2}/\lambda$, are converted by Fourier transform to the distribution of d_2. A single peak, whose position gives the layer thickness d_2, is obtained in the Fourier space.

As the number of layers is increased, the equation becomes rather complicated, but essentially the same analysis is possible. For example, for four-layered model, R is written :

$$R = (A \cos\gamma_2 + B \cos\gamma_3 + C \cos(\gamma_2+\gamma_3) + D \cos(\gamma_2-\gamma_3) + E) / F \qquad (2)$$

$$A = 2 F_{1,2} F_{2,3} (1 + F_{3,4}{}^2); \; B = 2 F_{2,3} F_{3,4} (1 + F_{1,2}{}^2); \; C = 2 F_{1,2} F_{3,4};$$
$$D = 2 F_{1,2} F_{2,3}{}^2 F_{3,4}; \; E = F_{1,2}{}^2 + F_{2,3}{}^2 + F_{3,4}{}^2 + F_{1,2}{}^2 F_{2,3}{}^2 F_{3,4}{}^2;$$
$$F = (1 - F_{1,2}{}^2) (1 - F_{2,3}{}^2) (1 - F_{3,4}{}^2)$$

In this equation, reflectivity oscillation is expressed as the sum of four cosine wave. The distribution of d_j is obtained by Fourier transform of the data plotted as a function of $(\theta^2-\theta c_j{}^2)^{1/2}/\lambda$.

Though the correction term for absorption should be included in γ_j, it is negligible when R is small, and therefore, equations (1) and (2) can be used for cases with absorption.

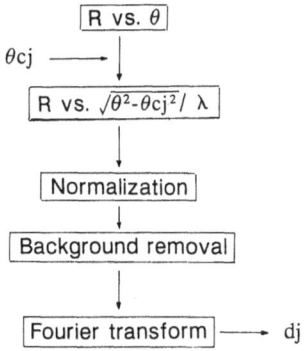

Figure 1
Analytical scheme for determining each layer thickness of multi-layered thin films.

ANALYTICAL SCHEME

The scheme for the analysis of interference oscillation was summarized in Fig.1. First, the experimental angular distribution of reflectivity is measured. Then, the data is re-plotted as a function of $(\theta^2-\theta c_i^2)^{1/2}/\lambda$. To compensate for the attenuation in the higher angle region, the data are normalized by the average curve, which is calculated in the logarithmic plot. After that, the oscillating part is extracted by subtracting the non-oscillating background. Finally, the data are Fourier transformed.

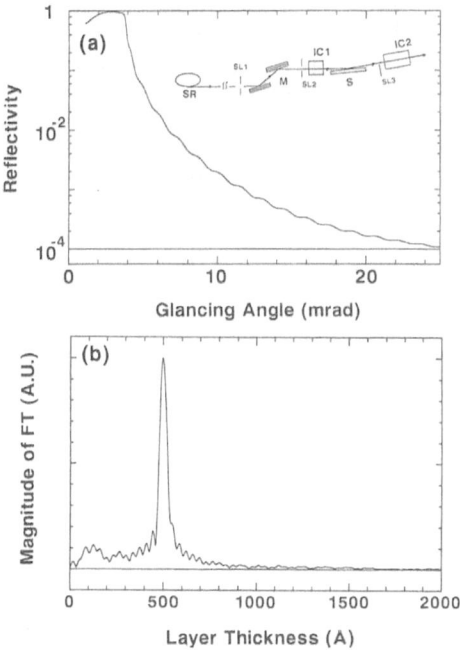

Figure 2
(a) Interference structure of $SiO_2[501Å]/Si$ in the 8 keV X-ray specular reflection. Schematic drawings of the experimental arrangement is shown in an inset; M: a monochromator, IC1, IC2: ionization chambers for incident and reflected X-rays, respectively, S: a sample, SL1, SL2, SL3: slits. (b) Magnitude of the Fourier transform of the oscillation extracted from Fig.2(a), after plotting as a function of $(\theta^2-\theta c_2^2)^{1/2}/\lambda$. The peak indicates the frequency component of the oscillation, and agrees with the film thickness.

When the sample has several layers, for determining each layer thickness, the critical angle θc_i should be determined for each layer. The data are plotted as a function of each $(\theta^2 - \theta c_i^2)^{1/2}/\lambda$ abscissa. Fourier analysis is done for each curve to determine each layer thickness d_j.

APPLICATION TO SiO$_2$/Si THIN FILMS [16]

The experiment was done using synchrotron X-rays on beam line 4A at the Photon Factory, in order to use tunable monochromatic X-rays with sufficiently high photon flux density. The apparatus for X-ray reflectivity measurements is shown in the inset of Fig.2(a), and is essentially the same as that described in our previous work.[7-9] Synchrotron radiation was monochromatized by a Si(111) double-crystal sagittal focusing monochromator. Reflectivity ranging from 1 to ~10^{-5} was measured by detecting the intensities of incident and reflected X-rays with two ionization chambers. Optical alignment was optimized by the translational and rotational motion of the sample stage. The measurement was made in air.

Figure 2(a) shows the experimental results of reflectivity of 8 keV X-rays for a SiO$_2$/Si sample, which were prepared by the conventional thermal oxidization process. The thickness of the SiO$_2$ layer obtained from ellipsometry was 501Å. The critical angle θc was 3.81 mrad, which was determined within 0.1 mrad by the direct comparison with the calculated reflectivity curve for SiO$_2$, and this accuracy is sufficient in this case.

The data were analyzed using the scheme shown in Fig.1. Figure 2(b) shows the magnitude of Fourier transform. A single sharp peak was clearly obtained, which gives the frequency of the oscillation observed in Fig.2(a). It is important that the peak position (499.5Å) agrees well with the thickness of the SiO$_2$ layer (501Å). The differences are less than 1 %.

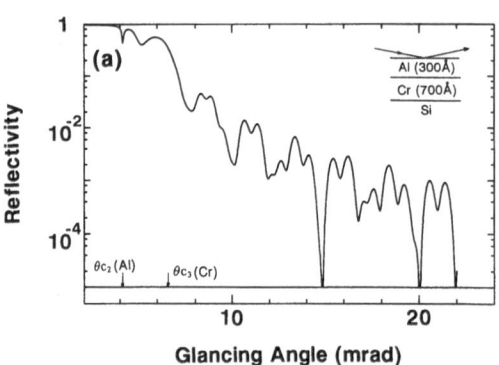

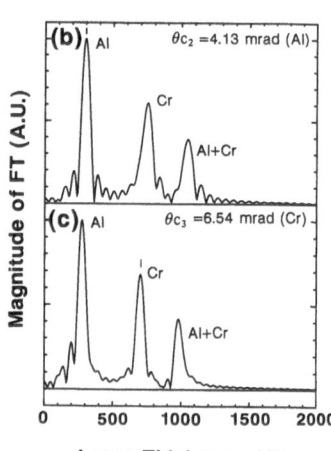

Figure 3
(a) Calculated X-ray reflectivity curve for a Al[300Å]/Cr[700Å]/Si sample at 8 keV. (b), (c) Magnitude of the Fourier transform of the oscillation extracted from Fig.3(a), after plotting as a function of $(\theta^2 - \theta c_i^2)^{1/2}/\lambda$; (b) for θc_2=4.13 mard, (c) for θc_3=6.54 mrad.

CALCULATED RESULTS FOR FOUR-LAYERED EXAMPLE

The advantages of this technique will become more apparent when applied to cases where the number of layers is increased. Figure 3(a) shows the calculated reflectivity of 8 keV X-rays for a Al[300Å]/Cr[700Å]/Si sample. The curve seems rather complicated owing to the mixture of several frequency components. The present technique is feasible for determining both aluminium and chromium layer thickness.

First, the data are plotted as a function of $(\theta^2-\theta c_2{}^2)^{1/2}/\lambda$ abscissa (θc_2, critical angle for aluminium layer is 4.13 mrad) and $(\theta^2-\theta c_3{}^2)^{1/2}/\lambda$ abscissa (θc_3, for chromium layer, is 6.54 mrad). Then, the oscillation part of each curve are extracted and Fourier transformed. Figures 3(b) and (c) show the Fourier transformed results for each curve, respectively. In each figure, three peaks are clearly seen. They qualitatively correspond to the thickness of the aluminium layer, chromium layer and their sum. A peak corresponding to their differences is too weak to be observed, since the coefficient D in eq.(2) is quartic. Accurate peak position corresponding to thickness d_2 (for aluminium layer) and d_3 (for chromium layer) is just given in the data obtained through the scheme using θc_2 and θc_3, respectively. Both d_2 and d_3 determined to be 301.3Å and 696.7Å from Fig.3(b) and (c), respectively, are in good agreement with the assumed thicknesses.

SUMMARY

In conclusion, Fourier transform is effective in the analysis of the interference oscillation of reflectivity observed in grazing incidence X-ray experiments. The peak position in the Fourier space was in good agreement with the layer thickness. The present technique is feasible for the analysis of complicated oscillating structure obtained from multi-layered thin films. Furthermore, combining Fourier analysis with least-squares curve fitting procedure is useful for attaining interface information, i.e., interface roughness or sharpness.

Information on the present computer programs (PASCAL codes) are available from the authors. They are designed to work on a personal computer (NEC PC-9801), but it is not so difficult to translate them for other machines.

ACKNOWLEDGMENTS
The authors would like to thank Dr. K. Saitoh and Dr. K. Honda for preparing samples and useful discussions. This work was performed under the approval of the Photon Factory Program Advisory Committee (Proposal No.91-074).

REFERENCES

1. L. G. Parratt, Phys. Rev. 95:359 (1954)
2. W. C. Marra, P. Eisenbergur and A. Y. Cho, J. Appl. Phys. 50:6927 (1979)
3. T. C. Huang, Adv. in X-Ray Anal. 33:91 (1990)
4. S. M. Heald, H. Chen and J. M. Tranquada, Phys. Rev. B38:1016 (1988)
5. W. B. Yun and J. M. Bloch, J. Appl. Phys. 68:1421 (1990)
6. M. J. Bedzyk, G. M. Bommarito, M. Caffrey and T. L. Penner, Science 248:52 (1990)
7. A. Iida, K. Sakurai, and Y. Gohshi, Adv. in X-Ray Anal. 31:487 (1988)
8. K. Sakurai and A. Iida, Adv. in X-Ray Anal. 33:205 (1990)
9. A. Iida, Adv. in X-Ray Anal. 34:23 (1991)

10. H. Kiessig, Ann. Physik 10:715 (1931)
11. M. Renninger, Z. Physik 100(1936)326.
12. N. Wainfan, N. J. Scott, and L. G. Parratt, J. Appl. Phys. 30:1604 (1959)
13. J. P. Sauro, J. Bindell, and N. Wainfan, Phys. Rev. 143:439 (1966)
14. A. Segmuller, Thin Solid Films, 18:287 (1973)
15. A. Segmuller, A.I.P. Conf. Proc., 53:78 (1979)
16. K. Sakurai and A. Iida, Jpn. J. Appl. Phys. 30 (1991) (in press)
17. M. Born and E. Wolf, "*Principles of Optics 6th ed.*", Pergamon, New York (1980)

X-RAY STUDIES OF CHROMIUM NITRIDE (Cr_xN_y) THIN FILMS

DEPOSITED BY REACTIVE MAGNETRON SPUTTERING

M. Charbonnier, M.Romand, A. Roche, J.P. Terrat*

Department of Chemistry and Chemical Engineering
(CNRS, URA 417), Université Claude Bernard - Lyon I
69622 Villeurbanne Cedex, FRANCE

* Centre Stéphanois de Recherches Mécaniques, Hydromécanique et
Frottement, 42160 Andrézieux Bouthéon Cedex, FRANCE

ABSTRACT

Chromium nitride hard coatings have been prepared by a plasma process by varying nitrogen partial pressure. The crystallographic structure of these samples has been investigated by XRD and their chemical composition and stoichiometry by Low Energy Electron Induced X-ray Spectrometry (LEEIXS).

INTRODUCTION

Chromium nitride, because of its remarkable properties (hardness, low friction coefficient, wear resitance, anticorrosion barrier) is widely used as hard coating for mechanical applications. Among the available deposition methods, plasma-based processes are the most suitable to coat any substrate. As film morphology, structure, composition and properties are linked to the deposition parameters (gas mixture composition, pressure, substrate temperature, source power, reactor geometry) it is of prime importance to fully characterize the deposited films and to know the influence of the multiple elaboration parameters on the final product. The films obtained in the present work were mainly studied as a function of nitrogen partial pressure in the preparation enclosure. The thickness was checked by X-Ray Fluorescence Spectrometry (XRFS), structure characterized by X-Ray Diffraction (XRD), stoichiometry and chemical composition evaluated by Low Energy Electron Induced X-ray Spectrometry (LEEIXS).

EXPERIMENTAL

Sample elaboration

Cr_xN_y films about 10 μm thick were deposited either on steel or glass substrates by reactive radio-frequency magnetron sputtering of a 99.9 % chromium target in a mixture of argon and nitrogen both 99.998 %, operating respectively as sputtering ions and reactive radicals.Power density was kept constant at 10 W. cm^{-2} with a negative bias of - 100 V on the cathode which was heated at 200 ± 30 ° C. The present study was performed only by varying

Table 1

Sample reference	a	b	c	d	e	f	g
nitrogen partial pressure (mbar)	1.10^{-5}	5.10^{-5}	2.10^{-4}	3.10^{-4}	4.10^{-4}	1.10^{-3}	3.10^{-3}

nitrogen partial pressure in the range 1.10^{-5} to 3.10^{-3} mbar to deposit different films of variable composition. Film thickness was monitored by a quartz crystal microbalance. Seven samples referred to as (a) to (g) were elaborated under the partial pressures indicated table 1.

Instrumentation

The film structure was investigated with a Philips diffractometer equipped with a cobalt anticathode tube. Film thickness was checked by XRFS with a Philips instrument using a gold anticathode tube, X-radiations being dispersed with a LiF (200) crystal . Finally, film composition and stoichiometry were determined by LEEIXS.This method less known than the preceding ones but widely presented elsewhere [1-4] will be briefly described here. It consists of bombarding the sample with a nonfocused electron beam supplied by a gas discharge tube working in the primary vacuum of the spectrometer and detecting the resulting X-rays with a conventional wavelength dispersive device. The depth probed (from 10 to 150 nm) depends on the primary electron energy (from 0.5 to 5 keV) and can therefore be chosen by selecting this energy according to the chemical nature of the sample and the radiation considered. Because of the large area analysed ($\sim$ 1 cm^{-2}) the method is very sensitive for detecting low energy radiations (particularly those of light elements : B, C, N, O, F) while using a low current density (0.2 mA.cm $^{-2)}$ to avoid sample damage.

RESULTS AND DISCUSSION

Tkickness control

During the film elaboration the samples were deposited on the 15 cm diameter cathode. To determine the zone in which homogeneous deposits were obtained, the substrates were distributed on the whole cathode, coated with pure chromium and the thickness of the films deposited checked by XRFS through the CrK$_\alpha$ radiation intensity. The results showed that only the samples within a 6 cm circle had identical and highest thicknesses. Beyond this zone, the film thicknesses decreased regularly up to the cathode limit.

Structural investigation

Many authors [5-12] have carried out structural studies using conventional or glancing incidence X-ray diffractometers on different nitride films deposited by plasma processes. These studies generally show a broadening of the diffraction lines and a preferred orientation. The results of our studies on Cr$_x$N$_y$ films are shown in figure 1. Spectra were recorded in the range 40 - 100° 2 Θ. The d values are indicated for each diffraction line. The spectra relative to CrN and Cr$_2$N were obtained from powders supplied by Johnson Mattey (purity announced : not given for CrN, 99.2 % for Cr$_2$N ; in fact, it appears that the latter is a mixture of CrN and Cr$_2$N). The Cr spectrum was recorded from a copper sample coated with an electrochemically deposited Cr layer. These 3 sample spectra show some thin and well defined diffraction lines all fitted (d values and relative intensities) to the data given by their corresponding ASTM card. On the other hand, the spectra of the Cr$_x$N$_y$ plasma films deposited on steel substrates are characteristic of poorly crystallized samples. Indeed they show, in the whole range probed, one, two or three wide or very wide lines. Spectrum (a) due to a film deposited by

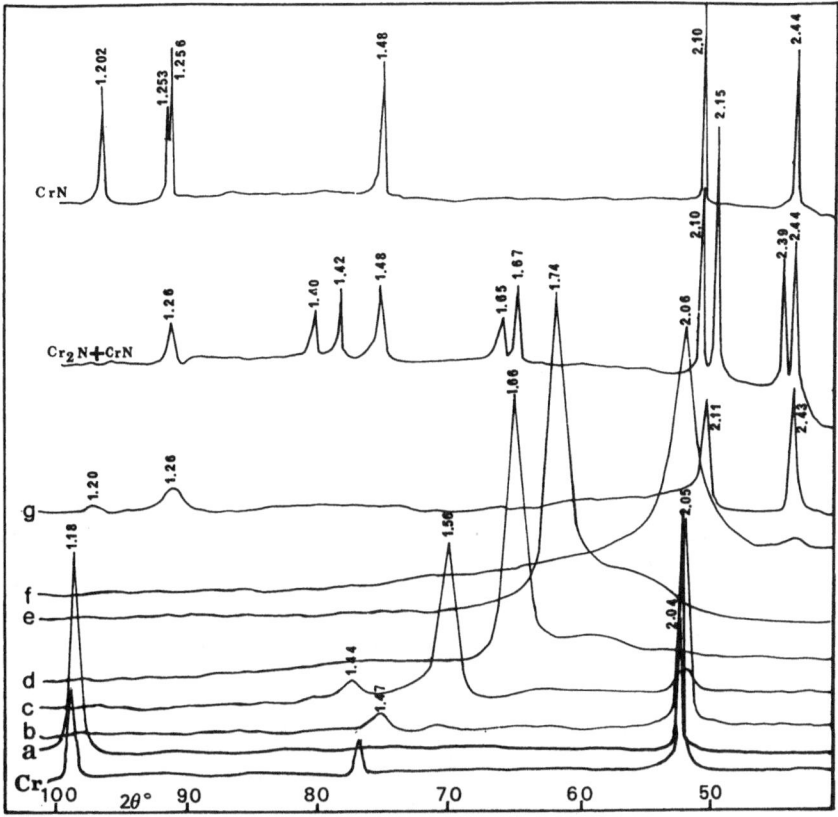

Fig. 1. Diffractograms of Cr, Cr$_2$N, CrN and "plasma coatings" (a) to (g).

the plasma method under the lowest nitrogen partial pressure corresponds practically to pure chromium and presents the least wide lines among all the "plasma films". The (200) line (d = 1.44 Å) is not observed and the intensity ratios of the others are different from those given by the ASTM card (table 2).   Furthermore, the spectrum recorded from the same film deposited on a glass substrate shows, for the main lines, intensity ratios not only different from those given by the ASTM card, but also different from those obtained from a film deposited on a steel substrate (table 2). All these deviations compared to well crystallized pure chromium

Table 2

Cr		relative intensities		
d (Å)	hkl	ASTM card (Cr)	nitride film/steel	nitride film/glass
2.0390	110	100	79	100
1.4419	200	16	/	/
1.1774	211	29	100	48

Table 3

CrN		relative intensities		
d (Å)	hkl	ASTM card (CrN)	nitride film/steel	nitride film/glass
2.394	111	80	100	100
2.068	200	100	71	92
1.463	220	80	/	/
1.249	311	60	21	26

data can be explained by a preferential orientation of the deposit induced by the substrate. Spectrum (g), relative to the film having received the maximal nitrogen content, corresponds approximately to the CrN one (obtained from the powder) , the (220) line (d = 1.48 Å) however missing and the thinnest ones being slightly shifted. As for spectrum (a), the relative intensities depend on the substrate nature and are different from those given by the ASTM card (table 3). The absence of the (220) line and the relative intensity variations observed on the others characterize an orientation effect (fiber texture) shown by scanning electron microscopy and already mentioned by FABIS et al. [4,5].   It is to be noted that none of the "plasma films" elaborated at increasing nitrogen pressure contains any chromium in either the metal form, or as Cr_2N or CrN. Spectra (a and b) on the one hand and (g) on the other, due to deposits realized at low and high nitrogen partial pressure respectively show thin lines characterizing well crystallized materials. The crystallographic structure of the samples (a and b) could derive from the Cr one (presence of the d = 2.05 Å line) and that of (g), from the CrN one as previously mentioned.

The other spectra (c), (d), (e), (f) show an intense and wide line with a diffuse structure more or less marked on the large d value side and one or two weak lines. The d value of the intense line increases with the nitrogen content. Figure 2 shows the variation of this interreticular distance as a function of the nitrogen content measured by LEEIXS. A linear variation is observed for the (c), (d), and (e) films which seems characteristic of a continuous solid solution in the corresponding limited composition range. Sample (b), represented by its weakest line (d = 1.47 Å) is outside the straight line which excludes it from the solid solution

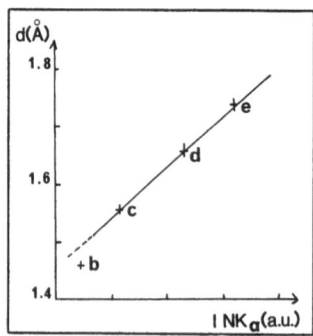

Fig. 2. Interreticular distance variation of the broadest and highest line of samples (b), (c), (d), (e) as a function of NK_α intensity.

previously mentioned. Sample (f), whose spectrum is characterized by an intense and wide line (d = 2.06 Å) and by a low intensity line beginning (d = 2.43 Å) both fitted to the most intense lines of CrN could be a solid solution derived from the CrN structure. The line width, particularly large for the (c), (d), (e), (f) deposits characterizes some low dimension crystallites and probably some high strain rates inside the material [4-7].

Chemical composition and stoichiometry

Composition and stoichiometry determination of nitride thin films is generally a major problem which confronts researchers who want to check the effect of the numerous plasma deposition experimental parameters. The most commonly used method seems to be Auger Electron Spectrometry (AES) [9-13] because of its sensitivity to nitrogen. However, it is little suited to determine the bulk composition of the films. Furthermore, AES quantification is often inaccurate and only a few well characterized standards are available. Other characterization methods such as X-ray Photoelectron Spectroscopy (XPS) [9,13], Rutherford Backscattering Spectroscopy (RBS) [9] and Electron Probe Micro Analysis (EPMA) equipped with a wavelength dispersive system [10] are also used. Among these methods, XPS, another common surface analysis technique, presents the same disadvantages as AES in the particular case investigated here and the other two methods which could be used because they can perform bulk analyses are not at all sensitive to nitrogen. Conversely, this difficult analytical problem of nitrogen determination may be quite well solved by LEEIXS.

In the above mentioned structural studies it was considered that the "plasma films" were essentially composed of chromium and nitrogen. This was justified by the fact that XRD is not sensitive to impurities. LEEIXS analyses have shown the presence of some traces of Ar and of small quantities of carbon and oxygen. The latter elements seem to be distributed partly on the

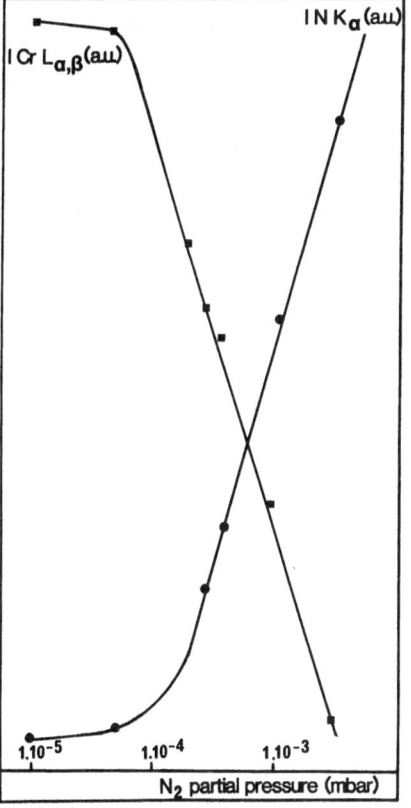

Fig. 3. CrL$_{\alpha,\beta}$ and NK$_\alpha$ intensity variation versus nitrogen partial pressure.

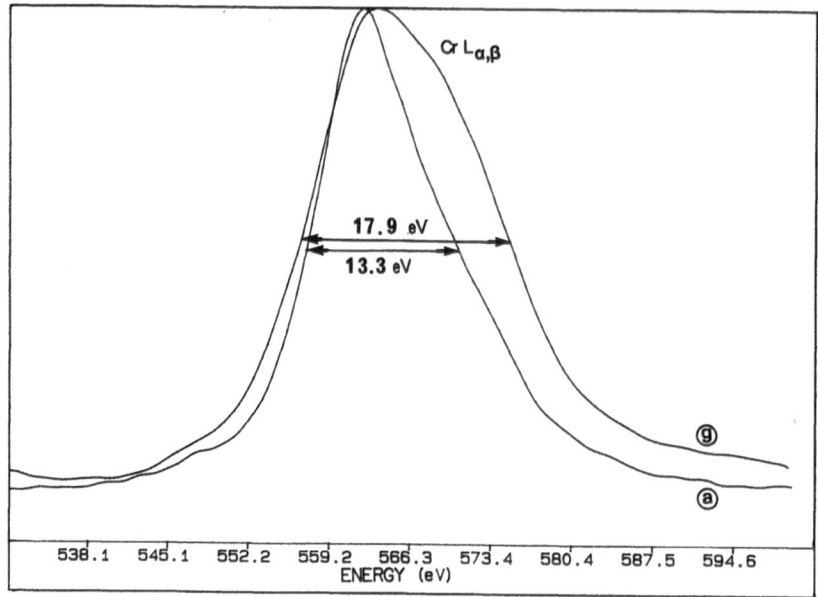

Fig. 4. CrL$_{\alpha,\beta}$ spectra of samples (a) and (g) recorded with an OV45 multilayer synthetic mirror.

surface (resulting from contamination in the laboratory atmosphere), partly inside the films for oxygen only (resulting from the residual oxygen in gas atmosphere inside the preparation enclosure). However, if we consider the high sensitivity of LEEIXS for detecting light elements, it can be assessed that the presence of these impurities (about 1 to 2 at. %) cannot affect the measurements realized on Cr - N bonding and on film stoichiometry.

As a first step, NK$_\alpha$ and CrL$_{\alpha,\beta}$ radiations generated respectively with an electron beam 3.1 and 3.2 keV in energy in order to probe the outer 100 nm of the samples, were dispersed with a 2d = 4.5 nm multilayer mirror (OV 45) for obtaining high count rates (for example net intensities of 980 cts. s^{-1} for NK$_\alpha$ and 4300 cts. s^{-1} for CrL$_{\alpha,\beta}$ on the CrN type sample (g)). Figure 3 represents the intensity variations of CrL$_{\alpha,\beta}$ and NK$_\alpha$ as a function of the nitrogen partial pressure . It shows that at low nitrogen pressures (less than 1. 10^{-4} mbar) , nitrogen incorporation into the coatings is very low, as already emphasized by X-ray diffraction measurements on (a) and (b) samples. Beyond 1. 10^{-4} mbar, nitrogen incorporation increases rapidly until CrN composition is reached. High count rates due to high reflecting power of the synthetic multilayer mirror are obtained to the detriment of resolution. Indeed, as shown in Figure 4 which represents the CrL$_{\alpha,\beta}$ spectra of the "plasma films" (a) close to Cr, and (g) close to CrN, CrL$_\alpha$ and CrL$_\beta$ are not at all separated, but the presence of CrL$_\beta$ widens the emission band towards the high energy side. Furthermore, the CrL$_{\alpha,\beta}$ emission band which is very sensitive to chemical state shows in Figure 4 a noticeable widening due to Cr - N bonding for spectrum (g) compared to spectrum (a) . This widening is observed on both sides of the band ; on the high energy side it is due to the increasing L$_\beta$ intensity versus the L$_\alpha$ one when going from metal to nitride and on the low energy side it is due to the raising of sub-bands directly associated to the formation of Cr - N bonds. The FWHM or normalized area under the CrL$_{\alpha,\beta}$ band values can therefore be used to characterize film stoichiometry. Figure 5 represents the logarithm of CrL$_{\alpha,\beta}$ / N K$_\alpha$ intensity ratio versus CrL$_{\alpha,\beta}$ FWHM values for different "plasma samples" and for CrN. The linear variation observed makes it easy to evaluate film stoichiometry, the line ends corresponding respectively to the ratio N/Cr = 0 for the pure Cr "plasma film" and to 1 for CrN.

In order to investigate more completely the fine structure of the CrL$_\alpha$ and L$_\beta$ emission bands in the non-stoichiometric chromium nitrides, we have also used a TlAP crystal

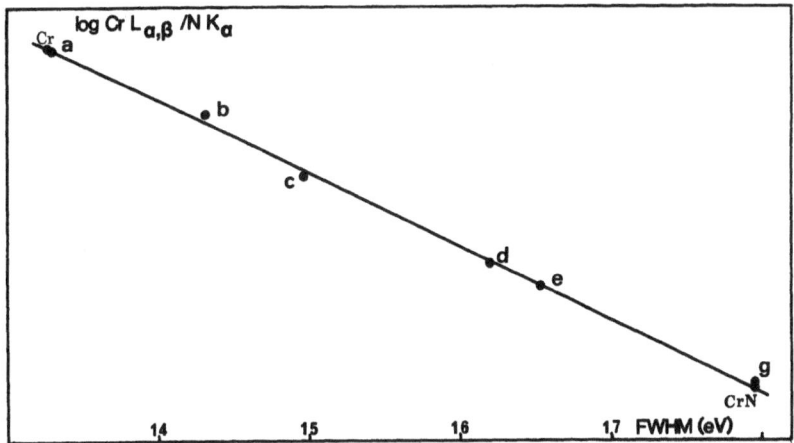

Fig. 5. Log CrL$_{\alpha,\beta}$ / N K$_\alpha$ as a function of CrL$_{\alpha,\beta}$ FWHM measured on spectra recorded with OV45.

which possesses a resolving power higher than OV 45 but a lower reflecting power. Indeed, as shown in Figure 6, CrL$_\alpha$ and CrL$_\beta$ emission bands (relative to Cr, Cr$_2$N, CrN and a "plasma film" referred to as CrN$_x$ and corresponding to sample (f)) are clearly distinguished instead of being completely separated. The different spectra show a progressive change in their form, in the CrL$_\alpha$ FWHM, and in the CrL$_\alpha$ / CrL$_\beta$ intensity ratio. The non resolved fine structures appearing on the low energy side participate in the L$_\alpha$ band widening and as already mentioned are characteristic of the chemical bonding between Cr and N. A shift of the band maximum (~ 0.66 eV) towards the high energy side is observed from Cr to CrN. For the "plasma film" CrN$_x$, as well as for all the other ones, the shift compared to chromium metal

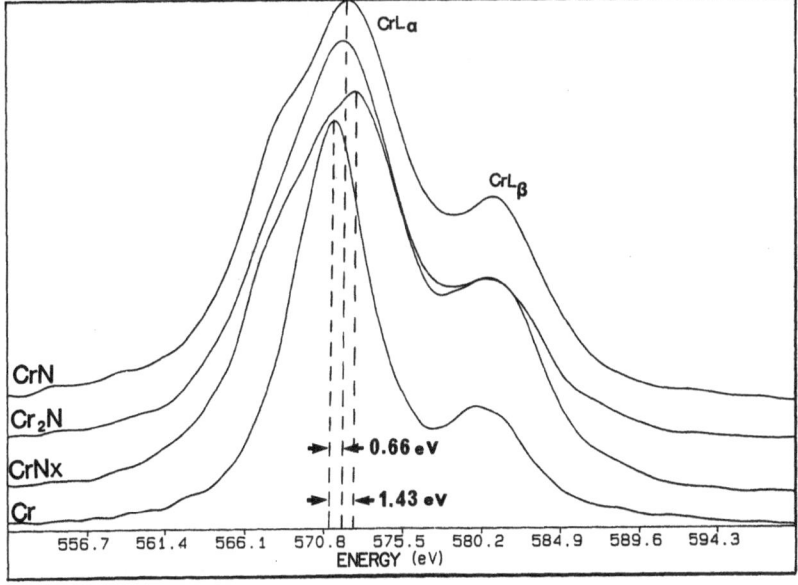

Fig. 6. CrL$_\alpha$ and CrL$_\beta$ spectra of Cr, Cr$_2$N, CrN and a "plasma film" (f) recorded with a TlAP crystal.

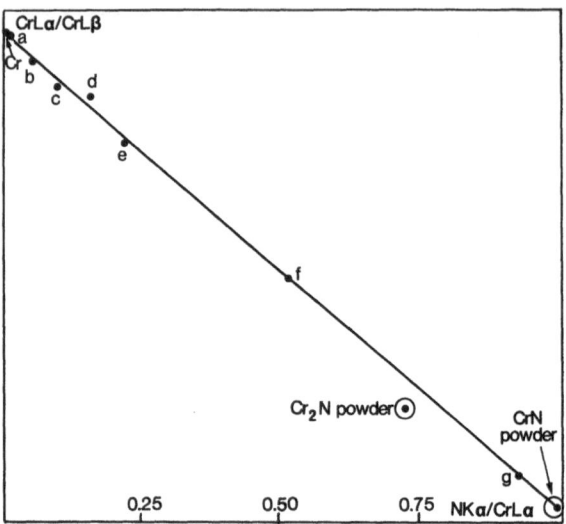

Fig. 7. CrL_α / CrL_β intensity ratios measured
with a TlAP crystal versus NK_α / CrL_α
intensity ratios for the "plasma films",
Cr_2N and CrN.

maximum is larger ($\sim$ 1.43 eV). On the other hand, the intensity ratio CrL_α / CrL_β is also characteristic of the chemical bonding and changes with stoichiometry. The largest widening of the L_α band and the lowest value of the CrL_α / CrL_β ratio are obtained for CrN which corresponds to the upper incorporation limit of nitrogen in chromium. The CrL_α / CrL_β intensity ratio measurement, easier than the CrL_α band FWHM one was considered to characterize Cr - N bonding in the "plasma" deposits. Figure 7 shows this ratio variation versus the nitrogen incorporation rate determined from the NK_α / CrL_α intensity ratio normalized at 1 for CrN. The experimental points are well distributed about a straight line so that the mere CrL_α / CrL_β ratio measurement allows to estimate the coating stoichiometry.

CONCLUSION

Experiments carried out in this work by using complementary techniques have shown the particular interest of studying Cr_xN_y chromium nitride films elaborated by plasma processes. XRFS, XRD and LEEIXS were respectively used for obtaining information on thickness, structure and chemical composition. More particularly this work has emphasized the interesting and unique capabilities of LEEIXS in nitrogen analysis. NK_α radiations were dispersed using a synthetic multilayer mirror and $CrL_{\alpha,\beta}$, using this same device or a conventional TlAP crystal. Stoichiometry of the nitride films was determined, the standardization being realized from Cr and CrN samples.

REFERENCES

1 . R. Bador, M. Romand, M. Charbonnier, and A. Roche, <u>Adv. in X-ray Anal.</u>,
    24 : 351 (1981).
2 . M Romand, R Bador, M. Charbonnier, and F Gaillard, <u>X-ray Spectrom.</u>, 16 : 7
    (1987).
3 . M. Romand, F. Gaillard, M. Charbonnier, and D.Urch, <u>Adv. in X-ray Anal</u>, 34 :
    (1991).

4 . M. Charbonnier, F. Gaillard, M. Romand, and D.Urch, <u>Adv. in X-ray Anal.</u>, 34 : 139 (1991).

5 . P. M. Fabis, R. A. Cooke, S. Mc.Donough, <u>J.Vac. Sci. Technol.</u>, A 8 (5) : 3809 (1990).

6 . P. M. Fabis, R. A. Cooke, S. Mc Donough, <u>J.Vac. Sci. Technol.</u>, A 8 (5) : 3819 (1990).

7 . K. K. Shih, D. B. Doue, J. R. Crowe, <u>J.Vac. Sci. Technol.</u>, A 4 (3) : 564 (1986).

8 . V. Valdova, R. Kuzel, Ir. and C. Cerny, <u>Thin Solid Films</u>, 156 : 53 (1988).

9 . R. C. Buschert, P. N. Gibsow, W. Gissler, J. Haupt, A. Manara, <u>Surf. Interf. Anal.</u>, 16 : 510 (1990).

10 . S. Komiya, S. Ono, N. Umezu, T. Narusawa, <u>Thin Solid Films</u>, 45 : 433 (1977).

11 . K. Kashiwagi, K. Kobayashi, A. Masuyama, Y. Murayama, <u>J.Vac. Sci. Technol.</u>, A 4 (2) : 210 (1986).

12 . H. A. Jehn, E. Grallath, I. Le Roux Strydom, S. Hofmann, <u>Surf. Interf. Anal.</u>, 16 : 540 (1990).

13 . S. Hofmann, <u>J.Vac. Sci. Technol.</u>, A 4 (6) : 2789 (1986).

MULTI-LAYER XRF CALCULATIONS

M. Sumini and J.E. Fernández[*]

Laboratorio di Ingegneria Nucleare di Montecuccolino
University of Bologna
Via dei Colli 16, 40136 Bologna, ITALY

ABSTRACT

An analytical framework for calculation of multiple scattering intensities emitted by a multi-layer sample are obtained with transport theory. The n-th order flux solution of the Boltzmann transport equation renders the iterative solution for multi-layers samples of thin or infinite thickness.
The first-order intensity is written for a generic sample with I layers and for the predominating interactions in the X-ray regime: the photoelectric effect, and the Compton and the Rayleigh scattering. As an example, the case of a thin deposit on an infinite thickness substrate is discussed.

INTRODUCTION

An interesting problem in X-Ray spectrometry is the analysis of the emission from a multilayer sample due to the prevailing interactions in the X-Ray regime: photoelectric effect, and Rayleigh and Compton scattering. The appropriate tool should allow the analytical study of the phenomenon enlightning the influence of the mathematical model for the different interactions and the geometry. The aim of this paper is to present a clear mathematical picture for the emitted X-ray intensity in terms of the phase-space variables and the characteristic parameters (layer thicknesses, compositions, etc.) for generic interactions.

---

[*] Member of CONICET, Buenos Aires, Argentina. On leave of absence from the Faculty of Mathematics, Astronomy and Physics, University of Córdoba, Argentina.

Advances in X-Ray Analysis, Vol. 35
Edited by C.S. Barrett *et al.*, Plenum Press, New York, 1992

THEORY

The most appropriate tool to analyze the multiple scattering diffusion problem in a general non-homogeneous media is the Boltzmann transport equation.[1] The integral form of the photon transport equation reads:

$$f(\vec{r},\vec{\omega},\lambda) = \int_0^\infty d\tau \left(\int_0^\infty d\lambda' \int_{4\pi} d\vec{\omega}' \; k(\vec{r}-\tau\vec{\omega},\vec{\omega},\lambda,\vec{\omega}',\lambda')f(\vec{r}-\tau\vec{\omega},\vec{\omega}',\lambda') + S(\vec{r}-\tau\vec{\omega},\vec{\omega},\lambda) \right) \times$$

$$\exp\left(-\int_0^\tau d\tau' \; \mu(\vec{r}-\tau'\vec{\omega},\lambda) \right), \tag{1}$$

where the notation has been previously defined (see Ref [1] and references therein). Note that τ is the space coordinate along flight direction $\vec{\omega}$.

We shall consider the simplified problem of an infinite domain characterized by a plane surface interface between a non interacting medium on the left side, and an infinite sequence of homogeneous slabs on the right. See Fig. 1. For such case, Eqn (1) becomes

$$f(z,\vec{\omega},\lambda) = \int_0^\infty d\tau \left(\int_0^\infty d\lambda' \int_{4\pi} d\vec{\omega}' \; k(\tau,\vec{\omega},\lambda,\vec{\omega}',\lambda')f(\tau,\vec{\omega}',\lambda') + S(\tau,\vec{\omega},\lambda) \right) \times$$

$$\frac{(1+\text{sgn}(\eta)\text{sgn}(z-\tau))}{2|\eta|}\exp\left(-\int_0^{\left|\frac{z-\tau}{\eta}\right|} \frac{d\tau'}{|\eta|}\mu(z-\tau',\vec{\omega},\lambda) \right). \tag{2}$$

In the present situation, the explicit space dependence for the interaction kernel and the absorption cross section results:

$$k(z,\vec{\omega},\lambda,\vec{\omega}',\lambda') = k_1(\vec{\omega},\lambda,\vec{\omega}',\lambda')(\mathcal{U}(z)-\mathcal{U}(z-d_1)) +$$

$$+ k_2(\vec{\omega},\lambda,\vec{\omega}',\lambda')(\mathcal{U}(z-d_1)-\mathcal{U}(z-d_2)) +$$

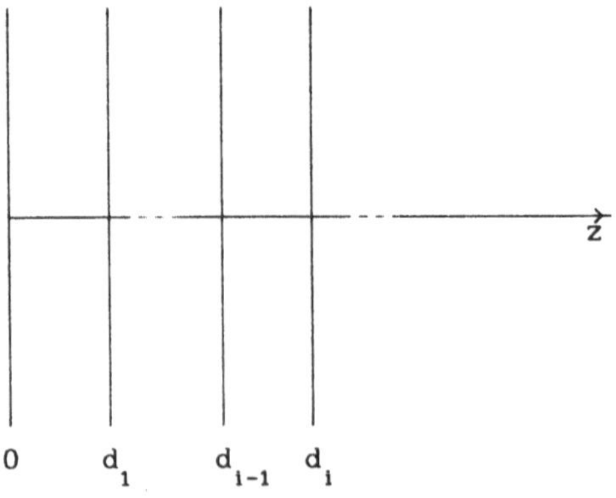

Fig. 1. Geometrical arrangement of the multilayer sample. d_i represents the i-th border.

$$+ k_3(\vec{\omega},\lambda,\vec{\omega}',\lambda')(\mathcal{U}(z-d_2)-\mathcal{U}(z-d_3)) + \ldots,$$

$$\mu(z,\lambda) = \mu_1(\lambda)(\mathcal{U}(z)-\mathcal{U}(z-d_1) + \mu_2(\lambda)(\mathcal{U}(z-d_1)-\mathcal{U}(z-d_2)) +$$

$$+ \mu_3(\lambda)(\mathcal{U}(z-d_2)-\mathcal{U}(z-d_3)) + \ldots,$$

or, rearranging the terms,

$$k(z,\vec{\omega},\lambda,\vec{\omega}',\lambda') = k_1(\vec{\omega},\lambda,\vec{\omega}',\lambda')\mathcal{U}(z) + (k_2(\vec{\omega},\lambda,\vec{\omega}',\lambda')-k_1(\vec{\omega},\lambda,\vec{\omega}',\lambda'))\times$$

$$\mathcal{U}(z-d_1) + (k_3(\vec{\omega},\lambda,\vec{\omega}',\lambda')-k_2(\vec{\omega},\lambda,\vec{\omega}',\lambda'))\mathcal{U}(z-d_2) + \ldots$$

$$= \sum_i (k_i(\vec{\omega},\lambda,\vec{\omega}',\lambda')-k_{i-1}(\vec{\omega},\lambda,\vec{\omega}',\lambda'))\mathcal{U}(z-d_{i-1}),$$

$$\mu(z,\lambda) = \mu_1(\lambda)\mathcal{U}(z) + (\mu_2(\lambda)-\mu_1(\lambda))\mathcal{U}(z-d_1) + (\mu_3(\lambda)-\mu_2(\lambda))\mathcal{U}(z-d_2) + \ldots$$

$$= \sum_i (\mu_i(\lambda)-\mu_{i-1}(\lambda))\mathcal{U}(z-d_{i-1}),$$

being $\mathcal{U}(z)$ the step Heaviside function. In operational form, defining the phase space operators:

$$\mathbb{L} = \sum_i \mathbb{L}_i ,\qquad (3)$$

$$\mathbb{L}_i = \int_{d_{i-1}}^{\infty} d\tau \int_0^{\infty} d\lambda' \int_{4\pi} d\vec{\omega}' \left\{ k_i(\vec{\omega},\lambda,\vec{\omega}',\lambda')-k_{i-1}(\vec{\omega},\lambda,\vec{\omega}',\lambda') \right\}\times$$

$$\frac{(1+\mathrm{sgn}(\eta)\mathrm{sgn}(z-\tau))}{2|\eta|} \exp\left(-\sum_i (\mu_i(\lambda)-\mu_{i-1}(\lambda)) \int_0^{\left|\frac{z-\tau}{\eta}\right|} \frac{d\tau'}{|\eta|}\mathcal{U}(z-\tau'-d_{i-1}) \right), \qquad (4)$$

and

$$f^{(0)}(z,\vec{\omega},\lambda) = \int_0^{\infty} d\tau \frac{(1+\mathrm{sgn}(\eta)\mathrm{sgn}(z-\tau))}{2|\eta|} \times$$

$$\exp\left(-\sum_i (\mu_i(\lambda)-\mu_{i-1}(\lambda)) \int_0^{\left|\frac{z-\tau}{\eta}\right|} \frac{d\tau'}{|\eta|}\mathcal{U}(z-\tau'-d_{i-1}) \right) S(\tau,\vec{\omega},\lambda), \qquad (5)$$

or, performing the integral for the computation of the optical path,

$$f^{(0)}(z,\vec{\omega},\lambda) = \int_0^\infty d\tau \frac{(1+\text{sgn}(\eta)\text{sgn}(z-\tau))}{2|\eta|} \exp\left[-\sum_i \frac{(\mu_i(\lambda)-\mu_{i-1}(\lambda))}{|\eta|} \times\right.$$

$$\left. \left(\frac{(|z-\tau|-z+d_{i-1})\text{sgn}(z-d_{i-1}-|z-\tau|)+|z-d_{i-1}|+|z-\tau|}{2}\right) \right] S(\tau,\vec{\omega},\lambda), \qquad (6)$$

as the uncollided flux term, we can write the original problem as:

$$f(z,\vec{\omega},\lambda) = \sum_i \mathbb{L}_i f(z,\vec{\omega},\lambda) + f^{(0)}(z,\vec{\omega},\lambda). \qquad (7)$$

Searching for the solution through a Neumann series, i.e. writing the total photon flux as a superposition of multiple scattering order terms, for an I-zone medium, we can write:

$$f = f^{(0)} + f^{(1)} + f^{(2)} + \ldots = f^{(0)} + \mathbb{L}f^{(0)} + \mathbb{L}(\mathbb{L}f^{(0)}) + \ldots$$

$$= f^{(0)} + \sum_{i=1}^{I} \mathbb{L}_i f^{(0)} + \sum_{i=1}^{I} \mathbb{L}_i \left(\sum_{i'=1}^{I} \mathbb{L}_{i'} f^{(0)}\right) + \ldots \qquad (8)$$

We can immediately verify that, in the hypothesis of a scattering kernel coming from J different interaction kinds, the n-th order contribution in the above series results from a superposition of $(J\times I)^n$ physically well characterized terms.

We report now some analytical developments for the simpler (but without loss of generality) problem of a layer of thickness d superimposed to a semi-infinite medium of different scattering properties (with kernels k_{a1}, k_{a2} for the first-order interactions and for media 1- and 2- respectively).

We consider a delta-shaped source like

$$S(z,\vec{\omega},\lambda) = I_0 \delta(z) \delta(\lambda-\lambda_0) \delta(\vec{\omega}-\vec{\omega}_0) \qquad (9)$$

so that the uncollided flux term results:

$$f^{(0)}(z,\vec{\omega},\lambda) = \frac{I_0}{2|\eta|} \delta(\vec{\omega}-\vec{\omega}_0) \delta(\lambda-\lambda_0) (1+\text{sgn}(\eta)\text{sgn}(z)) \exp\left[-\frac{\mu_1|z|}{|\eta|} -\right.$$

$$\left. - \frac{(\mu_2(\lambda)-\mu_1(\lambda))}{|\eta|} \left(\frac{(|z|-z+d)\text{sgn}(-|z|+z-d)+(|z-d|+|z|)}{2}\right) \right] \qquad (10)$$

Looking only for positive z, we get:

$$f^{(0)}(z,\vec{\omega},\lambda) = \frac{I_0}{2|\eta|} \delta(\vec{\omega}-\vec{\omega}_0) \, \delta(\lambda-\lambda_0) \, (1+\text{sgn}(\eta)) \, \exp\left[-\frac{\mu_1(\lambda)z}{|\eta|} - \frac{(\mu_2(\lambda)-\mu_1(\lambda))}{|\eta|} \left(\frac{z-d+|z-d|}{2}\right)\right],$$

(11)

and the general term of Eqn (8):

$$f^{(n)}(z,\vec{\omega},\lambda) = \frac{1}{|\eta|} \left[\frac{(1+\text{sgn}(\eta))}{2} \int_0^z d\tau \, \exp\left(-\frac{\mu_1}{|\eta|}(z-\tau) - \frac{\mu_2-\mu_1}{|\eta|} \times \right.\right.$$

$$\frac{((d-\tau)\text{sgn}(\tau-d)+|z-d|+z-\tau)}{2}\right) \int_0^\infty d\lambda' \int_{4\pi} d\vec{\omega}' \, k(\tau,\vec{\omega},\lambda,\vec{\omega}',\lambda') \, f^{(n-1)}(\tau,\vec{\omega}',\lambda') +$$

$$+ \frac{(1-\text{sgn}(\eta))}{2} \int_0^\infty d\tau \, \exp\left(-\frac{\mu_1}{|\eta|}\tau - \frac{\mu_2-\mu_1}{|\eta|}\left(\frac{|z-d|+(\tau-z+d)\text{sgn}(z-d-\tau)+\tau}{2}\right)\right) \times$$

$$\int_0^\infty d\lambda' \int_{4\pi} d\vec{\omega}' \, k(\tau+z,\vec{\omega},\lambda,\vec{\omega}',\lambda') \, f^{(n-1)}(\tau+z,\vec{\omega}',\lambda') \right]$$

(12)

The first order collision term results:

$$f^{(1)}(z,\vec{\omega},\lambda) = \frac{I_0}{|\eta||\eta_0|} \frac{(1+\text{sgn}(\eta_0))}{2} \left[\frac{(1+\text{sgn}(\eta))}{2} \times \right.$$

$$\exp\left(-\frac{\mu_1}{|\eta|}z - \frac{\mu_2-\mu_1}{|\eta|}z\,U(z-d)\right) \times$$

$$\int_0^z d\tau \, \exp\left(\frac{\mu_1}{|\eta|}\tau + \frac{\mu_2-\mu_1}{|\eta|}(\tau-d)U(\tau-d)\right) \exp\left(-\frac{\mu_{1,0}}{|\eta_0|}\tau - \frac{\mu_{2,0}-\mu_{1,0}}{|\eta_0|}(\tau-d)U(\tau-d)\right) \times$$

$$\left[k_{a1}(\vec{\omega},\lambda,\vec{\omega}_0,\lambda_0)U(\tau) + (k_{a2}(\vec{\omega},\lambda,\vec{\omega}_0,\lambda_0)-k_{a1}(\vec{\omega},\lambda,\vec{\omega}_0,\lambda_0))U(\tau-d)\right] +$$

$$+ \frac{(1-\text{sgn}(\eta))}{2} \int_0^\infty d\tau \, \exp\left(-\frac{\mu_1}{|\eta|}\tau - \frac{\mu_2-\mu_1}{|\eta|}U(z-d)((z-d)U(\tau+d-z)+\tau U(z-d-\tau))\right) \times$$

$$\exp\left(-\frac{\mu_{1,0}}{|\eta_0|}(z+\tau) - \frac{\mu_{2,0}-\mu_{1,0}}{|\eta_0|}(z+\tau-d)U(z+\tau-d)\right) \times$$

$$\left[k_{a1}(\vec{\omega},\lambda,\vec{\omega}_0,\lambda_0)\mathcal{U}(\tau+z) + (k_{a2}(\vec{\omega},\lambda,\vec{\omega}_0,\lambda_0)-k_{a1}(\vec{\omega},\lambda,\vec{\omega}_0,\lambda_0))\mathcal{U}(\tau+z-d)\right] \qquad (13)$$

From the albedo calculation we obtain for the first-order intensity:

$$I^{(1)}(\vec{\omega},\lambda) = \frac{I_0}{|\eta_0|} \frac{(1+\text{sgn}(\eta_0))}{2} \frac{(1-\text{sgn}(\eta))}{2} \times$$

$$\left\{ \frac{1 - \exp\left[-d\left(\frac{\mu_1}{|\eta|} + \frac{\mu_{1,0}}{|\eta_0|}\right)\right]}{\frac{\mu_{1,0}}{|\eta_0|} + \frac{\mu_1}{|\eta|}} k_{a1}(\vec{\omega},\lambda,\vec{\omega}_0,\lambda_0) \right.$$

$$\left. + \frac{\exp\left[-d\left(\frac{\mu_1}{|\eta|} + \frac{\mu_{1,0}}{|\eta_0|}\right)\right]}{\frac{\mu_{2,0}}{|\eta_0|} + \frac{\mu_1}{|\eta|}} k_{a2}(\vec{\omega},\lambda,\vec{\omega}_0,\lambda_0) \right\} \qquad (14)$$

Obviously, Eqn (14) coincides with the result for the semi-infinite homogeneous medium when $d\to\infty$, and with the one for a thin thickness homogeneous slab when $k_{a2}\to 0$. Higher-order scattering calculations or higher number of layers can be easily computed from Eqns (10) and (12).

CONCLUSIONS

The above procedure is able to give reference results for classical problems in X-Ray spectrometry like, for instance, the study of multiple scattering contributions from a light matrix substrate on a metallic coating. In this context, it looks promising also the improvement of the code SHAPE[2] including multi-layer calculations.

REFERENCES

1. J.E. Fernández and V.G. Molinari, X-Ray photon spectroscopy calculations, in: "Advances in Nuclear Science and Technology," Vol 22, M. Becker and J. Lewins, eds., Plenum Press, New York (1991).
2. J.E. Fernández and M. Sumini, SHAPE: a computer simulation of energy dispersive X-Ray spectra, X-Ray Spectrom. (1991), to be published.

MULTIPLE SCATTERING CONTRIBUTIONS OF THIN FILMS IN REFLECTION GEOMETRY

J.E. Fernández[*,**] and R. Sartori[*,***]

Laboratorio di Ingegneria Nucleare di Montecuccolino
University of Bologna
Via dei Colli 16, 40136 Bologna, ITALY

ABSTRACT

The multiple scattering contributions to the emitted intensity of a thin homogeneous sample under X-Ray excitation are studied with recourse to the Boltzmann transport theory. The corrective terms to the XRF characteristic line due to a second collision of either the photoelectric effect (secondary XRF), or the Compton, or the Rayleigh scattering, are deduced for reflection geometry. Analytical expressions for the intensities are given that allow their computation for variable incidence and take-off beam directions and source wavelength.

INTRODUCTION

Photons in the X-Ray regime have a large penetrating depth, and the probability of secondary interactions contributing to the XRF-intensity is not negligible, becoming a source of uncertainty for the computational methods of quantitative analysis.

The multiple scattering of photons in XRF has been studied with different approaches, from theoretical to Monte Carlo simulation. Traditional theoretical methods have shown severe limitations to study scattering processes having cross-sections with a certain complexity, and have mainly reduced to consider chains of pure photoelectric effects (having a simple isotropic cross-section). On the other way, Monte Carlo simulation provides numerical results of relative accuracy which are of difficult generalization when many variables are involved. This fact represent an obvious drawback to analyze problems where the excitation and detection geometry, the source energy, and the multi-element sample composition are free variables.

---

[*] On leave of absence from the Faculty of Mathematics, Astronomy and Physics, University of Córdoba, Argentina.
[**] Member of the CONICET, Buenos Aires, Argentina.
[***] Fellowship of CONICOR, Córdoba, Argentina.

Advances in X-Ray Analysis, Vol. 35
Edited by C.S. Barrett *et al.*, Plenum Press, New York, 1992

In contrast, transport theory provides the formal framework for studying appropriately photon transport in different geometries. An order-of-
interactions[1] scheme allows the solution of the one-dimensional equation
for plane geometry, making it possible to compute, analytically, the intensities due to chains of collisions using non approximated cross-sections.
Calculations[2-5] of such contributions have been recently performed for the
main three processes in the X-ray regime: photoelectric effect, Rayleigh
and Compton scattering, for an homogeneous sample of $infinite$ extent and
thickness. It is the aim of this work to extend the formalism to the case
of a $thin$ thickness homogeneous sample under X-ray excitation, in reflection geometry.

THEORY

        The flux of X-Rays is completely determined as solution of the photon
transport equation describing the balance between the numbers of photons of
given energy and direction entering and leaving an infinitesimal volume
element.

        We shall consider a plane and homogeneous sample of thickness t and
density ρ, with mass attenuation coefficient $\mu(\lambda)$, and a plane slant mono-
directional (along $\vec{\omega}_0$) and monochromatic (λ_0) source of intensity I_0 photons s^{-1} cm^{-2} in the beam direction that can be described by the one-
dimensional transport equation

$$\eta \, \frac{\partial f(z,\vec{\omega},\lambda)}{\partial z} = -\mu(\lambda) \, f(z,\vec{\omega},\lambda) +$$

$$\int_0^\infty d\lambda' \int_{4\pi} d\vec{\omega}' \; k(\vec{\omega},\lambda,\vec{\omega}',\lambda') \; \mathcal{U}(z)[1-\mathcal{U}(z-t)] \; f(z,\vec{\omega}',\lambda') +$$

$$I_0 \, \delta(z) \, \delta(\vec{\omega}-\vec{\omega}_0) \, \delta(\lambda-\lambda_0), \qquad (1)$$

where $\eta = \omega_z$, $\mathcal{U}(z)$ is the unitary-step Heaviside function, $f(z,\vec{\omega},\lambda)d\omega d\lambda$ is
the number of photons with wavelength between λ and $\lambda+d\lambda$, with direction
between $\vec{\omega}$ and $\vec{\omega}+d\vec{\omega}$, crossing a unit area at depth z per unit time, and
$k(\vec{\omega},\lambda,\vec{\omega}',\lambda')$ is the probability density, per unit pathlength, unit wavelength, and unit solid angle, that a photon change its phase-space coordi-
nates from $(\vec{\omega}',\lambda')$ to $(\vec{\omega}, \lambda)$. A sketch of the geometrical arrangement is
shown in Fig 1.

        We find an order-of-interactions[1] solution to describe the multi-
ple scattering contributions due to the prevailing effects in the X-ray
regime. We obtain the n-th order flux (n > 0) and (0 < z < t)

$$f^{(n)}(z,\vec{\omega},\lambda) = \frac{1}{|\eta|}\left[\frac{1+sgn(\eta)}{2} \, e^{-\alpha z}\int_0^z d\tau \; e^{\alpha\tau}\int_0^\infty d\lambda' \int_{4\pi} d\vec{\omega}' k(\vec{\omega},\lambda,\vec{\omega}',\lambda')f^{(n-1)}(\tau,\vec{\omega}',\lambda')\right.$$

$$\left. + \frac{1-sgn(\eta)}{2} \, e^{\alpha z}\int_z^t d\tau \; e^{-\alpha\tau}\int_0^\infty d\lambda' \int_{4\pi} d\vec{\omega}' k(\vec{\omega},\lambda,\vec{\omega}',\lambda')f^{(n-1)}(\tau,\vec{\omega}',\lambda')\right], \qquad (2)$$

with $\alpha = \frac{\mu(\lambda)}{|\eta|}$. For the zeroth-order we get

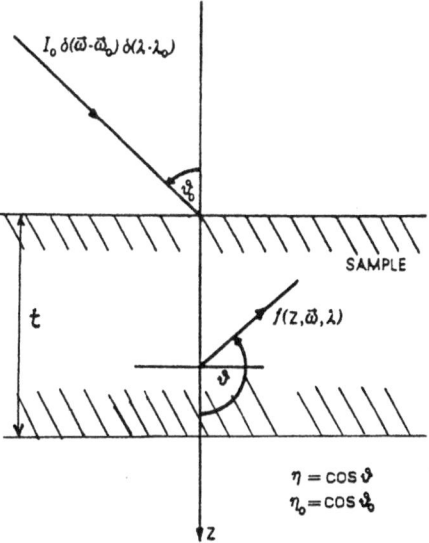

Fig. 1. Irradiation arrangement of an homogeneous sample of thickness t
under X-ray bombardment of a plane monochromatic X-ray source.

$$f^{(0)}(z,\vec{\omega},\lambda) = \frac{I_0}{2|\eta|}\, \delta(\vec{\omega}-\vec{\omega}_0)\, \delta(\lambda-\lambda_0)\, e^{-\alpha z}\, [1+\text{sgn }\eta \text{ sgn } z], \tag{3}$$

where $\vec{\omega}_0, \vec{\omega}$ are unitary vectors along the incident and take-off directions,
and λ_0 and λ denote the incident and emitted wavelength, respectively . As
expected, Eqn (2) approaches Eqn (33) of Ref. [2] when t→∞.

Solving Eqn (2) for n=1 and $\eta < 0$ we obtain the reflected flux
after the single collision a

$$f_a^{(1)}(z,\vec{\omega},\lambda) = \frac{I_0}{|\eta_0||\eta|}\, k_a(\vec{\omega},\lambda,\vec{\omega}_0,\lambda_0)\, e^{\alpha z}\, \frac{e^{-(\alpha+\alpha_0)z} - e^{-(\alpha+\alpha_0)t}}{\alpha + \alpha_0}. \tag{4}$$

The flux for photons scattered twice is represented by

$$f_{ab}^{(2)}(z,\vec{\omega},\lambda) = \frac{I_0}{|\eta_0||\eta|}\int_0^\infty d\lambda' \left[\iint_{\eta'<0} d\vec{\omega}'\, k_b(\vec{\omega},\lambda,\vec{\omega}',\lambda') k_a(\vec{\omega}',\lambda',\vec{\omega}_0,\lambda_0) R^-(z,\eta',\lambda') \right.$$

$$\left. + \int_{\eta'>0} d\vec{\omega}'\, k_b(\vec{\omega},\lambda,\vec{\omega}',\lambda')\, k_a(\vec{\omega}',\lambda',\vec{\omega}_0,\lambda_0)\, R^+(z,\eta',\lambda') \right], \tag{5}$$

where

$$R^+(z,\eta',\lambda') = \frac{1+\text{sgn }\eta'}{2}\, \frac{e^{\alpha z}}{|\eta'|\,(\alpha_0-\alpha')}$$

$$\left[\frac{e^{-(\alpha+\alpha')z} - e^{-(\alpha+\alpha')t}}{\alpha + \alpha'} - \frac{e^{-(\alpha_0+\alpha)z} - e^{-(\alpha_0+\alpha)t}}{\alpha_0 + \alpha}\right], \tag{6}$$

$$R^-(z,\eta',\lambda') = \frac{1-\text{sgn }\eta'}{2} \frac{e^{\alpha z}}{|\eta'|\ (\alpha'+\alpha_0)}$$

$$\left[\frac{e^{-(\alpha+\alpha_0)z} - e^{-(\alpha+\alpha_0)t}}{\alpha_0+\alpha} + e^{-(\alpha_0+\alpha')t}\ \frac{e^{-(\alpha-\alpha')t} - e^{-(\alpha-\alpha')z}}{\alpha - \alpha'}\right], \tag{7}$$

$$\alpha = \frac{\mu(\lambda)}{|\eta|}, \qquad \alpha' = \frac{\mu(\lambda')}{|\eta'|}, \qquad \text{and} \qquad \alpha_0 = \frac{\mu(\lambda_0)}{|\eta_0|}.$$

For $t\to\infty$ the above expression (5) approaches the corresponding term for the infinite thickness sample.[2]

MULTIPLE SCATTERING INTERACTIONS

In the X-ray regime, that is between 1 keV and 100 keV, there are three processes that prevail: the photoelectric effect, and the Rayleigh and Compton scattering. The kernels for these processes are respectively[2]

$$k_P(\vec{\omega},\lambda,\vec{\omega}',\lambda') = \frac{1}{4\pi} \sum_i Q_{\lambda_i}(\lambda')\ \delta(\lambda-\lambda_i)\ \left(1-U(\lambda'-\lambda_{e_i})\right), \tag{8}$$

$$k_C(\vec{\omega},\lambda,\vec{\omega}',\lambda') = \frac{\sigma}{\lambda_C}\ K_{kN}(\lambda,\lambda')\ S(\lambda',\vec{\omega}.\vec{\omega}',Z)\ \delta\left(1-\vec{\omega}.\vec{\omega}' + \frac{\lambda'-\lambda}{\lambda_C}\right), \tag{9}$$

and

$$k_R(\vec{\omega},\lambda,\vec{\omega}',\lambda') = \sigma\ \delta(\lambda-\lambda')\ \left(1+(\vec{\omega}.\vec{\omega}')^2\right)\ \frac{F^2(\lambda',\vec{\omega}.\vec{\omega}',Z)}{Z}, \tag{10}$$

where $Q_{\lambda_i}(\lambda')$ is the XRF emission probability for the λ_i line at λ', $\sigma = r_0^2 \rho N Z/(2A)$, being r_0 the classical radius of the electron, N the Avogadro's number, A the atomic weight, ρ the density, Z the atomic number, and λ_C the Compton wavelength. The factor $\sigma\ K_{kN}(\lambda,\lambda')$ represents the well known Klein-Nishina differential cross section, and $S(\lambda',\vec{\omega}.\vec{\omega}',Z)$ and $F(\lambda',\vec{\omega}.\vec{\omega}',Z)$ the atomic form factor and scattering function, respectively.

In order to calculate the emitted intensities for reflection geometry, we must evaluate the expressions (4) and (5) for z=0, and use the appropriate kernels for the interactions of interest.

Second-order contributions to a characteristic line are the chains (P,P), (P,C), (C,P), (P,R) and (R,P).[2] All of them, except the (P,C) chain give discrete spectra overlapping the line. In the (R,P) and (C,P) processes the scattered photon acts as source for photoelectric effect. In the (P,R) and (P,C) chains, the characteristic photon is scattered towards the detector. Note that the four scattering contributions to the XRF line are always present and that are an important correction for characteristic lines of pure elements.[4] In the (P,P) process a characteristic photon acts as source for producing a new photoelectric effect. Note that this process is possible not only for multicomponent samples but also for a pure sample where the interest line has been excited by another line whose energy is higher than the absorption edge of the interest line.[5]

To make feasible the comparison with the customary X-ray spectrometry intensity the following relationship must be used

$$I^{(n)}(\lambda,\vec{\omega}) = |\eta| \, f^{(n)}(0,\vec{\omega},\lambda).$$

Solving for the photoelectric, Compton and Rayleigh kernels we obtain the following relationships for the first- and second-order intensities:

Photoelectric: (P)

$$I^{(1)}_{(P)}(\lambda,\vec{\omega}) = \frac{I_0}{4\pi|\eta_0|} \sum_i Q_{\lambda_i}(\lambda_0) \, \delta(\lambda-\lambda_i) \left(1-\mathcal{U}(\lambda_0 - \lambda_{e_i})\right) \frac{1-e^{-(\alpha+\alpha_0)t}}{\alpha + \alpha_0} \qquad (11)$$

Photoelectric-Photoelectric: (P,P)

$$I^{(2)}_{(P,P)}(\vec{\omega},\lambda) = \frac{I_0}{8\pi|\eta_0|} \sum_i \sum_j Q_{\lambda_i}(\lambda_0) \, Q_{\lambda_j}(\lambda_i) \, \delta(\lambda -\lambda_i) \, [1-\mathcal{U}(\lambda_0-\lambda_{e_j})][1-\mathcal{U}(\lambda_j-\lambda_{e_i})]$$

$$\left\{ \frac{1}{\alpha + \alpha_0} \left[\frac{1}{\alpha_0} \ln\left[\frac{\mu_j + \alpha_0}{\mu_j}\right] + \frac{1}{\alpha} \ln\left[\frac{\mu_j + \alpha}{\mu_j}\right] \right] + \right.$$

$$\frac{e^{-(\alpha+\alpha_0)t}}{\alpha + \alpha_0} \left[\frac{1}{\alpha_0} \ln\left[\frac{\mu_j - \alpha_0}{\mu_j}\right] + \frac{1}{\alpha} \ln\left[\frac{\mu_j - \alpha}{\mu_j}\right] \right] +$$

$$\frac{1}{\alpha + \alpha_0} \left[\frac{E_1((\mu_j+\alpha_0)t)}{\alpha_0} + \frac{E_1((\mu_j+\alpha)t)}{\alpha} \right] +$$

$$\frac{e^{-(\alpha+\alpha_0)t}}{\alpha + \alpha_0} \left[\frac{E_1((\mu_j-\alpha_0)t)}{\alpha_0} + \frac{E_1((\mu_j+\alpha)t)}{\alpha} \right] -$$

$$\left. \frac{E_1(\mu_j t)}{\alpha \, \alpha_0} \left[e^{-\alpha t} + e^{-\alpha_0 t} \right] \right\} \qquad (12)$$

where $E_1(x)$ is the exponential-integral, defined as:

$$E_1(x) = \int_x^\infty \frac{e^{-t}}{t} \, dt$$

$E_1(x)$ goes to zero when $x\to\infty$, making Eqn (12) approaches the solution for the infinite sample (see Eqn (40) in Ref. 2).

Expression (12) for the $I_{(P,P)}^{(2)}$ intensity is identical to the one pre-viously calculated by de Boer[6] in a less general context.

Photoelectric-Rayleigh: (P,R)

$$I_{(P,R)}^{(2)}(\vec{\omega},\lambda) = \frac{I_0 \sigma}{4\pi |\eta_0| Z} \sum_i Q_{\lambda_i}(\lambda_0) \, \delta(\lambda-\lambda_i) \left(1-\mathcal{U}(\lambda_0-\lambda_{e_i})\right)$$

$$\int_{-1}^{1} d\eta' \; (R^+(0,\eta',\lambda_i) + R^-(0,\eta',\lambda_i)) \int_0^{2\pi} (1+(\vec{\omega}.\vec{\omega}'^2)) \; F^2(\lambda_i,\vec{\omega}.\vec{\omega}',Z) \; d\varphi' \qquad (13)$$

Rayleigh-Photoelectric: (R,P)

$$I_{(R,P)}^{(2)}(\vec{\omega},\lambda) = \frac{I_0 \sigma}{4\pi |\eta_0| Z} \sum_i Q_{\lambda_i}(\lambda_0) \, \delta(\lambda-\lambda_i) \left(1-\mathcal{U}(\lambda_0-\lambda_{e_i})\right)$$

$$\int_{-1}^{1} d\eta' \; (R^+(0,\eta',\lambda_0) + R^-(0,\eta',\lambda_0)) \int_0^{2\pi} (1+(\vec{\omega}_0.\vec{\omega}'^2)) \; F^2(\lambda_0,\vec{\omega}_0.\vec{\omega}',Z) \; d\varphi' \qquad (14)$$

Compton-Photoelectric: (C,P)

$$I_{(C,P)}^{(2)}(\vec{\omega},\lambda) = \frac{I_0 \sigma}{2\pi |\eta_0| \lambda_c} \sum_i \delta(\lambda-\lambda_i)$$

$$\int_{\lambda_0}^{\lambda_0+2\lambda_c} d\lambda' \; Q_{\lambda_i}(\lambda') \; K_{KN}(\lambda',\lambda_0) \; S(\lambda_0,a',Z) \left[1-\mathcal{U}(\lambda'-\lambda_{e_i})\right]$$

$$\int_{\eta_{min}}^{\eta_{max}} d\eta' \; \frac{R^+(0,\eta',\lambda') + R^-(0,\eta',\lambda')}{\sqrt{(1-\eta'^2)(1-\eta_0^2)-(a'-\eta'\eta_0)^2}}, \qquad (15)$$

where

$a' = 1+(\lambda_0-\lambda')/\lambda_c$, $\Delta=\sqrt{(1-a'^2)(1-\eta_0^2)}$, $\eta_{min}= a'\eta_0- \Delta$, and $\eta_{max}= a'\eta_0+ \Delta$.

Photoelectric-Compton: (P,C)

$$I_{(P,C)}^{(2)}(\vec{\omega},\lambda) = \frac{I_0 \sigma}{2\pi |\eta_0| \lambda_c} \sum_i Q_{\lambda_i}(\lambda_0) \left[1-\mathcal{U}(\lambda_0-\lambda_{e_i})\right] K_{KN}(\lambda,\lambda_i) \; S(\lambda_i,a_i,Z)$$

$$\int_{\eta_{min}}^{\eta_{max}} d\eta' \; \frac{R^+(0,\eta',\lambda_i) + R^-(0,\eta',\lambda_i)}{\sqrt{(1-\eta'^2)(1-\eta^2)-(a_i-\eta'\eta^2)}}, \qquad (16)$$

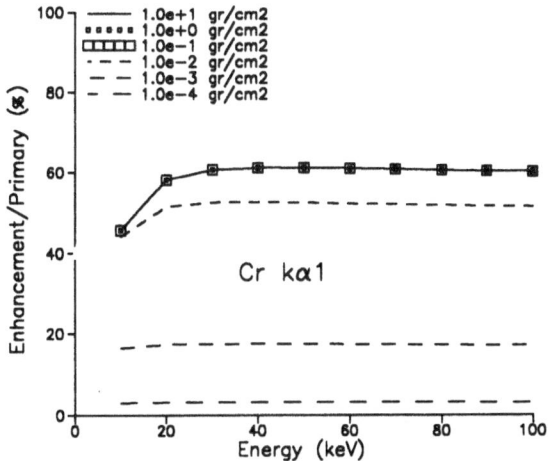

Fig. 2. The (P,P) contribution is plotted in units of the unmodified Cr Kα₁ line for a ternary alloy as a function of the excitation energy. Different sample thickness are considered. The alloy is Cr(25%)-Fe(60%)-Ni(15%).

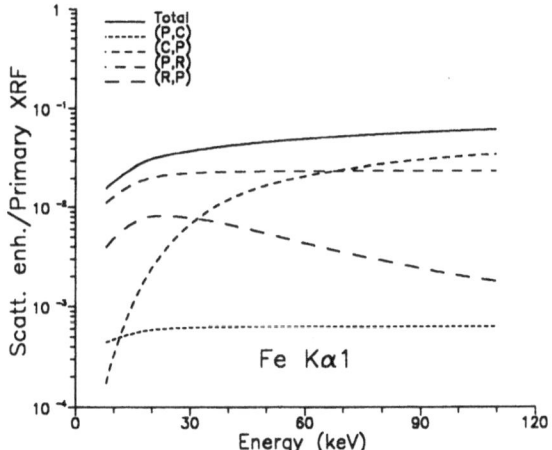

Fig. 3. The (P,R), (R,P), (P,C) and (C,P) contributions are plotted in units of the unmodified line intensity for a thin thickness sample. The (P,C) continuous spectrum is wavelength-integrated in order to be compared with the discrete contributions. The total contribution is given as the sum of the four terms. This graph can be compared with a similar one (Fig. 4(b) of Ref [2]) for infinite thickness.

where

$$a_i = 1+(\lambda_i-\lambda)/\lambda_c \;, \; \Delta=\sqrt{(1-a_i^2)(1-\eta^2)} \;, \; \eta_{min} = a_i\eta - \Delta, \text{ and } \eta_{max} = a_i\eta + \Delta.$$

RESULTS AND DISCUSSION

The behaviour of the above contributions as a function of the thickness was studied for two pure elements, Al and Fe, and for a ternary alloy, Cr(25%)-Fe(60%)-Ni(15%). An excitation-detection geometry with $\vartheta_0 = 45°$, $\vartheta = 135°$, $\varphi_0 = 0°$, and $\varphi = 0°$ was assumed. The innermost integrals of Eqn (14), (15), (16) and (17) were performed with recourse to a Romberg quadrature procedure.[7] Outer integrals when necessary were calculated with a trapezoidal algorithm. The evaluation of the exponential integral was done using a rational Chebyshev approximation.[8,9]

Fig. 2 shows the (P,P) contribution in units of the unmodified XRF-intensity for the Cr $K\alpha_1$ line of the alloy. Note that this contribution remains important (~20%) in spite of the very small sample thickness.

In Fig. 3 the scattering contributions are plotted as a function of the incident energy, for a thin thickness sample. It is shown that all the multiple scattering contributions decrease with the sample thickness. The (P,C) chain gives a continuous spectrum in the range $[\lambda_1, \lambda_1 + 2\lambda_c]$ that modifies the low energy tail of the peak as in the infinite thickness case.[4] This chain (wavelength-integrated) gives the weaker contribution of the four scattering chains. All of these contributions present azimuthal symmetry.

In Fig. 4 the definition of "infinite" thickness is evident. As the thickness grows, the primary XRF intensity and the enhancements contribu-

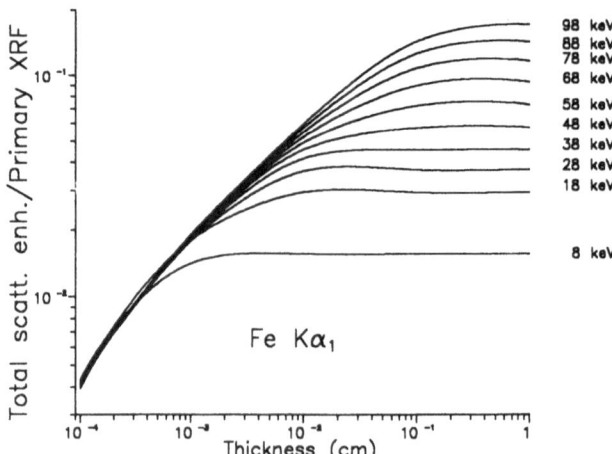

Fig. 4. The total scattering enhancement to the Fe $K\alpha_1$ line in units of the primary intensity, for different incident energies, as a function of the sample thickness.

tions do so. When for a certain thickness, all of them have reached the maximum value, we can say that the sample is an infinite one, because all the contributions produced in a deeper sample region cannot reach the surface. Graphics like this can be an useful tool in order to optimize the sample thickness in experimental work.

CONCLUSIONS

Transport theory permits us to acquire a deep insight in the knowledge of the multiple photoelectric, and scattering contributions to the XRF intensity, becoming an important tool for the analytical solution of a great variety of old problems that up to recently required strong simplifications.

Analitical relationships were obtained for all the second-order contributions that modify the primary XRF-line of a thin thickness target in terms of excitation-detection geometry, excitation wavelength, and sample composition and thickness.

REFERENCES

1. J.E. Fernández, V.G. Molinari and M. Sumini, Effect of X-Ray scattering anisotropy on the diffusion of photons in the frame of the transport theory, Nucl. Instr. and Meth. in Phys. Res. A 280:212 (1989).
2. J.E. Fernández and V.G. Molinari, X-Ray photon spectroscopy calculations, in: "Advances in Nuclear Science and Technology," Vol 22, 45-92, M. Becker and J. Lewins eds., Plenum Press, New York (1991).
3. J.E. Fernández, V.G. Molinari and M. Sumini, Corrections for the effect of scattering on XRF intensity, in: "Advances in X-Ray Analysis," Vol 33, C.S. Barrett at al. eds., Plenum Press, New York (1990).
4. J.E. Fernández, Rayleigh and Compton scattering contributions to the XRF intensity, X-Ray Spectrom. (1991), to be published.
5. J. E. Fernández, XRF Intensity in the frame of transport theory, X-Ray Spectrom. 18:271 (1989).
6. D.K.G. de Boer, Calculation of X-Ray fluorescence intensities from bulk and multilayer samples, X-Ray Spectrom. 19:145 (1990).
7. W.H. Press, B.P. Flannery, S.A. Teukolsky and W.T. Vetterling, "Numerical Recipes. The Art of Scientific Computing," Cambridge University Press, Cambridge (1986).
8. W.J. Cody and H.C. Thacher Jr., Chebyshev approximations for the exponential-integral Ei(x), Math. Comp. 22:289 (1968).
9. W.J. Cody and H.C. Thacher Jr., Rational Chebyshev approximations for the exponential-integral $E_1(x)$, Math. Comp. 22:641 (1968).

THE USE OF THE CONVENTIONAL ISOLATED ATOM MODEL FOR THE THEORETICAL

CALCULATION OF THE DEPENDENCE OF Lβ/Lα INTENSITY RATIO ON THE SAMPLE EXIT

ANGLE FOR UNOXIDIZED AND OXIDIZED TRANSITION METAL ALLOY THIN FILMS

Francis Fujiwara and George Andermann

Department of Chemistry, University of Hawaii
Honolulu, Hawaii   96822

INTRODUCTION

Variable sample exit angle x-ray fluorescence spectroscopy (VEA-XRF) employing the Lβ/Lα intensity ratio for transition metals and their oxides has been shown[1] to be useful for non-destructively studying transition metal surfaces and oxidation, as well as, superconductors. Thus the theoretical formulation of the Lβ/Lα intensity ratio dependence on the sample exit angle, θ_e, is of some interest and we develop it here. We also present methods of obtaining parameters needed in the formulation, such as Lβ absorption coefficients, which are not available in the literature, for wavelengths greater than about 12Å.

MODEL

To develop the Lβ/Lα equation we assume that the standard fluorescence intensity equation for the isolated atom[2] gives a valid accounting of the intensity of the Lβ and Lα lines. We specifically formulate the equation for the Lβ/Lα intensity ratio for a self-supporting sample of thickness t for the following two cases:

i)   a single compound of element A, and

ii)   a homogeneous mixture of two compounds of element A.

However, the formulation can be extended to more complex types of sample composition by considering the sample as consisting of consecutive layers of varying thicknesses of homogeneous mixtures of varying concentrations (see Figure 4, "b"). The sample geometry used in the formulation is as illustrated in Figure 1.

Advances in X-Ray Analysis, Vol. 35
Edited by C.S. Barrett *et al.*, Plenum Press, New York, 1992

THEORETICAL EQUATIONS

　　　Case i.　For the simple case of a single compound of element A, the equation for the Lβ/Lα intensity ratio follows directly from the ratioing of the standard fluorescence equations and is given by:

$$L\beta/L\alpha = K(\gamma,w,g)\ J(a(\Theta),b(\Theta),p,t) \tag{1}$$

where,

$$K = [w_\beta g_\beta(\gamma_\beta - 1)/\gamma_\beta]\ /\ [w_\alpha g_\alpha(\gamma_\alpha - 1)/\gamma_\alpha]$$

$$J = \{a[1 - \exp(-bpt)]\}\ /\ \{b[1 - \exp(-apt)]\}$$

$$a = \mu_i \csc\Theta_i + \mu_\alpha \csc\Theta_e \qquad\qquad b = \mu_i \csc\Theta_i + \mu_\beta \csc\Theta_e \tag{1a}$$

$$\mu_y = \Sigma_i\ F^j \mu^j \qquad (\text{sum is over elements j in compound}) \tag{1b}$$

and, y is "i", "α" or "β", w is the fluorescence yield, g is the degeneracy, γ is the jump ratio, p is the density of the compound, μ is the absorption coefficient (for the incident, i, or emitted, α or β, photon, and for the compound or element j), and F^j is the weight fraction of element j in the compound.

　　　Case ii.　For the case of a homogeneous mixture, "m", of two compounds of element A, "u" and "o", the following equation is obtained:

$$(L\beta/L\alpha)_m = [(C_u a_u + C_o a_o)\ /\ (C_u b_u + C_o b_o)]\ E_m\ K_m(C_u, C_o) \tag{2}$$

where,

$$E_m = [\ 1 - \exp(-b_m p_m t)]\ /\ [1 - \exp(-a_m p_m t)]$$

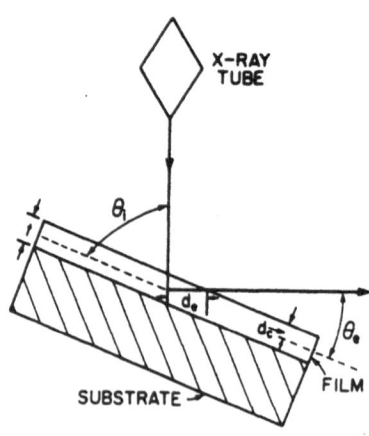

Fig. 1　Sample geometry.

$$K_m = (C_u F_u {}^A K_j {}^{\beta,A} + C_o F_o {}^A K_o {}^{\beta,A}) \, / \, (C_u F_u {}^A K_u {}^{\alpha,A} + C_o F_o {}^A K_o {}^{\alpha,A}) \qquad (2a)$$

$$a_x = \mu_i {}^x \csc\theta_i + \mu_\alpha {}^x \csc\theta_e \qquad\qquad b_x = \mu_i {}^x \csc\theta_i + \mu_\beta {}^x \csc\theta_e$$

$$\mu_i {}^m = C_u \mu_i {}^u + C_o \mu_i {}^o \qquad u_a {}^m = C_u \mu_\alpha {}^u + C_o \mu_\alpha {}^o \qquad \mu_\beta {}^m = C_u \mu_\beta {}^u + C_o \mu_\beta {}^o$$

$$\mu_y {}^x = \Sigma_j \, F_x {}^j \mu_y {}^j \quad \text{(sum is over elements j in compound x)}$$

$$K_x {}^{y,A} = w_x {}^{y,A} g_x {}^{y,A} (\gamma_x {}^{y,A} - 1) \, / \, \gamma_x {}^{y,A} \qquad (2b)$$

and, x is "u", "o" or "m", c_u is the weight fraction of "u" in "m", C_o, is the weight fraction of "o" in "m", $F_x {}^j$ is the weight fraction of element j in compound x, and the remaining parameters are the same as those for "Case i".

It should be noted that the following relation must hold:

$$C_u + C_o = 1.$$

This is useful because given all other terms, the weight fraction of compound "o", C_o, (or "u", C_u,) can be solved for by iteration, using equation 2.   (A particularly useful application of this is the determination of the extent of oxidation of a homogeneously, partially oxidized binary alloy.   Thus if "o" refers to the completely oxidized alloy and "u" refers to the unoxidized alloy, then C_o gives directly the extent of oxidation.)

METHODS: UNAVAILABLE PARAMETERS

Parameter: $\mu_\beta {}^j$.   While many of the absorption coefficients, including $\mu_\alpha {}^A$ of element A at the Lα wavelength of A, can be found in the literature[3], the $\mu_\beta {}^A$ of element A at the Lβ wavelength of element A are not readily available.   These $\mu_\beta {}^{A}$'s, of course, cannot be interpolated from available absorption coefficients because of the Lβ absorption edge of A.   In order to obtain the μ_β's needed in the above equations, we have used a ratioing method to experimentally determine them.

The ratio of interest is that of equation 1 for two different exit angles for the case of a pure bulk sample, (i.e., t = ∞), as shown below:

$$\{[L\beta/L\alpha](\theta_1)/[L\beta/L\alpha](\theta_2)\} = [a(\theta_1)/b(\theta_1)]/[a(\theta_2)/b(\theta_2)] \qquad (3)$$

The left side of equation 3 is experimentally determined for exit angle θ_1 and θ_2, and $\mu_\beta {}^A$, which enters in b(θ), (see equations 1a, and 1b), is

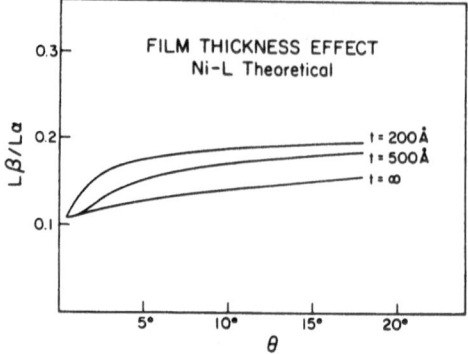

Fig. 2.   Film thickness dependence of Ni-L$_\beta$/L$_\alpha$ for Ni.

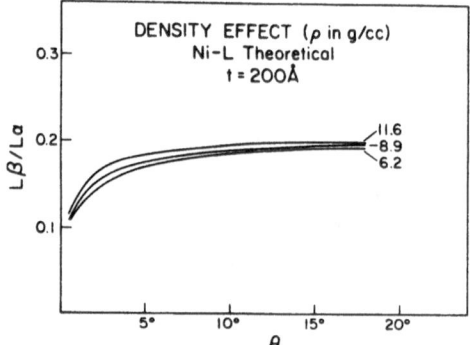

Fig. 3.   Density dependence of Ni-L$_\beta$/L$_\alpha$ for Ni.

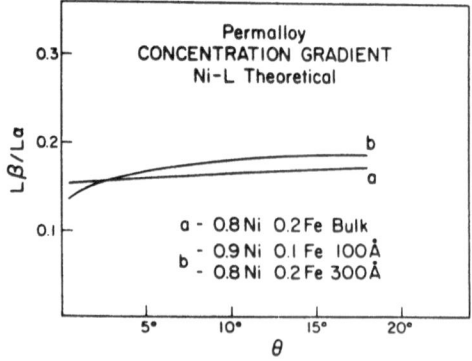

Fig. 4.   Concentration gradient effect on Ni-L$_\beta$/L$_\alpha$ for permalloy.

solved for by using the available values for all the other parameters. (Of course, it should be noted that because the magnitude of the left side of the equation may depend on such things as the instrumental broadening of the spectral resolution, the μ_β obtained should not be used independent of the Lβ/Lα experimental values from which it is derived unless proper corrections are made.)

Parameters: w, g and γ. The fluorescence yield, w, the degeneracy, g, and the jump ratio, γ in both compounds "u" and "o" are often not available in the literature. These parameters enter equation 2 via the $K_x^{y,A}$ terms. (See equations 2a and 2b.) Inspection of the form of equation 2a shows that the absolute magnitudes of the $K_x^{y,A}$ are not necessary but that normalized values of the following form can be used:

$$k_x^{y,A} = K_x^{y,A} / K_{ref}^{a,R}$$

where "ref" refers to a reference substance and "R" to the element R in the reference substance. These normalized values, $k_x^{y,A}$, can be determined experimentally by utilizing the following intensity ratio for bulk samples of compound x and reference substance. It is formulated using the same standard fluorescence intensity equation used above.

$$L_y^{x,A} / L_\alpha^{ref,R} = [(\mu_i^A F_x^A a_{ref}) / (\mu_i^R F_{ref}^R a_x)] k_x^{y,A} \tag{4}$$

where the "L"'s refer to the intensity of the L line specified by the subscripts and superscripts. The terms on the left side of equation 4 are measured under the same experimental conditions and the $k_x^{y,A}$ can be extracted by using the available values for the remaining parameters. It should be noted that in using this method to determine normalized $K_x^{y,A}$ values, extremely pure compounds u and o should be used since they serve as calibration standards, (i.e., see equation 2a,) in the determination of C_o (or C_u).

CALCULATIONS AND RESULTS

Case i. Equation 1 points to some interesting effects on Lβ/Lα due to film thickness, density and concentration changes. Such effects calculated from equation 1 are shown in Figures 2 - 4. However, as related in an earlier paper[4], optical effects may alter the behavior of Lβ/Lα at low θ_e angles.

Case ii. Plots of equation 2 applied to the case of bulk samples of varying degrees of uniform oxidation of the element A in the binary alloy AB are shown in Figure 5. As previously mentioned, optical effects may alter Lβ/Lα behavior at low θ_e values.

Equation 2 had also been applied to determine the extent of oxidation of permalloy thin films at a θ_e of 18°. These results which

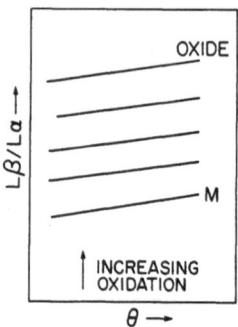

Fig. 5.   Calculated L_β/L_α Vi θ_e for different degrees of oxidation
of A in alloy AB.

have been previously reported[4], showed the iron to have mild oxidation
(2 and 4%) when ambiently oxidized and medium to extensive oxidation (36
and 91%) when thermally, air oxidized (200°C for 2 hours) for thin films
of thickness, 400 and 200 Å, respectively.

CONCLUSION

        In this paper, the standard fluorescence intensity equation
normally used for fixed angle bulk studies has been extended to variable
exit angle XRF for thin films and bulk.   In the process a simple
technique has been developed for obtaining $L\beta$ mass absorption coefficient
values and normalized fundamental K parameter values which deal with
fluorescence yield, degeneracy and jump ratio values.

REFERENCES

1.   G. Andermann, Appl. Surf. Sci. 31. 1 (1988).
2.   I. P. Bertin, Principles of X-Ray Spectrometric Analysis, 2nd. Ed.
        (Pleneum Press) New York (1975).
3.   B.L. Henke, P. Lee, T.J. Tanaka, R.L. Shimabukuro and B.K.
        Fujikawa, At. Data Nucl. Data Tables 27, 1 (1982).
4.   George Andermann, Francis Fujiwara, T.C. Huang, J.K. Howard and N.
        Staud, Ad. X-Ray Anal. 32 261 (1989).

X-RAY PHOTOELECTRON AND FLUORESCENCE SPECTRA OF
SEVERAL ZIRCONIUM OXIDE COMPOUNDS

Tatsuya Maruyama, Goh Sasaki*, Sei Fukushima, Kozo Kuchitsu*, and Naoto Koshizaki**

Analysis Research Center, Fuji Xerox Co., Ltd.
Kanagawa, JAPAN

*Department Chemistry, Nagaoka University of Technology
Niigata, JAPAN

**Composite Engineering Division, Structure Technology
Department, Industrial Products Research Institute
Ibaraki, JAPAN

ABSTRACT

IIa group metal - Zr perovskite type compounds ($MZrO_3$, M=Ba, Sr and Ca) and other oxides were studied by XPS to know the effects of M atoms to Zr atoms. And the feasibility of using XPS and high-resolution XRF for the coordination analysis of Zr atoms was examined for some kinds of stabilized zirconia. An almost linear relationship between the energy of Zr 3d peaks and the radii of IIa group metal ions was found, and, a simple model to interpret this relationship has been proposed. However, no relationship between the energy shifts of Zr 3d spectra or Zr $L\alpha_{1,2}$ spectra and the coordination numbers of Zr atoms in the several Zr oxide compounds was found. Thus, the coordination analysis using XPS or high-resolution XRF is not feasible.

1. INTRODUCTION

Stabilized zirconia is one of the new ceramic materials developed recently. As known well, zirconia (ZrO_2) has a very high melting point, but it shows very large volume change at the phase transition temperature. Thus, zirconia is hard to use as a heat-resistant structural material. On the contrary, the stabilized zirconia, which is a solid solution of zirconia containing about 7-10% CaO, MgO or Y_2O_3, shows very small volume change on heating, and can be used as a superior heat-resistant structural material.

The structural and thermodynamical stability of the stabilized zirconia is closely connected with its crystallographic structure, the coordination state of Zr atoms, for example. X-ray powder diffraction method (XRD) is used for the structural analysis in general, but there are difficulties in analyzing the complicated materials such as solid solutions or amorphous materials. For the coordination analysis of such materials, X-ray photoelectron spectroscopy (XPS) or high-resolution X-ray fluorescence spectroscopy (XRF) can be used. The coordination analysis of magnesium[1], aluminum[2], silicone[3] or germanium atoms[4] with their oxides were already studied by means of the measurement of the energy shifts of the characteristic X-rays, the so called "chemical shifts", in high-resolution XRF spectrum.

The oxidation state of zirconium atoms in the oxides can be regarded as the highest oxidation state (Zr(IV)), except ZrO. The principal factors which dominate the chemical states of Zr atoms are, therefore, the coordination number of Zr by O atoms and the effects to Zr-O clusters from the other cations in the compounds. The similar discussion for Ti oxide compounds was already reported[5], but the study about zirconium oxides has not be done.

In this report, IIa group metal - Zr perovskite type compounds ($MZrO_3$, M=Ba, Sr and Ca) and other oxides were studied by XPS to know the effects of M atoms to Zr atoms. And the feasibility of using XPS and high-resolution XRF for the coordination analysis of Zr atoms was examined for some kinds of stabilized zirconia.

2. EXPERIMENTAL

For the measurement of XPS spectra, a JEOL JPS-80 spectrometer with a magnesium anode X-ray source was used. The pressure in the analysis chamber was kept lower than 1×10^{-6} Pa. As a standard for the relative calibration of the effects of charge built up, C 1s of graphite powder was used. This powder was well baked out before use and then mixed uniformly into the samples. The photoelectron spectra of Zr 3d were chosen for analysis.

For the measurement of high-resolution XRF spectra, a two-crystal type spectrometer, made by RIGAKU, was used. Analyzer crystals were Ge(111)s (2d=6.532A). Samples were excited with X-rays generated by a chromium target tube. The spectrum of metallic Zr was measured just before and just after the measurement of each sample, in order to cancel the drift of signals in the measurement system. The fluorescence spectra of Zr $L\alpha_{1,2}$ were chosen for analysis.

All of the samples were fine powder and their structures were determined by XRD except stabilized zirconia. The coordination numbers of zirconium atoms in the stabilized zirconia were estimated from their chemical formulas.

3. RESULTS AND DISCUSSION

3.1 Zr 3d photoelectron spectra of perovskite type compounds

Fig. 1 shows the Zr 3d photoelectron spectra of several perovskite type compounds and zirconium oxide compounds. All the peaks were calibrated

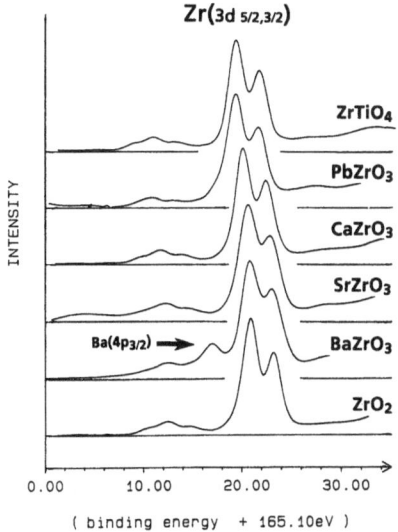

Fig.1. Zr 3d 5/2,3/2 photoelectron spectra of several perovskite type compounds and zirconia compounds.

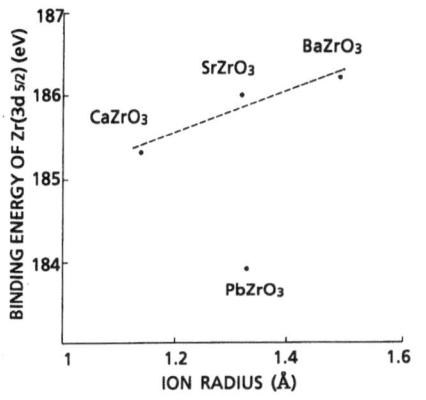

Fig.2. Peak positions of Zr 3d of perovskits vs. the ion radii of IIa group metals and Pb.

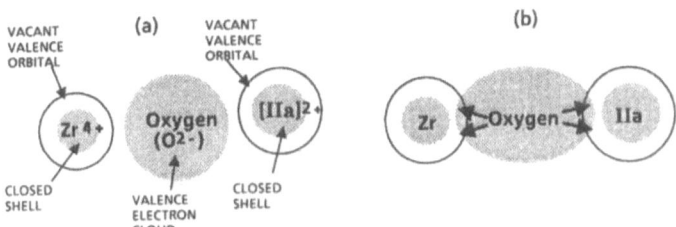

Fig.3. A simple model for explaining valence electron in Zr-O-IIa group metal bond.

relatively with the position of C 1s peak of the graphite powder mixed into
each sample.   Fig. 2 is the plot of binding   energy   of Zr 3d vs. ion radii
for   4 different perovskite type compounds . These metal ions   are IIa
group elements except lead.   It can be found that the energy of Zr 3d
increases   almost linearly against the radii of IIa group metal ions.

A simple model can be proposed to interpret this tendency of the
shifts of Zr 3d peaks shown in Fig. 2. As mentioned above, both   zirconium
ions   and IIa group metal ions in these   compounds take the highest oxidation state. If   these   ions   are in   free atomic state, they have closed
shells as in the case of noble gas atoms and their valence orbitals are
vacant, as shown in Fig. 3(a). Thus, it can be assumed that atoms in these
compounds are bonded only by valence electrons of the oxygen ions, which
flow into the vacant valence orbitals of Zr and other metal ions, as shown
in Fig. 3(b). In this model, it must be noted that total amount of valence
electrons at the bonds between zirconium, oxygen and other metal ions is
constant.

It is well known that   valence orbital   interaction depends upon their
spatial spread represented by ion radius. In the case of this study, as the
radius of IIa group metal ion increases, the interaction of valence orbital
of metal with oxygen becomes greater, and the electron density of the bond
of metal and oxygen atom also increases. This means that the electron
density of the bond between   zirconium and oxygen atoms decreases relatively, and the valence electron density of zirconium atom decreases. Consequently, Zr 3d peak shifts to a higher biding energy side. On the contrary,
as the radius of IIa metal ion decreases, the electron density of zirconium
atom increases, then Zr 3d peak shifts to a lower binding energy side. The
similar discussion for several Al-Si oxide compounds was developed   by West
et al. [6]

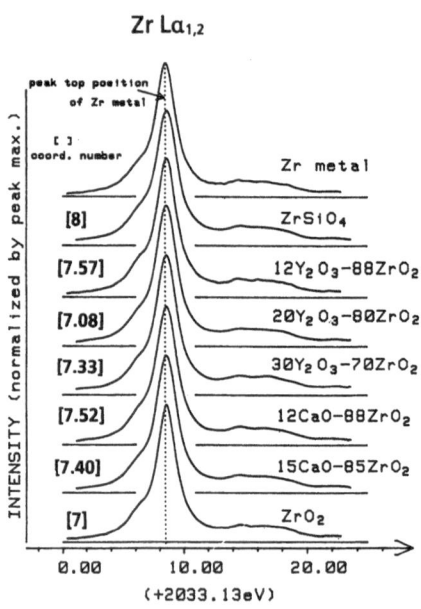

Fig.4.   High-resolution Zr L α1,2 spectra of several stabilized
         zirconia materials.

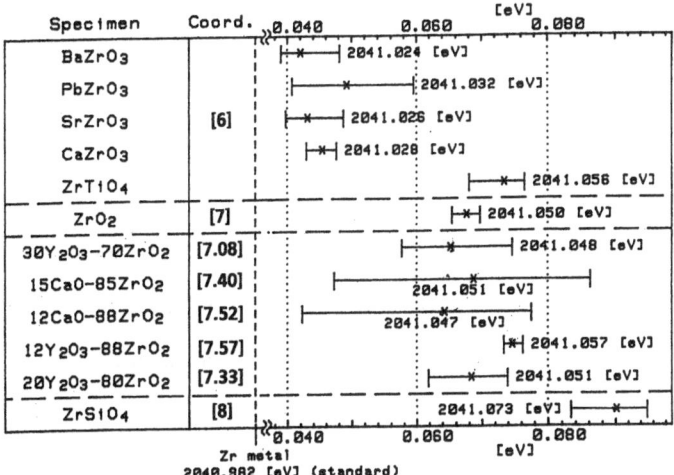

Fig.5. Summary of chemical shifts of Zr L α1,2 of several
zirconium oxide compounds.

The structure of valence band of lead ion (IVb group element) is
different from that of IIa group ones, and the nature of closed shell and
ion radius is also because lead ion has closed 4f atomic orbitals. Thus, the above discussion cannot be made with $PbZrO_3$.

3.2 Coordination state of Zr in stabilized zirconia

In Fig. 4, high-resolution Zr L $\alpha_{1,2}$ spectra of several stabilized
zirconia, ZrO_2 (7 coordination state), $ZrSiO_4$ (8 coordination state) and
metallic Zr are shown. Chemical shift of L $\alpha_{1,2}$ of each zirconium compound
from metallic zirconium is very small, and no obvious profile change is
observed. In Fig. 5, a summary of chemical shifts of L α_1 of zirconia compounds corresponding to coordination numbers of zirconium atoms is shown.
Error bars show the range of presence of L α_1 peaks, which are very wide

specimen	coord. No.	184.00	186.00	188.00
ZrTiO₄		•		
PbZrO₃		•		
CaZrO₃	6		•	
SrZrO₃			•	
BaZrO₃			•	
ZrO₂	7		•	
30Y-70Z	7.08			•
20Y-80Z	7.33			•
15C-85Z	7.40		•	
12C-88Z	7.52			•
12Y-88Z	7.57		•	
ZrSiO₄	8			•

Z = ZrO₂
Y = Y₂O₃ C = CaO   184.00   186.00   188.00
                         Binding energy of Zr 3d₅/₂ (eV)

Fig.6. Summary of Zr 3d5/2 photoelectron peak positions
of several zirconium oxide compounds.

because of   low spectral intensity . From this figure, no correlationship
between peak shifts and coordination number can   be found. This means that
the   coordination analysis for stabilized zirconia materials by high-resolution XRF is not possible.

A summary of peak positions of Zr 3d of several stabilized zirconia
and zirconia compounds is shown in Fig. 6. No obvious relationship **between**
peak positions and coordination numbers is found in this figure.

In addition to the above measurements, the differences between the
energy of Zr 3d and O 1s peaks were also examined against the coordination
numbers of Zr atoms, but, no correlation was found either. Thus, it can be
concluded that the coordination analysis by means of XPS or high-resolution
XRF for zirconium atoms in the stabilized zirconia is not feasible.

4. CONCLUSION

IIa group metal - Zr perovskite type compounds (MZrO₃, M=Ba, Sr and
Ca) were studied by XPS. An almost linear relationship between the energy
of Zr 3d peaks and the radii of IIa group metal ions was found, and, a
simple model to interpret this relationship has been proposed.

The feasibility of using XPS and high-resolution XRF for the coordination analysis of Zr atoms   was examined for some kinds of stabilized
zirconia.   However, no relationship between the energy shifts of Zr 3d
spectra or Zr L$\alpha_{1,2}$ spectra and the coordination numbers of Zr atoms in
the several Zr oxide compounds was found. Thus, the coordination analysis
using XPS or high-resolution XRF is not feasible.

ACKNOWLEDGMENTS

The authors thank Prof. Masanori Owari (Univ. of Tokyo) for his nice
cooperation for the improvement of the data acquisition system of XPS
spectrometer, and also thank to Dr. Kazuo Krosaki (RAC, Fuji Xerox Co.,
Ltd.) for his kind advises and discussion .

REFERENCES

1)   ex. K. Isozaki et al., J. Mat. Sci. **16** 2318 (1981)
2)   ex. S. Fukushima et al., Spectrochim. Acta **39B** 77 (1981)
3)   I. Okura et al., Spectrochim. Acta **45B** 711 (1990)
4)   S. Fukushima et al., J. Am. Ceram. Sic. **68** 490 (1985)
5)   M. Murata et al., J. Spectrosco., **6** 459 (1975)
6)   R. H. West et al., Surf. Interf. Anal. **4**, 68 (1982)

DEPTH PROFILING BY MEANS OF X-RAY PHOTOELECTRON SPECTROMETRY

M.F.Ebel and H.Ebel

Institut für Angewandte und Technische Physik
Technische Universität Wien, A 1040 Wien (Austria)

F.Olcaytug

Institut für Allgemeine Elektrotechnik und Elektronik
Technische Universität Wien, A 1040 Wien (Austria)

INTRODUCTION

The depth range d of x-ray photoelectron spectrometry (XPS) is determined by the inelastic mean free path λ of photoelectrons. The following considerations are dedicated to a correlation between d and λ. The contribution dn_i to the measured photoelectron signal of an element i of a sample, originating from a depth x and a volume with a thickness dx is given by

$$dn_i = \text{const} \cdot c_i \sigma_i \cdot \frac{dx}{\cos\gamma} S_i \varepsilon_i \left(1 - \beta_i \frac{3\cos^2\vartheta - 1}{4}\right) \cdot e^{-x/\lambda_{ic}\cos\gamma} \cdot e^{-t/\lambda_{il}\cos\gamma}$$

c_i...concentration of element i in atomic fractions
σ_i...photoabsorption cross-section
S_i...spectrometer function
ε_i...detector efficency
β_i...asymmetry parameter
ϑ....angle between x-rays and take-off of photoelectrons
$\lambda_{ic}, \lambda_{il}$..ineleastic mean free path of the photoelectrons in the sample and in the contamination layer
γ....angle between normal to the sample surface and take-off direction of photoelectrons
t....thickness of the contamination layer

For a sample thickness d and a homogeneous distribution of the chemical elements an integrated signal n_i is obtained.

$$n_i = \text{const} \cdot c_i \sigma_i S_i \varepsilon_i \lambda_{ic} \left(1 - \beta_i \frac{3\cos^2\vartheta - 1}{4}\right) \left(1 - e^{-d/\lambda_{ic}\cos\gamma}\right) \cdot e^{-t/\lambda_{il}\cos\gamma}$$

Thus, 1-1/e, or 63%, of the signal of a homogeneous thick sample are from a layer of a thickness $d = \lambda_{ic}\cos\gamma$. This thickness or depth is indicating the depth of analysis by XPS. With characteristic MgKα-radiation typical kinetic electron energies are found in the range from 100 to 1250eV and the corresponding values of λ are 1 to 4nm. This means that a variation of the depth of analysis can be performed from approximately 0.5nm to 4nm by a variation of the take-off angle γ. The present paper deals with depth profiling by means of a variation of the take-off angle γ. The concept is treated by a discussion of three different examples of application.

1. INVESTIGATION OF THIN GeC-FILMS

Thin films of $Ge_xC_yO_z$:H were prepared by rf-plasma deposition of tetraethyl germanium with Ar as carrier. The thicknesses of these films were in the range of 0.5μm to 1.5μm. For the application of these thin films as temperature sensors surface composition and depth profile are important.

On the outermost surface contaminations of adventitious hydrocarbons have to be expected. The positions of the C1s signals of the contamination and of $Ge_xC_yO_z$:H on the scale of kinetic energies are 0.4eV apart. Therefore, a variation of the take-off angle allows to distinguish between the two signals. An increase of the take-off provides a decrease of the C1s substrate signal from $Ge_xC_yO_z$:H and an increase of the C1s overlayer signal from the contamination.

Together with an algorithm[1] for quantitative analysis of homogeneous samples the angular variation gives an estimation of the depth responses of the composition. For this purpose, we assumed that the measured concentrations of the elements Ge, C and O are representative for an averaged composition in the depth of analysis. An estimation of the inelastic mean free path of the specimen material from Penn's tables[2] gives λ_{ic}=2.5nm. Thus, for take-off angles of 80°, 45° and 0° the analytical result represents these averaged compositions in depths d of 0.4nm, 1.8nm and 2.5nm.

Further investigations were carried out by depth profiling by means of a step-by-step Ar-ion etching of the outermost surface layers to a depth of 30nm. The experimental sequence was sputter etching, XPS-measurement, sputter etching, XPS-measurement, and so on. For these XPS-measurements the take-off angle was 0°. For the description of the depth profile of specimen composition we added the corresponding depth of analysis of 2.5nm to the actual values of sputter depths.

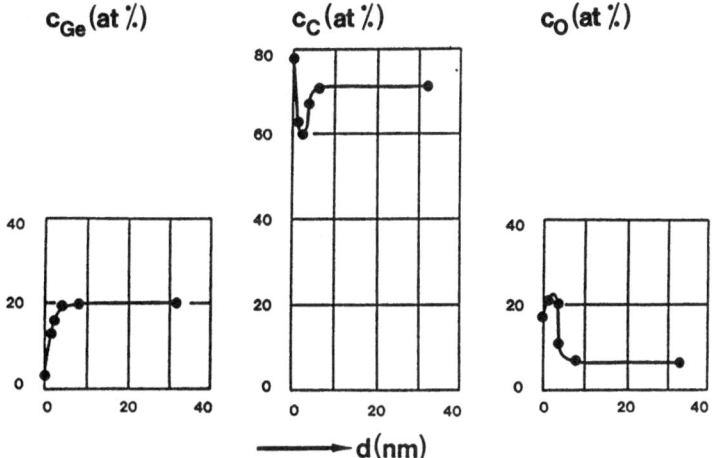

Fig.1   Typical ion etch XPS depth profiles of the concentrations of Ge, C and O in thin $Ge_xC_yO_z$:H-films.

A representative result of these investigations is depicted
in Fig.1. An enrichment of carbon on the outermost surface
confirms the existence of an adventitious hydrocarbon contamination. An oxygen rich interface between the contamination
and the bulk of the thin film specimen can be attributed to
GeO_2. It is formed at the end of the sputter deposition,
when the evacuated reactor chamber of the deposition system
is exposed to ambient atmosphere. For d>10nm the depth profile indicates a constant composition. Thus, the only exception to a homogeneous composition of the thin films is the
outermost surface.

This result could be confirmed by SIMS-investigations. The
SIMS depth profiles of Fig.2 confirm the homogeneous composi-

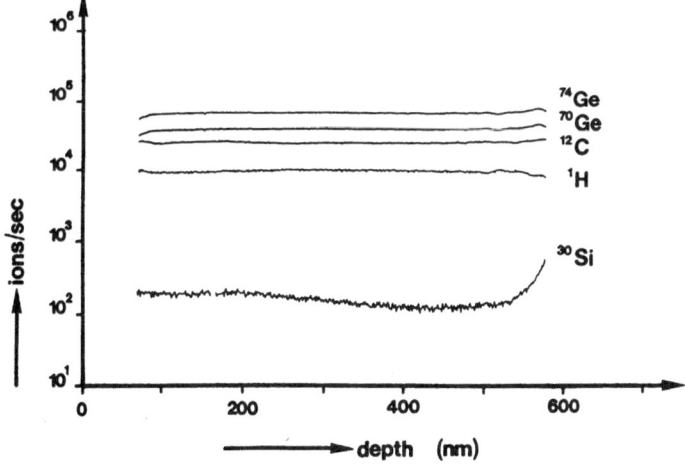

Fig.2   Typical depth profiles of Ge, C, H and Si measured by
SIMS with O^--ions.

tion. The end of the profile is indicated by the rapidly growing Si-signal from the glass substrate. Due to the use of oxygen ions, the depth response of oxygen cannot be given. From a comparison with reference specimens we were able to estimate the hydrogen content of the films to approximately 1at%. The remarkable oxygen content of 7at% in Fig.1 can be explained by the use of Ar and tetraethyl germanium of technical quality. We found no increase of the thickness of the GeO_2-interface with the time of exposure to air. The investigation of the electrical conductivity gave no evidence for ageing. From these results we conclude that the GeO_2-layer acts as a diffusion barrier for oxygen.

2. SiO_2-LAYERS ON SILICON WAFERS

We investigated thin dioxide films on monocrystalline silicon. The film thicknesses ranged from 2.7nm to 8.3nm and were determined by ellipsometry. A thickness of a few tenth of nm of the interface of Si^+, Si^{2+} and Si^{3+} between the substrate and the dioxide can be assumed[3]. As can be seen from Fig.3 the Si2p peaks from the substrate and from the dioxide film are well distinguishable.

The ratio r of the dioxide Si^{2+} signal and the substrate Si^0 signal is used for the determination of the film thicknesses. The usual evaluation of signal ratios r for the reduced thickness D/λ_{SiO_2} is taken from literature[4].

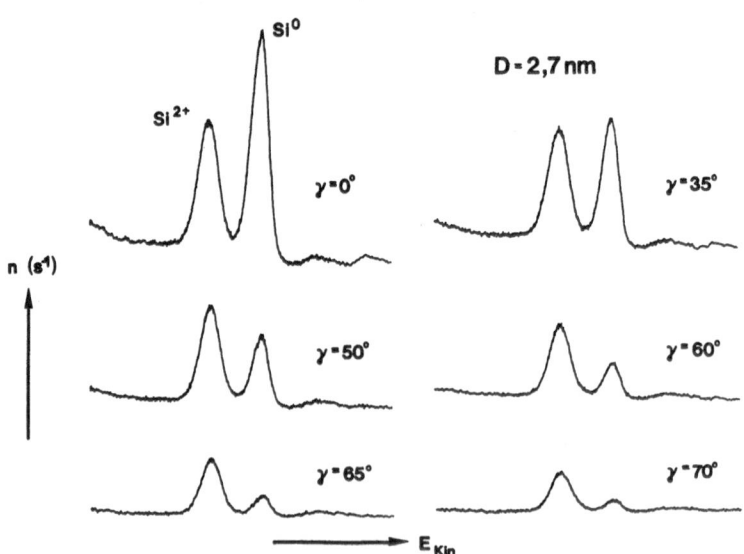

Fig.3 Si2p signals of a dioxide film with thickness D=2.7nm measured under take-off angles from $\gamma=0°$ to $\gamma=70°$.

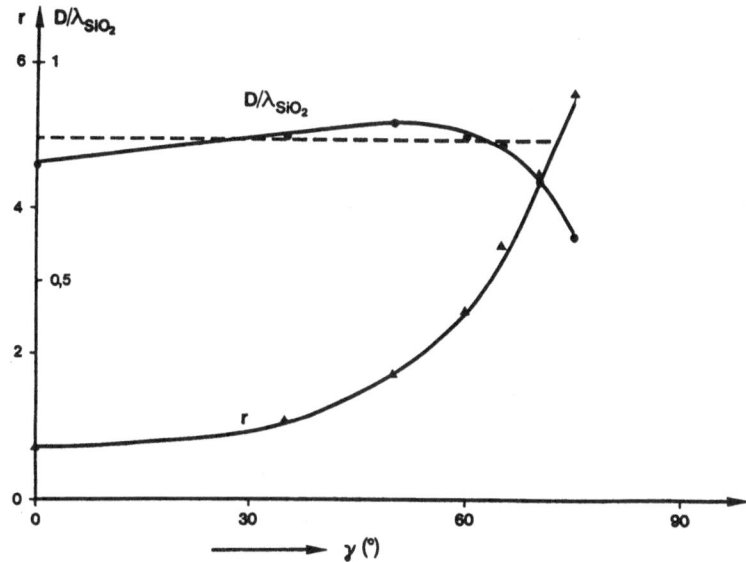

Fig.4 Signal ratio r of Si2p^{2+} from the dioxide and Si2p^{o} from the substrate and the corresponding reduced thicknesses in dependence on the take-off angle γ. The increase of the reduced thickness from $\gamma=0°$ to $\gamma=50°$ is caused by the neglection of the interface[8].

$$\frac{D}{\lambda_{SiO_2}} = \cos\gamma \cdot \ln(1+r/0.6)$$

Fig.4 depicts the responses of $r(\gamma)$ and the reduced thickness. Independent of the take-off angle a constant reduced thickness is expected. But this expectation only holds for a certain range of take-off angles. This has been attributed to the influences of surface roughness[5], the finite solid angle of electron detection[6], the influence of elastic electron scattering[7] and the neglection of the interface.[8]

The thickness D of the dioxide film cannot be determined employing the usual evaluation procedure of XPS-results. Only a reduced thickness in multiples of the inelastic mean free path of the photoelectrons in the film can be obtained. For this reason, the approach of depth profiling of thin films as used for $Ge_xC_yO_z$:H is modified and a more sophisticated concept is applied. This gives information on the depth profile of concentrations[9-13]. We assume seven different depth profiles and from our experimental Si2p results we calculate the characteristic quantities of each of the concentration responses. The quality of the least squares fit is expressed by the standard deviation of fitted and measured values. The depth profile showing the lowest standard deviation, gives the best description of the actual depth profile.

We apply the concept of depth profiling to a binary system
and a depth distribution $c_1(x)$ of the element 1. For the concentrations $c_1+c_2=1$ has to be met. Thus, $c_2(x)=1-c_1(x)$ results. The bulk concentration of the element 1 is $c_1(\infty)$. In
Fig.5 the selected depth profiles are depicted.

The signals n_1 and n_2 are obtained from:

$$n_1=\text{const}\frac{\sigma_1}{\cos\gamma}S_1\varepsilon_1\left(1-\beta_1\frac{3\cos^2\vartheta-1}{4}\right)e^{-\frac{t}{\lambda_{11}\cos\gamma}}\cdot\int_{x=0}^{\infty}c_1(x)\cdot e^{-\frac{x}{\lambda_{1c}\cos\gamma}}dx$$

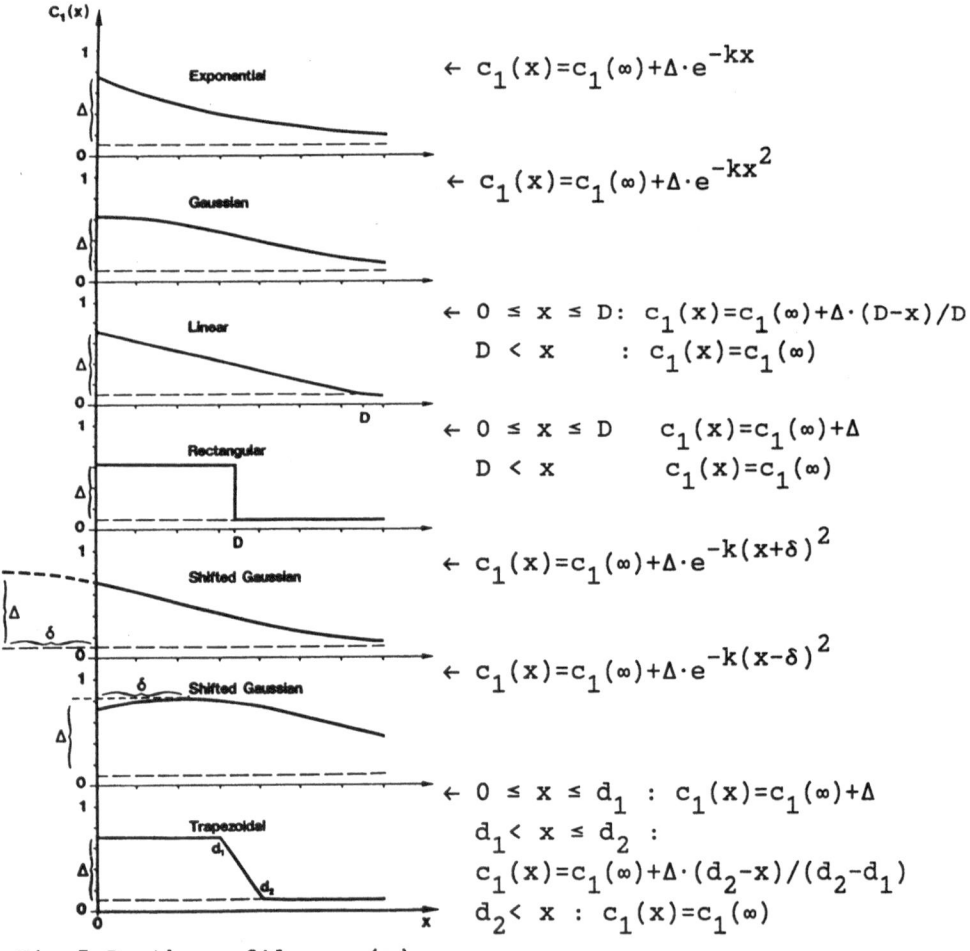

Fig.5 Depth profiles $c_1(x)$.

$$n_2 = \text{const} \frac{\sigma_2}{\cos\gamma} S_2 \varepsilon_2 \left(1 - \beta_2 \frac{3\cos^2\vartheta - 1}{4}\right) e^{-\frac{t}{\lambda_{21}\cos\gamma}} \cdot \int_{x=0}^{\infty} c_2(x) \cdot e^{-\frac{x}{\lambda_{2c}\cos\gamma}} dx$$

For dioxide films on silicon and a chosen depth profile $c_1(x)$ of $\text{Si}^{+2)}$ we obtain an expression for the signal ratio of the signals from the dioxide layer and the silicon substrate

$$r = \frac{n_{\text{Si}^{2+}}}{n_{\text{Si}^0}} = R \cdot \frac{\int_{x=0}^{\infty} c_{\text{Si}^{2+}}(x) \cdot e^{-\frac{x}{\lambda_{ic}\cos\gamma}} dx}{\int_{x=0}^{\infty} \left[1 - c_{\text{Si}^{2+}}(x)\right] \cdot e^{-\frac{x}{\lambda_{ic}\cos\gamma}} dx}$$

with

$$R = \frac{A_{\text{Si}}}{A_{\text{Si}} + 2A_0} \cdot \frac{\rho_{\text{oxide}}}{\rho_{\text{Si}}} \cdot \frac{\lambda_{i,\text{oxide}}}{\lambda_{i,\text{Si}}} .$$

A numerical value of R=0.6 has been reported by Fadley[4]. A_{Si} and A_0 are the atomic masses of silicon and oxygen, ρ_{Si} and ρ_{oxide} are the densities and $\lambda_{i,\text{Si}}$ and $\lambda_{i,\text{oxide}}$ are the inelastic mean free paths of Si2p photoelectrons in silicon and oxygen. For an evaluation of experimental r-values r_{meas} we need an error quantity $v=r-r_{\text{meas}}$. This error is valid for the specific depth profile $c_1(x)$. Thus, the standard deviation $\text{SDEV}(c_1(x))$ is

$$\text{SDEV}(c_1(x)) = \sqrt{\frac{1}{z-1} \cdot \sum_{m=1}^{z} v(c_1(x), \gamma_m)^2}$$

A minimum $\text{SDEV}_{\text{min}}(c_1(x))$ of $\text{SDEV}(c_1(x))$ is obtained by a systematic variation of the parameters of $r(c_1(x))$. This minimum is representative for the quality of the approximation of the true depth profile by the chosen depth profile $c_1(x)$. Following these considerations, the best fit on the true depth profile of the specimen is identical with the minimum of the seven different $\text{SDEV}_{\text{min}}(c_1(x))$-values of the specimen under investigation. Consequently, we expect for SiO_2 on Si the best fit with the rectangular depth profile, $\Delta=1$, $c_1(\infty)=0$ and $D=D_{\text{ell}}$. Another possibility is the trapezoidal response,

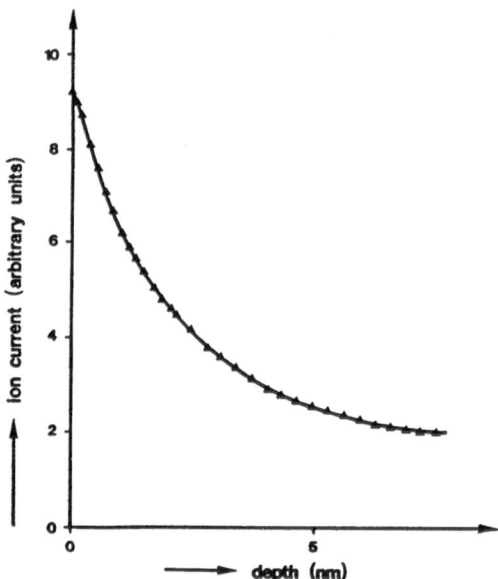

Fig.6 Depth dependence of the Li-ion current from a binary Al-Li alloy of 9.1at%Li after storage at room temperature for 48hours.

which allows to recognize the interface. The results of our 21 series of measurements performed on 11 dioxide films with different thicknesses confirm the rectangular depth profile. The Δ-values range from 0.98 to 1.02, the thicknesses are, in general, 0.5nm to 1nm larger than the ellipsometric results. It can be explained either by the neglection of the interface in the rectangular depth profile or by a systematic error of the ellipsometric thicknesses. Besides we obtained the numerical values of R = 0.61 ± 0.23 and of $\lambda_{ic} = \lambda_{i,oxide}$ = (3.3 ± 0.62)nm.

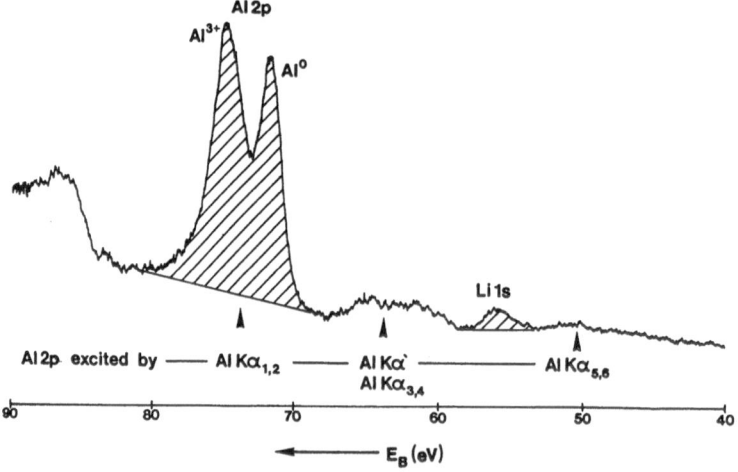

Fig.7 Photoelectron spectrum of an Al-Li alloy after extended heat treatment, for segregation. The Li concentration on the surface is 70at%Li.

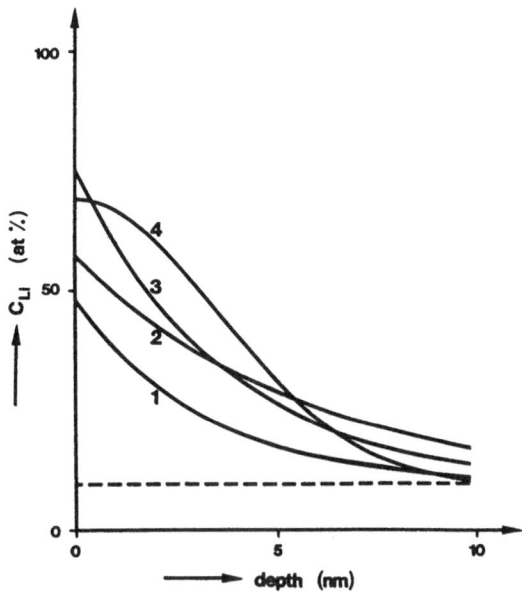

Fig.8 Segregation profiles of Li in an Al-Li alloy after heat treatment at 150°C.

It is worth noting that this approach of depth profiling enables the determination of the geometrical thickness D.

3. SEGREGATION OF Li IN AN Al-Li ALLOY

In Al-Li-alloys segregation causes a Li-enrichment on the outermost surface. SIMS investigations[14] have been conducted on an Al-Li alloy with 9.1at%Li, resulting in a well pronounced Li-enrichment after storage at room temperature for more than 48hours. Fig.6 depicts a SIMS depth response of the Li-signal. It allows a conclusion on the depth response of the Li-concentration. By Vonbank and Varga[14] this response has been assumed to be an exponential depth profile.

Our studies of the segregation profiles of an identical alloy after heat treatment at 150°C for 20, 80 and 140min were performed with XPS. A serious problem arises from the poor statistical significance of the Li1s signal, due to the much smaller photoabsorption coefficient of lithium. The spectrum given in Fig.7 has been measured after a heat treatment for 140min. In spite of the progress in segregation, we observe only a weak Li1s signal. As already mentioned, surface roughness claims restriction of the range of take-off angles. For this reason the maximum take-off angle has been 60°.

We modified the above given expression for the signal ratio into $r=n_{Li1s}/n_{Al2p}$ and evaluated the results of measurements performed under different take-off angles.

$$r = \frac{\sigma_{Li1s}}{\sigma_{Al2p}} \cdot \frac{1-\beta_{Li1s}\frac{3\cos^2\vartheta-1}{4}}{1-\beta_{Al2p}\frac{3\cos^2\vartheta-1}{4}} \cdot \frac{\int\limits_{x=0}^{\infty} c_{Li}(x)\cdot e^{-\frac{x}{\lambda_{ic}\cos\gamma}}\,dx}{\int\limits_{x=0}^{\infty}\left[1-c_{Li}(x)\right]e^{-\frac{x}{\lambda_{ic}\cos\gamma}}\,dx}$$

In spite of the poor statistical significance, the exponential response gave the smallest value of the standard deviation with the only exception for a specimen after exposure to ambient atmosphere. This evaluation asked for a description of the depth profile by a Gaussian distribution. It should be mentioned that Vonbank and Varga[14] made similar observations. In Fig.8 our results are depicted. The responses 1, 2 and 3 are from specimens after heat treatment for 20, 80 and 140min, respectively. The Gaussian response 4 is from the specimen after heat treatment for 140min, followed by an exposure to air. The depth of segregation is smaller than 10nm and comparable to the response of Fig.6.

ABSTRACT

A variation of the take-off angle of photoelectrons helps to extend the variety of possible applications of XPS to quantitative depth profiling. Three practical examples are discussed. First, the development of temperature sensors on the basis of amorphous thin films of $Ge_xC_yO_z$:H is reported. The evaluations of the experimental results are made by the assumption of a depth of analysis. Next, thin SiO_2 layers on silicon wafers have been chosen for their well established knowledge of the technique, the procedure of evaluation and the results which have to be expected. It is the aim of this example to investigate the validity of the mathematical approach for depth profiling and the reliability of the results. The measurements are evaluated by the assumption of seven different depth profiles. The evaluations give clear evidence for the existence of rectangular depth profiles. Finally, the power of depth profiling by XPS is demonstrated by the phenomenon of segregation of Li in a binary Al-Li alloy. The results of the experiments reveal depth profiles, confirmed by SIMS investigations.

ACKNOWLEDGEMENT

The research work has been supported by "Fonds zur Förderung der wissenschaftlichen Forschung", Projekt P 7012-PHY and Projekt Nr.P7234-TEC-Magnetron-PECVD".

REFERENCES

1 W.Hanke, H.Ebel, M.F.Ebel, A.Jablonski and K.Hirokawa,
   J. Electron Spectrosc. Related Phenom. 40, 241 (1986)
2 D.R.Penn, J.Electron Spectrosc.Relat.Phenom. 9, 29 (1976)
3 A.Ishizaka, S.Iwata and Y.Kamigaki, Surf.Sci. 84, 355
   (1979)
4 C.S.Fadley, in G.A.Somorjai and J.G.McCaldin (eds.),
   Progr.Solid State Chem., 11, 265 (1976)
5 M.F.Ebel, J.Electron Spectrosc.Relat.Phenom.14,287 (1978)
6 H.Ebel, M.F.Ebel, J.Wernisch and A.Jablonski, Surf. Interface Anal. 6, 140 (1984)
7 O.A.Baschenko and V.I.Nefedov, J. Electron Spectrosc.
   Relat. Phenom. 17, 405 (1979)
8 M.F.Ebel, G.Moser, H.Ebel, A.Jablonski and H.Oppolzer, J.
   Electron Spectrosc.Related Phenom. 42, 61 (1987)
9 V.I.Nefedov, V.V.Lunin and N.G.Chulkov: Surf.Interface
   Anal. 2, 179 (1980)
10 B.J.Tyler, D.G.Castner and B.D.Ratner, Surf. Interface
   Anal. 14, 443 (1989)
11 B.J.Tyler, D.G.Castner and B.D.Ratner, J.Vac.Sci.Technol.
   A7, 1646 (1989)
12 O.A.Baschenko and V.I.Nefedov, J. Electron Spectrosc.
   Relat. Phenom. 53, 1 (1990)
13 H.Ebel, M.F.Ebel, R.Svagera, E.Winklmayr and P.Varga, J.
   Electron Spectrosc. Relat. Phenom. (in press)
14 M.Vonbank and P.Varga, Vakuum Technik 37, 220 (1988)

THE USE OF X-RAY PHOTOELECTRON SPECTROSCOPY IN MATERIALS SCIENCE

James Castle

University of Surrey
Guildford, Surrey
GU2 5XH, England

ABSTRACT

This review will attempt to show how XPS now makes an important
contribution to Materials Science and to highlight the developments which
have brought it to this position. XPS is now a mature technique for
surface analysis but it has in addition a major role as a specialised tool,
being essential to studies in which derivitization methods are used to tag
surface groups.

The requirements of users in this field have led to the development
of X-ray sources which were not envisaged in the early development of the
spectroscopy. The usual sources of aluminium Kα and magnesium Kα have
limitations for those elements beyond magnesium in the periodic table which
would have the 1s line as the principal peak - aluminium, silicon, sulphur
and phosphorus for example. Higher energy sources such as silicon Kα[1] or
zirconium[2] and silver Lα[3] have made it possible to utilise the 1s lines up
to chlorine and have the additional advantage that a strong and well
resolved series of Auger lines also becomes available. The higher energy
radiations are thus particularly suited to the determination of relaxation
energies in materials by use of relative shifts between the photo- and
Auger lines of the spectrum. Such has been the utility of such relaxation
energies that use is often made of Auger lines derived from the
Bremmstrahlung component of the normal x-ray sources to make a similar
measurement[4]. This measurement is used in the study of insulating ceramics
in which electrostatic charging makes measurement of binding energies
uncertain.

Modern materials technology is particularly concerned with the
manufacture of composites; particulate, fibre and laminate composites are
all well known and the key to their success often lies within the interface
between the phases. Transfer of load across the interface places
particular requirements on adhesion at the phase boundary and an
understanding of the locus of failure during destructive testing is crucial
to the development of satisfactory bonding processes. In coated and
laminated products there is no problem in the use of XPS, with its
excellent chemical sensitivity but there is a problem of increasing
magnitude in fibre and particulate composites as the substructures become
finer. This stems, of course, from the difficulty of providing a focused
source of X-rays of sufficient magnitude. Imaging XPS is slowly becoming a
reality with several systems having a capability of 10 μm now available,
and one of the markets for such instruments is that of composite materials.

There are important areas of Materials Science in which XPS has been displaced by other techniques such as SIMS. One such area is that of polymer surface analysis. The selectivity of XPS for substituent groups in the surface region is not good. Derivitization methods have made an impact, enabling acidic or basic groups to be determined, but SIMS, which has the ability to detach molecular clusters, has obvious advantages which will become increasingly exploited as the problems of charging become solved. Until then however XPS will continue to find a role in polymer research and development.

INTRODUCTION

Materials Science has two distinct branches. Firstly, there is that branch which is concerned with the study of materials for structural use: that is for the manufacture of artifacts ranging in scale from the smallest to the largest objects of human endeavour. Secondly, there is the study of the intrinsic properties of materials such as superconductivity, optical and magnetic behaviour, or catalytic properties: that is of the properties which give rise to the phenomena which we exploit in creating devices. In discussing the impact of an analytical technique such as XPS on Materials Science it is necessary to be clear which area is of concern to the investigator. In this review I shall concentrate on the structural aspect but will make some comments on the latter towards the end.

The mechanical properties of materials have a great dependence on the microstructure which develops within them during processing and which is manipulated by the skilled producer. Processing to enhance mechanical properties has gone on hand-in-glove with technological progress since historic times. It has been driven by the need for better weapons, from swords to aircraft, for health care products in dentistry or prosthetics, and by the market for the artifacts of everyday life. Recently the development of sporting and leisure goods has become a driving force also for the development of advanced materials. A new emphasis in the development of materials, which is common to the majority of the examples cited above, is on mass. Thus a feature of the development of new materials is the specific property, ie. the mechanical property per unit mass. As a result lightweight materials and lightweight methods of assembly have become commonplace - placing in turn a focus on the joining and bonding of quite dissimilar materials. We have become familiar with honeycomb structures, with composite materials, and with glued joints in aircraft and cars as the drive for greater economy in fuel and more efficient use of materials gathers pace. Central to all developments of this nature is the transfer of load across an interface and hence on its chemical nature and on the characteristics of the surfaces which together form the interface.

Whilst it is self-evident that we cannot study the *made* interface by XPS we can determine the composition and bonding opportunities of the original surfaces; we can study the degradation of the surfaces prior to bonding and of the loci of failure of the interface; and we can use the technique to monitor deliberate modification of the interfaces for the enhancement of bond life. Materials of all types degrade in service and the study of surface blemishes or of more serious corrosion damage continues to provide much work for the surface analyst. Whilst this review will look in the main at the techniques which are used, it is also important to the successful application of the techniques that the potential user has some feel for the degree of difficulty of the projected investigation. Examples drawn from the examples above are therefore given to act as a guide for the user.

XPS: ITS PLACE IN ANALYSIS

The Venn diagram in Figure 1 is a convenient way of delineating the attributes of XPS. In this chart are shown the overlapping field of resolution in depth, position, chemical state, and atom or molecular microstructure. Some of the present advances in modern analytical techniques concern the overlapping regions of structure and depth, and much

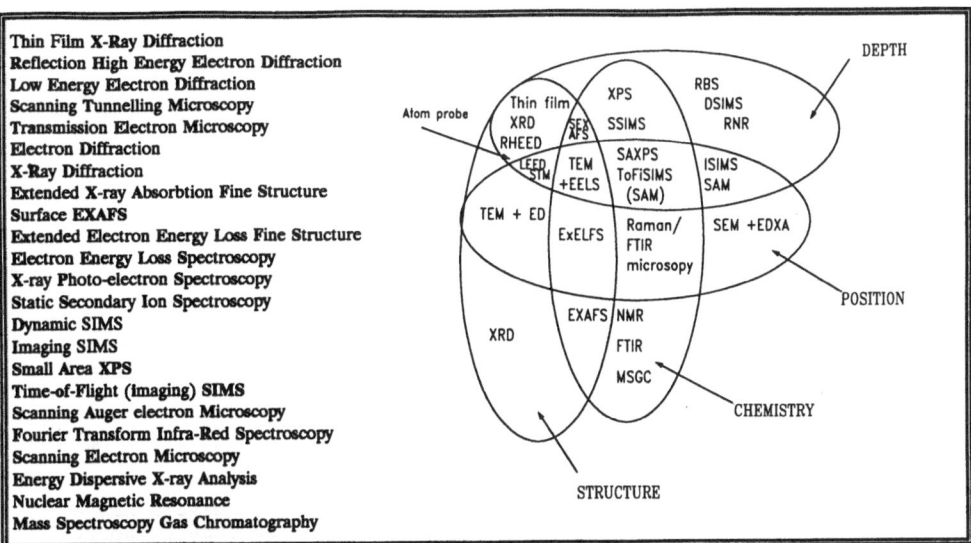

| Thin Film X-Ray Diffraction |
| Reflection High Energy Electron Diffraction |
| Low Energy Electron Diffraction |
| Scanning Tunnelling Microscopy |
| Transmission Electron Microscopy |
| Electron Diffraction |
| X-Ray Diffraction |
| Extended X-ray Absorbtion Fine Structure |
| Surface EXAFS |
| Extended Electron Energy Loss Fine Structure |
| Electron Energy Loss Spectroscopy |
| X-ray Photo-electron Spectroscopy |
| Static Secondary Ion Spectroscopy |
| Dynamic SIMS |
| Imaging SIMS |
| Small Area XPS |
| Time-of-Flight (Imaging) SIMS |
| Scanning Auger electron Microscopy |
| Fourier Transform Infra-Red Spectroscopy |
| Scanning Electron Microscopy |
| Energy Dispersive X-ray Analysis |
| Nuclear Magnetic Resonance |
| Mass Spectroscopy Gas Chromatography |

Figure 1. A Venn diagram illustrating the relationship between XPS and other forms of analysis used for the solid phase.

has been said of this topic in the present meeting, but XPS is concerned primarily with the overlap which gives depth resolution of chemical state. There are competing methods, notably Auger methods which give resolution in depth and position and sometimes chemical state; and secondary ion methods which again give a potential for resolution in all three modes. There are drawbacks to the use of these other methods, notably the sensitivity of samples to electrostatic charging in the former case and the surface damage which occurs in the latter. In the majority of surveys of the use of surface analytical techniques by consultant laboratories XPS is found to be that which is most often used.

XPS: WHAT SPATIAL RESOLUTION IS NECESSARY?

XPS, with is all-element chemical sensitivity, has not yet reached its limit for positional resolution. The ten years from 1975 to 1985 saw the limit fall from 1 mm to 0.15 mm and the limit has fallen yet again by a factor of ten to 0.01 mm[5]. There is a clear market pressure to see this improvement in spatial resolution made available to the Materials Science community. But there are offsetting considerations: cost, complexity and signal-to-noise ratio; so that over-specification may in fact may have an adverse effect on availability and reliability. It is therefore necessary to indicate the fields in which the needs might arise and this is attempted in the first Tables against a size scale which is appropriate for each type of XPS which has been developed.

In establishing the need for small or selected area surface chemical analysis it is worth recalling that for many obvious requirements only surface elemental analysis may be necessary. AES and EDX, for example, have proved perfectly satisfactory for the analysis of segregation to grain boundaries, giving embrittlement[6], or for the analysis of inclusions in steels[7]. In each case the elements are 'foreign' to the substrate and it is generally enough to have identified them as such. The need for an analysis which is both representative of a small area of the sample and chemically specific is more often concerned with the reaction pathway associated with small features. This might be of a destructive nature (e.g. electrochemical pitting associated with inclusions or the formation of brittle compounds at bonded interfaces); or of a constructive nature (e.g. the formation of chemically specific bonds between dissimilar phases in

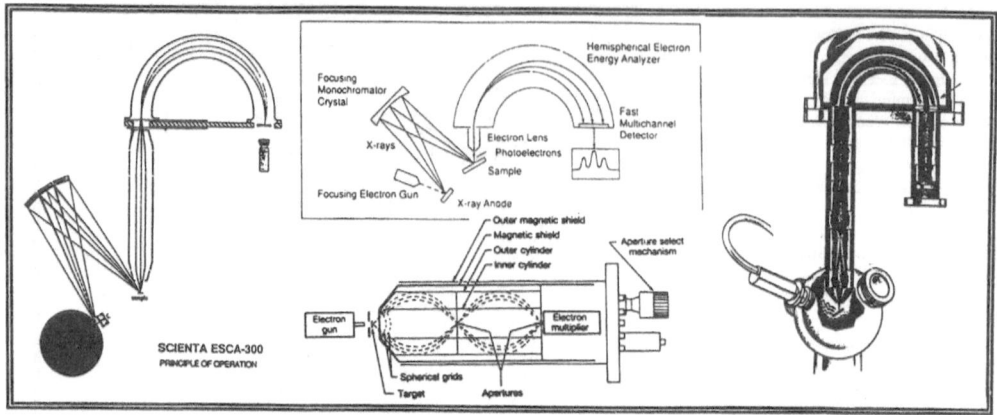

Figure 2.　Diagrams illustrating the differing approaches used to gain spacial resolution in XPS.

advanced materials). The size scale necessary for useful analysis has a range of several orders of magnitude. For example, in the case of bonded interfaces it might extend from millimetres (eg. acoustically damped steel), to sub-millimetre (eg. surface treatment of carbon fibre tow), to tens of micrometres (eg. examining interphase regions in tapered cross-sections), to micrometres (eg. pullout of individual fibres in failed composites), and to sub-micrometre (eg. surface of reinforcing whiskers or particles in MMCs, or active pigments in advanced coatings).

The final row, which is added to complete the picture, is the province of atom probe and analytical STM instrumentation and is unlikely to be accessible to XPS. The top row of this Table represents conventional XPS and the second row selected area XPS. The third row represents the subject areas which can be addressed by means of the imaging XPS systems which now represent the state of the art and the fourth row lists those topic areas of interest to scientists which will be outside the reach of current or foreseen instrumentation. This latter area may become possible by making use of National High Brightness Source of X-Rays as these become available. It is possible that certain of these topics could be addressed on the basis of an idea proposed by Cazaux some ten years ago which can be referred to as transmission XPS[8].

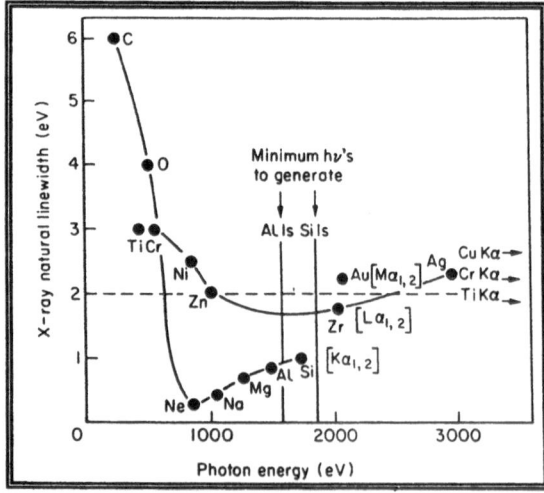

Figure 3.　The relationship between energy and resolution for X-rays of potential value in XPS.

Table 1. The Association of Topic Areas and Scale for XPS Analysis

	Size(m)	Bonding	Corrosion	Biocompat.	Biofouling
XPS	10^{-3}	Adhesion Tests	Surface Blemish	Implant Surfaces	Protein Adsorption
SAXPS	10^{-4}	Tapered Sections	Large Pit Electrochem.	Tablet Homogeneity	Diatom Surfaces
iXPS	10^{-5}	Fibre Pullout	Correlation of Potential & Composition in Pits	Drug Delivery Systems	Cell Adhesion
TXPS	10^{-6}	Microtomed Sections	Composition Gradients in Scales	Implant/ Bone Inter- Phase	Tissue Sections
STM Atom Probe	$<10^{-7}$	Polymer Molecules on Surfaces	Nuclei of Passive Films	DNA Helix	Adsorption Sites

The methodologies for microscale and imaging XPS are illustrated in Figure 2. Note that both local illumination of a selected area with a diffracted spot of X-rays and the use of defining apertures in a lens system are used to obtain selected areas of about 150 μm diameter. To go to diameters of 10 μm it is necessary to use a sophisticated lens design as in the VG ESCASCOPE or entrap the emerging photoelectrons in the field lines of a high field solenoid, as in the instrument designed by Turner[9]. A further reduction to 1 μm diameter is in principle possible if the transmitted geometry is used as suggested by Cazaux and by Hovland[10].

X-RAY SOURCES: CHEMICAL STATE RESOLUTION

Conventional XPS is now a mature technique for surface analysis but the requirements of users in some of the fields given in Tables 1 and 2

Table 2. The Association of Materials and Scale for XPS Analysis

	Size(m)	Composites	Ceramics	Semiconductors	Metals
XPS	10^{-3}	Tow Chemistry	Fracture Sections	Etch Profiles	Surface Segregation
SAXPS	10^{-4}	Fracture Surfaces	Wear Track Transfer Film	Surface Blemish	Powder Compacts
iXPS	10^{-5}	Interface Composit- ion	Toughening Mechanism	Etch Bevel Analysis	Oxide on Powder Particle
TXPS	10^{-6}	Whisker & Particle Surfaces	Phase Boundary Segregation	Bevel of Quantum Well Structure	Nuclei in Thin Sections
STM Atom Probe	$<10^{-7}$	Voids in Fibres	Active Centres in Catalysts	Quantum Well Structures	

have led to the development of X-ray sources which were not envisaged in
the early development of the spectroscopy. The usual sources of aluminium
Kα and magnesium Kα have limitations for those elements beyond magnesium in
the periodic table which would have the 1s line as the principal peak -
aluminium, silicon, sulphur and phosphorus for example.

Figure 3 illustrates the relationship between the natural line widths
and energy for the range of X-rays with potential for use in XPS. Higher
energy sources such as silicon Kα or zirconium and silver Lα have made it
possible to utilise the 1s lines up to chlorine and have the additional
advantage that a strong and well resolved series of Auger lines also
becomes available. For chemical state resolution to be retained the FWHM
of the line should be less than, say, 2 eV. Lines with widths greater than
this will need monochromatisation with the attendant loss in signal
strength.

Monochromatisation has been worthwhile in order to utilise the
advantages of AgLα which has an energy approaching 3 KeV. It can be argued
that this is as far as one should go in increasing photon energy since the
cross-section for the light elements decreases strongly with increase in
photon energy. Although the flux obtainable with the AgLα source is low,
there is some compensation by virtue of the high relative cross-sections of
the 1s lines of the elements beyond magnesium. Figure 4 gives a plot of
the cross-sections with, superposed, a spectrum for a sulphide.

Photo-peaks are sharp enough to reveal chemical state by their
position alone, the so-called, chemical shift which gives rise to the
possibility of distinguishing chemical state is still the best known
feature of XPS. Further effects are also helpful however: the shake-up
satellite on the divalent states of copper and other transition metals[11] is
a very effective indicator for this oxidation state and a satellite is also
generated by the π - π* transition in the carbon of aromatic compounds[12];

The XPS spectrum contains not only the photo-electron peaks generated
by the formation of the core hole but also the Auger electron peaks
generated by the filling of this hole. Different shifts on Auger and

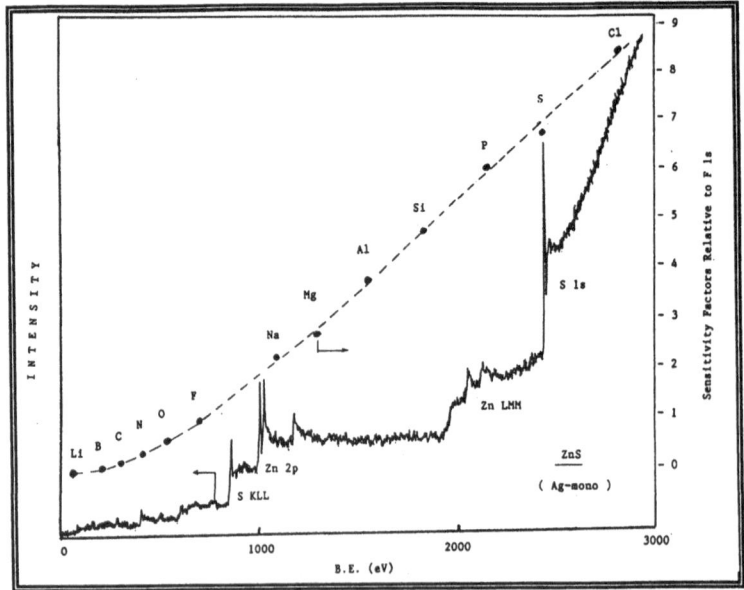

Figure 4. A spectrum for zinc sulphide in AgLα: this is superposed on a plot
of the photo-electron cross-sections of the 1s lines.

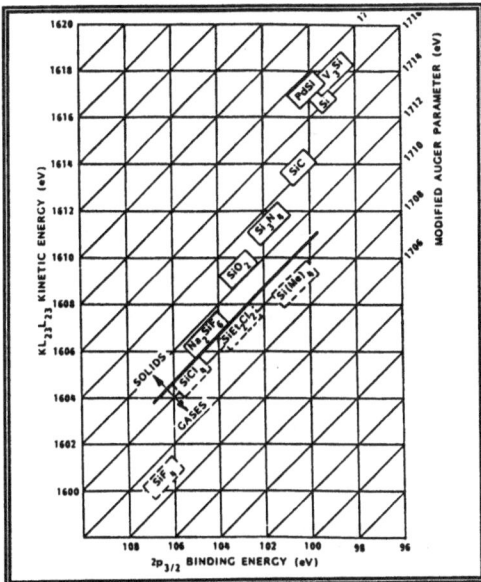

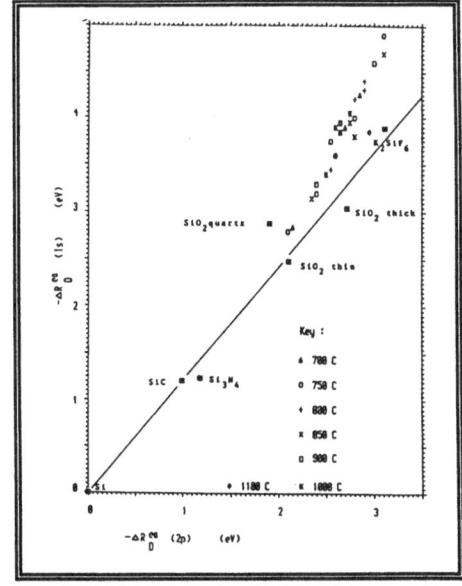

Figure 5.   A chemical state plot of the type introduced by Wagner.[14] The Auger Parameter is given by the position of the diagonal lines.

Figure 6. Thermal oxide on silicon: the extra-atomic relaxation energy depends on the temperature and duration of oxide growth.

photo-peaks arise from the greater influence of relaxation on the Auger process, which finishes up with two holes in the atom. Hence the difference in Auger and photoelectron shifts can be a useful indicator of chemical state as shown in figure 5. This measurement, introduced by Wagner and Biloen[13], is often called the Auger parameter[14]. Compilations of peak positions and Auger parameters may be found in the literature and an excellent database produced by Wagner may be obtained on disc from NIST[15].

The higher energy radiations are particularly suited to the determination of relaxation energies in materials using the Auger Parameter. Such has been the utility of relaxation energies that use is often made of Auger lines derived from the Bremmstrahlung component of the normal X-ray sources. This measurement is valuable in the study of insulating ceramics in which electrostatic charging makes measurement of binding energies uncertain. It has also become possible to distinguish, within the formal oxidation state, between ions in differing local environments and as a function of the strain within the oxide. Figure 6 shows how the relaxation energy associated with the silicon 1s peak depends on the temperature at which the oxide film on silicon is grown - probably because of thermal strain within the oxide[16].

DEPTH RESOLUTION: ANGULAR RESOLVED XPS

The signal obtained from an atomic layer within the material is attenuated exponentially by the overlying layers. The characteristic length is the inelastic mean free path(iemfp) of the electron which has a dependence on the kinetic energy and on the nature of the material. Seah and Dench[17] have provided a compilation from which the value of iemfp, lambda, can be obtained.

Certain combinations of the exponential relationship are particularly useful. Use is made of the fact that I_∞, the intensity from an infinitely thick, clean, substrate is, except at low angles, independent of collection angle. Then, if d_1 is the thickness of an outer absorbing layer and d_2 the thickness of an inner emitting layer, there are three basic cases:

$$I_{\infty d1\theta}/I_\infty = \exp(-d_1/\lambda \sin \theta) \tag{1}$$

representing the relative signal from an underlying substrate.

$$I_{d2\theta}/I_{\infty} = 1 - \exp(-d_2/\lambda \sin \theta) \qquad (2)$$

representing the relative signal from an overlying layer, and

$$I_{d2d1\theta}/I_{\infty} = \exp(-d_1/\lambda\sin \theta) - \exp[-(d_1+d_2)/\lambda\sin\theta] \qquad (3)$$

giving the relative signal from a thin overlying layer when covered with a second layer. This will be the usual case when examining a surface layer covered with adsorbed contamination.

A form which has been used by several authors occurs when the signal from the substrate (1) is attenuated by a layer of its own oxide, (2); d_1 then equals d_2 and

$$I_{\infty d\theta}/I_{d\theta} = \exp(-d/\lambda\sin \theta)/1 - \exp(-d/\lambda\sin \theta) \qquad (4)$$

When the collection angle is varied eq. (1) becomes

$$I_{\theta 1}/I_{\theta 2} = \exp(-d_1(\cosec \theta_1 - \cosec \theta_2)/\lambda) \qquad (5)$$

which describes the angular variation of the substrate signal in the presence of an overlayer. Equation (5) is particularly useful for confirming the order of a series of layers and this provides a basis by which a speculative concentration profile can be transformed into an angular profile and hence compared with that found in practise[18]. This has been much used by material scientists to study, for example, the formation of passivating layers on metals[19] or the segregation of organic material to the surface of a polymer[20].

DERIVITIZATION OR SURFACE TAGGING

The chemical shift is not always sufficient to distinguish closely related chemical states. An example occurs with the shifts of carbon in sites of differing polarity in organic compounds. It has thus become common to measure the concentration of differing substituent groups in the surface of polymers by forming a chemical derivative of the individual groups and measuring the surface concentration of this derivative. The technique has become known as derivitization and examples of its use are found in the following sections. Briggs[21] has listed a number of useful reagents for this purpose.

METALLIC CORROSION

The corrosion scientist is concerned above all else with the rate of reaction of a solid with its environment and especially with the prognosis for that rate of reaction. Mechanisms of reaction are explored at the atomic level and applied to problems on the engineering scale. Perhaps because of this range of endeavour, a wide range of techniques is employed in the work of the corrosion laboratory and will in general include XPS.

The special viewpoint of the corrosion scientist is that analysis should include a time dependency or perhaps a dependency on the pH of the solution, the electropotential of the surface or say, the oxygen potential of a gaseous atmosphere. It is the corrosion scientist or engineer who must supply the range of samples which enables the snapshot analysis provided from each sample to be assembled into a meaningful series, illustrating the approach to equilibrium or the rate law applicable to a given situation.

The fields in corrosion research having potential for use of electron spectroscopy have recently been summarised by the author[22]. Many examples exist in the literature of the application of XPS to problems in aqueous

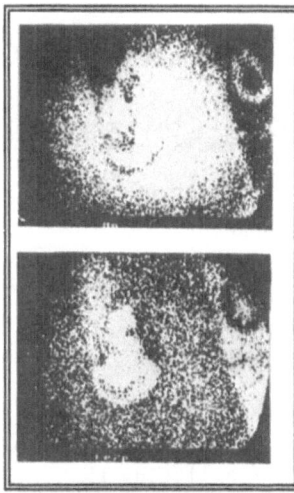

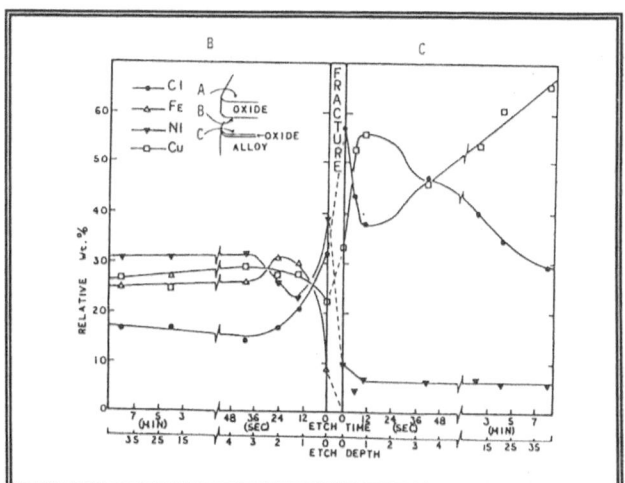

Figure 7. Imaging XPS can reveal the gradations of chemical states created by the potential field surrounding a large pit - here on an iron/manganese alloy.

Figure 8. The passivating layers on aluminium brass and cupronickel were revealed by removing superficial layer with adhesive tape - SiKa was used to determine Al in the present of Cu.

corrosion. However for consistency of study in a given field the work relating to the passivation and corrosion of the austenitic stainless steels is outstanding. One reason for this success is that chromium steels become very strongly passivated and thus the thickness of the films is ideal for XPS (about 2nm), not strongly dependant on time, and not altered greatly by subsequent removal from the electrochemcially controlled environment or exposure to vacuum.

Passivation of Stainless Steels

The first work with XPS concentrated on establishing the nature of the passive surfaces. Several authors[23-25] quickly showed by XPS that oxidised nickel did not appear in the film on stainless steel but that chromium(lll) did and did so at an enriched level. The fact that nickel did not appear in the oxide film enabled the intensity of the element to be used to measure the film thickness and to calibrate the ion gun used to provide etch profiles of thicker films formed at high temperatures[26]. In all of this early work great use was made of the depth relationships listed above, the stimulus undoubtedly being that both metal and oxide components could be seen in one and the same spectrum. This fact remains a stimulus to the present time.

When metals corrode in an aqueous medium, transport or loss of material in soluble form must be expected. Olefjord and also Asami were among workers who showed that the enrichment of chromium in the oxide does not occur because of selective oxidation of this element but, since the interface composition is identical to the bulk metal, arises from selective dissolution of iron and nickel. Recently, Castle and Qiu[27] have shown how ICPMS (Inductively Coupled Plasma Mass Spectroscopy) can be used to provide complete accountability for all the charge passing through the electrode during passivation of iron/chromium alloys. Kirchheim et Al[28]. have carried out similar studies on Fe/Cr and Fe/Cr/Mo alloys using on-line atomic absorption spectroscopy for the solution analysis.

Asami and Hashimoto showed by some elegant experiments in which the surface composition was measured as a function of potential and alloy composition, that the enrichment in chromium behaved as expected, being found only when passivation occurred. A further finding of Asami et al was that the enrichment of Cr(lll) was accompanied by Fe(ll), suggesting that air oxidation during transfer was not occurring with these passive films.

This has been verified by much later work in the form of a round robin examination of the passivation of iron/chromium alloys in which a number of laboratories produced closely similar results[29].

The breakdown of passivity gives rise to pits and it is this more than anything which dominates the corrosion of stainless steels. The pits are micrometer in scale and therefore studied by AES[30] but this does not mean that there is not a need for the chemical state resolution of XPS. A pit differentiates electrochemically, developing a central anodic region surrounded by a cathodic annulus: much of the detailed chemistry is associated with the valence states of ions found in one part or another and is inaccessible to SAM. Imaging XPS may help in the study of pitting and an image, Figure 7, taken with the ESCASCOPE of VG Scientific, of a pit in iron/manganese alloy illustrates the opportunities available.

Examination of thick oxide layers

The oxide layers formed as a result of corrosion are usually studied by electron excited X-ray analysis. XPS, however, has the advantage that the composition of phases formed at the various interfaces often present in thick layers can be identified by the simple expedient of splitting the layers by mechanical means. Figure 8 shows how the presence of a inner layer of copper (1) chloride[31] and of the mineral hydrotalcite[32], $Mg_6AL_2(OH)_{18}4H_2O$, have been identified on cupronickel and aluminium brass respectively. The thin layers identified by XPS can often be related directly to the thermodynamic stability diagrams, known as Pourbaix diagrams, which are important means of predicting corrosion in waters of differing pH values and at differing electropotentials.

ORGANIC COATINGS AND ADHESIVES

Cohesive and adhesive failure

It is a small step from consideration of the layers of corrosion products on metals to consideration of the adhesion of organic coatings, which are often used to protect against corrosion. The paint and coating technologist needs, above all else, an indication of the location of the weakest link in the bonds which hold the oxide - polymer system together. It is usual to think in terms of 'adhesive' failure when it is the heterogenous bond which fails, and of 'cohesive' failure when a homogenous phase on either one or other side of the bond provides the site of failure[33]. Such information can be provided by simple peel tests or by the failure in shear of an overlapping joint[34]. These basic tests can now be undertaken in the vacuum of the spectrometer but even when undertaken in air they have frequently showed that cohesive failure has actually occurred when the eye suggests otherwise[35].

The failure often occurs very close to the actual bond. The reconstruction of the composition gradients close to the region of failure using angle-resolved XPS, interpreted along the lines of equation 5 above, can show that poorly polymerised material can form the site of failure. This is the interpretation of the composition depth profiles obtained in this manner and illustrated in figure 9[36]. It can be seen that the carbon type associated with the epoxy bond actually lies beneath the plane of failure and is not present within the bonding zone.

Designer paints!

There have been great advances in painting technology in recent years and much of this stems from the pioneering work of Fowlkes[37]. He first demonstrated the importance of understanding the contribution of each component of a surface to the overall free energy, which is measured only as an average by parameters such as the contact angle. Best results are obtained when acidic bonds in the oxide or hydroxide of the metal surface match with basic bonds in the polymer - or vice versa. The acid/base character of a surface is readily monitored by use of XPS to measure the

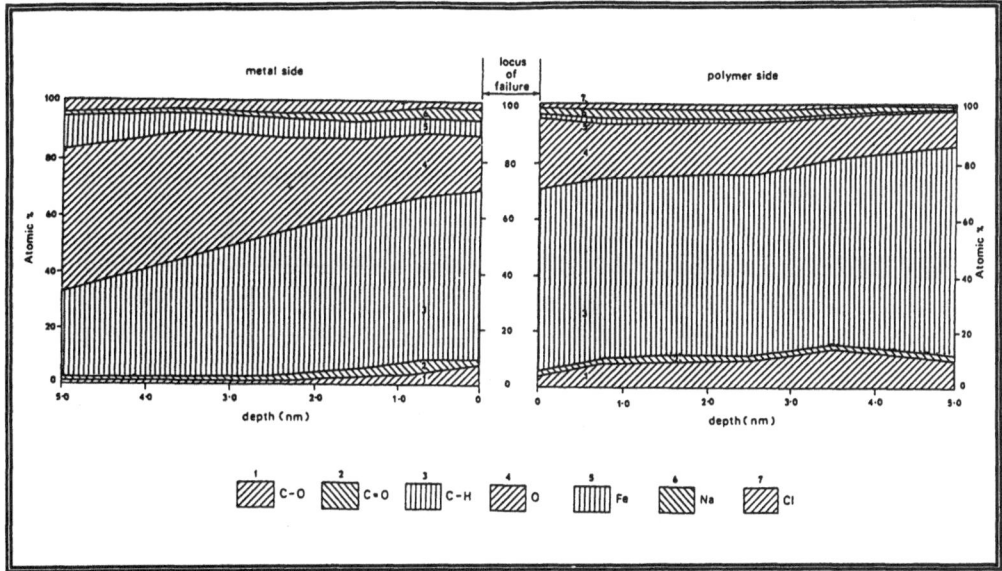

Figure 9   Angular resolved XPS used to show the nature of the locus of failure of an epoxide polymer on steel

uptake of a counter ion[38]. Calcium or barium ions are frequently used to determine the surface concentration of acidic groups on polymers. Watts and Kinlock have now shown how XPS measurement of carbon types in the surface can be related directly to the contact angle[39].

More information can be obtained if an adsorption isotherm is obtained: the uptake by the surface is plotted as a function of the concentration in solution and characteristic shapes of curve may be related to physical or chemical adsorption[40]. Watts has recently shown that the interaction between solvents having a range of acidities and a polymer can be used to assess the bonding opportunities[41].

COMPOSITE MATERIALS

The manufacture of composite materials has all the problems of adhesion found in coating technology, with the added disadvantage that the interacting surfaces are not planar but are in the form of a fibre and matrix. In the particular case of carbon fibre composites there is also a problem that bonding groups are not naturally found on the carbon surface[42]. These are generated by oxidative treatment which is usually carried out electrochemically. A relation has been found between the level of treatment and mechanical parameters such as the inter-laminar shear strength of the composite[43]. Much work has been carried out by means of XPS over the past twenty years with the aim of understanding and refining this surface treatment[44]. Since XPS cannot resolve detail from individual fibres, which have a diameter of ca. 6μm, bundles of fibres or 'tows' are used for this study. Both direct measurement of carbon types from curve-resolved spectra and labelling by derivitization have been used. Adsorption isotherms show the uptake of basic ions to be consistent with a reaction adsorption mechanism. The field has been reviewed by Castle and Watts[45] and this can be consulted for a wider spectrum of studies. However figure 10 provides a summary of the main sequence of studies in chronological order over the past fifteen or so years.

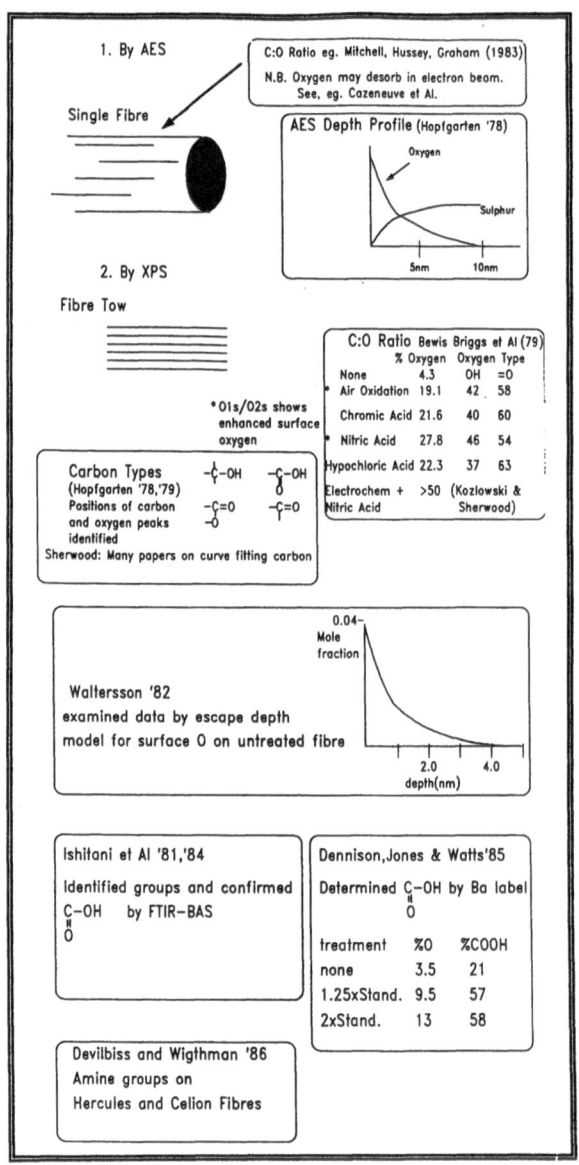

Figure 10.   Progress in the characterisation of carbon fibre. The references cited may be found in refs. 46 - 57 inclusive.

REFERENCES

J.E.Castle and R.H.West; J.Elec.Spec. and Rel.Phenom, **19**, 409-428,(1980)

J.E.Castle, L.B.Hazell and R.H.West; ibid,**16**, 97-106,(1979)

M.J.Edgel,R.W.Paynter, and J.E.Castle; ibid,**37**, 241-256,(1985)

J.E.Castle and R.H.West; ibid,**18**, 355-358,(1980)

U.Gelius et.Al; ibid,**52**, 747-785(1990), see also P.Coxon, same journal

M.P.Seah, pages 311-356 in 'Practical Surfaces Analysis' 2nd.Edn. D.Briggs and M.P.Seah Eds. John Wiley and Sons. Chichester Engand (1990)

J.E.Castle and R.Ke; Corros.Sci.**30**, 409-428,(1990)

J.Cazaux; Rev.Phys.Appl.**10**, 263-280,(1975)

G.Beamson,H.Porter, and D.Turner; Nature,**290**, 556,(1981)

C.Hovland; Appl.Phys.Letrs,**30**, 274,(1979)

J.E.Castle; Nature,(Phys Sci) **234**, 93,(1971)

J.A.Gardella, S.A.Ferguson and R.L.Chin, Appl.Spectros.,**40**, 224,(1986)

C.D.Wagner and P.Biloen; Surf.Sci.**35**, 82,(1973)

C.D.Wagner and A.Joshi; J.Elec.Spec and Rel.Phenom.**47**, 283-313,(1988)

NIST Database

M.J.Edgell, S.Mugford, J.E.Castle and N.A.Pirie; J.Electrochem.Soc.**137**, 201-206,(1990)

M.P.Seah and W.Dench; Surf.Interface Anal.**1**, 2-11,(1974)

R.W.Paynter; Surf.Interface Anal.**3**, 186,(1981)

H-H Streblow; Surf.Interface Anal.**12**, 363-379,(1988)

J-J.Pireaux; Proc.Int.Cong.Elec. Spectros.**4**, Honolulu, (1989)

D.Briggs; pages 450-454 in 'Practical Surfaces Analysis' 2nd.Edn. D.Briggs and M.P.Seah Eds. John Wiley and Sons. Chichester Engand (1990)

J.E.Castle; 'The Application of Surface Analytical Methods to Environmental/Material Interactions' D.R.Baer, C.R.Clayton and G.D.Davis Eds. The Electrochem. Soc.NJ. (1991) pp 1-21

I.Olefjord; and H. Fischmeister; Corros.Sci.**17**, 677-707,(1975)

K.Asami, K.Hashimoto, and S.Shimodaira; Corros.Sci.**18**, 153-160,(1978)

J.E.Castle and C.R.Clayton; Corros.Sci.**17**, 7-26,(1977)

C.R.Clayton and J.E.Castle; 'The Passivity of Metals' R.Frankenthal and J.Kruger, Eds. p714 (1978) The Electrochem.Soc.Princeton NJ.

J.E.Castle and J-H.Qiu; J.Electrochem Soc.**137**, 2031-2036,(1990)

R.Kircheim, B.Heine, H.Fischmeister, S.Hofmann, H.Knote and U.Stotz; Corros. Sci.**29**, 899-917,(1989)

P.Marcus and I.Olefjord; Corros.Sci.**28**, 589,(1988)

A.J.Sedriks, Int.Met.Rev.,**28**, 295-307,(1983)

C.Kato, H.W.Pickering and J.E.Castle; J.Electrochem.Soc.**131**, 1225-1229,(1984)

J.E.Castle D.C.Epler and D.B.Peplow; Corros.Sci.**16**, 145-157,(1976)

J.F.Watts; Surf.Interface Anal.**12**, 497,(1988)

J.E.Castle and J.F.Watts; J.Mats.Sci.**18**, 2987-3003,(1983)

J.E.Castle and J.F.Watts; I&EC Product Development,**24**, 361-369,(1985)

J.F.Watts, J.E.Castle and S.J.Ludlam; J.Mats.Sci.**21**, 2965-2971,(1986)

F.M.Fowlkes; J.Adhesion Sci.Tech.**4**, 669,(1990) see also ibid,**1**,7,(1987)

W.M.Riggs; Dupont Tech.Note (1974) 'The Use of Calcium Labelling for Acidic Groups on Polyethylene', see also W.M.Riggs and D.W.Dwight; J.Electron Spectros and Rel.Phenomena, **5**, 447,(1974)

A.J.Kinloch, G.Kadokian and J.F.Watts; Proc.Roy.Soc. to be published

S.J.Gregg and K.S.W.Sing; 'Adsorption, Surface Area and Porosity', 2nd Edition AP. London, (1982)

M.M.Chehimi, J.F.Watts, S.N.Jenkins and J.E.Castle; J.Mats.Chem. to be published

J. Harvey, C.Kozlowski and P. M A. Sherwood, J.Mats.Sci., **22**, 1585-1596,(1987)

C.A.Baillie, J.F.Watts, J.E.Castle, M.G.Bader, Proc. ICCMVIII, Honolulu, Ed. S. Tsai and G.S.Springer, Pub. SAMPE, Paper 11A, (1991)

W.Wright, Composite Polymers, **3** (1), Ed. P.Dickin, Part 1: 231, Part 2: 360,(1990)

J.E.Castle and J.F.Watts, 'Interfaces in Polymer Ceramic Metal Matrix Composites', H. Ishida Ed., Elsevier, 57-71 (1988)

46   D.F.Mitchell, R.J.Hussey and M.J.Graham, J.Vac.Sci.Tech.A,**1**, 1006, (1983)(1970)

47   C.Cazeneuve, J.E.Castle, J.F.Watts, J.Mats.Sci.,**25**, 1902-1908,(1990)

48   F.Hopfgarten, Fibre Sci.Tech.,**11**, 67-79,(1978)

49   F.Hopfgarten, Fibre Sci.Tech.,**12**, 283-294,(1979)

50   D.M.Brewis, J.Comyn, J.R.Fowler, D.Briggs, V.A.Gibson, Fibre Sci.Tech.,**12**, 41-52,(1979)

51   C.Kozlowski, P.M.A.Sherwood, Carbon,**24**, 357-363,(1986)

52   K.Waltersson, Fibre Sci.Tech.,**17**, 289-302,(1982)

53   A.Ishitani, Carbon,**19**, 269-275,(1981)

54   T.Takahagi, A.Ishitani, Carbon,**22**, 43-46,(1984)

55   P.Denison, F.R.Jones, J.F.Watts, J.Mater.Sci.,**20**, 4647-4656,(1985)

56   P.Denison, F.R.Jones, J.F.Watts, Surf.Interf.Anal.,**12**, 455-460,(1988)

57   T.A.DeVilbiss, D.L.Messick, D.J.Progar, J.P.Wightman, Composites,**16**, 207-219,(1985)

SURFACE AND THIN FILM ANALYSIS OF METALS AND SEMICONDUCTORS USING X-RAY PHOTOELECTRON SPECTROSCOPY

S. Hofmann

Max-Planck-Institut für Metallforschung
Institut für Werkstoffwissenschaft
Seestr. 92, D-7000 Stuttgart 1

ABSTRACT

X-ray excited Photoelectron Spectroscopy (XPS) has become an indispensable tool for the study of metals and semiconductors. Due to the small mean free path of the photoelectrons in solids of the order of a few nanometers for energies in the keV range, it is a surface analysis technique. Its capability of quantitative analysis of all elements except hydrogen and helium and their chemical bonding states has recently been combined with small area and imaging analysis to typical spatial resolutions of about 10 μm. After a brief survey of the basic capabilities and limitations of XPS, some illustrative examples in typical metals and semiconductor research areas are presented, such as surface contamination and failure analysis in microelectronics, oxidation and corrosion, segregation at surfaces and interfaces, oxide/metal and oxide/semiconductor interfaces, and thin film analysis using angle resolved XPS and sputter depth profiling. Recent developments emphasize improved data evaluation and quantification schemes as well as instrumental capabilities with respect to both high spatial and energy resolution, and high power excitation sources such as synchrotron radiation.

INTRODUCTION

During the past two decades, X-ray Photoelectron Spectroscopy (XPS) has become one of the most common methods of surface and thin film analysis of solids.

Advances in X-Ray Analysis, Vol. 35
Edited by C.S. Barrett *et al.*, Plenum Press, New York, 1992

This development was enhanced by the huge advances in ultra high vacuum techniques, electronics and data processing which resulted in a number of versatile commercial instruments with highly improved sensitivity, energy resolution and spatial resolution. In the early fifties Kai Siegbahn (Nobel prize 1981) and his coworkers[1] developed the technique and coined the acronym ESCA (Electron Spectroscopy for Chemical Analysis), emphasizing the main application then and now. Today, the more specific term XPS (X-ray Photoelectron Spectroscopy) is frequently used. XPS is an surface sensitive technique due to the small mean free path of the detected photoelectrons in the nanometer range for typical energies in the 1 keV region[2,3]. Therefore, the technique has become a major tool to study reaction products on surfaces as well as their distribution in thin films, which are of great importance in metallurgy and microelectronics, e.g. with respect to environmental control, oxidation and corrosion, wear, interface properties etc.[2,3]. The aim of the paper is to present a brief survey of the basic capabilities and limitations of XPS with respect to some typical applications in metals and semiconductor research. For a recent, detailed review, the reader is referred to the book by Briggs and Seah[3] and references therein.

2. CAPABILITIES AND LIMITATIONS OF XPS

Besides quantitative elemental analysis of surfaces, XPS is particularly useful for determination of the chemical bonding state based on characteristic energy shifts of the core level XPS peaks. Excitation by monoenergetic X-rays (usually MgK_α: $h\nu = 1253.6$ eV or AlK_α: $h\nu = 1486.6$ eV) is employed to induce emission of electrons from the core levels with kinetic energy E_{kin} which is analyzed. This process directly allows the determination of the electron binding energy E_b given by:

$$E_b = h\nu - E_{kin} - \emptyset_A \qquad (1)$$

where $\emptyset_A$ is the work function of the analyzer. The notation of the electron levels is that generally used in spectroscopy (for example 1s 1/2, 2s 1/2, 2p 1/2, 2p 3/2 etc.), i.e. by n l j with the principal (n), the orbital angular momentum (l= s, p, d for l = 1, 2, 3 ...) and the total spin momentum (j = l + s) quantum numbers, where s = ±1/2 denotes the electron spin. For an electron in the k level of an atom with N electrons, the binding energy is given by the energy difference between the total energy of the final state E_{tot}^f after electron emission and that of the initial state E_{tot}^i, i.e. $E_b = E_{tot}^f(N - 1,k) - E_{tot}^i(N,k)$. The binding energy depends on the atomic charge distribution and on the local potential caused by the surrounding atoms[2,3]. This causes a variation of the binding energy if an atom is in different bonding states (e.g. in compounds) and a so called chemical shift of the measured E_b with respect to a certain standard state. Handbooks[4] and data banks[5] are available where the binding energies of many compounds are tabulated. Besides the chemical shift, there are further features like plasmon loss peaks, shake up satellites and multiplett splitting which can be used for detailed characterization of the electronic bonding state. Curve fitting procedures allow a distinction of different bonding states which may occur in a surface layer. As a typical example, the 4f 5/2, 7/2 doublett of Ta in different oxidation states is shown in Fig. 1.

Quantitative XPS analysis depends on the total cross section for photoionization at the specific energy based on calculations by Scofield[6], the excitation–emission geometry and the transmission function of the analyzer[3]. Most often relative sensitivity factors are used in practice, preferably based on own standards. An advantage for quantification is that the relative sensitivities of the most intense XPS peaks of the elements are comparable within about one order of magnitude and that matrix effects are relatively small.

The surface sensitivity is given by the electron mean free path or attenuation length λ which depends on the matrix and on the kinetic energy ($\lambda \propto E^n$ with $n \approx 0.5$ ... 0.75)[7,8]. This fact affects quantification if the elemental distribution is not uniform in the first surface layers and in case of a contamination overlayer. On the other hand, variation of the take–off angle leads to different probing depths and can be used as a means to nondestructively determine elemental in–depth distribution.

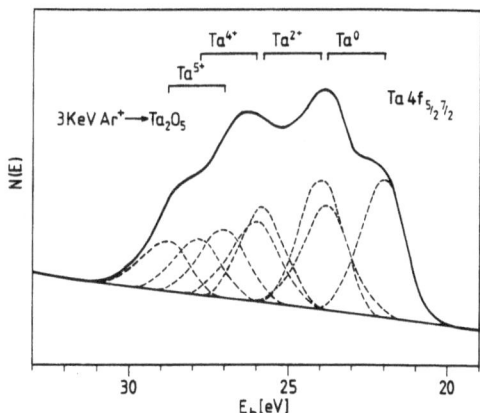

Fig. 1   Ta 4f 5/2, 7/2 XPS peak doublet obtained in sputter equilibrium of a Ta_2O_5 oxide layer, showing peak fitting of the different valence states of Ta corresponding to pure Ta (Ta^0), TaO (Ta^{2+}), TaO_2 (Ta^{4+}) and Ta_2O_5 (Ta^{5+}). Adapted from ref. 43.

The detection sensitivity of XPS is limited to typically about 0.1 at% (with respect to a monoatomic layer). The spatial resolution has been steadily improved and at present ranges from the mm region for conventional XPS down to about 10 μm for imaging instruments.

Excitation by X-rays inevitably leads to local distortions of the chemical bond, which for sensitive compounds, e.g. some oxides, halides and organic molecules may cause decomposition[9]. Most important is positive charging of insulating surfaces which distorts chemical shift determination but can usually be compensated by additional irradiation with low energy electrons[10].

A fundamental limitation of XPS is the necessity of ultra high vacuum ($< 10^{-7}$ Pa of reactive gas pressure) to prevent reactions during analysis. This precludes in-situ studies at higher pressures or in liquid media as well as the study of volatile surface constituents. In spite of the mentioned restrictions, the many advantages of XPS have lead to its widespread use particularly in the metals and semiconductor fields as illustrated for some typical examples in the following section. A recent review by Powell and Seah[11] shows the present state of the art in quantitative XPS.

3. APPLICATIONS OF XPS TO METALS AND SEMICONDUCTORS

Of the numerous reports on successful applications in metals and semiconductor research, only a few examples are chosen to show typical tasks and results.

3.1 Surface Contamination and Failure Analysis

Control of the surface composition prior to further treatment, as in the production of thin film structures in microelectronics, is one of the main applications of XPS[12,13]. Particularly useful is the capability of detecting various organic compounds identified by the chemical shift of the C1s peak[30]. The thickness and quantitative analysis of a contamination layer can be obtained by angle resolved measurements as shown e.g. by Sanz and Hofmann[14] (see also sect. 3.5). Failure analysis is facilitated by the new development of high spatial resolution XPS, which allows chemical mapping with a typical resolution of about 10 μm[15,16]. An example of the detection of residues after plasma etching of Si surfaces in semiconductor fabrication is shown in Fig. 2[13]. The different C1s binding energies allow the identification of several compounds. The F-containing plasma ($CHF_3 + O_2$) gives rise to the according surface compound formation.

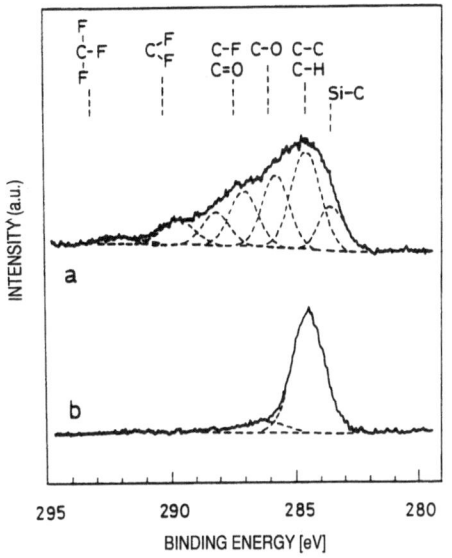

Fig. 2   C1s XPS spectra of (a) a polymer film on a Si surface after reactive ion etching of SiO_2 in a CHF_3/O_2 plasma; (b) after further treatment by Ar^+ ion etching. Adapted from Kolbesen and Pamler[13].

3.2 Oxidation and Corrosion

The broad field of oxidation and corrosion studies is particularly important for the detailed study of the reactions on metal and semiconductor surfaces in gaseous and aequeous environments[17-20].

At low pressures ($< 10^{-3}$ Pa), reactions with gases, e.g. oxygen, can be studied in situ and reveal the initial stages of oxide formation and oxide layer growth on surfaces in great detail[21-23]. For example, Fig. 3 shows the variation of the Ta 4f peak with increasing oxygen exposure of a Ta surface (1 L = 1 x 10^{-6} Torr · sec)[21]. These data can be quantitatively evaluated by the peak areas of the respective standards of pure metal and oxides. The peak shift of the Ta 4f doublett is about 1 eV per valency of Ta. As indicated in Fig. 3, the pure Ta peaks can be subtracted from the total spectrum and the remaining oxide peaks can be quantitatively fitted. The value of each component is plotted in Fig. 4 as a function of the oxygen exposure. It is clearly recognized, that the initial chemisorption of O_2 leads to the suboxides TaO and TaO_2, whereas the formation and growth of the pentoxide only starts at about 10 L. With further increasing oxygen exposure, only the pentoxide layer grows on top of the interfacial layer of about 2–3 monolayer thickness, which stays practically constant[21]. From the linear dependence of the Ta^{5+} intensity on the logarithm of the oxygen exposure and the according decrease of the Ta^{0-} intensity (Fig. 4), the logarithmic growth law of the pentoxide layer was quantitatively established[21].

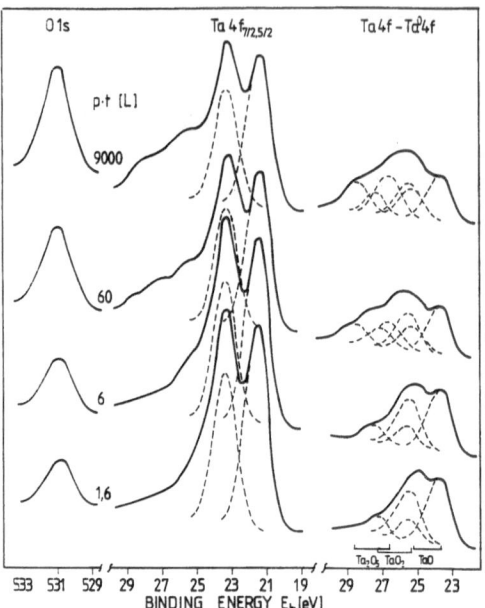

Fig. 3   Variation of the Ta 4f peak with increasing oxygen exposure (1 L = 10^{-6} Torr · sec), showing subtraction of the metallic peak and fitting of the remaining peak to reveal the different oxidation stages of Ta. From ref. 21.

Surface reactions in aequeous media, as in corrosion of metals, can only be studied ex–situ after certain reaction stages. Numerous publications on corrosion and passivation layers with XPS[17-20] have significantly contributed to a better understanding of the underlying mechanisms[24-26]. Again, the capability of resolving the different valence states of the respective elements is of decisive importance. For example, Fig. 5 shows the determination of Fe^{2+} and Fe^{3+} in the passivation layer of a ferritic stainless

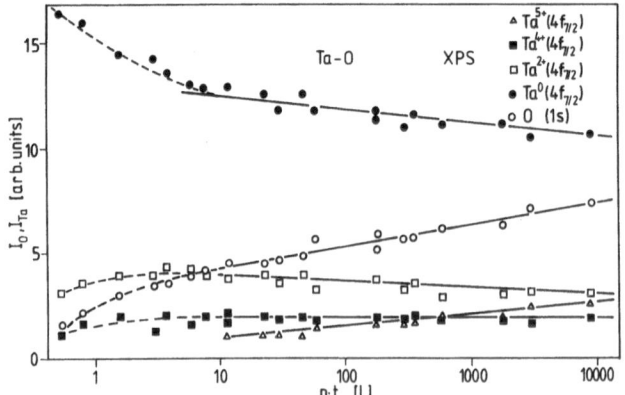

Fig. 4   Peak areas of the components in Fig. 3 as a function of the oxygen exposure. From ref. 21.

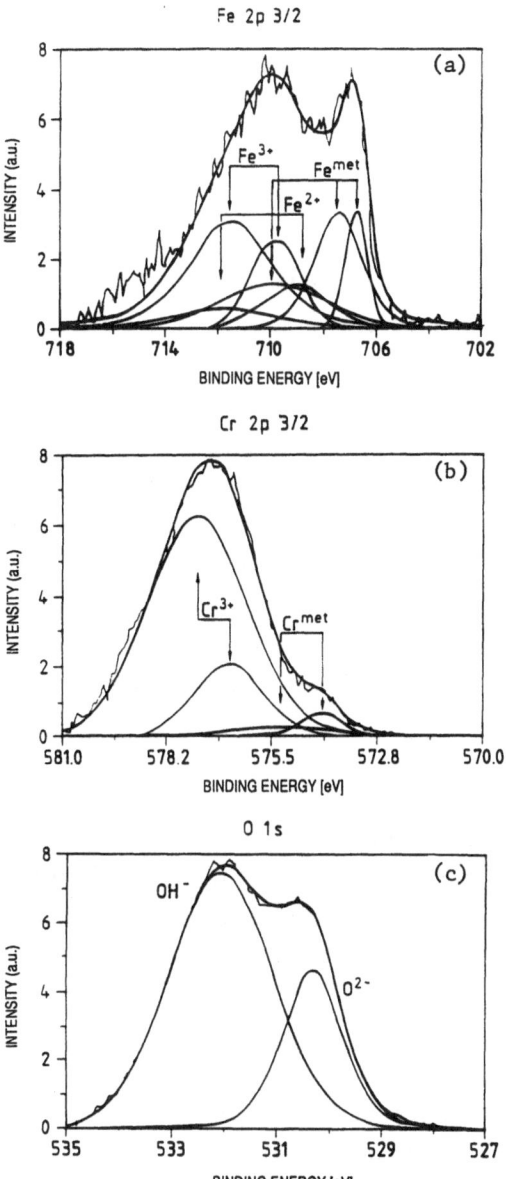

Fig. 5   XPS spectra of Fe, Cr and O for a passive layer on an Fe17.18Cr alloy after
1 h at +500 mV (SCE) in 0.5 mol/l H_2SO_4. (a) Fe 2p 3/2 decomposed by
standard peak fitting in Fe^{3+}, Fe^{2+} and Fe^{met} (metallic); (b) Cr 2p 3/2
decomposed in Cr^{3+} and Cr^{met}; (c) O1s with peak components for O^{2-} and
OH^-. After ref. 24.

steel Fe17.8Cr[24]. The ratio of the oxide to the metallic peak areas allows the determination of the oxide layer thickness (2.8 nm). Deconvolution of the oxygen 1s peak discloses the presence of hydroxides in the passive layer.

3.3 Segregation at Surfaces and Interfaces

Many alloys show a marked difference of the surface composition with respect to their bulk composition in thermodynamic equilibrium[26]. Furthermore, impurities dissolved in the crystal lattice are often enriched at the surface or at grain boundaries. These segregation phenomena are due to the lower chemical potential of certain constituents at surfaces and interfaces as compared to the bulk.

Segregation at surfaces as a function of sample temperature can be studied in situ. Segregation at grain boundaries is measured after fracture in the UHV chamber of an XPS instrument. Besides the elemental composition, the bonding state, revealed by the chemical shift of the relevant peaks, is particularly important for understanding segregation mechanisms and segregation enthalpies[26-31]. For example, Grabke and coworkers[29-31] were able to disclose the bonding of C segregated on a Fe (100) surface which shows a shift of –2 eV compared to graphite, indicating a strong polarity for the segregated carbon with electron transfer from iron to carbon atoms[29].

3.4 Interfacial Bonding

Adhesion of protective coatings on various substrates as well as electronic properties of metal/insulator interfaces are determined by the details of chemical bonding at interfaces, which have been frequently studied by electron spectroscopies including XPS[12,32-34]. An example of the determination of interfacial bonding states at the SiO_2/Si interface by the Si 2p XPS spectra from the work of Hattori et al.[34] is shown in Fig. 6. The different emission angles with respect to the surface plane reveal the in–depth distribution of the Si^{x+}, Si^{y+}, Si^{2+} states, whereas the relative peak areas represent the amount of those species with respect to Si^{4+} in the SiO_2 overlayer (Si^0 stems from the Si substrate). It was concluded that the transition interface layer contains SiO (Si^{2+}) and other suboxides (Si^{4+}) whereas the constant intensity of Si^{x+} is ascribed to the presence of Si-H bonds throughout the native oxide film of a few monolayer thickness.

3.5 Thin Film Analysis

The main application of surface analysis methods is in the analysis of thin films. Electron spectroscopies and in particular XPS provide several possible approaches to

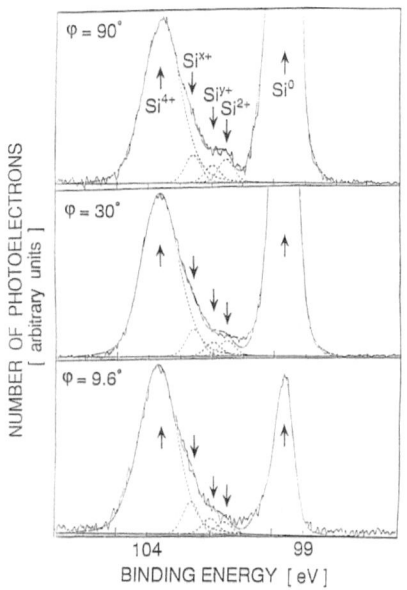

Fig. 6   Si 2p XPS spectra of a native oxide on Si (100) formed in hot HNO_3 for different take-off angles φ to the surface plane, showing suboxides (Si^{z+}, Si^{y+})) in the interface and Si–H-bonds (Si^{x+}) uniformly distributed in the oxide film. Adapted from Hattori[34], with permission of the author and Elsevier Sequoia.

achieve this goal. They can be categorized as non-destructive and destructive[3, 35]. Non-destructive methods make use of the dependence of the effective information depth on the photoelectron energy or on the angle of emission, whereas destructive methods apply combination with noble gas ion sputtering for depth profiling[35].

   3.5.1 Angle Resolved XPS. The advantage of the first, non-destructive category is the detection of the undistorted chemical state. However, the total probing depth is confined to about two to three times the electron attenuation length, i.e. to < 5 nm for conventional XPS[35-37]. For sputter depth profiling, there is no such confinement, but care must be taken with respect to chemical bonding since it is often changed by the energetic ion bombardment.

   The change of photoelectron energy is usually performed by applying synchrotron radiation with different X-ray energies. In conventional XPS with constant X-ray energy (MgK_α), the usual technique is tilting the sample perpendicular to the analyzer axis. An example is shown in Fig. 7 for an aluminium oxide layer of 2.3 nm thickness on Al[38]. The peak area I_{ox} of the Al 2p peak corresponding to Al_2O_3 (75.7 eV) is increasing relative to that of pure Al (72.3 eV) (I_{Me}) with increasing take-off angle φ. It can be shown that the thickness d of the Al_2O_3 layer is determined by[35-37]

$$d = \lambda \cdot \sin\varphi \cdot \ln(1 + k \cdot I_{ox}/I_{Me}) \qquad (2)$$

with λ the electron attenuation length (2.0 nm) and k a sensitivity correction factor of

ANGLE RESOLVED XPS

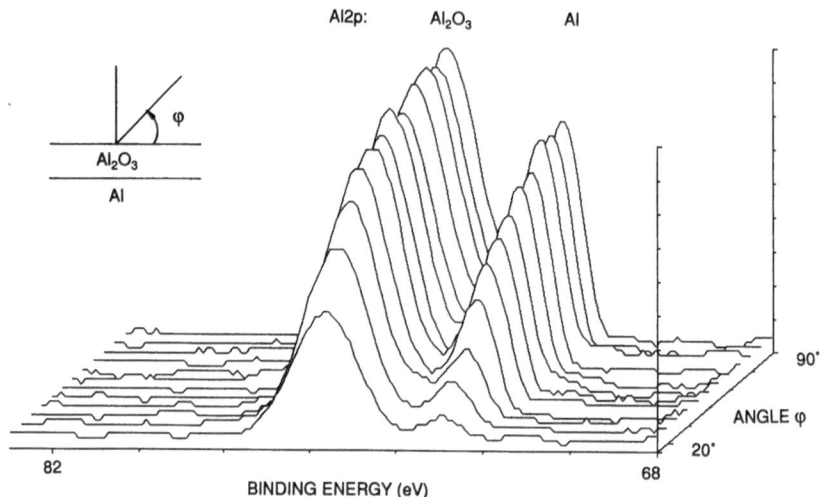

Fig. 7   Al 2p XPS spectra of an Al_2O_3 layer of 2.3 nm thickness on Al for different take-off angles φ to the surface plane.

about 1. A plot of the ratio I_{ox}/I_{Me} is given in Fig. 8 together with the theoretical curves. (Note that for high emission angles $\Theta = 90° - \varphi$, deviations may occur due to a finite acceptance angle of the instrument and/or a contamination layer). Using the general formulation of the angular dependence $I(\varphi)$ being a Laplace transform of $I(1/\lambda)$[14,39,40], more detailed layer profiles can be revealed, as e.g. shown by Bussing and Holloway[40] in the case of the altered layer of sputtered GaAs surfaces.

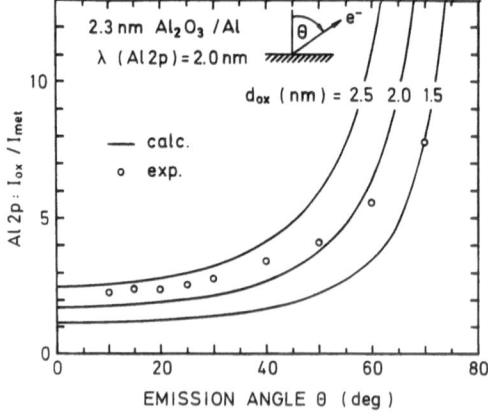

Fig. 8   Determination of the layer thickness of Al_2O_3 on Al from the peak area ratios in Fig. 7 according to eq. (2) with emission angle $\Theta = 90° - \varphi$.

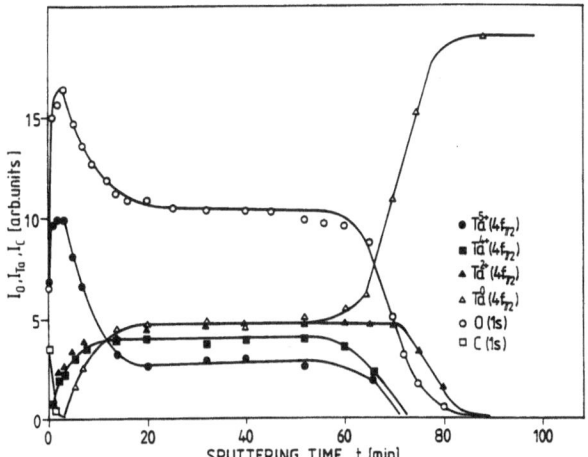

Fig. 9   XPS depth profile of a 30 nm thick Ta_2O_5 layer on Ta obtained with 3 keV
         Ar^+ sputtering. Deconvolution of the Ta 4f peaks show the decomposition of
         the pentoxide to lower oxides and to pure Ta due to preferential sputtering of
         oxygen. From ref. 43.

In recent years, Tougaard and coworkers[41,42] have consistently shown that the
background at the lower kinetic energy side of an XPS peak contains information about
the depth of origin of the respective photoelectrons and can therefore be used as a means
to determine in-depth distributions and layer thicknesses. Similar total depth
restrictions (< 5 nm) as in angle resolved XPS apply.

3.5.2 Sputter Depth Profiling: Decomposition and Compound Formation. The
limitations in both probed depth and obtainable depth resolution are in favor of depth
profiling by sputtering in combination with XPS. This technique discloses the in-depth
distribution of chemical compounds. In case of homogeneous sputtering, these are
practically unchanged due to ion bombardment, as e.g. in Ni, Fe and Cr oxides and
many nitrides[20]. In contrast, oxides of the heavier transition metals like Zr, Nb, Mo, Ta
etc. are decomposed due to preferential sputtering of oxygen[43,44]. Therefore, an XPS
depth profile of stoichiometric Ta_2O_5 on Ta gives direct evidence of the ion induced
oxide decomposition and the oxygen depletion, as shown in Fig. 9 for sputtering with
3 keV Ar^+ ions[43]. The formation of an altered layer of about 2.5 nm thickness consisting
of Ta, TaO and TaO_2 corresponding to a total oxygen depletion of about 20 at% could
be determined from results as in Fig. 9 and was confirmed by angle resolved XPS[43].

Surface engineering by means of laser and ion beam irradiation has become a
major activity in materials science and technology. Irradiation with a beam of energetic

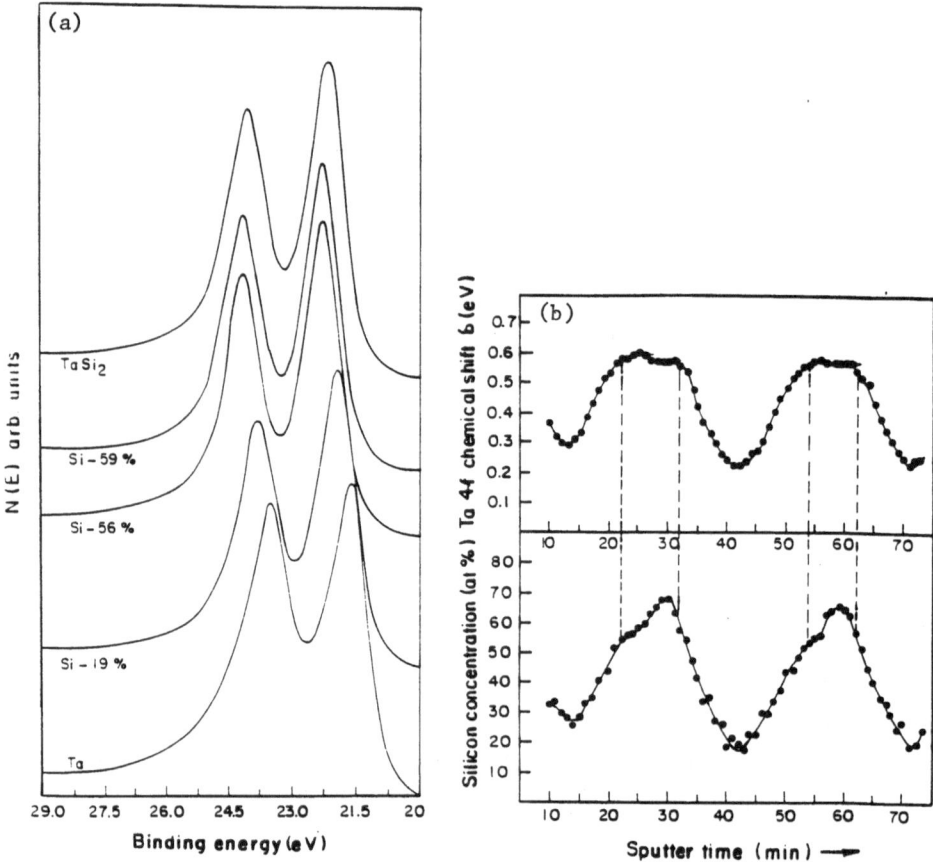

Fig. 10   Ion beam induced silicide formation at Ta/Si interfaces: (a) Ta 4f XPS
spectra showing silicidation at 56 at% and 59 at% Si from the atomic
concentration profile. (b) chemical shift δ of the Ta 4f 7/2 binding energy
(upper part) at different stages of the atomic concentration XPS profile
shown for Si in the lower part of the figure. From ref. 46.

noble gas ions produces a zone of intense ion mixing due to the momentum transfer of
the primary ions. The extension of the mixing zone is of the order of the ion range.
Recently for 3 keV Ar[+] ions about 4 nm were determined by TEM at Ta/Si interfaces in
accordance with LSS theory calculations[45]. Atomic mixing in the interface is capable of
producing compounds because the particle energies in the collisional cascade are well
above thermal equilibrium. Often stable compounds are formed. This process can be
monitored by XPS if the standard spectra for these compounds are known. An example
is shown in Fig. 10 for the in–depth profile of Si in profiling of Ta/Si multilayers (10.5
nm Si and 7.5 nm Ta, respectively)[46]. Fig. 10a shows the peak shift δ (eV) of the Ta 4f
XPS doublett peak together with the variation of the Si 2p signal intensity, calibrated in

at% Si (Fig. 10b). The value δ = 0.6 eV (δ = 0 refers to pure Ta) was determined for a TaSi$_2$ standard. It is clearly seen that this value coincides with the shoulder in the Si depth profile, indicating TaSi$_2$ formation with a lower sputtering yield at roughly the required stoichiometry.

3.6 Recent Developments

Recent progress in applied XPS mainly concerns improved data evaluation schemes and instrumental improvements.

Reliable quantification of XPS is based on the knowledge of the attenuation length of the photoelectrons and its dependence on energy and on matrix parameters. A summary of recent advances was given by Powell and Seah[11]. The techniques of background subtraction were considerably improved by using appropriate background functions[41,42] and corrections for plasmon loss features[41,47]. As shown by Tougaard, the electronic background can be used to determine the depth of origin of the XPS signal, and therefore to obtain the in-depth distribution of the measured species[41,42].

Progress in instrumentation development is mainly focussed on high energy resolution and on high spatial resolution. The former is achieved by improved X-ray monochromators, allowing an energy resolution of about 0.2 eV which enhances correct chemical sate determination. Different approaches to high spatial resolution comprise small area XPS by imaging a focussed electron beam impact spot on the anode through the monochromator on the sample or by selecting a small area of the sample by means of elaborate electrostatic lens systems in front of the analyzer[16]. New developments of imaging XPS systems use either monoenergetic imaging of a certain peak through the analyzer optics, or the energy dispersion in one axis (x) and the spatial image of a stripe region in the other axis (y) and subsequent mechanical scanning of the sample to generate elemental images of those of the chemical bonding state, both with resolutions of about 10 μm[15,16]. Recently, Fresnel lens systems are used to focus synchrotron radiation in spot diameters of less than 5 μm. With these developments, the application range of XPS is greatly enlarged.

4. CONCLUSIONS AND OUTLOOK

XPS has now gained a wide acceptance as one of the major characterization techniques of surfaces and thin films of metals and semiconductors. Its unique possibilities of quantitative determination of the elements and their chemical bonding state in thin surface layers, based on relatively straight forward interpretation of the chemical shifts have enhanced its application in various research and development areas, such as contamination and failure analysis in process control, oxidation and corrosion, adhesion, wear and surface engineering. Angle resolved measurements and combination with ion sputtering have opened the field of thin film and interface analysis which is of particular interest in coatings and semiconductor studies. New developments in both quantitative data evaluations and high spatial resolution imaging XPS will further enhance its widespread use, which is most efficient in conjunction with other surface analysis methods.

REFERENCES

1. K. Siegbahn et al., ESCA Applied to Free Molecules, North Holland Publ. Comp., Amsterdam (1969).
2. C.S. Fadley, in: Electron Spectroscopy - Theory, Techniques and Applications, C.R. Brundle and A.D. Baker, eds., Academic Press, London, Vol. 2 (1978).
3. D. Briggs and M.P. Seah, eds., Practical Surface Analysis Vol.1 1: Auger X-ray Photoelectron Spectroscopy, J. Wiley, Chichester (1990).
4. C.D. Wagner, W.M. Riggs, L.E. Davis, J.E. Moulder and G.E. Muilenberg, Handbook of XPS, Phys. Electronics Div., Perkin-Elmer Corp., Eden Prairie (USA) (19879).
5. NIST XPS Database, compiled by C.D. Wagner (1989).
6. J.H. Scofield, J. Electr. Spectr. Rel. Phen. 8:129 (1976).
7. M.P. Seah and W. Dench, Surf. Interface Anal. 1:2 (1979).
8. T. Tanuma, C.J. Powell and D.R. Penn, Surf. Interface Anal. 11:577 (1988).
9. J. Cazaux, Appl. Surf. Sci. 20:457 (1985).
10. C.P. Hunt, C.T.H. Stoddart and M.P. Seah, Surf. Interface Anal. 3:157 (1981).
11. C.J. Powell and M.P. Seah, J. Vac. Sci. Technol. A8:735 (1990).
12. J.H. Thomas III in: H Windawi and F.-L. Ho, eds., Applied ESCA, J. Wiley, New York (1982), p. 37.
13. B.O. Kolbesen and W. Pamler, Fresenius Z. Anal. Chem. 333:561 (1989).
14. S. Hofmann and J.M. Sanz, Surf. Interface Anal. 6:75 (1984)
15. N. Gurker, M.F. Ebel and H. Ebel, Surf. Interface Anal. 5:13 (1983)
16. R.L. Chaney, Surf. Interface Anal. 10:36 (1987).
17. D.R. Baer, C.R. Clayton and G.D. Davis, eds., Proc. of the Symp. on: The Application of Surface Analysis Methods to Environmental/Materials Interactions. The Electrochemical Society Inc., Pennington, N..J. (1991).
18. J.E. Castle, Surf. Interface Anal. 9:345 (1986).
19. D. Landolt, Surf. Interface Anal. 15:395 (1990).

20. S. McIntyre, chap. 10, p. 397 in ref. 3.

21. J.M. Sanz and S. Hofmann, J. Less Comm. Metals 92:317 (1983).

22. J. Steffen and S. Hofmann, Fres. Z. Anal. Chem. 329:250 (1987).

23. C. Morant, J.M. Sanz, L. Galán, L. Soriano and F. Rueda, Surf. Sci. 218:331 (1989).

24. H. Knote, S. Hofmann and H. Fischmeister, Fres. Z. Anal. Chem. 329:292 (1987).

25. P. Bruesch, K. Müller, A. Atrus and H. Neff, Appl. Phys. A38:1 (1985).

26. R. Kirchheim, B. Heine, S. Hofmann and H. Hofsäß, Corrosion Sci. 31:573 (1990).

27. S. Hofmann, Vacuum 40:9 (1990).

28. G. Hetzendorf and P. Varga, Nucl. Instr. Meth. Phys. Res. B18:501 (1987).

29. G. Panzner and W. Diekmann, Surf. Sci. 160:253 (1985).

30. W. Diekmann, G. Panzner and H.J. Grabke, Surf. Sci. 218:507 (1989).

31. B. Egert and G. Panzner, Surf. Sci. 118:345 (1982).

32. W.J. van Ooij, A. Sabata and A.D. Apelhans, Surf. Interface Anal. 17:403 (1991).

33. F.J. Grunthaner and P.J. Grunthaner, Mater. Sci. Rep. 1:65 (1986).

34. T. Hattori, Proc. ICMCTF91, San Diego, 22–26 April 1991, to be publ. in Thin Solid Films.

35. C.S. Fadley, Porgr. in Solid State Chemistry 11:265 (1976).

36. M.F. Ebel, J. Electron Spectr. Rel. Phen. 14:287 (1978).

37. M.F. Ebel, Surf. Interface Anal. 3:333 (1981).

38. I. Olefjord, H.J. Mathieu and P. Marcus, Surf. Interface Anal. 15:681 (1990).

39. S. Hofmann, Analusis 9:181 (1981).

40. T.D. Bussing and P.H. Holloway, J. Vac. Sci. Technol. A3:1973 (1985).

41. S. Tougaard, Surf. Interface Anal. 11:453 (1988).

42. S. Tougaard and H.S. Hansen, Surf. Interface Anal. 14:730 (1989).

43. S. Hofmann and J.M. Sanz, J. Trace Microprobe Techn. 1:213 (1982–83).

44. J.B. Malherbe, S. Hofmann and J.M. Sanz, Appl. Surf. Sci. 27:355 (1986).

45. S. Hofmann and W. Mader, Surf. Interface Anal. 15:794 (1990).

46. B.R. Chakraborty and S. Hofmann, to be publ. in Thin Solid Films.

47. I.LeR. Strydom and S. Hofmann, J. Electron Spectr. Rel. Phen. 56:85 (1981).

TRACE ELEMENT ANALYSIS USING TOTAL-REFLECTION
X-RAY FLUORESCENCE SPECTROMETRY

Andreas Prange* and Heinrich Schwenke

GKSS Research Centre, Institute for Physics, P.O. Box 1160
D-2054 Geesthacht, Germany

ABSTRACT

For the application of TXRF in trace element analysis, two characteristic features of the total reflection of X-rays are exploited. These are the high reflectivity on flat surfaces and the low penetration depth of the primary radiation. This allows the application of TXRF for both chemical trace and ultra-trace element analysis on the one hand and surface analysis on the other. For chemical trace element analysis, total reflection on a highly polished substrate is characterized by a *high reflectivity*, which leads to a drastic reduction of the spectral background. The sample to be analyzed is prepared *on* the substrate as a residue of small quantities by evaporation from solutions or fine-grained suspensions. Instrumental detection limits of a few pg or sub-ng/ml^{-1} are state of the art for commercially available equipment. Besides the high detection power, internal standardization is another important feature of TXRF, enabling very simple quantification of the detected elements. The small sample mass required enables especially ultra-micro analytical questions to be tackled. For trace element analysis *in* surfaces the *low penetration depth* of the primary radiation under the conditions of total reflection is exploited. The penetration depth is in the range of a few nanometers, hence, TXRF is intrinsically surface sensitive. Detection limits better than 10^{10} atoms/cm^2 are obtained for metal impurities on silicon wafers. For the examination of layered structures elemental composition, layer thickness, and density can be derived from the angle-dependent fluorescence intensities. The present paper describes the basic features of the total reflection of X-rays and gives some representative examples of the different uses of TXRF in trace element analysis.

INTRODUCTION

Energy dispersive X-ray fluorescence (EDXRF) has undergone its most significant progress in the past few years through the use of total reflection of the primary radiation on flat surfaces. Developments in the X-ray optics of EDXRF have led to a substantial improvement in the detection power.

---

* *Offprint requests to:* A. Prange

Table 1. 20 Years Total-Reflection X-Ray Fluorescence Spectrometry

1971	Initial idea to use total reflection of X-rays for trace analysis Detection limit: 1000 pg	*Kyushu University, Fukuoka Japan*
1974	Rediscovery of TXRF in Europe Detection limits: 1000 pg	*Atominstitut, Vienna Austria*
1978	Improvements in instrumental design and performance Detection limits: 30 pg	*GKSS, Geesthacht Germany*
1980	Introduction of cut-off mirror Detection limits: 5 pg	*GKSS, Geesthacht Germany*
1981	First commercially available instrument for chemical analysis	*Seifert, Ahrensburg Germany*
1985	Twin excitation module (Mo- and W-anode) Detection limit: 2 pg	*Seifert, Ahrensburg Germany*
1987/88	Starting surface analysis	*GKSS; Technos; Rigaku Germany and Japan*
1988	First commercially available instrument for surface analysis	*Atomika, München Germany*
1989/90	Use of rotating anodes Use of multilayers Detection limit: 0.2 pg	*Technos; Rigaku GKSS Japan and Germany*
1991	Character. of layered structures (composition, density and thickness)	*GKSS and Philips Research Germany and The Netherlands*

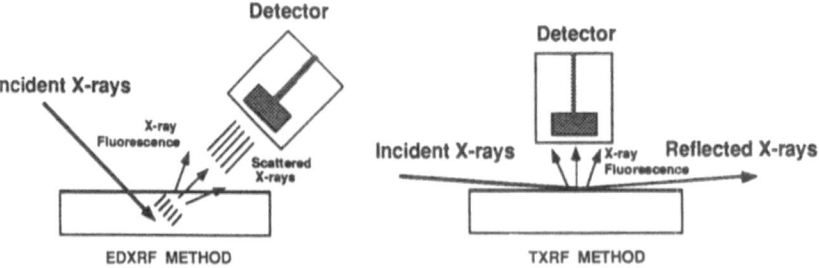

Figure 1. Conventional EDXRF excitation / detection geometry compared with total reflection geometry.

Classical EDXRF has relatively poor detection limits. This is because the fluorescence radiation of the elements to be determined and the scattered radiation from the matrix elements in the sample and from the sample support itself all strike the detector. The limits of detection of an analytical technique are known to be determined to a greater extend by the peak to background ratio. Because the spectral background is mainly produced by scattered radiation from the sample support using a ~ 45° excitation/detection geometry (see figure 1), YONEDA and HORIUCHI [1] suggested in 1971 that the background could be reduced by total reflection of the exciting radiation at an optically flat sample support. In this case the incident radiation has virtually no interaction with the sample carrier, resulting in a dramatic reduction of the scattered radiation and a substantial improvement in the peak to background ratio. Experimental detection limits of about 1 nanogram were reported, which were very impressive at that time, in particular for an X-ray fluorescence technique.

Table 1 gives an historical overview of developments over the past 20 years. It shows a dramatic development in instrumentation with the detection limits improving from 1000 pg down to below 1pg.

Inspired by the publications of AIGINGER and WOBRAUSCHEK [2], 1974, our laboratory began to work on TXRF in 1975 and succeeded in designing a compact, stable and easily adjustable total reflection module by the late seventies, increasing the performance of TXRF stepwise to detection limits better than 5 picogram for some metals [3, 4]. This was the decisive step towards the introduction of the method into the analytical practice of *chemical trace and ultra-trace element analysis* and a few years later for the commercialization of Total Reflection X-ray Fluorescence, where twin excitation using Mo- and W-anodes is now state of the art [5].

A new field of application became apparent in 1988 in that TXRF was also found to be an *inherently surface sensitive technique* due to the low penetration depth of X-rays in the total reflection regime. Instrumental improvements allow for accurate angle-adjustment of the incident X-ray beam [6]. Furthermore, the use of rotating anodes and the monochromatization of the exciting radiation results in improved detection limits down to below 1 picogram. Initially developed for the contamination control of silicon wafers, TXRF has recently been used for the quantitative characterization of near surface layers. Now, after twenty years, several companies such as Seifert and Atomika in Germany, Technos and Rigaku in Japan offer commercially available TXRF-instruments especially for wafer analysis.

TXRF FUNDAMENTALS

Total reflection occurs when radiation arrives from an optically denser medium and impinges on an optically thinner medium. As known, for X-rays any medium is optically thinner than a vacuum. The optical density of matter is described by the complex index of refraction n. For total reflection of X-rays n is smaller than unity:

$$n = 1 - \delta + i\beta$$

where δ is the dispersive term which is dependent on the wavelength λ or the energy of the incident radiation respectively, and also on material quantities such as atomic number Z, the atomic mass A and the density ρ of the reflector (i. e. sample support or material under investigation). β is the absorptive term which is dependent on the wavelength or on the energy, respectively, and is also dependent on the linear mass absorption coefficient. In the above formular β describes the attenuation of X-rays in matter and provides the damping which is an essential feature of the grazing incidence arrangement.

With δ the critical angle of total reflection is also defined, that is the angle below which total reflection occurs. The critical angle can be derived from Snell´s law to:

$$\varphi_c = \sqrt{2\delta}$$

Because δ is in the order of 10^{-6}, there is only a very small angle interval of a few millirad for the total reflection of X-rays. With δ the critical angle is proportional to the wavelength or energy of the exciting radiation, hence the range where total reflection occurs is less for Mo-K$_\alpha$ excitation (φ_c = 1.75 mrad) than for Cu-K$_\alpha$ excitation (φ_c = 3.86 mrad), when reflected from silicon. Furthermore, φ_c is proportional to material quantities, given by the reflecting medium.

With the terms of the complex index of refraction and the critical angle of total reflection two very important features can be derived with the help of the Fresnel equations. These are the reflectivity and the penetration depth.

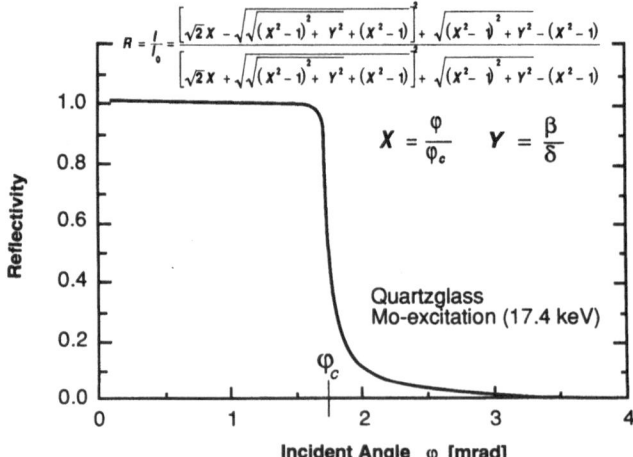

Figure 2. Reflectivity versus the angle of incidence for quartz glass material using a Mo-anode for excitation.

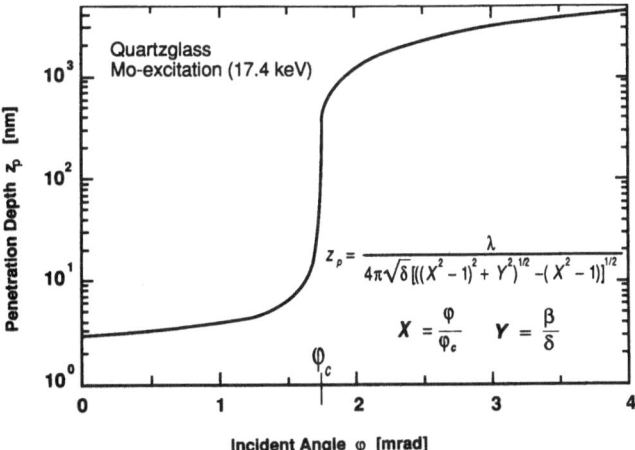

Figure 3. Variation of the penetration depth with increasing angle of incidence for quartz glass material using a Mo-anode for excitation.

Figure 2 shows the variation in the reflectivity R with the angle of incidence φ for quartz glass using a Mo-anode (17.4 keV) for excitation. The reflectivity R is defined as the ratio of the reflected intensity I to the incident intensity I_0. Below the critical angle, the reflectivity is near 1 (e.g. 0.999 at 1 mrad) and decreases rapidly after passing the critical angle. The formula-expression is the result obtained from the Fresnel equation [7]. The reflectivity as well as the penetration depth are basically described by the dependence of two quantities, namely X and Y, where X is the incident angle, given in units of the critical angle and Y is the ratio of the absorptive to the dispersive term, which are already known.

Figure 3 presents the penetration depth z_p versus the angle of incidence φ for quartz glass material using a Mo-anode (17.4 keV) for excitation, and shows the corresponding formula-expression [7], again containing the terms for X and Y. It can be seen that the penetration depth at angles greater than the critical angle is in the order of a few micrometers. Under total reflection conditions the penetration depth is drastically reduced to about 3 nm. At very low angles the penetration depth is nearly constant.

The very slight penetration at angles less than the critical angle makes this technique an inherently sensitive method for exploring thin surface layers. However, for this task additional developments in the instrumentation and quantification are required, compared to its chemical use.

CHEMICAL TRACE ANALYSIS

In the following section a closer look at the use of TXRF for chemical trace analysis is given, where the high reflectivity of appropriate sample carriers is exploited.

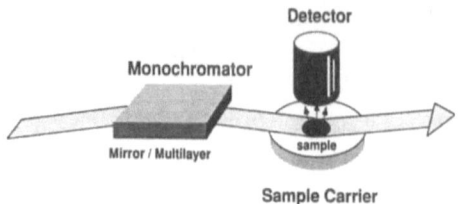

Figure 4. Schematic view of the total reflection excitation geometry.

Instrumental set-up. Figure 4 shows a schematic view of the instrumental set-up for chemical trace analysis. The sample to be analyzed is prepared *on* the sample carrier as a residue of small quantities by evaporation from solutions or fine-grained suspensions. It is excited under a solid angle of incidence of about 1 mrad using a collimated beam from a sealed fine focus X-ray tube with Mo- and W-anodes respectively. The fluorescence radiation, which is doubled in intensity because of the excitation of the sample by both the incident and the reflected beams, is detected by a Si(Li)- detector perpendicular to the support and is registered by a multichannel analyzer. A monochromator in front of the sample support leads to further improvements in the peak to background ratio. A quartz mirror which serves as a high energy cut-off filter, ensures that only exciting radiation below the critical

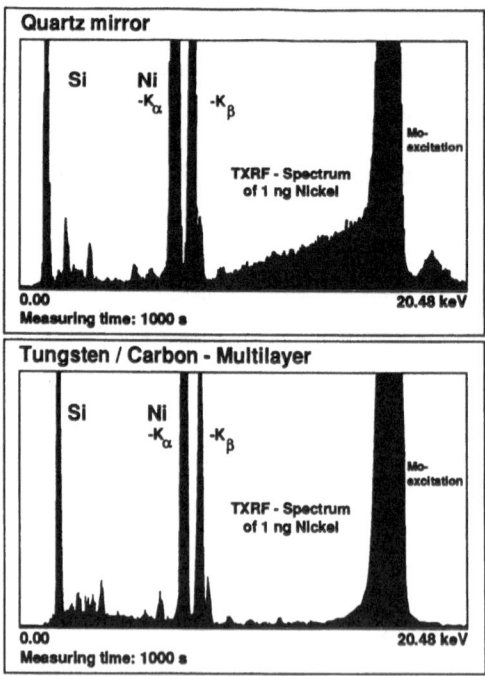

Figure 5. Comparison of TXRF spectra using a quartz mirror and a multilayer
mirror for monochromatization.

Table 2. Important characteristics of different sample carrier materials

	Quartz (Suprasil®)	Plexiglass	Glassy Carbon	Boron Nitride
Phys. characteristics for Mo-excitation:				
$R_{(1\ mrad)}$ [%]	0.997	0.999	0.999	0.999
Φ_C [mrad]	1.72	1.34	1.45	1.73
ρ [g/cm³]	2.20	1.20	1.55	2.25
Surface quality	very good	good	satisfactory	good
Purity	very good	good (Zn)	Fe, Cu, Zn	good (Zn)
Fluorescence Peaks	Si	none	none	none
Chem. inert against:				
Acids	good (not HF)	little	good	very good
Alkaline solutions	moderate	little	very good	very good
Organic solvents	very good	not good	good	very good
Mechanical Stability	good	good	good	good
Cleaning	easy	impossible	difficult	easy
Prize / carrier [$]	20	a few cents	20	~ 35

angle of total reflection for X-rays on quartz is reflected onto the sample carrier, thus eliminating high energy bremsstrahlung via diffuse scattering and absorption. Replacing the quartz mirror by a multilayer mirror (e.g. tungsten / carbon type), which recently became available, a further reduction of the background can be achieved due to the band-pass effect. This is especially true for higher energies next to the excitation radiation as demonstrated in figure 5 by the TXRF spectra of 1ng Ni using Mo-excitation.

Sample carriers. Aside from the excitation and monochromatization of the beam, a further essential part of the instrumentation are the sample carriers as the reflecting medium for TXRF. Up to now several sample carrier materials have been investigated and used for TXRF. There are some general requirements which the sample carriers must have to ensure an optimal use of TXRF. These include:

- a high reflectivity
- be chemically inert
- free of impurities
- no fluorescence peaks from the carrier material should occur over the considered energy scale

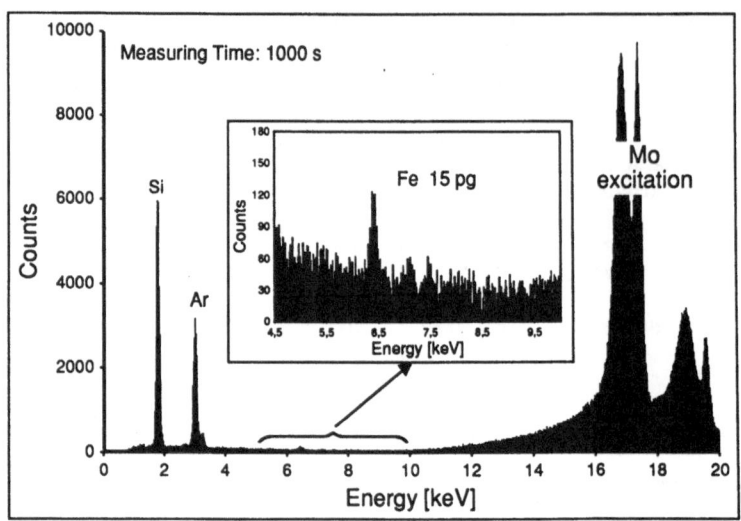

Figure 6. TXRF spectrum of a cleaned quartz glass carrier.

- be easy to clean
- and be cheap

Besides quartz glass carriers, plexiglass, glassy carbon and boron nitride have been used for our applications. Some important properties or characteristics are listed in table 2.

Figure 6 shows a TXRF spectrum of a cleaned quartz glass carrier. The Si-peak originates from the carrier, the Ar-peak from the air between carrier and detector and the peaks bet-

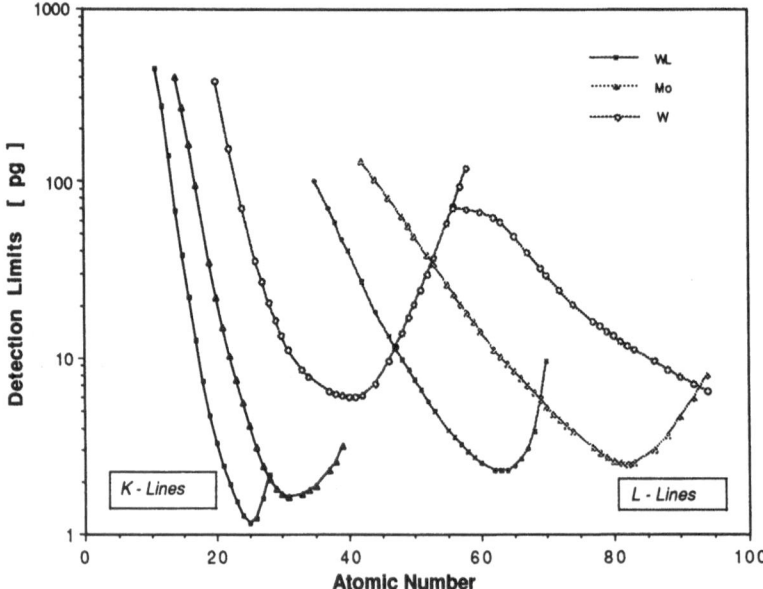

Figure 7. Instrumental detection limits for Mo- and W-excitation respectively.

ween 16 and 20 keV originate from the Mo-excitation. The residual Fe contamination, clearly visible in the zoomed out spectrum, refers to a mass of 15 pg and gives an idea of the excellent detection power of this technique.

Detection limits. Figure 7 shows the detection limits obtained by means of isolated element peaks measured on matrix free standard solutions. They are better than 10 pg for more than 60 elements for a counting time of 1000 s using a twin excitation source equipped with both a Mo- and W-anode. The W-anode can be used for excitation with the $W-L_{\alpha}$-line in order to better detect elements of low atomic number, or using a section of the bremsstrahlungs continuum for excitation of the Cd -K-line and neighboring elements. Both anode materials ideally complement each other, such that for practical situations the elements ranging in atomic number from 12 through the remainder of the periodic table are determinable.

The detectability of elements with atomic numbers below 12 is increasingly affected by a poor fluorescence yield, poor peak resolution and other detector influences. However, with the help of a windowless Si(Li)-detector operating in a high vacuum measuring chamber the determination of magnesium and even oxygen has been carried out by STRELI, WOBRAUSCHEK and AIGINGER. They have reported detection limits of about 10 ng and 0.8 ng for oxygen and magnesium, respectively [8].

Calibration and Quantification. Besides its high detection power the most important feature of TXRF is its ability to analyse very small sample quantities. Because of the use of low masses, matrix absorption or enhancement effects are avoided resulting in a constant relationship between the fluorescence yield and the atomic number. Once established, the calibration remains valid as long as major components such as the detector or tube anode remain unchanged. This calibration curve is independent on the properties of the sample matrix. Thus a quantification is very simply performed by internal one-element standardization; that is adding a single standard element before preparing the sample on the sample carrier. The concentration of the unknown elements are calculated by:

$$C_{unk} = C_{std} \cdot \frac{cps_{std}}{cps_{unk}} \cdot \varepsilon_{In}$$

where

C	= concentration	*std*	refers to internal standard
cps	= counts per second	*unk*	unknown for all elements/lines other than internal standard
ε_{In}	= relative sensitivity of the instrument		

The performance of any analytical technique is to a greater extent dependent on the instrument itself and its detection limits. Low detection limits are a necessary precondition for the

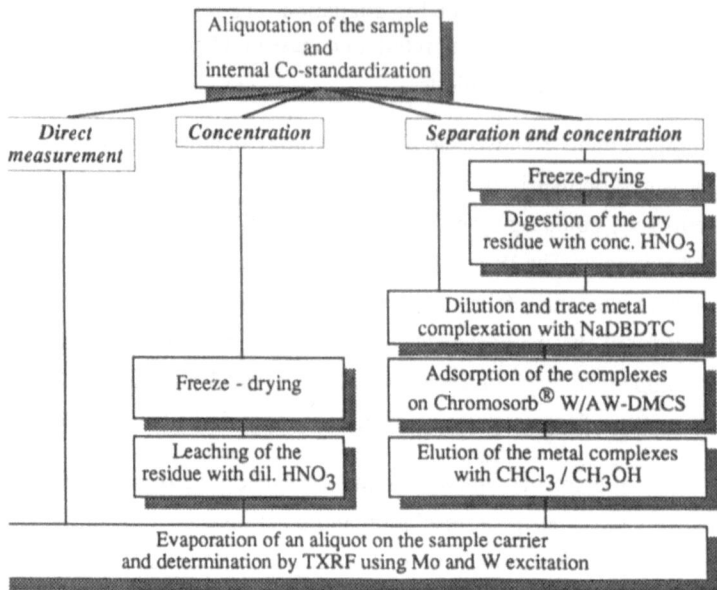

Figure 8. Sample preparation techniques for aqueous samples.

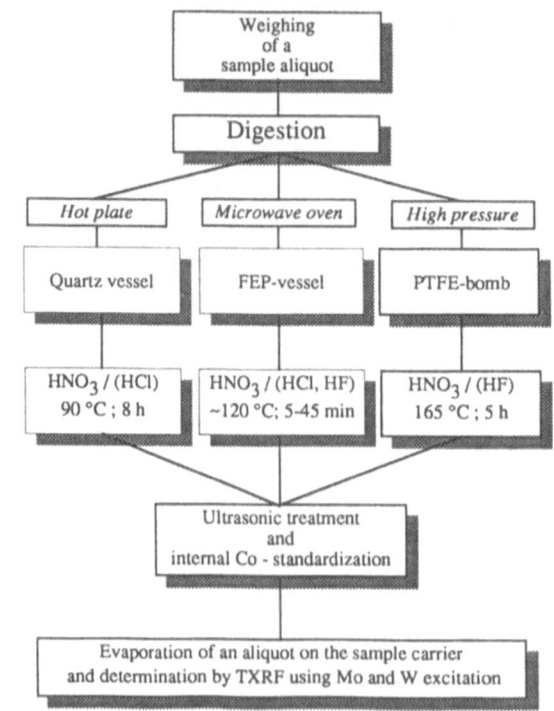

Figure 9. Sample preparation techniques for solid samples.

majority of applications, however, they alone are not sufficient in practice. Therefore, much work has been invested in the development of standardized sample preparation techniques for solid and aqueous samples of a variety of matrices, including tools for sample handling and sample preparation as well as the development of special devices for various practical problems [9, 10, 15]. It is precisely these developments which brings life into the instrument thus enabling it to be used in various fields of application.

Sample Preparation. For aqueous samples, very simple techniques such as direct measurement or physically concentrating the sample by freeze-drying are in many cases practicable. This is especially true for samples where no or only small amounts of matrix elements occur, as is the case, for example with rain- or riverwater. In the case of higher concentrations of matrix elements, separation and concentration procedures are mandatory. A reliable technique for the separation of alkaline and alkaline earth elements and for an enrichment of trace elements is the complexation of the trace elements with sodium dibenzyldithiocarbamate, chromatographic adsorption of the metal complexes on a hydrophobic reverse phase material and a subsequent elution of the complexes with organic solvents [10]. For natural waters with higher organic matrix content, digestion is additionally required prior to the separation and concentration step. This is shown in figure 8. This procedure looks more complicated as it actually is in practice. For instance, our laboratory and also the German authority for environmental monitoring of the North Sea are analyzing hundreds of sea water and estuarine water samples routinely using this technique adapted to TXRF.

For solid samples different digestion procedures are carried out on a hot plate, a microwave oven or, if more oxidation power is needed, using high pressure in PTFE bombs as is shown in figure 9. Favourable acids for digestion are nitric, hydrochloric or hydrofluoric acid. The resulting solution or fine grained suspension is then spiked with an internal standard element and prepared for measurement.

Application of TXRF in chemical trace analysis

In the following a few selected examples from the scope of TXRF-applications which are summarized in table 3 are presented. Some typical applications from the various fields such as the environment, mineralogy, medicine, biology and the purity control of ultrapure reagents are elucidated. Examples will include environmental samples with and without a high matrix content such as rainwater, seawater and sewage sludge. From the field of medicine examples of the microanalytical capability of TXRF for investigating human hair and tissue will be given and finally an example of the use of this technique for the purity control of ultrapure reagents such as HF and HNO_3 will be presented.

Rainwater. The more or less instrumental use of TXRF is exploited for the determination of trace elements in rainwater. The rainwater sample is spiked with a few microliters of a

Table 3. The scope of TXRF-applications

Environment	Mineralogy	Medicine	Biology	Ultrapure reagents
WATER	ROCKS	BLOOD	PLANT MATERIAL	ACIDS
Rain, River, Sea,	Obsidian	Whole, Serum	Fine roots, Leaves,	Nitric, Hydrochloric,
Drinking, Waste,			Needles, Wood, Hay	Hydrofluoric, Sulfuric
Susp. particulate matter	MINERALS	HAIR	powder, Wheat flower	
	Monazite, Betafite,			WATER
AIR	Columbite	URINE	MARINE	
Dusts, Aerosol particles,			Algae, Mussel, Fish	Ammonia Solution
Fly ashes	Trace Elements in SiO₂	TISSUES		
		Muscle, Lung,	FOOD	AMMONIA
SOILS	Rare Earth Elements	Liver, Kidney	Mushroom, Grapes	FLUORIDE
Sediments (marine, river)			Onions	
Light sandy soil				
Sewage Sludge				

standard solution, e.g. Co, for internal standardization. After a thorough mixing an aliquot of 25 µl of this solution is pipetted onto a siliconized quartz glass sample carrier, and is then allowed to dry for measurement. Figure 10 shows a section of the resulting TXRF spectrum of the rainwater sample excited using a Mo-anode and measured for 1500 s. The values are given in µg/l and Co was used as the internal standard. The concentrations are in the ppb and sub-ppb range. Altogether up to 20 elements can be determined in rainwater. These are S, K, Ca, Ti, V, Cr, Mn, Fe, (Co), Ni, Cu, Zn, As, Se, Rb, Sr, Mo, Cd, Ba and Pb [11].

Seawater. One of the great difficulties in trace analysis is when high matrix concentrations occur. This is especially the case for seawater. Therefore, it is necessary to separate the trace

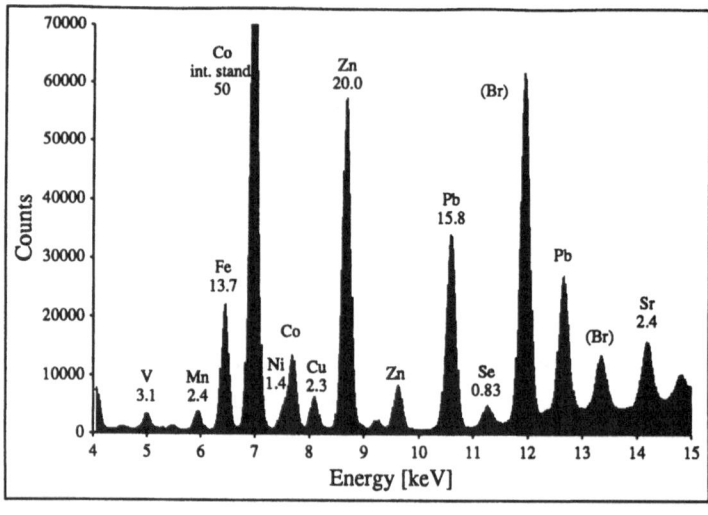

Figure 10. TXRF spectrum of a rainwater sample. Values are given in µg/l.

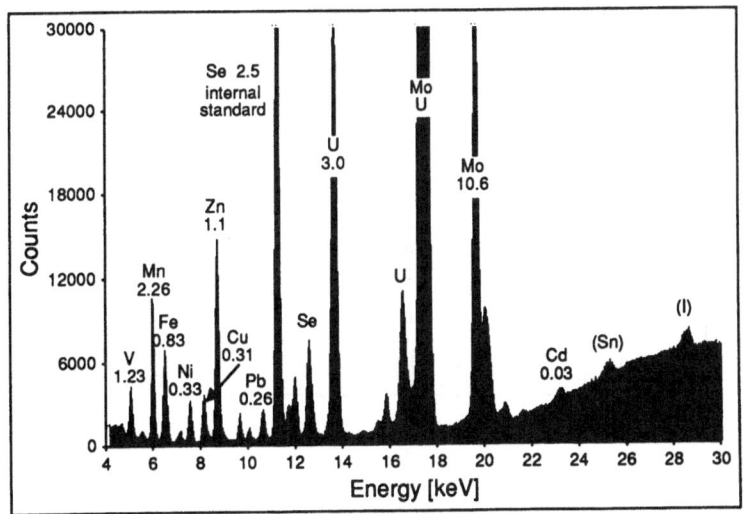

Figure 11. TXRF spectrum of the seawater reference sample CASS-1, excited
by a W-anode. The values are given in ng/l.

elements from the alkaline and alkaline earth bulk prior to measurement, as described earlier
on in the sample preparation techniques. Figure 11 presents a TXRF spectrum of the seawater reference sample CASS-1, exited by a W-anode. The values are given in µg/l and Se was
used as internal standard. The concentration ranges from several ppb down to the low ppt
level. Altogether 12 elements can be determined after separation and concentration. These
are V, Mn, Fe, Co, Ni, Cu, Zn, Se, Mo, Cd, U and Pb [10]. Table 4 gives the TXRF results
in comparison to the reference values, verified by several independent methods. As can be
seen, there is an excellent agreement even in the low ng/l range.

Sewage Sludge. A further application, which is quite different, deals with the determination of trace elements in sewage sludge, an example of a solid sample which had to be di-
gested by means of pressure digestion in PTFE vessels using a mixture of nitric and hydrofluoric acid. Figure 12 shows the resulting TXRF spectra using Mo- and W-excitation. Up to
30 elements have been determined and Sc was used as internal standard. Table 5 presents the
results for the sewage sludge reference material BCR 146. The concentrations given in this
table are in the ppm range and again show a very good agreement. For such samples TXRF
has the advantage of simple quantification by internal standardization.

Another field of application will be presented in the following, where the microanalytical
capability of TXRF is exploited.

Human Hair. This example deals with the trace element analysis of human hair. In this
case a TXRF spectrum of a piece of hair is shown (Figure 13), measured in a none-destruc-
tive manner (lined spectrum) and after oxygen plasma treatment (grey spectrum) which
yields somewhat better results due to the background reduction. Detection limits for a single

Table 4. Results of the seawater reference material CASS-1 [12]

Element	Concentration (µg/kg)					
	Reference values			TXRF (n = 4)		
V	-			1,23	±	0,150
Cr	0,118	±	0,021	-		
Mn	2,27	±	0,170	2,26	±	0,220
Fe	0,873	±	0,076	0,83	±	0,110
Co	0,023	±	0,004	0,025	±	0,001
Ni	0,29	±	0,031	0,33	±	0,035
Cu	0,291	±	0,027	0,315	±	0,027
Zn	0,98	±	0,099	1,1	±	0,010
As	1,04	±	0,070	-		
Pb	0,251	±	0,027	0,259	±	0,023
Mo	-			10,6	±	1,000
Cd	0,026	±	0,005	0,034	±	0,008
U	-			2,89	±	0,200

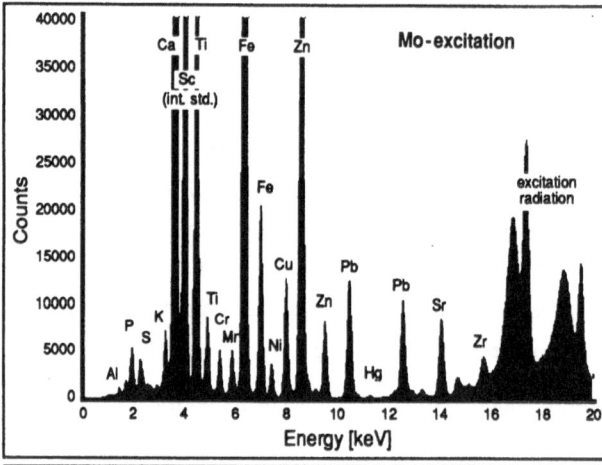

Table 5. Results of the Sewage Sludge reference material BCR 146

Elem.	Unit	TXRF		Ref. values	
S	mg/g	11.2	± 0.2		
K	mg/g	4.48	± 0.24	4.8	
Ca	mg/g	96.8	± 3.5	101.0	
Ti	mg/g	15.9	± 1.2	17.4	
V	µg/g	44.0	± 12.0		
Cr	µg/g	781.0	± 85.0	784.0 ± 37.0	
Mn	µg/g	594.0	± 20.0	588.0 ± 24.0	
Fe	mg/g	19.9	± 0.8	18.5	
Ni	µg/g	296.0	± 15.0	280.0 ± 18.0	
Cu	µg/g	963.0	± 30.0	934.0 ± 24.0	
Zn	mg/g	4.20	± 0.10	4.06 ± 0.09	
Ga	µg/g	8.0	± 2.0		
As	µg/g	19.0	± 5.0		
Rb	µg/g	25.0	± 1.0		
Sr	µg/g	289.0	± 10.0		
Y	µg/g	10.0	± 2.0		
Zr	µg/g	159.0	± 41.0		
Nb	µg/g	14.0	± 2.0		
Mo	µg/g	15.0	± 3.0		
Ag	µg/g	209.0	± 43.0		
Cd	µg/g	80.0	± 8.0	77.7 ± 2.6	
Sn	µg/g	167.0	± 17.0		
Sb	µg/g	112.0	± 18.0		
Ba	mg/g	2.10	± 0.32		
W	µg/g	16.0	± 10.0		
Hg	µg/g	6.2	± 2.0	9.49 ± 0.76	
Pb	mg/g	1.27	± 0.03	1.27 ± 0.03	
Bi	µg/g	36.0	± 4.0		
Th	µg/g	8.0	± 5.0		
U	µg/g	10.0	± 5.0		

Figure 12. TXRF spectra of the Sewage Sludge BCR 146.

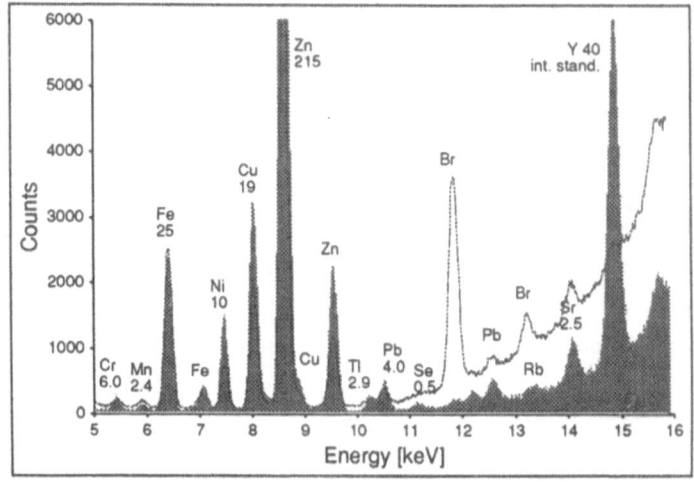

Figure 13. TXRF spectrum of a piece of a human hair (line: direct; grey: ashed).

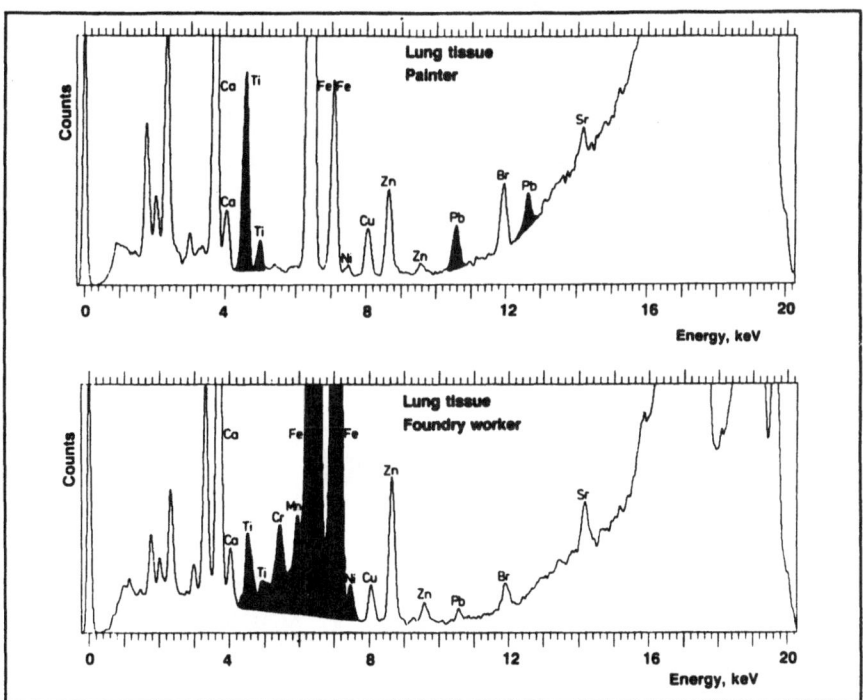

Figure 14. TXRF spectra obtained from lung tissue of a painter (upper spectrum) and from a foundry worker (lower spectrum) [13].

piece of hair of 3mm in length are in the order of 1 to 5 µg/g. Analyzing concentration profiles along a single hair enables the reconstruction of ingestion times and rates, and is thus a
very helpful tool in toxicological studies, clinical medicine or forensic science.

Lung Tissue. Another application, exploiting the microanalytical potential of TXRF, has
been developed by KLOCKENKÄMPER and VON BOHLEN [13,14]. They have investigated microtome sections of human lung tissues. These tissue samples (mostly fixed in solutions of formalin) were frozen and then cut by a microtome to a thickness of about 10 µm
and a diameter of less than 10 mm. The thin slices are prepared on a quartz glass carrier,
spiked with an internal standard, and are ready for measurement after drying under an infrared light. Figure 14 presents two TXRF spectra of human lung tissues of a man exposed

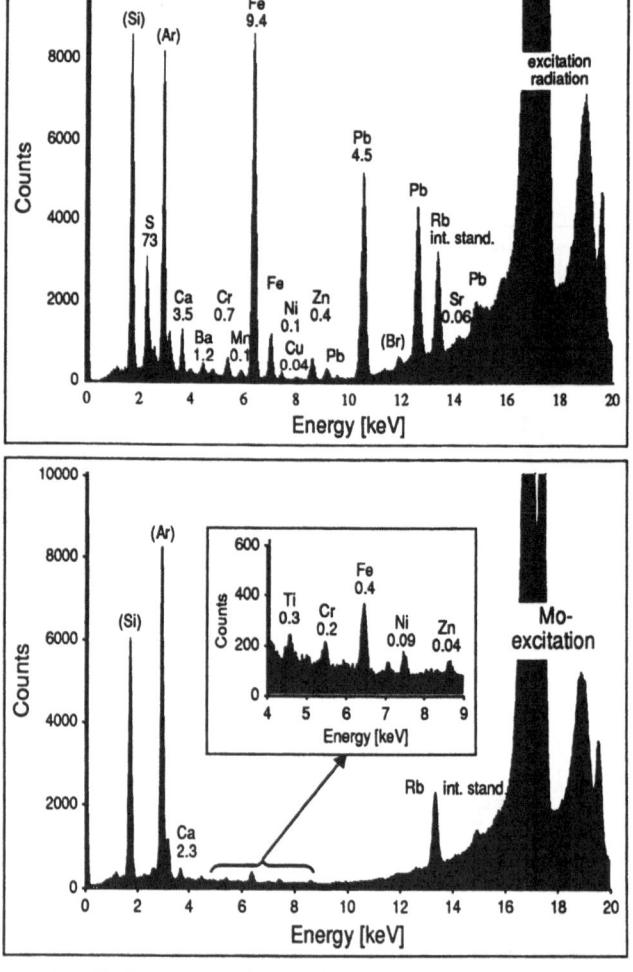

Figure 15. TXRF spectra of hydrofluoric acid before and after additional purification. Values are given in µg/l.

Table 6. Comparison of element concentrations determined in
nitric acid (65%) by ICP-MS and TXRF

Element	ICP - MS		TXRF			
	HNO_3 conc.		HNO_3 direct		HNO_3 enriched	
ng/ml	Mean	Std.Dev.	Mean	Std.Dev.	Mean	Std.Dev.
		n = 3		n = 6		n = 3
S	-		322,64 ± 9,38		334,45 ± 9,18	
K	-		3,92 ± 0,21		3,09 ± 0,07	
Ca	-		3,91 ± 0,34		4,34 ± 0,06	
Ti	0,90 ± 0,20		0,57 ± 0,12		0,37 ± 0,02	
V	-		< 0.08		0,04 ± 0,01	
Cr	0,95 ± 0,01		0,87 ± 0,04		1,00 ± 0,01	
Mn	0,26 ± 0,01		0,42 ± 0,02		0,39 ± 0,01	
Fe	28,60 ± 0,30		34,25 ± 0,36		34,96 ± 0,28	
Co	0,27 ± 0,01		0,34 ± 0,05		0,29 ± 0,04	
Ni	2,50 ± 0,06		2,02 ± 0,06		2,07 ± 0,01	
Cu	0,70 ± 0,04		0,85 ± 0,03		0,96 ± 0,01	
Zn	3,00 ± 0,05		4,40 ± 0,14		4,41 ± 0,03	
Ga	0,04 ± 0,01		< 0.05		< 0.01	
Sr	0,04 ± 0,01		0,05 ± 0,01		0,04 ± 0,01	
Zr	-		0,30 ± 0,09		0,13 ± 0,01	
Mo	-		-		0,15 ± 0,01	
Ag	0,01 ±		< 0.1		< 0.05	
Cd	0,05 ± 0,01		< 0.15		0,06 ± 0,02	
Ba	0,90 ± 0,02		1,04 ± 0,13		0,96 ± 0,04	
Tl	0,02 ±		< 0.05		< 0.02	
Pb	1,06 ± 0,01		1,16 ± 0,04		1,10 ± 0,01	
Bi	0,07 ± 0,01		0,05 ± 0,02		0,06 ± 0,01	

to dust. The upper spectrum was obtained from the lung tissue of a painter. During his professional live, he had probably inhaled large amounts of paint and varnish in the form of aerosol particles. Peaks of the significant components of many paints and varnishes, titanium and lead, are evident. The lower spectrum was obtained from the lung tissue of a foundry worker. An exogene siderosis could be diagnosed medically, based only on histological results. A further identification of the inhaled aerosol particles could be done by TXRF. Besides an excessive amount of iron in relation to the normal level of iron in the haemoglobin, a significant increase in titanium, chromium, manganese and nickel is also visible. Because of the very short time needed for this method, it could be a helpful tool in pathological and occupational medicine.

Ultrapure Acids. The determination of trace element impurities in ultrapure reagents is another, quite different application and the last example of the use of TXRF for chemical trace analysis [15]. TXRF spectra of a hydrofluoric acid sample are displayed in figure 15 before and after additional purification. The upper spectrum was obtained from a commercially available Suprapure® grade reagent (Merck, Darmstadt). Strong peaks are visible for S (73 ppb), Fe (9.4 ppb) and Pb (4.5 ppb) which are drastically reduced after sub-boiling destillation. Rubidium was used as internal standard. One can still evaluate small peaks in the lower spectrum, which is demonstrated in the window of the marked section. For instance the Ni-peak refers to a concentration of 90 ppt.

To ensure the reliability of the results obtained by TXRF, intercomparisons with ICP-MS have been performed. Table 6 presents the results from a sample of concentrated nitric acid. The ICP-MS analysis was performed with a VG PlasmaQuad 2 by Dr. Nachstedt from Rie-del-de Haën AG, Seelze. His group performs the analysis of acids routinely, particularly with the ICP-MS technique. For the TXRF results, two rows are displayed, relating to the di-rect measurement of a 100 μl aliquot and a determination after preconcentration by evapora-tion on a hot plate under a nitrogen stream. Both preparation techniques compare well. The results for most of the elements are also in good agreement with the ICP-MS results, espe-cially for the high mass elements. For elements such as Ti, Mn and Fe, differences might be explained by molecular ion interferences in the MS spectra which are difficult to correct for.

SURFACE AND THIN LAYER ANALYSIS

This latter example is a good transition point to the second part of this paper, namely the use of TXRF in surface and thin layer analysis, because ultrapure acids, such as HF and HNO_3, are also used in the purification of silicon wafers. Traces of metals in the surface layers of silicon wafers are known to seriously affect both performance and yield of integrat-ed circuits. For the processing of wafers it is of interest to know whether contaminants are located *on*, or embedded *in*, the surface of the wafer.

When discussing chemical trace element analysis in the section before, we only dealt with small quantities of residues - or expressed in another way: particulate type residues - on the surface of appropriate sample carriers utilizing their *high reflectivity*. As mentioned at the

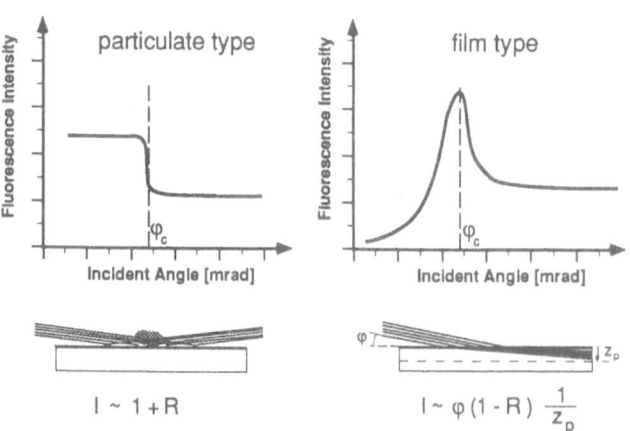

Figure 16. Angle dependent fluorescence intensity of atoms embedded in a near surface layer compared with atoms on the surface.

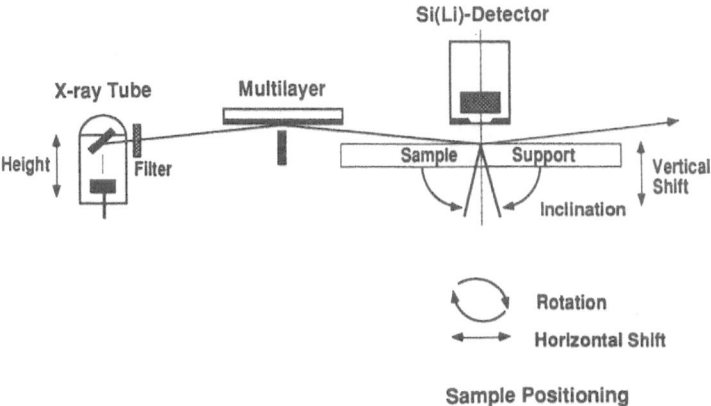

Figure 17. Schematic view of the instrumental set-up for surface analysis.

beginning of this paper, total reflection is also characterized by *low penetration depth*. The fluorescence intensity of atoms which are embedded in a near-surface layer is, however, completely different from the fluorescence radiation of atoms, which are obviously located on the surface. This is demonstrated in figure 16.

The incident intensity which excites particulate residues is proportional to $1 + R$, where R denotes the reflectivity. For angles below the critical angle, the reflectivity is near unity, hence the count rate is approximately doubled compared to angles above the critical angle. These atoms are excited by the incident as well as by the reflected beam. For homogeneously distributed atoms in surfaces a strong angular dependence is given, especially below the critical angle which is in contrast to when particulates are actually on the surfaces. Atoms in homogeneous surfaces are excited by the unreflected part $1 - R$, which is multiplied by two factors which take into account the refraction of the incident X-rays and also the angle dependent irradiance. Thus the primary intensity is nearly four times enhanced at the critical angle [16].

Instrumental set-up. In order to meet the requirements of an accurate angle setting for surface applications, TXRF instrumentation and the quantification software had to be further developed. In our apparatus, which is shown schematically in figure 17, the X-ray source is a 2 kW fine focus tube with a Mo-anode and a Cu-anode, respectively. Spectral purity is established by a multilayer mirror in the path of the incident X-rays. The sample is fixed onto the sample support which is mounted at three points to stepping motors, thus allowing for a vertical shift and inclination. The angle setting of an absolute accuracy of 0.05 mrad has been achieved by these stepping motors which are controlled by a software package using the fluorescence signals from the Si(Li)-detector. Thus a further detector is not required [17]. To reach every point of the surface, the sample support can be rotated and shifted horizontally in the beam direction.

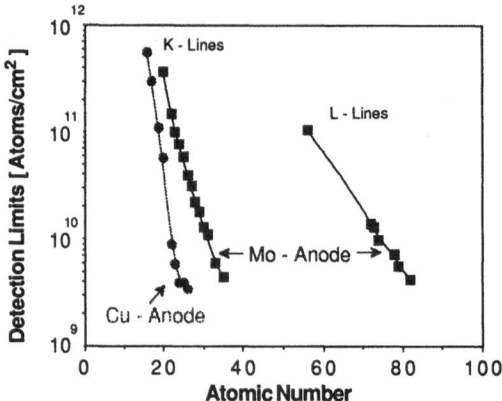

Figure 18. Detection limits for metal impurities on silicon wafers.
Measuring time 1000s.

Detection Limits. Figure 18 presents detection limits achieved by the improved instrumental set-up for metal impurities on silicon wafers. For elements of atomic number below Fe
(Z=26) a Cu-anode is used, otherwise a Mo-anode is used to attain detection limits better
than 10^{10} atoms/cm^2.

Quantification. For a quantification of wafer contaminants, residues of standard solutions
on a wafer are measured in a previous step, resulting in a particulate type curve. A comparison of this curve with the film type curve obtained from the contaminants of the wafer under
investigation leads to a quantification of the wafer contaminants.

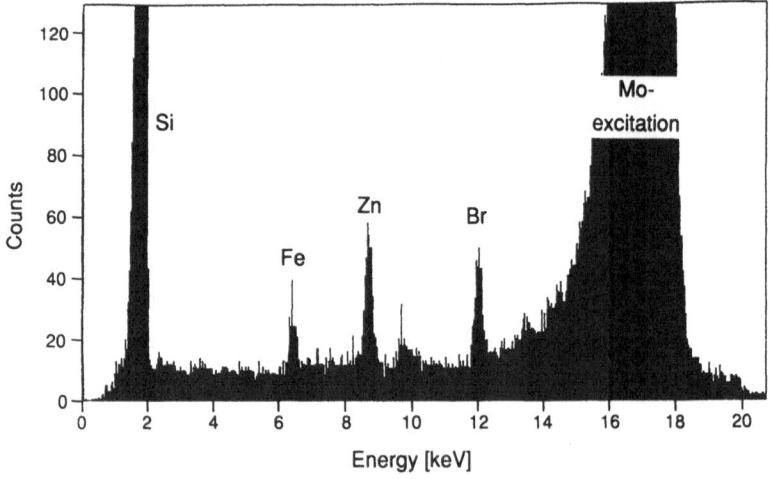

Figure 19. TXRF spectrum of a silicon wafer contaminated with Fe, Zn and Br.

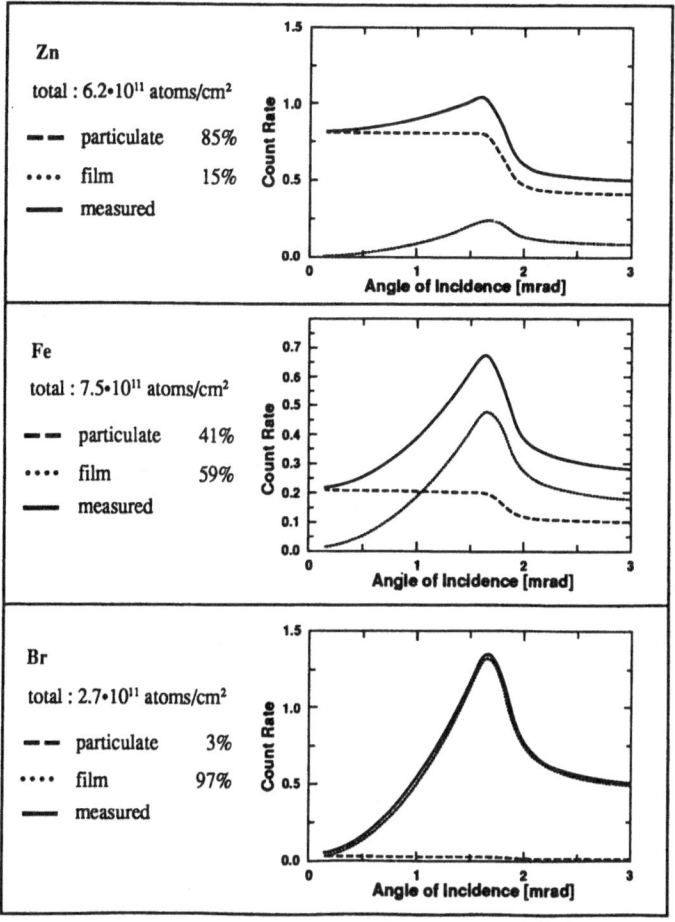

Figure 20. Different types of wafer contaminations embedded *in* the silicon surface and located *on* the surface.

Application of TXRF for surface analysis

Wafer contamination. The following example demonstrates the determination of contaminants on and in a silicon wafer, respectively. Figure 19 shows the TXRF spectrum of Fe, Zn and Br contaminations, measured at the grazing incidence of the primary beam as a function of the incident angle (in this case 1.3 mrad). At a first glance the single spectrum appears to show three similar element contaminations on a silicon wafer. However, using a specially developed program which is based on the Fresnel formalism and including regression procedures for quantitative estimates we are now able not only to quantify the total amount of atoms per area, but also differentiate between the different types of contamination mentioned above. The experimental data are displayed in figure 20 as a sum (solid line) of two components which represent the different types of contamination. The dashed line represents the particulate type (atoms located on the surface), and the dotted line the film type (atoms embedded in the surface). It has been determined that the total amount of for example $6.2 \cdot 10^{11}$ atoms/cm^2 for Zn consists of up to 85 % of a particulate and up to 15 % embedded contamination. For Fe only 41 % are particulate and 59 % are film type contaminations. Br is almost 100 % embedded. The 3% particulate component is due to the calculation and is within the analytical error of about 5 %.

Layered Structures. The last application example takes us away from the determination of trace elements. It deals with the analysis of layered structures which will be of increasing importance in the application of TXRF.

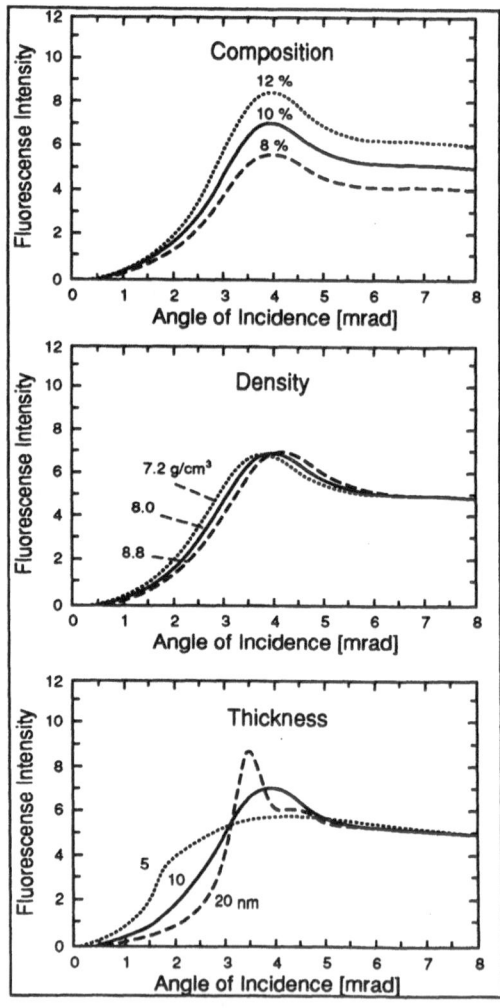

Figure 21. Influence of the angular dependence of the fluorescence intensity for layer composition, density and thickness.

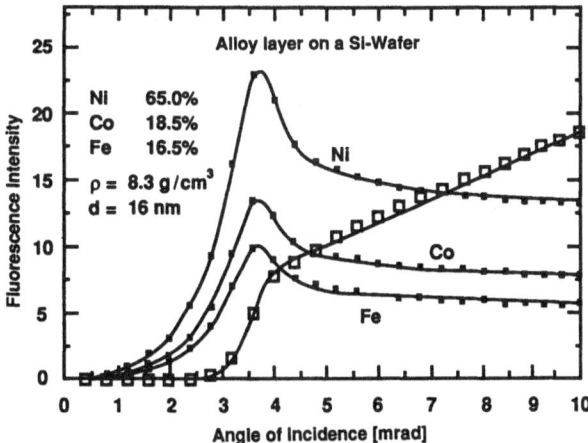

Figure 22. Alloy layer on a silicon wafer. The solid lines are the result of model calculations [18].

Layer parameters such as elemental composition, thickness and density have a strong influence on the angular dependence of the fluorescence yield. This is displayed in figure 21. The precise evaluation of parameters and the quantification of signals can only be achieved by model calculations. WEISBROD et al. [17, 18] as well as DE BOER [19] have developed software to calculate and predict fluorescence yields of arbitrarily composed stratified matter. The calculation procedure comprises the basic equations relating to the propagation of X-rays in absorbing media at grazing incidence. In addition it makes use of the fundamental parameter technique and includes interelement enhancement.

The last example describes an alloy layer on a silicon substrate (Figure 22). Calculated fluorescence yields (solid lines) are compared with the experimental data. Starting from an expected set of values, the layer parameters have been refined iteratively until the theoretical curve was in good agreement with the measured data. An absolute calibration, using a pure element standard, had been carried out to make the calculation comply with a correct quantitative analysis. The composition and mass fractions were determined to be Ni = 65.0%, Co= 18.5% and Fe = 16.5%. The density of the layer, basically deduced from the angular location of the bank of curve, was found to be ρ = 8.3 g/cm^3 with a thickness of 16 nm.

CONCLUSION

In conclusion it can be stated that total reflection X-ray fluorescence has greatly surpassed the respective capabilities of conventional EDXRF. For both fields of application, that is in chemical trace element analysis as well as in surface analysis, a substantial progress has been achieved. Totally reflecting sample carriers can be used in XRF to achieve detection limits in the pg- or ppt-range, respectively. The high detection power, the simple quantification by internal one-element standardization, the unique microanalytical potential and of course the multielement capability affords XRF a new, high rank in atomic spectroscopy. For various applications TXRF has clear advantages compared to ICP-MS.

Furthermore, in surface analysis ultra thin surface layers are accessible with TXRF with a probing depth in the nanometer range. The method is suitable for determining layer characteristics such as composition, density and thickness from the angular dependent fluorescence signal in a non-destructive manner. In the critical angle regime TXRF offers, for favourable cases, the advantage of combining surface sensitivity with depth analysis.

REFERENCES

1.   Yoneda, Y. and Horiuchi, T., Optical Flats for Use in X-ray Spectrochemical Micro-analysis, Rev. Sci. Instr. 42: 1069 (1971)

2.   Aiginger, H. and Wobrauschek, P., A Method for Quantitative X-ray Fluorescence Analysis in the Nanogram Region, Nucl. Instr. Methods 114: 157 (1974)

3.   Knoth, J. and Schwenke, H., An X-ray Flourescence Spectrometer with Totally Reflecting Sample Support for Trace Analysis at the ppb Level, Fresenius Z. Anal. Chem. 291: 200 (1978)

4.   Schwenke, H. and Knoth, J., A Highly Sensitive Energy-Dispersive X-Ray Spectrometer with Multiple Total Reflection of the Exciting Beam, Nucl. Instr. Methods 193: 239 (1982)

5.   Freitag, K., Energy Dispersive X-Ray Fluorescence Analysis with Multiple Total Reflection - An Improvement of Detection Limits, in: B. Sansoni (Ed.) Instrumentelle Multielementanalyse, VCH Verlagsgesellschaft, Weinheim (1985) pp. 257 - 268

6.   Eichinger, P., Rath, H.J. and Schwenke, H., Application of total reflection X-ray fluorescence analysis for metallic trace impurities on silicon wafer surfaces, ASTM Spec. Tech. Publ. 990: 305 (1989)

7.   Underwood, J. H., Glancing Incidence Optics in Astronomy, Space Sci. Instr. 1: 289 (1975)

8.   Streli, C., Wobrauschek, P. and Aiginger, H., Light Element Analysis with TXRF, this volume

9.   Prange, A., Total reflection X-ray spectrometry: method and applications, Spectrochim. Acta 44B: 437 (1989)

10. Prange, A., Knöchel, A. and Michaelis, W., Multi-Element Determination of Dissolved Heavy Metal Traces in Sea Water by Total-Reflection X-Ray Fluorescence Spectrometry, Anal. Chim. Acta 172: 79 (1985)

11. Stößel, R.-P. and Prange A., Determination of Trace Elements in Rainwater by Total-Reflection X- Ray Fluorescence, Anal. Chem. 57: 2880 (1985)

12. Prange, A., Knoth, J., Stößel, R.-P., Böddeker, H. and Kramer, K., Determination of Trace Elements in the Water Cycle by Total-Reflection X-Ray Fluorescence Spectrometry, Anal. Chim. Acta 195: 275 (1987)

13. von Bohlen, A., Klockenkämper, R., Otto, H., Tölg, G. and Wiecken, B., Qualitative survey analysis of thin layers of tissue samples - Heavy metal traces in human lung tissue, Int. Arch. Occup. Environ. Health 59: 403 (1987)

14. Klockenkämper, R., von Bohlen, A. and Wiecken, B., Quantification in total reflection X-ray fluorescence analysis of microtome sections, Spectrochim. Acta 44B: 511 (1989)

15. Prange A., Kramer K. and Reus U., Determination of trace element impurities in ultra-pure reagents by total reflection X-ray spectrometry, Spectrochim. Acta 46B: in press (1991)

16. Iida, A., Sakurai, K., Yoshinaga, A. and Gohshi, Y., Grazing Incidence X-ray Fluorescence Analysis, Nucl. Instr. Methods A246: 736 (1986)

17. Weisbrod, U., Gutschke, R., Knoth, J. and Schwenke, H., Total Reflection X-ray Fluorescence Spectrometry for Quantitative Surface and Layer Analysis, Appl. Phys. A53: 2031 (1991)

18. Weisbrod, U., Gutschke, R., Knoth, J. and Schwenke, H., X-ray induced fluorescence spectrometry at grazing incidence for quantitative surface and layer analysis, Fresenius Z. Anal. Chem. 341: 83 (1991)

19. de Boer, D.K.G. and Leenaers, A., XRF and TXRF of (Multiple) Thin Films, X-Ray Spectrometry, in press (1991)

TXRF WITH VARIOUS EXCITATION SOURCES

Peter Wobrauschek, Peter Kregsamer, Christina Streli,
Robert Rieder and Hannes Aiginger

Atominstitut der Österreichischen Universitäten
Schüttelstraße 115, A 1020 Wien, Austria

ABSTRACT

Improving the detection limits in TXRF by optimizing the excitation
conditions is the goal of this work. The properties of the exciting radiation
due to spectral distribution, polarisation, intensity and energy are
investigated and compared to find best conditions. Results are given from
experiments performed with synchrotron radiation, Bragg polarized monoenergetic
x-rays, high energy cut-off reflector in the primary beam path of a high power
x-ray tube and several geometries for the sample reflector.

INTRODUCTION

The features and the working principle of Total Reflection X-Ray
Fluorescence Analysis (TXRF) have been described in many papers.[1-7] In order
to improve further the lower limits of detection (LLD) in TXRF, still some
physical and technical ideas can be introduced and realized.

Collimation system

Optimization of the experimental setup can be done by minimizing the
distances between source - sample - detector. The minimum distance from source
to sample is limited because the collimation system requires a certain length
which is determined by the slit width and the allowed divergence of the beam.
As some typical values, 0.05 mm slit width and 100 mm distance between the two
slits are adequate. Double reflector[2] and double plate[1] collimator systems were
tested and results are compared later. The distance from sample to detector is
already as close as possible in TXRF, about 1 mm.

Sources of excitation

Due to the dimensions of the collimation system a line shape of the beam
is required to pass with minimum losses through the collimator. The ideal
source is undoubtedly synchrotron radiation.

All desired features of an ideal source are combined: high brilliance, natural collimation, linear polarisation and optionally the continuous spectral distribution can be monochromatized by Bragg crystals, of course at the cost of intensity. The only disadvantage is the restricted availability of beam time for research groups so routine analysis can not be performed.

Laboratory scale solutions have been developed with compromises to the ideal properties of a source. Increase in brilliance from x-ray tubes can be achieved by using rotating anodes instead of fixed anodes. For Mo and W targets 625 W/mm^2 are the maximum load for fixed anodes whereas up to 35000 W/mm^2 can be dissipated with rotating anodes[8]. The choice of the anode material with respect to the elements to be analysed is straightforward, but sometimes a compromise like excitation with W-L lines, instead of Cu K lines, leads to acceptably good results and versatility for other elements as well. The use of a Au anode and operating this tube with 100 kV leads to new aspects in excitation conditions, because the K-series of higher Z elements can be excited. This leads to improved conditions for the evaluation of the spectral data, because the severe line overlap in the case of L-lines is avoided.[1]

The use of a windowless tube with Cu anode and standard Cr anode tube for the improved excitation of low Z elements has been succesfully tested and is described in a separate contribution in this volume.

A monoenergetic beam produced by Bragg reflection on single crystals[9], or only suppression of high energy photons by using a cut-off reflector[1], leads to improved background conditions. Single crystals like LiF have an integrated reflectivity R in the order of 10^{-3} to 10^{-4}, respectively peak diffraction coefficients of 0.1 - 0.5. Multilayer synthetic microstructures of e.g. W-Si or W-C have comparatively higher peak diffraction coefficients in the range of 0.8 increasing the excitation intensity of the sample thus leading to improved LLD[10,11].

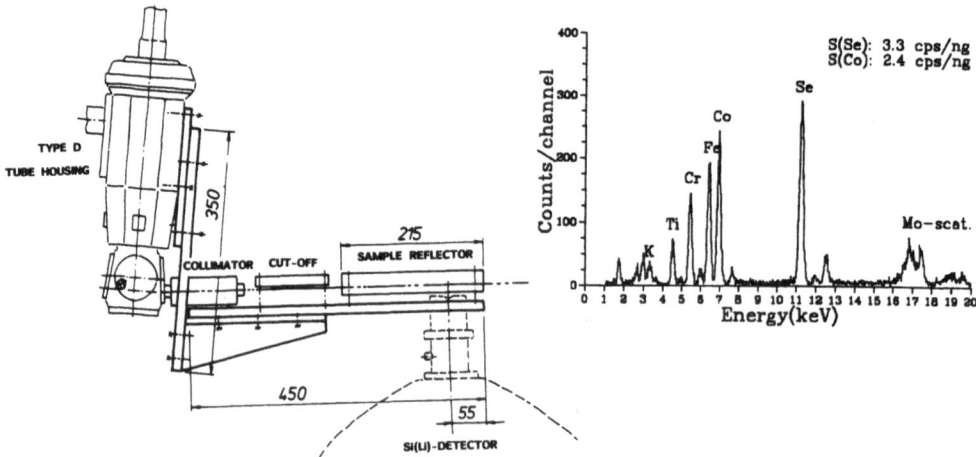

Fig.1 Scheme of the TXRF module and a spectrum of a multielement standard. The sample was 1 μl aqueous solution with 10 ppm of each of the elements K,Ti,Cr,Fe,Co,Se. Excitation conditions: Mo-anode / 45 kV / 15 mA, measuring time: 100 s.

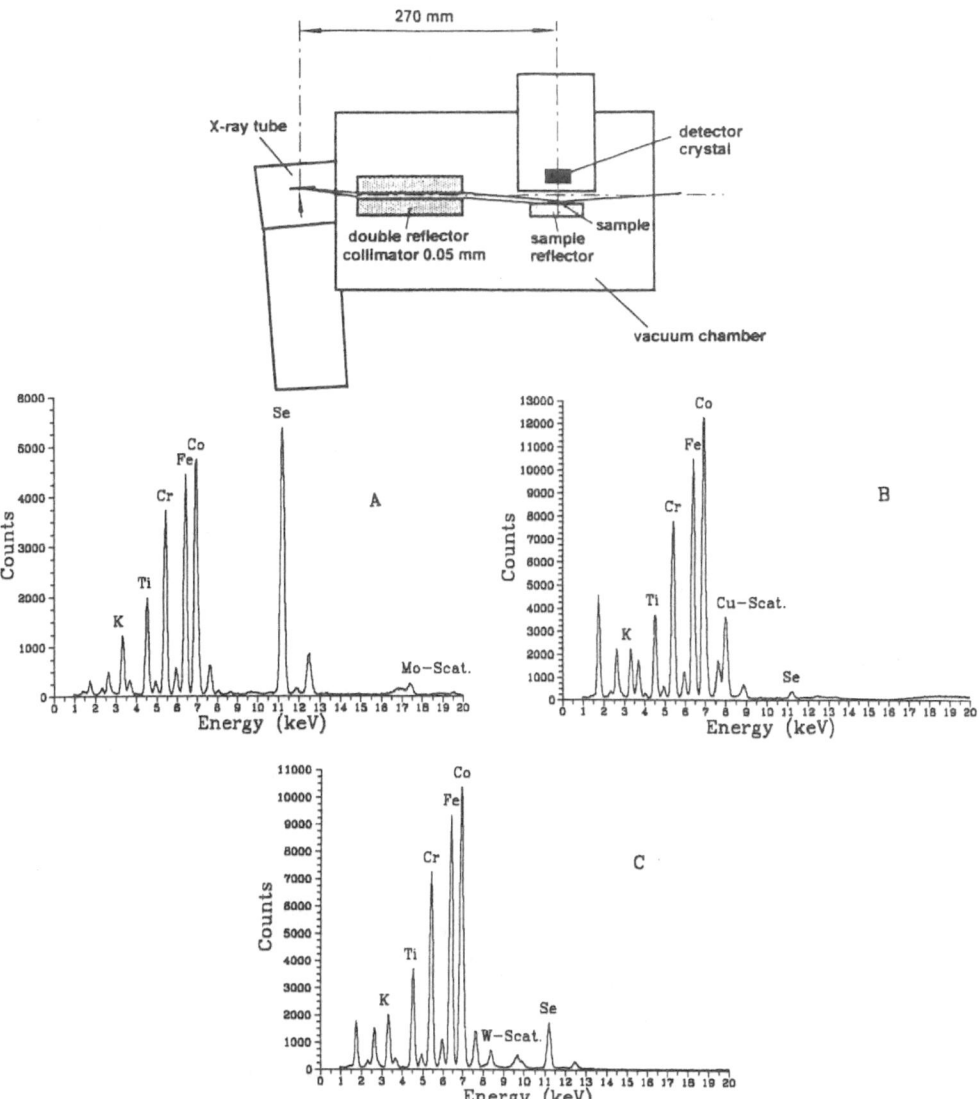

Fig.2 Scheme of the double reflector collimator and spectra of a multielement standard excited with A) Mo-, B) Cu-, C) W-anode tubes. In all cases the cut off energy was 20 keV and operating conditions 45 kV, 10mA, 250 s. Sample composition: 5 μl, 10 ppm (50 ng) of each of the elements K,Ti,Cr,Fe,Co,Se.

EXPERIMENTAL

A compact module which can be attached to standard tube housings for high power diffraction tubes has been developed[2]. The schematic experimental setup and a typical spectrum obtained with that system is given in Fig. 1 where also the sensitivity for some elements can be read.

The use of two reflector plates for collimating the beam and producing the cut off effect has been described and published[4]. The scheme and spectra

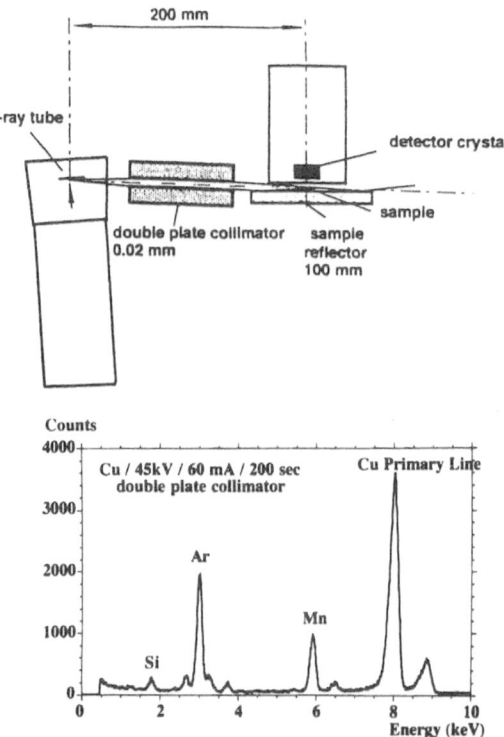

Fig.3 Scheme of the double plate collimator and a spectrum from a Mn sample where extrapolated 1000s LLD of 2 pg were obtained.

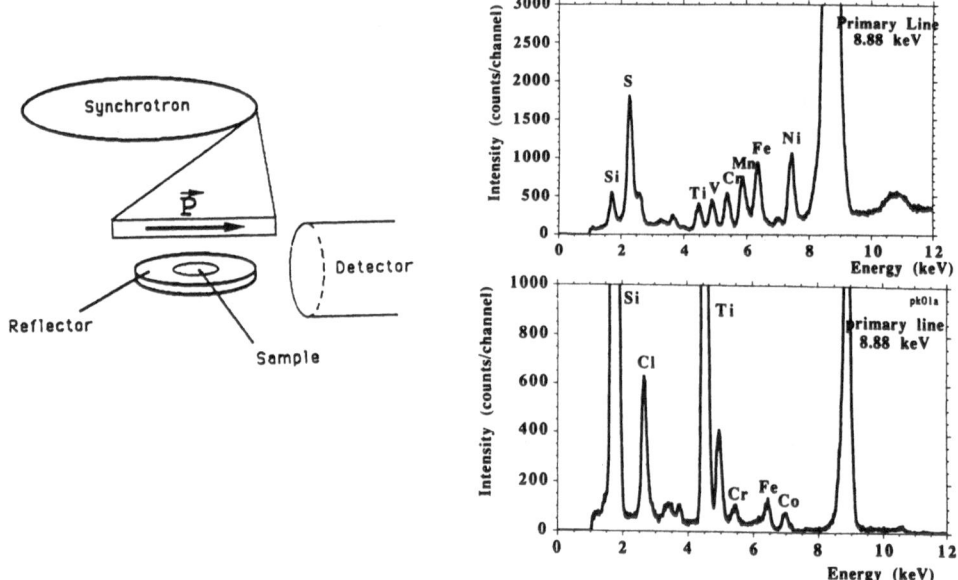

Fig.4a Scheme of excitation with synchrotron radiation, horizontal reflector and spectra from a multielement oil standard (30 ppm of each element) and a triple element standard with 10 ppm Ti (0.1 ppm Cr, Fe and Co) as internal reference.

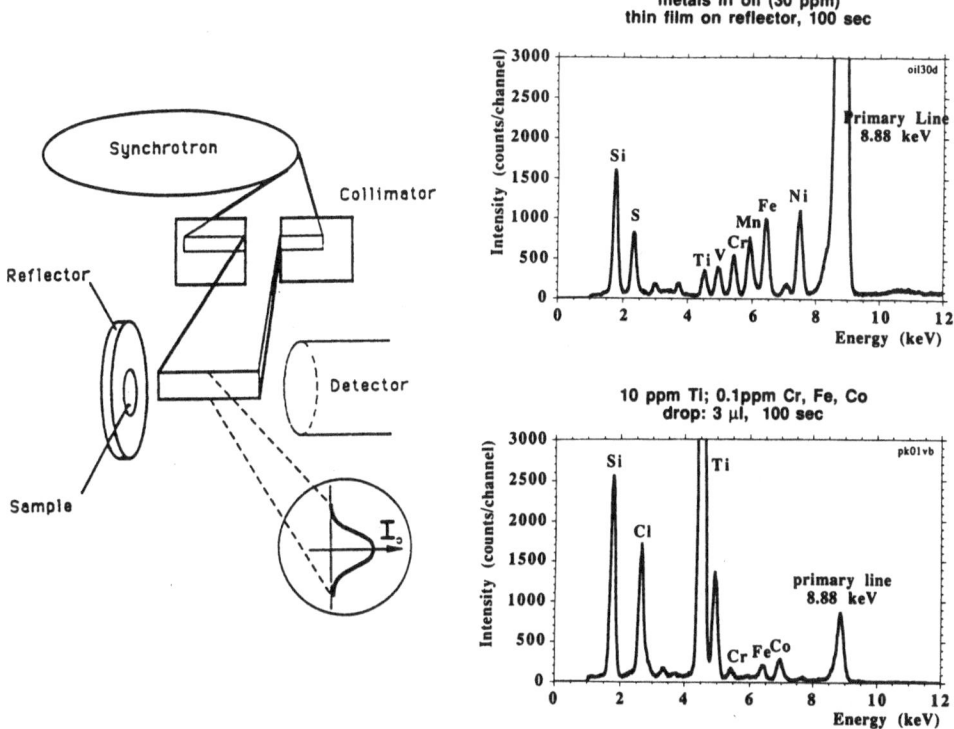

Fig.4b Scheme of excitation with synchrotron radiation, vertical reflector and spectra from a multielement oil standard (30 ppm of each element) and a triple element standard with 10 ppm Ti (0.1 ppm of Cr, Fe and Co) as internal reference.

collected with three different anode materials, Cu, Mo and W are shown in fig. 2.

Replacing the double reflector collimator by a double plate collimator with Ta slits defining the beam divergence leads to the experimental setup given in Fig. 3 where also a spectrum is shown. Details of the setup where described in Reference 1. It is worth noting that the geometry of the reflector was changed from of a disc of 30 mm diameter to a rectangle 100 mm x 30 mm.

In a series of experiments with synchrotron radiation two excitation geometries under total reflection condition were investigated. Two types of samples were analyzed: Aqueous or acidic solutions where matrix removal is no problem and samples with remaining matrix like oil. In Fig.4 the experimental scheme can be seen. The detector is always placed in that way to fully use the polarization effect, thus lowest scatter from sample and substrate is expected. Spectra collected with the SSRL - beam line 10 - 2 at Stanford are shown in the same figure. All experiments were performed in a recently designed vacuum chamber with fully remote control of all necessary motions to adjust total reflection on translation and rotation stages. Details on that system will be published elsewhere.

Tab. 1   Extrapolated LLD for different excitation modes

Collimator type	Operating rate	LLD 1000 s (pg)
Double reflector	Cu, 40 kV, 60 mA	Mn 5
Double reflector	Mo, 55 kV, 32 mA	Sr 22
Double plate	Cu, 45 kV, 60 mA	Mn 2
TXRF module	Mo, 45 kV, 20 mA	Sr 30

Tab. 2   (Sensitivities extrapolated to 1000 s and I = 50 mA)

DETECTOR	REFLECTOR	LLD for Ni in oil ng/g	SENSITIVITY cps/(μg/g)
horizontal	horizontal	50	12
horizontal	vertical	150	4

(Sample consists of 30 ppm of metals present in oil. No matrix removal. Thin film placed on reflector.)

DETECTOR	REFLECTOR	LLD for Co in water pg	SENSITIVITY cps/ng
horizontal	horizontal	5	50
horizontal	vertical	3	100

(Sample consists of 0.1 ppm of metals present in water. Matrix removal by evaporation. 3 μl drop placed on reflector.)

RESULTS

The LLD of the different experimental configurations described and displayed above are given in Table 1.

The LLD from the experiments with SSRL - beam line 10-2 are summarized in Table 2.

CONCLUSIONS

The use of synchrotron radiation gives impressive results, especially if due to the sample structure, special preparation techniques can not be applied, so the matrix is suppressed either by the polarisation effect for low Z elements or by the sharp tunability to an energy value just below the respective absorption edge of the high Z matrix. As an interesting remark the use of Bragg polarized x-rays produced with a Cu anode tube operated at 2400 W for the analysis of trace elements in oil yields LLD which are only worse by a factor of 5 compared to synchrotron radiation[1]. TXRF can be seen as an analysis technique well-suited for almost all elements of the periodic system, where, at optimized conditions, LLD in the sub picogram range or in concentration levels of sub ppb can be reached. In many applications the advantage of a required sample volume of a few microliters is appreciated.

ACKNOWLEDGEMENT

Parts of the work were supported by grants of the "FONDS zur FÖRDERUNG der Wissenschaftlichen Forschung" projects   P 7115 and P 8490. The hospitality of the staff in Stanford at SSRL beam line 10 - 2 is gratefully acknowledged.

REFERENCES

1   P.Wobrauschek, P.Kregsamer, C.Streli, H.Aiginger
    Adv.X-Ray Anal.34, 1 (1991)
2   P.Wobrauschek, P.Kregsamer, Spectrochim.Acta 44B, 458 (1989)
3   A.Prange, Spectrochim.Acta 44B, 437 (1989)
4   H.Schwenke,W.Berneike,J.Knoth,U.Weisbrod Adv.X-Ray Anal.32, 105(1989)
5   P.Wobrauschek,P.Kregsamer,C.Streli,H.Aiginger
    X-Ray Spectrom.20, 23 (1991)
6   A.Iida, Adv.X-Ray Anal.34, 23 (1991)
7   A.Knöchel, Fresenius J.Anal.Chem.337, 614 (1990)
8   Rigaku Catalogue
9   P.Wobrauschek, H.Aiginger, Adv.X-Ray Anal. 28, 69 (1985)
10 M.Schuster Adv.X-Ray Anal. 34, 71 (1991)
11 D.H.Bilderback, B.M.Lairson, T.W.Barbee,Jr., G.E.Ice, C.J.Sparks,Jr.,
    Nucl. Instr. & Meth. 208, 251 (1983)

INSTRUMENTATION FOR TOTAL REFLECTION FLUORESCENT X-RAY SPECTROMETRY

Tadashi Utaka and Tomoya Arai

Rigaku Industrial Corporation
Takatsuki, Osaka, Japan

ABSTRACT

This article describes the instrumentation for a total reflection fluorescent x-ray spectrometer. The reflecting intensity and the angular divergence were studied with respect to various kinds of monochromators. Using silicon wafers, the angular divergence effect of the incident beam, surface roughness influences and the smoothing of background x-ray intensity for the improvement of the lower limit of detection were investigated.

INTRODUCTION

Total reflection x-ray fluorescence (TRXRF) analysis offers several advantages: better sensitivity to trace-element analysis and detection of surface contamination, nondestructive analysis, and short measuring time in analytical chemistry and material-science applications.

The equipment used in this experiment ensures high sensitivity in several ways. A high power rotating-anode x-ray generator is used to increase the x-ray output for fluorescence excitation. For the sake of adoption of the limited wavelength-range of excited x-rays and reduction of background, monochromatized x-rays were used which were selected by using an analyzer. To reduce the background x-rays caused by scattering from the collimating, detecting or positioning devices, the position and the size of the slit located between the monochromator and the source must be determined so that the incident beam strikes only the analyzing area of a sample.

A laser-beam height gauge is used to control the incident angle precisely. The role of the gauge is also to maintain the relationship of the incident angle and sample height during sample translation, allowing identical measurements on different parts of a surface. In order to keep the sample surface clean during x-ray measurement, the analyzing chamber, sample holder, etc., must be assembled in a clean production room free from particle contamination.

In this paper, we report the analysis of the constituents of mono-chromatized x-rays with regard to various monochromators, and examine the relationship between the intensity of fluorescent x-rays and incident angle

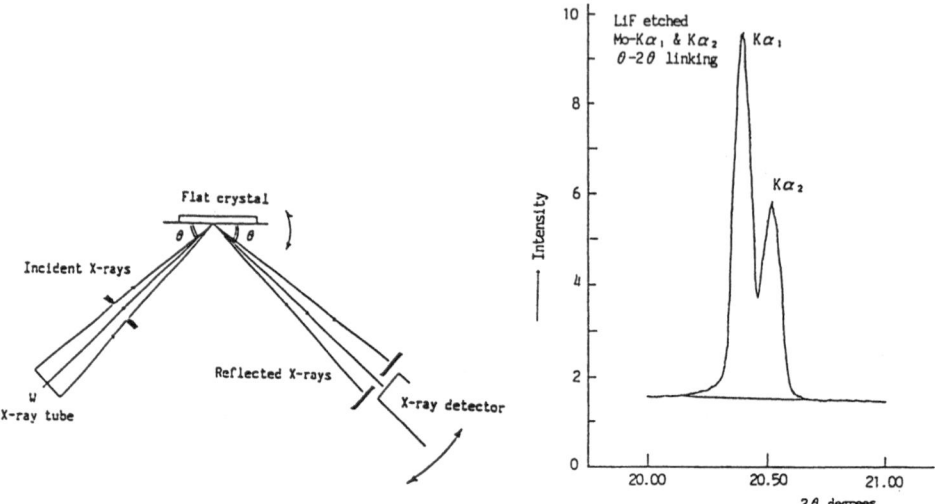

Fig.(1) X-ray optics for estimation
of angular divergence of monochromatic x-rays and their intensities

Fig.(2)-a Angular separation of Mo-Kα₁
and Kα₂ x-rays using an etched LiF
under the condition of θ-2θ linked
scanning operation of a goniometer

for W-Lβ₁ radiation, and the angular divergence obtained from various
surfaces. The effects of surface roughness on fluorescent x-rays are also
described, and the importance of data smoothing is discussed.

MONOCHROMATOR MEASUREMENTS AND EVALUATION

Since the incident angle in grazing-angle measurements is below one
degree, and the angular dispersion of the incident beam should be less than
half of the critical angle for total reflection, a reasonable angular
dispersion of the incident beam for a typical sample would be 0.1 degrees or
less. The distortion of a total reflection curve which shows the dependence

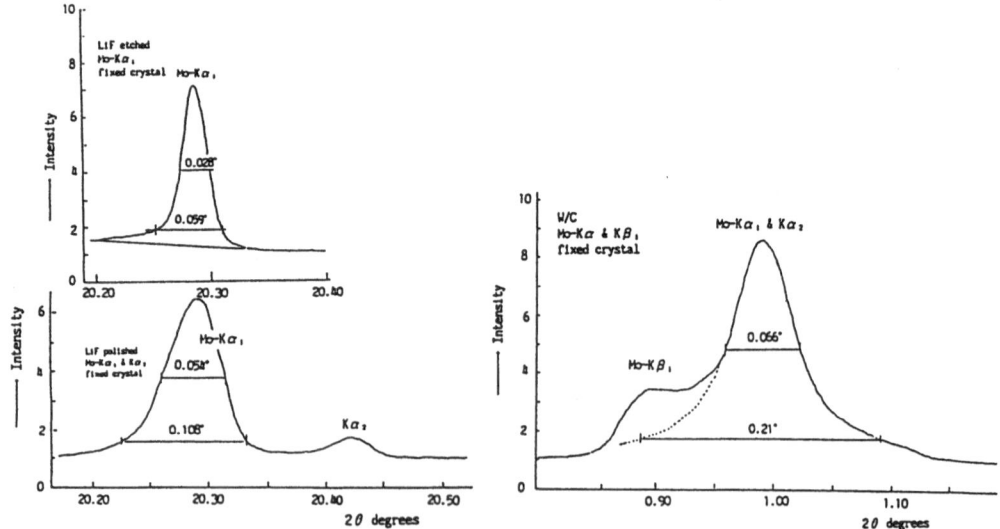

Fig.(2)-b Angular divergence of Mo-Kα₁ and Kα₂ x-rays from a fixed
monochromator by means of counter arm scanning mode

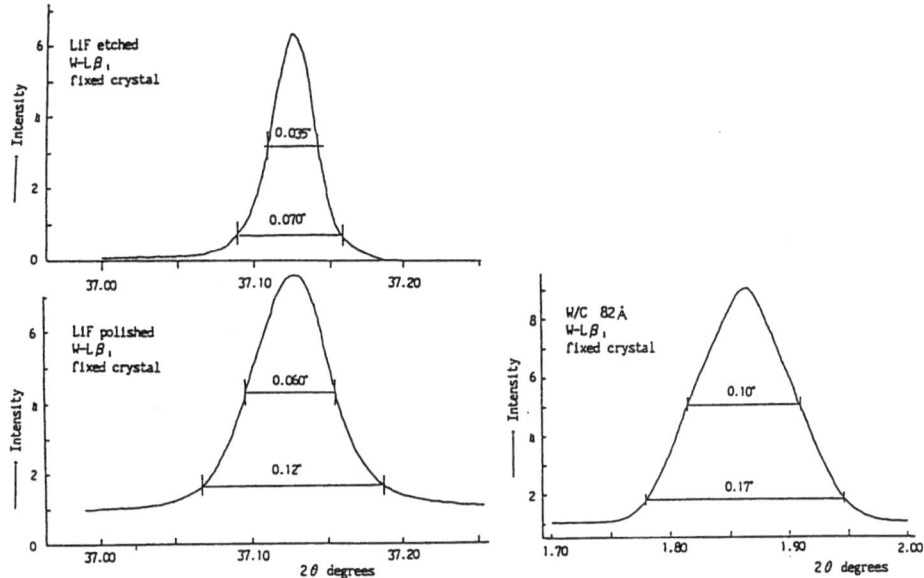

Fig.(2)-c   Angular divergence of W-Lβ_1 x-rays from a fixed monochromator
by means of counter arm scanning mode

of totally reflected x-ray intensity on grazing incident angle depends on the
angular divergence and the spread of the x-ray wavelength in the incident
beam.   These two factors depend on the source size of the x-ray tube, the
monochromator crystal and its geometry, the natural width and separation of
the characteristic x-rays of the anode material, and the x-ray path length to
the sample.   To study these effects, measurements of the angular divergence of
radiation from a monochromator were made with the experimental arrangement
shown in Fig. (1).   Fine focus sealed-off x-ray tubes (with molybdenum and
tungsten targets) were used.   The small spot source of these tubes, combined
with a 0.05 mm slit in front of the detector, causes a geometrical divergence
of 0.015° 2θ full width.   To show the separation of the Mo-Kα_1 and Mo-Kα_2 x-
rays, a θ-2θ linked scanning profile on a cleaved and etched nearly perfect
crystal, is shown in Fig. (2)-a.   To make the measurements for this
instrumentation, first the incident angle to the monochromator crystal was
adjusted to maximize the intensity of the diffracted characteristic x-rays of
interest, and then fixed.   A radiation detector arm with a slit was rotated
around the fixed monochromator to measure the angular distribution of the
diffracted x-rays.   Figs. (2)-b and (2)-c show peak profiles using an etched
LiF crystal, mosaic LiF crystal, and a synthetic W/C multi-layer.

Table (1) lists the measured FWHM's of outgoing beams and the corrected
FWHM's for several crystals, for both Mo-Kα and W-Lβ radiations.   The
corrected angular divergence was calculated taking into account the focal spot
size of the x-ray tube, the natural width of the characteristic x-rays, and
the width of the narrow receiving slit in front of the detector.   For the peak
intensity ratio, the results with the etched LiF was adopted as a standard for
intensity comparison.   The distribution of diffracted x-rays is seen to be
broadened in the mosaic LiF, the graphite, and the synthetic W/C multi-layer.
Because the monochromators featuring a large spacing reduce the angle
separation between Mo-Kα_1 and Mo-Kα_2 x-rays, one can obtain a high intensity
beam sharply defined in energy and angle.   As the natural width of W-Lβ_1 x-
rays is approximately 2.7 times larger than that of Mo-Kα_1 and Mo-Kα_2 x-rays,
it cannot be neglected when using a tungsten target x-ray tube.

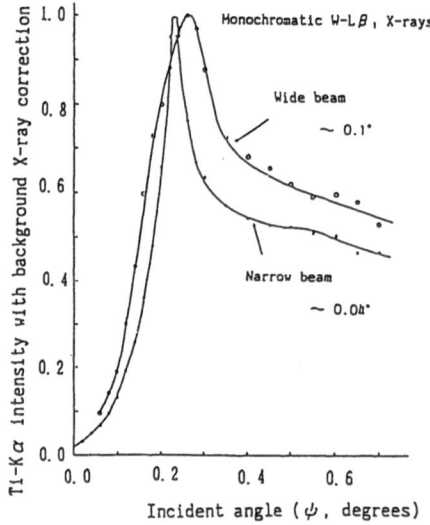

Fig.(3)   Comparison between narrowly
          and widely divergent beams
          against incident angle

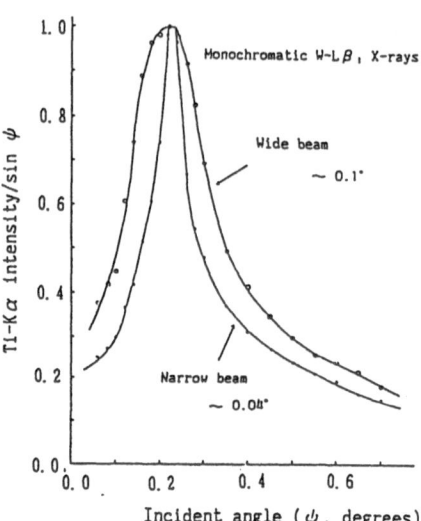

Fig.(4)   Relationship between
          intensity/sinψ and
          incident angle

        The measured divergence FWHM's tabulated in Table (1) represent roughly
the values of incident beam divergence that would be found in a TRXRF
spectrometer arrangement.   The x-ray source selection should be made
considering both the divergence and the peak intensity of the monochromator
crystal.

CASE STUDIES USING THE TRXRF TECHNIQUE

Ultra-thin Metal Films on Silicon Wafers
        Fluorescent x-rays from a titanium evaporated-metal film of 2.5 nm
thickness on a silicon wafer was measured with the excitation of mono-

Table (1)   Relationship among various factors of angular
            divergence for monochromatized x-rays

Crystals	2d Å		Mo-Kα_1 & Kα_2						W-Lβ_1				
				$\Delta(\theta)$ degrees						$\Delta(\theta)$ degrees			
		Bragg angle	α_1-α_1	Kα_1 natural width	50% width		Peak intensity ratio		Bragg angle	W-Lβ_1 natural width	50% width		Peak intensity ratio
					measured	corrected					measured	corrected	
LiF(200) (P)	4.0267	10.15	0.06	0.0039	0.014	0.005	1.0		18.56	0.014	0.018	−0.005	1.0
LiF(200) (M)	4.0267	10.15	0.06	0.0039	0.027	0.023	6.3		18.56	0.014	0.029	0.025	4.3
Ge(111)	6.533	6.24	0.035	0.0034	0.012	0.005	2.1		11.32	0.009	0.014	0.006	3.6
Graphite	6.708	6.08	0.035	0.0033					11.02	0.008	0.042	0.04	3.6
PET	8.742	4.61	0.025	0.0025					8.42	0.007	0.017	0.013	2.8
Mica	20.0	2.05	0.01	0.0011	0.014	0.005	2.8		3.68	0.003	0.010	−0.005	3.0
TAP	25.763	1.08	0.01	0.0009					2.85	0.002	0.012	0.008	6.6
W/C	82.	0.50	0.003	0.0003	0.033	0.025	11.		0.90	0.0007	0.05	−0.05	26.

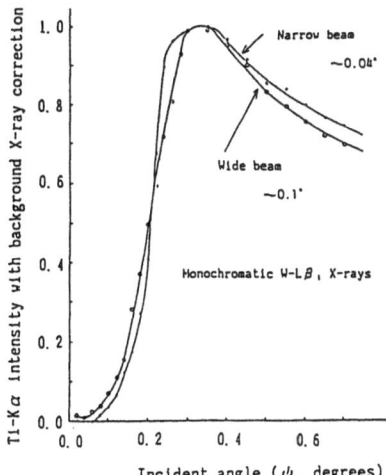

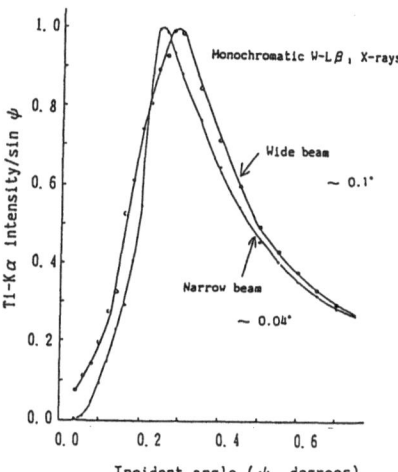

Fig.(5) Comparison between narrowly and widely divergent beams against incident angle

Fig.(6) Relationship between intensity/sinψ and incident angle

chromatized W-Lβ_1 radiation, with results shown in Fig. (3). A changeable slit located between the monochromator and the sample was used to collimate the incident beam to either 0.04 or 0.10 degrees FWHM. These two collimations will henceforth be called narrowly divergent (ND) and widely divergent (WD), respectively. As seen in the ND case in Fig. (3), the intensity maximum is shifted relative to the WD beam, and the peak profile is sharper. Fig. (4) shows the measured intensity divided by sin ψ in order to eliminate the expansion of the irradiating area corresponding to the incident angle reduction. With this correction, the WD intensity maximum is shifted to the lower angle side, now coinciding with the ND maximum. A titanium evaporated-metal film 10 nm thick on a silicon wafer was also measured with both collimations, with results shown in Fig. (5). The measured intensity divided by sin ψ is shown in Fig. (6). The differences between the ND and WD results shown in Figs. (5) and (6) are clearly smaller than those observed with the 2.5 nm titanium film sample.

Contaminant Levels on Silicon Wafers

The levels of artificially deposited zinc, copper, nickel and iron contamination (spin coated at 5×10^{13} atoms/cm^2) on a silicon wafer were measured by ND and WD beams, with the results shown in Fig. (7). The fluorescent x-ray intensity maximum in the ND case is shifted to the lower angle side, and the profile is sharper than that of the WD x-ray beam. Clearly the contaminants are easily detected with either beam. More detailed analysis such as depth profiling could, however, require the sharper peak profiles obtained with the ND.

Surface Roughness

Surface roughness effects on similarly contaminated silicon wafers were investigated by measuring the dependence on the incident angle of scattered W-Lβ_1 radiation, Si-Kα fluorescent x-rays, and fluorescent x-rays from the contaminating elements. Each wafer had a rough and a smooth surface, both coated with contaminants. The results are shown in Figs. (8) and (9). Fig. (8) shows the normalized intensities of the W-Lβ_1 scattered radiation and the Si-Kα fluorescent x-rays for several wafers. Below the critical angle, both intensities show a strong dependence on surface quality; for the smooth

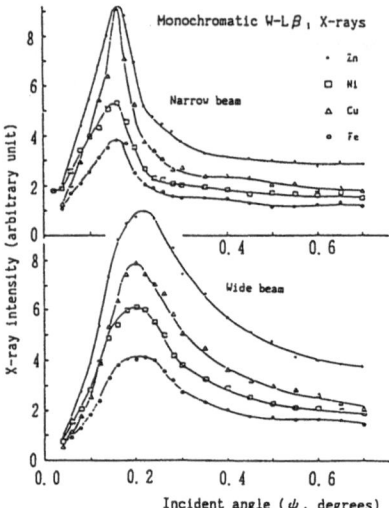

Fig.(7)   Effects of narrowly and
widely divergent x-ray beams
of contamination elements on
a silicon wafer

surface, they drop off quickly below the critical angle; total reflection
greatly reduces interaction with the bulk of the sample.   The intensities
trailed off more gradually for rougher surfaces, with much higher levels of
scattered and silicon fluorescent x-rays below the critical angle.

Fig. (9) shows contaminant fluorescent x-rays for a single wafer.   The fluorescent x-ray curves show a strong peak profile for the smooth surface, peaking
close to the critical angle.   The rough surface data show no comparably
distinct peaks.   TRXRF data clearly indicates the presence or absence of total
reflection from a surface, providing a distinctive measure of surface quality.

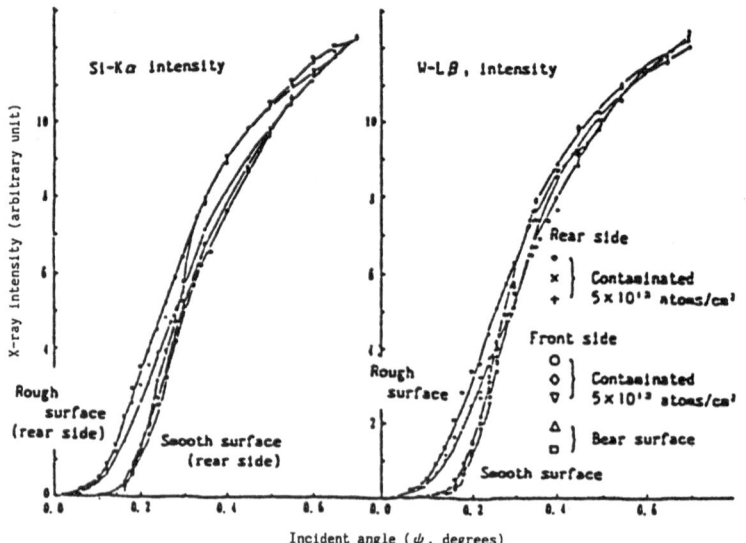

Fig.(8)   Effects of surface roughness
of silicon wafers

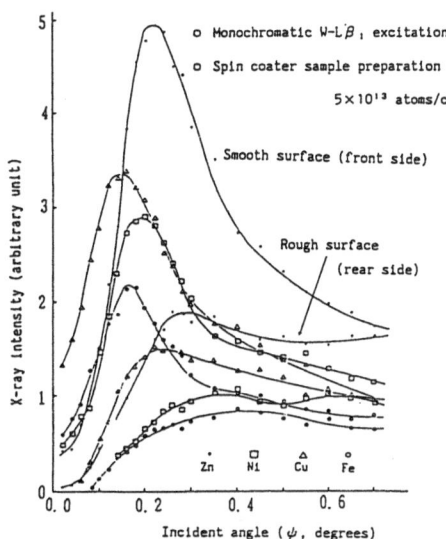

Fig.(9) Surface roughness effects on analyzing elements on silicon wafers

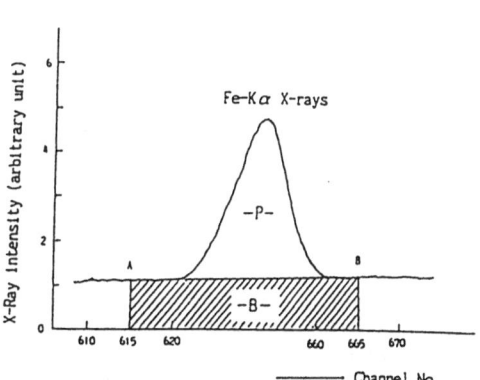

Fig.(10) Example for the determination of LLD and background x-ray intensity in the detection of 5×10^{10} iron atoms/cm² contamination on a silicon wafer

Enhancement of Trace Element Detection through Data Smoothing

The error caused by the background x-ray intensity governs the lower limit of detection (LLD) in trace element analysis. The LLD is defined as three times the standard deviation of the integrated background x-ray intensity, which should be equivalent to simple statistical counting error.[3] Data smoothing improves the LLD by lowering the standard deviation of the background x-ray intensity. Generally, this allows detection of lower concentrations of elements on a surface.

Table (2) shows a typical example of results from a silicon wafer with 5×10^{10} atoms/cm² of iron contamination (Fig. 10). Fluorescent x-ray data were

Table (2) Relationship of peak and background x-ray intensity measurements and calculated errors

| | X-ray intensities rerated to errors | | | | |
| | A position | B position | Integrated Background | Integrated peak | L.L.D. * |
Channen No.	610 – 620 th	660 – 670 th	615 – 665 th	615–665th	atoms/cm²
Raw background	7.46 at 615th $\sigma =3.04$	6.54 at 665th $\sigma =2.35$	357 $\sigma =19.4$	434 $\sigma =34$	0.67×10^{10}
Smoothed background	6.94 at 615th $\sigma =1.24$	6.46 at 665th $\sigma =1.30$	342 $\sigma =9.1$	450 $\sigma =30$	0.30×10^{10}
Difference between raw and smooth background	0.52	0.08	15.2	16.	

* $3 \times \sigma_{Background} \times$ Contamination gradient W-Lβ_1 excitation, 30kV, 400mA, 1000sec

accumulated for 1000 sec using W-Lβ_1 x-rays from a rotating anode x-ray tube source set at 30kV, 400 mA. The results show that smoothing the data improved the counting statistics, in this case reducing the LLD by more than a factor of 2.

CONCLUSION

The angular dispersion properties of various monochromator crystals have been investigated for total-reflection fluorescent x-ray measurements. The values of the angular dispersions and the intensities obtained from different crystals with both molybdenum and tungsten x-ray sources have been tabulated. In the case of a molybdenum x-ray source, the monochromator with a large spacing produced narrow x-ray beams and was found to provide brighter and less divergent beams.

The investigations of the fluorescent x-rays from titanium films of different thickness and the detection of contamination on silicon surfaces were performed. The studies were carried out with beams of different divergence; the wider beam, less collimated and more intense, allowed detection of surface contaminants more quickly than the narrow beam. The narrow beam method offers sharp angular profiles useful for depth analysis.

Similar fluorescent x-ray studies of contaminated silicon wafers on smooth and rough surfaces found that the intensity of scattered incident beam radiation and bulk fluorescent x-rays below the critical angle depends strongly on the quality of the surface.

Data analysis on a typical contaminant detection measurement shows the importance of data smoothing to improve the lower limit of detection for trace element analysis.

REFERENCES

1)   Iida and Gohshi; Japanese Journal Applied Physics Vol. 23 [11], 1543 (1984), Nuclear Instruments and Methods in Physics Research 556 (1985) and ibid, A2461, 736 (1986). Special issue on total reflection x-ray analysis; Spectrochimica Acta Vol. 44B [5] (1989). Arai; Advances in X-Ray Analysis Vol. 32, 131-139 (1989).

2)   Brunel and Gilles; Colloque De Physique, Colloque C7 supplement au n° 10 Tome 50 Octobre 85-96 (1989).

3)   Leland, Bilbrey, Leyden, Wobrauschek and Aiginger; Anal. Chem. Vol. 59, 1911 (1987). Curie; Anal. Chem. Vol. 40, 586 (1968).

CHARACTERIZATION OF NEAR SURFACE LAYERS BY MEANS OF

TOTAL REFLECTION X-RAY FLUORESCENCE SPECTROMETRY

H. Schwenke, R. Gutschke and J. Knoth

GKSS-Forschungszentrum, Institut für Physik

Postfach 1160

W-2054 Geesthacht, Germany

ABSTRACT

Total Reflection X-ray Fluorescence Spectrometry (TXRF) has been used for the characterization of a 20 nm thick Ni/Fe/Cr-layer on a silicon substrate. Instrumental aspects of the technique as well as the data evaluation procedure on the basis of modelling calculations are outlined in this paper. The effect of standing waves is discussed by means of the selected example. This particular layer serves also as an illustration of the capabilities and limitations of TXRF. At least three surface parameters are covered by the technique, elemental composition, density and layer thickness.

INTRODUCTION

Total Reflection X-Ray Fluorescence Spectrometry (TXRF) has not only been used for trace element analysis and the examination of metallic contamination in and on silicon wafer surfaces[1,2], but also for the characterization of a variety of surface layers ranging from a few nm to approximately 500 nm in depth[3].

The particular suitability of TXRF for surfaces is based on the low penetration depth of X-rays in the total reflection mode. Contrary to normal X-ray fluorescence spectroscopy, which is not appropriate for the examination of near-surface layers, TXRF is an inherently surface sensitive technique. In addition, the formation of standing waves, which occur as a result of interference between reflected and refracted portions of the primary beam, can be exploited for data evaluation. This is why the fluorescence intensities, measured as a function of

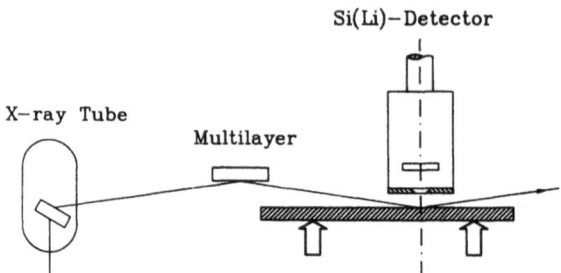

Fig.1 Design of the instrument.

the grazing incident angle, provide information about the properties of single layers or a layered surface structure.

We first outline briefly the instrumentation and the modelling
formalism, which is used for data analysis. Specific features
of the method are exemplified by means of experimental and
simulated results obtained from a metallic layer on a silicon
substrate.

INSTRUMENTATION AND MEASURING PROCEDURE

The basic features of the instrument used are displayed in
Fig.1. The X-ray beam from a fine focus anode is directed via
a monochromatizing multilayer mirror onto the surface target.
A Si(Li)-detector is positioned closely above the sample in
order to ensure an optimal take-off angle range. The sample is
precisely adjusted with respect to the incident beam by means
of stepper motors controlled by the fluorescence signals from
the surface. Other stepper motors set the lateral position.
The measuring procedure consists of recording the fluorescence

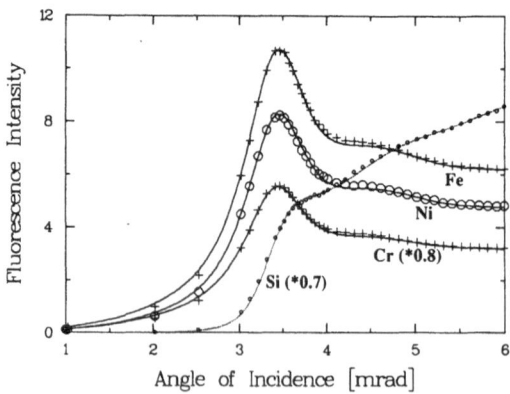

Fig.2. Measured data (symbols) and best fit of an alloy layer
on a silicon substrate. For analysis results see Tab.1.

Tab.1. Analysis results of the examined layer. Values in pa-
       rentheses are not independently determined but supposed
       to be already known (composition and density of the
       silicon substrate) or complemented to 100% (Oxygen)

Layer	Element concentration (weight percent)					Density (g/cm^3)	Thickness (nm)
	NI	Fe	Cr	Si	O		
Oxide layer	7.4	23.5	39.0	-	(30.1)	3.7	2.5
Ni/Fe/Cr-layer	20.9	39.8	39.3	-	-	7.4	17.5
Substrate	-	-	-	(100)	-	(2.33)	∞

spectra as a function of the incident angle starting from an
angle close to zero up to approximately two times the critical
angle for total reflection (s. Fig.2).

DATA ANALYSIS

The raw data, i.e. the fluorescence intensities of each constituent element as a function of the incident angle (symbols in
Fig. 2) are analyzed by means of a fitting procedure using
modelling calculations which describe the propagation of X-
rays in absorbing media at grazing incidence. For reasons of
clarity we neglect here the corrections for absorption and enhancement effects which have to be applied in surface TXRF similar to normal XRF and which are considered elsewhere[3].Basical-
ly the fluorescence intensity I_i of the element i depends on
rather few parameters in the corresponding modelling formalism:

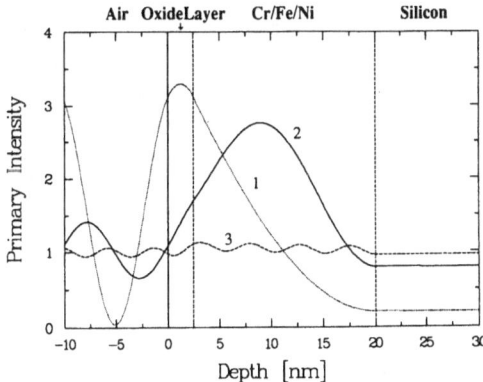

Fig.3. Formation of standing waves: primary intensity vs. dis-
       tance from the interface for various incident angles Ø
       (curve 1 → $Ø_1$ = 3.1 mrad, curve 2 → $Ø_2$ = 3.5 mrad,
       curve 3 → $Ø_3$ = 8.0 mrad), sample as described in
       Tab.1, excitation energy is 17.5 keV.

$$I_i = I_i(\{ C_{i1}, \rho_1, D_1\}, \ldots, \{ C_{ij}, \rho_j, D_j\}, \ldots \varepsilon_i, \phi)$$

where the variables denote:

C_{ij} = concentration of element i in layer j

ρ_j = density of layer j

D_j = thickness of layer j

ε_i = instrumental sensitivity for element i

ϕ = angle of incidence

After calibration of the instrument using only thick monoele-
ment layers or bulk samples for the determination of ε_i, we apply these modelling calculations to fit the measured data
(Fig.2). The data analysis results in the determination of the
free parameters of the model, that are C_{ij}, ρ_j and D_j.

EFFECTS OF STANDING WAVES

The capacity of TXRF for the determination of composition,
density and thickness of near surface layers, which comprises
a certain capability for depth profiling, is based on the formation of standing waves in layered structures due to interference of reflected and refracted portions of the primary
beam. Fig.3 illustrates what happens in the case of the selected example (Tab.1).For a low grazing angle of e.g. 3.1
mrad the primary intensity in the air (or vacuum) above the
interface (displayed as negative depth in Fig.3) varies
strongly between almost zero and more than three times the un-
affected incident intensity and reaches a maximum in the oxide
layer. Within the material the intensity decreases approximately exponentially for this particular angle, showing a penetration depth of about 15 nm. A small change in the incident
angle, however, e.g to 3.5 mrad yields a substantially different course of the intensity. Compared to curve 1 (in Fig.3),
curve 2 shows just minor variations above the interface but in
the oxide layer a steep increase of the intensity takes place
which continues even further into the middle of the alloy
layer before it likewise decreases. Curve 3 displays the intensity beyond the total reflection regime at a comparatively
large angle of 8 mrad. The residual, unspecific oscillations
around unity are the only indicators of total reflection effects in this angular range. The intensity in the silicon substrate is, in any case, no more influenced by standing waves.
The deviations from unity in the substrate are caused by absorption, depending on the path of the primary X-rays through
the metal layer which increases with decreasing angle.Owing to
their specific angular dependence the standing waves turned
out to be a sensitive probe for surfaces. This may be demonstrated by the following studies.

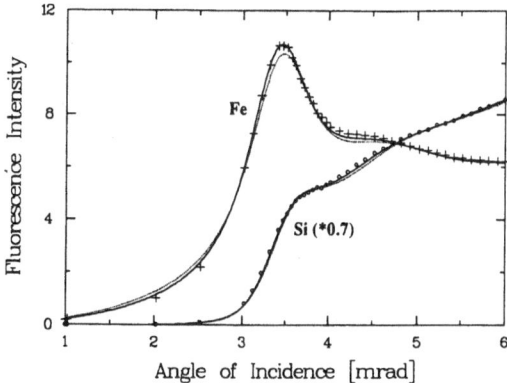

Fig.4. Comparison of the final fit for Fe and Si (solid li-
        nes) using the data of Tab.1 with a preliminary fit
        (dotted lines) based on the following data: no oxide
        layer, thickness of the Ni/Fe/Cr-layer 18.2 nm, con-
        centration of Ni, Fe, and Cr as in Tab.1.

Fig.4 demonstrates a certain capability of TXRF for depth profiling. It shows a small but significant measuring effect
which points to the existence of a thin oxide layer. Without
the assumption of an additional top layer characterized by reduced density and metal concentrations the agreement between
calculated and measured data could not be further optimized.
The dotted curve represents the best fit achieved with the
single layer model. The thickness of the alloy layer was replaced by 18.2 instead of 17.5 nm in order to fit the high angle range which yields the measured overall areal density[4].

Fig.5 shows the effect of a variation of the layer density. An
alteration of the best fit density of 7.4 g/cm^3 by 10 % results

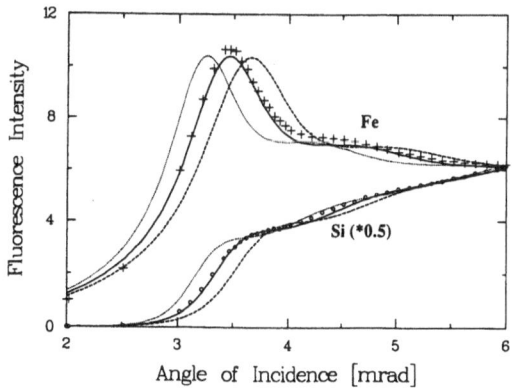

Fig.5. Effect of a 10 % density variation on the Fe and Si
        signals of the layer under investigation (s. Tab.1).
        The solid curve represents a density of 7.4 g/cm^3
        (dotted curve 6.7 g/cm^3, dashed curve 8.1 g/cm^3).

in an angle shift of approximately 0.15 mrad. This is beyond the uncertainty of the instrumentation used, which was found to be better than 0.05 mrad in absolute terms. Therefore, we assess the accuracy of the density determination to about 5 %.

CONCLUSION

TXRF is, as far as we know, the only technique available, which is able to determine elemental composition, thickness and density of near surface layers. In addition, it works in a technically simple and even nondestructive manner. Though TXRF needs no standards with the exception of pure elements it provides strictly quantitative results. At the present stage of development, however the method requires a skilled operator for data analysis. Work is in progress to facilitate a nearly automatic or at least strongly supported data inversion procedure.

REFERENCES

1. A. Prange, "Total Reflection X-Ray Spectrometry: Method and Applications ", Spectrochim. Acta, 44B: 437 (1989).

2. P. Eichinger, H.J. Rath and H. Schwenke, "Application of Total Reflection X-Ray Fluorescence Analysis for Metallic Trace Impurities on Silicon Wafer Surfaces", 5. International Symposium on Semiconductor Processing, Santa Clara, CA (USA),(1988).

3. U. Weisbrod, R. Gutschke, J. Knoth and H. Schwenke, "Total Reflection X-Ray Spectrometry for Quantitative Surface and Layer Analysis", Appl. Phys. A, in press.

4. T.C. Huang, "Thin-Film Characterization by X-Ray Fluorescence" X-Ray Spectrometry, 20: 29 (1991).

LIGHT ELEMENT ANALYSIS WITH TXRF

Christina Streli, Peter Wobrauschek,
Hannes Aiginger

Atominstitut der Österreichischen Universitäten
Schüttelstraße 115, A-1020 Wien, Austria

ABSTRACT

Total Reflection X-Ray Fluorescence Analysis (TXRF) has become a powerful analytical tool for trace element analysis.[1-4] Because of its advantages in excitation and background reduction TXRF has been applied for the analysis of light elements (C,O,F,Na,...). A special Ge(HP) detector offering an ultra thin window in combination with a spectrometer specially designed for the requirements of light element analysis was used. Also a new windowless X-ray tube for efficient excitation of the light elements was tested. The system was checked with standard aqueous solutions; detection limits in the ng range (7 ng for O) are obtained.

INTRODUCTION - PROBLEMS OF LIGHT ELEMENT ANALYSIS WITH EDXRF

Photoninduced energy dispersive X-ray fluorescence analysis of light elements (C,O,F,Na,...) is connected with several problems. The energy of the emitted fluorescence radiation of the light elements is very low, in the range of 1 keV and below and therefore absorption plays an important role in the sample and in all barriers before the detector crystal The intensity of the fluorescence radiation $I_{K\alpha}$ of a measured element is determined by the following factorsbesides the sample mass :

$$I_{K\alpha} \propto I_0(E) \cdot \tau_K(E) \cdot \omega \cdot \varepsilon \cdot \Omega/4\pi \tag{1}$$

The fluorescence yield ω is very low for light elements and therefore a natural restriction. For standard solid state detectors the detector efficiency ε for low energy radiation is poor due to absorption in the entrance window, the contact layer and the dead layer. Electronic noise influences the energy resolution and limits the detectable elemental range. Using standard 45° geometry the solid angle of the fluorescence radiation $\Omega/4\pi$ is low. Using standard X-ray tubes the fluorescence cross-section $\tau_K(E) = \tau(E) \cdot (1-1/S_K) \cdot p_\alpha$ (τ is the photoelectric mass absorption coefficient, S_K is the jump ratio and p_α is the probability of the K_α radiation). This parameter is low in comparison to medium Z elements. The primary intensity $I_0(E)$ depends on the tube power, the distance from the source to the sample and the geometry between source, collimator and sample size. Therefore the intensity of the fluorescence signal is low. But this weak fluorescence signal is superimposed upon a background caused by scattering of the primary radiation by the sample and sample carrier. Especially high energy photons primary radiation by the sample and sample carrier. Especially high energy photons cause an increase of the low energy background due to Compton backscattering in the detector crystal, which produces a Compton electron that is scattered in the active volume of the detector.

The advantages of TXRF have been well described [1,5 - 8]

EVOLUTION OF THE SPECTROMETER

In 1985 the first TXRF spectrometer using a conventional Si(Li) detector where the Be-window had been removed operating in a high vacuum chamber and using a standard fine focus Cr-tube was built[9]. Routine measurements were not possible because residual gases in the chamber condensated on the detector crystal producing an ice-layer that led to efficiency losses and a non operating diode. An ultra thin carbon-foil was installed to prevent the crystal from icing[10]. This foil was thin enough to allow the low energy fluorescence radiation to pass, but not strong enough to stand mechanical shocks or vibrations. It was replaced by an aluminized Polypropylen-window, which offered enough mechanical stability but attenuated e.g.O-K-radiation about 50%[11]. The detector used did not offer the necessary low noise conditions and therefore O was the last detectable element. Sensitivity to grounding problems and electronic spikes prevented stable measuring conditions. The detector efficiency was not really adequate but allowed the collection of analytical data including Oxygen.

During recent years the spectrometer was equipped with a cut-off filter to eliminate the high energy photons, which limproved the low energy background conditions.[10] Then the beam path was evacuated to minimize attenuation of the exciting radiation by the air.[11] A disadvantage was complicated angle adjustment under vacuum conditions because the angle adjustment was done by rotating and translating the X-ray tube. A second disadvantage was the large distance between anode and sample (350 mm).

NEW SPECTROMETER - IMPROVEMENT IN DETECTION AND EXCITATION

Due to the disadvantages mentioned above a new spectrometer was designed and built under the aspects of optimization of excitation and detection.

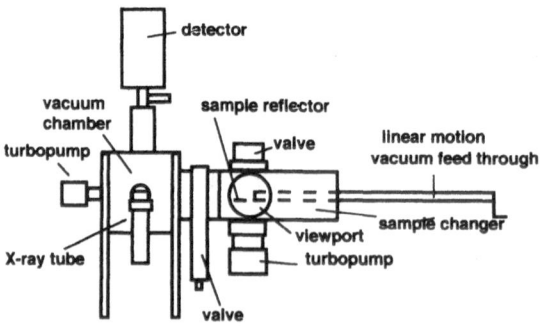

Fig.1.Schematic of the new Low-Z-spectrometer

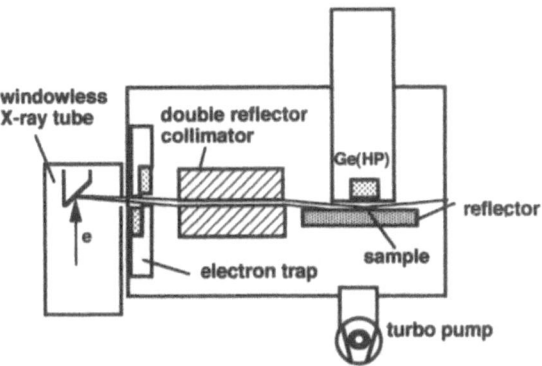

Fig.2.Schematic scetch of the measuring chamber

New detector special for light element detection

A new commercially available Ge(HP) detector was used, with an ultrathin (0.4 µm thick) window (Diamond-window), which offers high transmission of the low energy fluorescence radiation e.g. 85% for O-K-radiation [13] and the necessary low noise contribution leading to 125 eV FWHM for Mn-Kα and start of noise peak at 140 eV. Ge as detector material has the advantage that a larger number of charge carriers are produced by the incoming radiation and that the crystal has nearly no dead layer. No grounding problems or microphony was observed. There were also no problems with low energy tailing or Ge-L-escape peaks. The crystal size is 30 mm^2 and the distance between window and crystal 2 mm.

Special X-ray tube for efficient excitation of light elements

To excite the light elements efficiently an X-ray source that emits a large number of low energy photons is required. A special X-ray tube without a Be-window (to reduce the absorption) was designed and constructed. However in this case the absorption takes place in the anode. It depends on the electron incidence and photon take-off angle. The larger the take-off angle, the shorter the path length of the photons in the anode and therefore the smaller the absorption. But the large take-off angle reduces the brilliance, and high brilliance is very important when collimation is necessary because a maximum number of photons should pass the collimation system to hit the sample. A 45° take-off angle was used producing a focal spot of 1 x 10 mm, but minimizes the absorption in the target. Cu was used as anode material offering the advantage of its K and L radiation The Cu-L lines have energies of 0.96 keV and are therefore well suited for the excitation of O and C.

The spectrometer consists of a vacuum chamber to which the windowless tube was connected by a flange without a valve to minimize the distance between source and sample (now 130 mm). Therefore it was necessary to install a sample changer that could be evacuated. When the vacuum in the changer is good enough, a valve between measuring chamber and sample changer is opened and the sample moved into the measuring chamber by a linear vacuum feedthrough. Thus, the X-ray tube and the measuring chamber are never flushed during sample changing. The vacuum is produced by turbopumps and is in the range of 6.10^{-6} mbar , which is only necessary for the operation of the windowless tube. Using a standard X-ray tube high vacuum is not required because the new detector is not connected to the measuring vacuum circuit.

Fig.1 shows the schematic of the spectrometer.

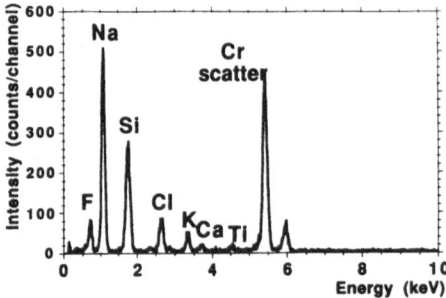 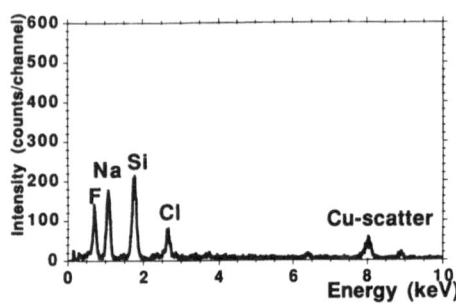

Fig.3 NaF sample (125 ng Na, 100 ng F) measured with
a. Cr-tube standard 6° take-off 40 kV 12 mA 100 sec
b. Cu-tube windowless 45° take-off 20 kV 25 mA 100 sec

Fig.2.shows a schematic sketch of the measuring chamber.The backscattered electrons from the X-ray tube are derected to the walls by a special diaphragm and a magnetic trap. In the measuring chamber there is a double reflector collimator[14] with a length of 5 cm and 0.1 mm spacers, which can be adjusted to the beam by steel bellow tightened micrometer screws. Also the sample reflector can be adjusted from outside the chamber by micrometer screws. The sample reflector is pressed against three steel balls in a copper block in which is a central opening for the detector.This block can be moved with micrometer screws. The detector can be moved very close to the sample after having made the correct adjustment. Because the adjustment can be done from outside under vacuum there are no problems with reproducability of the geometry. The rotation centre of angle adjustment is very close to the sample so that angle scanning can be performed. The sample reflector is pressed against the steel balls by a bellow, which can be withdrawn from outside to change the sample. There is also a beam exit window where the beam can be controlled. The detector is shielded by an aluminum collimator (0.5 mm) with an opening of 8 mm.

RESULTS

The new spectrometer was tested with two different X-ray tubes. First a standard Cr-tube with 6°take-off angle and a focal spot size of 0.1 x 10 mm was used. Second the new windowless Cu-tube with 45° take-off angle and a focal spot size of 1 x 10 mm was used in combination with the sample changer. Fig.3 shows spectra of a NaF

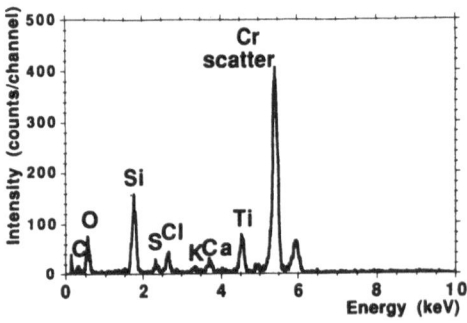 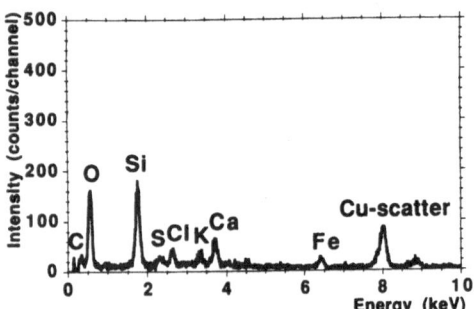

Fig.4.Li$_2$CO$_3$ sample (500 ng O and 125 ng C) measured with.:
a. Cr-tube standard 6° take-off 40 kV 12 mA 100 sec
b. Cu-tube windowless 45° take-off 20 kV 25 mA 100 sec

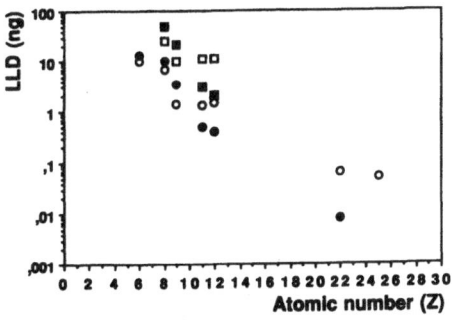 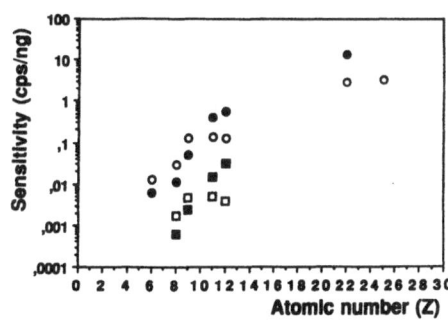

Fig.5 comparison of detection limits (ng) and sensitivity (cps/ng) for
●.Cr-tube standard 6° take-off 40 kV 12 mA - new spectrometer
○.Cu-tube windowless 45° take-off 20 kV 25 mA - new spectrometer.
■.Cu-tube windowless 6° take-off 25kV 20 mA - old spectrometer.
▢.Cu-tube windowless 45° take-off 25 kV 20 mA - old spectrometer.

sample measured with the two tubes. The intensity of F has been increased by a factor of two using the windowless Cu-tube in comparison to the Cr-tube. The reason is the Cu-L-line which has an energy of 0.96 keV and excites very well F and elements below. But the Na intensity has been decreased by a factor of 2.5 due to the loss of brilliance using the large focal spot of the Cu-tube.

Fig. 4 shows the comparison for a Li_2CO_3 sample. The spectrum measured with the windowless tube shows a 2 times higher O peak.

Fig.5 shows the comparison of detection limits (extrapolated to 1000 sec) and sensitivity obtained with the different spectrometers and tube types. The increase in sensitivity and the reduction of the detection limits obtained by the new spectrometer can be seen. Using the windowless tube there was a further improvement for O and C. The detection limits for O are now 7 ng (or 1 ppm concentration in an aqueous solution) and the sensitivity 0.03 cps/ng.

Fig.6. shows spectra of a Carbon-film and of a Mineralwater sample.

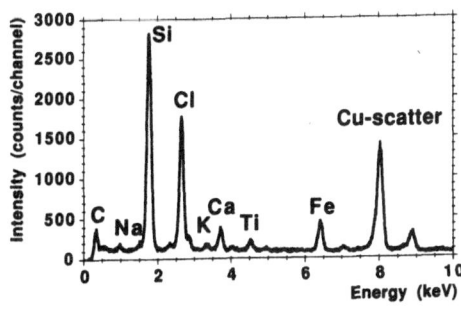 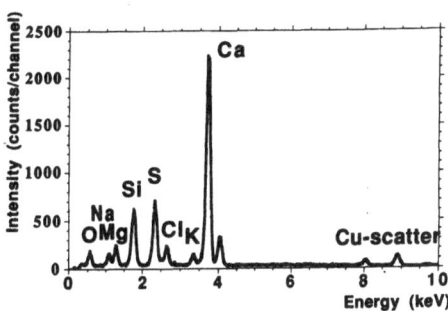

Fig.6 Spectra of Carbon film and mineralwater
Cu-tube windowless 45° take-off 20 kV 25 mA - new spectrometer.
a. 40 $\mu g/cm^2$ Carbon-film
b. 5 μl mineralwater

CONCLUSIONS

With this new TXRF spectrometer for light element analysis it has been shown that TXRF is a method for the analysis of low Z elements with improved detection limits. For the future new anode materials will be tested and the applicability of the spectrometer will be checked..

ACKNOWLEDGEMENT

We would like to thank the "Fonds zur Förderung der wissenschaftlichen Forschung" for the financial support of this project (P7120), Dr.Ernst Unfried for the helpful discussions and Walter Drabek for the help during construction and the technical realisation of the mechanical components.

REFERENCES

1   H.Aiginger,P.Wobrauschek,Nucl.Instr.Meth.114,457 (1974)
2   A.Prange and H.Schwenke, Adv.X-Ray Anal.32, 211 (1989)
3   R.Klockenkämper, Spectroscopy 5,26 (1990)
4   P.Wobrauschek, P.Kregsamer, C.Streli, H.Aiginger
     Adv.X-Ray Anal.34 ,1 (1991)
5   E.Schnabel, R.Hosemann , B.Röde, J.Appl.Phys.43,3237 (1972)
6   H.Schwenke, J.Knoth Nucl.Instr.193,239 (1982)
7   H.Aiginger, P.Wobrauschek, Adv.X-Ray Anal.28,1 (1985)
8   H.Schwenke, W.Berneike, J.Knoth, U.Weisbrod
     Adv.X-Ray Anal.32,105 (1989)
9   G.Lurf, doctoral thesis, Technical University Vienna (1984)
10   C.Streli, P.Wobrauschek, H.Aiginger,
     Spectrochim. Acta 44B,491 (1990)
11   C.Streli, P.Wobrauschek, H.Aiginger,
     Spectrochim. Acta 46B (accepted for publication)
12   C.Streli, doctoral thesis, Technical Univ.Vienna 1989
13   TRACOR technical information (1989)
14   P.Wobrauschek, P.Kregsamer, Ch.Streli, H.Aiginger,
     X-Ray Spectrom.20, 23 (1991).

APPLICATION OF TXRF IN ENVIRONMENTAL RESEARCH

Walfried Michaelis, Rudolf Pepelnik and Andreas Prange

Institute of Physics
GKSS Research Centre Geesthacht
D-2054 Geesthacht, Germany

ABSTRACT

In the framework of interdisciplinary research projects TXRF has been applied for multielement analysis of rainwater, size-fractionated airborne particulates, throughfall samples and litter fall in forests, soil solutions, seepage water, river water filtrates and suspended particulate matter. Essential aspects of sample preparation techniques are briefly described. Some results are given on trace element fluxes in a forest ecosystem and on contaminant transport phenomena in estuaries in order to demonstrate the outstanding capabilities of TXRF in demanding environmental research projects.

INTRODUCTION

TXRF has been used with great success in various fields of environmental research. This paper concentrates on (i) studies of the transfer of atmospheric pollutants into a forest ecosystem and the associated element fluxes within the water cycle, (ii) on research with regard to the transport of contaminants in estuaries. During these investigations TXRF has proved to be a powerful analytical tool, in particular owing to the pronounced multielement characteristics, the high sensitivity and the minute sample masses required.

MATERIALS AND METHODS

A measuring station comprising a 48 m tower with extensive sampling and meteorological instrumentation has been set up in a North German forest ecosystem, about 40 km north-east of Hamburg (Michaelis et al., 1989). The site is characterized by a mixed stand consisting predominantly of spruce with some pine and beech. The aim of the research is to investigate, among others, the wet and dry deposition of ecologically relevant trace elements and their fluxes in the throughfall, the litter fall and the seepage water under environmental stress conditions. Rainwater samples are collected by means of an automatic device which is controlled by a moisture sensor. Undissolved constituents are strained off and analysed separately. Samples of size-fractionated airborne particulates are obtained using a high-volume sampler equipped with a 5-stage slotted cascade impactor. The flux onto the canopy is determined by means of the concentration method. Throughfall samples are collected with the aid of radially arranged teflon-coated grooves. Parallel sets of lysimeter suction tubes deliver soil solutions from different depths down to 165 cm. The flux via the seepage water is determined by using chloride as an inert tracer. Analysis of liquid samples is preferably achieved by both a direct measurement and freeze-drying for preconcentration. Particulates are analysed using digestion with nitric acid in teflon bombs (Michaelis and Prange, 1988). 20 to 25 elements can be determined.

Estuaries are characterized by pronounced spatial heterogeneities and temporal variabilities in the composition of the water body. These features make water quality assessment and flux measurements a difficult task. The research at GKSS aims at establishing the necessary fundamentals for optimum instrumentation and strategies. Examples are: (i) investigations of the representativeness of moored stationary platforms, (ii) validation of numerical transport models, (iii) studies of the contaminant fluxes under different hydrological conditions, and (iv) quantification of the long-term discharge over selected cross-sections. For performing research of this kind, a methodology, the so-called BILEX concept, has been developed which combines in an effective way hydrographic measurements with numerical transport models and trace analytical procedures (Michaelis, 1990). For the analysis of filtrates two techniques are used: (i) direct measurement and (ii) nitric acid digestion of the freeze-drying residue with subsequent matrix separation and enrichment of the trace elements. A simple and convenient sample preparation for suspended particulate matter is nitric acid digestion in quartz vessels (Prange et al., 1990). Between 15 and 20 elements are detectable in filtrates and suspended matter.

RESULTS

Element Fluxes in a Forest Ecosystem

Table 1 summarizes the results for the wet and dry deposition of some selected trace elements in $\mu g/m^2 d$. The data cover the period from April 1987 to January 1990. In the case of dry deposition intervals are quoted in order to take account of the relatively large uncertainties which are mainly due to possible errors in the deposition velocity. The evaluation is based on values of this quantity which have been obtained using natural aerosols (Hertlein, 1990; Waraghai and Gravenhorst, 1989; see also literature cited there). If deposition velocities measured with artificial aerosols are taken as a basis, the results for the dry deposition are in part markedly higher (Michaelis et al., 1989). In contrast to grass surfaces, the dry deposition clearly exceeds the wet flux in the case of forest canopies.

The comparison of atmospheric input and throughfall (Fig. 1) allows estimates of the retain and wash-out processes in the crown compartment (Table 2). Three groups of elements may be distinguished: (i) Elements which are retained in the tree-tops. Examples are heavy metals like Zn and Pb. The same behaviour is observed for hydrogen ions and rainwater. (ii) Elements which on a long-term average prevail in the throughfall. These elements are obviously released in the crown compartment. They must

Table 1.   Long-term averages of the daily wet and dry deposition of trace
           elements in µg/m²d

Element	Wet deposition		Dry deposition	
	dissolved	particulate		
S	2750	40	5790	– 10000
K	167	52	599	– 1080
Ca	690	35	1070	– 2030
V	1,5	0,5	11	– 19
Cr	0,43	0,65	5,1	– 9,4
Mn	9,1	2,5	27	– 49
Fe	36	172	710	– 1340
Ni	1,4	0,45	8,0	– 14
Cu	3,1	1,0	15	– 28
Zn	69	3,4	103	– 184
As	1,2	0,12	7,0	– 12
Se	0,52	0,02	1,9	– 3,3
Cd	0,91	0,04	2,0	– 3,4
Pb	11	2,1	98	– 168

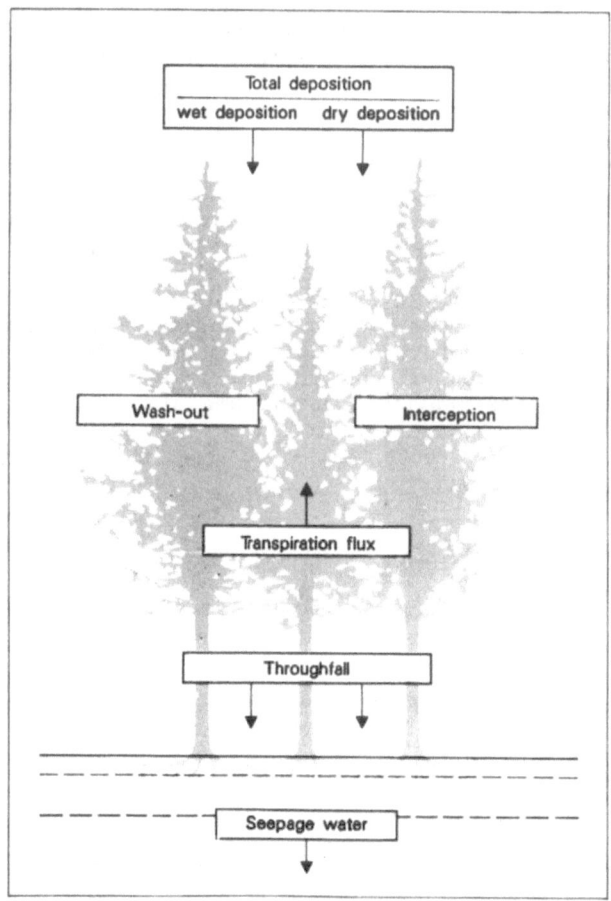

Fig. 1.   Fluxes and processes in a forest ecosystem

Table 2.   Fluxes of some selected elements in a forest ecosystem.
           All data in mg/m²d

Element	K	Ca	Mn	Pb
Total deposition	0,8 - 1,4	1,7 - 2,9	0,04 - 0,07	0,10 - 0,18
Wash-out	5,1 - 7,2	2,6 - 5,0	0,4 - 0,6	(-0,16)-(-0,08)
Throughfall[*)	7,3	6,1	0,55	0,022
Seepage water 5 cm	3,3	8,5	1,6	0,010
Seepage water 45 cm	1,0	5,4	1,1	0,002

[*) Estimated error ± 10 %

be replenished via uptake in the root system. Exponents of this category are mobile nutritional elements like K and Mn. (iii) Finally, other elements which show an indifferent behaviour.

Retention of trace elements in the crowns is ultimately balanced by litter fall. The wash-out of nutrient elements is, among others, due to the neutralization of hydrogen ions deposited. This buffer mechanism only operates as long as there is sufficient compensation with nutritional cations from the soil solution via an efficient fine root system. During the latter process the buffered amount of acid is transferred to the soil. There it is buffered again with temporal delay. In the case of many years' acid deposition the bonding sites in the humus and soil constituents, which initially carry in particular nutrients at disposal for the plants, are occupied by ions that are physiologically ineffective. As a consequence, nutrients can no more be sufficiently retained and thus get partially lost for the ecosystem via the seepage water (Table 2). This will finally result in a deficient supply of the plant organisms. Toxic elements such as Cd and Pb increase the harmful effects. For example, Pb has been fixed in the humus layer for decades without physiological effectiveness. With decreasing pH value it was mobilized and became disposable for the plants. The results are increased concentrations in the soil solution (0.07 - 0.25 µg/ml) and in the fine roots (100 - 300 µg/g) as well as the danger of migration into the ground water.

Transport Phenomena in Estuaries

From the hydrographic data and the trace analytical results for the liquid and the particulate phase, the transport rates during a BILEX campaign can be calculated. Moreover, the experimental data obtained on one of the open boundaries of the river section investigated may be used to simulate the fluxes at the other cross-section by means of numerical models, which in particular have to take into account interactions at the sediment-water interface by sedimentation and resuspension, but also meteorological boundary conditions. The temporal variation of the trace metal concentration and thus the time-integrated fluxes as well strongly depend on the partitioning between the liquid and the particulate phase of the element considered (Michaelis, 1990; Fanger et al., 1990). This means that the transport behaviour differs from element to element.

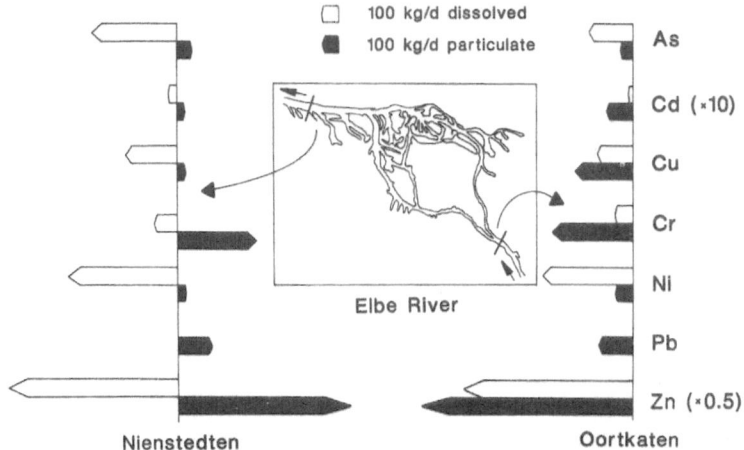

Fig. 2.   Balance of the heavy metal transport through the Hamburg Harbour

        The complexity of the transport phenomena may be well illustrated by
Fig. 2 in the light of the heavy metal transport through the Hamburg
Harbour during a six tides' campaign. Data for the liquid and particulate
phase are shown separately by open and full arrows. Both phases exhibit
clearly different rates upstream and downstream from the harbour. A striking result is the upstream transport of suspended particulate matter and
associated heavy metals at the harbour exit during the period considered.
Obviously, the harbour area acts as a sink for this material from both
upstream and downstream. Depending on the specific partitioning between the
phases, as a whole either the downstream (Ni, Cu, As) or the upstream
transport (Cr, Cd, Pb) prevails at the exit. Probably, this phenomenon is
caused by the combined action of local, meteorological and hydrological
factors.

        The multielement characteristics of TXRF allow the efficient applica-
tion of chemometric methods which might give deeper insight into the con-
fusing diversity of the experimental data. For example, Fig. 3 shows a den-

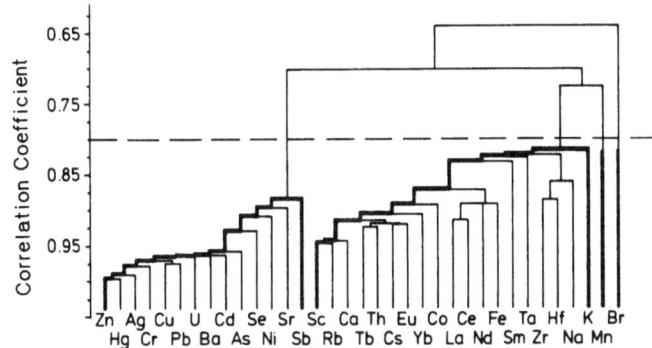

Fig. 3.   Dendrogram of the elements in suspended particulate matter in the
           Elbe River

dogram of suspended material, as obtained by statistical cluster analysis. In this case additional elements have been determined by neutron activation. Four groups of elements can be identified, the first two of them being of particular importance. They comprise elements of anthropogenic and geogenic origin, respectively, each with a characteristic behaviour in spatial concentration distribution and temporal variabilities.

REFERENCES

Fanger, H.-U., Kappenberg, J., and Männing, V., 1990, A study of the transport of dissolved and particulate matter through the Hamburg Harbour, in: "Estuarine Water Quality Management", W. Michaelis, ed., Coastal and Estuarine Studies, Vol. 36, Springer Berlin - Heidelberg - New York, 127-134.

Hertlein, F., 1990, Untersuchungen zur Anwendung der Gradientenmethode auf luftgetragene Partikel, Diplomarbeit (GKSS), Fachhochschule Lübeck.

Michaelis, W., 1990, The BILEX concept - research supporting public water quality surveillance in estuaries, in: "Estuarine Water Quality Management", W. Michaelis, ed., Coastal and Estuarine Studies, Vol. 36, Springer Berlin - Heidelberg - New York, 79-88.

Michaelis, W., and Prange, A., 1988, Trace analysis of geological and environmental samples by total reflection X-ray fluorescence, Nucl. Geophys., 2, No. 4: 231.

Michaelis, W., Schönburg, M., and Stößel, R.-P., 1989, Deposition of atmospheric pollutants into a North-German forest ecosystem, in: "Mechanisms and Effects of Pollutant-Transfer into Forests", H.-W Georgii, ed., Kluwer Academic Publishers, Dordrecht, 3-12.

Prange, A., Niedergesäß, R., and Schnier, C., 1990, Multielement determination of trace elements in estuarine waters by TXRF and INAA, in: "Estuarine Water Quality Management", W. Michaelis, ed., Coastal and Estuarine Studies, Vol. 36, Springer Berlin - Heidelberg - New York, 429-436.

Waraghai, A., and Gravenhorst, G., 1989, Dry deposition of atmospheric particles to an old spruce stand, in: "Mechanisms and Effects of Pollutant-Transfer into Forests", H.-W Georgii, ed., Kluwer Academic Publishers, Dordrecht, 77-86.

DETERMINATION OF HEAVY METALS IN ENVIRONMENTAL WATER BY TOTAL REFLECTION
X-RAY FLUORESCENCE METHOD USING OPTIMIZED ROENTGEN OPTICS CUT-OFF FILTER

A. I. Egorov, L. P. Kabina, I. A. Kondurov, E. M. Korotkikh,
V. V. Martynov, A. F. Shchebetov, P. A. Sushkov

Neutron Research Division
Leningrad Nuclear Physics Institute
Gatchina, 188350 U.S.S.R.

INTRODUCTION

The total reflection x-ray fluorescence (TXRF) method of analyzing
elemental contents is based on the small angle irradiation of thin samples
placed on a total reflecting backing with a narrow photon beam. Two
instrumental problems are to be solved here. The first is to form the narrow
beam with a small angular deviation. The usual way to solve this problem is
to use collimators with small solid angles.[1,2] These angles must be less
than the critical angle for x-ray total reflection, which, in the energy range
10 - 20 keV has an order of magnitude around 10^{-3} rad.

To reach a sufficient intensity of the beam one must use powerful x-ray
generators producing also high energy bremsstrahlung. The total reflection
critical angle for the tail of the high energy exciting spectrum is very
small. The result is elastic and inelastic scattering of this part of the
spectrum causing an increase of the background. To reject this high energy
tail one can use a cut-off filter which improves the spectral distribution of
the exciting radiation by filtering out the unwanted high-energy components of
the bremsstrahlung. However such a filter works well only when the angular
deviation of the imaging beam is much less than the glancing angle, the first
has an order of magnitude around 10^{-4} rad. This causes an additional decrease
in the intensity of the incident beam.

NEW ROENTGEN OPTICS COLLIMATING CUT-OFF FILTER

A new roentgen optic cut-off adjustable filter for total-reflection x-
rays has been developed at LNPI. The main advantage of the filter is its
rather large input solid angle, 7-10 times as large as found in existing
instruments.[1,2] This became possible because of the use of a bent reflecting
mirror instead of a plane mirror usually used. The law of bending is a
logarithmic spiral. In this case all the beams beginning at the focus of the
spiral are falling on the mirror surface at the same angle. This angle in our
case is just the critical angle for total reflection.

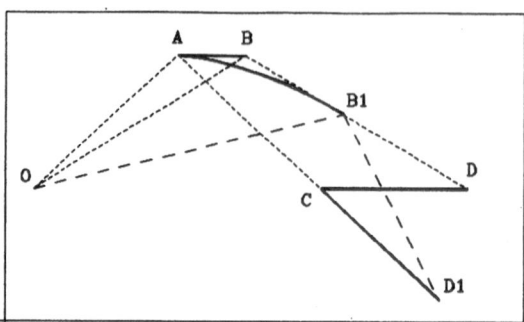

Fig. 1   Comparison of plane (A-B) and bent (A-B1) reflectors
         at equal length of the samples (C-D = C-D1).

From Fig. 1 one can see that the bent mirror reflects beams from a
broader solid angle than the plane one does. As a result the total flux of
irradiating x-rays is higher at the same length (and area) of the sample.

The bent mirror works also as an effective cut-off filter because x-rays
with energy higher than the critical angle (corresponding to the total reflection critical angle) are not reflected but are scattered when falling on the
mirror surface and never reach the sample. The critical cut-off energy can be
changed with variation of the logarithmic spiral curvature.

EXPERIMENTAL SET-UP

The filter produced has a mirror of 123 mm length bent as described
above. The mirror is made of polished glass covered with an evaporated layer
of nickel. The distance between the x-ray source position and the center of
the sample equals 280 mm; the input solid angle is 16.5 min of arc; the output
angle is 5.5 min of arc. The input slit of the filter is 10 x 0.01 mm^2. The
cross section of the output beam is 20 x 0.18 mm^2. The backing of the sample

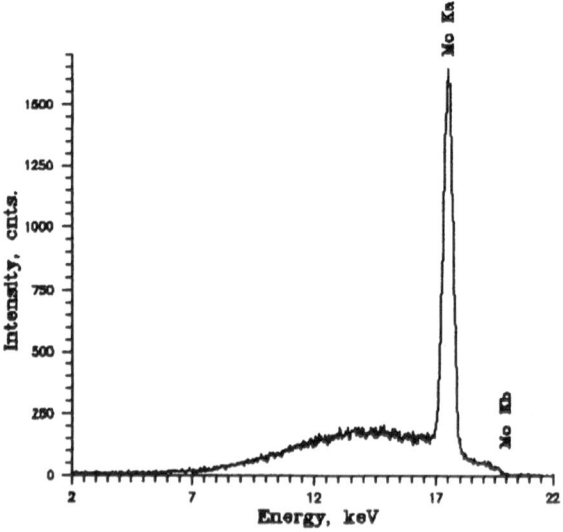

Fig. 2.   The spectrum of the beam
          exciting the fluorescence

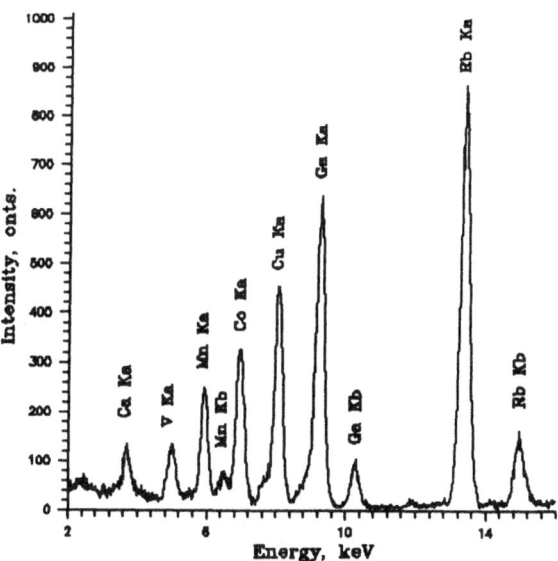

Fig. 3.   The spectrum of the
standard sample.

is made of the same glass but without nickel coating.   The glass used was
originally prepared for producing car windows.

On the basis of the filter an instrument is designed for the determination of concentration of heavy metals in environmental waters.   Copper and
molybdenum x-ray tubes (30 kV, 10 milliamp) with linear 11 x 1.2 mm anodes
were used as an x-ray source.   To make the exciting x-ray spectrum more
monochro- matic a thin additional passive filter made of niobium foil is
placed in front of the cut-off filter.   The main aim of this filter is to cut
off the low-energy part of the spectrum.   One can see on Fig. 2 the spectrum
of the outgoing x-rays exciting fluorescent emission in the sample.

The measurements made with the new roentgen optics filters show a good
perspective of using such a filter in TXRF.   In the near future we plan to
decrease the linear dimension of the filter to increase the x-ray flux at the
sample and also add one more roentgen optics element.   We hope to create an
excitation beam with angular deviation of 1 min of arc and 0.04 mm width.

SAMPLE PREPARATION

Microsamples are prepared by evaporating several calibrated drops of
water of interest on the surface of reflecting glass in the form of a spot of
15 mm diameter.

Special standard samples were prepared for calibration of the system.
They were made from a standard solution containing chemical elements in the
range of atomic numbers 19 through 39 with equal concentrations of 1 μg/l.

MEASUREMENTS

In Fig. 3 one can see a spectrum of a standard sample measured with a
Si(Li) semiconductor spectrometer placed at a distance of 1 mm above the
surface of the substrate glass.

Table 1.   Behavior of Q with Z (normalized to Ga)

Element	Q/Q_{Ga}	Element	Q/Q_{Ga}
Ti-19	1.190 ± 0.040	Ni-28	1.061 ± 0.016
V-23	1.050 ± 0.024	Cu-29	1.062 ± 0.014
Cr-24	1.025 ± 0.025	Zn-30	0.887 ± 0.018
Mn-25	1.021 ± 0.015	Ga-31	1
Fe-26	1.060 ± 0.021	Rb-37	0.883 ± 0.015
Co-27	1.019 ± 0.014	Y-39	0.635 ± 0.013

The peaks on the spectrum show a regular behavior increasing with atomic number. The area of each peak can be expressed as

$$A^{z}_{q,i} = Q \; \Omega \cdot \frac{S^{z}_{q} - 1}{S^{z}_{q}} \cdot \tau^{z}_{q} \cdot \omega_{q,i} \cdot \varepsilon(E^{z}_{q,i}) \cdot C^{z} \tag{1}$$

where: Q   - a constant proportional to the intensity of x-ray source;

Ω - geometric factor;

S^{z}_{q} - q-shell edge step ratio;

τ^{z}_{q} - total cross section for q-shell of element

$\omega_{q,i}$- fluorescence yield for q-shell;

$\varepsilon(E)$- energy dependence of the detector efficiency

C_{z} - concentration of the element Z.

An efficiency curve was measured using calibrated γ-ray sources. All the atomic data was taken from the X-Ray Data File XRDF.[3]

The constant Q must be independent of Z if the mathematical model of the measuring process is right and there is no loss of the chemical elements during sample preparation. In Table 1 the calculated values of Q for several elements are presented. All the values are normalized relative to Q for Ga as an internal standard.

Table 2.   Contents of metals in waste water samples in mg/l

Element	Sample SL3	Sample SL2
Mn	0.023 ± 0.010	0.028 ± 0.015
Fe	0.050 ± 0.008	0.174 ± 0.012
Ni	0.033 ± 0.008	0.053 ± 0.008
Cu	0.025 ± 0.007	0.055 ± 0.007
Zn	≤ 0.021	0.061 ± 0.006
Ga	1.0	1.0
Br	0.052 ± 0.004	0.029 ± 0.003
Sr	0.042 ± 0.004	0.118 ± 0.008

Table 2 presents, as an example, the results of measuring the contents of metals in a waste water. The sensitivity of such measurements is of the order of μg/l. The actual sensitivity can be determined after essential improvement of optical components of the system.

REFERENCES

1. H. Aiginger and P. Wobrauschek, A Method for Quantitative X-Ray Fluorescence Analysis in the Nanogram Region, Nucl. Instr. & Meth. 114, 157-158 (1974).
2. H. Schwenke and J. Knoth, A Highly Sensitive Energy-Dispersive X-Ray Spectrometer with Multiple Total Reflection of the Exciting Beam, Nucl. Instr. & Meth. 193, 239-243 (1982).
3. I. A. Kondurov, P. A. Sushkov, T. M. Tukavina, G. I. Shulyak, Using *a priori* Information in EDXRF Analysis of Complex Samples (published elsewhere in these proceedings).

URANIUM CONCENTRATION MEASUREMENT IN WATER SAMPLES WITH TXRF

F.Hegedus and P.Winkler

Paul Scherrer Institute
CH-5232 Villigen Switzerland

ABSTRACT

The Total Reflection X–Ray Fluorescence method was used to detect low level (ppb) uranium concentration in water.

INTRODUCTION

The Total Reflection X–Ray Spectrometer of the Paul Scherrer Institute (detailed description is given in Ref.1) was used to measure uranium in water sediments. The optimum condition for detection was obtained by using a Mo X-ray tube as the primary X–ray source. The beam was collimated by means of two 0.05 mm slits. The Mo Ka (E=17.5 keV) beam can excite the U–LIII shell (16.7 keV), but can not excite the U–LII (21.0 keV) or the U–LI (21.7 kev) shells. Therefore only La and Lb lines are induced. Only the La lines (13.62 and 13.44 keV) can be detected because the Lb lines (16.5-17.5 keV) are interfering with the Mo Ka line (17.4 keV) of the scattered primary beam. The intensity of the U–La line is high because the primary beam energy (17.5 keV) is ideal to excite the U–LIII shell.

SAMPLE PREPARATION

The water sample was concentrated 50 times by slow evaporation. Ten times 5 μl of water was pipetted on the quartz sample holder. By that means 50 X 10 X 5 μl = 2.5 ml of sample sediment was deposited on the quartz glass. The diameter of the deposit was 3 to 4 mm.

MEASUREMENT AND RESULTS

The U–La1 peak intensity vs. U weight calibration was performed in the range of 0.1 ng to 50 ng. The resulting curve was linear. Fig.1. shows the spectrum of a sample containing 1.0 ng of uranium.

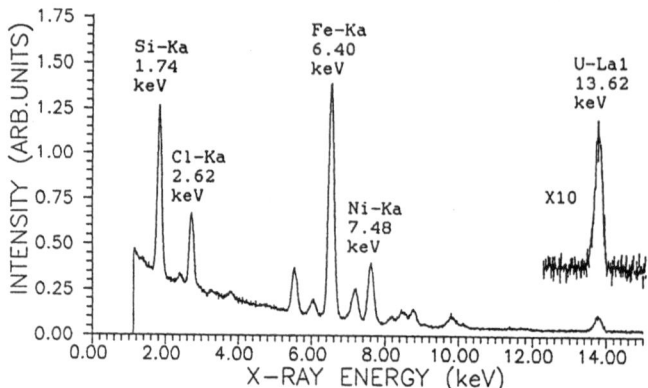

Fig.1.   X-ray intensity vs. energy.   Sample: 1.0 ng U dissolved
         in 5 µl water.

In most of the water samples uranium was not detected. Fig.2. shows a
sample in which 0.5 ng (0.2 ppb) of uranium was measured.

DISCUSSION OF THE RESULTS

Fig.2. shows that many elements were detected in the natural water
samples. The U-Lal (13.62 keV) line is between the Rb-Ka (13.40
keV) and the Sr-Ka (14.17 keV) lines. For that reason a sophisticated evaluation program was used to evaluate the peak intensity
of the U-Lal line.   Under these bad conditions the 3σ detection
limit, with 1500 s counting time, was as high as 300 pg, corresponding
to 0.1 ppb. In water samples with low concentrations of Rb and Sr,
the detection limit will be at least ten times lower.

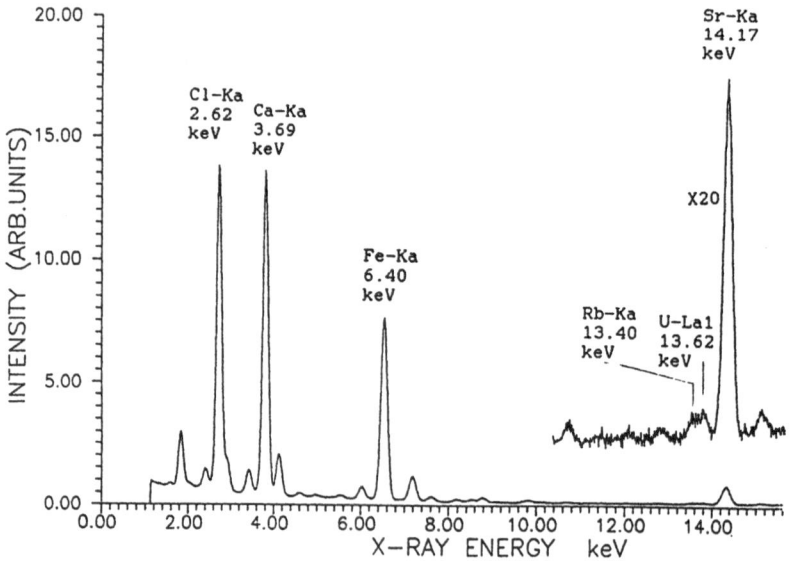

Fig.2.   X-ray intensity vs. energy.   Sample: 2.5 ml natural water.

REFERENCE

1. F.Hegedus, P.Winkler, P.Wobrauschek and C.Streli: TXRF Spectrometer
   for Trace Element Detection, Advances in X-Ray Analysis, Vol.33.

APPLICATION OF TOTAL REFLECTION X-RAY FLUORESCENCE SPECTROMETRY TO DRUG ANALYSIS

S.Nomura, T.Ninomiya[*], K.Taniguchi and S.Ikeda[**]

Osaka Electro-Communication University, Osaka, Japan

[*] Hyogo Pref. Police H.Q., Hyogo, Japan

[**] Ryukoku University, Shiga, Japan

ABSTRACT

A new procedure for indirect determination of pharmaceutical drugs is presented. The procedure consists of extracting ion pairs between organic basic compounds, that is, pharmaceutical drugs and cobalt tetrathiocyanate and of determining Co contents in the organic extraction phase using total reflection X-ray fluorescence spectrometry. Quinine, papaverine and pilocarpine are used as pharmaceutical drugs and 1,2-dichloroethane is adopted as extraction medium. Quinine cobalt tetrathiocyanate complex was isolated and was analyzed by FT-IR (Fourier Transform Infra-Red spectroscopy) suggested that approximately 4 ng of quinine could be detected by 1 μl of sampling of organic phase under ideal conditions using total reflection X-ray fluorescence spectrometry. These drugs have their own optimal pH for extraction. This technique can be applied to cocaine analysis.

INTRODUCTION

In the field of forensic science, drug analysis takes an important role in chemical identification. TLC (thin layer chromatography), FT-IR, GC (gaschromatography) and GC-MS (gaschromatograph-massspectrometry) are often used as useful methods for drug analysis. These methods, however, require specific pretreatment on specimens respectively. On the other hand, total reflection X-ray fluorescence spectrometry (TXRF) is a rather new technique for trace elemental analysis. Yoneda and Horiuchi had reported TXRF as a microanalytical technique.[1] The application of TXRF to forensic specimens has been reported by our group.[2-5] Then it seemed to be impossible that TXRF would apply to organic drug analysis because many organic pharmaceutical drugs have no heavy elements as ingredients. However it is known that some alkaloids or drugs can form ion pairs with cobalt tetrathiocyanate and the ion paired complexes being formed can be extracted into organic medium. On the basis of this principle, C.Nerin et al. reported the indirect drug analysis using AAS

Table 1 Measuring Conditions

X-ray target	Mo
Exciting voltage	20kV
Tube current	30mA
Glancing angle	0.03°
Area of Excitation	10mm × 10mm
Detector	SSD=Si(Li)
Area of detector	30mm²
Sample volume	1μl
Sample holder	Pyrex glass
	(26mm × 19mm × 1mm)

(atomic absorption spectrometry).[6]   So that it may be possible to determine indirectly some drugs by detecting Co element using TXRF when those drugs could form ion paired complexes with cobalt tetrathiocyanate.

In this paper, indirect drug analysis using TXRF has been reported on three drugs on the basis of forming drug-ion paired complexes with cobalt tetrathiocyanate.

EXPERIMENTAL

Instrumentation

The schematic diagram of a prototype TXRF spectrometer is shown in Fig.1.   The measuring conditions are summarized in Table 1.   The specimens were dropped on the Pyrex glass reflector.   The angle of the incident X-ray beam on the reflector is adjusted to 0.03 degree. Standards used to determine sensitivity factors for Co element were counted for 200 seconds real time.   Data for drug determination were obtained with 600 seconds real time measurement.

A Horiba F-14 pH meter was used for pH measurements, and a Nicolet 5SXC FT-IR spectrometer was used having a Spectratech microanalysis system.

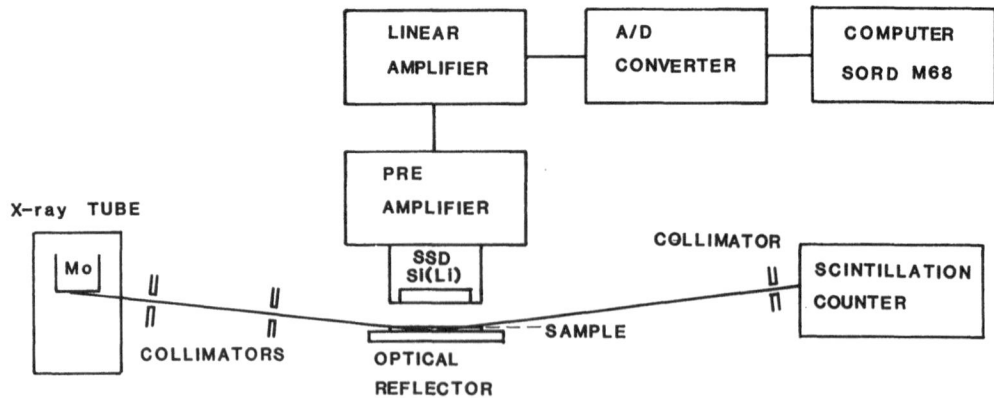

Fig. 1 Schematic diagram of TXRF spectrometer.

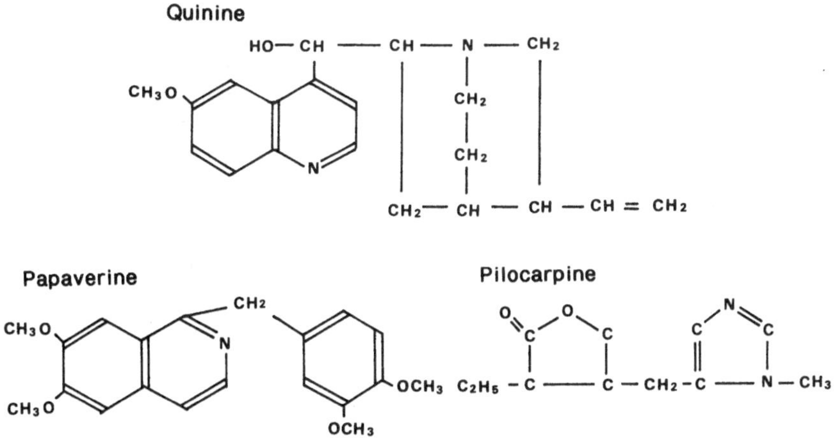

Quinine

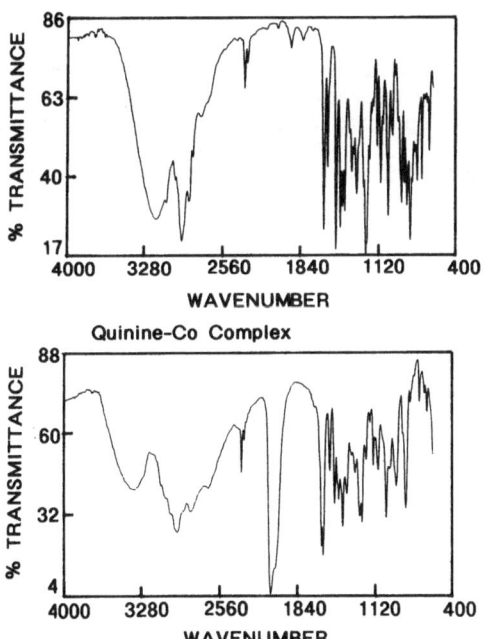

Fig. 2 Formulas of quinine, papaverine and pilocarpine.

Solutions

Co standard solutions were prepared from commercial standard solution (Wako pure chemical Co. Ltd.). Standard tetrathiocyanate cobalt(II) solution was prepared by dissolving 571 g of NH_4SCN and 437 g of $Co(NO_3)_2 \cdot 6H_2O$ in distilled water to give 1500 ml of solution. Aqueous solutions of three alkaloids were used: quinine (Wako Pure Chemical), papaverine hydrochloride (Sigma), and pilocarpine hydrochloride (Sigma).

Fig. 3 FT-IR spectra of quinine(upper) and of
        Co complex of quinine(lower).

Procedures

In separatory funnels of 100 ml volume, a fixed amount of alkaloid (5 mg for papaverine hydrochloride, 5 mg for quinine and 3 mg for pilcarpine hydrochloride), 10 ml stock solution of $Co(SCN)_4^{2-}$ (20 ml stock solution of $Co(SCN)_4^{2-}$ for quinine), solution of 0.1 molar HCl or NaOH up to a determined pH and distilled water up to 50 ml are added.   Ten milliliters of 1,2-dichloroethane is added, then the mixture is shaken for five minutes and the organic extract is filtered through 9 cm P/S filter paper.   And 1 µl of the organic extract was spotted on the Pyrex glass and the extractingsolvent was dried off.

The intensities of Co Kα in the organic phase were measured.   To examine optimal pH for those drugs, the same procedure was carried out with different value in the range of pH 1-7.

RESULTS AND DISCUSSION

The formulas of three drugs are shown in Fig. 2.   In the case of quinine, cobalt tetrathiocyanate complex of quinine was isolated.   Its FT-IR spectrum was shown in comparison with that of quinine only in Fig. 3.   This spectrum indicates the coexistence of both SCN group and quinine skeleton.   C.Nerin et al. mentioned that the weak complex cobalt tetrathiocyanate ion should be stabilized by the formation of an ion pair with alkaloids as $(alkaloid)_2Co(SCN)_4$.[6] The result from FT-IR spectrum supports the suggestion by C.Nerin et al.[6]

The working curve for Co standard is shown in Fig. 4.   By use of the calibration curve approximately 0.4 ng of Co element can be detected as low detection limit of Co so that 4.4 ng of quinine can be calculated on the basis of the molecular formula $(quinine)_2 -Co(SCN)_4$ complex in principle.

Figure 5 shows the results obtained with three drugs studied against pH change for extraction and with blank without any drugs.   It has been proved that the extract in case of blank solution containing no drugs, could not pick up any Co compounds.   The result indicates that 1,2-dichloroethane is a good solvent for extracting cobalt tetrathiocyanate complex of drugs as mentioned by C.Nerin et al.[6]   Also

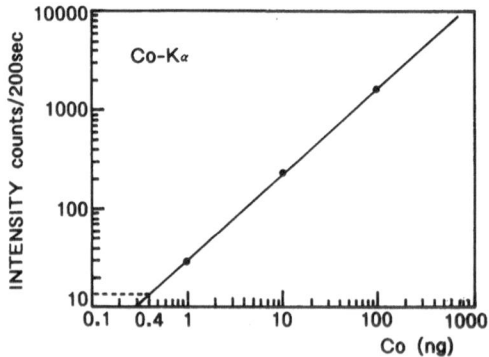

Fig. 4 Working curve for Co element.

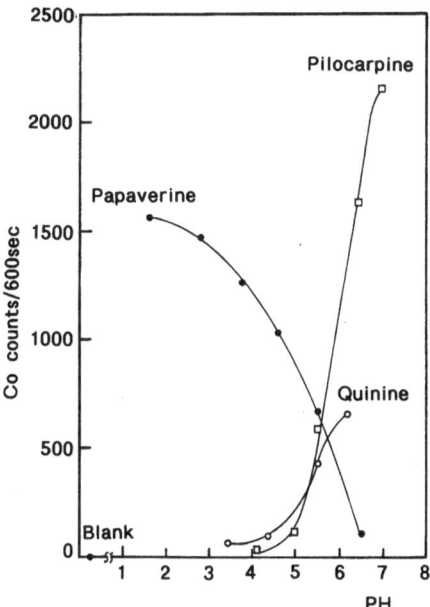

Fig. 5   Influence of pH in the determination of
          drugs by TXRF.

it can be seen that the extraction of cobalt tetrathiocyanates of three drugs were strongly affected by the pH of the solution.   On the basis of this result, at least papaverine could be determined selectively against pilocarpine at nearly pH 2.   In the case of quinine, extraction efficiency of $Co(SCN)_4^{2-}$ complex of quinine became fairly lower when amounts of $Co(SCN)_4^{2-}$ solution was 10 ml which was half of the case shown in Fig. 5.   Therefore it can be expected that pilocarpine could be determined differentially agaist quinine at near pH 7 under the condition of 10 ml of $Co(SCN)^{2-}$ solution.

As conclusions[4], indirect determination of drugs by TXRF technique requires only trace amounts of drugs while AAS method by C.Nerin et al.[6] needed mg order of drugs sample and also in case of TXRF method drugs could be recovered while AAS method should consume drugs perfectly. Moreover it can be expected that a certain drug could be determined differentially among other drugs by selecting optimal conditions,for example, pH, amounts of $Co(SCN)_4^{2-}$ solution and so on.   In fact, it has been reported by Inoue et al. that Scott reagent as a screening reagent should be specific for cocaine among 105 kinds of drugs examined.[7]   Scott reagent is a modified cobalt thiocyanate solution so that this TXRF method with Scott reagent can also be applied to cocaine analysis.

REFERENCES

1. Y.Yoneda and T.Horiuchi, Optical Flats for Use in X-Ray spectrochemical
   Microanalysis, Rev.Sci.Instrum., 42, 1069(1971)
2. T.Ninomiya, S.Nomura and K.Taniguchi, Application of Total Reflection
   X-Ray Fluorescence Analysis to Forensic Model Samples, Memoirs of
   Osaka Electro-Communication University, 22, 51,(1986)

3. S.Nomura, T.Ninomiya and K.Taniguchi, Trace Elemental Analysis of Titanium Oxide Pigments using Total Reflection X-Ray Analysis, Advances in X-Ray Chemical Analysis Japan, 19, 217(1988)
4. T.Ninomiya, S.Nomura and K.Taniguchi, Elemental Analysis of Trace Plastic Residuals using Total Reflection X-Ray Fluorescence Analysis, Advances in X-Ray Chemical Analysis Japan, 19, 227(1988)
5. T.Ninomiya, S.Nomura, K.Taniguchi and S.Ikeda, Quantitative Analysis of Arsenic Element in a Trace of Water Using Total Reflection X-Ray Fluorescence Spectrometry, Adv. X-Ray Anal.,32:199(1989)
6. C.Nerin and A.Garnica, Indirect Determination of Alkaloids and Drugs by Atomic Absorption Spectrometry, Anal. Chem., 57, 34(1985)
7. T.Inoue, K.Tanaka and T.Niwase, Field Tests for Cocaine, REPORTS of the National Research Institute of Police Science, 41, 225(1988)

AN INTEGRATED X-RAY TUBE - POLARIZER EDXRF-SPECTROMETER

IN CARTESIAN GEOMETRY

Robert Rieder, Peter Wobrauschek and
Hannes Aiginger

Atominstitut der Österreichischen Universitäten
Schüttelstraße 115, A-1020 Wien, Austria

ABSTRACT

Linear polarized x-rays are used as exciting radiation to reduce the background and to improve the detection limits in XRF. A combination of an x-ray tube and a measuring module realizing a compact cartesian geometry was constructed. An on-line adjustable Bragg-polarizer, a Barkla-polarizer or a secondary target can be used optionally. The anode material of the tube can be changed easily.

INTRODUCTION AND THEORY

Energy-dispersive x-ray fluorescence (EDXRF) is a powerful method for elemental analysis of various samples. With standard detectors elements from $Z=11$ upwards can be detected. To improve the detection limits (DL), the use of polarized radiation produced by scattering at an angle of 90° has been shown to be an appropriate excitation source [1-9].

Generally the DL of an XRF method can be improved by increasing the counting rate (e.g. by using a higher tube current). In EDXRF usually the radiation source is not the limiting point, it is the detection system which is determining the maximum counting rate. The counting rate depends on two contributions: the background and the counts belonging to the signal measured within a certain time. Using direct excitation, the background resulting from scatter processes is mainly responsible for the counting rate - especially when analyzing infinitely thick samples with a low Z matrix. Therefore various special methods for background reduction in XRF have been developed. One of those is XRF with polarized radiation. An electromagnetic wave propagating parallel to the x-axis of an cartesian system and scattered elastically into the y-direction will be linear polarized with a polarization vector parallel to the z-axis. In the energy range used in XRF this fact is also valid in good approximation for inelastic scatter. Now a second scattering process will produce no intensity into the z-direction. In XRF using polarized radiation in a cartesian geometry, the second scatterer is the sample and to reduce the background the detector is located in the z-direction. The first scatterer can be either a Barkla-polarizer or a Bragg-polarizer.

The most suitable material for a Barkla-polarizer is a low Z material with a high density [10]. Boroncarbide (B_4C) fulfills these requirements quite good and therefore it is the most used polarizer. Just for high energies the polarizer should be a material with a higher atomic number (e.g. Al (Z=13)). Since a Barkla-polarizer scatters the whole primary spectrum, it will be preferred for bulk analysis. The excitation conditions are sufficient over a wide energy range of the elements to be analyzed. Another point which makes the Barkla-polarizer attractive is, that there is no fine-adjustment of the polarizer with respect to the x-ray beam necessary. The disadvantages of a Barkla-polarizer are, that the unpolarized fraction of the exciting radiation (mainly caused by multiple scattering in polarizer and sample) increases the background in the full range of the spectrum and that the scatter intensity per solid angle unit is rather low.

When using a Bragg-polarizer, this disadvantages of a Barkla-polarizer can be overcome. It has been shown, that a crystal with a lattice spacing, which is suitable to the characteristic energy of the anode material of the tube for Bragg-reflection at $2\theta=90°$, is an excellent polarizer [3-7,11,12]. The excitation radiation will be either monoenergetic or it will consist of several energies if there are harmonics. So good excitation only can be expected for the elements with an absorption edge immediately under the Bragg-reflected line(s). On the other hand, the monoenergetic excitation spectrum simplifies a quantitative analysis performed by the fundamental parameter method [13].
It should be noted, that the characterization Barkla-polarizer and Bragg-polarizer represents a simplified image of the reality. A polarizer will always work simultaneously as a secondary target. In the case of a Barkla-polarizer, which is a low Z material, this fact won´t matter. But a Bragg-polarizer sometimes is a material, which might be excited excellent by the primary radiation, thus emitting its characteristic radiation and so increases the non-polarized fraction of the radiation exciting the sample dramatically. If one chooses a low Z crystal as Bragg-polarizer, it simultaneously works as Barkla-polarizer.

An important fact which has to be considered in designing an XRF spectrometer with cartesian geometry is, that the detection limits will decrease if the solid angles of the collimators are increased [10,14,15]. Enlarging the collimator divergences of course will cause a loss in the polarization factor, but this disadvantage is more than compensated by the increase of intensity. The sometimes suggested optimization between the degree of polarization and intensity won´t achieve its purpose. So one can expect the best DL with collimator diameters, which lead to the maximum counting rate of the detection system (under the precondition, that the x-ray tube is already operated at full power). Compared to direct excitation maintaining the same counting rate, now the peak-to-background ratio is much better.

As a result of this considerations an XRF module realizing the cartesian geometry with shortest distances, high flexibility and comfortable adjustment was constructed. Since the for this purpose available x-ray tubes have the inherent disadvantage of a rather large focus-window distance, a special tube, which can be coupled easy to the module, has been designed.

EXPERIMENTAL EQUIPMENT - DESIGN

The constructed x-ray tube is a so called open tube, which means, that it can be disassembled by easy removing several screws. The design is shown in Figure 1. Part (1) is the neck of the x-ray tube. On the upper side it has to be mounted to the high voltage connector (that device, which connects the high voltage cable with the cathode inside the tube). (2) and (11) are the drillings for the turbomolecular pump and the gauge to measure the vacuum. The position of the tungsten filament is show in (3).

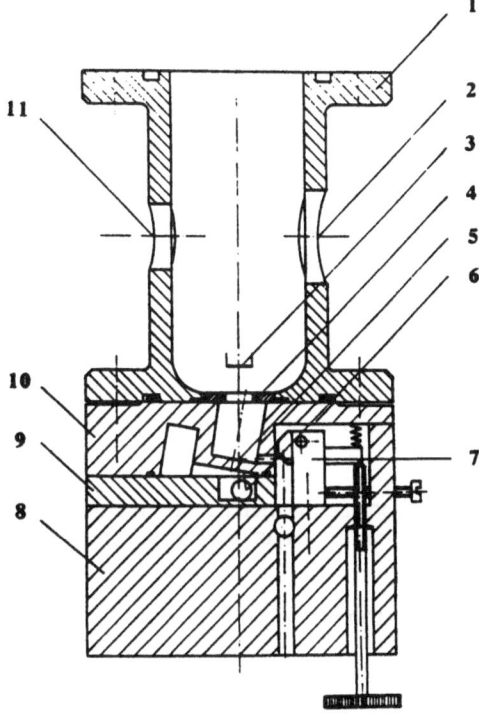

Figure 1

On the other side of part (1) the anode block (10) is fixed. Between the neck of the tube and the anode block is an exchangeable electron diaphragm to define the electron focus on the anode (4). In the center of the block is the anode plate (5) and right of the plate is the x-ray outlet sealed by a 30 μm thick aluminium window. The anode plate is tilted to give a take-off angle of 10°. Since the focus on the anode is about 1.5 mm * 8 mm, an optical size of about 1.5 mm * 1.4 mm is achieved. In the cooling-cap (9) at the backside of the anode block there is a nozzle with a profile according in size and position to the focus on the anode. Approximately 6 liters cooling water per minute flow over the backside of the anode (6) and later on around the entire anode chamber. The tube has already been operated at 40 kV / 25 mA with a vacuum at about $5*10^{-6}$ mbar.

In the present x-ray tube there is a molybdenum disk as anode material soldered in the anode block. If there is manufactured a set of anode blocks with different anode disks, the XRF spectrometer in combination with several polarisators will be a highly flexible analytical tool optimizable for various samples.

The XRF module (8) is fixed by a fitting bolt and a screw to the x-ray tube. It contains a kind of on-line controllable goniometer (7) to adjust the polarisator in the case of Bragg-reflection. Each piece of the set of polarisators has been glued on an aluminium base, which can be inserted into the goniometer and fixed by a screw. The drilling for the polarized beam to the sample passes the sample chamber and so enables the measurement the spectrum of the secondary radiation. All three drillings defining the cartesian geometry are coated with molybdenum sheets to guarantee an acceptable blank

Table 1.Dimensions of the cartesian geometry

beam path from - to	collimator diameter in mm	distance in mm
focus (center) - polarizer	2.5	14
polarizer - sample	5	20
sample - detector window	6	16

measurement. If the entrance for the polarized beam into the XRF module is sealed by capton foil, it is possible to evacuate the beam path between the sample and the detector.

Compared to existing cartesian geometries the distances could be reduced drastically. This was achieved due to the special design of the tube, which has focus (center) - window distance of only 11 mm. All relevant distances along the three axis and the collimator diameters are given in table 1. Of course this distances are not equal to the length of the collimators.

The set-up of the fully assembled spectrometer is so, that the inserted sample carrier cup is in horizontal position. As a benefit of this fact, measuring of liquid or powder samples is enabled.

The materials used for the spectrometer are brass for the XRF module and copper for the parts (1), (4), (9) and (10).

Up to now measurements with two different polarizers have been made: Mo (611) and highly oriented pyrolythic graphite HOP-C (00.14).

RESULTS

Test measurements have been performed with three types of samples: single elements in aqueous dilution, NBS standard Orchard Leaves SRM 1571 (pressed pellet)

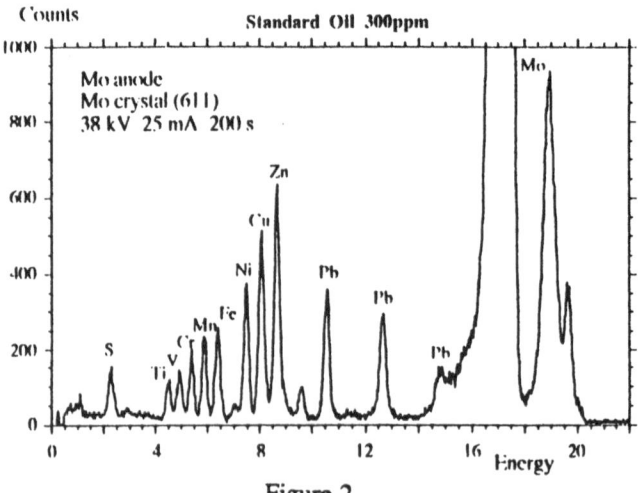

Figure 2

Table 2. Detection limits for various samples, the polarizer is Mo (611)

sample	DL
Se in aqueous dilution	0.65 ppm
Rb (Orchard Leaves)	0.51 ppm
Fe (Oil, 300 ppm standard)	3.90 ppm

Table 3. Detection limits for various samples, the polarizer is HOP-C (00.14)

sample	DL
Fe in aqueous dilution	0.59 ppm
Rb (Orchard Leaves)	0.67 ppm
Fe (Oil, 300 ppm standard)	0.83 ppm

and metals in oil. For all samples both types of polarizers - Mo (611) and HOP-C (00.14).- have been used. The first one acts as a Bragg-polarizer and secondary target simultaneously, the second acts as Bragg- and Barkla-polarizer. The x-ray tube has been operated at 38 kV / 25 mA and the time for collecting the samples was 200 seconds.

Under this conditions the dead time indicated by the MCA was in the range of 10% to 15%. The figures 2 and 3 show the spectra of oil measured with the two polarizers described above. The concentration of the metals present in oil is 300 ppm.

The achieved detection limits with the both polarizers are presented in table 2 and 3. They are extrapolated to 1000 seconds measuring time.

In the case of monoenergetic excitation the DL are just for elements with an absorption edge close to the primary line in the sub-ppm range. When using the HOP-C polarizer, also elements with lower energies reach this range.

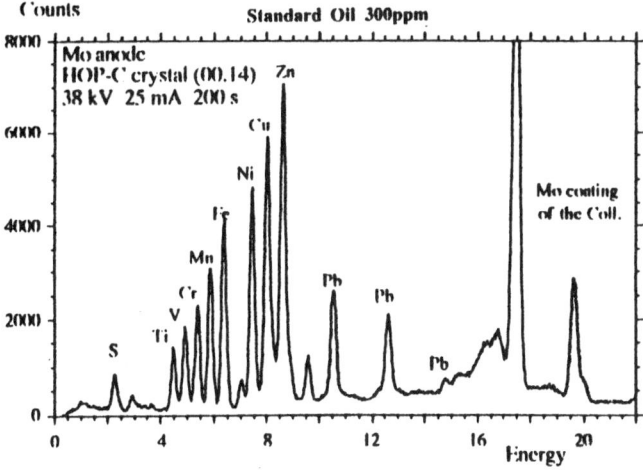

Figure 3

SUMMARY

A high flexible XRF spectrometer for EDXRF with polarized excitation radiation has been constructed. It is designed as an analytical tool for elements from Z=11 to Z=40 in samples with a low Z matrix. Anode material and polarizer can be chosen to optimize the system according to the samples for lowest detection limits. The existing spectrometer has a Mo-anode and Mo (611) or HOP-C (00.14) as polarizer. Detection limits almost down to 500 ppb can be achieved. If one is interested in elements in the range from Z=11 to Z=27, the combination Cu-anode and Cu (113) as polarizer should deliver even better results due to the high reflectivity of the Cu (113) lattice planes.

REFERENCES

[1]   T.G. DZUBAY, B.V. JARRET and J.M. JAKLEVIC, Nucl. Instr. Meth. 115, 297 (1974)

[2]   R.H. HOWELL, W.L. PICKLES and J.L. CATE Jr., Adv. X-Ray Anal. 18, 265 (1974)

[3]   H. AIGINGER, P. WOBRAUSCHEK and C. BRAUNER, Nucl. Instr. Meth. 120, 541 (1974)

[4]   P. WOBRAUSCHEK and H. AIGINGER, X-Ray Spectrom. 9, 57 (1980)

[5]   P. WOBRAUSCHEK and H. AIGINGER, X-Ray Spectrom. 12, 72 (1983)

[6]   P. WOBRAUSCHEK and H. AIGINGER, Adv. X-Ray Anal. 28, 69 (1985)

[7]   R.W. RYON, J.D. ZAHRT, P. WOBRAUSCHEK and H. AIGINGER, Adv. X-Ray Anal. 25, 63 (1982)

[8]   W.E. MADDOX, Adv. X-Ray Anal. 27, 519 (1984)

[9]   W.E. MADDOX and W.C. KELLIHER, Adv. X-Ray Anal. 29, 497 (1986)

[10]   R.W. RYON and J.D. ZAHRT, Adv. X-Ray Anal. 22, 453 (1979)

[11]   J.D. ZAHRT, Adv. X-Ray Anal. 26, 331 (1983)

[12]   J.D. ZAHRT, Nucl. Instr. Meth. A242, 558 (1986)

[13]   R.M. ROUSSEAU, X-Ray Spectrom. 13, 115 (1984)

[14]   R.W. RYON, Adv. X-Ray Anal. 20, 575 (1977)

[15]   R.W. RYON and J.D. ZAHRT, UCRL Report 102936, Lawrence Livermore Lab. (1990)

VARIABILITY OF CRYSTAL PERFORMANCE IN X-RAY FLUORESCENCE

SPECTROMETERS

Jules V.Dubrawski and Ken E.Turner

BHP Research Newcastle Laboratories
PO Box 188, Wallsend, NSW 2287, Australia

ABSTRACT

The intensity performance of various common XRF crystals (LiF200, Ge, PET and LSM) has been investigated under fixed experimental conditions. Measured were the raw intensities of elements particularly suitable to a given crystal, and peak-to-background ratios at different elemental concentrations. Peak profiles for each crystal type were also recorded and analysed in terms of apparent crystallite size and mosaic spread. In general the LiF crystals showed least variability in performance, whereas the PET crystals demonstrated the most.

INTRODUCTION

The performance of spectrometer crystals has a significant bearing upon the quality of results produced by any XRF instrument. A variety of crystals have been commonly employed[1,2] such as LiF200, Ge111, PET, KAP and T/AP. More recently layered synthetic microstructure crystals, LSM, have become available. LSM are large d-spacing crystals suitable for light element analysis. Such crystals with d-spacings of approximately 55Å are suitable replacements for the acid phthalates (eg.T/AP). In this study the performance of a range of crystal types including two specimens of LSM has been evaluated.

EXPERIMENTAL

Evaluated were four pairs of crystals, situated in two Siemens SRS-1 spectrometers designated XRF#1 and #2. The crystal pairs consisted of the types LiF200, Ge111, PET and LSM. Using conventional fusion techniques glass beads were prepared containing elements suitably sensitive to each crystal. High and low concentrations of each element were prepared. In each spectrometer the beads were irradiated by the same chromium X-ray source operated at 50kV and 55mA. Peak and background intensity measurements were made using a 40sec counting period for Fe and 100sec period for P, Si and Mg. Kα lines were used for each element. Concentrations were expressed relative to the sample flux system used. Measurements were also made with coarse (0.4º) and fine (0.15º) collimators. Intensity profiles for each crystal were obtained and the apparent average crystallite size determined from the Scherrer equation[3],

$$L = K\lambda/B \, Cos \, \theta$$

where the shape constant K, was assumed to be 0.9.

RESULTS AND DISCUSSION

The intensity data obtained from the two sets of crystals in different spectrometers are shown in Tables 1 and 2. As expected the intensities in all cases are higher when coarse rather than fine collimators are used. The performance of each crystal was determined from the peak to background ratio (PTB) obtained in each spectrometer. The results calculated for each crystal set are given in Tables 3 and 4. Two effects were considered. Firstly the performance of the crystal themselves within the two sets and secondly, the influence of the spectrometers.

Table 1   Intensity Data for Crystal Set 1 in Different Spectrometers

Crystal	XRF #1 Pk	Bgd	XRF #2 Pk	Bgd	Collimator	% Conc.
LiF200	54774	394	50406	404	C	(Fe) 56.4
	25819	61	17106	61	F	"
	2339	291	2033	293	C	2.3
	1092	48	5672	38	F	"
Ge111	7672	84	6291	86	C	(P) 15.1
	2467	21	1296	19	F	"
	500	46	425	48	C	0.9
	162	12	89	10	F	"
PET	3935	29	3500	35	C	(Si) 46.5
	1280	7	671	8	F	"
	141	26	126	33	C	1.2
	44	6	24	7	F	"
LSM	4852	150	5620	205	C	(Mg) 58.3
	1415	36	1530	41	F	"
	273	112	372	185	C	2.2
	77	30	90	37	F	"

Table 2   Intensity Data for Crystal Set 2 in Different Spectrometers

Crystal	XRF #1 Pk	Bgd	XRF #2 Pk	Bgd	Collimator	% Conc.
LiF200	39529	292	41662	305	C	(Fe) 56.4
	13907	66	17197	53	F	"
	1725	206	1733	222	C	2.3
	604	44	695	34	F	"
Ge111	5746	59	6185	59	C	(P) 15.1
	1669	15	1813	13	F	"
	372	20	398	20	C	0.9
	107	6	117	5	F	"
PET	6753	18	6326	19	C	(Si) 46.5
	2599	5	1702	4	F	"
	198	10	180	11	C	1.2
	75	3	47	2	F	"
LSM	7078	167	6754	196	C	(Mg) 58.3
	2258	41	1805	40	F	"
	349	112	403	170	C	2.2
	107	30	101	35	F	"

Table 3   Peak to Background Ratios for Crystal Sets in Different
Spectrometers (Coarse Collimators)

Crystal	% Conc.	XRF#1		XRF#2	
		Set#1 Pk/Bgd	Set#2 Pk/Bgd	Set#1 Pk/Bgd	Set#2 Pk/Bgd
LiF200	56.4	139	135	125	137
	2.3	8.0	8.4	6.9	7.8
Ge111	15.1	91	97	73	105
	0.9	11	19	8.8	20
PET	46.5	136	375	100	332
	1.2	5.4	20	3.8	16
LSM	58.3	32	42	27	35
	2.2	2.4	3.1	2.0	2.4

Table 4   Peak to Background Ratios for Crystal Sets in Different
Spectrometers (Fine Collimators)

Crystal	% Conc.	XRF#1		XRF#2	
		Set#1 Pk/Bgd	Set#2 Pk/Bgd	Set#1 Pk/Bgd	Set#2 Pk/Bgd
LiF200	56.4	369	211	280	325
	2.3	23	14	18	20
Ge111	15.1	118	111	68	140
	0.9	14	18	9	23
PET	46.5	183	520	84	426
	1.2	7.3	25	3.4	24
LSM	58.3	39	55	37	45
	2.2	2.6	3.6	2.4	2.9

From the PTB values obtained using coarse collimators, the commonly used LiF crystals were found to perform similarly in both spectrometers and at both concentration levels. However some variability was observed using fine collimators.

The Ge crystals produced some variation in PTB values which suggested that the crystal in set 2 performed marginally better. With fine collimators a marked difference was observed for the crystal in set 1 in XRF#1 and #2.

The largest discrepancy in performance, however occurred with the PET crystals. Clearly the PET of set 2 yielded far superior PTB values, irrespective of the spectrometer and collimators used. It was noted that the crystal of set 1 was much darker in appearance and suspected of radiation damage due to a history of more exposure to the X-ray beam.

From the results of Tables 3 and 4 it is apparent that the PTB values for both LSM crystals are signficiantly lower than other crystals, due to high backgrounds in relation to peak intensities[4]. However, significant differences in PTB between LSM crystals were observed, and the performance of crystal 2 was generally superior.

Crystal performance was also influenced by the spectrometer and in this respect XRF spectrometer #2 produced a detrimental effect. Almost without exception the PTB values for

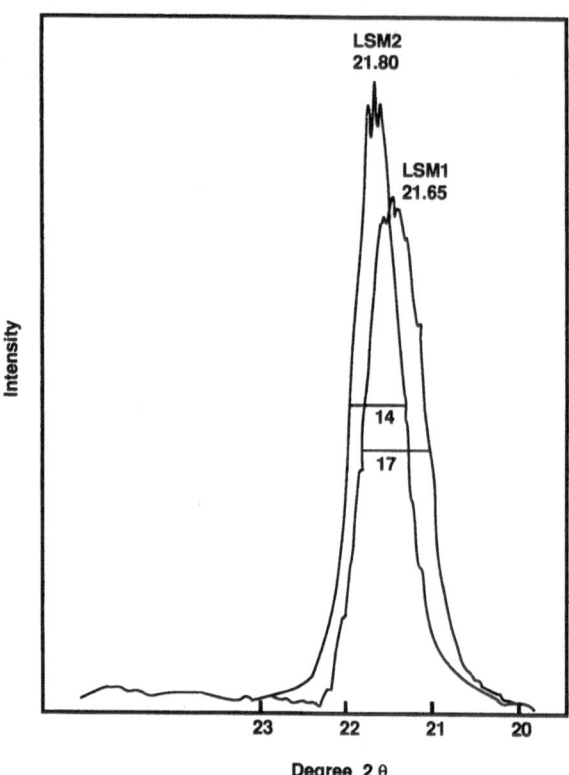

Figure 1. Intensity profile of LSM crystals

all crystals decreased when switched from XRF#1 to #2. The drop in PTB occurred for both fine and coarse collimators, and was quite marked in some cases (eg. PET and Ge of set 1). The influence of spectrometers may stem from differences in alignment or possibly X-ray tube sample coupling.

The results indicate that analyses requiring fine collimation are best carried out using LiF from set 1 and Ge, PET and LSM from set 2, in XRF#1. The same arrangement would apply for coarse collimation and the analysis of less sensitive elements.

The intensity profile for each crystal was plotted as shown for the LSM crystals in Figure 1. From each was determined the apparent crystallite size L, using the Scherrer equation (Table 5). The largest variation occurred with LSM crystals. In addition, significant differences in d-spacing values resulted in the peaks being separated by 0.15° 2-theta. This is not uncommon with LSM crystals, resulting from the manufacturing process that controls layer

Table 5   Apparent Crystallite Size

Crystal	L(Å) Set 1/XRF#1	Set 2/XRF#2	% Variation
LiF200	241	222	8.2
Ge	241	218	10
PET	134	144	7.2
LSM	68	88	26

thickness. The higher L value of crystal in set 2, consistent with larger crystallites within the mosaic, corresponded with the higher PTB ratio.

The similarity of L values for LiF was consistent with similar PTB values, excluding spectrometer effects, while for Ge PTB ratios were more divergent, in agreement with their L values. No comparison can be drawn, however between PET crystals since radiation damage appeared to be the dominant factor affecting the performance of these crystals.

CONCLUSIONS

(1)   LiF200 crystals were found to perform similarly. Ge crystals were also generally similar.

(2)   A large difference in PTB was observed between PET crystals. The poorer specimen was suspected of radiation damage.

(3)   LSM crystals showed significant variability in performance. Measurement of apparent crystallite size indicated that a larger L value corresponded with higher PTB values and better performance.

(4)   Of the two spectrometers, XRF#2, was found to have the poorest sensitivity for all crystals. This may be due to small differences in spectrometer alignment and/or X-ray tube sample coupling.

ACKNOWLEDGEMENTS

The authors wish to acknowledge the assistance of Ms C. Ferguson in this study and the support of the Broken Hill Proprietary Company Ltd.

REFERENCES

1.   R. Jenkins, "An Introduction to X-ray Spectrometry", Heyden-Son, New York, 88, (1974).

2.   K.G. Carr-Brion and K.W. Payne, X-ray Fluorescence Analysis, The Analyst, 95, 977, (1970).

3.   H.P. Klug and L.E. Alexander, "X-ray Diffraction Procedures for Polycrystalline and Amorphous Materials", 2nd Ed. John Wiley & Sons, New York, 656, (1974).

4.   G.J. Mulheron, AXAA-88 Conf. Uni. WA, Perth, Australia, 14-19 Aug., 455, (1988).

AN X-RAY SPECTROMETER FOR PIXEL ANALYSIS OF ART OBJECTS

M.Mantler, M.Schreiner[*], F.Weber, R.Ebner,
F.Mairinger[*]

Institute of Applied and Technical Physics
Technical University, Vienna, Austria

[*]Institute of Chemistry
Academy of Fine Arts, Vienna, Austria

ABSTRACT

An x-ray spectrometer has been designed for pixel by pixel analysis along lines or across selected areas of paintings and other art-objects. Characteristic technical data are: $0.8mm^2$ pixel size, 800mm (vert.) by 1000mm (horiz.) by 200mm (perpendicularly to object) motion distances, $\pm 20\mu m$ precision in positioning the system, 2x3m maximum object size (mounted vertically); 2.8kW x-ray tube; Si(Li)detector. PC's are used for instrument control and new, complex data evaluation software.

INTRODUCTION

Chemical analysis of the materials used in old art-objects reveals important information pertaining to the history of the object, the circumstances of its creation, and the artist. It is also valuable and sometimes indispensable for restoration and conservation projects. Many methods of chemical analysis are routinely applied to this purpose, such as wet chemistry and various spectroscopical methods including electron microprobe analysis and XRFA. Additional information related to the structure of a painting and to hidden (overpainted) features is achieved by x-ray radiography and photography at optical, infrared and ultraviolet wavelengths.

Most of the methods of chemical analysis are destructive and require small samples to be taken from the object. This is, if at all, acceptable only from normally invisible areas and is

Advances in X-Ray Analysis, Vol. 35
Edited by C.S. Barrett *et al.*, Plenum Press, New York, 1992

a severely limiting factor. Only XRFA can prove to be a sufficiently nodestructive method of direct elemental analysis, if
used with care with respect to possible damages due to excessive radiative doses.

The presented paper describes an x-ray spectrometer designed for nondestructive analysis of small individual pixels,
selected areas, or scans along lines on large art objects, such
as paintings with sizes of up to 2x3 meters.

GENERAL DESIGN

Mechanical support system

X-ray tube, detector and additional devices are mounted on
a support system, which can be moved in three dimensions in
order to select the pixel to be measured and to adjust the distance between measuring system and object. The maximum motion
distances are 800x1000mm parallel and 200mm perpendicular to
the object.

A scheme of the assembly is shown in figure 1. The top
support desk is mounted on 4 vertical linear ball bearings on a
second desk, which in turn moves to and from the object. The
whole system rests on a third desk, which moves horizontally
along the object. All bearings are of high precision quality
and run on burnished shafts of about 25mm diameter. Provisions
on the top desk allow for additional equipment to be mounted at
defined distances to the x-ray beam, such as microscopes and
photographic cameras. The object is mounted in vertical position on a frame, which runs on rails along the instrument for
convenient placement to the measuring position.

A number of protective measures have been taken in order
to minimize the danger of any destruction of the often extremely valuable objects being examined. The moving parts are
mechanically and electronically blocked from running into the
object, in case of malfunctioning of control logic. In addition, highly touch-sensitive micro-switches are mounted in
front of the tube/detector assembly, which immediately shut
down the system when activated, bypassing all electronic
control circuits.

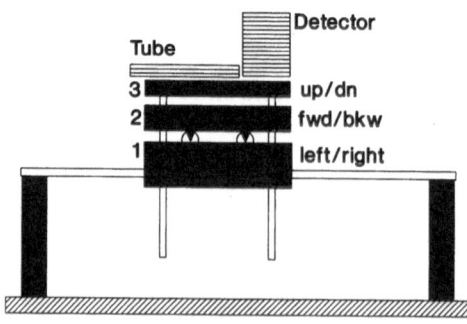

Figure 1. Mechanical Set-Up.

Excitation and detection system

The primary x-ray source is a conventional spectroscopy tube (Siemens AG60, W-target, max. 60kV/2.8kW), powered by a 3kW generator (Siemens Kristalloflex). The apparent target size observed perpendicularly to the electron beam is 6x7mm. A collimator and shutter unit is attached to the tube defining an effective beam diameter of 1mm by a pinhole at a target-distance of 181 mm. The collimator-object distance is normally less than 10mm and the illuminated area on the object is therefor practically equal to the area of the collimator-opening (0.8mm^2). Collimators for other beam diameters can easily be mounted. A Si(Li) detector (Tracor-Northern) is mounted at an angle of 30° to the primary beam with its entrance window at a distance of (normally) 25mm from the illuminated spot.

A beam trap behind the object runs synchronously with the tube/detector assembly and absorbs the primary beam penetrating the object. A set of lead-plated panels is mounted around all unshielded areas of the instrument for radiation protection.

Electronic Control System

The system is fully controlled by a personal computer (NCR, 80286) equipped with a MCA-card (Canberra, connected to an external Canberra amplifier/ADC and a Tracor Si(Li) detector); an IEEE488 card for 96 digital I/O lines (I/Otech; 80 bit TTL + 16bit open collector, used for various logical controls and switches as well as for 6x8bit parallel I/O of the position control system); an IEEE488 card for 4 analog output lines (I/Otech, 10bits, ±5V, 3 lines used for DC-motor controls for the mechanical positioning system); and a RS232 device for an external stepper motor controller for the beam trap system).

The output signals from the computer are fed into an interface designed to meet the electrical requirements of the various external devices. Its main components are the driver circuits for the three DC-motors (300Watts,2x70Watts), drivers and control logics for the solenoid of the shutter unit and safety circuits, digital displays showing the actual position of the x-ray system, and an interface for external joysticks for manual positioning and pixel selection.

Measurement of positions and distances. Optically encoded glass rulers with a resolution of 10µm (Heidenhain, Germany) are employed for the determination of the current position of the tube/detector assembly relative to the base support system (i.e. to the object). Computer controlled positioning is precise within ±2 resolution-units (±20µm) in each coordinate. A motion distance of 20mm is required for an initial self calibration procedure after a system-restart.

An additional laser-based device is used to determine the actual distance between the collimator of the x-ray tube and the object. This distance is kept constant within 0.1mm during any motion along the object. Such adjustments are necessary in the case of objects with uneven surfaces, as for example bent

wooden boards, in order to protect the object from collisions
with the collimator and to provide identical x-ray geometries
for the measurement of all pixels.

Selection of pixels, lines and areas to be measured. A
joystick device is mounted on a long cable and is provided for
manual control of the positioning system. Visual selection of
pixels is supported by a pointing laser which is mounted next
to the x-ray collimator with its beam parallel to the x-ray
beam and a known offset. A push-button selects the lit pixel
for measurement and commands the computer to store its coordinates. Optionally, the computer program combines the selected
points to lines (for line scans) or to closed polygons to
define areas for area scans. The density of the grid-points for
line and area scans can be arbitrarily selected by the
operator.

SOFTWARE

The software package developed for the system consists of
three parts: instrument control, primary evaluation of spectral
data, and advanced evaluation of spectral data. Instrument
control routines and the primary evaluation package are installed at the computer mentioned above. A second computer
(NCR, 80386) is used for data evaluation and has the two data
evaluation packages installed. The programs were mainly written
in Fortran, except for the device control routines and graphics
routines, which have been programmed in assembly language.

Instrument control. These routines allow to access all
instrument functions individually ("manually"), to set up a
measurement task, and to automatically carry out a complete set
of measurements. Accessable instrument functions include self
calibration of the positioning system, individual positioning
to user defined coordinates, and control of the shutter, of the
pointing laser, of the 'keep distance-to-object constant'-
option, and of the joystick-mode.

After set-up, all selected coordinates are stored in
ASCII-files and can be recalled and/or edited, if required.
Measurements can be repeated with the same absolute coordinates, as long as the object has not been removed from the system; interrupted measurement sequences continue at the point
of interruption even after complete shut-downs.

Data evaluation.

Primary evaluation of spectral data. The basic acquired
information are the individual spectra of each pixel and their
coordinates. All basic operations, such as graphics display,
scaling/zooming, comparison of two spectra, element markers,
and background modelling using polynomials are provided. Background subtraction and the determination of the resulting net
intensity for a line can be modelled for one selected spectrum
and the procedure is automatically repeated for all spectra of
a set.

Advanced evaluation of spectral data. This package is used
for two main purposes: to graphically display or plot data from
line scans and area scans (see the example below representing
element distributions over a measured area), and to use fundamental parameter methods to model a multiple layer structure
representing the different layers of paint and to compare such
data with experiments.

Problems of fundamental parameter approaches. The fundamental parameter routines are based on the theory of multiple
thin film analysis and computer programs for pertinent analyses, as developed by Mantler[2,3]. The assumptions made there
differ, however, from the 'real situation' in the case of
paints by the fact, that the individual (around 5) layers of
paint-structures are relatively thick (in classical paintings
several micrometers, in some cases even up to 100 micrometers
or more), and that they are not necessarily homogeneous. Each
layer consists of finely ground, generally inorganic pigment,
mostly from minerals, in an organic environment (dried oil).
The grain-size of the pigment varies by its kind and manufacturing process, and is almost always below 1μm.

The pigments are, in x-ray terms, of 'intermediate size'
where they can be neither seen as bulk nor as sufficiently
small for linear approaches of intensity versus sample mass. In
addition, paints (i.e. pigments) may have been mixed, but not
thoroughly intermingled, which adds a second kind of inhomogeneity. Results from fundamental parameter models are currently still of limited accuracy, but the only possibility for
estimates of the composition of invisible, overpainted areas.

RESULTS

The following example is discussed from a technical point of
view. A paper by Schreiner[1] published in this proceedings is
dedicated to interprete further results from measurements using
this spectrometer, in terms of paint-chemistry and art-history.
The original used for this example is an Indian miniature
painting dating from the 17th century. A section with an area
of 14x18cm is shown in figure 2. The headdress of the man
(marked area) was measured by using the area-scan feature. 5
points were selected to define the area and a grid density of
2.5mm was chosen, resulting in 205 pixels within this area. The
spectra of the 205 pixels were measured for 30 seconds each.
The system requires additional 25-35 seconds per pixel for
positioning and data manipulation, resulting in a total time of
3½-4 hours. An energy window was set for each elemental line of
interest and background regions were defined for subtraction by
using 3rd-order polynomial fits. The procedure, as defined
once, was automatically repeated for all 204 additional spectra
by the primary-data-evaluation program. The result is a data-
file containing 205 net-intensities for each selected element,
and the pixel-coordinates.

The advanced evaluation program was used to display the element-maps on the screen. Among the features of the program are

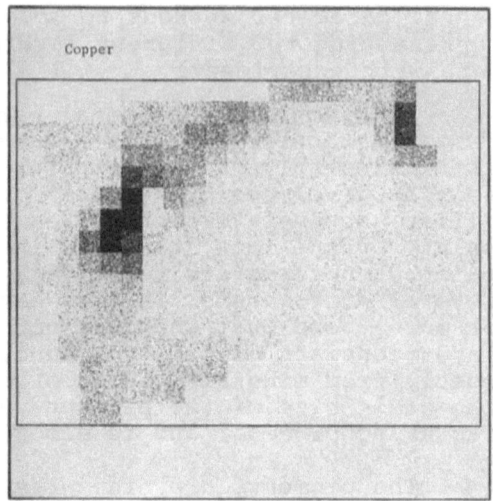
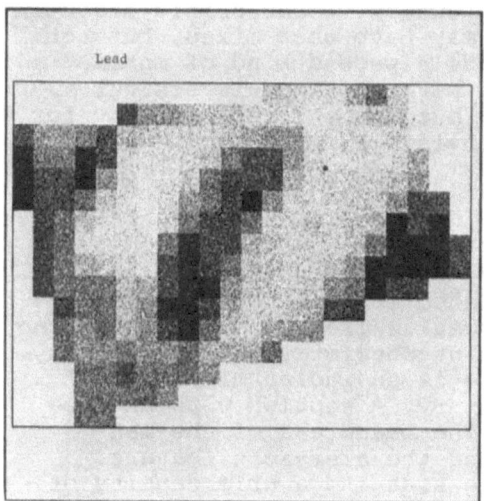
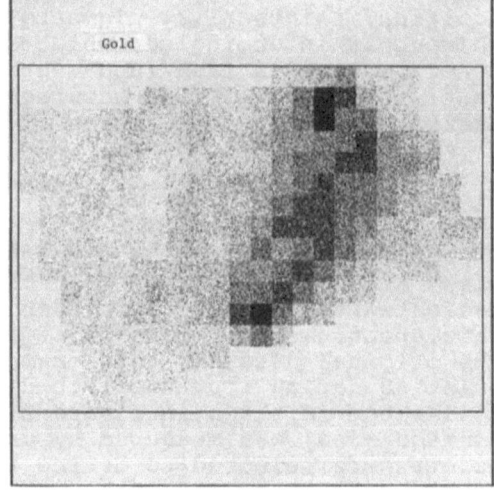

Figure 2. Section of an Indian miniature (17th century). 205
pixels within the area of the headdress (≈ 50mm width) have
been measured. The colors in the squared area are gold and red;
the 3 thick stripes and the (dark) feature at the back are
green, the thin stripes red and the 8 white pompoms also white
in the original. The background is blueish with reads and
whites. Selected element-maps show Cu, Pb, and Au. The hard
copy-maps were obtained by a screen-dump function with
randomized gray-shading.

contrast- and brightness control (effectively a gray-level discriminator setting upper and lower levels for display), a normalization function which divides the intensity value for each pixel by that of the corresponding pixel of a reference element-map, a mouse-controlled cursor to aid displaying the numerical intensity-value and coordinate data of a pixel, a cross-section function for x/y plots of intensities vs. position along a selected line, and routines to plot the displayed maps. Examples of resulting element-maps are shown in figure 2.

LITERATURE

1. M.Schreiner, M.Mantler, F.Weber, R.Ebner, F.Mairinger:
   Adv. X-ray Analysis vol.35 (1992).
2. M.Mantler: Analytica Chimica Acta, 188(1986), pp.25
3. M.Mantler: Adv. X-ray Analysis, vol.27 (1984),pp 433.

MICRO-X-RAY FLUORESCENCE ANALYSIS ON A SYNCHROTRON RADIATION

WIGGLER BEAM LINE

J.V. Gilfrich[*], E.F. Skelton, S.B. Qadri, N.E. Moulton[#]
and D.J. Nagel

Naval Research Laboratory
Washington, DC 20375-5000, USA

J.Z. Hu

Carnegie Institute of Washington
Washington, DC

ABSTRACT

It has been well established over recent years that synchrotron radiation possesses some unique features as a source of primary x-rays for x-ray fluorescence analysis. Advantage has been taken of the high intensity emanating from the bending magnets of storage rings to develop x-ray microprobes utilizing apertures or focussing optics, or both, to provide a beam spot at the specimen of the order of micrometers. The use of insertion devices, wigglers and undulators, can further increase the available intensity, especially for the high energy photons. Beam Line X-17C at the National Synchrotron Light Source (NSLS) at Brookhaven National Laboratory, accepts the unmodified continuum radiation from a superconducting wiggler in the storage ring. Some initial XRF measurements have been made on this beam line using apertures in the 10 to 100 micrometer range. The fluorescent radiation was measured by an intrinsic Ge detector having an energy resolution of 300 eV at 15 kev, and located at 90° to the incident beam in the plane of the electron orbit. In samples containing many elements, detection limits of a few ppm were achieved with 100 μm beams.

INTRODUCTION

Micro-X-Ray Fluorescence Analysis (μ-XRF)is becoming important as a materials characterization technique. The use of a small beam to excite x-ray emission from local areas on a sample provides spatial resolution enabling, for example, the analysis of single grains within a polycrystaline specimen. When micrometer spatial resolution has been required, it has become common practice to use electrons as the excitation quanta, as in scanning electron microscopes or electron probe microanalyzers. X-ray

---

* Also: SFA, Inc., Landover, MD
# Office of Naval Technology Postdoctoral Fellow at NRL

excitation possesses several advantages over electrons, including no necessity for electrical conductivity, less energy deposition, less surface sensitivity, and the fact that x-ray photons do not generate a continuous spectrum to increase the background.

Over the last decade, several instruments have been developed to take advantage of x-ray excitation, while providing some level of spatial resolution, particularly that necessary to attack a specific problem. A new instrument was reported at the 1985 Denver Conference[1], designed to operate with a rotating anode source and permit simultaneous XRF and XRD on a spatial region only 10 μm in dimension. The impact of that development was such that it received an IR100 Award in 1986, and led one of the instrument manufacturers to develop a similar device, using a much lower power x-ray tube (for XRF only)[2]. This commercial instrument is normally provided with 100 μm and 2 mm apertures[3].

Two other instruments, representing the extremes of spatial resolution for those which might be considered in this present context, need to be mentioned. At Colorado State University, an x-ray analyzer, using apertures to define a sub-millimeter beam, was constructed[4] to evaluate changes in the bulk composition of fine-grained mineral aggregates. And at the other end of the scale, the "X-Ray Microprobe" was implemented on Beam Line X-26 at the National Synchrotron Light Source (NSLS), where combinations of focussing optics (mirrors, multilayers and/or crystals) and apertures achieved synchrotron radiation (SR) beam sizes in the micrometer range[5]. It is an interesting sidelight to note that one of the first XRF experiments using SR excitation (looking for "super-heavy elements" in monazite inclusions[6]) used a curved crystal to monochromatize and focus the primary x-rays onto the small areas of interest on the specimen. A most interesting concept has been put forth recently, taking advantage of total reflection on the inside of a capillary tube to intercept a larger solid angle from the source, while confining the beam to a small size[7]. This has been carried a significant step farther, by packaging a number (upwards of 1000) of these capillaries into a bundle for the so-called "Kumakhov Lens"[8], to focus high intensity beams into a small area.

These are only a few examples of the work in this area that has been done up to now. They represent a wide range of experimental conditions, sizes ranging from a few micrometers to a fraction of a millimeter, and sources from low-power x-ray tubes, through rotating anodes, to SR, but, heretofore only "bending magnet" SR. Wiggler SR, which results from the

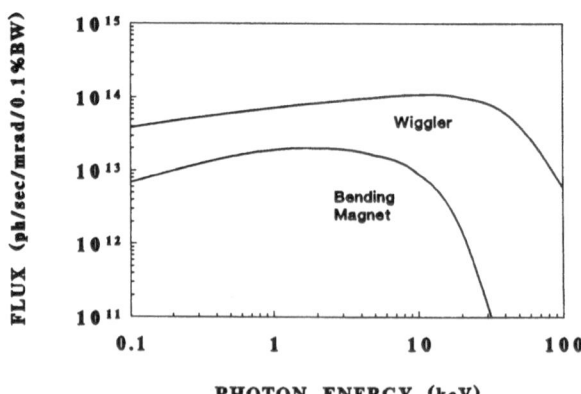

Fig. 1. Comparison of bending magnet & wiggler spectra at NSLS.

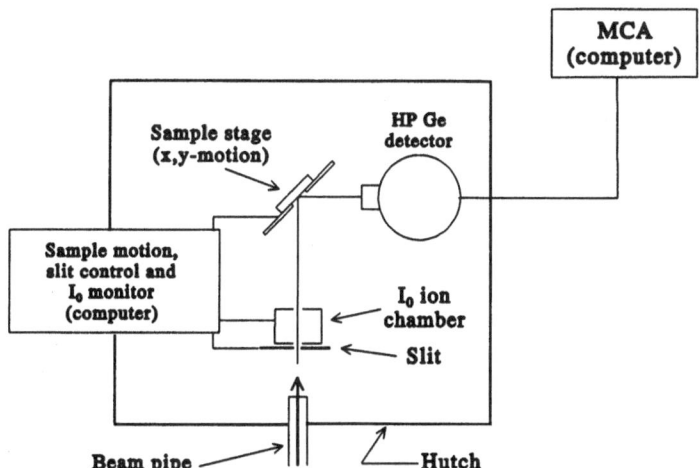

Fig. 2. Experimental arrangement at NSLS.

interaction of the electron beam in the storage ring with the field of a multiple-pole magnet inserted in the ring (the insertion device), is much more intense than that from the bending magnets, and it extends to higher energy. Figure 1 is a comparison of the bending magnet spectrum at NSLS with that from the superconducting wiggler providing x-rays to Beam Line X-17C, both for a storage ring current of 300 mA. The increase in intensity at photon energies higher than 30 keV provides the ability to analyze for high atomic number elements using their K-radiation, eliminating the overlap problem which occurs when analyzing these high-Z elements using L-lines, in the presence of high intensities of lower-Z K-lines. The higher intensity should lead to better detection limits, for energy-dispersion (ED), at least with "thin" samples, where the detector counting-rate limit is not compromised, and for wavelength dispersion (WD) even with bulk samples.

EXPERIMENTAL

Preliminary experiments were performed to evaluate the potential of this high energy radiation for μ-XRF. The immediate observation from these experiments was the increased difficulties brought about by the high intensity. The environment inside the radiation proof enclosure (the "hutch") was very hostile. Air scattering of the primary beam caused all the materials inside the hutch to fluoresce, creating a very high background. The high energy compounds the problem, making it more difficult to shield the detector from the extraneous radiation inside the hutch. However, appropriate shielding was employed to permit measurement of low (50 ppm) concentrations from small areas.

Figure 2 is a schematic diagram of the experimental arrangement on Beam Line X-17C at NSLS. The size of the opening in the collimating slit and the sample motion are controlled, and the current in the I_0 ion chamber measured, by a computer. Data were collected by the multichannel analyzer (MCA) card in the computer. Most often, the detector was HPGe, to take advantage of its higher quantum efficiency for the high energy portion of the spectra. The energy resolution of that detector was 300 eV at 15 keV. Some measurements were made using a Si(Li) detector, but the poor efficiency at high energies made it unacceptable.

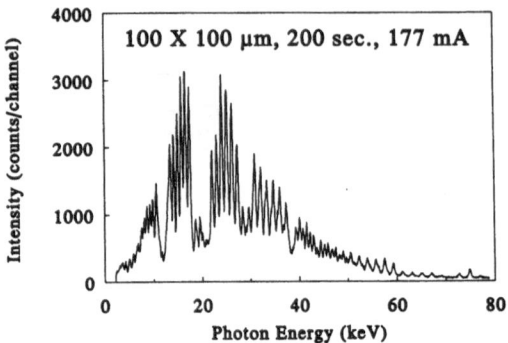

Fig. 3.   Spectrum of NIST 610, 0 to 80 keV.

RESULTS

        The primary specimens used for these measurements were the NIST Trace
Elements in Glass Standard Reference Materials.   With the glass having
nominal composition of 72% SiO_2, 12% CaO, 14% Na_2O and 2% Al_2O_3, there are as
many as 50 trace elements present in SRM's 610 and 612, at nominal values
of 500 and 50 ppm, respectively.   The concentrations of only 7 or 8 of these
elements were certified, with information values given for another 9 or 18.
Figure 3 shows the 200 second spectrum of the 500 ppm glass, with the beam-
defining aperture set at 100 X 100 μm and the storage ring current at 177
mA.   It is easy to see that the energy region from 12 to 30 keV is the most
efficiently excited, with the Kα lines of all elements from Rb to Mo, and
from Ag to Te, all having intensities within a factor of two of each other,
even though only Rb and Sr are certified, with values of 425.7 and 515.5
ppm, respectively.   The information value for Ag is 254 ppm; none of the
other 9 elements in that range have any values given at all.   Figure 4
expands the scale of the spectrum.

        From this measurement, a conservative value for the 3σ detection limit
of Sr can be determined as 15 ppm, conservative because immediate neighbors
on each side cause an artificial increase in the background which would not
be present in a sample which did not contain such interfering lines. The Rb
peak is overlapped by the PbLβ on its low energy side, making it impractical
to attempt any quantification.   It should be observed that RbKβ and PbLγ_1 do
not contribute intensity to the Sr line because they both occur very near

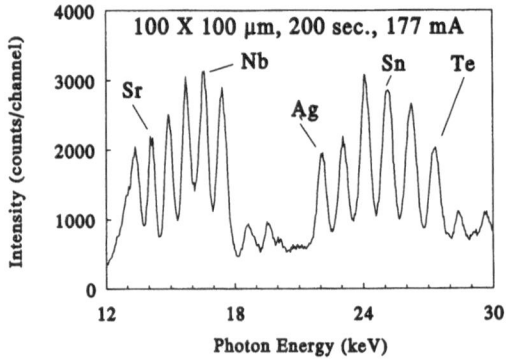

Fig. 4.   Spectrum of NIST 610, 12 to 30 keV.

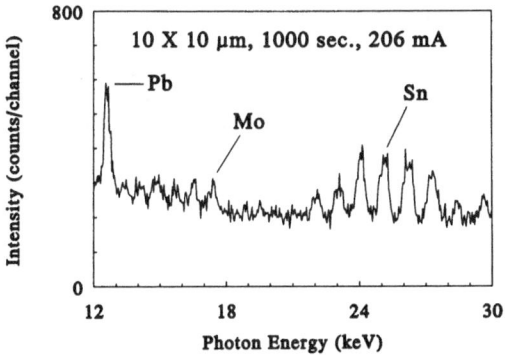

Fig. 5.   10 x 10 μm spectrum of NIST 610.

15 keV, well outside the limits of integration for the Sr peak.   Applying
all the caveats listed above, and assuming that the information value for
Ag is a reasonable estimate, the conservative detection limit for that
element is determined to be 8 ppm, better than for Sr because no overlapping
peak is on the low energy side.   Reducing the size of the aperture to 10 X
10 μm, decreases the intensity quite dramatically.   It was necessary to
increase the counting time to 1000 seconds to achieve a reasonable spectrum.
Figure 5 shows the same energy range as Figure 4, for the smaller aperture.

The 500 second, 12 to 30 keV, spectrum of the lower concentration
glass (NIST 612), using the 100 X 100 μm aperture, is shown in Figure 6.
The same selection of elements can be seen as in NIST 610, although the
relative intensities of some of the lines is somewhat different, because the
relative concentrations are not the same.   Detection limits from this
measurement are 8.2 and 4.5 ppm for Sr and Ag, respectively.

CONCLUSION

The results of these measurements illustrate that it is relatively
straightforward to achieve detection limits of a few ppm using beam sizes
of 100 μm, even for bulk samples emitting many overlapping x-ray lines.
With beam sizes of 10 μm, comparable counting times lead to detection limits
of a few tens of ppm.

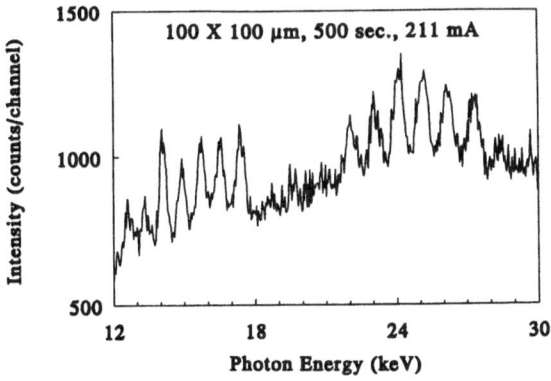

Fig. 6.   Spectrum of NIST 612.

Further experiments are planned on Beam Line X-17C at NSLS in preparation for work with the Advanced Photon Source being constructed at Argonne National Laboratory. The fluxes which should be available from the insertion devices being planned for that facility (wigglers and undulators) indicate that it should be possible to achieve ppm detection limits with micrometer-sized beams.

REFERENCES

1. M. C. Nichols and R. W. Ryon, Adv. in X-Ray Anal. 29:423 (1986).
2. M. C. Nichols, D. R. Boehme, R. W. Ryon, D. Wherry, B. Cross and G. Aden, Adv. in X-Ray Anal. 30:45 (1987).
3. Kevex Instruments, 355 Shoreway Road, P.O. Box 3008, San Carlos, CA 94070-1308, U.S.A., OMICRON brochure, 1990.
4. N. L. Gilfrich, D. E. Leyden and E. A. Erslev, Adv. in X-Ray Anal. 33:593 (1990).
5. M. L. Rivers, S. R. Sutton and K. W. Jones, Synchrotron Radiation News 4:23 (1991).
6. C. J. Sparks, Jr., S. Raman, E. Ricci, R. V. Gentry and M. O. Krause, Phys. Rev. Lett. 38:205 (1977).
7. A. Ringby, P. Engstrom, S. Larsson and B. Stocklassa, X-Ray Spectrom. 18:109 (1989).
8. M. A. Kumakhov, Nucl. Instrum. Meth. Phys. Res. B48:283 (1990).

THE COMPARISON OF THREE EXCITATION MODES IN THE ENERGY DISPERSIVE X-RAY

FLUORESCENCE ANALYSIS

Birgit Kanngießer, Burkhard Beckhoff, Jens Scheer,
and Walter Swoboda

University of Bremen, Dept. of Physics
Bremen, Germany

ABSTRACT

With a new irradiation chamber a comparison of three excitation modes
under same conditions and with the same material has been done. The results
of this comparison are presented in the following manner:
- Barkla scattering (graphite) versus Bragg reflection (HOPG)
- Bragg reflection (Mo) versus secondary target fluorescence (Mo)
- secondary target fluorescence (Sn) versus Barkla scattering (graphite)
Excitation spectra and detection sensitivities of a NBS standard will be
discussed for the different modes. The Barkla scattering was found to be the
best excitation mode for a wide elemental range.

INTRODUCTION

In EDXRF with X-ray tubes, the following three excitation modes are
discussed controversely with respect to their advantages and ranges of
application.
- The fluorescence radiation of a secondary target. Here the exciting
radiation is nearly monochromatic and unpolarized.
- The Barkla scattering of the primary spectrum at a matter of low Z: here
the exciting radiation under 90° is polychromatic and polarized.
- The Bragg reflection of the characteristic line of the primary spectrum at
a single crystal: here the exciting radiation under 90° is monochromatic and
polarized.

A new irradiation chamber with Cartesian xyz geometry was developed for
the investigations of these three excitation modes[1]. Its principle of design
allows comparisons under identical experimental conditions. In the following
the results are presented in the form of three comparisons:
- comparison of Barkla scattering and Bragg reflection
- comparison of Bragg reflection and secondary target fluorescence
- comparison of secondary target fluorescence and Barkla scattering

Advances in X-Ray Analysis, Vol. 35
Edited by C.S. Barrett *et al*., Plenum Press, New York, 1992

CRITERIUM FOR JUDGEMENT

The different excitation modes were judged using the following criterium:

As a comparable figure for orientation, we apply the detection sensitivity (DS) for an element or - in case of overlapping elements - for an energy, which for a fixed measuring live time - usually 1000 seconds - is defined as:
$$DS := N \; / \; 3 \; \sqrt{B} \geq 1 \quad (N := \text{Net intensity}, \; B := \text{Background}).$$

The value of unity means reaching the minimum detection limit according to the definition of the minimum detection limit.

Earlier investigations have shown that the decisive factor for an increase in the detection sensitivity is a high net intensity of the fluorescence line, rather than a lowering of the background by polarization effects[2]. Therefore, for the purpose of comparison we opened the collimators far enough so that a maximum count rate could be handled by the detection electronics, keeping the energy resolution at a maximum.

COMPARISON OF BARKLA SCATTERING AND BRAGG REFLECTION

The comparative measurements for Barkla scattering and Bragg reflection were performed using the same material, namely Carbon. As Barkla scatterer we used polycrystalline graphite. For Bragg reflection, a HOPG - highly oriented pyrolytic graphite - crystal was at our disposal. A comparative measurement with the NBS standard Orchard Leaves (1571) is shown in figure 1. The collimator openings and the tube current were chosen so that in both measurements about the same count rates were achieved. The parameters of the measurements were: live time 1000s, anode Mo, total count rate ≈4000 cps.

Because of its higher density the HOPG showed a higher scattering effectivity. However, this caused also a larger background, as can be seen from the figure.

DETECTION SENSITIVITIES

Figure 2 shows the ratio of the detection sensitivities for both excitation modes. Within the range of errors one finds only slightly higher detection sensitivities for the HOPG. Measurements with other specimen types, thin standards as well as thick liquid standards, corroborate this behavior of detection sensitivities for both excitation modes[3]. Reports[4,5] on an considerable increase of detection sensitivities by means of Bragg reflection on HOPG with a molybdenum tube could, therefore, not be confirmed.

The poor effectivity of highly polarized Bragg reflexes with HOPG can be explained by collimation effects. Figure 3 shows the exciting radiation spectra obtained with the HOPG and with the ordinary graphite target, observed at the site of the specimen and obtained with a narrower collimation (primary collimator Ø5mm, secondary collimator Ø6mm+lead diaphragm Ø0.3mm) than the collimation for the NBS standard measurement. The increase of the MoKα line due to Bragg reflection is here only a factor of 1.5. The intensity of the Compton Peak related to the MoKα line is higher than the MoKα itself.

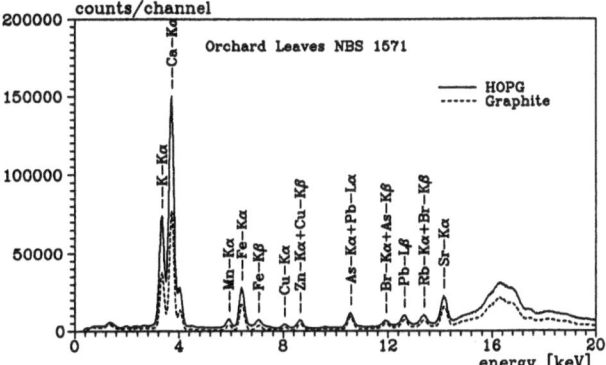

Fig. 1 Comparative measurement, NBS 1571

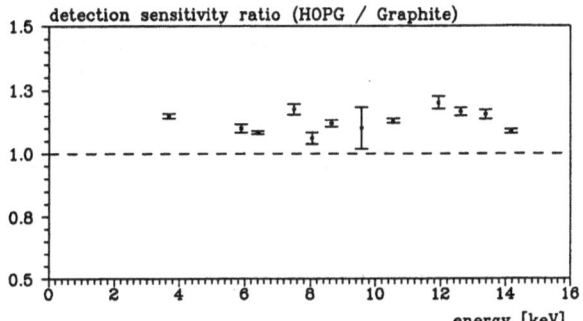

Fig. 2 Detection sensitivity ratio

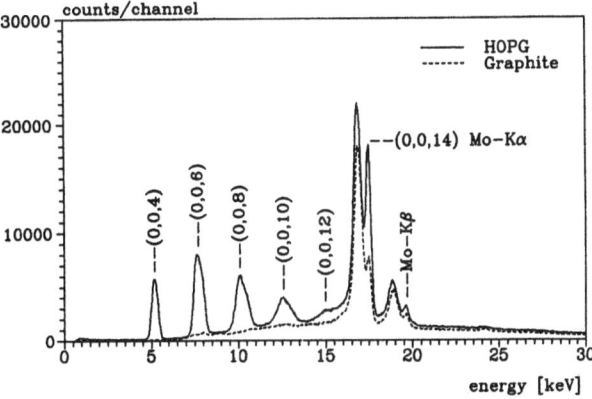

Fig. 3 Exciting radiation spectra

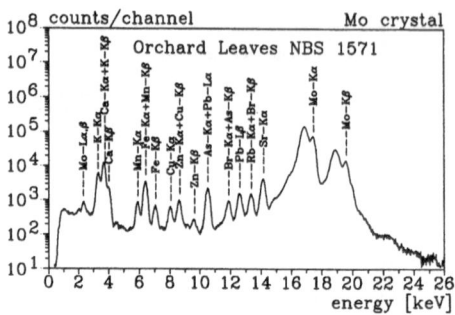

**Fig. 4 Spectrum with excitation by the
molybdenum crystal**

From this one may conclude that with still wider collimator openings,
as is necessary for practical EDXRF analytical applications, the increase of
the MoKα line due to Bragg reflection will be even less. In regard to the
total excitation spectrum, the Bragg contribution appears truly minimal.

This collimation effect is due to the different angular ranges in which
Bragg reflection and Barkla scattering occurs: while Bragg reflection occurs
only in a narrow angular range around the Bragg angle, for the Barkla scattering there is no angular limitation. Therefore, the larger the acceptance
angle of the secondary collimator, the more Barkla scattered radiation is
passed relative to Bragg reflexes. In other words, under these conditions
the HOPG acts mainly as a simple Barkla scatterer.

COMPARISON OF BRAGG REFLECTION AND SECONDARY TARGET FLUORESCENCE

These comparative measurements were also performed with the same
specimen material - NBS Orchard Leaves. For the Bragg reflection of the MoKα
line we used a molybdenum crystal; for the excitation by means of a secondary target we used a metal sheet of molybdenum. The spectra, obtained from
the same specimen with the same collimation and count rates appear almost
congruent! Figure 4 shows the fluorescence spectrum with excitation by the
molybdenum crystal.

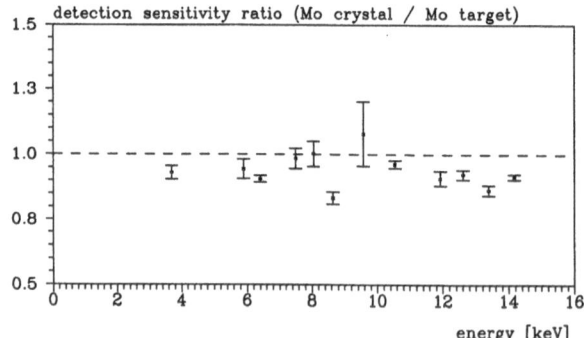

Fig. 5 Detection sensitivity ratio

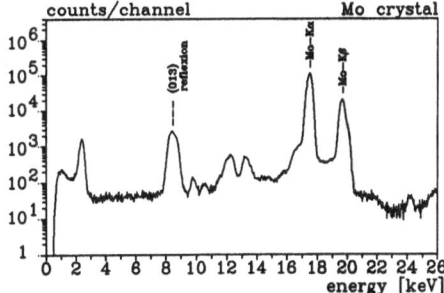

Fig. 6 Excitation spectrum

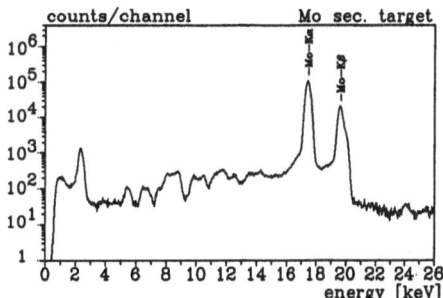

Fig. 7 Excitation spectrum

DETECTION SENSITIVITIES

Figure 5 shows the ratio of the detection sensitivities for both excitation modes. As it is expected from the almost congruent spectra one finds
within the range of errors approximately the same detection sensitivities.
Investigation of the exciting spectra at the specimen site also yielded an
explanation of this phenomenon. Figures 6 and 7 show the excitation spectra
from the Mo crystal and from the secondary target as observed at the
specimen site. The collimation was the same as was used in the investigation
of the graphite excitation spectra with wide collimation shown before. The
comparison shows that the intensity of the Bragg reflex is less than 10% of
the intensity of the fluorescence from the molybdenum secondary target.

The MoKα Bragg reflection contributes, therefore, only minutely to the
excitation of the specimen and can, therefore, only marginally improve the
detection sensitivity - in spite of its high degree of polarization. In
other words: The molybdenum crystal acts mainly as a secondary target in
exciting the specimen!

DISCUSSION

In our device, the tube radiation originates from an extended anode
area, which, seen from the direction of outgoing radiation, appears to have
the dimensions of 13mm x 8mm. One might argue that the use of a fine focus

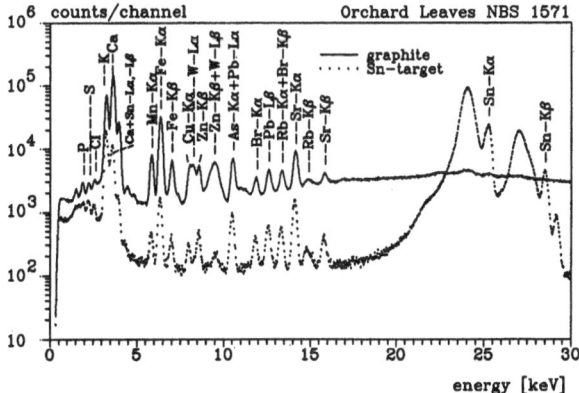

Fig. 8 Comparative measurement, NBS 1571

Table 1 Minimum detection limits

z	el.	Cu target MDL [ppm]			Sn target MDL [ppm]			graphite MDL [ppm]		
25	Mn	0.87	±	0.04	8.27	±	0.55	1.11	±	0.05
26	Fe	0.73	±	0.05	4.86	±	0.33	0.87	±	0.06
35	Br		-		0.85	±	0.09	0.51	±	0.05
37	Rb		-		0.62	±	0.06	0.43	±	0.04
38	Sr		-		0.53	±	0.02	0.45	±	0.01
82	Pb		-		1.96	±	0.14	1.46	±	0.10

tube with its much increased specific density of radiation could change the situation sufficiently to justify the use of a Bragg polarizer. This will require further investigation, which would, however, encounter several severe problems. Among them are the following:
- even with an ideal point focus because of the unavoidable distances and the divergence of the X-rays, the irradiated areas on the polarizer will be of the order of millimeters, hence not orders of magnitudes smaller than with the large focal area.
- the usual line foci of diffraction tubes present particular problems: if one uses appropriately shaped rectangular collimators both as primary and secondary collimators, there will be small scattering angles in one plane, but extremely large ones in the orthogonal plane. This means that, at any rate there will be a very large contribution of Barkla scattering or unpolarized fluorescence from the crystal material . From our measurements one may conclude that this Barkla component will completely overrule the Bragg contribution, even if the line focus might be very narrow.
- a further effect for low Z crystals reduces the amount of radiation available for excitation even further:
The primary radiation penetrates deeply into the crystal, because of its low Z. For instance, with the MoKα radiation, the half length is about 2mm in the HOPG. Therefore, the Bragg reflected radiation originates from a region several millimeters wide. A narrow collimation, required to ensure Bragg predominance, will therefore only accept a very small fraction of Bragg reflected radiation. Opening the angle will quickly lead to the Barkla predominance.

COMPARISON OF BARKLA SCATTERING AND SECONDARY TARGET FLUORESCENCE:

As low Z crystals behave mainly like Barkla scatterers and high Z crystals as secondary target, a comparison between secondary target fluo-

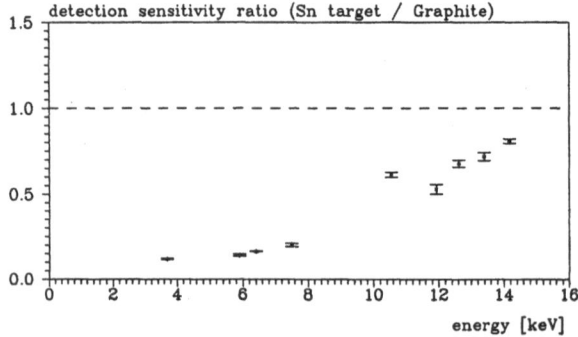

Fig. 9 Detection sensitivity ratio

rescence and Barkla scattering will have the more fundamental importance for the methodological comparison. These comparative measurements were also per-formed with the NBS Orchard Leaves specimen with the same count rates, using a Tungsten tube (figure 8). As Barkla scatterer we again used polycrystal-line graphite, and as secondary target, Tin. It is evident that specimen excitation by Barkla scattering yields an increase of background in spite of its polarization. Furthermore, and contrary to the Tin excitation, there is a considerable disturbance due to the characterictic lines of the Tungsten tube. In the upper elemental range the signal-to-background ratio is clearly worse in the case of the graphite excitation. Only in the lower elemental range, which is energetically far away from the Tin excitation line, the signal-to-background ratio is better with graphite excitation.

DETECTION SENSITIVITIES

Figure 9 shows the ratio of detection sensitivities determined from the comparative measurements as shown before. The general impression is that the graphite excitation yields higher detection sensitivities in the presented elemental range. In the upper elemental range the Tin secondary target yiel-ded similar, though slightly lower detection sensitivities. In the lower elemental range, an additional comparative measurement with a Copper secon-dary target yielded somewhat higher detection sensitvities than did the graphite measurement. In table 1 there are some selected minimum detection limits from the same measurement.

CONCLUSION

In conclusion, one may say that the graphite excitation yields a wider elemental range with lower detection limits. As there are also Barkla scatterers with higher scattering effectivity than graphite, the Barkla scattering is to be prefered over the secondary target fluorescence. Furthermore, the lower detection limits are about 25% lower than the corres-ponding values with Tin excitation. The lower detection limits were achieved in spite of the poorer peak-to-background ratios. The reason for this is the absence of the strong (Rayleigh and Compton) scattering peaks in the case of secondary target excitation, which aggravate the electronics without contri-buting to the analysis.

ACKNOWLEDGEMENTS

The author wishes to thank the DFG for financial support for her work and for the PICXAM conference.

REFERENCES

1. W. Swoboda, B. Kanngießer, B. Beckhoff, K. Begemann, H. Neuhaus, and J. Scheer, "A New Device for Energy Dispersive X-Ray Fluorescence (EDXRF)," accepted in Rev.Sci.Instrum..

2. R. W. Ryon, "Polarized Radiation Produced by Scatter for Energy Dispersive X-Ray Fluorescence Trace Analysis," Advances in X-Ray Analysis 22:575-590 (1979).

GEOMETRIC CONSIDERATIONS IN EDXRF TO INCREASE FLUORESCENCE

INTENSITIES AND REDUCE BACKGROUND

Igor Tolokonnikoff

Nuclear Radiometric Methods
Moscow Geologic Prospecting Institute
23 Miklukho-Maclay Str., GSP-7
117873 Moscow, U.S.S.R.

ABSTRACT

Two methods for improving the sensitivity of EDXRF will be presented. One method is to use a spherical geometry for the measurements. The analyzed specimen is made as a spherical layer (i.e., "orange rind"), with the exciting x-ray source and the detector being located in the inner surface of the specimen at opposite points on the diameter. The source radiation scattered into the detector is minimized by the 90° scattering angle at all points on the specimen. It has been shown that the sensitivity is improved by several times while carring out EDXRF measurements under conditions of spherical geometry as compared to the conventional flat arrangement.

The other method we have investigated makes use of polarized radiation. The source radiation is polarized by scattering from a thin, low atomic number material. Higher energy radiation from the x-ray tube passes through the polarizer, where it impinges on a secondary fluorescer. This secondary radiation is then polarized by the same scatterer as the primary radiation. A quasi-monoenergetic polarized source is produced with maximized intensity by these means. It has been shown experimentally that the use of this polarized source for excitation leads to detection limits which are several times lower in the whole range of moderate energy. It has further been shown that the use of the easily added secondary target to the standard orthogonal polarization geometry increases the polarized beam intensity by 10-15%.

A SPHERICAL GEOMETRY OF MEASUREMENTS

In the specimen spectrum with a light-weight matrix the main source of the background is known to be a scattered radiation. This radiation is minimum when the angle between the exciting and detecting radiation is 90°. A spherical geometry of measurements [1,2] enables

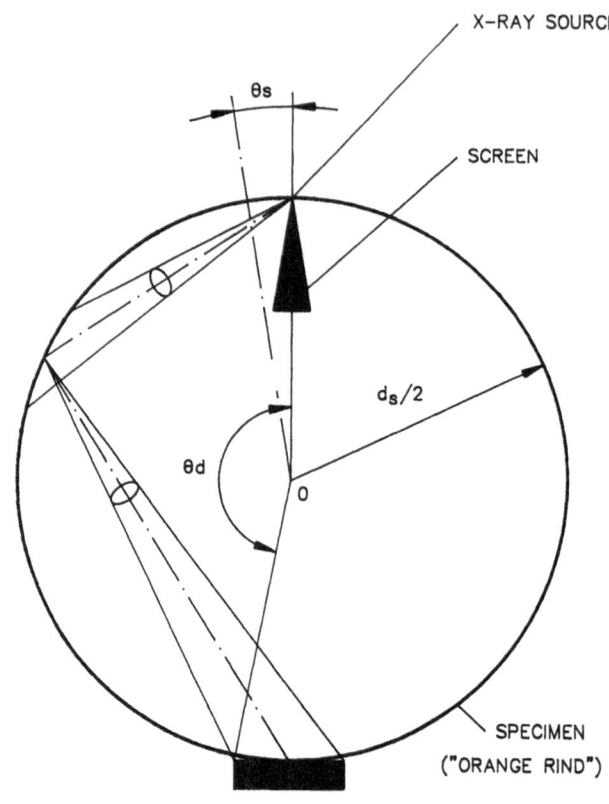

Fig. 1. Energy-dispersive x-ray fluorescence analysis
under conditions of spherical geometry
of measurements.

one to provide a required angle for all the elementary parts of the
specimen surface.

By a spherical geometry of measurements we mean that the analyzed
specimen is made as a spherical layer ("orange rind"), the exciting
x-ray radiation source and detector being located in the inner surface
of specimen at diametrically opposite points (Figure 1).

The effectiveness of spherical geometry of measurements as
compared to conventional flat one (specimen is made in the form of
flat layer) can be estimated by comparing the fluxes of the
characteristic x-ray radiation (CXR) of the atoms of the analyzed
element which hit the detector in both cases. For simplicity of
calculation we take the following assumptions: exciting radiation
source is point one; analysis is carried out in thin layers; in case
of flat specimen the source and detector are at the same point. Taking
the preceding assumptions into account we derive the following
relation:

$$\frac{I_s}{I_f} = \frac{2(\theta_d - \theta_s)}{d_{sa}^2 \, 2\left(\frac{1}{a^4} - \frac{1}{[a^2 + (d_f/2)^2]^2}\right)} \tag{1}$$

where θ_d is the angular size of detector in spherical coordinates, θ_s is the angular size of conical screen in spherical coordinates, d_s is the specimen diameter in the form of spherical layer, d_f is the flat specimen diameter, a is the distance between the source (detector) and the flat specimen. In terms of the formula obtained we make a numerical estimate assuming the diameters of the spherical and flat specimens are equal to a double distance between the source (detector) and the flat specimen ($d_s = d_f = 2a$). Moreover, we assume that $\theta_d \sim \pi$, $\theta_s \sim 0$. After inserting the values into formula we finally obtain: $I_s/I_f \sim 2.09$.

For experimental verification of the effectiveness of the use of spherical geometry we have mounted a setup. The main element of the setup is a cuvette made of two concentrically located polymeric film hemispheres with the distance between each other 4 mm. The cavity between hemispheres was filled with uranyl nitrate solution (U content is 90 ppm). The ^{109}Cd source was placed on the inner surface of hemisphere. A semiconductor detector was placed in the truncated part of the cuvette so that the Si(Li) crystal would be on the extension of generator of the inner hemisphere at the point diametrically opposite to the source location. To protect a detector from the source radiation a conical multilayer (W, Cd) screen was placed between them. In our experiments the intensity ratio of analytical line $UL_{\alpha1}$ for two geometries is $I_s/I_f \sim 2.5$. A discrepancy between the calculated and experimental results is related, obviously, to those assumptions which were taken in obtaining expression (1). The background value in the region of analytical peak turned out to be equal both for flat and spherical geometry, i.e. the background increase for the latter geometry due to increase of the square of the specimen surface is compensated by smaller yield of scattered radiation, since for all parts of the specimen surface the scattering angle is ~90°.

THE USE OF QUASI-MONOENERGETIC POLARIZED RADIATION

The use of exciting polarized radiation has become recently one of the ways in increasing the EDXRF sensitivity. There are different ways in producing under laboratory conditions both wideband[3] and monoenergetic[4] polarized radiation. It is known that while exciting x-ray fluorescence by monoenergetic radiation the signal/background ratio is higher than wideband one. As was noted[5], in the range of moderate energy (MoK_α) the problem of producing monoenergetic polarized radiation is still unsolved.

Papers[6-8] presented one way of solving this problem: a crystal-less method of obtaining quasi-monoenergetic polarized radiation in the moderate energy range. Figure 2 shows two variants of a schematic drawing of the setup for accomplishing the EDXRF at the excitation of x-ray fluorescence by quasi-monoenergetic polarized radiation. The exciting quasi-monoenergetic polarized radiation is produced both due to formation of the spectrum of the x-ray tube and application of secondary targets where the wideband bremsstrahlung, having passed through a polarizer, is transformed to a monoline.

The effectiveness of use of the target was evaluated by comparing the values of the flux $I_{K_{\alpha+\beta}}^a$ of the K-series of the CXR of the atoms of the anode material of the x-ray tube, and the flux $I_{K_{\alpha+\beta}}^t$ of the

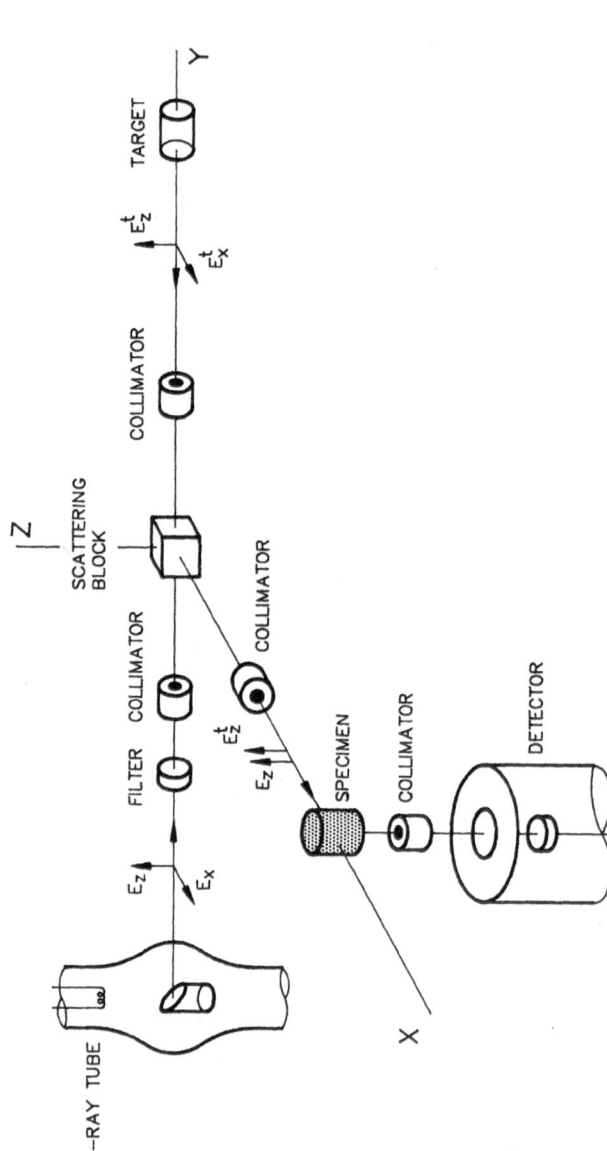

Fig. 2a. A schematic drawing of EDXRF setup for carrying out measurements at the excitation of x-ray fluorescence by quasi-monoenergetic polarized radiation. Variant 1.

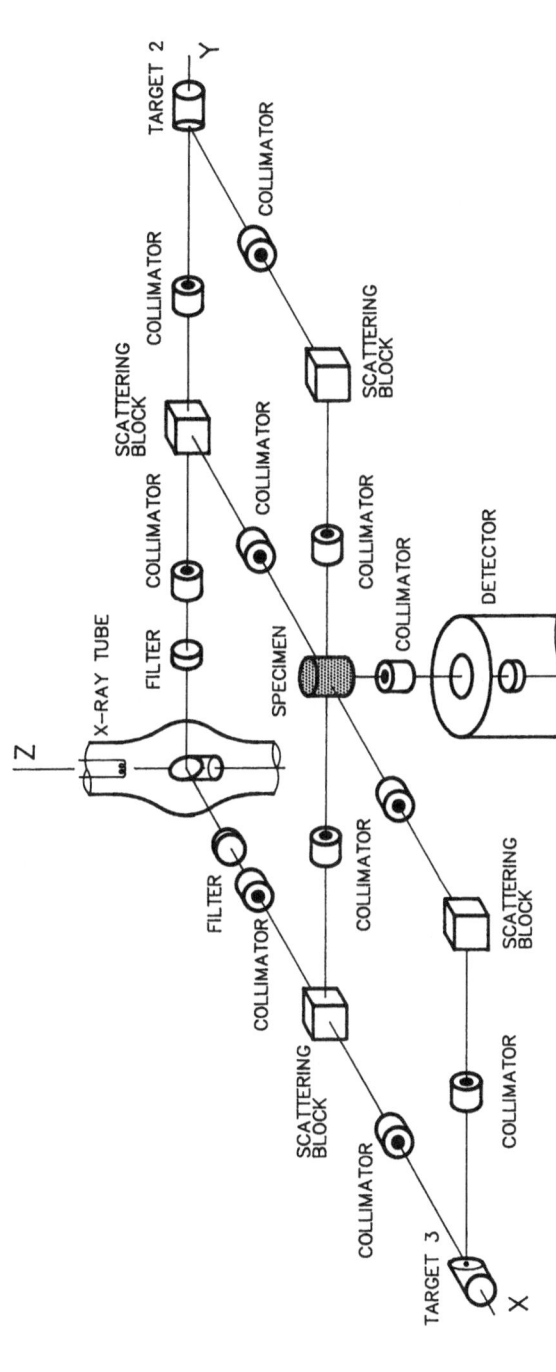

Fig. 2b. A schematic drawing of EDXRF setup for carrying out measurements at the excitation of x-ray fluorescence by quasi-monoenergetic polarized radiation. Variant 2.

Table 1. Experimental and calculated values of the ratio
          of the flux of the K-series of CXR of the atoms of
          the target material to the flux of the K-series of
          CXR of the atoms of the anode material

High voltage,kV	Target effectiveness: $I^t_{k_{\alpha+\beta}} / I^a_{k_{\alpha+\beta}}$, %	
	Experiment	Theory
75	16.6 ± 13.2	13.9
100	15.6 ± 11.4	16.0
130	13.6 ± 6.61	18.2

K-series of the CXR of the atoms of the target material at the location
of the scattering block (polarizer). For simplicity of calculation
we made the following assumptions: the thickness of the polarizer is
much less than the distances from the anode to the polarizer R_1 and
from the polarizer to the target R_2; the "bell" of bremsstrahlung can
be approximated by an equilateral triangle with base extanding from
$1/3E_0$ to E_0, and the height of this triangle is m times less than the
height of the peak of the $K_{\alpha 1}$ line of the CXR of the atoms of anode
material; we neglect the attenuation of the flux density of the
high-frequency component of bremsstrahlung owing to dispersion and
absoption in the material of the polarizer. With these assumptions
we obtained the formula:

$$\frac{I^t_{k_{\alpha+\beta}}}{I^a_{k_{\alpha+\beta}}} = \frac{E_0 \; \eta \; D^2 R_1^2}{48 \; m \; \Delta E^a_{k_{\alpha 1}} \; (1+p+q \;) \; (R_1+R_2 \;)^2 \; R_2^2} \tag{2}$$

where E_0 is the maximum voltage across the x-ray tube, eV; η is the
relative yield of radiation of the K-series for secondary targets;
D is the diameter of the collimators; $\Delta E^a_{k_{\alpha 1}}$ is the natural width of
the $K_{\alpha 1}$ line of CXR of the atoms of anode material; p is the flux
density of the $K_{\alpha 2}$ line in relative units with respect to the flux
density of the $K_{\alpha 1}$ line; q is the flux density of $K_{\beta 1 \beta 2}$ lines in
relative units with respect to the flux density of the $K_{\alpha 1}$ line.

    Papers[9,10] present the experimental data on verification of the
theoretical estimate. There was registered in this case the x-ray
tube and target radiation, scattered by a polarizer at the angle of
90° . For this the Si(Li) semiconductor detector was placed at the
location of specimen (see drawing in Fig.2a) so that the detector
axis would coincide with the X direction. A 1 mm - thick perspex plate
served as a scattering block. A lead collimator with the hole diameter
50 μm was placed between the polarizer and detector. The collimators
diameter of the setup was 5 mm. The distance from the anode to the
scattering block was 35 mm, and from the scattering block to the
target - 10 mm. Table 1 lists the experimental and calculated values
of the ratio of $I^t_{k_{\alpha+\beta}} / I^a_{k_{\alpha+\beta}}$ for molybdenum anode and target. The

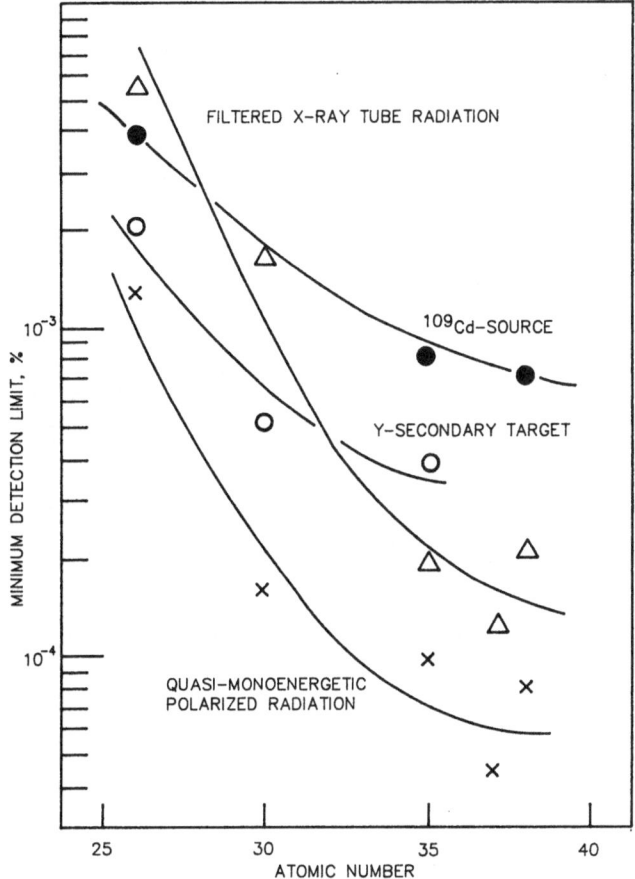

Fig. 3. The MDL versus the atomic number of element under
different methods of excitation of the x-ray
fluorescence.

calculation is made in the assumption that the value of parameter m,
which has not been measured experimentally, is 100. The large experimental error arises from the instability of the operation of
high-voltage generator employed.

Moreover, we performed experiments on comparing the different
methods of excitation of x-ray fluorescence. The standard SBMT-01
specimen (organic matrix) spectra produced with the help of four
methods of excitation of the x-ray fluorescence: quasi-monoenergetic
polarized radiation, direct filtered radiation of the x-ray tube,
[109]Cd source and direct yttrium secondary target radiation. The total
count rates and the times of measurements (1 hour) were equal. On the
basis of the data obtained we built a graph (Figure 3) of the MDL as
a function of atomic number.

CONCLUSION

As compared to flat geometry, spherical geometry of measurements
provides the increase of the EDXRF sensitivity approximately by 2.5

times. The background level of scattered radiation in spherical and flat geometry is approximately equal when the Compton scattering peak "smearing" for spherical geometry is less. The application of spherical geometry of measurements is most advantageous while analyzing specimens with the elements whose CXR lines are arranged on the "tail" of Compton scattering peak of exciting radiation. Construction features of the employed detector (Si(Li) crystal position relative to the input Be-window, the presence of the collimator ahead of Si(Li) crystal) made it possible to use a cuvette only as a hemisphere. The main disadvantage of the present method is the specimen preparation. This refers both to fabrication of cuvettes and pressed specimen in the form of hemispherical layers.

It is experimentally shown that the use of secondary target makes additional contribution (10-15%) to the intensity of exciting quasi-monoenergetic polarized radiation. More proper experimental verification of the effectiveness of using polarized radiation of the secondary targets is possible only at the presence of high-voltage generator with more stable parameters. So the results obtained are preliminary. The use of quasi-monoenergetic polarized radiation as compared with the other sources of exciting radiation leads to the MDL decrease by several times in the whole range of moderate energy. The total count rate obtained was more than two orders of magnitude less than the one obtained in the present time in the semiconductor x-ray spectrometry. So a future progress in applying the EDXRF of quasi-monoenergetic polarized radiation arises both from the use of the higher-power x-ray tubes (5-10 kW) and the construction of the setup in a close-coupled geometry by the scheme presented in Figure 2b.

ACKNOWLEDGEMENTS

I wish to thank Dr. Richard W. Ryon for very useful discussions and encouragement. L.P.Ignatyeff and A.V.Erastoff were also particularly helpful in the preparation of this manuscript.

REFERENCES

1. Yu.N.Burmistenko and I.A.Tolokonnikoff, The use of spherical geometry of measurements in x-ray analysis, Izv. VUZov Geologiya i razvedka Dep. N°2847-84 (in Russian).

2. Yu.N.Burmistenko, I.A.Tolokonnikoff and O.N.Chernobrivets, On possibility of improving the metrological characteristics of the x-ray analysis with the use of spherical geometry of measurements, Atomnaya energiya 60(3):218 (1986) (in Russian).

3. T.G.Dzubey, B.V.Jarrett and J.M.Jaklevic, Background reduction in x-ray fluorescence spectra using polarization, Nuclear Instruments and Methods 115:297 (1974).

4. P.Wobrauschek and H.Aiginger, X-ray fluorescence analysis using intensive linear polarized monochromatic x-ray after Bragg reflection, X-Ray Spectrometry 9(2):57 (1980).

5. R.W.Ryon, J.D.Zahrt, P.Wobrauschek, e.a., The use of polarized x-ray for improved detection limits in energy dispersive x-ray spectrometry, Advances in X-Ray Analysis 25:63 (1982).

6. Yu.N.Burmistenko and I.A.Tolokonnikoff, The apparatus for x-ray analysis of the content of substance, Inventor's Certificate USSR N⁰ 1224689 B.I. N⁰ 14 (1986).

7. I.A.Tolokonnikoff and Yu.N.Burmistenko, The apparatus for x-ray analysis of substance, Inventor's Certificate USSR N⁰ 1300353 B.I. N⁰ 12 (1987).

8. I.A.Tolokonnikoff, On the possibility of obtaining quasi-monoenergetic polarized x-ray radiation in the moderate energy range, Atomnaya energiya 61(3):224 (1986) (in Russian).

9. I.A.Tolokonnikoff, To the problem on effectiveness of using the polarized secondary target radiation in energy-dispersive x-ray analysis, Izv. VUZov Geologiya i razvedka 3:123 (1990) (in Russian).

10. I.A.Tolokonnikoff, O.V.Gorbatyuk and K.I.Schekin, The use of exciting quasi-monoenergetic polarized radiation in energy-dispersive x-ray analysis, Atomnaya energiya 69(2):115 (1990) (in Russian).

X-RAY MICROBEAM SPECTROSCOPY WITH THE USE OF
CAPILLARY OPTICS

S. Larsson and P. Engström

Chalmers University of Technology, Dept of Physics
S-41296 Göteborg, Sweden

I   INTRODUCTION

X-ray micro analysis had suffered from the limited intensity of conventional X-ray sources like X-ray tubes etc. However, with the optical technique described in this paper, the so called "X-ray Capillary Optics" (XCO), it is possible to make X-Ray Fluorescence (XRF) analysis in micro-scale, down to a few μm^2 still with conventional sources[1,2].

With the development of more intensive sources such as synchrotron radiation light sources, the interest for XRF micro-spectroscopy has increased. With modern detectors and electronic equipment it is possible to collect a great amount of analytical information in a very short time from different kind of samples. It is shown here that within the framework of modern technology and the XCO technique it is possible to achieve **submicron** X-ray beams and then to do X-ray analysis down to sub-microscopic level.

II   CAPILLARIES

Due to total external reflection of X-rays, capillaries can be used as waveguides for X-rays in analogy with the optical fibre technique. X-ray microbeams can easily be obtained with this kind of technique. It is important to point out that the capillary is not a one to one imaging system, so there will not be a focal plane. The beam will have its smallest cross section at the capillary end and will then diverge with the maximum divergency not exceeding the critical angle. In a way the capillary "collects" the X-ray beam close to the source and then transfers it to the other end of the capillary. If the angle of incidence is below the critical angle the losses will be very small. If the capillary is conical with the large end facing the source, the X-ray beam will be "squeezed" to a smaller size with a correspondingly higher intensity compared to the primary beam. However, for each reflection the glancing angle will increase with two times the "cone" angle, and if the glancing angle is greater than the critical angle, then the reflectivity will drop rapidly. Due to the energy dependence of the critical angle, capillaries can be effective to suppress high energy X-rays[1].

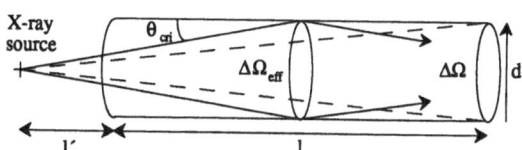

Fig 1. The capillary gain factor (I_g) is defined by the ratio between $\Delta\Omega_{eff}$ and $\Delta\Omega$.

Advances in X-Ray Analysis, Vol. 35
Edited by C.S. Barrett *et al.*, Plenum Press, New York, 1992

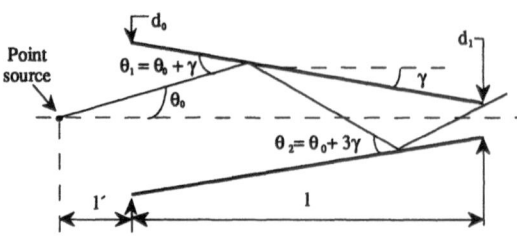

Fig 2. Conical capillary geometry.

For a straight capillary the gain factor I_g can be defined as the ratio between the effective solid angle $\Delta\Omega_{eff}$ seen by the X-ray source (corresponding to the critical angle) and the solid angle when no reflections occur $\Delta\Omega$ (defined by the exit end), assuming the reflectivity to be unity up to the critical angle θ_{cri} and not taking the absorption into account

$$I_g = \Delta\Omega_{eff}/\Delta\Omega \qquad (1)$$

The gain factor shows the behaviour of a low-pass filter where $\Delta\Omega_{eff}$ is proportional to $1/E^2$ as θ_{cri} will decrease for higher energies see fig 3 and 5.

When calculating the transmittance of X-rays through the capillary one has to know the gancing angle of incidence and the number of reflections. For a straight capillary the angle of incidence for each reflection will be the same and hence the reflectivity. This is not the case for a conical (tapered) capillary, where the angle will increase for each reflection - see fig 2, and one has to calculate the reflectivity for each reflection.

The total reflection coefficient can be written as the product of each individual reflection coefficient given by Fresnel formula for small angles

$$R_{tot}(\theta_0,E) = \prod_{i=1}^{I} R(\theta_i,E) \qquad (2)$$

The intensity gain factor[3], which is defined as the ratio of the intensity with and without reflection, can be written as

$$I_g = \frac{1}{\Delta\Omega} \int_{\theta_{min}}^{\theta_{max}} R_{tot}(\theta_0,E) \cdot 2\pi \cdot \sin(\theta_0) \cdot d\theta_0 + 1 \qquad (3)$$

where $\Delta\Omega$ is the solid angle seen by the X-ray source when no reflections occur, corresponding to the far end of

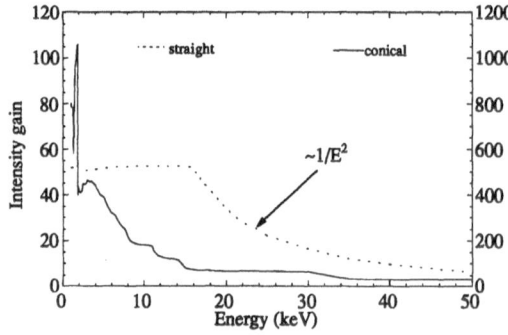

Fig 3. Calculated gain factor for a straight and a conical capillary. Input data where l=19 cm, l'=3 cm, d_0=98 μm, d_1=98 μm (straight, left hand scale) and d_1=12 μm (conical, right hand scale). This conical capillary was used in the experiments described below.

the capillary. The limits of integration, θ_{min}, θ_{max}, are the minimum and the maximum glancing angles respectively for the X-rays inside the capillary as defined by the geometric dimensions of the capillary and the X-ray source-capillary distance, see fig 2. The calculations in fig 3 are made for two different capillaries, one straight 19 cm long and with a diameter of 98 μm, and the conical one has an exit end of 12 μm and with the other dimensions identical to the straight one. The graphs shows clearly the low pass effect of the capillary.

III   CALCULATION OF CONCENTRATIONS

Normally the fluorescence radiation intensity from an element depend on several different parameters related to the element of interest, the other elements in the sample and the geometry of the spectrometer. This relation is usually written as

$$I_i = I_0 \cdot \Omega \cdot \mu_{ik} \cdot \varepsilon_i \cdot \omega_i \cdot m_i \cdot C_i \qquad (4)$$

where;
I_i	intensity from element i in the sample.
I_0	intensity of the primary radiation.
Ω	detector solid angle (i.e efficient constant).
μ_{ik}	photoelectric cross-section for element i.
ε_i	relative intensity ratio for element i.
ω_i	fluorescence yield.
m_i	mass of element i in g/cm².
C_i	correction factor for the attenuation of the primary and secondary radiation in the sample which depends on the beam-sample-detector geometry and sample composition.

This relation (eq. 4), in which all parameters have a definite physical significance, is normally recognized as the "fundamental parameter method". This means that the calculation can be performed with only one single calibration data point valid for all types of samples. If light elements (elements which are not detectable) are abundant in the sample, some estimates of the quantity of these elements have to be done.

In conventional XRF spectroscopy the detector is usually placed perpendicular to the primary beam direction and with the sample positioned in 45 degrees direction in relation to the primary beam. However, in the micro beam spectrometer described here and in ref 2 the sample is placed perpendicular to the primary beam direction, as shown in figure 4. The sample position has been selected in order to take full advantages of the capillary technique. This position enables the operator to have a micrograph image of the sample and the capillary cross-section without changing the optical microscope position (see ref 2). The specific geometry of the beam-sample-detector has to be taken into account when calculating the attenuation correction factor.

This correction factor C_i is usually called the self attenuation factor and with the geometry used in the micro beam spectrometer (i.e sin90 ≈ 1) C_i can be divided into two factors

$$C_i = c_1 \cdot c_i \qquad (5)$$

where c_1 is related to the attenuation of the primary beam and can be written as

$$c_1 = \frac{1 - \exp(-A)}{A} \qquad \text{where} \qquad A = \sum_j \rho_j \cdot d \cdot \mu_j^{att}(E_0) \qquad (6)$$

and c_i is related to the attenuation of the characteristic radiation from element i in the sample matrix. c_i can be written as

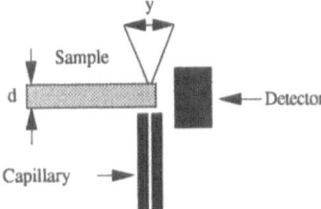

Fig.4a. Shows the the sample, the capillary and the detector geometry from above.

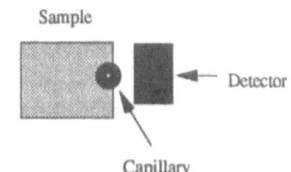

Fig.4b. Shows the same as a/, but from the view by the primary beam.

$$c_i = \frac{1 - \exp(-B_i)}{B_i} \quad \text{where} \quad B_i = \sum_j \rho_j \cdot y \cdot \mu_j^{att}(E_i) \tag{7}$$

ρ_j the density of elements j in the sample.

d   sample thickness.

$\mu_j^{att}(E_0)$ attenuation coefficient for element j at the primary beam energy E_0.

y   average pathway for the characteristic X-rays through the sample.

$\mu_j^{att}(E_i)$ attenuation coefficient for element j in the sample at the characteristic energy i.

The relation between the scattered intensity and the total sample mass can be written as

$$I_{sc} = I_0 \cdot \Omega \frac{d\mu_{sc}}{d\Omega} m_{tot} \cdot c_1 \cdot c_{sc} \tag{8}$$

where c_{sc} is the self attenuation factor for the scattered radiation

$$c_{sc} = \frac{1 - \exp(-A_{sc})}{A_{sc}} \quad \text{where} \quad A_{sc} = \sum_j \rho_j \cdot d \cdot \mu_j^{att}(E_{sc}) \tag{9}$$

where: I_{sc} measured scattered intensity.

I_0 primary radiation intensity.

Ω detector solid angle.

$d\mu_{sc}/d\Omega$ differential scattering cross-section[4] for the primary radiation.

m_{tot} total sample mass (g/cm^2).

$\mu^{att}_j(E_{sc})$ attenuation coefficient for element j in the sample at the energy of the scattered radiation.

Thus, the ratio between the intensity of characteristic and scattered radiation will be independent of the attenuation of the primary beam and therefore also independent of the sample thickness.

$$Conc_i = \frac{I_i}{I_{sc}} \frac{d\mu_{sc}}{d\Omega} \frac{4\pi \cdot c_{sc}}{\varepsilon_i \cdot \omega_i \cdot \mu_{ik} \cdot c_i} \tag{10}$$

A very fine point occurs when the beam diameter is small and the impact is just at the edge of the sample. Then the self attenuation factor for the characteristic - and the scattered radiation will be close to unity, thus, correction for the attenuation needs not be made in first order, i.e c_{sc}/c_i can bee set to unity in equation 10. The ratio between characteristic and scattered radiation will then be directly proportional to the elemental concentration. In that case neither the sample thickness (or chemical composition), primary beam intensity or detector solid angle have to be taken into account.

The peak areas can easily be determined and due to the monochromatic properties of the primary beam the excitation efficiency can easily be calculated.

IV X-RAY TUBE EXPERIMENT

The desktop microbeam X-ray spectrometer described in ref 2 has been improved by new translational tables for the sample. The new sample holder can be moved perpendicular to the primary beam both vertical and horizontal, and in direction along the beam by computer controlled dc-motors with a step-increment of 0.06 μm. The sample can also be rotated with an angular step-increment of $5.14 \; 10^{-5}$ degree. These new motorized tables have improved the accuracy of the beam position versus the sample.

A capillary with a length of 19 cm and with an outcoming beam of 12 μm in diameter, a 10 mm^2 Si(Li) detector (PGT, model PO-12) and a long-fine focus diffraction Cr-tube (Philips) as X-ray source, has been used in these measurements.

The spectrometer has been used for analysing two NBS glass standards in order to test equation (10). Good agreement was obtained (see table 1), and a spectrum is shown in fig 5.   Later samples from a normal newspaper were analysed, which had an average thickness of 65 μm, to obtain the concentration detection limits (DL) and the minimum detectable amount (MDA) for some elements, which also are presented in table 1.

The concentration detection limits are defined as $3\sqrt{background}$ and are in the range of conventional XRF spectrometers but the MDA values are very low. Due to the small size of the primary beam the radiated volume

Tab 1 Present status of the XRF micro-beam spectrometer at our institute.

Element	NBS (float glass) 1830		NBS (container glass) 621		Samples from a Newspaper		
	Meas conc	Ref conc	Meas conc	Ref conc	Conc (ppm)	DL (1000s)	MDA (1000s)
Si	31 %	34.1 %	31 %	33.2 %			
Cl					1520 ppm	14 ppm	72 fg
K			1.6 %	1.67 %	716 ppm	2.9 ppm	16 fg
Ca	6,5 %	6.1 %	7.6 %	7.65 %	287 ppm	2.2 ppm	12 fg
Ti	69 ppm	66 ppm					

will become small, in cubic μm range, therefore the MDA will reach a level comparable to XRF with a synchrotron light source.

V SYNCHROTRON RADIATION EXPERIMENT

The X-ray synchrotron radiation spectrum in fig 6. is recorded at the F3 beamline HASYLAB in Hamburg[5], with a beam from a sub-micron focusing capillary impinging a solid brass sample represented by the solid line spectrum. The dotted spectrum corresponds to the direct beam. The two spectra are normalized to each other at the Cu K_α peak. Observe the different spectral distributions in the high energy region which indicates the

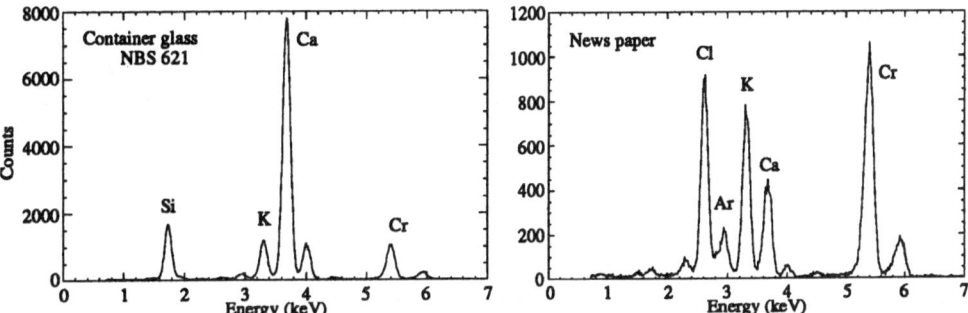

Fig. 5 Spectra recorded 1000s with a Cr X-ray tube.

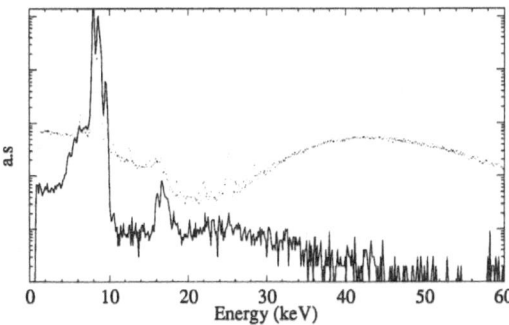

Fig. 6. X-ray spectra recorded, at the HASYLAB synchrotron facility, with a submicron (with a diameter of about 0.1 μm) focused "capillary" beam impinging a solid brass sample (solid line). and the dotted line represents the direct beam. The two spectra are normalized to each other at the Cu K_α peak, largest peak.

Tab 2 Present status of X-ray optics used in Synchrotron X-ray fluorescence microscopy, with the exception of HASYLAB data which are taken from ref 6. For the latter the band width data should be interpreted so that 0.1% corresponds to the intensity data and the beam is white. * KBML, spherical Kirkpatrick - Baez mirrors coated with multi-layers. ** EMC, Monochromator +ellipsoid mirror + collimator. *** EC, Ellipsoid crystal. ****extrapolated values.

Source	Optics	Size [μm]	Energy [keV]	Band width ΔE	Intensity [ph/μm²*s*mA*0.1%bw]	MDA (Mn) [fg]
NSLS	KBML*	6*6	6-14	10%	40000	3 (60 s)
Photon factory	Monocr+Wolter	1.6*34	8	2eV	4000	160 (100 s)
Photon factory	EMC**	10*10	8-10	2eV	5600	20 (100s)[****]
SRS	EC***	10*20	15	300eV	5000	30 (300s)[****]
HASY-lab	Capillary	1	5-30	white(0.1%)	200000	-
HASY-lab	Capillary	~0.1	5-30	white(0.1%)	1000000	-
X-ray tube	Capillary	12	5.411	Cr-line width	230	40 (100s) Ca

low-pass property of the capillary. Beside the fluorescence lines from copper and zinc at about 8 - 10 keV "pile-up" lines appear at about 16 - 17 keV. The increased intensity at the direct beam spectra in the low energy part of the spectrum compared to the capillary spectrum, corresponds to the detector effect called Compton escape, which originates from the high energy components in the spectrum.

VI CAPABILITY OF CAPILLARIES

As an example of what is achievable today with X-ray fluorescence (XRF) microscopy[6], one can mention the micro beam set-up in use at synchrotron facilities like: the National Synchrotron Light Source (NSLS) in Brookhaven, USA; the Photon Factory[7] in Tsukuba, Japan; the Synchrotron Radiation Source (SRS) in Daresbury, UK; and the HASYLAB at DESY in Hamburg, Germany. The data presented in table 2 are mostly taken from Rivers et al (ref 6) and the data in the table shows: the type of optics used, the beam cross section, the X-ray energy, the bandwidth, the intensity at 8 keV except SRS and the X-ray tube, and finally the MDA of manganese with the beam used on a thin organic sample as well as the time of analysis, are given.

VII CONCLUSIONS

From the experimental data given above it is evident that capillary optics provide an efficient and a convenient method to generate micro beams of X-rays from both X-ray tubes and synchrotron radiation sources. With a normal X-ray tube such a table top micro beam spectrometer can be made quite easily with relatively low cost. This capillary technique will probably give a new era in XRF analysis, the possibility to investigate samples in a microscopic scale (micrometer scale) by making maps of elements of interest and micro-scale density determination[8].

Compared to other micro-baem techniques the X-ray technique is very useful in it's simplicity and penetration ability and clean spectra are achieved which facilitate the quantitative analysis.

By making capillaries in a conical form and using them for "focusing" hard synchrotron radiation submicron beams can be achieved. A summary of synchrotron micro beams made by Jones and Gordon[9] states "synchrotron radiation... opens new opportunities for exploiting the XRF method". In an excellent review, given by Rivers et al. ref 6 several different synchrotron micro beam set-ups have been described using different types of conventional X-ray optics. From the beam data given it is obvious that capillary optics can provide not only a superior spatial resolution but also that the intensity of the capillary beam can be several orders of magnitude higher compared to other X-ray focusing techniques.

VII ACKNOWLEDGEMENT

The authors would like to express their thanks to Dr A. Rindby, P. Voglis, B. Stocklassa and Dr G. Nilsson for their helpful cooperation.

REFERENCES

1   N.Shakir, S.Larsson, P.Engström and A.Rindby. *Nucl Instrum Meth* **B52** 194 (1990).
       S.Larsson. X-ray microbeam spetroscopy. (thesis) ISBN 91-7032-579-0 (1991)
       P.Engström. Development of capillary optics for X-ray focusing. (thesis) ISBN 91-7032-584-7 (1991)
       D.A.Carpenter, M.A.Taylor, C. E.Holcombe. *Adv X-ray Anal* **v32** 115 (1989)
2   S.Larsson, P.Engström, A.Rindby, B.Stocklassa. *Adv X-ray Anal* **13** 623 (1990).
3   P.Engström, S.Larsson, A.Rindby, B.Stocklassa. "2nd European Conference on Progress in X-Ray
       Synchrotron Radiation Research" *Conference proceedings* vol. **25** 283 (1990).
4   J.H Hubbel, Wm.J.Veigele, E.A.Briggs, R.T.Brown, D.T.Cromer and R.J.Howerton.
       J. of Physical and Chemical Reference Data, Vol. 4, No. 3, p. 471 (1975)
5   P.Engström, S.Larsson, A.Rindby, A.Buttkewitz, S.Garbe, G.Gaul, A.Knöchel, F.Lechtenberg.
       Nucl Instrum Meth **A302** 547 (1991)
6   M.L.Rivers, S.R.Sutton, K.W.Jones. *Synch Rad News* **4** **2** 23 (1991)
7   S.Hayakawa, Y.Gohshi, A.Iida, S.Aoki, M.Ishikawa. *Nucl Instr Meth* **B49** 555 (1990)
8   A.C.Thompson, J.H.Underwodd, Y.Wu, R.D.Giauque, K.W.Jones, M.L.Rivers.
       Nucl Instr Meth **A266** 318 (1988)
9   K.W.Jones, B.M.Gordon. *Anal Chem* **61** **5** 361A (1989)

SYNCHROTRON RADIATION X-RAY FLUORESCENCE ANALYSIS

WITH A CRYSTAL SPECTROMETER

Kazutaka Ohashi, Mamoru Takahashi and **Yohichi Gohshi**
Department of Industrial Chemistry, Faculty of Engineering
The University of Tokyo, Hongo, Bunkyoku, Tokyo 113, JAPAN

Atsuo Iida and **Shunji Kishimoto**
Photon Factory, National Laboratory for High Energy Physics
O-ho, Tsukubashi, Ibaraki 305, JAPAN

ABSTRACT

A wavelength dispersive spectrometer which consists of a flat crystal analyzer and a position sensitive proportional counter has been developed for X-ray fluorescence analysis using synchrotron radiation. The advantages of this spectrometer are high energy resolution, multielemental nature, and high efficiency, and these match well with the high brightness synchrotron X-ray source. The minimum detection limits are of the order of ppm or pg. An application to elemental mapping has also been demonstrated. The present system is useful for practical analysis of small samples or small regions.

INTRODUCTION

X-ray fluorescence (XRF) analysis using synchrotron radiation (SR) is a powerful technique for microbeam analysis[1,2], chemical state analysis[3], surface analysis[4] and trace element analysis[5]. SR-XRF analysis usually employs energy dispersive (ED) type detection in which a Si(Li) detector is almost exclusively used because of its high efficiency. However, the ED measurement is not suitable for those samples in which line interference is dominant because of its poor energy resolution.

Wavelength dispersive (WD) XRF analysis has higher energy resolution than ED analysis. Several types of WDXRF system for use with SR have been developed ; a flat crystal with a solar slit[6,7,8], which is the standard for the laboratory WDXRF system, a flat crystal spectrometer with a position sensitive detector[9,10], and a curved crystal spectrometer of high energy resolution[11]. Among these, a spectrometer consisting of a flat crystal with a position sensitive proportional counter (PSPC) is suitable for elemental analysis of practical samples because of the high brightness of SR. Furthermore, its high efficiency and multielemental nature are promising for both qualitative and quantitative analysis.

In this work, analytical aspects of SR-WDXRF with a PSPC are reported emphasizing the use of a flat crystal analyzer. The experimental parameters which affect the energy resolution, the signal intensity and the signal to background ratio (S/B) were studied. An application to XRF imaging,i.e. elemental mapping, is also presented.

EXPERIMENTAL

The experiment was performed at the Photon Factory on beam line 4A. The experimental arrangement is schematically shown in Fig.1. A flat crystal and a linear PSPC(50mm effective length, 0.15mm spatial resolution) were mounted on the θ-2θ goniometer. Analyzer crystals used were a perfect Ge (111) crystal mirror polished and mosaic crystals of a Ge (111) lapped with SiC powder and a pyrolytic graphite. Several excitation modes were studied experimentally. The continuous spectrum of SR has high intensity, but scattered radiation results in high background. A reflecting mirror (Pt/SiO$_2$) before the sample is used to reduce radiation damage to the sample. A synthetic multilayer (W/Si,2d=61.2Å) was used for monochromatic excitation.

A pair of slits were used for adjusting the irradiation area of the sample. The sample was positioned at 45° to the incident beam. Samples used were reference materials from NIES(National institute for environmental studies) and NIST(National institute of standards and technology).

RESULTS and DISCUSSION

Energy Resolution

When a perfect crystal is used as an analyzer, the energy resolution is determined by (1) the horizontal beamsize at the sample (X-ray source size for the spectrometer), (2) the sample-PSPC distance, (3) the spatial resolution of the PSPC, (4) the Bragg angle and (5) the crystal diffraction width. Usually the first two parameters can be varied experimentally. A smaller beam size and a longer sample-PSPC distance give higher energy resolution. Fig.2 shows the dependence of the energy resolution on the horizontal beam size for a powder sample. For conventional analysis, a beam size of more than 0.1 mm, Ge 111 reflection and the sample-PSPC distance of 460 mm were used to optimize the signal intensity and energy resolution.

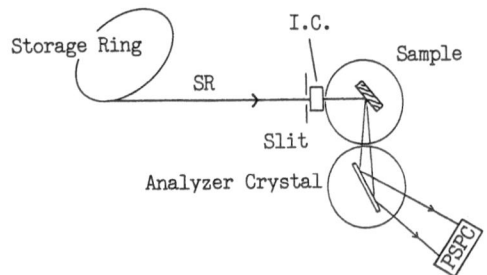

Fig. 1 Experimental arrangement. I.C : He filled ionization chamber, PSPC : position sensitive proportional counter

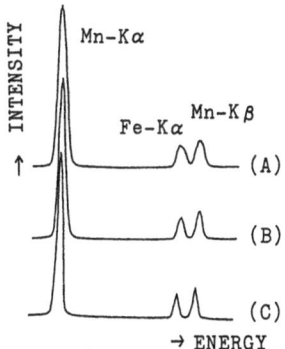

Fig.2 Dependence of the energy
resolution on the beam size.
Horizontal beam size are
1.0mm(A), 0.5mm(B) and 0.1mm(C).

The spectra obtained with the WDXRF and the EDXRF are shown in Figs.3
(a) and (b) respectively.   The WD spectrometer resolves the adjacent Kα
and Kβ lines from the transition metal elements and the Lα lines from the
rare earth elements. Thus, WDXRF is effective for analysis of those samples
in which line interference is dominant.

A CuKα high energy resolution spectrum is shown in Fig.4 that was
obtained by Ge 333˙ reflection, a sample-PSPC distance of 1380 mm and a
beam size of 50μm. The FWHM (full width at half maxima) obtained was 3.3 eV
while the natural width is 2.6 eV [12].

When a mosaic crystal, a lapped Ge(111) crystal or a pyrolytic graphite
is used as an analyzer, due to the focusing effect of the mosaic crystal,
the longer sample to analyzer distance degrades the energy resolution.

Sensitivity

Table 1 compares the signal intensity and S/B for excitation modes
and analyzer types. The continuous spectrum of SR is about 100 times more
efficient than monochromatic excitation. The efficiency of the analyzer
crystal is highest for pyrolytic graphite.

In our experiment, the crystal spectrometer was set at 90° to the
incident beam so as to minimize scattered radiation from the sample using
SR polarization. To verify the polarization effect on the S/B, the
intensities of incident and scattered X-rays from the sample were
measured as a function of SR vertical angular divergence and is shown in
Fig.5. The scattered X-ray intensity shows a minimum at the center of the
X-ray beam and maxima at the off center corresponding to the vertically
polarized component of SR.   The MDL for the practical sample obtained at
the center of X-ray beams was improved by a factor of two compared with
that at off center. The linearly polarized nature of SR is effective in
reducing the background of WDXRF analysis just as it has been for EDXRF[13].

To achieve higher S/B, excitation with linearly polarized
monochromatic X-rays, which is obtained with a monochromator and a narrow
slit, is favorable. Polarized continuum SR is also useful because of the

Table 1   Comparison of the signal intensity and the S/B
          for excitation modes and analyzer crystals
                                    Sample : Pepperbush (NIES No.1)

Excitation Mode	Analyzer crystal	Relative signal intensity of MnKα	S/B for MnKα
Continuous SR	Ge(111)Lapped	1	183
Monochromatic SR	Ge(111)Lapped	0.01	320
Monochromatic SR	Graphite C(0002)	0.08	394

extremely high photon flux density. The graphite analyzer has the highest
signal intensity, while the energy resolution can be improved to around 1eV
with a perfect crystal analyzer.

Table 2. shows the minimum detection limits (MDLs) of those elements
that cannot be detected with the Si(Li) detector due to overlapping peaks.
These MDLs were obtained with SR continuum excitation and Ge (111) analyzer
with a collection time of about 1000 sec. The MDLs were less than ppm. Due
to the small beam size, the MDL in absolute amount is of the order of pg
and   is improved by a factor of a few hundreds compared to that obtained
with a conventional X-ray tube crystal spectrometer.

Imaging

One of the advantage of WDXRF utilizing SR is the high sensitivity for
a small samples or in a small regions.   XRF imaging is a promising
application of SR-WDXRF analysis. The elemental mapping of a rock sample
(peridotite) with 100 μm spatial resolution is shown in Fig.6.   The high
and low concentration regions of Cr correspond to chromite and olivine
respectively[14]. The Co concentration distribution can not be obtained by
EDXRF because of overlapping peaks.

Table 2   Minimum Detection Limits (MDLs) in reference materials

Sample	Element	MDL/ppm	Concentration/ppm (Certified Value)
Pepperbush (NIES #1)	Cr	0.2	1.3
	Co	0.4	23
	Ni	1.5	8.7
Pond Sediment (NIES #2)	Co	0.9	27
Mussel (NIES #6)	Mn	0.2	16.3
NIST SRM #612 (Glass,Trace element 50 ppm)	Fe	0.9	51

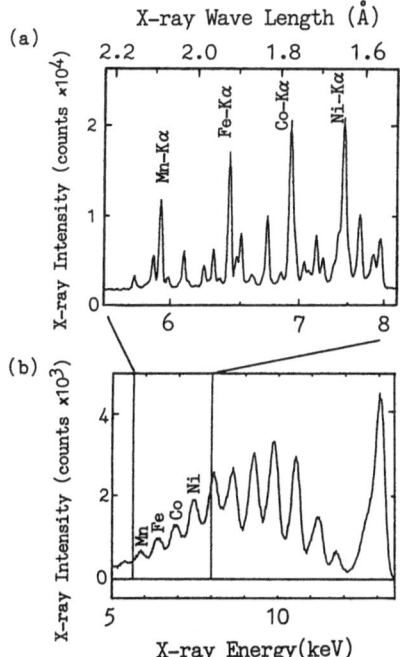

Fig. 3   WD (a) and ED (b) spectra from NIST No.610 (Trace elements <500ppm> in glass).
(a) 2000 second measurement time. 0.5mm*1.0mm beam size. Continuous SR excitation.
(b) 200 second measurement time. 0.3mm*1.3mm beam size. Si (111) double crystal monochromator was used for monochromatization of excitation X-rays.
Excitation energy was 13keV.

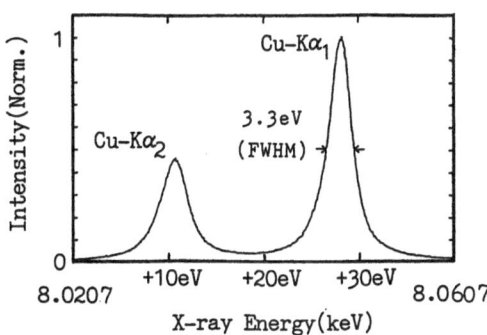

Fig. 4   Cu Kα high energy resolution spectrum obtained with Ge 333 reflection.

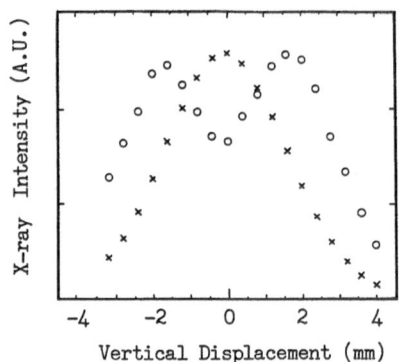

Fig.5 Measured intensities of scattered radiation from a polymer plate (o) and the incident SR beam (×) as a function of the vertical displacement of the horizontal slit before the sample. The slit width was 400 μm. The SR source to the slit distance was 14m.

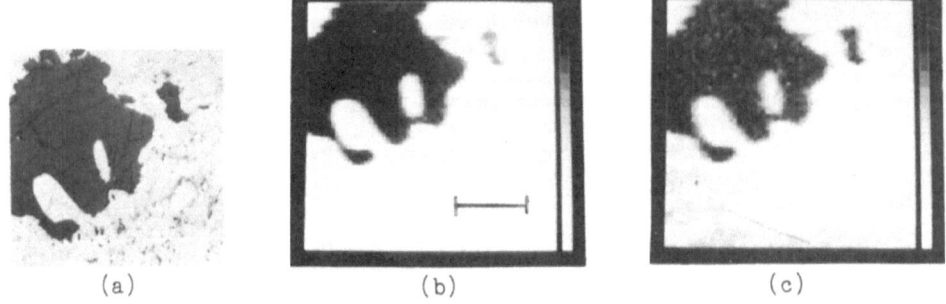

(a)   (b)   (c)

Fig.6 (a) Optical micrograph of the peridotite and WDXRF images of CrKβ (b)and CoKα (c). Continuous SR excitation. 100*100 μm^2 beam size. 60*60 pixels. 7sec. collection time for each pixel. Scale mark corresponds to 1 mm.

ACKNOWLEDGMENTS

We would like to thank Prof. Y. Amemiya for stimulating discussions, Dr. K. Sato for providing the rock sample and Dr. T. Scimeca for critically reading the manuscript. We also wish to thank the PF staff for their help during the experiments. This work has been performed under the approval of Photon Factory Program Advisory Committee (Proposal No.89-171).

REFERENCES

1. K.W. Jones and B.M. Gordon, Anal. Chem., 61, 341A (1989)
2. S. Hayakawa, A. Iida, S. Aoki and Y. Gohshi, Rev. Sci. Instr., 60, 2452 (1989)
3. K. Sakurai, A. Iida and Y. Gohshi, Jpn. J.Appl.Phys., 26, 1937 (1987)
4. A. Iida, Adv. in X-ray Anal., 34, 23 (1991)
5. A. Iida and Y. Gohshi, in " Handbook on synchrotron radiation " vol.4, p.307, eds. S. Ebashi, M. Koch and E. Rubenstein (North-Holland, Amsterdam, 1991)
6. J.V. Gilfrich, E.F. Skelton, S.B. Quadri, J.P.Kirkland and D.J. Nagel, Anal. Chem., 55, 187 (1983).
7. M. Prins, W. Dries, W. Lenglet, S.T. Davies and K. Bowen, Nucl.Instrum and Method, B10/11, 299 (1985).
8. A. Iida, Y. Gohshi and H. Maezawa, Adv. in X-ray Anal., 29, 427 (1986)
9. P. Chevallier, C. Jehanno, M. Maurette, S. R. Sutton and J. Wang, J. Geophys. Res., 92, E649 (1987).
10. P. Chevallier, J. de Physique, C9, 39 (1987)
11. S. Brennan, P. L. Cowan, R. D. Deslattes, A. Henins, D. W. Lindle and B.A. Karlin, Rev. Sci. Instr., 60, 2243 (1989)
12. M. Deutsch and M. Hart, Phy. Rev.,B26, 5558 (1982)
13. K. Sakurai, A. Iida and Y. Gohshi, Anal. Sci., 4, 3 (1988).
14. S. Hayakawa, Y. Gohshi, A. Iida, S. Aoki and K. Sato, Rev. Sci. Instr., in press.

A NEW USER ORIENTED INTELLIGENT XRF SPECTROMETER SYSTEM

Y. Kataoka and N. Masukawa

Rigaku Industrial Corporation
Takatsuki, Osaka, Japan

K. Toda

Rigaku/USA, Inc.
Danvers, Massachusetts USA

INTRODUCTION

The semi-quantitative analysis, which is called 'Standardless Analysis', plays a major role in X-ray fluorescent analysis, especially in the field of research and development. The main feature of the semi-quantitative analysis is the fact that the composition of a sample can be obtained directly from a qualitative scan without any prior knowledge of the sample.

This paper describes a new wavelength dispersive X-ray fluorescence spectrometer, RIX 3000, which has full multi-tasking and multi-window software, and gives maximum flexibility in operations with a user friendly interface. Through rapid qualitative scan and simultaneous data processing, the total analysis time from qualitative scan measurements, to the output of the semi-quantitative analysis is within 5 minutes. The function of conventional semi-quantitative analysis software was limited to bulk analysis a new software package has been expanded to thin film analysis. A new algorithm for automatic element identification for light element range and peak separation techniques demonstrate accurate analysis results on semi-quantitative analysis.

This paper describes the principles and the results concerning the new techniques for the qualitative and semi-quantitative analysis.

EXPERIMENTAL

A brief configuration of the spectrometer and data processing system used in this study is shown in Table 1. An end-window dual target X-ray tube gives optimum sensitivities for all elements merely by switching the target. The tube is fully controlled by the computer so that the target selection is

Table 1 Spectrometer and data processing system configuration

Spectrometer	Rigaku sequential X-ray spectrometer RIX 3000	
	X-ray tube	: End-window type Rh target, Rh/Cr or Rh/W dual target
	Analytical range	: $_5B - _{92}U$
	Analyzing crystal	: LIF, PET, Ge, TAP, Synthetic multi-layers Total reflection devices (10 positions)
	Detector	: SC, F-PC
	Slit	: Coarse, Fine, Ultra coarse (3 positions)
	Attenuator	: 1/1, 1/10
	Measuring area	: 5-35 mm dia.(6 positions)
Data Processing	Software	: Rigaku Dataflex 3000
	Computor	: IBM PS/2
	Operating System	: IBM OS/2 (Multi-task, Multi-window)

programmable for each element. A 10 position crystal changer and 3 position Soller slit changer allows selection of a combination to give the highest sensitivities for the ultra light elements and the highest resolution for the heavy element range.

The data processing system connected to the spectrometer is an IBM PS/2. The software supports full multi-tasking and multi-window functions. The function of the multi-tasking is not limited, so any combination of programs can be run simultaneously. The multi-window and mouse operation give a user friendly interface.

The programmable system control function enables an operator to program automatic power on and off operation. An example of a program sequence is: The system shuts off the X-ray after completion of a set analyses and then turns on the X-ray at a preset date and time. Then it executes tube aging, adjusting pulse height for the detectors, and runs a check program which may include drift corrections and check analyses. All these operations can be done automatically.

RAPID QUALITATIVE AND SEMI-QUANTITATIVE ANALYSES

Short turn around time of analysis can be crucial in laboratories involved in trouble shooting production problems. The total run time from qualitative scan measurements to output of semi-quantitative analysis for the element range $F_9 - U_{92}$ can be within 5 minutes using this system. It is achieved by fast data collection and simultaneous data processing. Figure 1 shows a time chart of an analysis. The measurements include the time for preliminary measurements for automatic intensity scaling of chart output. The scanning speed can be preset up to 120°/min for a continuous scan. As shown in the figure, all processing is done on-line and the simultaneous processing minimizes the run time. The qualitative scan, using the conditions in Figure 1, is shown in Figure 2 with the automatic identification result. Tables 2 and 3 are semi-quantitative analysis results obtained by using same conditions as above. As the chart shows, the peaks

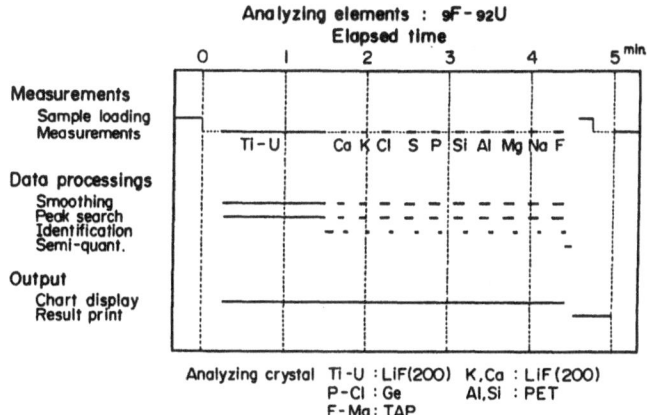

Figure 1   Time chart of rapid qualitative and
semi-quantitative analyses

of trace elements are clearly detected.   Table 2 exhibits the
semi-quantitative analysis results of low allow steel NBS 1168
and stainless steel JSS 655-11.   Table 3 lists the results on
brass NBS1105 and soda lime glass NBS620 which is a light element
based sample.   The sensitivity for each element in the semi-
quantitative analysis was obtained from a sensitivity library[1],[2]
which was set up by measuring typical pure materials.   The
sensitivity library gives sensitivity for any element from F to
U.   As the results show, in general, that the relative accuracies
are less than 5% for major elements and the lower analytical
range is 0.01 - 0.05%.

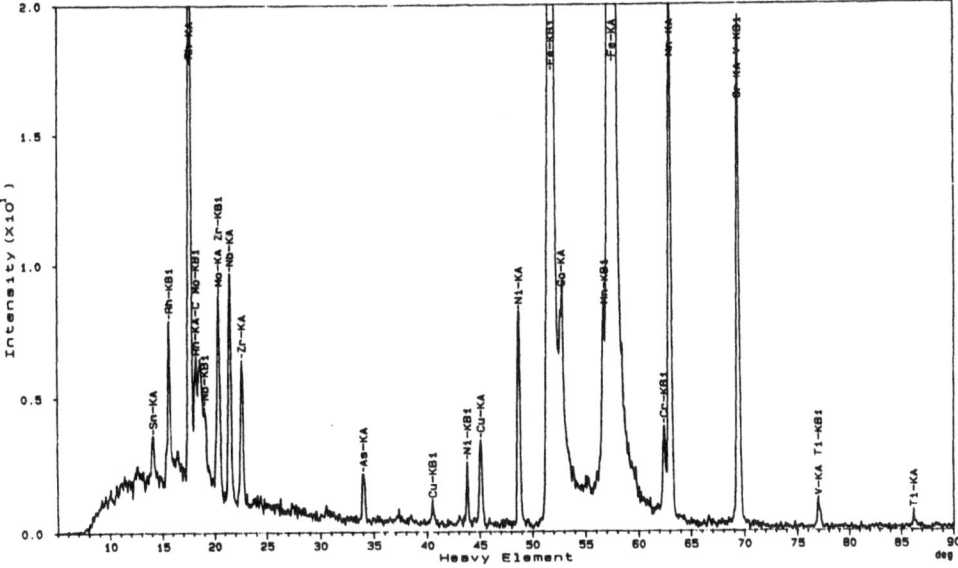

Figure 2   Qualitative chart and the identification result
by rapid qualitative scan   (sample : low alloy steel)

Table 2   Semi-Quantitative analysis results by
          rapid qualitative scan -(1)

Low alloy steel (NBS 1168)

unit : %

Element	Chemical	XRF	Deviation
Si	0.075	0.086	0.011
Mn	0.47	0.49	0.02
P	0.023	0.020	-0.003
Ni	1.03	1.05	0.02
Cr	0.54	0.57	0.03
Mo	0.20	0.21	0.01
Cu	0.26	0.27	0.01
V	0.17	0.22	0.05
Co	0.16	0.17	0.01
W	0.077	0.090	0.013
Al	0.042	0.065	0.023

Stainless steel (JSS 651-11)

unit:%

Element	Chemical	XRF	Deviation
Si	0.42	0.42	0.00
Mn	1.70	1.79	0.09
P	0.035	0.035	0.000
Ni	10.11	10.51	0.40
Cr	18.47	18.73	0.26
Mo	0.16	0.17	0.01
Cu	0.39	0.43	0.04
Co	0.17	0.17	0.00

Table 3   Semi-Quantitative analysis results by
          rapid qualitative scan -(2)

Brass (NBS 1105)

unit : %

Element	Chemical	XRF	Deviation
Cu	63.72	62.82	-0.90
Zn	34.00	35.07	1.07
Pb	2.01	1.78	-0.23
Fe	0.044	0.060	0.016
Sn	0.21	0.20	-0.01
Ni	0.043	0.063	0.020

Soda lime glass (NBS 620)

unit : %

Component	Chemical	XRF	Deviation
SiO_2	72.80	71.76	-1.04
Al_2O_3	1.80	1.74	-0.06
Fe_2O_3	0.043	0.042	-0.001
CaO	7.11	6.35	-0.76
MgO	3.69	4.16	0.47
K_2O	0.41	0.41	0.00
Na_2O	14.39	15.05	0.66
As_2O_3	0.056	0.077	0.021
SO_3	0.28	0.33	0.05

SEMI-QUANTITATIVE ANALYSIS FOR THIN FILMS

     The functions of the semi-quantitative analysis have been
expanded to thin film applications. Figure 3 illustrates a flow
chart of the semi-quantitative analysis for thin films. The film
to be analyzed is single layer and the substrate may be included
in the calculation.   After qualitative data processing, the
program selects analytical lines for the fundamental parameter
calculation using the sensitivity library as mentioned above.
Figure 4 exhibits typical analyzed results of metal films coated
on PET film. The result can also be output as the area density
of each component instead of concentration. This output may be
used for the analysis of airborne particulates on filter.

ELEMENTAL IDENTIFICATION

     Accurate elemental identification is essential for automated
data processing from qualitative data collection to semi-
quantitative analysis. Figure 5 illustrates the flow chart of
the new elemental identification method. For heavy elements,

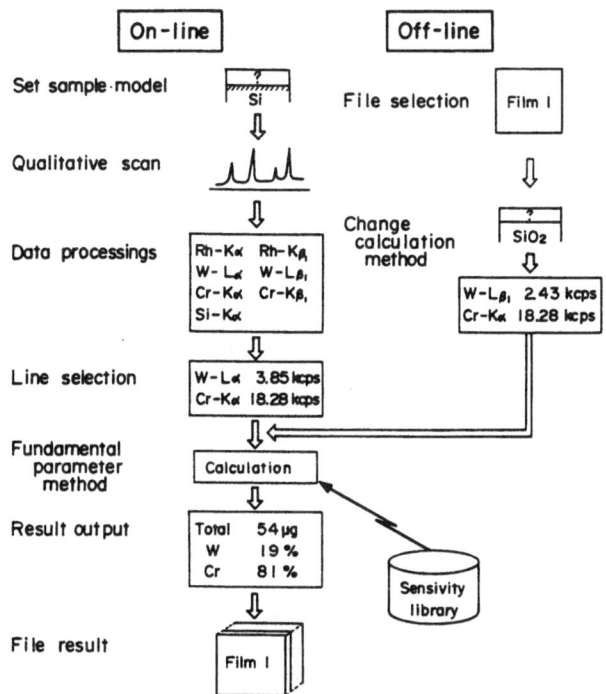

Figure 3   Semi-quantitative analysis of thin film

the qualitative scan range to be taken is commonly wide and is started from the shortest wavelength. The range covers from Ti_{22} to U_{92} using a LiF (300) crystal. Therefore, accurate identification can be dome logically by utilizing the intensity ratios among lines of $K\alpha$, $K\beta_1$, $L\alpha$, $L\beta_1$, $L\beta_2$...and the intensity ratios of the first order to higher order lines. The algorithm was included in the conventional Rigaku elemental identification software and the software gave successful results[1]. The new algorithm provides excellent results in both heavy and light element ranges. In the light element range, the scanning range is usually taken within a short region around the analyte line. Accordingly, a single peak of L, M series line, or higher order line of a heavy element often appears. To find the most probable line of the peak, the program searches all lines of heavier elements which have been identified from an X-ray line data base.

CoCr			
PET			
Component	Chemical	XRF	Deviation
CoCr (µg/cm²)	51	54	3
Co (%)	80.6	81.3	0.7
Cr (%)	19.9	18.7	-0.7

CoCrW			
PET			
Component	Chemical	XRF	Deviation
CoCrW(µg/cm²)	176	164	-12
Co (%)	76.4	77.2	0.8
Cr (%)	18.8	17.9	-0.9
W (%)	4.8	4.9	0.1

Figure 4   Semi-quantitative analysis results of thin films

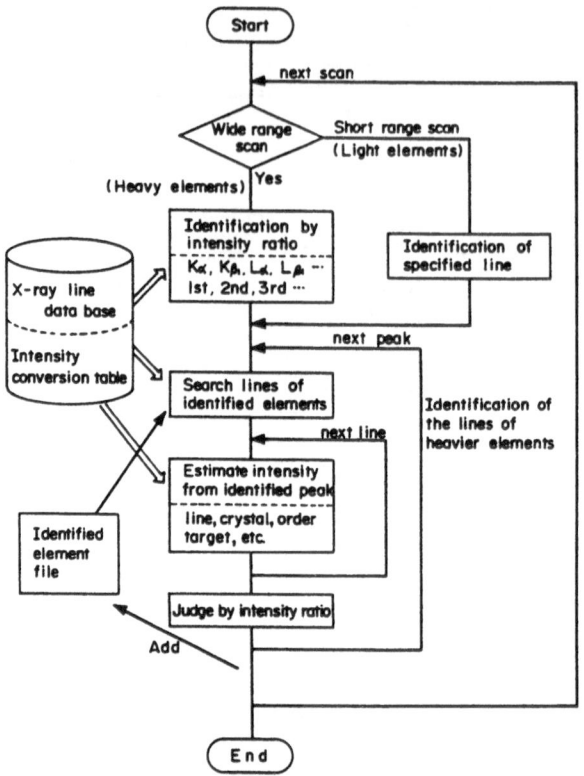

Figure 5   Flow chart of elemental identification procedure

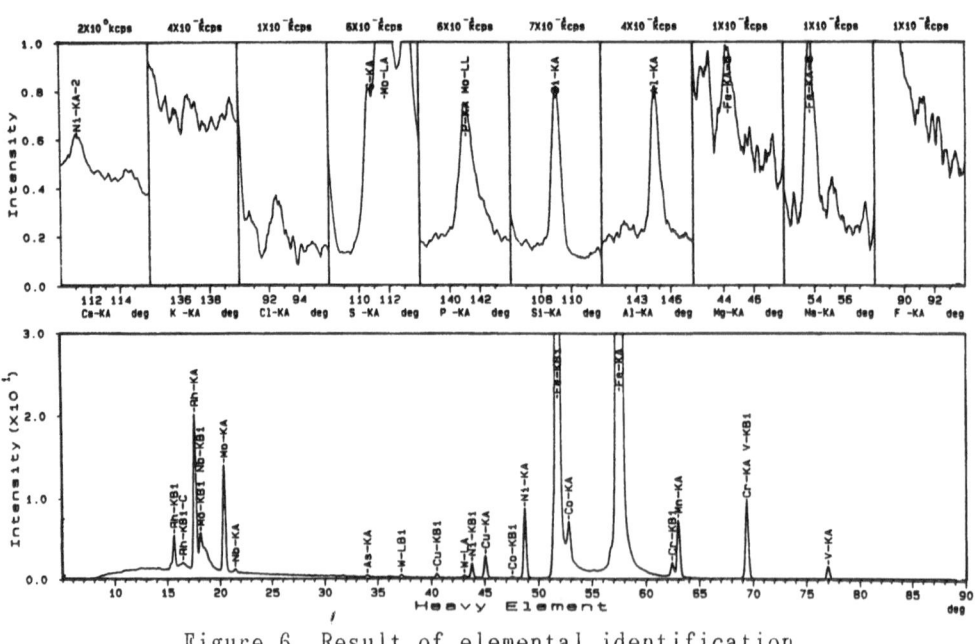

Figure 6   Result of elemental identification
(sample : low alloy steel)

Table 4   Peak separation method

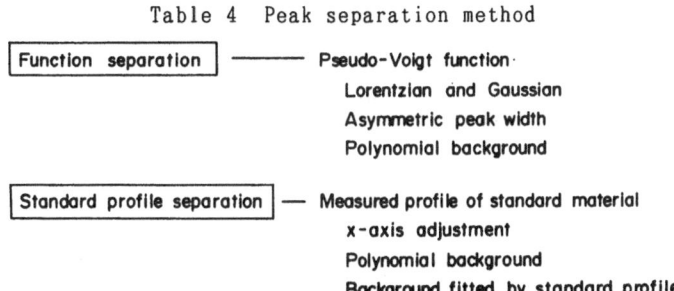

It predicts the possible peak intensity of each line searched from the identified result in the heavier element ranges, considering the differences in measuring conditions and lines. Then the program makes a final judgement by using the intensity ratio of an observed peak intensity to the predicted intensity.

Figure 6 is the identification result of a steel sample obtained by the automatic identification program. As shown in the figure, the higher order lines of Ni and Fe-Kα lines, whose first order lines were identified in the heavy element scan, were identified. Lines of Mo are also L identified as possible lines.

PEAK SEPARATION

Peak separation techniques have not been commonly used in wavelength dispersive types of spectrometers (WDXRF). This is because the intensities of only peak and background positions are measured in conventional quantitative analysis and line overlappings are corrected by using line overlapping coefficients.

The peak separation techniques developed are used to correct line overlaps in qualitative scan data for semi-quantitative analysis. There are two kinds of peak separation methods developed, as shown in Table 4. The function separation, pseudo-Voigt function, consists of both Gaussian and Lorentzian components and provides good fitting for WDXRF[1] peak profiles. The standard profile separation uses standard profiles obtained by measuring standard materials instead of using a mathematical function. Table 5 shows the formulae for the peak separation. A profile for the separation is fitted with the summation of individual profiles and background. The background is calculated using a polynomial function. In the function separation, an individual peak is treated as asymmetric and the widths for the left and right sides from the peak are calculated independently. All coefficients are calculated by a least squares fit. In the standard profile separation, the fitted profile is obtained with a scaling factor α_k and 2θ adjustment β_k.

Figure 7 illustrates the principle of the standard profile separation using an example of Mn-Kα and Cr-Kβ_1 overlapping. The samples used for the standard profiles are pure Cr and Mn metals. All measured standard profiles can be saved in a file and they can be used when required. The backgrounds of the

Table 5   Peak separation formulas

$$I(2\theta) = \sum_{k}^{k} f_k(2\theta) + B(2\theta)$$

$f_k(2\theta)$: Profile of peak k
$B(2\theta)$: Background function

Function separation (Pseudo-Voigt function)

$$f_k(2\theta) = h_k \{ r_k f_G(2\theta) + (1-r_k) f_L(2\theta) \}$$

$f_G(2\theta)$: Gaussian $f_L(2\theta)$: Lorentzian

$$f_G(2\theta) = \exp\left\{ -\ln 2 \frac{(2\theta - 2\theta_P)^2}{W^2} \right\}$$

$$f_L(2\theta) = \frac{1}{1 + \frac{(2\theta - 2\theta_P)^2}{W^2}}$$

h_k : Peak intensity
r_k : Gauss content
W : Half of 50% width
$2\theta_P$: Peak 2θ

Standard profile separation

$$f_k(2\theta) = \alpha_k S_k(2\theta + \beta_k)$$

$S_k(2\theta)$: Measured standard profile for peak k
α_k : Peak intensity ratio
β_k : Adjusted 2θ

measured standard profiles can be removed using a background removing program so that only peak profile can be saved in the file.   Peak profile of a X-ray fluorescence line is generally consistent.   Strictly speaking, peak profiles of a line can slightly vary with the chemical structure of the compound. Accordingly, standard profiles obtained from a sample having the same chemical structure as an actual sample should give the best fitting.

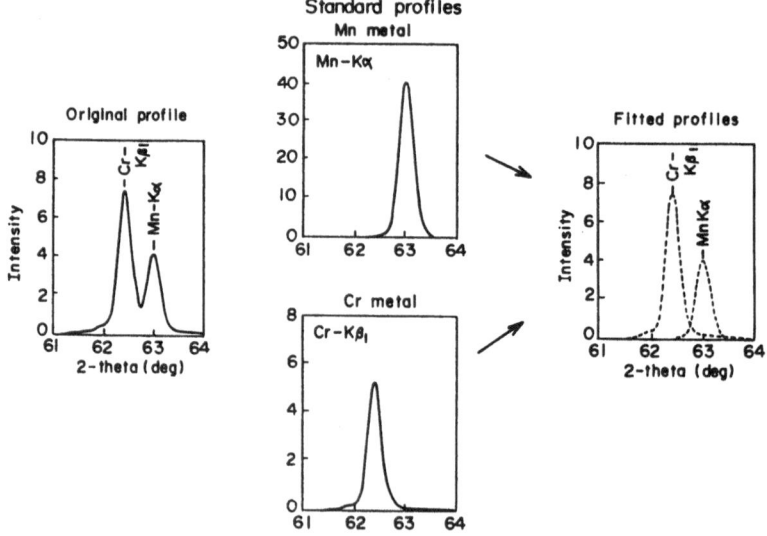

Figure 7   Principle of standard profile separation

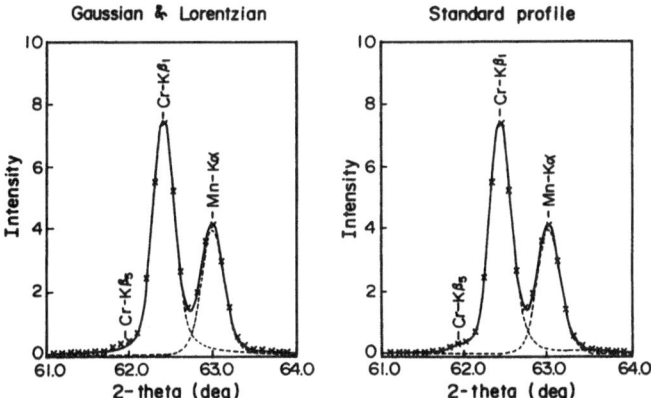

Figure 8   Comparison of peak separation methods
(sample : stainless steel)

Figure 8 shows the comparison of the peak separation of both methods for the Mn-Kα and Cr-Kβ₁ overlap in a stainless steel. In the figure, the "x" mark is a measured data point. The broken line is an individual fitted profile with the calculated background and the solid line is a synthetic fitted profile. As the results show, both methods gave excellent fits and the difference between the two separations is the fit around 62°. In the function separation, the Cr-Kβ₅ line is calculated as part of the Cr-Kβ₁ line. On the other hand, the Cr-Kβ₅ line is included in the standard profile of the Cr-Kβ₁ peak profile in the standard profile separation and thus gave a perfect fit.

Figure 9 shows the peak separation results of V-Kα and Cr-Kα lines in a titanium alloy using the standard profile separation. The background was calculated as a linear line. As shown in the figure, the fits are very good even for serious overlaps. Table 6 lists the semi-quantitative analysis results of the titanium alloy NBS 654a, which is a 6A1-4V titanium alloy, using the peak

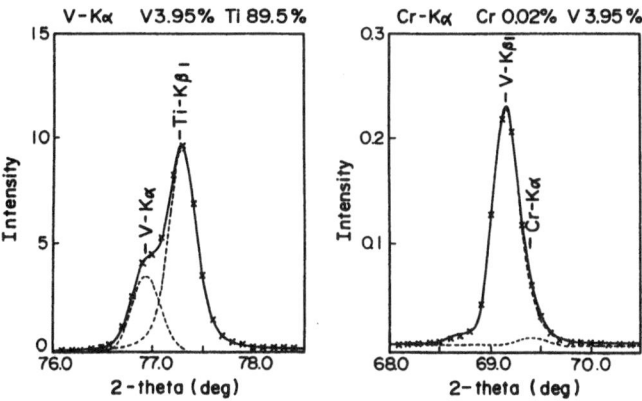

Figure 9   Peak separation in titanium alloy
(standard profile separation)

Table 6   Semi-quantitative analysis of titanium alloy
using peak separation

Sample : NBS 654a (6Al-4V)

unit : %

Element	Line	Chemical	with separation	Deviation	without separation	Deviation
Cr	Cr-Kα	0.02	0.02	0.00	0.45	0.43
Fe	Fe-Kα	0.20	0.22	0.02	0.21	0.01
Al	Al-Kα	6.34	6.16	-0.18	6.10	-0.24
V	V-Kα	3.95	4.00	0.05	4.76	0.81

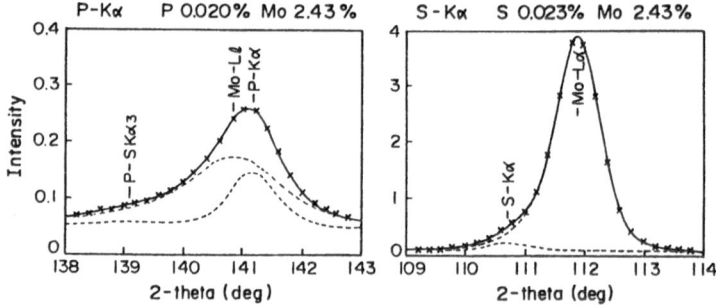

Figure 10   Peak separation in stainless steel
(standard profile separation)

separation results shown in Figure 9.   The analyzed results of
V and Cr without peak separation gave higher concentrations than
the chemical values due to the line overlap. The results with
the peak separation are well matched with the chemical values.

Figure 10 and Table 7 show another application of the peak
separation method. The sample is stainless steel BAS SS66. The
peak separation is applied to P and S determinations.   The

Table 7   Semi-quantitative analysis of stainless steel
using peak separation

Sample : BAS SS66

unit : %

Element	Line	Chemical	with separation	Deviation	without separation	Deviation
Si	Si-Kα	0.51	0.51	0.00	0.51	0.00
Mn	Mn-Kα	0.81	0.83	0.02	0.83	0.02
P	P-Kα	0.020	0.020	0.000	0.041	0.021
S	S-Kα	0.023	0.028	0.005	Not detected	-0.023
Ni	Ni-Kα	9.48	9.55	0.07	9.55	0.07
Cr	Cr-Kα	17.60	17.35	-0.25	17.35	-0.25
Mo	Mo-Kα	2.43	2.62	0.19	2.62	0.19
Co	Co-Kα	0.063	0.062	-0.001	0.062	-0.001

Table 8 Background subtraction formulas

$$I_{net} = I_P - I_B$$

One point	$I_B = b_1 I_{B1}$
Two points	$I_B = b_1 I_{B1} + b_2 I_{B2}$

Multi-point fitting

Polynomial function $I_B = b_1 + b_2(2\theta) + b_3(2\theta)^2 + b_4(2\theta)^3$

Lorentz function $I_B = b_1 + b_2(2\theta) + \dfrac{b_5}{\{(2\theta) + b_3\}^2 + b_4}$

Hyperbolic function $I_B = b_1 + b_2(2\theta) + \dfrac{b_4}{(2\theta) + b_3}$

concentration of Mo is 2.43% which is about 100 times higher than the concentrations of P and S. In the separation of P-Kα and Mo-L$_1$ lines, the peak positions of the two lines are close to each other. However, the difference of the peak widths makes the separation possible. In the figure of the S-Kα, the S-Kα peak is completely hidden in the Mo-L$_1$ peak. The S-Kα peak appeared after the separation. Table 7 lists the analyzed results of the sample by semi-quantitative analysis using the peak separation. The concentration of P without separation is 0.041% which is about double of the chemical value. The results of P and S with the separation show good agreement with the chemical values.

BACKGROUND FITTING

In most cases, one or two point measurements give good estimation of background intensity assuming straight background. However, when a small peak is located near a large peak, the assumption of a straight background can give an error in the background estimation for curved backgrounds. In this case, the background should be calculated as a curved line. The formulae used for the background subtraction are shown in Table 8. The Lorentz or hyperbolic function gives a good fit for curved backgrounds. The multi-point background fitting can be used for either qualitative scan data or quantitative analysis in the software.

Figure 11 illustrates the results of the background fitting for Bi determination in solder samples. The function used is

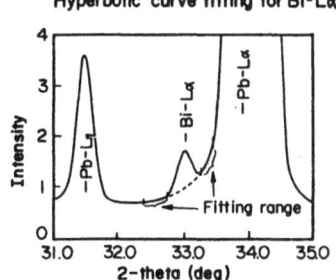

Hyperbolic curve fitting for Bi-Lα

Semi-quantitative analysis result of Bi

unit : %

Sample	Chemical	with BG fitting	without BG fitting
MBH40-1	0.05	0.05	0.03
MBH40-3	0.25	0.26	0.22
MBH50-1	0.06	0.06	0.03
MBH50-2	0.16	0.15	0.12

Figure 11

hyperbolic. The data points for the fit calculation were selected before and after the peak position. The concentrations of Bi were determined by the semi-quantitative analysis. The results without fitting are obtained assuming straight background and are about 0.03% lower than the chemical values. The results with fitting showed good agreement with the chemical values.

CONCLUSION

Rapid qualitative scan and simultaneous data processing provide a total analysis from qualitative measurements to output of semi-quantitative analysis results within 5 minutes. The lower analytical range is about 0.01 - 0.05% in general. The new algorithm for automatic elemental identification provides excellent results in both heavy and light element ranges. The peak separation techniques show accurate estimation of net intensities and the results prove that they are powerful tools for the semi-quantitative analysis. The function of the semi-quantitative analysis have been expanded to thin film applications.

REFERENCES

1.   H. Kohno, M. Murata, Y. Kataoka, T. Arai: Jpn Adv. X-Ray Anal., 19, 307 (1988)

2.   Y. Kataoka: The Rigaku Journal, 33, 6 (1989)

3.   T. C. Huang, G. Lim, X-Ray Spectrom., 15, 245 (1986)

A HIGH RESOLUTION PORTABLE XRF HgI$_2$ SPECTROMETER FOR FIELD

SCREENING OF HAZARDOUS METAL WASTES

M. Bernick[1], P. F. Berry and G. R. Voots[2], G. Prince[3], J. B. Ashe[4],
J. Patel and P. Gupta[1].

1 Roy F. Weston, Inc./REAC, Edison, New Jersey, 08837
2 TN Technologies, Inc. Round Rock, Texas, 78664
3 USEPA/Environmental Response Team, Edison, New Jersey, 08837
4 Ashe Analytics, Butte, Montana, 59711

ABSTRACT

The use of a field portable XRF analyzer incorporating a semiconductor, mercuric iodide, energy dispersive spectrometer is described with emphasis on the benefits of high resolution x-ray detection for rapid screening of hazardous metallic wastes. Results are presented of "in-situ" and "prepared sample" soil measurement for different sites to show the potential of Fundamental Parameter analysis to obtain acceptable quality data with minimum calibration effort, obviating the need for site-specific standards.

INTRODUCTION

Many of the U.S.EPA Hazardous Waste Site Evaluation/Removal programs now use field portable x-ray fluorescence (FPXRF) instruments to provide rapid, low cost, on-site analysis of soils and sediments for potentially-toxic metal contaminations from past industrial activity. Application, within the guidelines of the EPA Field Analytical Screening Program[1], has shown FPXRF to be well suited to the program's objectives in delivering acceptable quality results for most metal contaminants at the level of concern and, since the measurements are performed at the site, immediate results are available to assist in the ongoing evaluation/removal activities. The real-time nondestructive nature of x-ray analysis coupled with the application versatility of a remote probe measurement also promote rapid data turnaround and reduction in project cost. Sampled materials need simply to be dry and homogenous and the preparation of moderately dry soils for multi-point in-situ assay may only entail the removal of rocks and vegetation. In-situ measurement promises to be one of the most cost-effective methods of FPXRF application both in rapidly delineating the zones of high contamination and in greatly reducing the number of samples to be taken for the expensive and time-delayed Contract Laboratory analysis[2].

As the use of FPXRF continues to develop for environmental application, as does the impetus to improve the analytical performance and ease-of-use by non-technical personnel. Improvement in the x-ray spectrometer resolution is a step in this direction and the topic of our paper.

INSTRUMENT DEVELOPMENTS

Present instruments use "enhanced-resolution", gas filled proportional detectors and achieve a typical energy resolution of about 850 eV at FWHM for Mn K x-rays. At this resolution the x-rays of adjacent atomic number elements can not be resolved resulting in significant multielement spectrum overlaps. These instruments have been successful in field analysis provided the chemical characteristics of the site have been predetermined and the instrument calibrated accordingly[3]. Spectrum overlap factors are derived by measurement on pure elements representing both target elements and those expected to cause spectral interference. Calibration requires measurement of "site-specific" chemically analyzed samples. Multivariate regression techniques are used to develop and optimize the analytical equations.

All XRF instruments require calibration of the fluorescent x-ray response by measurement on a known elemental standard. Interelement effects within the analyzed material also influence the measured fluorescence and hence the need for a multielement analysis algorithm. Many laboratory-grade instruments now use Fundamental Parameter (FP) methods to quantify the element interaction coefficients and thus in principle may only require standardization of the instrument's pure element response. For soil application this could eliminate the need and expense of a site-specific calibration. Practical application of the FP-method requires that the spectrum resolution be sufficient to identify and separate all of the principal element components.

High resolution x-ray measurement is usually associated with semiconductor detectors such as Si(Li) or intrinsic-Ge. Their need for cryogenic operation is a major drawback for field use. Of the other semiconductor-like detector materials Mercuric iodide (HgI_2) can offer both high resolution capability and non-cryogenic operation. Commercial development has produced at least one instrument design based on this detector that is now widely used for on-site verification of alloy materials[4]. Its use on soils with an appropriate FP-analysis algorithm has been reported with good results for the analysis of soil material from different sites[5]. Application on other site materials and in-situ measurement will be described here.

Mercuric Iodide Detector Instrumentation

The main features of the instrument design and its field utility are illustrated in Figure 1. The weight, with batteries, is about 17 lb (9kg). A hand-size probe can be used either with a base-mount for sample assay or arranged in a free-standing inverted mode for in-situ measurement. Push-button controls and menu-style instructions on a large area LCD make for ease-of-use. Following a measurement the results are presented on the LCD in a table that includes the element symbols, assay values, and calculated statistical errors. Over 100 assay reports may be stored in the on-board memory for later retrieval. A 2000-channel x-ray spectrum for each of two excitation conditions can be viewed on the LCD or also stored, with up to 30 entries, for later review.

Two safely shielded radioisotope excitation sources (Am^{241} & Cd^{109}) each of ~3mCi are used in the probe and a low-power Peltier cooler maintains the detector at an optimum temperature. The energy resolution is typically < 300 eV (FWHM on Mn K x-rays) that provides good multielement separation, as seen in the spectra of Figure 2. The uniformly-high detection efficiency over a wide energy range is also advantageous in collecting information on the higher energy scattered x-rays to supplement the analysis algorithm.

The instrument for this work was configured to analyze for the following 19 elements - Ca:Cr:Fe:Co:Ni:Cu:Zn:As:Se:Sr:Mo:Cd:Sn:Sb:Ba:Hg:Pb:Bi:U. Spectral data for all these elements are internally processed to obtain standardized relative x-ray intensities for input to a multielement FP-analysis algorithm of a type previously described[5]. The output of the algorithm is element concentration, presently in % wt units.

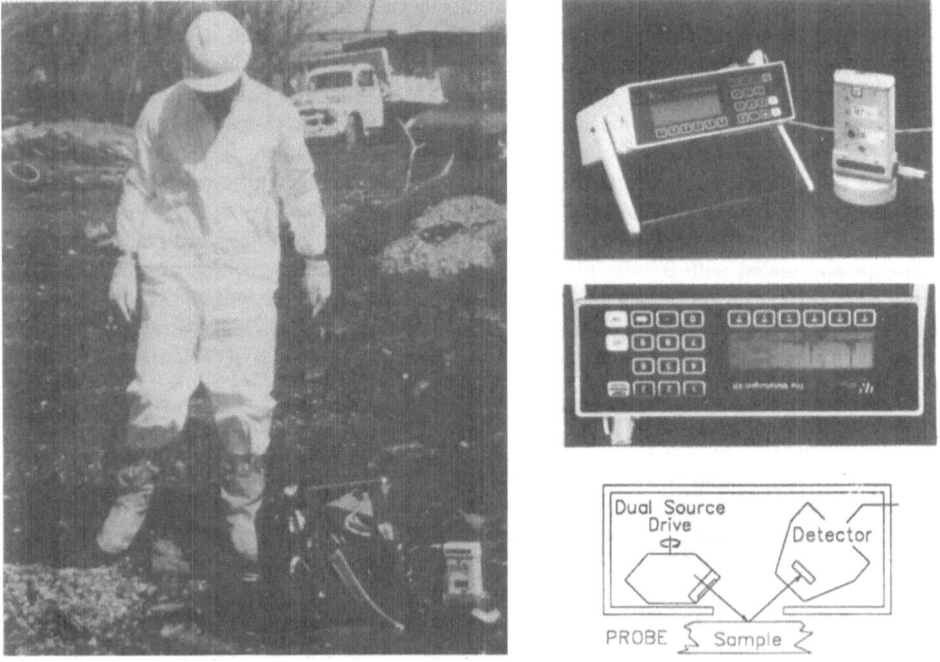

Figure 1. HgI$_2$ Instrument Design for Field Application

INSTRUMENT APPLICATION

Field measurements were performed on soils at an abandoned metal-recycling site, now undergoing removal and on samples collected from an inactive Cr-plating facility. The metals-site presented mostly Cu:Zn:Pb contaminants but a variety of chemical waste had also been dumped on the site, over the years.

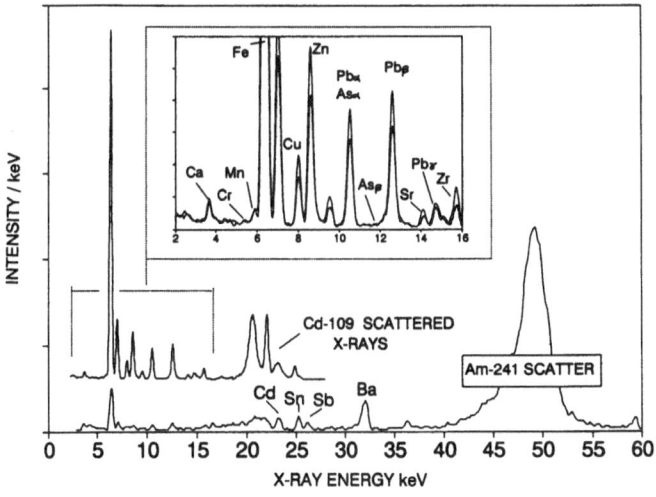

Figure 2. Example Fluorescence Spectra for Am^{241} & Cd^{109} Excitation of a Soil Sample

In-situ measurement called for raking the surface at the sampling point to break up the soil, to remove large "non-soil" objects and to homogenize the soil to a 1 inch depth over a ~1 x 1 ft area. Moisture conditions were not recorded but there had been recent rain. X-ray measurements were made with the probe in contact with the soil at three separated points within the prepared zone. This was repeated at several locations around the site. Results and spectra at each point were stored in the instrument memory for recall off-site. After measurement ~1 kgm of soil in the mixed zone was collected for further preparation and assay, along with other samples taken at various depths from both sites. Collected samples were dried, broken-up and screened through a 20 mesh sieve (0.8mm). Portions of the homogenized undersize material were placed in 30mm sample cups and sealed with 0.2mm plastic film for x-ray measurement. Subsequent laboratory analyses were performed on ~1gm of the soil from each cup using chemical digestion followed by either Atomic Absorption or ICP analysis. Some triplicate preparations were included. The results of both x-ray and lab assay on these samples showed less than a 10% variation about the average.

For the results that follow, no site-specific calibration was performed or was it necessary to restandardize over the 2-week test period. Adjustments for any gain drift were automatically performed in the software by peak-verification on the acquired spectrum. The precision of the measurement was checked using several standards during the tests. These results showed less than a 10% relative deviation in the element assays at concentrations of ~1000 mg/kg.

An example of **in-situ, prepared-sample**, and **lab**, assays associated with the Figure 2 spectra is shown in Table 1. The report format is illustrated by cols 1-3. The Date, Time-of-day and Sample No. are always part of the report. STD denotes the assay standard deviation computed for each measurement. Assay values less than the STD are presently excluded from the report (denoted by the "*"). In this example, 7 elements - Cd & Co ($<$80mg/kg); Hg & Bi ($<$25mg/kg); and Mo, U & Se ($<$10mg/kg) - were never reported: Cd may be seen in the spectrum of Figure 2 but it is a background feature of the current instrument design. Background and spectrum strip errors are combined with those of the multielement FP-analysis calculations to obtain the STD on each measurement. Assay times mainly dictate the error. In-situ assays (cols 2,4,5) used a 10:60 sec (Am:Cd) measurement sequence; prepared sample assays (col 7) used a 40:240 sec sequence. Comparative data (cols 9-10) are expressed in the customary lab assay units of mg/kg.

Table 1. Example Results for the Soil Sample of Figure 2

SAMPLE No 1481			1482	1483		:	1480		:		
DATE 3 Apr 91			3 Apr	3 Apr		:	5 Apr 91		:	**SAMPLE CUP**	
TIME 13:50			13:52	13:54		:	13:17		:	**COMPARATIVE DATA**	
COMP:	(%)	(%)	(%)	(%)	(%)	:	(%)	(%)	:	**(mg/kg)**	
ELM	CONC	STD	CONC	CONC	CONC	:	CONC	STD	:	XRF	CHEM
Sb	*	.004	*	0.012	0.007	:	0.013	.002	:	130	na
Ba	0.046	.006	0.036	0.038	0.040	:	0.062	.003	:	620	na
Cr	0.142	.08	0.102	0.268	0.170	:	0.123	.045	:	1,230	280
Sr	0.007	.001	0.006	0.005	0.006	:	0.010	.001	:	100	na
Fe	17.02	.2	21.55	17.17	18.58	:	26.45	.12	:	264,450	230,000
Ni	0.057	.01	0.057	0.051	0.055	:	0.068	.007	:	680	na
Cu	0.546	.02	0.649	0.588	0.594	:	1.006	.01	:	10,060	8,200
Zn	1.018	.02	1.251	1.032	1.100	:	1.904	.012	:	19,040	11,000
Ca	8.995	1.00	11.06	11.43	10.48	:	11.59	.48	:	115,900	na
As	0.046	.01	*	0.048	0.035	:	0.077	.006	:	770	na
Sn	0.060	.016	0.042	0.067	0.056	:	0.024	.007	:	240	na
Pb	0.490	.0095	0.672	0.551	0.571	:	0.976	.006	:	9,760	7,700
		3-POINT IN-SITU DATA				:	**SAMPLE CUP DATA**		:	**na = not analyzed**	

COMPARATIVE RESULTS

Results for comparative analysis were obtained at 50 in-situ locations and on 72 prepared samples from the metal recycling site. 32 prepared samples were assayed for the Cr-site. Graphed data for the elements Zn, Pb, Cu, Fe, & Cr are shown in Figure 3 along with a summary of the main parameters of the correlation (Table 2). In the graphs, in-situ results are vertically matched with the prepared sample result as noted in the graph legend. The balance of the prepared sample results are identified by the legend "OTHER". All Cr results are for prepared samples. For Cu, it may be noted that part of the scale has been adjusted (x3) to bring one of the points into the graph. The correlation summary results were obtained by linear regression analysis using the lab

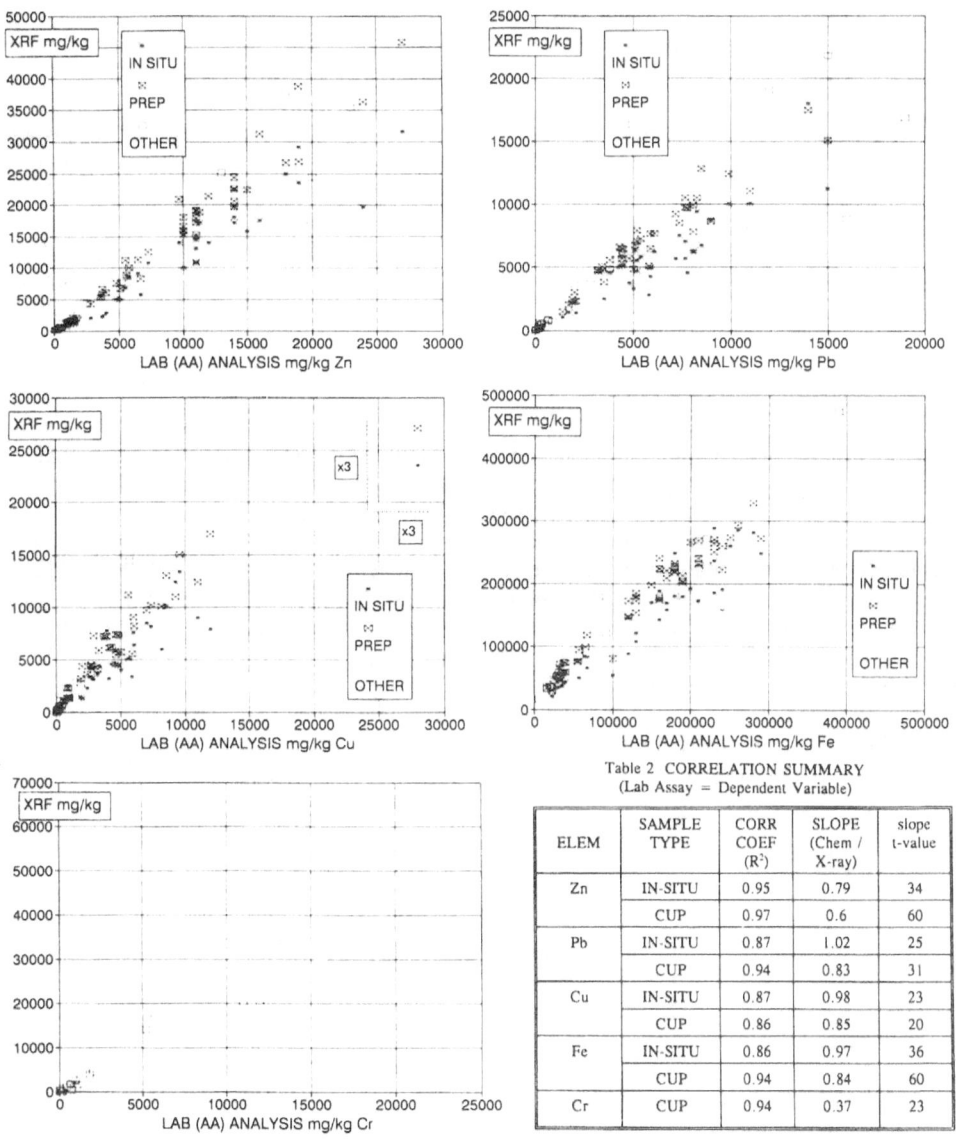

Table 2 CORRELATION SUMMARY
(Lab Assay = Dependent Variable)

ELEM	SAMPLE TYPE	CORR COEF (R^2)	SLOPE (Chem / X-ray)	slope t-value
Zn	IN-SITU	0.95	0.79	34
	CUP	0.97	0.6	60
Pb	IN-SITU	0.87	1.02	25
	CUP	0.94	0.83	31
Cu	IN-SITU	0.87	0.98	23
	CUP	0.86	0.85	20
Fe	IN-SITU	0.86	0.97	36
	CUP	0.94	0.84	60
Cr	CUP	0.94	0.37	23

Figure 3 Comparative Assay Results for Site 1 (Zn Pb Cu Fe) and Site 2 (Cr).

assay values as the dependent variable. Thus the derived "Slope" coefficient represents a factor by which the x-ray assay would need to be multiplied to best fit the lab assay. All of the graphed data were included in the regression and the coefficients reflect the full range of the variables.

DISCUSSION

It is not unusual that the comparative analysis results show the degree of scatter exhibited in the graphs considering the highly heterogeneous nature of most soils and the relatively coarse particle size condition of the measured materials on the scale of the x-ray penetration. The correlation coefficients (R^2) are seen to well exceed the value of 0.7 that is presently accepted for screening application[3]. The "slope-factor" results indicate the need for some adjustment in the FP-analysis algorithm for future application to the soils at these sites. With this adjustment the comparative analyses are found to agree well within the range of $\pm$ 50% that is acceptable for field application.

The need for a calibration adjustment for these sites compared to previous experience[5], where results showed less than a 10% deviation from the FP-derived sensitivity, is attributed to a combination of effects, some particle size related and some to the more extreme range of Fe in these materials. Operationally this is not a drawback since most site application work does not commence until the initial lab assays are available; giving time to make any necessary adjustments. Further refinement of the FP-algorithm to reduce the site-dependency of the calibration are planned.

CONCLUSIONS

Many benefits of using high resolution x-ray detection and a Fundamental Parameter analysis method for field screening application were demonstrated in this work with a relatively new FPXRF instrument. Using only a pure element calibration, acceptable screening-quality results were obtained on sites with highly variable metal content. In addition to the ease-of-use, other notable advantages included;- automatic 19 element coverage; multi-spectrum displays for element peak verification; and the convenience of on-board storage of data and spectra. Due to the good inter-element resolution, spectrum overlap conditions were minimal and there was a low incidence of "false negatives". The main shortcoming common to all present FPXRF instrumentation is the relatively high limit of detection for Cr. The conventionally-defined "3-sigma" limit in the soils investigated (containing up to 30% Fe) was ~ 500 mg/kg. Reduction to the level of ~ 100 mg/kg is desirable and is expected to be achieved in a future design. The detection limits for most other elements, as seen in Table 1, are in the 50-200 mg/kg range previously reported [5].

REFERENCES

1.   H. M. Fribush and J. F. Fisk, "Field Analytical Methods for Superfund," Proceedings of 2nd International Symposium on Field Screening Methods for Hazardous Wastes and Toxic Chemicals, Feb 12-14, 1991, Las Vegas, pp. 25-29.
2.   G. A. Rabb, C. A. Kuharic, and W. H. Cole III, "The Use of Field-Portable X-Ray Fluorescence Technology in the Hazardous Waste Industry," Advance in X-Ray Analysis, Vol. 33, 629-637, 1990.
3.   "Field Portable X-Ray Fluorescence," U.S.EPA Quality Assurance Technical Information Bulletin, Vol.1, No.4, May, 1991.

4.   P. F. Berry and G. R. Voots, "On-Site Verification of Alloy Materials with a New Field Portable XRF Analyzer Based on a High-Resolution HgI$_2$ Semiconductor X-Ray Detector," Proceedings of 12th World Conference on Non-Destructive Testing, Apr 23-28, 1989, Amsterdam, pp 737-742, Elsevier Science Pub.

5.   J. B. Ashe, P. F. Berry, G. R. Voots, M. Bernick, and G. Prince, "A High Resolution Portable XRF HgI$_2$ Spectrometer for Field Screening of Hazardous Metal Wastes," Proceedings of 2nd International Symposium on Field Screening Methods for Hazardous Wastes and Toxic Chemicals, Feb 12-14, 1991, Las Vegas, pp. 507-515.

EXPERIMENTAL XRF CALCULATION METHOD WITH CORRECTION

FOR A POLYDISPERSE MATERIAL PARTICLE SIZE

V. V. Zagorodny, V. I. Karmanov

E. O. Paton Electric Welding Institute
of the Ukr.SSR Academy of Sciences, Kiev, U.S.S.R.

ABSTRACT

A new experimental calculation method for polydisperse (i.e. heterogeneous) multicomponent material analysis has been developed using the dependence of element fluorescence intensity on the particle size and its distribution in the specimen. It is shown that correction of the influence of matrix particle size is possible using this experimental calculation method. For its application, the information on particle size distribution for each of the components is sufficient. Sample preparation includes only the pelleting of specimens under standard conditions. The efficiency of the method proposed is demonstrated by the analysis of the multicomponent mixtures of welding materials.

INTRODUCTION

The existing theoretical methods that take into account the effect of particle size on analytical line intensity of elements of multicomponent polydisperse mixtures, based on physical models of interaction between x-ray irradiation and heterogeneous materials are characterized by approximations and limitations. They are not sufficiently perfect and cannot be used for quantitative analysis.[1,2,3,4]

For practical application, methods with correction or compensation of heterogeneity effect are required, based on the reliably determined experimental regularities of correlation between the radiating sample particle size and the fluorescence intensity of the elements determined.

EXPERIMENTAL CALCULATION METHOD FOR ANALYSIS OF ONE-COMPONENT MATERIALS

Analytical line intensity of elements of one-component polydisperse materials depends on the nature of the material, density of radiating sample and conditions of its preparation.[5] It was found experimentally that, at fixed extrusion pressure, a certain intensity of the analytical line of the determined element A corresponds to each class of particle size (Fig. 1).

Thus in a polydisperse material, the intensity of fluorescence of the element A can be given in the form of the additive sum of products of

Table 1. Comparison of the Calculated (1) and the Measured (2)
          Intensities of Manganese and Iron K_α-line in Ferro-
          Manganese Samples of Various Particle Sizes

element	sample number	1	2	3	4	5	6	7
Mn	1	1.08	1.06	1.02	1.02	0.99	0.97	0.96
	2	1.09	1.08	1.04	1.03	1.01	0.99	0.98
Fe	1	1.17	1.14	1.06	1.04	1.00	0.98	0.96
	2	1.20	1.17	1.09	1.07	1.03	0.99	0.97

fluorescence intensity from the particles of each size class and the particle
fractions in the material, corresponding to these classes, using the following
relation:

$$I_{2i}^A = a_1 (I_{2i}^A)_1 + a_2 (I_{2i}^A)_2 + \ldots + a_n (I_{2i}^A)_n = \sum_j^n a_j (I_{2i}^A)_j \tag{1}$$

where $(I_{2i}^A)_j$ is the analytical line intensity of the element A in the radiating
sample with particle size of j-class; a_j is a share of j-class particles in the
polydisperse radiating sample, n is the number of size classes. For a_j values
the correction condition should be fulfilled:

$$\sum a_j = 1 \tag{2}$$

A_j coefficients characterize the density of the particle size distribution
function that can be found experimentally. Thus, having measured with the
instrument the intensity of element A fluorescence in a polydisperse sample and

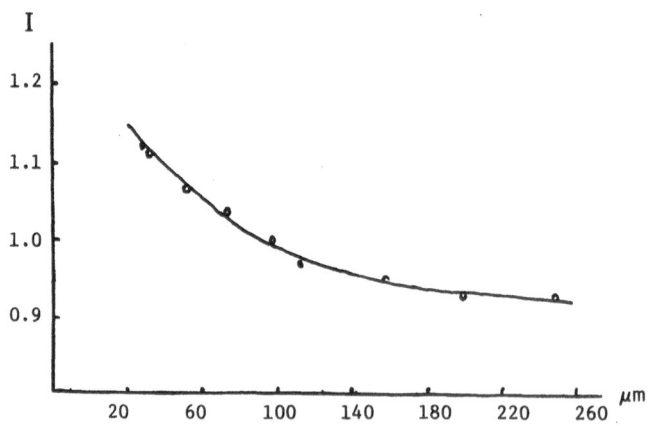

Fig. 1. Dependence of Relative Intensity of Manganese
         K_α-line on the Particle Size of Ferromanganese

Table 2. Comparison of the Measured (1) and Calculated (2) Intensities
of Iron K_α-line in Multicomponent Polydisperse Materials

Sample number	1	2
1	1.15	1.17
2	1.18	1.19
3	0.88	0.90
4	0.93	0.90
5	1.04	1.01

knowing the fractions of particles of individual size classes in the material,
as well as the dependence of analytical line intensity on particle size c (Fig.
1), it is possible to calculate by the relation (1) the intensity of element A
in this radiating sample. The experimental tests using ferromanganese powder
with various particle size distribution showed that a good agreement can be
reached between the calculated and experimental values. The results are given
in Table 1.

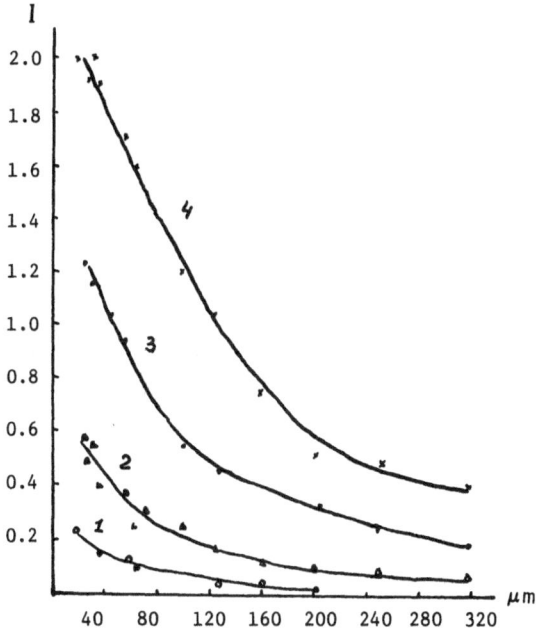

Fig. 2. Dependence of Intensities of Iron K_α-lines on Particle Size
of Iron Powder in Mixtures with Constant Particle Size Matrices:
(1) 10% Fe + 90% TiO_2; (2) 50% Fe + 50% TiO_2; (3) 10% Fe + 90% MgO;
(4) 50% Fe + 50% MgO

When using this method, the sample preparation consists only of making pellets, which are the radiators of the fluorescence spectrum.

EFFECT OF ABSORPTION CHARACTERISTICS OF CONSTANT PARTICLE SIZE MATRIX

When mixing the analyzed polydisperse materials with a filler of the constant granulometric composition, the analytical line intensity of the determined elements in the samples with variable particle size depends on absorption characteristics of the matrix and the ratio of matrix and the material analyzed. Thus, it was found that the intensity of K_α-line in a mixture of iron powder with magnesium and titanium oxides, in proportions 1:9 and 1:5, is increasing with the decrease in particle size of iron powder. As is shown in Fig. 2, the intensity of iron K_α-lines in mixtures with magnesium oxide is higher than in mixtures with titanium oxides. Therefore, for the use of the experimental calculation method, the dependence of the determined element's intensity on the particle size of the material analyzed, corrected to the mixture composition, is required to be plotted on the basis of experimental data.

The application of equation (1) for the analysis of flux mixtures of electrodes and flux-cored wires, containing iron powder, will be considered below. In such materials, all the mineral components can be ground except for the iron powder. The electrode coating mixture that contained 10% iron powder was analyzed. The matrix consisted of the mixture of rutile, magnesite, mica and ferromanganese. Sample preparation consisted of making the pellets (irradiating samples) at the constant pressure. The comparison of the relative intensities of iron fluorescence measured by the instrument or calculated is given in Table 2.

The differences between the calculated and measured intensities do not exceed 2-3% in various iron powder particle sizes in the mixtures. Hence the errors in iron determination are not more than ± 0.5%, despite the variations of K_α-line iron intensity within the limits 15-20% (Table 2).

Thus, the experimental calculation method can be used for the correction of the variable granulometric composition of the material, contained in the mixture of components with constant particle size.

EFFECT OF MATRIX PARTICLE SIZE ON ANALYTICAL LINE INTENSITY
OF ELEMENTS DETERMINED

When using the x-ray fluorescence analysis for the control of flux mixtures used in production of welding consumables, where the granulometric composition of all the components can be changed simultaneously within a wide range, improvement in accuracy can be attained by taking into account the effect of matrix material dispersion on the intensity of analytical lines of elements determined.

The effect of the matrix granulometric composition can be considerable. For instance, it was found in the work[6] that the variation of silicon K_α-line depending on the marble particle size in the mixtures with 85% marble, and 15% quartz for samples with different quartz particle size, can reach 40-60%.

The experiments with the use of the two-component mixtures, consisting of marble (85%) with variable granulometric composition and ferroalloys (ferro-titanium and iron powder) with constant particle size showed that with the decrease of marble particle size, the intensity of analytical lines of iron and titanium is also decreased. This is caused by the fact that with the decrease of size and with the rise of the specific surface of marble particles, the

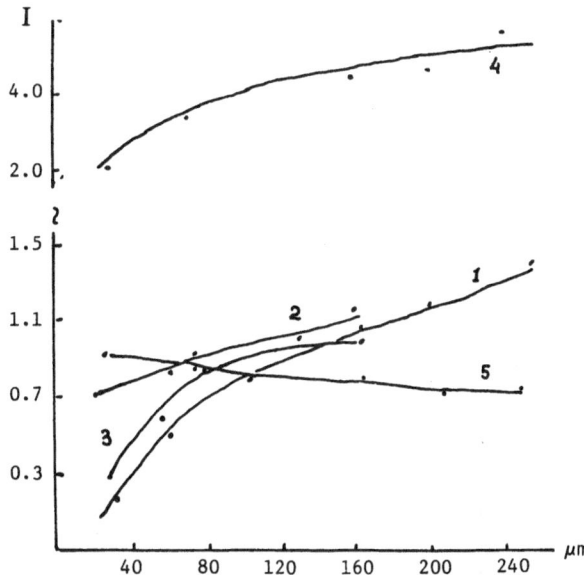

Fig. 3.   Dependence of Intensity of K_α-lines for Elements in the Mixtures of
85% Marble and 15% Ferroalloys on the Particle Size of Marble:
(1) iron powder - 250 μm;   (2) iron powder - 125 μm;
(3) ferrotitanium - 160 μm;   (4) ferrotitanium - 32 μm;   (5) marble

contact surface is increasing and the number of contacts between marble and
ferroalloys is growing, as well as the areas of autohesive interaction of
particles.   Due to this increase in contacts, the effective irradiating surface
of ferroalloy particles is decreased.   In this irradiation layer the shielding
of fluorescence irradiation of ferroalloy atoms is growing, which causes the
decrease in the intensity of characteristic lines of iron and titanium in the
mixture considered, as is shown in Fig. 3.

A similar process, associated with the variation of particle size of the
components, occurs in multicomponent compositions of ores and minerals, such as
flux mixtures of electrode coatings: at the simultaneous variation of particle
size of all the components, the intensity of the determined element A, present
in any material, depends on particle size of this material and on particle size
of other components due to the increase in particle contacts and to the
absorption characteristics of the matrix.

THE EXPERIMENTAL CALCULATION METHOD FOR THE ANALYSIS OF MULTICOMPONENT
POLYDISPERSE MATERIALS

The increase in particle contacts is caused by the matrix particle size.
In the supposition that the shielding by the matrix particles is similar for all
size fractions of the polydisperse material analyzed, the effect of matrix
particle size variation on the intensity of the K_α-line of an element determined,
can be presented in the form:

$$I_{2i}^{A} = \prod [\ \sum_{j=1} b_j (I_{2i}^{A})_j]_1 \tag{3}$$

Table 3.   Relative Errors in Determination of Components
           of Experimental Flux Mixtures, %:
           1. Without Corrections
           2. With Corrections by Relation (5)

sample	ferrotitanium		ferromanganese		silicon		iron	
number	1	2	1	2	1	2	1	2
1	0.57	0.20	0.59	0.0	0.63	0.01	0.49	0.05
2	0.81	0.03	0.53	0.15	0.22	0.20	0.42	0.05
7	0.30	0.10	0.52	0.26	0.85	0.13	0.26	0.10
8	0.55	0.20	0.58	0.05	0.42	0.14	0.46	0.02
9	0.06	0.02	0.67	0.17	0.15	0.02	0.22	0.08
11	0.66	0.01	0.02	0.01	0.71	0.05	0.19	0.20

where: b_j is the share of j-class size particles of the 1th component of the
matrix; $(I_{2i}{}^A)$ is given by the relation (1).

Then the coefficient, correcting differences in granulometric composition
of the analyzed and the reference samples can be given in the form:

$$Q_i^j = \Pi \frac{[\sum_j b_j \, (I_{2i}^A)_j]_1^{r.s.}}{[\sum_j b_j \, (I_{2i}^A)_j]_1} \qquad (4)$$

where the expression in the numerator is related to the reference specimen K and
that in the denominator - to the sample analyzed.

The relation (4) was used for correction in the analysis of multicomponent
flux mixtures of electrode coatings.   The polydisperse mixtures of variable
elemental and granulometric composition were prepared.   The variation in
elemental composition did not exceed several percent (from 3 to 10%).   The
granulometric composition was varied at two levels.   The components with small
particles were not more than 71 μm and those with large particles contained 99%
of particles not less than 71 μm.   The pellets-radiators were prepared after
mixing.   For calculation of element concentrations the following relation was
used as a reference sample:

$$C_x^A = C_{r.s.}^A \frac{I_x^A}{I^{r.s.}} \prod_{j=1}^{n} Q_j^A \qquad (5)$$

where $C_X{}^A$ and $C_{r.s.}{}^A$ are the concentrations of element (A) in the samples analyzed and in the reference specimen; the $Q_J{}^A$ coefficients were calculated using relation (4).

The results of the analysis of flux mixtures are given in Table 3.

CONCLUSION

From the comparison of the results, it can be concluded that the corrections raise the accuracy of flux component analysis considerably. The existing differences can be attributed to the incomplete consideration of the interelemental effects and to the insufficient representation of individual samples.

Thus, the analytical experimental calculation method developed enables the accuracy of x-ray fluorescence analysis of multicomponent polydisperse materials to be increased when the information on dispersed composition of the components is available. For its application the experimental dependences of analytical line intensity of the elements determined on both the particle size of the materials analyzed and of the matrix are required.

REFERENCES

1. A. Ya. Shpalansky and Sh. I. Duimakaev, Effect of sample hete-
      rogeneity in X-ray fluorescence spectroscopy. Review.,
      Rostov State University, Rostov-on-the Don, 54 (1982).
2. P. F. Berry, T. Furuta, and I. K. Phodes, (1968) Adv. X-Ray
      Anal., 12:612.
3. P. F. Berry, Particle Size Heterogeneity Phenomena in X-Ray
      Analysis, in: "Application of Low Energy X- and Gamma Rays",
      London, (1972)
4. C. B. Hunter and I.R. Phodes, (1972) X-Ray Spectrometry, 3:107.
5. V. V. Zagorodny and V.I. Karmanov, Fluorescence intensity of
      monodisperse materials, in: "Methods of X-Ray Analysis",
      Nauka, Novosibirsk, 33 (1986).
6. F. Bernstein, (1962) Adv. X-Ray Anal., 6:436.

FABRICATION AND SELECTED APPLICATIONS OF A NIST

X-RAY MICROFLUORESCENCE SPECTROMETER

P.A. Pella

National Institute of Standards and Technology
Gaithersburg, MD   20899

L. Feng

Kevex Instruments, Inc.
San Carlos, CA

INTRODUCTION

An x-ray microfluorescence (XRMF) spectrometer has been designed and fabricated at NIST for multi-point compositional analysis of small samples with x-ray beam sizes on the order of 50 micrometers or greater. This system was developed as part of an industrial cooperative research agreement with Kevex Instruments, Inc., San Carlos, CA., and consists of commercially available components incorporated in an aluminum vacuum chamber (see Figs. 1 and 2).

Key features include a low-power, Mo anode x-ray tube (50 kV, 1 mA), which provides a small focused x-ray beam (0.25 x 0.25 mm), a single aperture collimator, and a close coupled source-sample-detector geometry for optimum count rate. A motorized X-Y stage permits automated, programmable X-Y scans of samples with step sizes as small as one micrometer. A manual Z axis adjustment is provided to optimize positioning of the sample relative to the x-ray source and Si(Li) detector. The spectrometer is capable of operation in vacuum for analysis of low atomic number elements. A PC computer, equipped with an appropriate hardware interface and Toolbox software (Kevex Instr.)*, controls spectral data acquisition for up to seven elements

---

*Certain commercial equipment, instruments, or materials are identified in this report to specify adequately the experimental procedure. Such identification does not imply recommendation or endorsement by the National Institute of Standards and Technology, nor does it imply that the materials or equipment identified are necessarily the best available for this purpose.

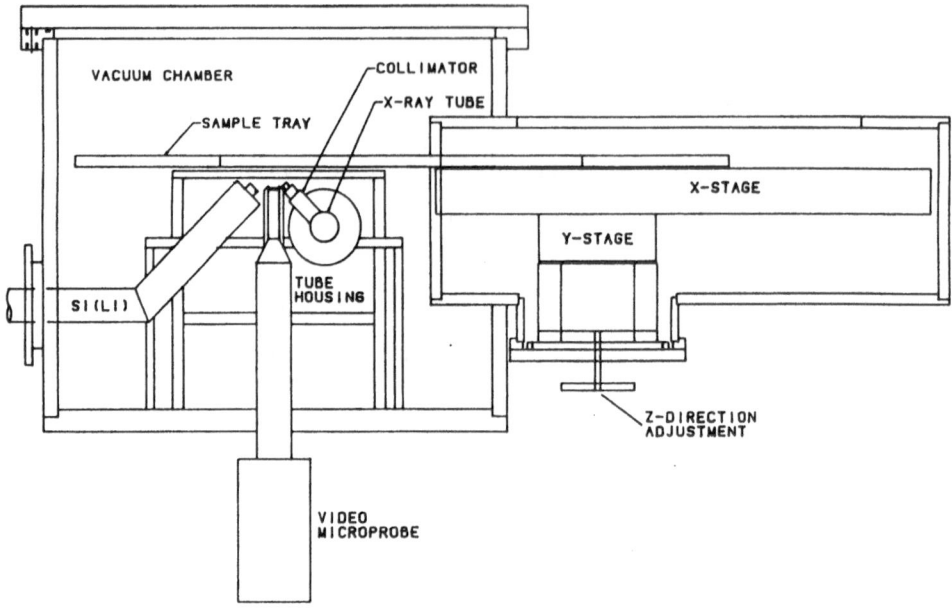

Fig. 1.   Spectrometer (Side View)

simultaneously, controls movement of the stage, and keeps track of the X-Y
stage positions during an area scan.   Other components include a color
video camera for continuous viewing of the sample at normal incidence,
using either direct (90°) illumination of the sample or side illumination
at 45° for viewing highly reflective samples.

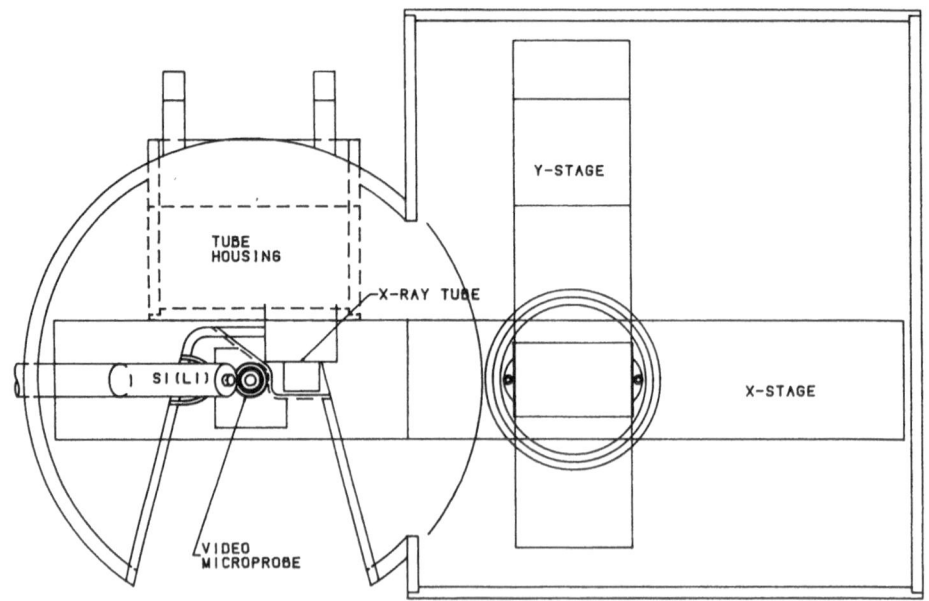

Fig. 2.   Spectrometer (Top View)

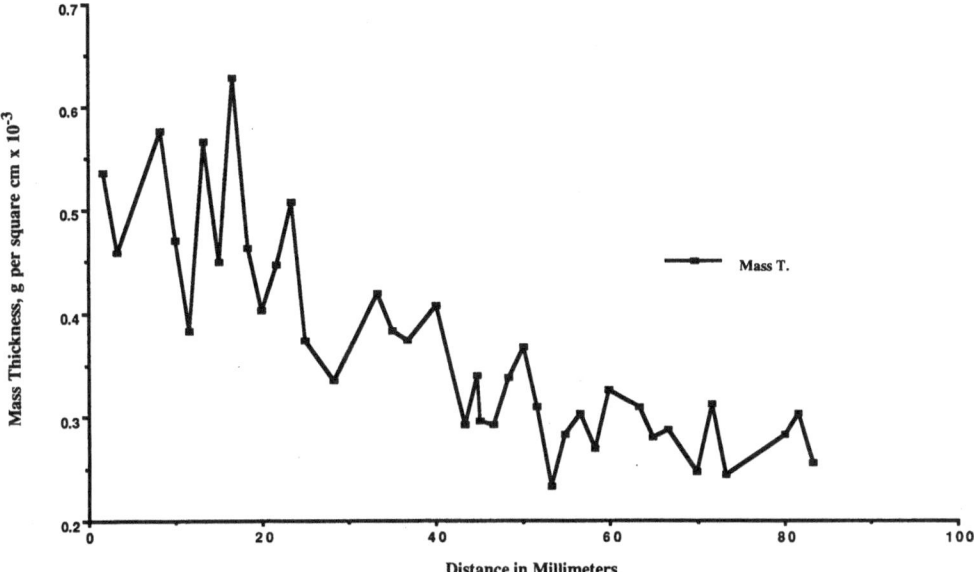

Fig. 3.   Thickness of Manganese Oxide Layer Along Catalyst (One Channel)

Measurement of the x-ray beam size was performed by using a 70 μm beam collimator, a 4 mm dia. detector collimator, and by step scanning across 50 μm diameter copper wires in both the X and Y directions. The beam at the sample surface was found to be elliptical with the long axis about 120 μm, and the short axis about 90 μm.

Some selected applications using the above experimental conditions include measurement of the mass thickness of a Mn_3O_4 layer deposited over the wash-coat surface of an automotive catalyst exposed to the fuel additive MMT (1), and quantitative analysis of major constituents in NIST-SRM stainless steel chips.

A cylindrical automotive catalyst (courtesy of F. Kunz, Ford Motor Co.) having dimensions of 14 cm long x 9 cm wide (O.D.) was sliced in half along its length exposing two flat surfaces containing approximately 1 mm wide channels coated with a Mn_3O_4 layers. XRMF of the mass thickness of manganese along the length of one channel on one piece is shown in Figure 3. The layer was found to be quite heterogeneous with the thickest

Table 1.   Standard Deviation of C1152 Alloy Chip Measurements
(900s each pt, n=10)

Element	Conc. %	Counts/s	S.D. (c/s)
Cr	17.81	11.73	0.11
Mn	0.96	0.64	0.11
Fe	68.35	34.12	0.21
Ni	10.88	3.55	0.08

Table 2. X-Ray Microfluorescence Analysis of NIST/SRM Stainless Steel Chips
(900s counting time each pt.)

Element	SRM C1151 Calibration STD C.V.	SRM 1246 Found	C.V.	SRM C1152 Found	C.V.	SRM C1153 Found	C.V.	SRM C1154 Found	C.V.
Cr	22.79±.08	19.9	20.1±.1	17.67	17.81±.12	16.70	16.69±.09	18.92	19.06±.11
Mn	2.50±.08	0.81	0.91±.02	0.87	0.96±.03	0.45	0.50±.01	1.19	1.42±.04
Fe	65.48 NC	46.1	46.2±.1	68.48	68.35 NC	71.68	71.64 NC	64.98	64.50 NC
Ni	7.29±.05	31.0	30.8±.1	10.63	10.88±.06	9.09	8.77±.04	12.78	12.92±.09
Elements Added As Fixed Concentrations									
Co	0.032 C.V.	0.01		0.01		0.00		0.00	
Cu	0.418 C.V.	0.20		0.10		0.20		0.50	
Al	0.10	0.20		0.30 C.V.		0.001		0.01	
Si	0.38 C.V.	0.20		0.80 C.V.		1.07 C.V.		0.50 C.V.	
Nb	0.20	0.60		0.4		0.01		0.50	
Ti	0.10	0.50		0.5		0.5		0.40	
Mo	0.80 C.V.	0.60		0.4		0.5		0.40	
Total	100.00	100.11		100.16		100.23		100.19	

C.V. = Certified Values.
NC = Not certified but obtained by difference.
Uncertainties are from NIST/SRM Certificates of Analysis.

portion on the inlet side of the catalyst, and about a factor of two less toward the outlet side. These results are generally in agreement with those reported by Hurley et al. (1) using bulk XRF analysis of inlet, middle, and outlet sections of the catalyst.

Alloy chips cut from NIST-SRM stainless steels (2 mm x 4 mm x 1 mm thick) were fixed in epoxy in a stainless steel ring holder (25 mm I.D.) and ground flat and polished in a manner similar to samples prepared for electron-probe microanalysis. At least ten X-Y measurements on each sample were made at random points and averaged. The standard deviation (counts/s) of a single intensity measurement after spectral deconvolution is shown in Table 1. The results of quantitative analysis of major constituents using fundamental parameter interelement corrections are summarized in Table 2. Only one standard sample was used for spectrometer calibration. The concentrations of minor constituents that were not measured were fixed in the program so that a more complete interelement correction could be made and the sum total is 100%. In most cases the concentrations of these minors were estimated assuming the same type material as the standard (C1151).

The agreement between the XRF values found and the NIST certified values is good and is typical of large area EDXRF analysis of bulk samples when the counting statistics and spectral resolution are comparable. The differences between the values found and the certified values were usually within the errors due to counting statistics as shown in Table 1. For measurement of elements such as manganese, the relatively large errors are due to low count rates with a 100 μm x-ray beam at these concentration levels combined with deconvolution errors contributed by the CrK_β peak.

CONCLUSION

X-ray microfluorescence analysis with x-ray beams less than 100 μm combined with an x-ray imaging capability is a rapidly growing XRF technique. The development of sophisticated software for low cost personal computers provides convenient instrument control coupled with effective graphic display. We expect that we can improve the spatial resolution of this particular system through continued development of higher x-ray intensity microfocus sources coupled with appropriate collimators such as glass capillary waveguides. We are currently investigating x-ray microfluorescence of particles in the 50-200 μm range for geological applications.

REFERENCES

1. Hurley, R.G., Watkins, W.L.H., and Griffis, R.C., Ford Motor Co., SAE Technical Paper Series No. 890582, presented at International Congress and Exposition, Detroit, Michigan, 1989.

PROBLEMS IN THE USE OF MULTILAYERS FOR SOFT X-RAY SPECTROSCOPY AND ANALYSIS: A COMPARISON OF THEORETICALLY AND EXPERIMENTALLY DETERMINED REFRACTION EFFECTS

E. Martins and D.S. Urch

Chemistry Department
Queen Mary and Westfield College
University of London
Mile End Road, London E1 4NS, UK

ABSTRACT

Refraction effects cause the effective 2d spacing of multilayers to vary so that the simple Bragg equation is not applicable when such devices are used for soft X-ray spectroscopy. This paper describes how the refraction term can be determined experimentally. These results compare well with dispersion and absorption terms calculated from atomic scattering factors.

1.   INTRODUCTION

The advent of multilayer devices with large 2d spacings and high reflectivity has revolutionised soft X-ray spectroscopy.[1] It is now possible to consider,[2] quite seriously, the extension of the range of elements that can be analysed by X-ray fluorescence (XRF) spectroscopy to encompass all the light elements up to and including beryllium.   As this extension includes the biochemically important elements carbon, nitrogen and oxygen, the potential for XRF analysis has expanded dramatically.   There are, of course, some attendant problems. As the wavelength of the characteristic X-ray increases so the thickness of the layer from which they are emitted decreases. Furthermore the 'escape depth' for soft X-rays is greatly affected by the chemical composition of the surface layers.[3]   Contamination, corrosion and other chemical changes at, or near, the surface can thus modify the analytical results.   Another problem, if problem it be, is that characteristic soft X-rays usually exhibit striking chemical effects which cause peak shifts and changes in peak profiles.[4] This 'problem' can, however, be turned to analytic advantage by providing, not only information about the concentrations of elements present in sample surfaces (analysed depths will vary from tens to thousands of nanometers), but

Advances in X-Ray Analysis, Vol. 35
Edited by C.S. Barrett *et al.*, Plenum Press, New York, 1992

also the chemical state of those elements.[5] Realisation of this potential in the soft X-ray region is restricted when multilayers are used because of their low resolving power (typically $\lambda/\Delta\lambda \sim 50$ or less). Chemical effects in multilayer soft X-ray spectra are not usually apparent at wavelengths of less than 5nm. Problems associated with the efficient excitation of soft X-rays (as far as possible given the natural preference for core ionised light elements to relax by Auger electron emission) can be overcome by the use of direct bombardment with low energy electrons or by the use of an open window gas-discharge X-ray tube.[6]

There remains, however, one major problem associated with the use of multilayers for spectroscopy at long X-ray wavelengths and that is the problem of refraction.[7] At wavelengths of greater than 2nm or so dispersion and, to a lesser extent, absorption, effects become increasingly important and cause X-rays to be reflected at angles that differ significantly from those calculated by using the simple Bragg relationship.

Preliminary experiments,[8] in which the Bragg equation was used to calculate the 2d spacing of a multilayer by observing the angle (θ) for first order reflection of X-rays of known wavelength (λ), gave results that varied by $\pm 5\%$ for the range 15nm $> \lambda >$ 1.5nm; the apparent 2d spacing increased and then decreased as the X-ray wavelength grew longer. When higher order reflections were investigated it was found that $2d_n$ (the effective 2d spacing for the nth order) increased with n, and approached a limiting value $2d_\infty$. These results show that, despite the low resolving power of multilayers, the application of the simple Bragg equation in the soft X-ray region will give erroneous and quite misleading results. In order that multilayers can be used for soft X-ray spectroscopy it is essential that they be fully characterised for refraction effects.

In the absence of refraction the Bragg equation can be written

$n.\lambda = 2d_\infty.\sin\theta_B,$

so that if the angle at which reflection is actually observed is θ and $\theta - \theta_B = \Delta\theta$ then

$n.\lambda = 2d_\infty.\sin(\theta - \Delta\theta).$

Thus,

$$n.\lambda = 2d_\infty.\sin\theta.\left(1 - \frac{\cos\theta.\sin\theta.\Delta\theta}{\sin^2\theta}\right) \qquad (1)$$

if $\Delta\theta$ is small.

The term $\cos\theta.\sin\theta.\Delta\theta$ can be related[9,10] to refraction within the multilayer (see below) and is dominated by dispersion effects which are not order (n) dependent. Thus, for a given X-ray wavelength, $\cos\theta.\sin\theta.\Delta\theta$ can be replaced by a constant A_λ and equation (1) rewritten as

$$\frac{n.\lambda}{\sin\theta} = 2d_\infty - \frac{A_\lambda.2d_\infty}{\sin^2\theta} \qquad (2)$$

In this form[8] experimental results for the values of θ at which the different orders of reflection were observed can be plotted as $(n\lambda/\sin\theta)$ *versus* $(1/\sin^2\theta)$ to give a straight line graph. The intercept of this graph enables the limiting value of the 2d spacing, $2d_\infty$ to be found whilst $A\lambda$ can be determined from the slope $(A\lambda.2d_\infty)$. This procedure has also been described by Henke et al.[11]

It is the purpose of this paper to show how this method can be used to determine $2d_\infty$ and $A\lambda$ values for multilayers and then to compare these experimental $A\lambda$'s with calculated values of the dispersion constant (δ_λ). To this end two multilayers were examined, with 2d spacings of ~ 20nm and ~ 30nm and made from different pairs of materials, Mo/B_4C and Pt/Si respectively.

2.   EXPERIMENTAL

All the experiments were carried out using a Philips PW 1410 X-ray fluorescence spectrometer. Characteristic X-rays were excited by means of a Compagnie Général de Radiation Elent-10 open window X-ray tube in which a discharge is struck between tungsten and aluminium electrodes in air at about 10 Pa. After passing the main Soller slit collimator the X-rays were reflected from a plane crystal or from the multilayer under test. After passing the main Soller slit collimator the X-rays were reflected from a plane crystal or from the multilayer under test. Detection was by means of a gas-flow proportional counter (P-10 gas: 90% Ar + 10% CH_4, ~10^5Pa) fitted with a one micron aluminised polypropylene window. Harwell 2000 series equipment was used for pulse amplification, selection and counting. The ratemeter output was registered using a chart recorder. The spectrometer was fitted with a five position crystal holder into which a suite of crystals of known 2d spacings could be loaded together with the test multilayer(s). This enabled the optimum conditions (amplification, pulse amplitude selection etc) to be established for each characteristic X-ray line before the multilayer was examined. The crystals[12] used were; $T\ell AP$ (thallium acid phthalate) for F $K\alpha$ and O $K\alpha$, OHM (octadecyl hydrogen maleate) for N $K\alpha$ and C $K\alpha$ and OAO (dioctadecyl adipate) for B $K\alpha$. The samples used as sources of suitable X-rays included

    potassium fluoride (F $K\alpha$)
    silica (O $K\alpha$, Si L)
    boron nitride (N $K\alpha$, B $K\alpha$)
    graphite (C Ka)
    yttrium oxide (Y $M\zeta$)
    zirconium-niobium alloy (Zr and Nb $M\zeta$)

For each sample the spectrometer was first adjusted to a characteristic wavelength using the appropriate crystal. The crystal was then replaced by a multilayer and, starting at a low angle, all possible orders were recorded.

Two multilayers were investigated. The first, OV-200H, was manufactured by the Ovonic Synthetic Materials Co., Inc. (Troy, MI, 48084, USA) with alternate layers of molybdenum and boron carbide and an intended d spacing of 10nm (relative thicknesses of the layers not specified). The second multilayer was prepared by Dr. B. Evans (Physics Dept., Reading University, Reading, UK) with layers of platinum (4.5nm) and amorphous silicon (10.5nm) to have a d spacing of 15nm and a γ of 0.3.

3.   CALCULATIONS

The index of refraction is equal to $1 - \delta - i.\beta$. The terms δ and β vary with both the wavelength of the radiation and the material through which the radiation is passing. In a multilayer composed of alternate layers of substances 1 and 2 with thicknesses t_1 and t_2 the average values for δ and β will be given by

$$\bar{\delta} = \gamma.\delta_1 + (1 - \gamma).\delta_2$$

and

$$\bar{\beta} = \gamma.\beta_1 + (1 - \gamma).\beta_2,$$

where $\gamma = t_1/(t_1 + t_2)$ and the values of δ and β for the different layers are denoted by the subscripts 1 and 2. Also,

$$\Delta\delta = \delta_1 - \delta_2 \text{ and}$$
$$\Delta\beta = \beta_1 - \beta_2.$$

Refraction within a multilayer will cause the angle at which X-ray reflection occurs to be displaced from the 'Bragg' angle, θ_B, by Δθ. It can be shown[10] that Δθ, δ and β are related as follows,

$$\Delta\theta.\cos \theta_B.\sin \theta_B = \bar{\delta} - \frac{\Delta\delta.\Delta\beta.\sin^2(n.\pi.\gamma).P^2(\theta)}{n^2.\pi^2.\bar{\beta}} \tag{3}$$

where the first term is associated with dispersion and the second with absorption effects. The function $P^2(\theta)$ is determined by the polarisation of the incident radiation and has been set equal to one in this paper. δ and β can be calculated[13] from the atomic scattering factors f_1 and f_2 respectively;

$$\delta = \frac{r_0}{2\pi} \cdot \frac{\rho}{RAM} \cdot L \cdot \lambda^2 \cdot f_1$$

$$\beta = \frac{r_0}{2\pi} \cdot \frac{\rho}{RAM} \cdot L \cdot \lambda^2 \cdot f_2 \tag{4}$$

where

$$r_0 \quad = \quad \text{classical electron radius} = 2.81794 \times 10^{-15}m$$
$$\rho \quad = \quad \text{density}$$
$$L \quad = \quad \text{Avogadro's number} = 6.02 \times 10^{23}$$
$$RAM \quad = \quad \text{relative atomic mass.}$$

The values of f_1 and f_2 have been tabulated[13] for all elements in the soft X-ray region. It is thus possible to calculate Δθ for any multilayer at any desired wavelength provided the physical structure (i.e. γ) is known. For the multilayers considered in this work γ was known for Pt/Si but not for OV-200H. Preliminary calculations were made with γ set equal to 0.5 but better agreement with experimental results (see below) was obtained with a gamma value of one third. The results of calculations of $\Delta\theta$ using γ (OV-200H) = 0.333 and γ (Pt/Si) = 0.3 for many different wavelengths for the two multilayers are presented in tables 1 and 2. These tables clearly show that the absorption correction (derived from the second term in equation 3) is usually an order of magnitude less important than that due to dispersion (calculated from the first term in equation 3) and also, as might be expected from equation 3, rapidly becomes negligible as the order of reflection increases.

4.   RESULTS AND DISCUSSION

The angles at which first and higher order reflections of specific X-ray lines were observed have been included in tables 1 and 2 to permit a direct comparison between theory and experiment. In the case of the Pt/Si multilayer a very high background was found at low angles. This made the

TABLE 1

Observed and calculated angles for first and higher order
reflections from OV-200H

$(2d_\infty = 21.05nm : \gamma = 0.333)$

	λ	n	θ_B	$\Delta\theta$ dispersion	$\Delta\theta$ absorption	θ corrected	θ observed	θ_K
O Kα	2.36	1	6.44	+1.26	-0.014	7.69	7.3	7.01
		2	12.96	+0.53	-0.002	13.49	13.4	13.26
		3	19.65	+0.36	0	20.01	19.95	19.86
		4	26.64	+0.29	-2×10^{-4}	26.93	26.85	26.82
		5	34.10	+0.25	nc	34.35	34.2	34.25
		6	42.27	+0.23	nc	42.50	42.45	42.43
N Kα	3.16	1	8.63	+1.11	-0.004	9.74	9.9	9.41
		2	17.47	+0.57	-6×10^{-4}	18.04	17.9	17.87
		3	26.77	+0.41	0	27.18	not obs.	27.05
		4	36.90	+0.34	-1×10^{-4}	37.24	37.15	37.14
		5	48.64	+0.33	nc	48.97	48.75	48.87
C Kα	4.47	1	12.26	+0.92	+0.003	13.18	13.0	13.36
		2	25.13	+0.50	$+3\times10^{-4}$	25.63	25.45	25.73
		3	39.57	+0.39	0	39.96	40.0	40.04
		4	58.19	+0.43	$+1\times10^{-4}$	58.62	58.7	58.66
B Kα	6.76	1	18.73	+1.25	-0.042	19.94	19.7	20.46
		2	39.96	+0.77	-0.007	40.72	40.7	41.03
Nb Mζ	7.22	1	20.06	+1.63	-0.115	21.58	21.5	21.94
Zr Mζ	8.21	1	22.96	+2.13	-0.096	24.99	24.9	25.12
Y Mζ	9.36	1	26.40	+2.63	-0.07	28.96	29.0	28.95
Be Kα	11.4	1	32.79	+3.61	-0.03	36.37	38.0*	36.13

*BeSO$_4$ not beryllium metal

θ_K : corrected values of θ using equation (5)

TABLE 2

Observed and calculated angles for first and higher order
reflections from the Pt/Si multilayer

$(2d_\infty = 33.6nm : \gamma = 0.3)$

	λ	n	θ_B	$\Delta\theta$ dispersion	$\Delta\theta$ absorption	θ corrected	θ observed		
F Kα	1.83	1	3.12	+1.97	-0.58	4.51	-		
		2	6.25	+0.99	-0.10	7.14	7.45		
		3	9.40	+0.67	-0.003	10.07	10.15		
		4	12.58	+0.50	-5×10^{-3}	13.08	13.3		
		5	15.80	+0.41	nc	16.21	16.2		
		6	19.07	+0.35	nc	19.42	20.35		
		7	22.41	+0.31	nc	22.72	23.7		
		8	25.80	+0.27	nc	26.07	26.75		
O Kα	2.36	1	4.02	+2.19	-0.48	5.74	-		
		2	8.08	+1.10	-0.08	9.10	9.4		
		3	12.16	+0.75	-2.7×10^{-3}	12.91	13.05		
		4	16.32	+0.57	nc	16.88	17.0		
		5	20.56	+0.47	nc	21.03	21.25		
N Kα	3.16	1	5.40	+2.51	-0.38	7.53	-		
		2	10.84	+1.27	-0.07	12.04	12.5		
		3	16.39	+0.87	-2.1×10^{-3}	17.26	17.6		
		4	22.1	+0.68	nc	22.78	23.0		
C Kα	4.47	1	7.65	+2.85	-0.18	10.32	-		
		2	15.43	+1.46	-0.03	16.86	17.5		
		3	23.52	+1.03	-1.1×10^{-3}	24.55	25.1		
		4	32.15	+0.84	nc	33.00	33.0		
		5	41.70	+0.77	nc	42.47	42.8		
		6	52.96	+0.81	nc	53.77	55.0		
B Kα	6.76	1	11.61	+3.00	-0.07	14.54	-		
								B$_4$C	BN
		2	23.73	+1.61	-0.01	25.33		25.5	26.0
		3	37.13	+1.24	-0.5×10^{-3}	38.37		39.3	39.5
		4	53.59	+1.28	nc	54.87		54.5	55.5

nc = not calculated

observation of discrete peaks at low angle impossible. As discussed below reasonable agreement with theoretical calculations was only possible if it was assumed that the first order peaks were in fact lost in the background at low angle.

The experimental data were plotted using equation (2) and the results are shown in figures 1(a) and 1(b). These diagrams are particularly important because they enable $2d\infty$ to be found. It is these experimental values of $2d\infty$ that have been used to calculate the θ_B's listed in the tables. The use of the 'physical' 2d values (20nm for OV-200H and 30nm for Pt/Si) gave 'Bragg' angles which, after correction for dispersion and absorption effects, were in poor agreement with the angles observed spectroscopically.

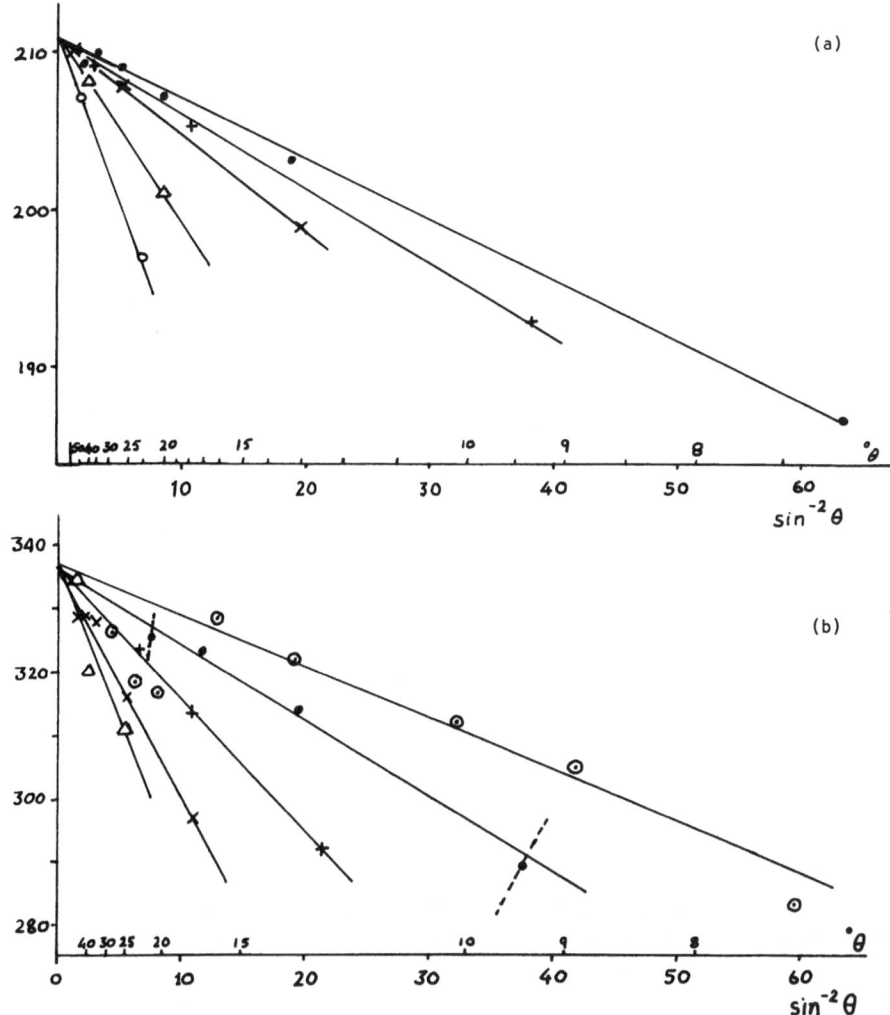

Figure 1. The function $(n\lambda/\sin\theta)$ — vertical scale, plotted versus $(\sin^{-2}\theta)$ — horizontal scale, for different X-rays using (a) the OV-200H and (b) the Pt/Si multilayers. The X-ray lines are distinguished as follows: F Kα, $\odot$; O Kα, $\bullet$; N Kα, +; C Kα, x; B Kα, $\triangle$; S L$_{2,3}$M, o. The displacement of a data point that would be caused by an error of $\pm 0.25°$ is shown as ----- for two points on the oxygen graph in (b).

Theoretical and experimental values of $\bar{\delta}_\lambda$ and A$_\lambda$ are compared in Table 3. Whilst the overall agreement is quite good it should be noted that the data from the Pt/Si multilayer has been calculated assuming that the first peak to be clearly observed was in fact due to a second order reflection. When the data are plotted assuming this peak to be first order much higher values of A$_\lambda$ are found (between five and ten times larger). The lack of accord with theoretical predictions in this case suggests that it is more reasonable to assume that the first order peaks were not, in fact, observed.

TABLE 3

Observed and calculated values for dispersion and absorption terms

All values listed are $\times 10^5$

		OV-200H			Pt/Si			
	λ n m calc.	A_λ obs.	$\bar{\delta}$ calc.	$\bar{\beta}$ calc.	A_λ obs.		$\bar{\delta}$ calc.	$\bar{\beta}$
F	1.83				240	(850)	190	94
O	2.36	180	201 (245)	123	360	(1340)	273	163
N	3.16	220	287 (333)	217	610	(2380)	418	313
C	4.47	290	333 (357)	304	1060	(4520)	667	578
B	6.76	550	664 (918)	124	1410	(8830)	1050	1240
Nb	7.22	860	920 (1186)	139				
Zr	8.34	1220	1388 (1672)	nc				
Y	9.36	1900	1829 (2272)	237				
Be	11.4	3000	2866 (3551)	363				

The values in the first column in brackets are calculated using $\gamma = 0.5$.

The second column in brackets shows the values of A_λ that are obtained if it is assumed that the lowest angle peak is first order.

It might be supposed that the differences between $\bar{\delta}_\lambda$ and A_λ could be ascribed to absorption effects but as can be seen from tables 1 and 2 this effect is pronounced only for first order reflections. This would be seen as a displacement of the 'first-order' point in the figures. But such displacements are not apparent in figure 1(a) and first order peaks were not observed for the Pt/Si multilayer. In any case the calculated magnitudes of $\Delta\theta$ (absorption) are similar to the experimental error ($\pm 0.5°$ 2θ) in determining angles of reflection using multilayers. It can therefore be concluded that absorption effects will not normally be detected using the experimental procedures described in this paper and that, within experimental error, A_λ should be equated with $\bar{\delta}_\lambda$.

5.   REFRACTION EFFECTS: A SIMPLIFIED EQUATION

It is apparent from the tables of atomic scattering factors[12] that f_1 rarely changes by more than an order of magnitude for the range of wavelengths normally covered by a particular multilayer in spectroscopic use; indeed the change in f_1 is often only a factor of two or three. Thus the major cause of variation is $\bar{\delta}_\lambda$ is the λ^2 term in equation 4. This prompts the idea that it might be interesting to set $\bar{\delta}_\lambda \approx K.\lambda^2$ where K would be a constant for a given multilayer and independent of wavelength.

Equation 2 could then be written

$$n.\lambda = 2d_\infty \sin\theta[1 - K\lambda^2/\sin^2\theta],$$

from which it can be shown (if $K.\lambda\ 2d_\infty/n$ is small) that,

$$\sin\theta = \frac{n\lambda}{2d_\infty} + \frac{K.2d_\infty.\lambda}{n} \tag{5}$$

($K.2d_\infty$ would be a constant for a given multilayer.)

For OV-200H it would appear that the best value for K is about 2.0×10^{14}. Values of θ corrected in this way and using this value of K have been included in Table 1 as θ_K. The most critical comparison is in the values calculated and found for first order reflections. Here it can be seen that some deviations are as large as 0.76° (excluding Be where chemical effects obscure comparison), although most are 0.3° or less. Clearly the agreement with experiment is not as good as that based on the complete calculations of equation 3. Table 1 shows that using this equation no differences greater than 0.4° were found (excluding Be) and that the agreement is usually better than ±0.2°. Whether the advantage of simplicity inherent in (5) will outweigh the greater error in the calculated value of θ can only be determined by applying the approximate equation to many other multilayers and seeing if the results are of practical value. 'The proof of the pudding is in the eating.'

6.   CONCLUSIONS

Using the tabulated values of atomic scattering factors, corrections in the angle of soft X-ray reflection (θ) from a multilayer due to refraction have been calculated. Experimentally the 'no-refraction' value of 2d and the values of the 'dispersion constant' at various wavelengths have been measured for two multilayers. From a comparison of the theoretical and experimental results it can be concluded that acceptable agreement is found, both for the dispersion constants (δ) and for angles of reflection. In an approximate equation to take account of refraction effects, the wavelength dependant dispersion constant (δ_λ) is replaced by a constant. The application of this equation to one multilayer gave fair agreement with the experimental values of θ.

7.   ACKNOWLEDGEMENTS

The authors thank the Royal Society, the Science and Engineering Research Council and the Central Research Fund of the University of London for grants for the purchase of equipment. They also gratefully acknowledge support from the European Economic Community (DGXII-BCR) for a bursary (EM) and from the Royal Society for a travel grant (DSU)

REFERENCES

1.   T. Arai, T. Shoji and R.W. Ryan, *Advan. X-ray Anal.* **28**: 137 (1985)
2.   S. Luck and D.S. Urch, *Physica Scripta* **41**: 749 (1990)
3.   M. Charbonnier, M. Romand and F. Gaillard, *Analysis*, **16** (supp. to No. 9-10): 17 (1988)
4.   D.S. Urch in *X-ray Spectroscopy in Atomic and Solid State Physics*, Ed. J.G. Ferreira and M.T. Ramos. Pub., Plenum Press, New York, USA 155 (1988)

5.   A.K. Gyani, P. McClusky, D.S. Urch, M. Charbonnier, F. Gaillard and M. Romand, *Adv. X-ray Anal.* **33**: 247 (1990)

6.   S. Luck and D.S. Urch, *J. de Phys.* **48-C9**: 63 (1987)

7.   J.H. Underwood and T.W. Barbee Jr., *Proc. Topical Conf. on Low-Energy X-ray Diagnostics,* pub., Amer. Inst. Phys. – Conf. Proc. **75**: 170 (1981)

8.   S. Luck, D.S. Urch and D.H. Zheng, *X-ray Spectrometry* in press (1991)

9.   B.L. Evans and B.J. Kent, *App. Optics* **26**: 4491 (1987)

10. A.E. Rosenbluth and P. Lee, *Appl. Phys. Lett.* **40**: 466 (1982)

11. B.L. Henke, E.M. Gullikson, J. Kerner, A.L. Oren and R.L. Blake, *J. X-ray Sci. and Tech.* **2**: 17 (1990)

12. J.M. Arber, P. Norman, D.S. Urch and D. Bloor, *J. Cryst. Growth* **84**: 145 (1987)

13. B.L. Henke, P. Lee, T.J. Tanaka, R.L. Shimabukuro and B.K. Fujikawa, *At. Data and Nuc. Data Tables* **27**: 1 (1982)

A METHOD FOR IN-SITU CALIBRATION OF SEMICONDUCTOR DETECTORS

J.Wernisch, H.J.August and A.Lindner-Schönthaler

Institut für Angewandte und Technische Physik
Technische Universität Wien
Wiedner Hauptstraße 8-10, A 1040 Vienna, Austria

ABSTRACT

A method of semiconductor detector calibration by variation of the incidence angle of the x-radiation is discussed.

INTRODUCTION

Semiconductor detectors are widely used in quantitative
analysis by electron probe microanalysis. The spectral response of the detector efficiency depends, in the low energy
range of x-ray photons, on the thickness of the layers the
radiation has to pass before arriving at the detectors active
zone, viz., the Be-window, the Au-contact layer and the Si-
dead layer. Baker et al (1) described an approach for cali-
brating the detector by simply varying the incidence angle of
radiation from radioactive standard sources. The angular variation causes a variation of the pathlengths in the mentioned layers and thus, an increasing absorption of the incident
radiation in the layers with increasing angle of incidence.
In the present paper we describe the application of this
principle to an in-situ calibration of the detector in an
electron probe microanalyzer.

THEORY

We assume a sequence of 3 layers of Be, Au and inactive
Si. Corresponding to literature (2) the Au-layer has an estimated thickness of 15 to 20 nm. The measured x-ray signal n_m,
in dependence on the incidence angle ε, follows from the un-
attenuated photon flux n_o according to

$$n_m = n_o \cdot \exp\left(-\sum_{i=1}^{3} \tau_i \rho_i t_i / \cos\varepsilon\right)$$

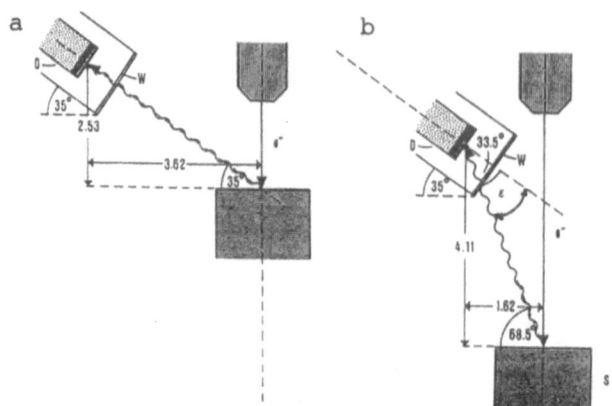

Fig.1a,b: Two geometries of detector, electron gun and specimen in the electron probe microanalyzer (distances in cm).

From the ratio $r_{1,2} = n_m(\varepsilon_1)/n_m(\varepsilon_2)$ of two n_m-values measured
under ε_1 and ε_2 the unknown quantity D

$$D = \sum_{i=1}^{3} \tau_i \rho_i t_i = (\ln(r_{1,2}))/(1/\cos\varepsilon_2 - 1/\cos\varepsilon_1)$$

is obtained. D depends on the photon energy E and can be
evaluated by means of a least squares fit for the unknown
layer thicknesses t_i. Photoabsorption coefficients τ_i and
densities ρ_i are known from literature.

EXPERIMENTAL

The determination of the incidence angle can be seen from
Fig.1. The spectrum processing asks for no special deconvolution or background subtraction routine. This is due to the
fact that for our purposes it is unimportant whether the radiation of a given energy E is continuous or characteristic

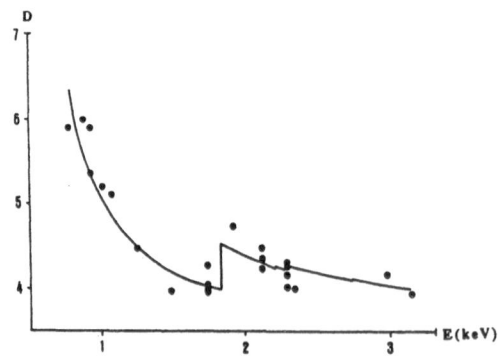

Fig.2 Experimental results and least sqares fit.

radiation. Two serious problems arise from statistical errors of the data in the range of white radiation and the restricted angular variation within $0° < \varepsilon < 33.5°$. This explains the scatter of D(E) in Fig.2. But, in spite of the scatter of measured data the results of Fig.2 are a valuable contribution to the description of the detector efficiency.

ACKNOWLEDGEMENT: The research work has been supported by Fonds zur Förderung der wissenschaftlichen Forschung in Österreich (P7336 PHY) and by Jubiläumsfonds der österreichischen Nationalbank (3264)

LITERATURE

1. C.A.Baker, C.J.Batty and S.Sakamoto, Nucl.Instr.Meth., A259, 501 (1987)
2. N.J.Zaluzek, in: Introduction to Analytical Electron Microscopy, J.J.Hren, J.I.Goldstein and D.C.Joy, eds., Plenum Press, New York, 121 (1979)

THE USE OF BRAGG REFLECTION ON SINGLE CRYSTALS FOR THE PRODUCTION

OF POLARIZED EXCITATION RADIATION IN THE EDXRF

Burkhard Beckhoff, Birgit Kanngießer, Jens Scheer, Walter Swoboda

University of Bremen, Department of Physics
P.O. 330 440
D–2800 Bremen 33, Germany

ABSTRACT

A monochromatic and linear polarized X-ray beam obtained by a Bragg reflection of a characteristic tube line under 90^0 on single crystals can be used as a EDXRF excitation source in a triaxial beam geometry in order to profit by the polarization effects reducing the scattering background in the fluorescence spectrum. A systematic selection procedure for 55 single crystals suitable for Bragg reflection was established. The selection procedure takes the following conditions and criteria into account. First of all the existence of appropriate lattice planes $h\,k\,l$ fulfilling the Bragg reflection condition and having low Miller indices as well as a Bragg angle close to 45^0 ensures a high degree of polarization. Furthermore the integrated reflectivity R_i for these lattice planes should be as high as possible. For the validity of the formula for R_i of an ideal mosaic crystal, the effects of primary and secondary extinction as well as of true absorption have to be considered. The critical limits of these parameters are calculated. The results of this systematic crystal selection procedure for the characteristic $K_\alpha-$ lines of Cr, Cu, Mo and Ag tubes are presented.

INTRODUCTION

The use of Bragg reflection on single crystals with characteristic X−ray tube radiation is a well known method in the EDXRF to obtain a monochromatic and linear polarized excitation in a geometry having an angle of $2\theta_B = 90$ degrees between the incident tube radiation and the Bragg reflected beam. Placing the detector in the z-direction of a triaxial geometry one can profit by the polarization effects reducing that part of the background in the fluorescence spectrum which is due to Compton and Rayleigh scattering of the exciting radiation at the specimen. On the other hand reasonable improvements of minimum detection limits are strongly linked with a high absolute intensity of the radition exciting the specimen. This requires a Bragg-reflected beam at least having approximately the same intensity as e.g. the unpolarized exciting radiation of a comparable ordinary secondary target.

In the recent years a few single crystals have been tested experimentally [1,2,3]. But a more general conclusion about the performance of this methological approach is still missing, maybe

Advances in X-Ray Analysis, Vol. 35
Edited by C.S. Barrett *et al.*, Plenum Press, New York, 1992

because of the argument [2], that crystals having a higher integrated reflectivity than the tested ones are still to be found, especially for shorter wavelengths like the wavelength of $Mo-K_\alpha$ radiation. The subject of the present investigation is a contribution to solve one of the basic problems of this EDXRF excitation mode, which is to find theoretically appropriate crystals ensuring experimentally a high intensity Bragg reflected beam.

THEORY

According to the theory [4] for the X-ray diffraction the integral or integrated reflectivity R_i of an ideal mosaic crystal may be considered as a figure of merit for the intensity of Bragg reflected X-ray beam. Under certain conditions given below a real crystal can be treated as such an ideal mosaic crystal for a specific reflection hkl. In the *symmetrical Bragg case* the formula of R_i has the following form:

$$R_i = r_0^2 \frac{1 + \cos^2(2\theta_B)}{2\sin(2\theta_B)} |F_{hkl}|^2 \frac{\lambda^3}{V^2} \frac{1}{2\mu_0} \exp\left(-2B\frac{\sin^2\theta_B}{\lambda^2}\right) \tag{1}$$

with $r_0 = \frac{e^2}{mc^2}$ as classical electron radius

θ_B as Bragg angle

$|F_{hkl}|^2$ as square of the structure factor of the lattice plane hkl

λ as the wavelength of the radiation

V as volume of the unit cell

μ_0 as linear mass absorption coefficient

$B = B_{0K} + B_{293K}$ as thermal Debye parameter at room temperature.

This formula for the integrated reflectivity R_i is valid for a particular Bragg reflection hkl if absorption and extinction effects within the mosaic crystal are negligible because these effects reduce the intensity of the Bragg reflected beam in comparison to an ideal mosaic crystal. For the judgement whether the real crystal can be considered as an ideal mosaic crystal for the concrete reflexion hkl one may introduce three critical values for these absorption and extinction effects. For the *primary extinction* effect the corresponding critical value is:

$$t_{critical, primary\ extinction} = \frac{2V\cos\theta_B}{r_0(1 + |\cos 2\theta_B|)\lambda |F_{hkl}| \exp\left(-B\frac{\sin^2\theta_B}{\lambda^2}\right)} \tag{2}$$

for the *secondary extinction* effect it is:

$$\eta_{critical, secondary\ extinction} = r_0^2 \frac{1 + \cos^2(2\theta_B)}{2\sin(2\theta_B)} |F_{hkl}|^2 \frac{\lambda^3}{V^2} \frac{1}{\sqrt{2\pi}\mu_0} \exp\left(-2B\frac{\sin^2\theta_B}{\lambda^2}\right) \tag{3}$$

and for the *true absorption* effect it has the form:

$$t_{critical, true\ absorption} = \frac{\cos\theta_B}{\mu_0} \tag{4}$$

The *range of validity* of the above formula for the integrated reflectivity R_i may then be expressed in terms of the following *three conditions* for the mosaic block thickness t_0 and for the mosaic spread η of the crystal:

$$t_0 \ll t_{critical, primary\ extinction} \tag{5}$$

$$\eta \gg \eta_{critical, secondary\ extinction} \tag{6}$$

$$t_0 < t_{critical, true\ absorption} \tag{7}$$

with η being the standard deviation of the gaussian shaped distribution of the mosaic block orientations.

CRYSTAL SELECTION

For the characteristic $K_\alpha-$ lines of Cr, Cu, Mo and Ag tubes a systematic selection code for a set of 55 single crystals suitable for Bragg reflection was established. The selection code was written in the programing language Turbo Pascal (version 5.0). This automatic crystal selection procedure has to take the following conditions and criteria into account:

1. Existence of appropriate lattice planes $h\,k\,l$ fulfilling the Bragg reflection condition $2d_{hkl}\sin\theta_B = \lambda$ and having low Miller indices as well as a Bragg angle θ_B close to 45 degrees as so to ensure a high degree of linear polarization.

2. Besides the existence of appropriate lattice planes $h\,k\,l$, which fulfill for a given wavelength λ the Bragg condition, their Miller indices $h\,k\,l$ should be as small as possible as so to reduce the number of undesired low order reflexes accompanying the $h\,k\,l$ reflection.

3. The integrated or integral reflectivity R_i for these lattice planes $h\,k\,l$ should be as high as possible in order to ensure a strong Bragg reflex for the specimen excitation. This is mainly a condition for the structure factor F_{hkl} and only in the second place the thermal Debye parameter B is required as small as possible for that.

4. The assumption that the real crystal acts for the calculated Bragg reflection $h\,k\,l$ as an ideal mosaic crystal and with it the validity of the formula given above for the integrated reflectivity R_i should be always controlled. For that purpose the three calculated critical extinction and absorption values of the reflection $h\,k\,l$ are compared according to the conditions (5), (6) and (7) with the typical values for the mosaic block thickness t_0 and the mosaic spread η, as given in the literature for real crystals. These typical literature values range from 10^{-5} through 10^{-4} cm for the mosaic block thickness t_0 and are around $2\cdot10^{-4}$ radians for the mosaic spread η. As this comparison, namely (5), (6) and (7) , fails to yield a clear tendency one must assume that the actual integrated reflectivity may be considerably smaller than the one calculated according to (1).

5. The energy of the fluorescence radiation characteristic for the crystal material should not be situated in the measuring range of interest.

DATA BASIS OF THE CRYSTAL SELECTION

Concerning the calculation of the interplanar spacing d_{hkl} the necessary crystal data like the structure types and the unit cell edges were taken from Wyckhoff [5]. All other data like the crystal density, atomic scattaring factors, mass absorption coefficients and thermal Debye parameters were taken from the International Tables for X-rays Crystallography [6]. The values of the thermal Debye parameter of HOPGraphite are given by Kanngießer [7]. In particular the thermal Debye parameter B is often indicated with two different values due to its determination in different experiments. Therefore the calculated parameters which are dependent on the thermal Debye parameter are presented in the corresponding Tables 1, 2 and 3, with two values, if two different thermal Debye parameters are given in litterature [6,7].

RESULTS

The reflecting lattice planes $h\,k\,l$, the Bragg angles θ_B, the integrated reflectivities R_i and the three critical absorption and extinction values were calculated for the four characteristic $K_\alpha-$lines of Cr, Cu, Mo and Ag tubes. The $Mo-K_\alpha$ characteristic tube radiation was of particular interest from an experimental and application orientated point of view. The results of these values are presented in the Tables 1 and 2. The tables are ordered by the atomic

Table 1. Lattice planes, Bragg angles, integrated reflectivities and critical absorption and extinction values of the systematic crystal selection for Mo-Kα characteristic tube radiation in the atomic number range Z = 3 upto 42

Z	crystal	structure type	thermal Debye parameter $B = B_{0K} + B_{293K}$ [Å²]	(2nd value)	lattice plane h k l	Bragg angle θ_B [degree]	integrated reflectivity R_i [radians]	(2nd value)	t crit., true absorption [cm]	t crit., primary extinction [cm]	(2nd value)	η crit., secondary extinction [radians]	(2nd value)
3	Lithium	bcc	6.48	(2.77)	4 4 4	44.551	5.24E-11	(7.24E-08)	6.82E+00	2.53E+01	(6.80E-01)	4.18E-11	(5.77E-08)
4	Beryllium	hcp	0.48	(0.29)	2 2 4	47.511	2.93E-05	(4.41E-05)	1.50E+00	1.45E-02	(1.18E-02)	2.34E-05	(3.52E-05)
5	Boron	tetragonal	0.28		0 0 10	44.948	4.05E-05		8.76E-01	1.04E-02		3.23E-05	
6	Carbon	diamond	0.15	(0.11)	4 4 4	43.649	1.17E-04	(1.26E-04)	3.84E-01	3.94E-03	(3.79E-03)	9.30E-05	(1.00E-04)
6	HOPGraphite	hexagonal	0.63	(0.37)	0 0 14	47.758	2.13E-05	(3.74E-05)	5.56E-01	1.03E-02	(7.75E-03)	1.70E-05	(2.98E-05)
11	Sodium	bcc	6.92	(4.57)	0 6 6	44.649	6.90E-12	(6.84E-10)	2.50E-01	1.34E+01	(1.34E+00)	5.51E-12	(5.45E-10)
12	Magnesium	hcp	1.69	(1.22)	2 2 8	44.648	2.93E-07	(7.36E-07)	1.03E-01	4.16E-02	(2.63E-02)	2.34E-07	(5.87E-07)
13	Aluminium	fcc	0.92	(0.72)	0 0 8	44.588	1.77E-06	(2.61E-06)	5.23E-02	1.21E-02	(9.93E-03)	1.41E-06	(2.08E-06)
19	Potassium	bcc	8.63	(3.27)	1 3 10	45.259	7.43E-14	(3.32E-09)	4.99E-02	5.75E+01	(2.72E-01)	5.93E-14	(2.65E-09)
20	Calcium	fcc	1.76	(1.61)	1 1 11	44.974	1.09E-07	(1.46E-07)	2.42E-02	3.34E-02	(2.88E-02)	8.68E-08	(1.17E-07)
22	Titanium	hcp	0.62	(0.40)	4 1 4	44.902	2.14E-06	(3.31E-06)	6.77E-03	3.97E-03	(3.20E-03)	1.71E-06	(2.64E-06)
23	Vanadium	bcc	0.76	(0.40)	0 0 6	44.834	1.96E-06	(3.97E-06)	4.61E-03	3.42E-03	(2.40E-03)	1.56E-06	(3.17E-06)
24	Chromium	bcc	0.41	(0.29)	0 0 4	44.189	4.20E-06	(5.29E-06)	3.41E-03	1.98E-03	(1.76E-03)	3.35E-06	(4.22E-06)
25	α-Manganese	cubic	0.51	(0.38)	0 5 17	45.071	5.64E-07	(7.30E-07)	2.99E-03	5.14E-03	(4.52E-03)	4.50E-07	(5.82E-07)
26	α-Iron	bcc	0.49	(0.29)	0 0 4	44.528	2.91E-06	(4.29E-06)	2.40E-03	2.01E-03	(1.66E-03)	2.32E-06	(3.43E-06)
27	α-Cobalt	hcp	0.39		0 0 8	44.324	3.48E-06		1.96E-03	1.65E-03		2.78E-06	
28	Nickel	fcc	0.42	(0.27)	4 4 4	44.318	3.11E-06	(4.15E-06)	1.70E-03	1.63E-03	(1.41E-03)	2.48E-06	(3.31E-06)
29	Copper	fcc	0.59	(0.47)	0 4 6	45.141	1.85E-06	(2.34E-06)	1.60E-03	2.08E-03	(1.84E-03)	1.47E-06	(1.87E-06)
30	Zinc	hcp	1.30	(0.56)	4 1 0	44.880	3.34E-07	(1.43E-06)	1.79E-03	5.18E-03	(2.50E-03)	2.66E-07	(1.14E-06)
31	Gallium	orthorhombic	3.10	(0.85)	2 6 10	42.919	1.15E-08	(7.16E-07)	2.18E-03	2.94E-02	(3.73E-03)	9.17E-09	(5.71E-07)
33	Arsenic	rhombohedral	0.91	(0.56)	6 6 8	45.119	4.64E-07	(9.30E-07)	1.87E-03	4.47E-03	(3.16E-03)	3.70E-07	(7.42E-07)
34	Selenium	hexagonal	2.35		6 0 5	42.039	3.96E-08		2.24E-03	1.58E-02		3.16E-08	
37	Rubidium	bcc	11.73	(5.46)	1 2 11	45.369	4.63E-17	(1.34E-11)	5.53E-03	7.63E+02	(1.42E+00)	3.69E-17	(1.07E-11)
38	Strontium	fcc	1.76	(1.54)	1 5 11	45.078	3.51E-08	(5.43E-08)	3.08E-03	2.09E-02	(1.68E-02)	2.80E-08	(4.33E-08)
40	α-Zirconium	hcp	0.59	(0.46)	2 2 8	44.913	5.61E-06	(7.26E-06)	6.77E-03	2.46E-03	(2.16E-03)	4.48E-06	(5.79E-06)
41	Niobium	bcc	0.58	(0.41)	2 2 2	45.577	7.17E-06	(1.01E-05)	4.82E-03	1.79E-03	(1.51E-03)	5.72E-06	(8.07E-06)
42	Molybdenum	bcc	0.28	(0.23)	0 2 6	45.568	1.47E-05	(1.63E-05)	3.72E-03	1.10E-03	(1.05E-03)	1.17E-05	(1.30E-05)

Mo Kα

Table 2. Results of the systematic crystal selection for Mo-Kα radiation in the range Z = 44 up to 92 (Cont. of Table 1.)

Mo Kα

Z	crystal	structure type	thermal Debye parameter B = B0K + B293K [Å²]	(2nd value)	lattice plane h	k	l	Bragg angle θB [degree]	integrated reflectivity Ri [radians]	(2nd value)	t crit., true absorption [cm]	t crit., primary extinction [cm]	(2nd value)	η crit., secondary extinction [radians]	(2nd value)
44	Ruthenium	hcp	0.22	(0.18)	1	1	8	45.568	1.85E-05	(2.00E-05)	2.69E-03	8.35E-04	(8.02E-04)	1.47E-05	(1.60E-05)
45	Rhodium	fcc	0.34	(0.24)	2	4	6	44.364	1.46E-05	(1.78E-05)	2.49E-03	9.11E-04	(8.27E-04)	1.17E-05	(1.42E-05)
46	Palladium	fcc	0.46	(0.41)	1	3	7	44.564	1.03E-05	(1.13E-05)	2.40E-03	1.07E-03	(1.02E-03)	8.18E-06	(9.02E-06)
47	Silver	fcc	0.70	(0.62)	3	3	7	45.384	5.08E-06	(5.96E-06)	2.54E-03	1.56E-03	(1.44E-03)	4.05E-06	(4.76E-06)
48	Cadmium	hcp	2.08	(1.02)	3	3	0	45.704	2.29E-07	(1.96E-06)	2.91E-03	7.76E-03	(2.65E-03)	1.82E-07	(1.57E-06)
49	Indium	tetragonal	4.83	(1.77)	2	6	2	45.039	8.96E-10	(3.86E-07)	3.33E-03	1.36E-01	(6.57E-03)	7.15E-10	(3.08E-07)
50	Tin(grey)	diamond	0.38		2	8	10	45.199	4.26E-06		3.90E-03	2.13E-03		3.40E-06	
50	Tin(white)	tetragonal	1.08	(0.43)	4	8	4	44.926	1.39E-06	(5.02E-06)	3.11E-03	3.34E-03	(1.76E-03)	1.11E-06	(4.01E-06)
51	Antimony	rhombohedral	1.42	(0.70)	1	4	8	45.199	5.65E-07	(2.37E-06)	3.20E-03	5.29E-03	(2.58E-03)	4.51E-07	(1.89E-06)
52	Tellurium	hexagonal	1.84		6	0	7	44.046	2.42E-07		3.40E-03	8.20E-03		1.93E-07	
53	Iodine(I2)	orthorhombic	1.20		1	13	7	44.688	5.85E-07		3.97E-03	5.79E-03		4.67E-07	
55	Cesium	bcc	14.01	(8.69)	1	1	12	45.049	2.06E-18	(7.87E-14)	9.19E-03	4.73E+03	(2.42E+01)	1.64E-18	(6.28E-14)
56	Barium	bcc	1.86		0	0	10	45.005	9.83E-08		4.66E-03	1.54E-02		7.85E-08	
57	α-Lanthanum	double hcp	1.39	(1.05)	0	0	24	44.540	4.26E-07	(8.27E-07)	2.54E-03	5.41E-03	(3.89E-03)	3.40E-07	(6.60E-07)
64	Gadolinium	hcp	0.93		5	2	0	44.809	5.58E-07		8.51E-04	2.75E-03		4.45E-07	
72	Hafnium	hcp	0.41		0	0	10	44.634	3.40E-06		6.21E-04	9.48E-04		2.72E-06	
73	Tantalum	bcc	0.36	(0.31)	2	2	6	45.482	4.42E-06	(4.89E-06)	4.72E-04	7.17E-04	(6.82E-04)	3.53E-06	(3.90E-06)
74	Tungsten	bcc	0.26	(0.13)	0	2	6	45.248	5.98E-06	(7.75E-06)	3.81E-04	5.60E-04	(4.92E-04)	4.77E-06	(6.18E-06)
75	Rhenium	hcp	0.24	(0.19)	0	2	2	44.644	6.92E-06	(7.63E-06)	3.43E-04	4.94E-04	(4.71E-04)	5.52E-06	(6.09E-06)
76	Osmium	hcp	0.29	(0.27)	4	1	2	44.981	6.56E-06	(6.82E-06)	3.14E-04	4.90E-04	(4.80E-04)	5.23E-06	(5.45E-06)
77	Iridium	fcc	0.22		1	3	7	45.310	7.38E-06		3.04E-04	4.49E-04		5.88E-06	
78	Platinum	fcc	0.34	(0.28)	1	3	7	44.087	5.69E-06	(6.38E-06)	3.08E-04	5.10E-04	(4.82E-04)	4.54E-06	(5.09E-06)
79	Gold	fcc	0.71	(0.47)	3	3	7	45.497	2.24E-06	(3.63E-06)	3.27E-04	8.39E-04	(6.59E-04)	1.78E-06	(2.89E-06)
81	Thallium	hcp	1.78	(1.65)	6	0	0	45.428	1.51E-07	(1.96E-07)	4.96E-04	3.99E-03	(3.50E-03)	1.21E-07	(1.57E-07)
82	Lead	fcc	3.42	(1.48)	4	4	8	44.693	6.39E-09	(2.85E-07)	5.10E-04	1.99E-02	(2.97E-03)	5.10E-09	(2.28E-07)
83	Bismuth	rhombohedral	4.20	(1.12)	5	7	9	44.629	1.19E-09	(4.90E-07)	5.77E-04	4.89E-02	(2.41E-03)	9.50E-10	(3.91E-07)
90	Thorium	fcc	0.69	(0.50)	0	2	10	45.460	1.70E-06	(2.49E-06)	6.13E-04	1.32E-03	(1.09E-03)	1.36E-06	(1.99E-06)
92	α-Uranium	orthorhombic	0.35		5	5	2	44.988	5.66E-06		3.84E-04	5.84E-04		4.52E-06	

Table 3. Best results of the systematic crystal selection for Cr-Kα, Cu-Kα and Ag-Kα characteristic tube radiation

Cr Kα

Z	crystal	structure type	thermal Debye parameter $B = B_{0K} + B_{293K}$ [Å²] (2nd value)	lattice plane h k l	Bragg angle θ_B [degree] (2nd value)	integrated reflectivity R_i [radians] (2nd value)	t crit., true absorption [cm]	t crit., primary extinction [cm] (2nd value)	η crit., secondary extinction [radians] (2nd value)
4	Beryllium	hcp	0.48 (0.29)	1 0 1	41.368	1.30E-03 (1.34E-03)	1.28E-01	1.17E-03 (1.15E-03)	1.03E-03 (1.07E-03)
6	HOPGraphite	hexagonal	0.63 (0.37)	0 0 4	42.986	5.36E-04 (5.62E-04)	2.24E-02	7.84E-04 (7.66E-04)	4.28E-04 (4.48E-04)
33	Arsenic	rhombohedral	0.91 (0.56)	0 0 2	47.396	1.49E-04 (1.60E-04)	5.27E-04	2.17E-04 (2.09E-04)	1.19E-04 (1.28E-04)
42	Molybdenum	bcc	0.28 (0.23)	0 0 2	46.710	1.53E-04 (1.55E-04)	1.47E-04	1.16E-04 (1.16E-04)	1.22E-04 (1.23E-04)
57	α-Lanthanum	double hcp	1.39 (1.05)	1 1 4	45.645	1.40E-04 (1.50E-04)	5.08E-04	2.35E-04 (2.28E-04)	1.12E-04 (1.20E-04)
72	Hafnium	hcp	0.41	1 1 0	45.778	2.11E-04	1.18E-04	9.20E-05	1.69E-04
73	Tantalum	bcc	0.36 (0.31)	0 0 2	43.868	2.72E-04 (2.74E-04)	9.55E-05	7.34E-05 (7.30E-05)	2.17E-04 (2.19E-04)
74	Tungsten	bcc	0:26 (0.13)	0 0 2	46.377	2.96E-04 (3.04E-04)	7.60E-05	6.08E-05 (6.01E-05)	2.36E-04 (2.43E-04)

Cu Kα

Z	crystal	structure type	thermal Debye parameter $B = B_{0K} + B_{293K}$ [Å²] (2nd value)	lattice plane h k l	Bragg angle θ_B [degree] (2nd value)	integrated reflectivity R_i [radians] (2nd value)	t crit., true absorption [cm]	t crit., primary extinction [cm] (2nd value)	η crit., secondary extinction [radians] (2nd value)
4	Beryllium	hcp	0.48 (0.29)	1 1 0	42.398	1.08E-03 (1.16E-03)	3.99E-01	1.90E-03 (1.83E-03)	8.58E-04 (9.23E-04)
6	Carbon	diamond	0.15 (0.11)	1 1 3	45.793	2.21E-04 (2.25E-04)	4.69E-02	1.47E-03 (1.46E-03)	1.76E-04 (1.79E-04)
6	HOPGraphite	hexagonal	0.63 (0.37)	0 0 6	43.496	2.49E-04 (2.76E-04)	7.61E-02	1.76E-03 (1.67E-03)	1.98E-04 (2.20E-04)
28	Nickel	fcc	0.42 (0.27)	1 1 3	46.516	2.18E-04 (2.33E-04)	1.58E-03	2.64E-04 (2.56E-04)	1.74E-04 (1.86E-04)
29	Copper	fcc	0.59 (0.47)	1 1 3	45.014	1.93E-04 (2.03E-04)	1.54E-03	2.94E-04 (2.87E-04)	1.54E-04 (1.62E-04)
74	Tungsten	bcc	0.26 (0.13)	0 2 2	43.550	1.72E-04 (1.81E-04)	2.20E-04	1.14E-04 (1.11E-04)	1.37E-04 (1.45E-04)
75	Rhenium	hcp	0.24 (0.19)	0 0 4	43.762	1.81E-04 (1.85E-04)	1.93E-04	1.04E-04 (1.03E-04)	1.44E-04 (1.47E-04)
77	Iridium	fcc	0.22	2 2 2	44.071	1.80E-04	1.67E-04	9.81E-05	1.44E-04

Ag Kα

Z	crystal	structure type	thermal Debye parameter $B = B_{0K} + B_{293K}$ [Å²] (2nd value)	lattice plane h k l	Bragg angle θ_B [degree] (2nd value)	integrated reflectivity R_i [radians] (2nd value)	t crit., true absorption [cm]	t crit., primary extinction [cm] (2nd value)	η crit., secondary extinction [radians] (2nd value)
4	Beryllium	hcp	0.48 (0.29)	4 1 4	46.090	4.15E-06 (7.77E-06)	1.84E+00	4.01E-02 (2.93E-02)	3.31E-06 (6.20E-06)
5	Boron	tetragonal	0.28	1 21 3	44.021	5.61E-06	1.20E+00	2.84E-02	4.47E-06
6	Carbon	diamond	0.15 (0.11)	0 4 8	44.680	4.08E-05 (4.63E-05)	5.71E-01	7.38E-03 (6.93E-03)	3.26E-05 (3.70E-05)
6	HOPGraphite	hexagonal	0.63 (0.37)	0 0 16	41.884	7.60E-06 (1.59E-05)	9.31E-01	2.06E-02 (1.42E-02)	6.06E-06 (1.27E-05)
44	Ruthenium	hcp	0.22 (0.18)	6 0 0	45.928	7.60E-06 (8.67E-06)	5.18E-03	1.58E-03 (1.48E-03)	6.06E-06 (6.92E-06)
45	Rhodium	fcc	0.34 (0.24)	1 3 9	44.695	5.42E-06 (7.43E-06)	4.80E-03	1.86E-03 (1.59E-03)	4.33E-06 (5.93E-06)
46	Palladium	fcc	0.46 (0.41)	4 4 8	44.934	3.33E-06 (3.91E-06)	4.63E-03	2.34E-03 (2.16E-03)	2.66E-06 (3.12E-06)
75	Rhenium	hcp	0.24 (0.19)	6 0 0	44.721	3.00E-06 (3.51E-06)	6.40E-04	9.14E-04 (8.45E-04)	2.39E-06 (2.80E-06)

number Z of the crystal material. Table 1 contains the results for the range $Z = 3$ up to 42 and Table 2 for the range $Z = 44$ up to 92. For $Cr-K_\alpha$, $Cu-K_\alpha$ and $Ag-K_\alpha$ tube lines the best results, i.e. the crystals having the highest calculated integrated reflectivity R_i, are shown in Table 3.

Apart from some singular maximum R_i values the orders of magnitude of the largest integrated reflectivities are for $Cr-K_\alpha$ radiation $3\cdot10^{-4}$ radians, for $Cu-K_\alpha$ radiation $2\cdot10^{-4}$ radians, for $Mo-K_\alpha$ radiation $2\cdot10^{-5}$ radians and for $Ag-K_\alpha$ radiation 10^{-5} radians.

As the best candidates of the selected crystals for the $Mo-K_\alpha$ tube line the HOPGraphite crystal [7,8], the Molybdenum crystal [8,9] and the Ruthenium crystal [9] were investigated experimentally in regard of their use for EDXRF in a new set-up [10]. The ratios of the measured intensities of their respective Bragg-reflected $Mo-K_\alpha$ beams confirm their calculated integrated reflectivities R_i and critical absorption and extinction values.

REFERENCES

1. Ryon, R.W., Zahrt, J. D., Wobrauschek, P., Aiginger, H., 'The Use of Polarized X-Rays for Improved Detection Limits in Energy Dispersive X-Ray Spectrometry', Advances in X-Ray Analysis 25 (1982).

2. P. Wobrauschek, H. Aiginger, 'X-Ray Fluorescence Analysis with a Linear Polarized Beam after Bragg Reflection from a Flat or Curved Single Crystal', X-Ray Spectrometry 12:72-78 (1983).

3. P. Wobrauschek, H. Aiginger, 'The Application of Linear Polarized X-Rays after Bragg Reflection for X-Ray fluorescence Analysis', Advances in X-Ray Analysis 28:69-74 (1985).

4. Zachariasen, W.H., 'Theory of X-Ray Diffraction in Crystals', Dover Publications, New York (1967).

5. Wyckoff, R.W.G., 'Crystal Structures', Volume 1, Interscience Publishers, John Wiley & Sons, New York (1965).

6. International Tables for X-Ray Crystallography, Vol. II - IV, Kynoch Press, Birmingham (1969).

7. B. Kanngießer, 'Die Anwendung von HOPG-Kristallen in der energiedispersiven Röntgenfluoreszenzanalyse', Diploma thesis, University of Bremen (1990).

8. B. Kanngießer, B. Beckhoff, J. Scheer, W. Swoboda, 'The Comparison of three excitation modes in the Energy Dispersive X-Ray Fluorescence', to be published in Advances in X-Ray Analysis 35.

9. B. Beckhoff, 'Untersuchung der Einsatzmöglichkeiten der Bragg-Reflexion an Einkristallen zur Erzeugung polarisierter Anregungsstrahlung für die energiedispersive Röntgenfluoreszenzanalyse', Diploma thesis, University of Bremen (1990).

10. W. Swoboda, B. Kanngießer, B. Beckhoff, K. Begemann, H. Neuhaus, and J. Scheer, 'A New Device for Energy Dispersive X-Ray Fluorescence (EDXRF)', accepted in Rev.Sci.Instrum.

X-SPECTRUM DETERMINATION APPLIED TO FLASH RADIOGRAPHY

Loïc LE DAIN and Jean-Marc DINTEN

Centre d'Etudes de Vaujours. BP N^0 7
77181 - Courtry - France

ABSTRACT

We propose a method, by absorption [1,2,3,4], of determination of the X-ray spectrum emitted by a flash X-ray radiography generator. This approach leads to an ill posed inverse problem very sensitive to the measures precision.

We improve the robustness of the method by introduction of a smoothness constraint on the X-ray spectrum to reconstruct. We also propose an automatic choice of the weight between the measurement and the constraint by using a cross-validation technique.

We present an application on an unusual flash X-ray radiography generator - i.e. 75 ns of duration pulse, 6.5 MeV of maximum energy, 350 Rad for dose - for which an instrument of direct measure has not yet been found.

1]PRINCIPLE OF THE METHOD

We observe on the detector (figure 1) :

$$D_j = \int_{Emin}^{Emax} I_0(E) \cdot \exp(\frac{\mu}{\rho}_j(E) \cdot \rho_j \cdot l_j) dE$$

where $\frac{\mu}{\rho}_j(E)$ is the mass attenuation coefficient, ρ_j the density and l_j the object length.

A discretization on the energy leads to the system : $D = A.I_0$, where D is the observed dose vector, A the kernel matrix and I_0 the X-ray spectrum that we want to reconstruct.

The resolution of this system by a generalized inverse method ($I_0 = A^{-1}D$) is very sensitive to the precision on the measurement. On figure 2, we present a reconstruction from data disturbed by a very low noise ($\frac{\Delta D}{D} \leq 1\%$).

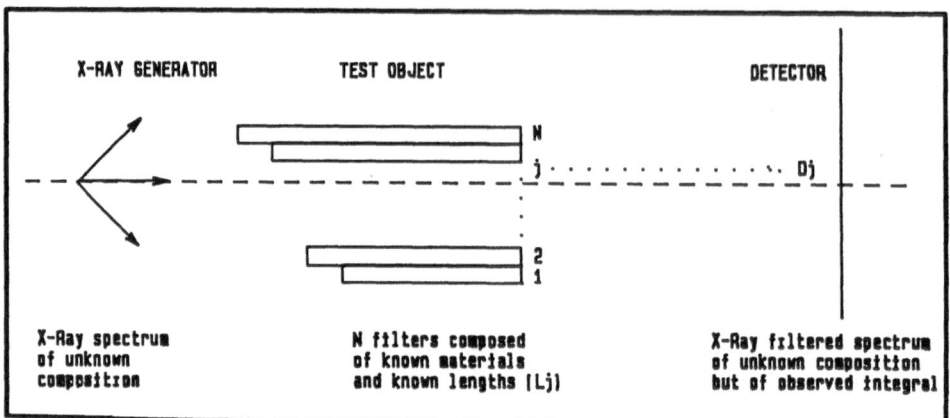

FIGURE 1

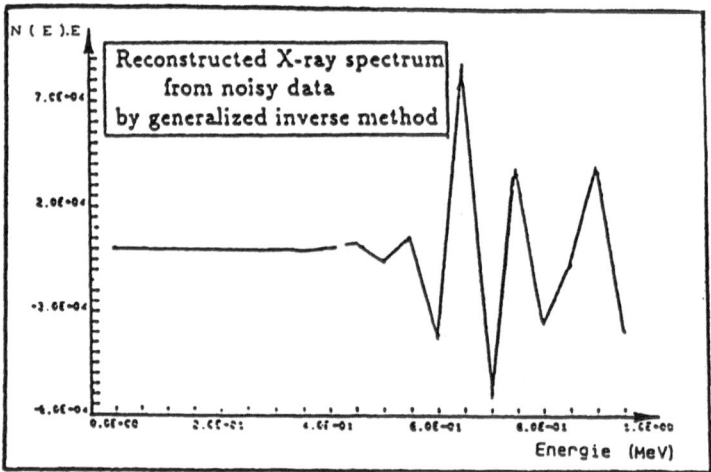

FIGURE 2

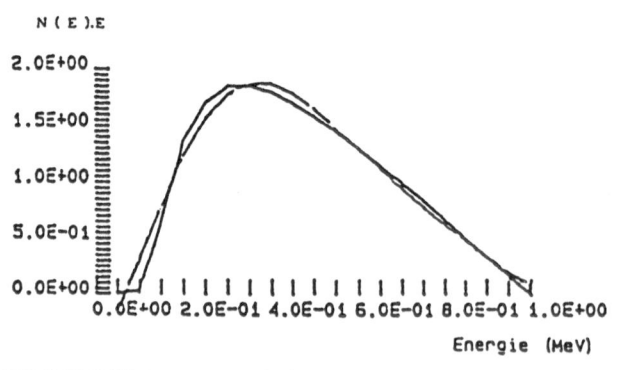

FIGURE 3

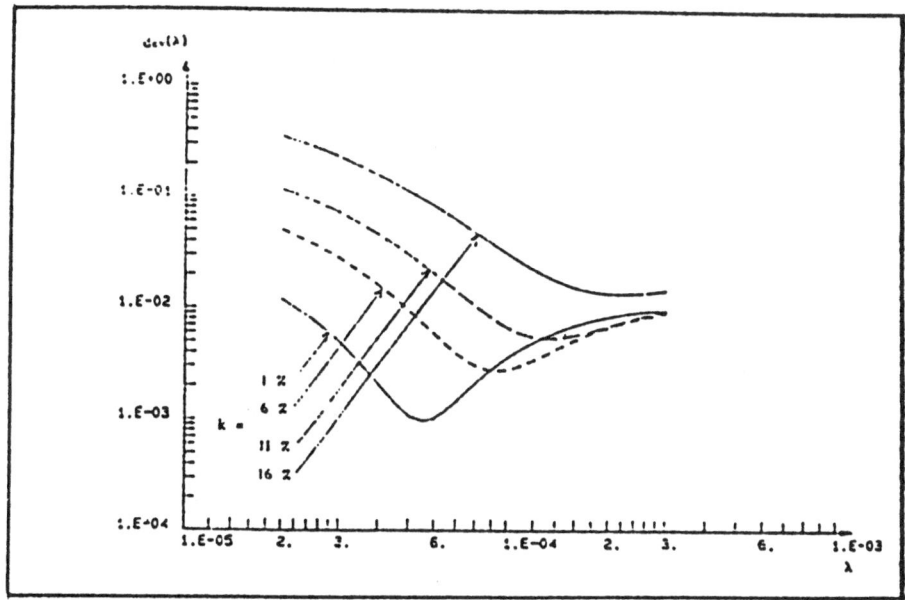

FIGURE 4

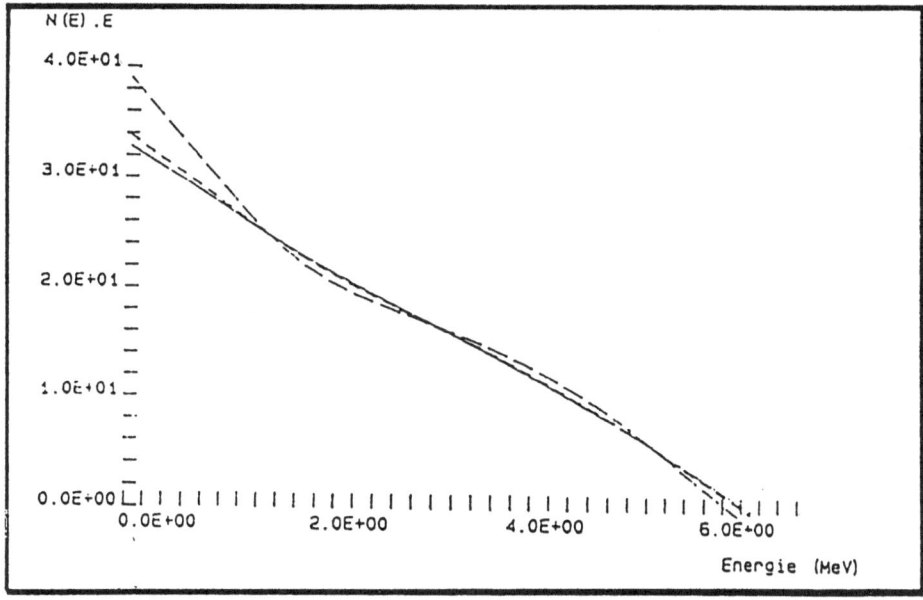

FIGURE 5

2]REGULARIZATION OF THE METHOD

To regularize [5,6], we introduce a constraint of smoothness on the spectrum to reconstruct : a small second derivative.

The solution then becomes $I_0 = (A^t A + \lambda \cdot V^t V)^{-1} A^t D$ with λ the regularization parameter expressing the weight between the data and the constraint, V the second derivative operator. With an optimal choice of λ, from the previous data, we obtain (figure 3) .

3]AUTOMATIC DETERMINATION OF THE REGULARIZATION PARAMETER

The choice of the optimal regularization parameter is done thanks to a cross-validation technique [7,8].

If $\hat{I}_0^{(i)}$ is the reconstructed spectrum from the data vector minus observation (i), we define $\hat{D}(i) = [A.\hat{I}_0^{(i)}]_i$. The optimal λ is then the argument of the minimum of the distance $d_{cv}(\lambda) = \sum_{i=1}^{N} [\frac{D(i) - \hat{D}(i)}{D(i)}]^2$. Let us notice that the value of optimal λ is determined from only the observations.

We test (figure 4) the performance of this technique by comparaison with the optimal λ value which minimizes the distance to original spectrum distribution (which we know in our simulation).

4]EXPERIMENTAL VALIDATION

We have applied this technique on a flash radiography X-ray generator - i.e. 75 ns of duration pulse, 6.5 MeV of maximum energy, 350 Rad for dose. The test-object was made from 8 differents materials (Z=5,...,92). The detector was a linear screen-film radiography behind a multi-hole collimator in order to reduce the scattering.

We obtain (figure 5) this result for "bremsstrahlung spectrum" for screen-film like Pb/DEF/Pb.

CONCLUSIONS

This experimental example shows the performance of this automatic method that we have constructed. We foresee to carry on this study by testing our technique on a X-ray flash radiography generator for which we could face our result, with the ones obtained by instrumental techniques.

REFERENCES

1)J.R. Grenning, The determination of X ray energy distributions by the absorption method, Physics Departement, Westminster Hospital, London, 1947.

2)D.E.A. Jones, The determination from absorption data of the distribution of X ray intensity in the continuous X ray spectrum, Physics Departement, Hammersmith and Lambeth Hospital, London country council, 1940.

3) R.T. Mainardy and Raul Barrea, "X-Ray Spectrum Determination by successive Modifications of Beam intensity",Nuclear Instruments and Methods in Physics Research, <u>A280</u>, 387-391 (1989).

4) M. Rubio and R.T. Mainardi, "Determination of X-Ray Spectra using Characteristic Line Intensities from Attenuation Data", Physics of Medical Biology, <u>29</u>, 1371-1376 (1984).

5) D.L.Phillips,-A technique for the numerical solution of certain integral equations of the first kind, Argonne National Laboratory, Argonne June 1961, PP 84,97.

6) S. Twomey,- On the numerical solution of fredholm integral equations of the first kind by the inversion of the linear system produced by quadrature, U.S Weather Bureau, Washington, D.C. september 1962, pp 97,101.

7) G.Wahba,-Practical approximate solutions to linear operator equations when the data are noisy, S.I.A.M., J.Numer. Anal., Vol 14, pp 651,677, Numero 4, Sept1977.

8) G.Wahba,-Constrained regularization for ill-posed linear operator equations, with applications in meteorology and medecine. statistical decision theory and related topics 3, vol 2, pp 383,417, 1982, Academic Press.

ANALYTICAL X-RAY ANALYSIS TECHNIQUES: A PANORAMA OF SOME OF THE APPLICATIONS IN LATIN AMERICA

C. Vazquez, D.V. de Leyt, J.J. LaBrecque[*]

Departamento de Química Analítica
Comisión de Energía Atómica
Buenos Aires, Argentina

[*]Atomic and Nuclear Spectroscopy Laboratory
Apartado 21827, Caracas 1020A, Venezuela

In general Latin America countries function differently in respect to the rest of the world because of their geographical locations, economic and social situations. The areas of science and technologies are not excluded from this general rule. Thus, the following question arises: How has the area of analytical X-ray analysis techniques developed in Latin America in respect to the developed countries?

A rough estimate of the number of X-ray "fluorescence" spectrometers in Latin America based on information from a questionaire sent to our contacts in the different countries is approximately four hundred (400). The distribution is shown in figure 1. The ratio of wavelength dispersive versus energy dispersive X-ray spectrometers is about four to one,

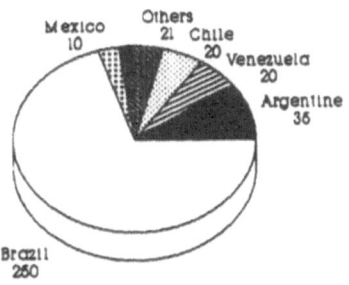

Fig.1. X-Ray Fluorescence Spectrometers in Latin America.

similar to developed countries. About 72% of the reported
X-ray spectrometers are employed for analytical services,
mostly in industrial and geological applications.

In general, the wavelength-dispersive spectrometers
are commerical integrated instruments while, the energy-
dispersive systems consists of various modular components.
The International Atomic Energy Agency (IAEA) has
maintained an X-ray fluorescence program since the early
1980´s in Latin America, in which it has supplied
equipment and training to those member countries which
have requested it. In general it started with
radioisotope excitation, followed by secondary target
excitation with conventional X-ray diffraction X-ray tube
systems to total reflection X-ray fluorescence with the
same X-ray diffraction X-ray tube systems.

The first "Seminario de Análisis por técnicos de
Rayos-X (SARX; The Seminar on X-ray Analysis Techniques)
was held in 1977 as an Argentine national meeting and has
now sucessfully completed seven conferences,
approximately one every two years, with a national
participation of more than one hundred (100) professionals
from research centers, universites, goverment and
industry. The last three meetings in 1985, 1987 and 1990
have also included the first three Latin American
Conferences on X-ray Analysis, but the participation from
members of others Latin America countries were very
limited, mostly from neighboring countries, because of
their geographical location, communication, economic and
social situations. But, the last meeting had the pleasure
of having a few experts from Europe and North America
(Canada and the United States of America)

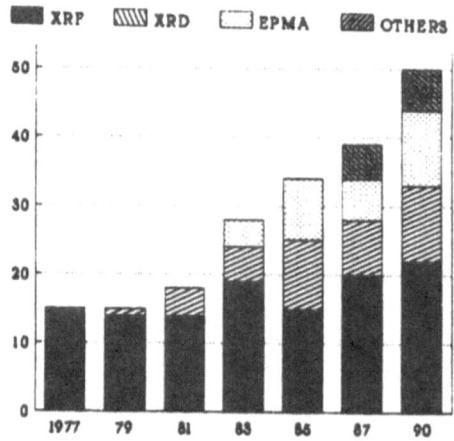

Fig.2. Papers presented in the SARX meetings classified by
       the different analytical X-ray techniques.

SARX for Latin America (at least the southern region) is similar to the Denver X-ray conference for the USA, the Annual X-ray Analysis Group meetings in Japan and the bi-annual meeting of the Australian X-ray Analytical Association in that it meets periodically in a region that is somewhat separated from the others as well as publishing the proceedings in its native language (Spanish). Finally, a histogram of the total number of papers presented in the SARX meetings discriminated by the different analytical techniques is presented in figure 2 in order to give some idea of how the different areas are developing.

BACKSCATTER/FUNDAMENTAL-PARAMETERS ANALYSIS OF UNWEIGHED SAMPLES USING

MULTI-TARGET, MULTI-CRYSTAL REGIONS OF INTEREST FROM WDXRF AND EDXRF

Richard J. Arthur and Ronald W. Sanders

Pacific Northwest Laboratory (PNL)
Richland, WA 99352

ABSTRACT

A method has been developed to simultaneously compute matrix corrections from a composite spectrum of multi-target energy-dispersive (EDXRF) and multi-crystal wavelength-dispersive (WDXRF) x-ray fluorescence systems. A serial line installed between the WDXRF and EDXRF spectrometers via a PDP 11/34a computer allows acquired wavelength data to be digitally transformed into an energy spectrum. The low-energy x-ray information from the WDXRF unit is then coupled with the backscatter coherent/incoherent information from the EDXRF unit, enabling enhanced quantitative analysis for low-atomic-number (low-Z) elements. The peak resolution obtainable from the WDXRF spectra often removes the necessity for peak-overlap corrections.

Backscatter intensities obtained from the EDXRF unit are used to provide information on total sample mass and to correct for matrix effects. The resulting backscatter fundamental-parameter (BFP) calculations generally provide an accurate analysis of samples without prior knowledge of the sample matrix. Such an approach is particularly useful for samples in which quantities of carbon, oxygen, and other low-Z constituents cannot be explicitly determined.

Regions of interest (ROI) are created by the computer code "PREP" and processed by the BFP code "MSAP" an extension of the "SAP3" computer program for quantitative multielement analysis by energy-dispersive x-ray fluorescence.[1]

INTRODUCTION

Some typical requirements for a BFP analysis are that all backscatter must originate from the sample, that the sample must be reasonably homogeneous, and that the coherent and incoherent backscatter peaks must be resolvable. Spectral peaks with no associated discrete coherent/incoherent backscatter peaks, such as those arising from WDXRF and from Ti fluorescence in EDXRF, are candidates for BFP analysis through formation of a composite spectrum.

Advances in X-Ray Analysis, Vol. 35
Edited by C.S. Barrett *et al.*, Plenum Press, New York, 1992

The BFP code "SAP3" has been used with success in evaluating multi-target
EDXRF spectra[2] to determine matrix correction, sample mass, and elemental
concentrations. Figures 1a and 1b illustrate the methodology involved in the
composite dual-spectra approach in which 2 EDXRF spectra are combined that
have been acquired from the same sample but that have different excitation
sources, Ti and Zr. Since the Ti contribution has no resolvable coherent-
incoherent backscatter peaks, "SAP3" uses only the scatter cross-sections
associated with the Zr backscatter peaks for the BFP calculations. The
combined results take advantage of features which are common to both analyses:
identical sample thickness and identical matrix composition, resulting in
similar x-ray absorption characteristics. Based on the success of the
composite dual-spectra approach, the BFP technique was extended to a composite
multi-spectra approach in which incoherent and coherent backscatter
intensities from a single excitation source (Zr K_α) define matrix corrections
for elements determined by WDXRF, using 4 crystals, and by EDXRF, using 3 or
more secondary targets.

The extended approach is intended for direct, multielement analysis of samples
without similar standards and is based on the previously reported BFP
method[1,2,3] for performing matrix corrections. Observed elemental masses (per
unit area) are first estimated from characteristic emission intensities. Then
the scatter contributions from the masses to the backscatter intensities are
determined. These contributions are subtracted from the scatter intensities,
and the difference is used to characterize the unobserved, light-element
(low-Z) component of the sample (H, C, N, O, etc.) Two representative light
elements are chosen from low-Z scatter intensity ratios, and their masses are
estimated by relating intensities to scatter cross sections. Matrix
corrections for self-absorption and enhancements are then computed from both
light- and heavy-element masses and are applied to emission and scatter
intensities for each element. The process of light-element selection and
matrix correction is repeated with the corrected intensities until no further
significant changes are obtained. Elemental concentrations and total sample
mass are then computed from the corrected light- and heavy-element masses.

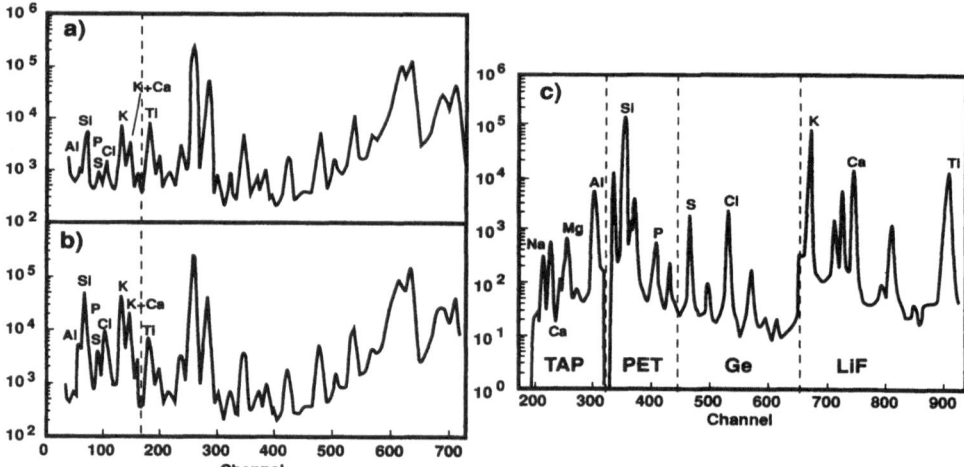

Figure 1. a) EDXRF spectrum of MESS-1 acquired under Zr secondary-source
excitation (0.025 keV/channel). b) Dual spectrum formed by combining the
spectrum in 1a) with the EDXRF spectrum acquired under Ti secondary-source
excitation. c) WDXRF spectrum of MESS-1 acquired using four dispersion
crystals: TAP, PET, Ge, LiF. The wavelength data has been digitally
transformed to form a high-resolution energy spectrum (0.005 keV/channel).

The BFP program "MSAP" was originally intended for EDXRF analyses employing Ti, Zr, Ag, and Gd targets. Although "MSAP" was partially developed in late 1983, it was never fully implemented, due, in part, to the success and versatility of the "SAP3" code. The "SAP3" approach, with the exception of the Ti-Zr composite spectrum, employs backscatter intensities from the actual secondary source providing the excitation energy. The "MSAP" code uses the backscatter intensities only from the Zr spectrum and operates against the ROI file prepared by the program "PREP" using a library, MULTSP.LIB, which provides the parameters to set gain and integrate peak areas. The ROI file is processed by "MSAP," using MULTSP.LIB, to provide matrix corrections and to determine corrected element concentrations. The reemergence of the code "MSAP" resulted from the acquisition of a wavelength spectrometer. Rate meter (2Θ) scans demonstrate the exceptional peak-resolution and intensities for low-Z elements when compared to EDXRF, as shown by comparing Figure 1c to 1a. The wavelength spectrometer was originally acquired for stand-alone development; however, the success of the "SAP3" dual-spectrum approach encouraged a limited "MSAP"-BFP evaluation by combining EDXRF spectra with a pseudo-energy spectrum composed of digitally transformed wavelength spectra.

The "MSAP" approach operates similarly to the dual-spectrum approach of "SAP3" in that two backscatter peaks from one analysis (Zr target) are used in the matrix corrections for all analyses. Also, elemental masses from observed characteristic peaks in all spectra are used to define scattering contributions to the backscatter peaks of the Zr secondary target. The main difference between single- and multisource fluorescence is the use of different excitation energies in computing the matrix corrections for characteristic emission peaks excited by targets other than Zr. The "MSAP" approach, like that of dual spectra, is of particular interest in extending the BFP method to element peaks that would not normally be candidates for BFP analysis.

EXPERIMENTAL

This paper evaluates a limited "MSAP" application of multi-crystal, multi-target, ROI-spectrum BFP matrix correction methodology with the analysis of four geological standard reference materials: USGS rock (BCR-1), USGS rock (G-2),[4] NRCC marine sediment (MESS-1) and NRCC marine sediment (BCSS-1).[5] Four dispersive crystals (PET, TAP, Ge, and LiF) are used for analysis of the light elements Na, Mg, Al, Si, P, S, Cl, K and Ca. The Ti peak is also acquired to set gain. Secondary targets for the EDXRF contributions are Zr, Ag, and Gd.

The samples are analyzed as unweighed, self-supporting pressed pellets. The dimensions of the prepared samples are 3.18 cm in diameter with an approximate thickness of 62 mg per cm². The samples are ground in a high density alumina mortar and pestle to a size of approximately 300 mesh. The resultant powder is pressed in a hardened carbon-steel die at 25000 kg. The prepared samples are placed in the four-position sample changer of the WDXRF excitation system and the sample chamber is placed under vacuum for spectrum acquisition. The samples are removed from the WDXRF system, mounted on 35-mm slides, and placed in the sample chamber of the EDXRF system under vacuum for excitation and spectrum acquisition.

EDXRF excitation utilizes Ti K_α/K_β (used in the "SAP3" comparison), Zr K_α/K_β, Ag K_α/K_β, and Gd K_α/K_β x rays from secondary sources excited by a tungsten x-ray tube. Acquisition is under control of the PNL-developed codes "XRF" and "MCA" using the parameters provided in Table 1. The x-ray detector is a 30-mm² Si(Li) diode with a resolution of 180 eV full-width at half maximum

Table 1.   EDXRF Acquisition Parameters

Secondary Source	Tube High Voltage (kV)	Tube Current (mA)	Acquisition Time (s)
Ti	40	10	375
Zr	40	10	3000
Gd	60	10	750
Ag	45	10	2000

Table 2.   WDXRF Acquisition Parameters

Analyzer Crystal	Detector High Voltage (kV)	PHS Setting (E_ℓ)	ΔE	2Θ Scan
TAP	1666	0.25	1.35	$60°$ – $34°$
PET	1554	0.25	1.35	$120°$ – $55°$
Ge	1542	0.20	1.60	$145°$ – $50°$
LiF	1464	0.20	1.45	$140°$ – $72°$

(FWHM) at 6.4 keV.   A multielement, thin-film sensitivity curve[1,3] was used as the primary elemental calibration.   The Zr backscatter peaks are both calibrated with sensitivity factors (intensity per unit mass per unit scatter cross section) as described elsewhere.[1]

WDXRF excitation uses a Cr x-ray tube operating at 30 kV and 20 mA.   The detector is a gas flow proportional counter.   The acquisition from each dispersion crystal is at a 2Θ scan rate of 0.05 with a readout setting of 2 s. Acquisition parameters are provided in Table 2.   A complete thin-film calibration was not performed due to time limitations so that much of the calibration curve is based upon the USGS BCR-1 standard.   Some thin-film standards were processed to evaluate higher order overlaps, such as third-order Ca on Mg.   The 2Θ versus intensity data is transfered via serial line using the PNL code "RDO" to create a temporary 2Θ spectrum on the storage disk of the dedicated EDXRF computer.   Each temporary spectrum is processed by the PNL code "DNO," which converts the 2Θ versus intensity spectrum to an energy-versus-intensity spectrum at a user-selected gain.

The library MULTSP.LIB used by the peak analysis program "PREP" and the BFP program "MSAP" was originally developed to process four different secondary excitation sources.   To facilitate data processing, a composite spectrum is formed using the WDXRF spectra.   The resulting multicrystal spectrum is used in place of the spectrum normally obtained by analyzing the sample using a Ti secondary source.   The splicing program "WXA" reads the four WDXRF spectra from disk and outputs selected regions to a single spectrum representing the desired contributions from each crystal.   The "TST" code reads the user-specified energy spectrum from disk and directs the peak analysis program "PREP" to use the library MULTSP.LIB to form a ROI file based on the associated excitation energy.   The resulting ROI file incorporates 1) number of counts observed in each element peak and its associated background integral, 2) live time, 3) real time, 4) tube current and voltage, and 5) values representing the coherent and incoherent scatter for both K_α and K_β excitation peaks.   Scatter cross sections contained in the library for observed and unobserved elements are for Zr excitation, since only the Zr backscatter intensities are used in computing the BFP matrix corrections.   The Zr K_β coherent/incoherent ROI are used for matrix definition to achieve slightly better resolution of the scatter intensities than is available for the Zr K_α coherent/incoherent ROI.   All scatter cross sections, photoelectric cross sections, and jump ratios are from McMaster et al.[6] and fluorescent yields are from Bambynek et al.[7]

Table 3. Summary of MSAP and SAP3 Results for Rock and Soils Samples with Reference Concentrations

REFERENCE SAMPLE		Na %	Mg %	Al %	Si %	P %	S ppm	Cl ppm	K %	Ca %	Ti %	V ppm	Cr ppm	Mn ppm	Fe %
USGS BCR-1	MSAP-BFP	2.5	2.05	7.65	25.8	0.21	507	105	1.47	4.96	1.33	386	<34	1441	9.5
USGS BCR-1	SAP3-BFP			8.41	26.2	<0.07	<200	<90	1.42	4.92	1.33	376	<33	1397	9.2
Ref. Conc.		2.5	2.09	7.20	25.5	0.17	570	99	1.41	4.95	1.32	399	7.6	1400	9.4
NRCC MESS1	MSAP-BFP	1.8	0.83	6.00	28.6	604	7125	8235	1.80	0.51	0.47	72	74	482	3.0
NRCC MESS1	SAP3-BFP			6.58	29.6	<500	8190	10740	1.75	0.48	0.49	67	67	486	3.0
Ref. Conc.		1.9	0.86	5.84	31.5	630	7200	8200	1.86	0.48	0.54	72.4	71	513	3.1
NRCC BCESS-1	MSAP-BFP	1.9	1.29	6.04	28.1	756	7096	10540	1.80	0.55	0.39	43	138	217	3.3
NRCC BCESS-1	SAP3-BFP			6.37	29.1	<470	4380	12550	1.65	0.50	0.45	105	139	255	3.6
Ref. Conc.		2.0	1.47	6.26	30.9	672	3600	11200	1.80	0.54	0.44	93.4	123	229	3.3
USGS G-2	MSAP-BFP	3.0	0.45	7.76	33.0	462	134	123	3.71	1.51	0.30	<50	<18	252	2.0
USGS G-2	SAP3-BFP			7.74	31.6	<500	<130	<90	3.26	1.42	0.31	<49	<19	252	1.9
Ref. Conc.		3.1	0.46	8.10	32.3	640	250	99	3.74	1.43	0.30	35.4	7	260	1.9

REFERENCE SAMPLE		Ni ppm	Cu ppm	Zn ppm	Ga ppm	Pb ppm	As ppm	Br ppm	Th ppm	Rb ppm	Sr ppm	Y ppm	Zr ppm	Nb ppm	Ba ppm	La ppm	Ce ppm
USGS BCR-1	MSAP-BFP	14	19	119	20	13	26.3	<.90	3.6	46.8	352	38	202	11	676	28	52
USGS BCR-1	SAP3-BFP	23	20	121	21	12	26.0	<1.0		47.1	328	38	190	13	673	<19	67
Ref. Conc.		16	18	120	20	18	0.79	0.15	6.0	46.6	330	37	190	14	675	26	54
NRCC MESS1	MSAP-BFP	29	24	168	11	37	9.1	54.4	3.9	91.7	82.9	33	524	14	233	29	60
NRCC MESS1	SAP3-BFP	32	26	178	12	35	10.6	93.7		93.7	87.3	39	582	17	361	32	74
Ref. Conc.		30	25	191		34	10.6				(89)						
NRCC BCESS-1	MSAP-BFP	59	18	101	9.9	24	10.4	75	3.6	72.0	110	20	250	14	252	25	54
NRCC BCESS-1	SAP3-BFP	65	22	117	12	23	12.1	101		85.2	99	25	244	12	291	<20	61
Ref. Conc.		55	19	119		23	11.1				(96)						
USGS G-2	MSAP-BFP	6.8	12	87	22	37	<1.6	<.64	14	173	430	8.8	291	8.9	1444	51	126
USGS G-2	SAP3-BFP	6.5	12	89	23	34	<1.6	<.78		174	458	11	320	10	1813	72	155
Ref. Conc.		5.1	12	85	22	31		0.3	12	168	479	12	300	14	1870	96	150

RESULTS AND DISCUSSION

Results for the rock and soils samples are presented in Table 3 in terms of analytical element concentration from "MSAP" and "SAP3" for comparison with the reference concentrations. The results for an non-optimized analysis are encouraging. Many of the heavier elements from the Gd target are below detection level. However, the low-Z elements obtained by wavelength dispersion, (Na, Al, P, S, K, and Cl) are intense and adequately resolved. Mg has a large third-order Ca interference and requires severe overlap corrections. Most elemental concentrations, however, agree within the statistical error of the peak analysis and the error associated with the reference material. Some notable exceptions to agreement are Si on the soil samples for both "MSAP" and "SAP3" and elements determined by "MSAP" using Ag and Gd excitation. The low Si values in the soil samples is probably explained by heterogeneity: examination of the samples under 5-power magnification reveals small reflective particles, probably associated with mica. The bias in Ag- and Gd-excited elements for "MSAP" is presently undetermined. The As value in the BCR-1 sample is attributed to another experiment not connected with this study and should be compared with the "SAP3" data only.

In summary, the performance for the WDXRF analysis is favorable and certainly demonstrates improved low-Z analysis over the limited range of this application. Thickness calculations over the small range evaluated are satisfactory. The MSAP-EDXRF analysis based solely on the incoherent and coherent backscatter intensities from Zr, unexpectedly, did not perform as well as "SAP3" analyses using the backscatter information obtained directly from the Ag and Gd targets. More evaluation is required for the technique (or variations of the technique) to be implemented routinely, including extending the analysis to significant weight variation of the sample material and to the analytical determination of tissue samples[3] and metals.[8]

ACKNOWLEDGMENT

The authors express appreciation to K. K. Nielson for implementing major portions of the "MSAP" code, to D. P. Brown for developing the electronic data transfer hardware, and to J. M. Deal for assisting in the review of the manuscript. Appreciation is also expressed to the PNL Analytical Chemistry Laboratory (XRF Center) for financial support. This work was supported by the U.S. Department of Energy under Contract DE-AC06-76RLO 1830. Pacific Northwest Laboratory is operated for the DOE by Battelle Memorial Institute.

REFERENCES

1. K. K. Nielson and R. W. Sanders, "The SAP3 Computer Program for Quantitative Multielement Analysis by Energy Dispersive X-Ray Fluorescence," PNL-4173, Richland, Washington (1982).
2. R. W. Sanders, et al. Anal. Chem. 55:1911 (1983).
3. K. K. Nielson and R. W. Sanders, Adv. X-Ray Anal. 26:385 (1983).
4. F. J. Flanagan, Geochim. Cosmochim. Acta 37:1189 (1973).
5. National Research Council of Canada, Marine Anal. Chem. Std. Prog. (1987).
6. W. H. McMasters, et al. "Compilation of X-ray Cross Sections," UCR-50174, Sec. II, Rev. 1 (1969).
7. W. Bambynek, et al. Rev. Mod. Phys. 44:716 (1972).
8. K. K. Nielson, R. W. Sanders, J. C. Evans, Anal. Chem. 54:1782 (1982).

L X-RAY LINE SHAPE OF COPPER(II) COMPOUNDS AND THEIR COVALENCY

Jun Kawai, Katsumi Nakajima,* Kuniko Maeda, and
Yohichi Gohshi*

*RIKEN (The Institute of Physical and Chemical Research)
Wako, Saitama, 351-01, Japan*

**Department of Industrial Chemistry, University of Tokyo
Hongo, Bunkyo-ku, Tokyo 113, Japan*

ABSTRACT——Profile changes of copper Lα (L$_3$-V) and Lβ (L$_2$-V) X-ray fluorescence spectra are interpreted. It is shown that the intensity of the high energy shoulder of the Cu L X-ray spectra has a relation to the intensity of the so-called 'shake-up' satellite in 2p X-ray photoelectron spectra; the high energy shoulder of the divalent copper Lα is weak for covalent compounds and strong for ionic compounds. It is also shown that the Lβ/Lα intensity ratio changes with the change of copper concentration of the analyte as well as the change of chemical state of the analyte.

RELATION BETWEEN Cu Lα X-RAY FLUORESCENCE SPECTRA AND Cu 2p X-RAY PHOTOELECTRON SPECTRA

It is believed that the copper Lα (L$_3$-V) X-ray fluorescence spectra (XRF) are related to the local (Cu atom) and partial (3d) densities of states (DOS) of the compounds. Figure 1 compares the measured Lα X-ray emission spectrum[1] of a copper oxide with that calculated by the local and partial DOS.[2] Since the DOS calculation by Redinger et al.[2] is quite reliable, the disagreement between theory and experiment in Fig.1 is mostly due to the fact that the Lα X-ray emission is not directly related to the local and partial DOS. Therefore we have proposed a theory[3] to interpret the Cu Lα line shape of various copper compounds based on the theory of X-ray photoelectron spectra (XPS) of van der Laan et al.[4] as follows.

The Cu 2p$_{3/2}$ X-ray photoelectron spectra (XPS) of divalent copper compounds (nominally 3d^9 electron configuration) have generally one main peak (at 933.6 eV) and one higher binding energy satellite (at 943 eV, so-called 'shake-up' satellite, since this satellite was believed to originate from electron

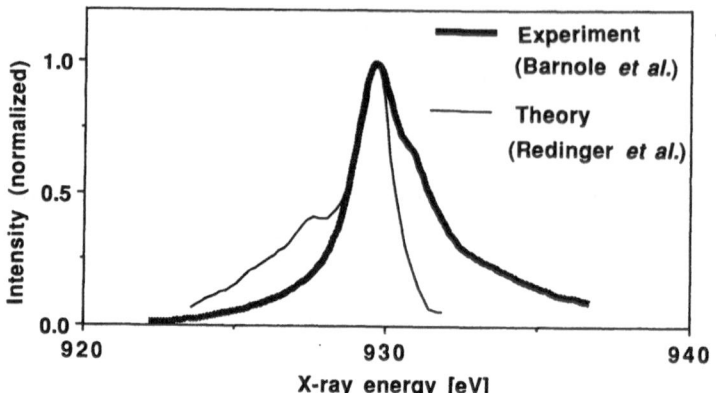

Fig.1 Comparison of Lα line shape of La₂CuO₄ between experiment (Barnole *et al.*[1]) and theory which has been calculated from the local and partial DOS (Redinger *et al.*[2]).

shake-up). After the study of van der Laan *et al.*,[4] it has been revealed that the main peak is, in fact, a charge-transfer satellite. That is to say, an electron is transferred from one of the ligands to the central Cu^{2+} ion at the moment ($<10^{-14}$ sec) of the 2p electron photoionization. On the other hand, the so-called 'shake-up satellite' is, in fact, the main line, *i.e.*, one of the 2p electrons is photoionized and the rest of the electrons in the vicinity of the photoionized Cu ion remains still.[5] The final states of the $2p_{3/2}$ XPS are the initial states of the Lα X-ray emission, thus we must include this charge-transfer effect for the line shape analysis of Cu Lα spectra. Figure 2 shows the relation between the 2p XPS and Lα XRF, which we have proposed in the previous paper.[3] The Lα main line (930 eV) of XRF corresponds to the main line of $2p_{3/2}$ XPS; the high energy shoulder (932-935 eV) of XRF to the so-called

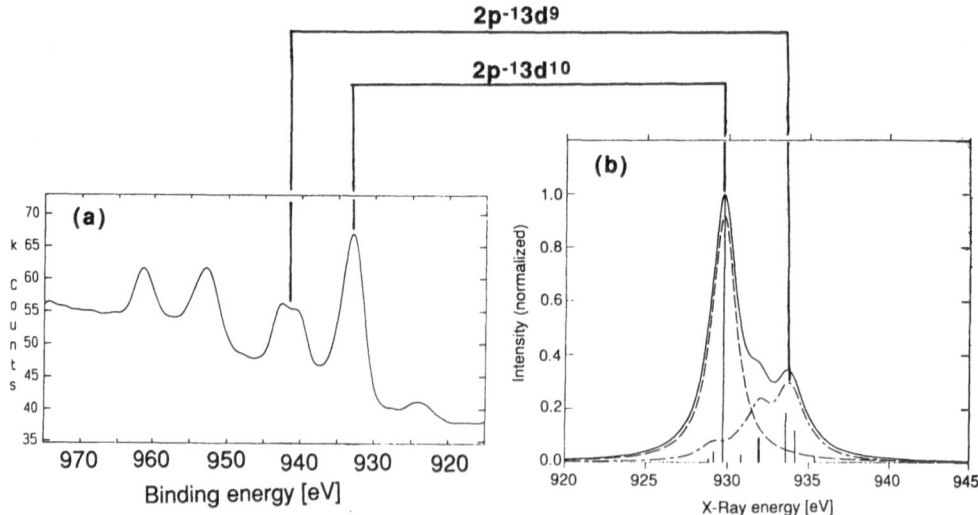

Fig.2 Relation between XPS and XRF. (a) Measured Cu 2p XPS spectrum of CuO. (b) Calculated Cu Lα XRF spectrum of CuO.

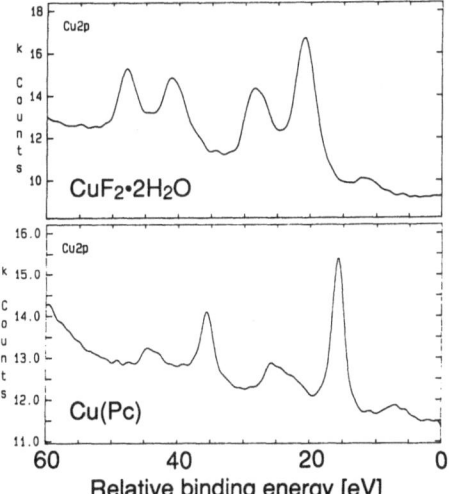

Fig.3 Measured Cu 2p XPS spectra of $CuF_2 \cdot 2H_2O$ and Cu-phthalocyanine. Though CuF_2 (nonhydrate) was also measured, the line shape was the same as that of $CuF_2 \cdot 2H_2O$, since the surface of the powder sample was hydrate. The spectra were smoothed by Savitzky-Golay method, but shifts due to sample charging were not corrected.

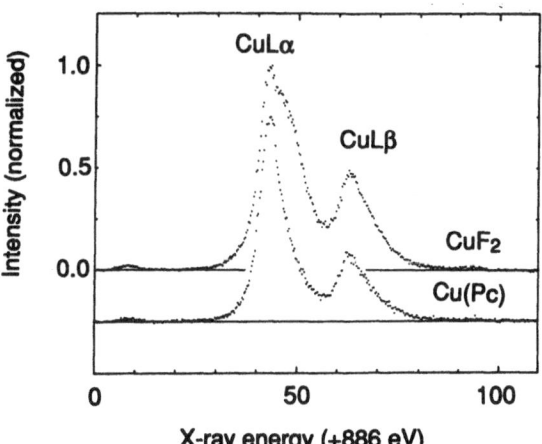

Fig.4 Measured Cu Lα,β XRF spectra of CuF_2 and Cu-phthalocyanine. Spectrum of CuF_2 (nonhydrate) was significantly different from that of $CuF_2 \cdot 2H_2O$; the high energy shoulder of Lα was much stronger for CuF_2 than for $CuF_2 \cdot 2H_2O$.

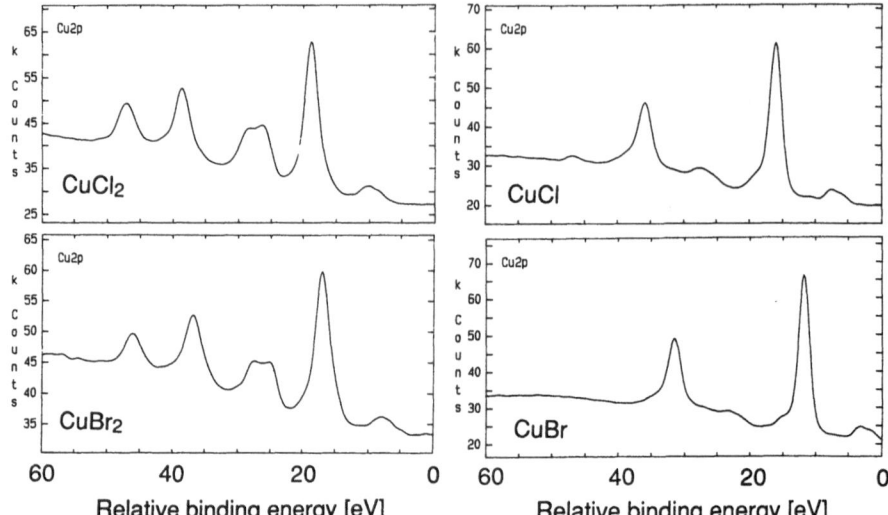

Fig.5   Measured Cu 2p XPS spectra of divalent and monovalent copper compounds. The samples were powder form. The spectra were smoothed by Savitzky-Golay method, but shifts due to sample charging were not corrected.

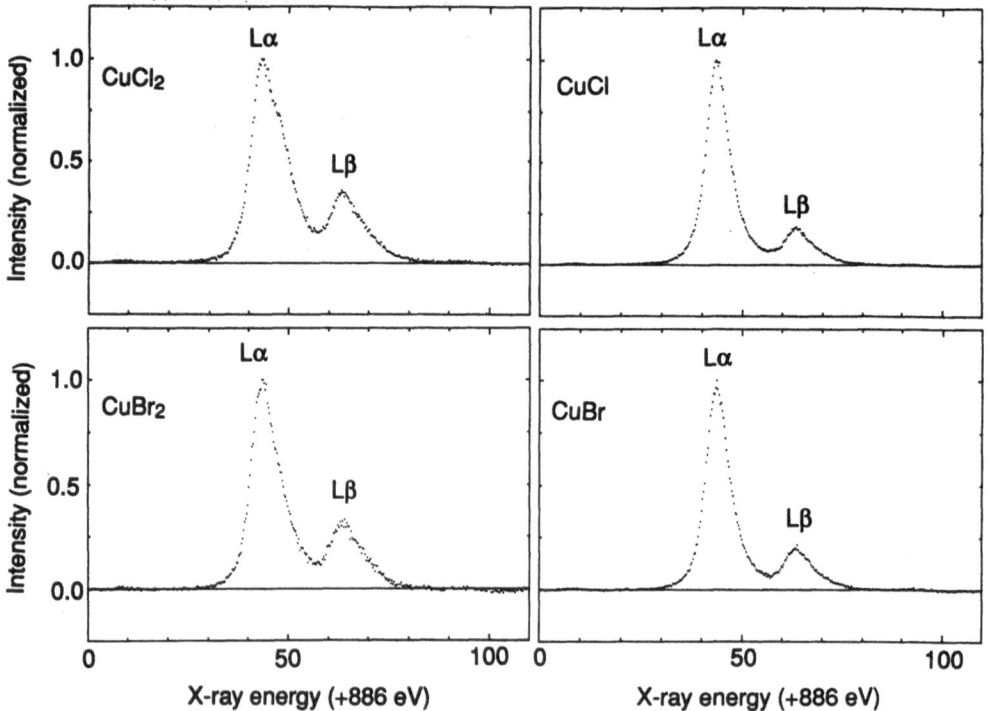

Fig.6 Measured Cu Lα,β spectra of divalent and monovalent copper compounds.

'shake-up' satellite of $2p_{3/2}$ XPS. It has been also predicted that the Cu Lα XRF spectra of ionic compounds (e.g. CuF_2) have a strong higher energy shoulder at 932-935 eV; the Lα spectra of covalent compounds [e.g. copper oxide superconductors and Cu-macrocyclic complexes such as copper phthalocyanine=Cu(Pc)] have a weak high energy shoulder, because the XPS satellite is strong for ionic compounds and weak for covalent compounds.

To check the above theory shown in Fig.2, we have measured both XPS and XRF spectra of divalent copper compounds (Figs.3-4). The XPS spectra were measured with a VG ESCA LAB MK II with Mg anode X-ray tube (15 kV, 10-20 mA). The XRF spectra were measured by a RIGAKU X-ray fluorescence spectrometer [TAP(100) (2d=25.76 Å) analyzing crystal and Rh anode X-ray tube with the applied power of 50 kV and 50 mA]. The samples were powder form.

It is found from Figs.3-4 that when the so-called 'shake-up' satellite is strong (this is the case for ionic compounds, see CuF_2 in Fig.3), then the high energy shoulder of Cu Lα XRF is also strong (see CuF_2 in Fig.4) and that when the so-called 'shake-up' satellite is weak [this is the case for covalent compounds, see Cu(Pc) in Fig.3], then the high energy shoulder of Cu Lα XRF is also weak [see Cu(Pc) in Fig.4].

We have measured both XPS and XRF spectra of mono- and divalent copper compounds as shown in Figs.5-6 to check again the validity of the theoretical prediction of Fig.2. It is found that the XPS spectra of monovalent copper compounds do not have the so-called 'shake-up' satellites as shown in Fig.5, and that the high energy shoulder of the Lα XRF does not exist for monovalent compounds, and vice versa. This again supports the validity of our theory shown in Fig.2.

Though the high energy shoulder also emerges due to an L_3V double hole initial state which is created by an $L_{1,2}L_3V$ Coster-Kronig transition,[6,7] this effect is negligible compared with the charge-transfer effect. Therefore since the X-ray fluorescence measurements are hardly affected by surface contaminations, the intensity ratio of the high energy shoulder (932-935 eV) relative to the main Lα peak (930 eV) provides information on the covalency, viz., similar information as XPS, of copper(II) compounds such as high T_c oxide superconductors.[8,9] It should be noted that Tanaka[10] and his coworkers[11,12] calculated the Lα line shape of copper oxides with a different method but reached a similar conclusion as described here, and that Hague et al.[13] characterized local structures of superconductors comparing the measured Lα line shape with that calculated by Tanaka et al.[12]

EFFECTS ASSOCIATED WITH SELF-ABSORPTION

Liefeld,[6] Holliday,[14] and Koster[15] studied self-absorption effects about 20 years ago. The Cu L_3 absorption maximum exists at the high energy shoulder of the Lα peak, therefore after the correction of the self-absorption effects, the spectral line shape is completely changed as well as the Lα peak maximum is shifted.

The spectra shown in Fig.6 are Cu Lα,β spectra of copper chlorides and copper bromides of both divalent and monovalent chemical states. The Lβ intensity ratio is stronger for divalent than for monovalent compounds. Therefore we can determine the valency by measuring the Lβ intensity ratio.

The X-ray absorption near edge structure (XANES) of Cu L3 is different among the monovalent, divalent, and metallic copper compounds.[16] The L3 XANES of the divalent copper compounds is not an absorption edge but a peak. This is due to 2p to 3d electron transition, since the 3d shell of divalent copper compounds has a hole. Therefore the relative absorption coefficient at the energy of the Lβ peak is lower for divalent compounds than that for the monovalent or metallic copper compounds. This is the reason why the Lβ intensity is weaker for monovalent copper compounds than for divalent copper compounds as shown in Fig.6.

We measured Cu Lα XRF spectra of copper alloys as shown in Fig.7 to study the concentration dependent self-absorption effect. The copper concentration was from 70 % to 100 % in weight. We find from Fig.7 that if the copper concentration is higher, then the relative Lβ intensity is lower. This is due to the self-absorption effect. If we measure an absorption-free spectrum, then the theory predicts the Lβ intensity to be one-half of the Lα intensity.

We have mentioned the effects associated with the self-absorption. These effects sometimes interfere with measuring the true line shape, but sometimes help us to characterize the copper compounds by line shape analysis.

CONCLUSIONS

In this paper, we have described, from both experiment and theory, and from both X-ray fluorescence spectra and X-ray photoelectron spectra, the implications of the Cu Lα and Lβ line shapes.

(i) The Cu Lα XRF spectra of Cu(II) compounds have a shoulder (932-935 eV) on the higher energy side of the Lα main line (930 eV). The Cu Lα XRF spectra of Cu(I) compounds do not have the high energy shoulder.

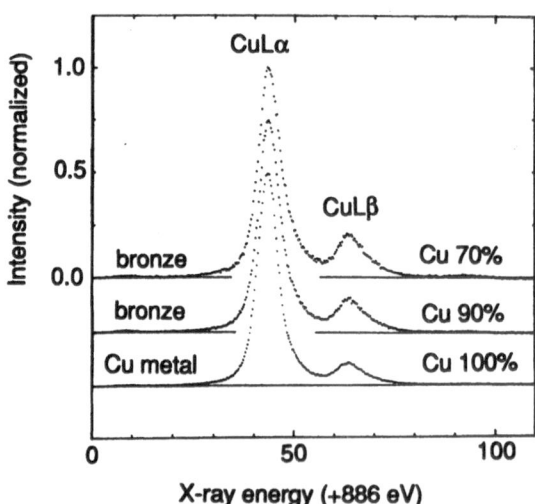

Fig.7 Measured Cu Lα,β spectra of bronze (Cu-Sn alloy) and copper metal. Bronzes were powder form.

(ii) The intensity of the high energy shoulder in XRF is roughly equal to the intensity of the so-called 'shake-up' satellite in $2p_{2/3}$ XPS.

(iii) The high energy shoulder in XRF is strong for ionic Cu(II) compounds, e.g., CuF_2, CuO, and $CuCl_2$; weak for covalent Cu(II) compounds, e.g., high T_c oxide superconductors and macrocyclic compounds such as phthalocyanine which has Cu-N bond. The intensity of the shoulder relative to the $L\alpha$ main peak is a measure of covalency of divalent copper compounds.

(iv) The Cu $L\alpha$ line width of Cu(II) compounds is wider for ionic compounds; narrower for covalent compounds. The Cu $L\alpha$ line width of Cu(I) compounds and metallic copper alloys is very narrow.

The following four are due to the self-absorption effects.

(v) The intensity ratio of $L\beta/L\alpha$ is different from 1/2.

(vi) The intensity ratio of $L\beta/L\alpha$ is large when the copper concentration is low; $L\beta/L\alpha$ small for copper concentration high.

(vii) The intensity ratio of $L\beta/L\alpha$ is large for Cu(II) compounds, whereas small for Cu(I) compounds.

(viii) The Cu $L\alpha$ and $L\beta$ line shapes are modified due to self-absorption since the absorption coefficient is jumping on the high energy shoulder of both $L\alpha$ and $L\beta$ lines.

ACKNOWLEDGEMENTS——Thanks are due to Dr. C. F. Hague, Prof. A. Kotani and his group, and Dr. S. Tanaka for fruitful discussions as well as sending us results prior to publication, and due to Ms. A. Nakao for technical support for the ESCA measurements. A part of the present work was done by Special Researchers' Basic Science Program, RIKEN.

REFERENCES

1. V. Barnole, J. -M. Mariot, C. F. Hague, C. Michel, and B. Raveau, *Phys. Rev.*, **B41**, 4262 (1990).
2. J. Redinger, J. Yu, A. J. Freeman, and P. Weinberger, *Phys. Lett.*, **A124**, 463 (1987).
3. J. Kawai and K. Maeda, *Spectrochim. Acta*, **46B**, 1243 (1991).
4. G. van der Laan, C. Westra, C. Haas, and G. A. Sawatzky, *Phys. Rev.*, **B23**, 4369 (1981).
5. J. Kawai, *Adv. X-Ray Anal.* **34**, 91 (1991).
6. R. Liefeld, *Soft X-Ray Band Spectra and the Electronic Structure of Metals and Materials*, Ed. D. J. Fabian, p.133, Academic, London (1968).
7. J. Kawai and Y. Gohshi, *Spectrochim. Acta*, **41B**, 265 (1986).
8. J. Kawai and K. Maeda, *Physica C* (Proc. M^2S-HTSC III), (in press).
9. K. Nakajima, J. Kawai, and Y. Gohshi, *ibid* (in press).
10. S. Tanaka, Doctoral Thesis, Tohoku University (1991).
11 S. Tanaka, K. Okada, and A. Kotani, *J. Phys. Soc. Jpn.*, **58**, 813 (1989).
12. S. Tanaka, K. Okada, and A. Kotani, *Physica C* (Proc. M^2S-HTSC III), (in press).
13. C. F. Hague, J. -M. Mariot, V. Barnole, and C. Michel, *ibid* (in press).

14. J. E. Holliday, *Adv. X-Ray Anal.*, **14**, 243 (1971).
15. A. S. Koster, *Mol. Phys.*, **26**, 625 (1973).
16. M. Grioni, J. B. Goedkoop, R. Schoorl, F. M. F. de Groot, J. C. Fuggle, F. Schäfers, E. E. Koch, G. Rossi, J. -M. Estava, and R. C. Karnatak, *Phys. Rev.*, **B39**, 1541 (1989).

A XANES STUDY OF SQUARE COPPER(II) COMPLEXES

Hisanobu WAKITA[1], Toshio YAMAGUCHI[1], Hirohiko ADACHI[2],
Manabu FUJIWARA[3],and Seiichi YAMASHITA[4],

[1]Department of Chemistry, Fukuoka University, Fukuoka, Japan
[2]Hyogo University of Teacher Education, Hyogo, Japan
[3]Department of Material Chemistry, Ryukoku University, Shiga
Japan
[4]Asahi Chemical Industry Co., Ltd., Shizuoka, Japan

ABSTRACT

The XANES (X-ray absorption near edge structure) spectra of
copper(II) ions in solid state and in solution of the square-planar
copper(II) complexes with tetraaza macrocycles were measured. The
peaks in the measured XANES spectra shifted to lower energy side with
increasing the electron density of central copper(II) ions. The
molecular orbital calculations for the complexes were carried out by the
DV-Xα method, and the theoretical XANES spectra were estimated. The
clear chemical shift obtained by this XANES study is evaluated and leads
to a new concept of π-back donation between the copper(II) complexes and
counter anion in aqueous solution.

Key words: XANES, tetraaza macrocycles, copper(II) complexes, DV-Xα
calculation, chemical shift

INTRODUCTION

At the rising part of the X-ray absorption edge, X-ray absorption
spectrum can be affected by the electronic state of the absorption atom
in the compounds. In the present report the coorelation between the
XANES spectra of the complexes and the electron-donating abilities of
the macrocyclic ligands in used complexes was studied. In addition,
some peaks in XANES spectra were assigned on the basis of the results of
the molecular orbital calculation using the DV-Xα method. The compounds
in this paper (Fig. 1 and 3) have four-coordinated, stable square-planar
structure in both crystal and solution. The chemical shift obtained by
this XANES study was evaluated and applied to get a new concept of π-

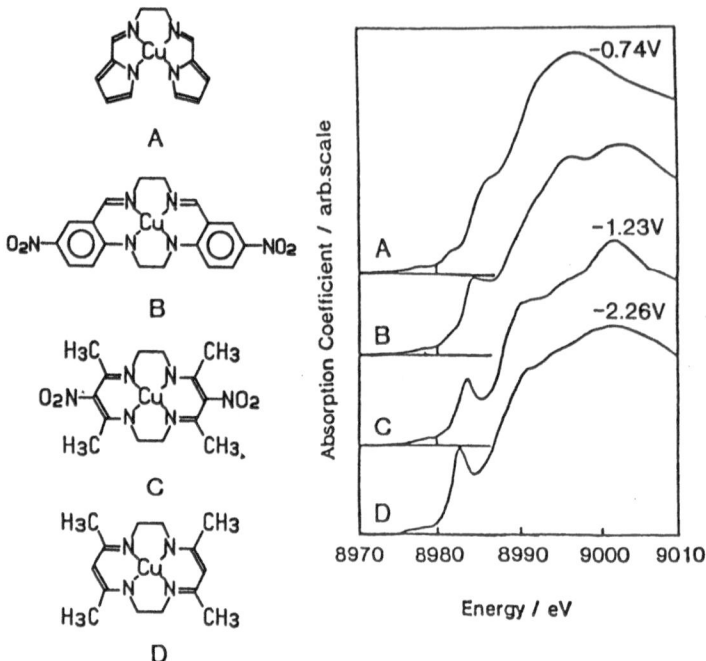

Fig. 1.   Tetraaza macrocyclic Cu(II) complexes, the reduction potential
values, and the XANES spectra in solid state.

back donation between the copper(II) complexes and counter anion in
solution.

EXPERIMENTAL

X-Ray Absorption Measurements   X-Ray absorption measurements for
solid samples were performed at the BL-7C station of the Synchrotron
Radiation(the Photon Factory) in the National Laboratory for High Energy
Physics (KEK).[1]   In case of solution samples the measurements were
performed by a laboratory EXAFS spectrometer made by us.[2]   In both
cases X-Ray absorption data were collected in the range from ca.9020 eV
around copper K-edge (8981.83 eV).   The time to collect X-ray photons
was 2 seconds at each measuring point.

Reduction Potential Measurements   The reduction potentials of
complexes A, C, and D in Fig. 1 and all complexes in Fig. 3 were
measured.   These values were obtained in DMF for A, C, and D of Fig. 1
and in water for the samples of Fig. 3 by the polarographic method with
a dropping mercury electrode and a mercury pool as the working and
counter electrodes at 25℃.

Molecular Orbital Calculations   Molecular orbital calculation by
using the DV-Xα method for the tetraaza macrocyclic copper(II) complexes
in Fig. 1 were performed with an FACOM M-780/10S computer.   The

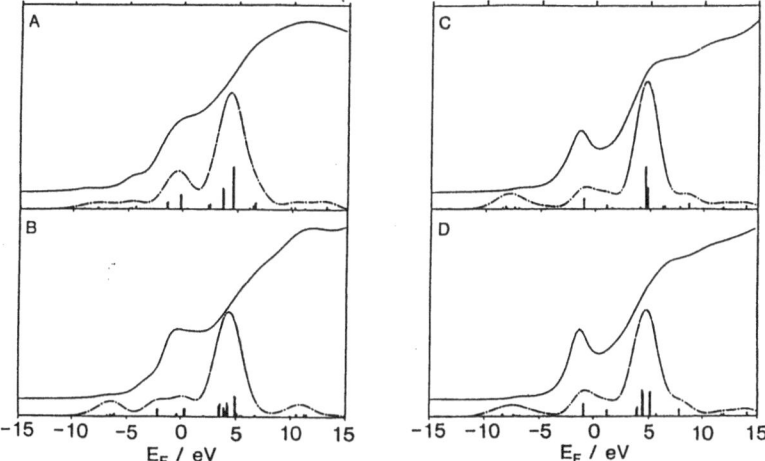

Fig. 2.   Theoretical(dashed and dotted line) and measured(solid line)
XANES spectra of the Cu(II) complexes in Fig. 1.

reliable crystal parameters of the molecular structures of the
copper(II) complexes have not yet been reported.   Therefore, their
atomic positions of the macrocyclic ligands were estimated from similar
structures in reported analogous complexes.[1]   For the Cu-N bond
distances within the copper(II) complexes was used an average value (1.98
Å) obtained from XAFS analysis for the macrocyclic copper(II) complexes.

The point group symmetries are C for complexes A and B,and D for
complexes C and D (Fig. 1).   Numerical atomic orbitals taken into
account in the calculations are 1s to 4p orbitals for Cu atom, 1s to 2p
orbitals for C, N, and O atoms, and 1s orbital for H atom.   The DV-Xα
calculations were repeated until the difference of charge densities
became less than 0.01 electron, where the charge densities were
calculated from the Mulliken population analysis.   The values of the
ionization energies of 2p orbitals for N and Cu atoms were calculated
from the DV-Xα calculation and compared with those measured by XPS
measurements.   The calculated values agreed well with the observed
values(Fig. 2).

RESULTS AND DISCUSSION

For the tetraazacopper(II) complexes, the reduction potentials of
complexes A, C, and D are given in Fig. 1.   As is seen, all
values are negative and become to more negative in order of A>C>D; these
results indicate that the electron donating ability of the ligands to a
central copper(II) ion is stronger in the order of A<C<D.   The XANES
spectra of complexes A, B, C, and D are also shown in Fig. 1.   Three
peaks are recognized in all spectra.   The energy values of the first
peak are almost the same.   For the second peak, however, the peak
intensities become smaller in the order of A<B<C<D, and the peak

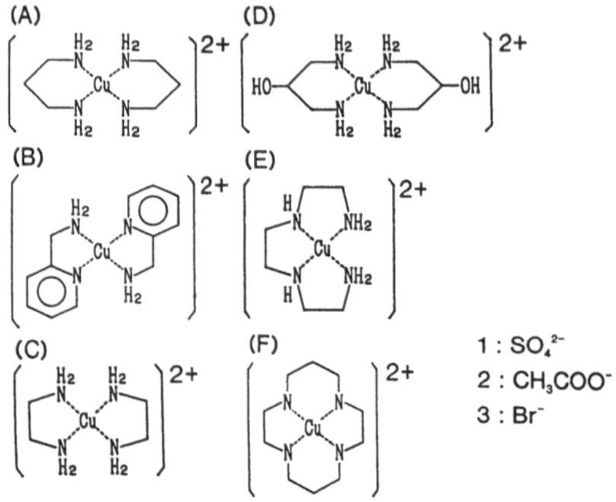

Fig. 3.   Tetraaza macrocyclic Cu(II) complexes for solution samples

positions of A, B, C, and D shift (called the XANES shift in this work)
to the lower energy side from A to D.   The third peak of A, B, C and D
also shifts to lower energy from A to D and the peaks become sharper.
This tendency corresponds to the result of the electron donating ability
of the ligand to its central copper in complexes from A to D.   The
third peaks in the XANES spectra shift to lower energy region with
increasing electron donating ability of the ligands, that is, with
increasing the electron density of a central copper(II) ion.

     To perform more detailed invesitigation for the XANES peaks,
molecular orbital calculations of the complex molecules were carried
out, and the first, second and third peaks in the XANES spectra were
assigned by comparing them with the corresponding calculated XANES
spectra.   From the results of DV-Xα calculations of the complexes, some
energy levels of calculated molecular orbitals and their atomic orbital
components are obtained.   To estimate the probabilities of the 1s
electron transition by an electronic dipole theory, theoretical XANES
spectra could be produced using molecular orbital calculations shown in
Fig. 2.   The peak positions in the theoretical XANES spectra were
fitted well with those of observed ones in all cases.   The peaks of the
measured XANES spectra could be assigned to the electron transition
mainly from Cu1s orbital to Cu4p and/or Cu4s orbitals from comparison
with the theoretical spectra.   The values of the atomic-charge
densities of the copper atoms and of the bond-charge densities of the
Cu-N bonds in the complexes were also calculated and are consistent well
with their peak positions in the measured XANES spectra.

     Comparing to this XANES shift, the spectral peaks measured by the
fluorescent X-ray and ESCA methods for the central copper(II) ion in
these complexes did not show any noticeable shift.   This shows that the
XANES spectroscopy is expected to be used as an analytical method for
the characterization of the chemical state of the X-ray absorbing atom.

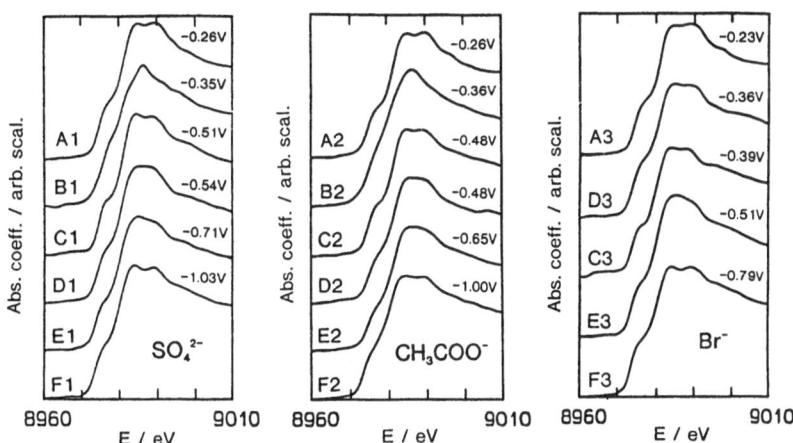

Fig. 4. Tetraaza macrocyclic Cu(II) complexes, the reduction potential values, and the XANES spectra in aqueous solution.

In case of solution samples in Fig. 3, we also measured XANES spectra and reduction potentials using each aqueous solution sample which has three different counter anion, bromide, acetate, and sulfate anions(Fig. 4). The values of the reduction potentials obtained(Fig. 4) are negative and become   more negative in the order of A to F for any case; these results indicate that the electron donating ability of the ligands to a central copper(II) ion is stronger in the order of A to F. The XANES spectra of complexes are also shown in Fig. 4. Three peaks are recognized in all spectra. The energy values of the first peak are almost the same. For the second peak, however, the peak intensities become smaller in the order of A to F, and the peak positions shift (called the XANES shift in this work) to the lower energy side from A to F. The third peak of the complexes also shifts to lower energy from A to F and the peaks become sharper. This

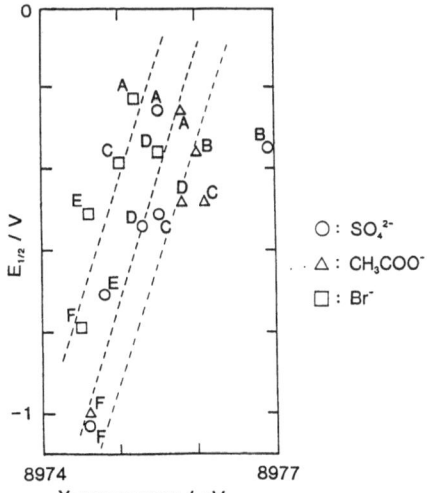

Fig. 5   The relationship between the XANES peak positions and the reduction potential values

tendency corresponds to the result of the electron donating ability of the ligand to its central copper in complexes from A to F.   The third peaks in the XANES spectra shift to lower energy region with increasing electron donating ability of the ligands, that is, with increasing the electron density of a central copper(II) ion.

The second XANES peak positions in the XANES spectra of solution samples were potted for their reduction potential values in Fig. 5.   As seen in Fig. 5, the second peaks shift to lower energy region with increasing electron densities of the central copper(II) ions, especially, with increasing electron densitites of the axially directed orbitals, p orbitals because of the MO analysis mentioned above.   This phenomenon is called as ?-back donation.   But interestingly, in case of the complexes countered by bromide anions electrons are going out from the axially directed orbital, ?-back donating from the orbital comparing to the complexes which have another counter anions.   This phenomenon is also called as ?-back donation, but there is no report that the ?-back donation is observed between cationic and anionic species.   This phenomenon was observed by precise measurements of XANES spectra.

REFERENCES

1. S.Yamashita, Y.Yanase, T.Yamaguchi, and H.Wakita, Bull. Chem. Soc. Jpn., 62, 2902-2907(1989).
2. S.Yamashita, K.Taniguchi, S.Nomoto, T.Yamaguchi, and H.Wakita, X-Ray Spectroscopy, in press.

COMPARISON OF ELEMENTAL SENSITIVITIES INDUCED BY RADIOISOTOPE AND SECONDARY TARGET EXCITATION FOR SIMULTANEOUS MULTIELEMENT X-RAY ANALYSIS

J.J. LaBrecque*

Atomic and Nuclear Spectroscopy Laboratory
Instituto Venezolano de Investigaciones
Cientificas-IVIC
Apartado 21827, Caracas 1020A, Venezuela

SUMMARY

The major advantage of secondary-target X-ray fluorescence analysis is the enchancement of sensitivity and detection limit of an element or small group of elements with similiar atomic numbers by selecting a target material which produces X-rays slightly higher than the absorption edges of the analyte of interest. Secondary target excitation systems are, however, being employed for simultaneous determinations of a large range of elements. Thus, it was decided to compare the application of a simple secondary target X-ray fluorescence system with excitation from a Cd-109 annular radioisotope source.

A simple secondary target X-ray fluorescence system was constructed and optimized which operates at less than 800 kilowatts. The "relative" elemental sensitivities of this system and a radioisotope system with a Cd-109 annular source were compared by the analysis of the 1-2-3-multi-spectral standard from Chemplex which contains 1.23% of 53 elements. The characteristic X-rays were collected

*Partial Travel funds to assist the Pacific-International Congress on X-Ray Analytical Methods (PICXAM) was granted from the Consejo Nacional de Investigaciones Científicas y Tecnológicas (The Venezuelan National Science and Technological Foundation).

with the same Si(Li) detector and analyzed with the same
Apple IIe microprocessor with a Nucleus ADC/interface
card. Finally, the secondary target system described
herein with a molybdenum X-ray tube and zirconium or
cadmium target produced similiar elemental sensitivities
as a Cd-109 annular source of about 7.5 mCi and 5.0 mCi
intensity respectively.

INTRODUCTION

     Secondary target X-ray fluorescence (STXRF) not to be
confused with source-tuned X-ray fluorescence (1) can
greatly increase selected elemental sensitivities and
detection limits by maximizing the element´s fluorescence
by employing an optimium excitation energy of secondary
X-rays. This is accomplished by selecting the appropiate
elemental secondary target, more specifically by choosing
one in which the secondary target X-ray energies are
slightly higher than the absorption edges of the elements
of interests. For example, the determination of sodium and
magnesium employing an aluminium secondary target. Even
the high energy K- X-rays of Cs, Ba, La, Ce and Pr can be
excited and measured when the high energy K- X-rays from a
gadolinium secondary target are employed, this has
important applications in the analysis of catalysts,
glasses and marine sediments.

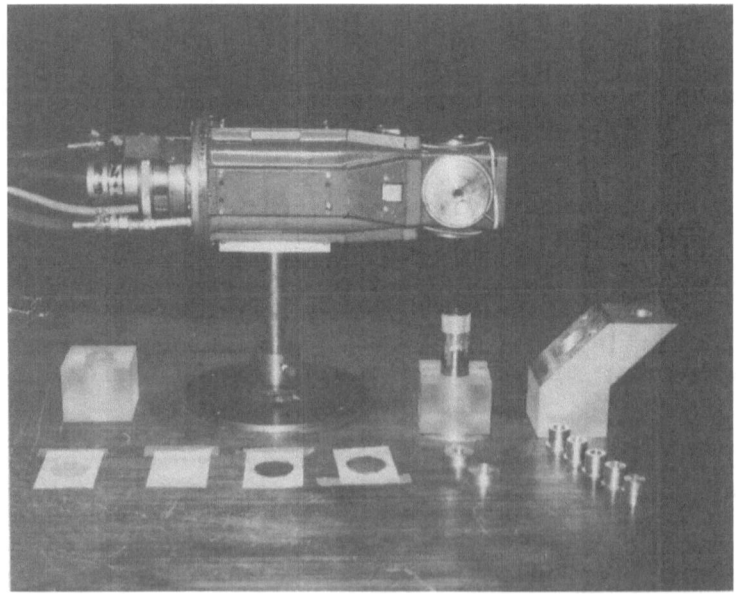

     Fig. 1. A composite photo of the X-ray coupler
               with its collimators, the specimens, the
               X-ray tube and Si(Li) detector head.

Unlike, other forms of elemental analysis such as wet chemical methods, atomic absorption and emission spectroscopy, the sample is not required to be decomposed to a liquid form and is also based on the analysis of all the elemental X-rays in the spectrum, in other words it is not only a multielemental technique but is also a simultaneous analysis. Finally, this technique involves little if any sample preparation and can even be truly non-destructive and employed in in-vitro analysis, such as lead in human bones (2).

Actually, STXRF is similiar to radioisotope excitation (3) in which the background is almost eliminated completely but the exciting X-rays can be selected from any of the elemental target rather than a few selected radionuclides such as, Fe-55, Cd-109, Am-241.

In this work, we have designed and studied the optimization of a simple, economic apparatus for coupling the X-ray tube, the secondary target and the detector for STXRF as well as comparing the relative elemental sensitivities of this system with a conventional Cd-109 annular source.

EXPERIMENTAL

The coupler for STXRF

The coupler was machined from commercial aluminium not pure aluminium. The housing was manufactured from a 75 mm x 75 mm bar approximately 120 mm long, cut into two pieces at 53 mm with a 45o angle. The collimators were also fabricated from commercial aluminium rods. The diameters for the holes (canals) between the secondary target and sample (C1) were 8, 10, 12 and 15 mm, while those of the collimator (C2) between the sample and the detector were 12 and 14 mm. Finally, the distance between the sample and the detector face was varied by using three plexglass blocks with a 30 mm diameter aperture in which the detector head was placed; the blocks were located between the coupler's base and the table which surrounds the detector head (see figure 1).

The excitation system

The X-ray generator was a Philips PW 1010/80 model with currents up to 54 kV at 1000 W, previously used for X-ray diffraction analysis. The X-ray tube had a molybdenum anode and was housed in a Phillip PW 1316 tube shield and supported horizontally as seen in figure 1. It was operated at 40 kV and 16mA for both the zirconium and cadmium secondary target.

A weak Cd-109 annular source (about 2 mCi) was employed for the radioisotope excitation with the distance between the sample and detector face similiar to those in the STXRF coupler.

The detection system

The characteristic X-rays from the sample were collected by a Si(Li) detector with a resolution of about 155 eV at 5.9 keV. The signals from the amplifier were processed by an Apple IIe microprocessor with a Nucleus interface ADC card. This detection system has been fully described elsewhere (4). The same Apple IIe microprocessor was also used for the data analysis with an X-ray software package supplied by Dapple Systems.

The X-ray measurements

The 1-2-3 standard from Spex industries which contains 1.23 wt. % of 53 different elements was used to determine the relative sensitivities for the different combinations of collimators and the distance between the sample and detector face. This standard was spread smoothly between pieces of Scotch Magic Tape supported by a 50 mm x 50 mm piece of card with a 35 mm aperture. Similiarly, a blank was prepared and measured. All measurements were for 250 seconds of live time.

RESULTS AND DISCUSSION

After measuring the blank for the forty-five different combinations for collimator (C1): between the secondary target and sample, collimator (C2): between the sample and the detector, and the distance between the sample and detector face, only eight combinations were found acceptable in which no characteristics X-rays from the coupler or scattered X-rays from the anode of the X-ray tube or secondary targets were seen in the spectra. The relative sensitivities for these eight combinations for Rb (the largest Kα peak) from the zirconium target excitation is given in Table 1, as well as those from the cadmium secondary target. The peak intensity for rubidium from the cadmium target excitation was only 66.8% compared to the zirconium target excitation giving an idea how the elemental fluorescence can be increase by employing a secondary target that produces X-rays more closer to the absorption edge of the element of interest. It can readily be seen that the optimium combination for the sensitivity for Rb as well as the other elements for both the Zr and Cd secondary target excitation is when the collimator (C1) is 12 mm, the collimator (C2) is 12 mm and the distance between the face of the detector and sample is 200 mm. As the distance between the sample and detector face is increased to 250 mm, the relative sensitivities are reduced by about 50% while they are slightly reduced when the distance is decreased to 150 mm.

Table 1. The relative sensitivities for Rb from the Zr and
Cd secondary target excitation for the acceptable
combinations of collimators (C1) and (C2) and the
distance (D3)

Collimator (C1) (mm)	Collimator (C2) (mm)	Distance (D3) (mm)	Relative Sensitivities for Rb %	
			Zr-target	Cd-target
8	12	150	51.4	54.6
10	12	150	62.1	68.3
10	14	150	75.2	82.3
8	12	200	53.2	60.9
10	12	200	76.2	82.0
12	12	200	100	100 (66.8)*
8	12	250	34.1	39.9
10	12	250	51.8	59.7

* Normalized to the Rb sensitivity for the Zr-secondary target
excitation.

The area of the secondary target that is excited by
the X-ray beam from the X-ray tube was measured to be an
ellipse of 10 mm x 15 mm by placing an X-ray film in the
position of the secondary target. Similiarly by placing
other films in the sample position it was found out that
the ellipse area of the excited part of the sample was
13 mm x 18 mm, when the (C1) collimator was 8 mm;
16 mm x 12 mm when C1 was 10 mm and; 20 mm x 27 mm when
C1 was 12 mm. The total sample area available for
excitation is an ellipse of 30 mm x 40 mm.

The relative sensitivities for the other elements
from the excitation of the Zr and Cd secondary target are
presented in Table II for C1=12 mm, C2=12 mm and D3=200
mm. They were similiar for the other acceptable
combination too. For Rb and the lower Z-elements the
sensitivities for the Cd secondary target excitation was
less than those of the Zr-secondary target excitation,
which is suppected since the Cd K-rays are further away
from the absorption edges of these elements than the
Zr-X-rays. Finally, a comparison of the relative sensitivities for a Cd-109 radioisotope source excitation is
given too in Table II. The relative sensitivity the
Rb K-X-rays from the Zr and Cd secondary targets is
comparable to a Cd-109 radioisotope source of about
7.5 mCi and 5.0 mCi respectively for this simple secondary
target X-ray fluorescence system.

Table 2. Relative sensitivities for the other elements in the 1-2-3 standard for the Zr and Cd secondary target excitation for Cl=12 mm, C2=12 mm and D3=200 mm combination and a Cd-109 radioisotope source with D3=200 mm

Element	Relative Sensitivities		
	Zr-target (%)	Cd-target (%)	Cd-109 (%)
V	4.34	5.7	-
Cr	6.97	8.4	6.5
Mn	7.28	8.7	7.4
Fe	12.2	14.8	15.9
Co	20.6	18.6	16.0
Ni	19.8	19.7	18.0
Cu	27.9	30.2	25.8
Zn	37.6	37.3	32.5
Ga	48.0	47.5	38.9
Ge	80.9	66.6	63.3
As	59.1	64.0	61.8
Se	53.1	51.1	55.6
Br	93.7	87.5	80.6
Rb	100	100*	100
Sr	-	99.4	88.1
Y	-	28.0	25.7
Zr	-	109.2	104.1
Nb	-	117.4	110.3
Mo	-	160.2	141.8

* Again it should be noted the Rb sensitivity relative to the Zr target is only 66.8%

REFERENCES

1) K. Stehr and J. Bogert, "Source-Tuned XRF spectrometry with light element analysis", American Laboratory, March, 1988.

2) P.A. Pella and C.G. Soares, "Secondary Target X-ray Excitation for In Vivo Measurement of Lead in Bone", presented at the 39th Annual Denver Conference on Applications of X-ray Analysis, July 30-August 3, 1990, Steamboard Springs, Colorado, USA.

3) J.J. LaBrecque, "Radiosiotope induced X-ray fluorescence", Progress in Analytical Atomic Spectroscopy, 4, (1981) 191-217.

4) K. Borowski, I.L. Priess, J.J. LaBrecque and C. Pauley, "Use of an Apple II plus microcomputer as a multichannel analyzer for X-ray fluorescence spectrometry", Computer Enhanced Spectroscopy, 1, (1983) 99-105.

THE STUDY OF VALUABLE MEDALS USING X-RAY ANALYSIS

Dudley Creagh

Department of Physics, University College, The University of New South
Wales, Australian Defence Force Academy, CAMPBELL ACT 2600
AUSTRALIA

ABSTRACT

The exact compositions of the Victoria Crosses held in custody by the
Australian War Memorial were needed so that its conservation staff could
formulate a proper strategy for their conservation. This analytical study not only
provided the required information but it resolved important ambiguities in the
historical record.

INTRODUCTION

Medals and coins are collectable items and as such can have considerable
monetary value. A large number of private collectors and museums vie for the
acquisition of important items and it is essential therefore that these can be
authenticated in some way. Once authenticated and acquired it is equally as
important that they are preserved in a suitable manner: proper conservation
measures must be undertaken to prevent their deterioration.

Authentication of the provenance of metallic objects can often be performed
using x-ray analytical techniques once independent historical provenance has been
established for objects of the same class. Hence it is possible to date, for example,
Imperial Roman bronze coins through their zinc content[1] and Indian
subcontinent bronze objects through their zinc-to-tin ratio[2].

Medals acquire a value, often much greater than the value they would have
by virtue of age or metal content, through their association with outstanding feats
of heroism. Medal collectors collect as much for the association with a specific act
of valour as for the rarity of a type of medal.

Although there are many private collectors, a considerable proportion of
medals of value are held in museum collections. One such collection is the

collection of Victoria Crosses which is held by the Australian War Memorial, Canberra, Australia.

The Victoria Cross (Figure 1) is the highest award for bravery in the British Commonwealth and the Australian War Memorial holds 52 of the 1352 of the medals awarded since the creation of the award by Queen Victoria in 1856. This is the largest collection of these medals in the world.

Because of the value of the collection it was thought necessary that a proper metallurgical investigation of the medals be undertaken so that a proper conservation strategy be put into effect. As with all scientific investigations of museum artefacts it is folly not to relate the scientific investigation with historical and stylistic information.

In every respect the Victoria Cross is unique among medals. Medals are usually cast or struck using materials which are easily worked. Medal materials range from gold alloys, through silver alloys to bronzes and cupro-nickels. Because these materials are similar in composition to coinage the production of medals has frequently been the responsibility of mints in their country of origin and over time there are usually a number of changes in manufacturing agencies. The design of the medal is often the work of an important sculptor or silver-smith.

Not so the Victoria Cross: the material from which they are fabricated has its origins in cannon captured from the Russians by the English at Sebastapol during the Crimean War. The design is reputed to be largely the work of Prince Albert, Queen Victoria's Consort. Certainly Queen Victoria had a significant role in the design of the medal, which of course, bears her name. Finally the jeweller responsible for its production has been the same since its first award, namely, Hancocks and Company of London.

It would seem that the production of such an important medal would be well documented. Unfortunately this is not the case and conflicting reports are given in what are considered authoritative texts[3,4,5]. What is agreed is that the material is indeed from cannon captured at Sebastapol. What is uncertain is the origin of the cannon and how many cannon were involved. It is agreed that the

(a) Face   (b) Obverse

Figure 1 The Victoria Cross

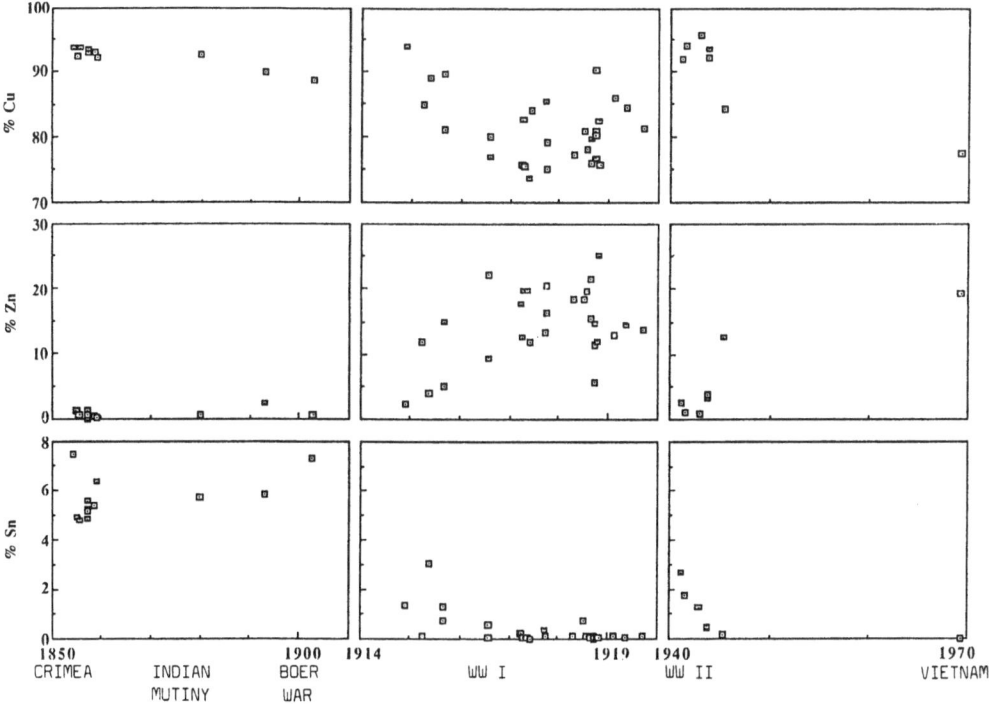

Figure 2   The major elements present in Victoria Crosses

material taken for manufacturing the medals comes from the cascabel, the rounded knob at the end opposite the muzzle of a muzzle-loading cannon. The cascabel is used as an anchoring point for ropes and pulley systems used to manoeuvre the cannon into position, in particular to run out the cannon to its gun port, and to aim the cannon.

Starting with this general lack of information was problem enough without the added restriction that one was not allowed to harm the medal in any way, a restriction which limited the type of analytical procedure which could be used.

EXPERIMENTAL DETAILS

Given the shape of the medal and the necessity to sample the largest area of its surface possible it was decided to make the examination on the circular area on the obverse side of the medal (Figure 1b). This is not a first-rate surface for x-ray fluorescence analysis and x-ray powder diffraction, because it is not absolutely flat, is covered with lacquer, and is engraved with the date and place of the action for which the medal was awarded. But it is the only area on which measurements can be made with any confidence of achieving reproducibility from one medal to another. Modifications had to be made to a holder of the Rigaku 3134 x-ray fluorescence spectrometer to hold the Victoria Cross. The other five holders contained standards, usually NBS brasses and bronzes. Spacers had to be fabricated to ensure that the surfaces of the standards were level with the surface of the Victoria Cross.

Table 1 Comparison of d-spacing and intensity ratio for two typical medals
belonging to the "bronze" and the "gun-metal" series

"BRONZE"		"GUN METAL"	
d-spacing (Å)	Intensity (rel)	d-spacing (Å)	Intensity (rel)
2.130	100	2.125	99
1.843	40	1.836	100
1.297	23	1.290	41
1.104	21	1.099	41
1.055	14	1.055	23

Some preliminary experiments were undertaken to establish the effect
lacquers and varnishes which are present on the medal would have on the x-ray
count rate of the fluorescent x-rays. This gives some of the information necessary
to compensate for the varnished surfaces of the medals.

For x-ray diffraction studies a Sietronics 112 modification to the Philips
PW1050 diffractometer enabled control of the diffractometer and the acquisition of
data by a NEC Powermate Plus personal computer. All data were taken using an
AMR diffracted beam monochromator mounted on the detector arm.

Care was taken in mounting the medals to ensure that the surface under
investigation lay on the rotation axis of the θ–2θ system. The medals were always
inserted in a fixed orientation relative to the incident beam.

(a)   (b)

Figure 3   The two cannon from which the cascabels were removed
for the production of the Victoria Crosses

RESULTS

The major elements in the medals under investigation are copper, zinc and tin. Figure 2 shows the percentage composition of these major components of the medals plotted as a function of the date of the action for which the medal was awarded. Note that there may have been a delay of almost a year between the date of award and the date of presentation. Surprisingly, information concerning the dates of presentation of a medal is not readily available.

For medals awarded for actions in the Crimean War, the Indian Mutiny, the Boer War, and the early part of World War I, the composition is characteristic of a bronze. Thereafter the composition changes sharply and is consistent with that of a gun metal. This situation remains consistent through World War II to the Vietnam War, with the exception of three medals awarded to Royal Australian Air Force personnel, all of which have compositions characteristic of a bronze.

As expected the XRD results confirm that significant differences exist between the medals, and, in particular between the two metal series. In Table 1 the d-spacing and intensity ratio of two typical medals, one issued in 1941 and the other in 1945. One belongs to the "bronze" series and the other to the "gun-metal" series.

A detailed examination of the XRD data is in progress.

DISCUSSION

The existence of two distinct classes of alloy led to further a historical investigation of the problem. There are, indeed, two separate sources of material for the medals because cascabels were removed from two cannon brought back to England from Sebastapol. That these cannon are dissimilar can be seen in Figure 3. It has been thought by most historians that these were Russian cannon: after all, the cannon had been captured from the Russians. However, the compositions of medals are not consistent with European gun bronzes and gun metals of the nineteenth century. Close investigation of one of the cannon showed inscriptions in archaic Chinese and it seems reasonable to believe now that the cannon are of Chinese origin, and were probably cast in the eighteenth century.

Further indirect evidence that these are Chinese lies with the casting techniques which were apparently used. It is believed that the Chinese cast their cannon with the muzzle down whereas the European technique was to cast with the muzzle up. The cascabels usually exhibit compositional inhomogeneities because of their mode of production. This is likely to be one of the prime reasons why there is so much scatter in the data. The remaining piece of cascabel from the VC cannon is held at the Central Army Ordnance Depot at Donnington, England.

CONCLUSION

Prior to this study little was known about the way in which the Victoria Crosses were manufactured. The little information available was conflicting in almost every detail. The use of x-ray fluorescence analysis has helped to resolve many of these discrepancies, and taking the XRF data in conjunction with stylistic and historical evidence, new directions for future historical study have opened up. As well, the conservation staff of the Australian War Memorial now know how best to preserve the medals and the curatorial staff can now accurately identify each medal. Not only is each medal unique in composition but because of the method of casting with the metal from the cannon each medal has its own unique set of imperfections.

If seems fitting that a medal that commemorates a unique act of valour be, in its own way, unique.

ACKNOWLEDGEMENTS

The author is indebted to the Australian War Memorial for the award of its Conservation Award to a Fellow. Without the active support of John Ashton and Anne I'Ons of the conservation staff of the Australian War memorial much of this work could not have been done. As well, Robert Smith, the Head of Conservation of the Royal Armouries at the Tower of London, has given invaluable assistance to the project.

REFERENCES

1.   R. Reece, World Archaeology, 6: 298 [1975].

2.   O. Werner, "Spektralanalytische und Metallurgische Untersuchungen und Indischen Bronzen", Brill, Leiden [1963].

3.   M.J. Crook, "The Evolution of the Victoria Cross", Midas, London [1975].

4.   L. Wigmore, "They Dared Mightily", Australian War Memorial, Canberra [1963].

5.   D.C. Creagh, Sabertache, XXXII: 15 [1991].

X-RAY FLUORESCENCE ANALYSIS OF OXIDE MAGNETIC TAPE

USING THIN LAYER FUNDAMENTAL PARAMETER ANALYSIS

Ko Kimura, Hideaki Wakamatsu, Takeshi Kitamura,
Ryozo Maeda, and Kunihiro Fujiwara

Analytical Center, Konica Corporation
No.1 Sakura-machi, Hino-shi, Tokyo 191, Japan

INTRODUCTION

In recent years, for the development of magnetic tape, it has become increasingly important that the elemental content be determined accurately and rapidly. In the past, the elemental content of magnetic tape was determined by calibration analysis using a wavelength dispersive X-ray fluorescence spectrometer(WDXRF). For calibration analysis many standard samples and large amount of sample were needed. Preparation of sample for calibration analysis, as well as for inductively coupled plasma emission spectral analysis(ICP) was difficult. On the other hand, for thin layer fundamental parameter analysis(we call TLFP) using a WDXRF, a few standard samples and less sample are needed and preparation of the sample is easy. The superiority of TLFP analysis can be seen in comparison with calibration analysis and ICP analysis in Table 1. For this reason, this study has established a method for accurate and rapid determination of the elemental composition of oxide magnetic tape by TLFP analysis using a WDXRF.

THEORY

Basic factors of TLFP analysis[1]

As used in our study, TLFP analysis determines elemental composition using the functional form Eq. (1) with the several factors.

$$I_T = f(W, P, K, T) \qquad (1)$$

where I_T is theoretical fluorescence X-ray intensity, W is chemical composition, P are physical constants, K are instrumental sensitivity and instrumental factors and T is sample weight per unit area.

Implementing TLFP analysis

TLFP analysis is performed a shown in Fig. 1[1,2].

Table 1. The superiority of thin layer fundamental parameter analysis

METHOD	WDXRF[a] TLFP	WDXRF CALIBRATION	ICP[b]
Preparation of sample	Easy	Difficult	Difficult
Number of standard samples	1 to 3 pieces	20 pieces	2 or 3 pieces
Weight of samples	0.15 grams	3 to 5 grams	0.1 grams
Matrix effect correction	Mathematical	By measurement	None
Calibration curve	None	Needed for each element	Needed for each element
Instrumental sensitivity	Needed for each element	None	None
EASE OF ANALYSIS	EASY	DIFFICULT	DIFFICULT

[a]WDXRF: wavelength dispersive X-ray fluorescence spectrometer
[b]ICP: inductively coupled plasma emission spectral analysis

EXPERIMENTAL

Equipment

The following equipment was essential to our study of TLFP analysis.
The WDXRF spectrometer was System 3080 of Rigaku Industrial Corporation
with a Machlett Rh-target X-ray tube. The computer was NEC PC-9801 RA.
The software was FP Method Ver.2.1 of Technos Corporation. The
spectrometer specifications are shown in Table 2.

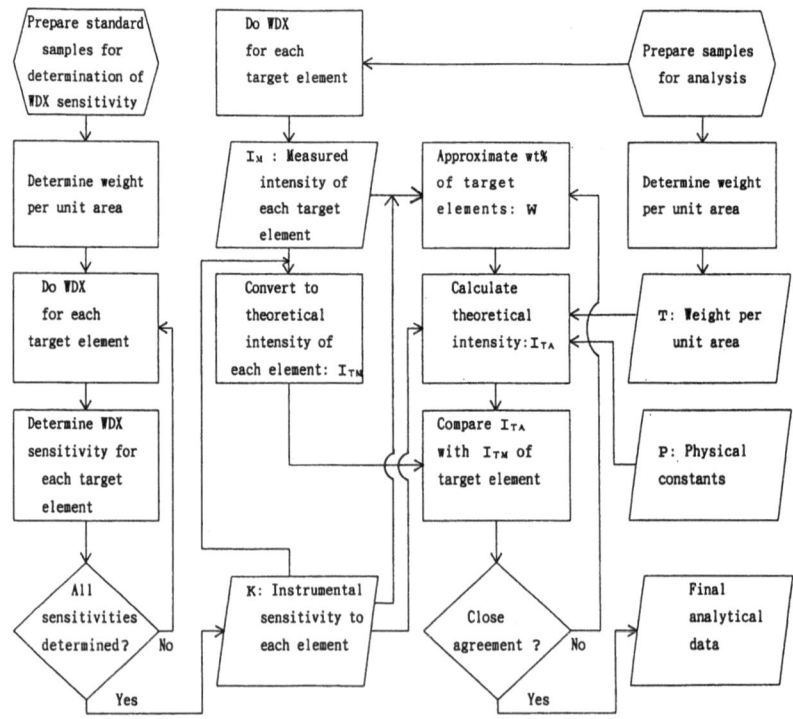

Fig. 1. Diagram of TLFP analysis

Table 2. WDXRF specifications

X-ray potential	50 kV	Crystals	Elements	Detector
X-ray current	50 mA	PET	Al	PC[a]
Absorber	1/1 (1/10 for Fe only)	RX-4	Si	PC
Counting peak	40 seconds	Ge	P, S, Cl	PC
Background 1	40 seconds	LiF3	Ca	PC
Background 2	40 seconds	LiF1	Heavy elements	SC[b]

[a]PC: proportional counter
[b]SC: scintillation counter

Preparation of oxide tape samples

Oxide tape samples were prepared using the following steps. The
mixtures were dispersed by a sandgrinder with ceramic beads and then were
coated on base with a knifecoater. The oxide tape sample was dried for 15
seconds at 333.15 to 343.15 K. No calender was used. The thickness of the
magnetic layer was approximately 4 μm.

Weight of magnetic layer per unit area

The weight of magnetic layer per unit area was determined in the
following way. The weight of the magnetic layer and base(0.5 inch x 39.37
inches) was measured by a balance.The magnetic layer was removed by acetone.
The weight of base was measured by a balance and the weight of base was
subtracted from the weight of magnetic layer and base. The weight of
magnetic layer was divided by the area(0.5 inch x 39.37 inches).

Establishing instrumental sensitivity

The sensitivity of the WDXRF to each for the target elements is the
constant to convert measured intensity to theoretical intensity. Its
dimension is 1/counts. It was established using the standard samples
shown in Table 3.

Table 3. The standard samples

	Sample 1	Sample 2	Sample 3
High purity compounds:			
Al_2O_3	0.87	5.28	9.77
SiO_2	0.42	0.97	1.53
FeP	0.27	0.72	1.03
$BaSO_4$	0.65	0.85	1.05
Cl(Binder)[a]	(4.02)	(4.02)	(4.02)
$CaCO_3$	0.13	0.23	0.44
TiO_2	0.80	2.35	3.63
Cr_2O_3	2.65	5.28	9.07
MnO_2	0.10	0.14	0.34
Fe_2O_3	73.77	59.70	42.65
Co_2O_3	2.56	5.17	8.02
NiO	0.01	0.01	0.03
ZnO	0.52	0.91	1.70
ZrO_2	1.40	2.54	4.89
Lubricant	1.92	1.92	1.92
Dispersant	1.58	1.58	1.58
Binder	12.35	12.35	12.35
Total	100.00	100.00	100.00

[a]Cl(Binder) was Cl in binder.

Table 4. TLFP models

Model	A: Pre-TLFP	B: TLFP	C: Post-TLFP
1	Weight of magnetic layer	Al-Zn content	Binder content
2	Weight of magnetic layer Cl content	Al-Zn content	Binder content
3	Weight of magnetic layer Binder content	Al-Zn content (Except Fe)	Fe_3O_4 content
4	Weight of magnetic layer Cl content Binder content	Al-Zn content (Except Fe)	Fe_3O_4 content

SELECTION OF THE TLFP MODEL

TLFP models

Four TLFP models were considered and shown in Table 4. In each model, factors in Column A are determined prior to TLFP itself. Factors in Column B are then determined by TLFP analysis. Finally, the material remaining in Column C is calculated.

Accuracies of TLFP models

Accuracies of TLFP models were shown in Table 5. To determine the accuracies of the four models, the models were applied to samples of known content. Among the data, wt% for the binder in Models 3 and 4 and for Cl in Models 2 and 4 were taken as known values.

Table 5. Accuracies of TLFP models

	Known value [wt%]	Model 1 [wt%]	Model 2 [wt%]	Model 3 [wt%]	Model 4 [wt%]
Binder	12.48	9.85	5.30	12.48	12.48
Al	0.00	0.00	0.00	0.00	0.01
Si	0.23	0.25	0.25	0.25	0.24
P	0.14	0.11	0.12	0.12	0.11
S	0.18	0.09	0.09	0.09	0.09
Cl	4.24	0.34	4.24	0.29	4.24
Ca	0.02	0.03	0.03	0.03	0.03
Ti	0.00	0.00	0.00	0.00	0.00
Cr	0.00	0.00	0.00	0.00	0.00
Mn	0.14	0.18	0.19	0.18	0.18
Fe(Fe_3O_4)	78.89	85.18	85.80	82.56	78.62
Co	2.75	2.82	2.84	2.82	2.83
Ni	0.02	0.00	0.00	0.00	0.00
Zn	0.91	1.14	1.15	1.14	1.14
Average accuracy[a] [wt%]	0.00	2.10	2.66	1.44	0.10

[a] Average accuracy $= \sqrt{\dfrac{\Sigma(W_i - X_i)^2}{(n-2)}}$

W_i: Known value [wt%]
X_i: Analytical value [wt%]
n: Number of elements and compounds

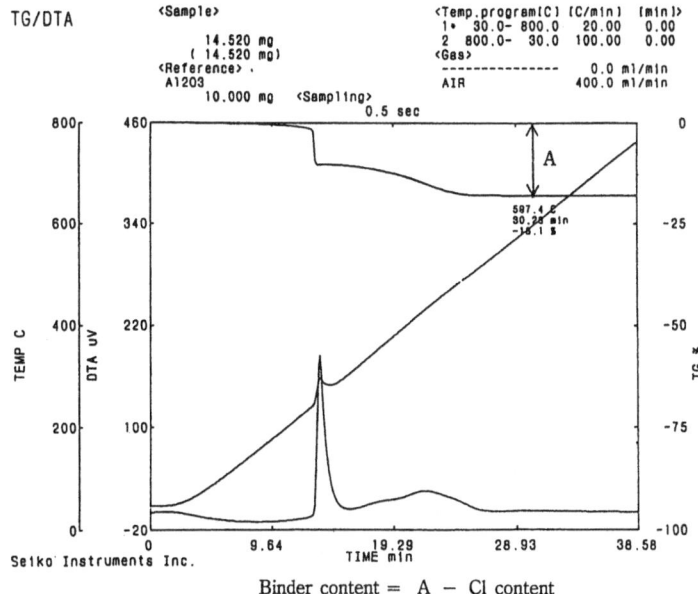

Binder content = A − Cl content

Fig. 2. Typical data of binder content determined by TG analysis

Model 4 was selected for its high accuracy. Three factors explain this. Since Fe is by far the element of highest concentration in oxide magnetic tapes, and since counting loss renders it the element least able to be accurately estimated by TLFP, the Fe content was chosen to be calculated as the remainder after actual TLFP analysis. Since Cl is destroyed by X-ray irradiation, thus leading to underestimation in TLFP analysis, the Cl content was chosen to be determined by flask combustion. Since the binder content cannot be determined through TLFP analysis, and

Table 6. The results of actual application

	Sample 1		Sample 2		Sample 3	
	Known value [wt%]	Analytical value [wt%]	Known value [wt%]	Analytical value [wt%]	Known value [wt%]	Analytical value [wt%]
Binder	12.48	11.43	16.15	15.50	16.12	15.47
Al	0.00	0.01	4.90	4.72	2.58	2.46
Si	0.23	0.24	0.21	0.21	0.64	0.42
P	0.14	0.11	0.12	0.06	0.40	0.31
S	0.18	0.09	0.16	0.09	0.13	0.03
Cl	4.24	4.73	3.76	4.38	3.60	4.19
Ca	0.02	0.03	0.02	0.03	0.10	0.08
Ti	0.00	0.00	1.11	1.24	0.00	0.00
Cr	0.00	0.01	0.00	0.00	6.93	7.60
Mn	0.14	0.19	0.12	0.14	0.19	0.16
Fe(Fe_3O_4)	78.89	79.19	70.19	70.36	64.21	65.56
Co	2.75	2.83	2.44	2.38	5.07	3.69
Ni	0.02	0.00	0.01	0.01	0.03	0.01
Zn	0.91	1.14	0.81	0.88	0.00	0.02
Average accuracy [wt%]	0.33		0.25		0.60	

since the Fe content was already chosen for calculation as a remainder, the binder content was chosen to be determined by thermogravimetric analysis.

DETERMINING CL AND BINDER CONTENT

Cl content

The flask combustion method used to determine Cl content is found in JIS K 6388 Testing Methods for Synthetic Rubber CR[3], and follows these steps. The weight of sample was measured by a micro balance. Sample was burned in O_2. The Cl gas was absorbed in pure water. After adding some alcohol and N/10 HNO_3 the solution was adjusted to pH=3. The solution was titrated by N/100 $AgNO_3$.

Binder content

Binder content was determined by thermogravimetric(TG) analysis. Typical data supplied by thermogravimetric analysis to determine binder content is shown in Fig. 2. The amount A in Fig.2 is the total organic content in the magnetic layer. The content of Cl is subtracted from the total organic content to approximate the binder concentration.

TLFP ANALYSIS APPLIED

The results of actual application of the TLFP model are given in Table 6. Samples of known content were used in order to determine the accuracy of the method. Accuracy is displayed at the bottom.

For each sample, TLFP showed high accuracy. Note, however, Sample 3. Here, average accuracy was lowered by inaccurate estimations of Cr and Co content. This is because oxides of these elements settle heterogeneously within the binder once they have reached a certain concentration, with a consequently sharp reduction of their rates of secondary emission.

CONCLUSION

By first determining Cl content by flask combustion and binder content by TG analysis, and by leaving Fe, the element of highest concentration, to be calculated as a last step, TLFP analysis proves to be a highly accurate method of determining the elemental composition of oxide magnetic tape.

REFERENCES

1. H. Kohno, M. Murata, Y. Kataoka, and T. Arai, Automation of X-Ray Fluorescence Analysis, Advances in X-RAY Chemical Analysis, JAPAN, 19:307 (1988).
2. J. Yoshitomi, S. Nakahama, H. Naganuma, and H. Oguro, XRF analysis of magnetic thin film conducted by a fundamental parameter method, BUNSEKI KAGAKU, 38:T160 (1988).
3. Japanese Industrial Committee, JIS K6388-1977.

XRF SPECTROMETRIC DETERMINATION OF YBCO SYSTEM, BPSCCO SYSTEM, AND TBCCO SYSTEM HIGH Tc OXIDE SUPERCONDUCTORS

Keiji KANEKO, Hiroko KANEKO, Hideo IHARA, Masayuki HIRABAYASHI, Norio TERADA and Masatoshi Jho

Electrotechnical Laboratory, 1-1-4, Umezono, Tuskuba-shi, Ibaraki 305 Japan

INTRODUCTION

Since Bednortz and Muller[1] discovered high Tc oxide superconductors in 1986, many oxide superconductors are synthesized by many workers. These are YBCO system[2] (Tc 90 K), BPSCCO system[3] (110 K) and TBCCO system[4] (125 K). Even now, the study of preparing higher Tc superconductors has been continued. The starting materials, in which the compositions are delicatly varied, are mixed and high Tc oxide superconductors are prepared. Therefore, high sensitivity and accurate composition analysis is required to an accuracy of ± 1 %. In general, ICP-MAS method[5], EPMX method[6] and electrochemical method[7] have been used as analysis methods. We have studied the non-destructive X-ray methods. As the first step, the X-ray method with polypropylene films and standard solutions has been studied.

EXPERIMENTS

Instruments and reagents

X-ray instrument used is semi automatic X-ray fluorescence spectrometer, type 3134, made by Rigaku Dcnki Industrial Corporation.

Yttrium, copper, bismuth, lead and thallium oxides, barium, strontium and calcium carbonates (99.99-99.999 % high purity reagents, Rare Metallic Co. Ltd.) and barium, strontium and calcium peroxides (chemical reagent grade, Kanto, or Showa Chemical Co. Ltd.) are used for starting materials and atandard solutions. Other reagents, polyvinyl alcohol, nitric acid and hydrochloric acid (analytical reagent grade, Kanto, or Wako Chemical Co. Ltd.) are used for standard solutions.

Preparation of high Tc oxide superconductors

1) YBCuO system : Y_2O_3, $BaCO_3$ and CuO reagents are weighted at 0.5:
2:3 molar ratio, respectivity and are mixed in an agate mortar. The mixed
samples (1-2 g) are pressed to pellets of 20 mm diameter by a press. The
pellets are calcined in an electric furnace at 900 ℃ for 15 hours in
oxygen (0.3 L/min) and after the pellets are powdered in an agate mortar
and pressed again, the pellets are burned at 930 ℃ for 15 hours in oxygen
(0.3 L/min).

2) BPSCCO system : Bi_2O_3, PbO, SrO_2, CaO_2 and CuO reagents are weiht-
ed at 0.92:0.34:2:2:3.06 molar ratio, respectivity and are mixed in an
agate mortar and then pressed into pellets. The pellets (1-2 g, 20 mm in
diameter) are wrapped in gold sheets and calcined in the electric furnace
at 830 ℃ for 24 hours. After the pellets are powdered and pressed, repeat-
ly, they are sintered in the electric furnace at 850 ℃ for 210 hours.

Preparation of standards and XRF measurements

Standard solutions for YBCO systems, BPSCCO systems and TBCCO systems
are prepared by dissolving Y_2O_3, $BaCO_3$ and CuO (0.5:2:3 molar ratio),
Bi_2O_3, PbO, $SrCO_3$, $CaCO_3$ and CuO (0.5:0.2:1:1:1 molar ratio), and Tl_2O_3,
$BaCO_3$, $CaCO_3$ and CuO (0.5:1:1:2 molar ratio) in 0.1 M nitric acid. 0.1 %
PVA solution is prepared by dissolving polyvinylalcohol (0.1 g) in 100
ml of pure water.

20 μ l of 0.1 % PVA solution are pipetted on each of the eleven cover
glass plates or magnesium oxide plates (18 x 18 mm, 0.12-0.17 mm thickness), and the eleven polypropylene films (52 mm diameter, 6 μ m thick-
nss) in 10 mm diameter, and 0,10, 20, 30, – – – 100 μ l of the standard
solutions are pipetted on the eleven plates and the films, respectivily.
Each of the eleven plates and the films are dried at room temperature.
Schematic of sample assembly is shown in Fig. 1.

The x-ray measuring conditions are as follows: Cr target x-ray tube
(50 kV – 40 mA), the crystals are lithium fluoride (220) for Y Kα, Ba
Lβ_2, Cu Kα, Bi Lα, Pb Lα, Sr Kα, Tl Lα, and germanium (111) for Ca

X-rays

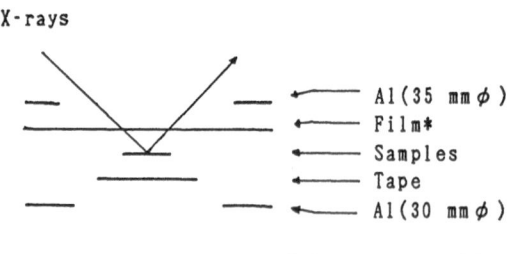

Al(35 mm ϕ)
Film*
Samples
Tape
Al(30 mm ϕ)

* Polypropylene film

Fig. 1.   Schematic of sample assembly

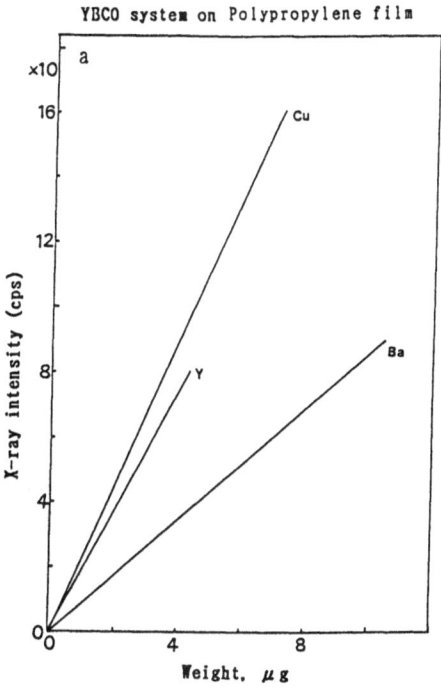

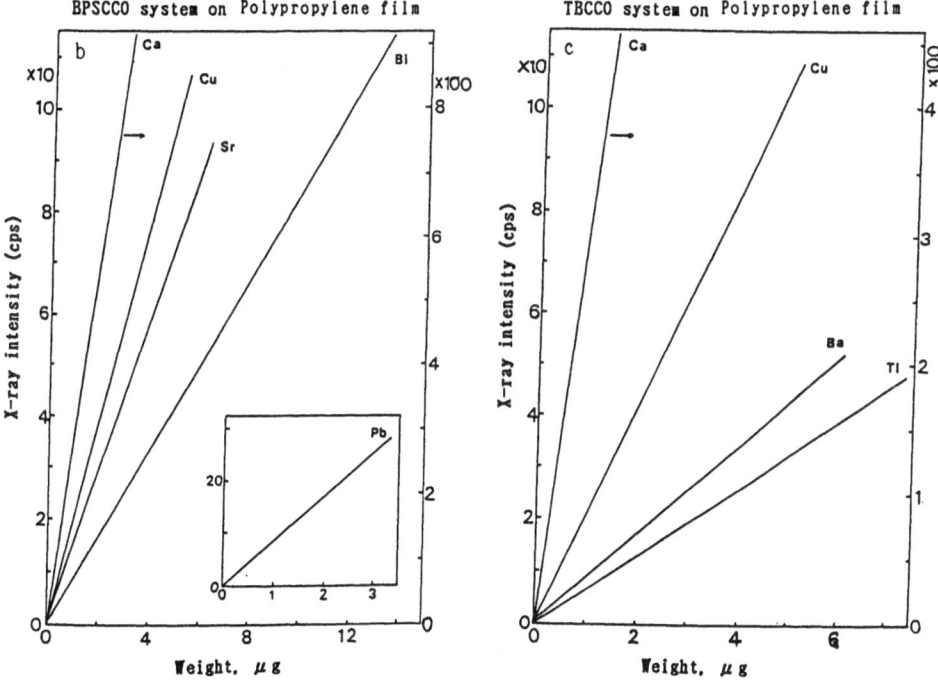

Fig. 2.   Calibration curves for composition elements in YBCO, BPSCCO
          and TBCCO systrms

Table 1 Analytical data for the YBCO, BPSCCO and TBCCO system
standards on polypropylene films

E	Anal. range (μ g)	Equation	α	ρ
Y	0.18~2.67	Log C = 1.06 Log I -1.37	0.062	0.990
Ba	0.55~9.61	Log C = 1.08 Log I -1.06	0.037	0.995
Cu	0.38~6.67	Log C = 0.95 Log I -1.26	0.033	0.996
Bi	0.42~9.40	Log C = 1.04 Log I -0.98	0.036	0.996
Pb	0.42~2.49	Log C = 1.08 Log I -1.04	0.018	0.997
Sr	0.18~5.26	Log C = 0.98 Log I -1.15	0.025	0.998
Ca	0.08~1.20	Log C = 0.98 Log I -2.41	0.047	0.991
Cu	0.13~2.86	Log C = 1.03 Log I -1.35	0.028	0.998
Tl	0.82~7.15	Log C = 0.96 Log I -0.76	0.014	0.999
Ba	0.28~4.12	Log C = 1.08 Log I -1.05	0.019	0.999
Ca	0.16~1.42	Log C = 0.91 Log I -2.44	0.018	0.998
Cu	0.26~4.45	Log C = 0.98 Log I -1.30	0.020	0.998

C:Concentration(μ g), I:X-ray intensity(cps), α :Standard
variation, ρ :Coefficient of correlation

Table 2 Analytical data for the YBCO, BPSCCO and TBCCO system
standards on cover glass

E	Anal. range (μ g)	Equation	α	ρ
Y	0.53~1.78	Log C = 1.78 Log I -1.44	0.021	0.992
Ba	1.10~4.40	Log C = 1.05 Log I -1.38	0.025	0.992
Cu	0.38~3.81	Log C = 0.93 Log I -1.19	0.010	0.999
Bi	6.27~104.5*	Log C = 1.06 Log I -0.65	0.025	0.996
Pb	1.24~24.86*	Log C = 0.95 Log I -0.38	0.063	0.991
Sr	2.63~43.81*	Log C = 1.09 Log I -1.03	0.017	0.999
Ca	1.20~20.04*	Log C = 1.30 Log I -2.29	0.047	0.996
Cu	1.91~31.77*	Log C = 0.99 Log I -0.44	0.033	0.997
Tl	0.21~20.44	Log C = 1.02 Log I -0.86	0.042	0.996
Ba	0.96~12.36	Log C = 1.11 Log I -1.03	0.044	0.994
Ca	0.08~4.01	Log C = 0.92 Log I -2.00	0.054	0.995
Cu	0.38~12.70	Log C = 0.94 Log I -1.18	0.028	0.998

C:Concentration(μ g), I:X-ray intensity(cps), α :Standard
variation, ρ :Coefficient of correlation, *:In the case of much
sampling

Table 3　　Analytical Results of YBCO Samples

No	Preparation Method	Condition	Analytical Method	Results Y	Ba	Cu	O	Tc K
				(Atomic ratio)				
1.			Starting[#]	1.0	2.0	3.0	6.5	
	Coprecipi-tation	600 ℃,1.5h	IP	0.95	1.95	3.00[*]		
		900 ℃,15h	XRF	1.02	2.08	3.00[**]	6.70[+]	77[@]-88[&]
2.			Starting	1.0	2.0	3.0	6.5	
	Oxide mixture	900 ℃,15h	IP	0.99	1.98	3.00[*]		
		930 ℃,15h	ICP	1.02	2.07	3.00[***]		
			XRF	0.97	1.97	3.00[**]	6.77[+]	91[@]-92[&]
3.			Starting	1.0	2.0	3.0	6.5	
	Coprecipi-tation	930 ℃,15h	IP	0.92	1.92	3.00[**]		
		as sintered	XRF	0.93	1.97	3.00[**]	6.74[+]	74[@]-80[&]

* 2 times average, ** 5 times average, *** 10 times average,
# Starting composition ($Y_{1.0}Ba_{2.0}Cu_{3.0}O_{6.5}$), + Gas analysis,
IP Isotachophoresis method, @ Endpoint, & Onset

Table 4　　Analytical Results of BPSCCO Samples

No	Preparation condition	Anal. method	Bi	Pb	Sr	Ca	Cu	O Calcu.[*] Mea.[+]		Tc K
			(Atomic ratio)					(Atomic ratio)		
1.	Oxide mixture									
	Starting[#]		1.85	0.35	1.9	2.1	3.0	14.13		
	830 ℃, 24h	ICP	1.78	0.20	2.01	2.09	3.00[**]			
	850 ℃,210h	XRF	1.73	0.17	2.04	2.16	3.00[**]	9.97	9.76	101 K
2.	Oxide mixture									
	Starting		1.82	0.48	2.0	2.0	3.2	14.41		
	830 ℃, 24h	ICP	1.82	0.25	2.08	2.04	3.20[**]			
	850 ℃,210h	XRF	1.84	0.21	2.18	2.12	3.20[**]	10.47	10.20	106 K
3.	Oxide mixture									
	Starting		1.84	0.34	2.0	1.94	3.05	13.23		
	830 ℃, 24h	ICP	1.74	0.19	2.03	2.00	3.05[**]			
	850 ℃,210h	XRF	1.68	0.16	2.11	2.09	3.05[**]	9.93	10.07	107 K

Starting composition (For example $Bi_{1.85}Pb_{0.35}Sr_{1.9}Ca_{2.1}Cu_{3.0}O_{14.13}$),
** 5 times average, * Calculating results, + Measuring results (Gas analysis).

Kα, the detectors are scintillation counter and proportional counter, and counting time is 40-100 sec.

RESULTS AND DISCUSSIONS

Calibration curves

Calibration curves for YBCO, BPSCCO and TBCCO systems are shown in Fig. 2, a, b, c. In the calibration curves for Ca and Ba, the former is more sensitive but the later is less by using Ba Lβ_2 X-ray line. The calibration curves on the polypropylene film substrates have the best sensitive compared to the other curves. As blanks (X-ray counts per sec for 0 μg) of calcium, copper for cover glass, magnesium oxide, and of calcium for the polypropylene films are larger, it is necessary to make accurate corrections. All the calibration curves are proportional to the concentration of each element.

Tables 1-2 show calibration curve equations, standard variations and coefficients of correlations.

Applications to actual samples

Ten to 100 mg samples of high Tc oxide superconductors prepared by the above methods are dissolved in 10 ml of pure water containing 0.1-1.0 ml of nitric acid. The sample solutions are pipetted on cover glasses or polypropylene films with 20 μl of 0.1 % PVA solution, are evapolated at room temperature and X-ray intensity is measured. The results are shown in Table 3-4. The results from XRF measurements agree with the results from ICP-AES method and isotachophoresis method[7]. For examples, as the limits of the determination by this method are Y 0.53, Ba 0.55, Cu 0.38 μg$\diagup$10 mm diameter, the films are contain Y 3.5 x 10^{15}, Ba 2.4 x 10^{15}, Cu 3.5 x 10^{15} atoms.

If the standards are prepared on the same substrates as actual film samples, the composition analysis of thinner high Tc oxide superconductor films would become possible. Therefore, we are now aiming at this target.

CONCLUSION

Microgram amounts of high Tc oxide superconductors have been determined by cover glass method or polypropylene film method with standard solutions and surfactant solution (0.1 % PVA solution). Analytical range, standard variation and coefficient of correlation for Yttrium systems, Bismuth systems, and Thallium systems are 0.08 - 20.44 μg, 0.014 - 0.062, and 0.990 - 0.999, respectively. Analytical results of actual bulk samples agreed with those from ICP-AES method and isotachophoresis method within ± 4 %.

REFERENCES

1) J. G. Bednorz and K. A. Muller : Z. Phys., B **64**, 189 (1986).
2) M. K. Wu, J. R. Ashburn, C. J. Torng, P. H. Hor, R. L. Meng, L. Gao, Z. J. Huang, Y. Q. Wang and C. W. Chu : Phys. Rev. Lett. **58**, 908 (1987).

3) H. Maeda, Y. Tanaka, M. Fukutomi and T. Asano : Jpn. J. Apl. Phys. **27**, L 209 (1988).

4) Z. Z. Sheng and A. M. Hermann : Nature **332**, 55 (1988).

5) K. Kurosawa and C. Homma : Proc. of the 36th Annual Meeting of the Japan Society for Analytical Chemistly, p. 747 (1987)

6) O. Michiami, H. Asano, Y. Kotoh, S. Kubo and K. Tanabe : Extended Abstracts of the 48th Autumn Meeting of the Japan Society of Applied Physics. I. p. 81 (1987).

7) K. Kaneko, H. Kaneko, H. Ihara, T. Hoshino, T. Okai, M. Hirabayashi, N. Terada M. Joh, H. Unoki : Mol. Cryst. Liq. Cryst. **184**, 297 (1990).

MEASUREMENTS OF LOW CONCENTRATION COMPONENTS

IN IRON ORES USING FUSION METHOD

K. Yamada, H. Kohno and T. Arai

Rigaku Industrial Corporation
Takatsuki, Osaka, Japan

Abstract

The problem of background radiation in X-ray fluorescence trace element analysis of fused-glass iron ore samples is addressed. A first-order model of coherent and Compton scattering with primary absorption is presented and used to correct measurements. Overlap coefficients for elements in iron ores are presented. The importance of these corrections is demonstrated. The accuracy achieved with X-ray measurements after background corrections compares well with the accuracy of chemical analysis.

Introduction

The X-ray fluorescence analysis of rocks, minerals, and mixed materials such as iron ores and cement, can be often be facilitated by melting and mixing the sample with a glass. This technique, called the fusion method, provides a homogeneous sample, which is necessary for reliable results; however, the fusing agent can interfere with X-ray fluorescence measurements. Since the measured volume

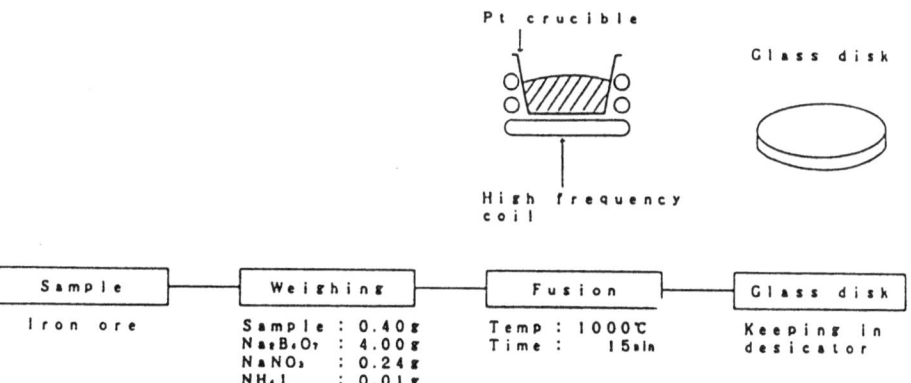

Fig. 1. Glass disk preparation using high-frequency induction furnace.

cannot be varied, the measurable quantity of analyzed elements decreases as the
ratio of fusing agent to sample material increases. This dilution effect hinders
detection of low-concentration and trace elements, and cannot be circumvented.

Another problem particular to fused glass specimens is background radiation,
caused by scattering off of the elements in the glass and sample. This background
radiation raises the threshold of sensitivity to trace elements, and distorts the
measured fluorescence spectrum.

The latter problem can be overcome by removing the expected background, thus
recovering the undistorted fluorescence spectrum. For this purpose the expected
background intensity has been calculated taking into account coherent and Compton
scattering of the primary beam (i.e. first-order scattering) as well as absorption
attenuation.[1]

This report describes the results of studies of background intensities in
an X-ray fluorescent spectrometer arrangement using samples of glass and iron ore.
A background correction method for such measurements is also introduced.

Experimentation

A Rigaku sequential spectrometer (RIX3000) equipped with a Rh-target end
window X-ray tube (OEG-76H, 50kV, 50mA) was employed for the first measurement.
The incident angle of the beam, ψ_1, was 65° with the analyzed angle, ψ_2, set at
a 40° take off angle. Measurements were made in a vacuum path, using a combination
of primary coarse Soller slit and a secondary Soller slit, a flat LiF(200) mosaic
analyzing crystal, and a NaI(Tl) scintillation counter detector, allowing analysis
of fluorescence lines in the 0.5 ~ 3.5 Angstrom range. Pulse height discrimination
was used to eliminate stray X-rays and other noise.

The quantitative studies were performed on a Rigaku simultaneous multi-channel
spectrometer (Simultix) with an identical tube ψ_1 for the measurements was 90°
and ψ_2 was 35°. Sealed-off proportional counters were used with pulse height
analyzers to make the measurements. A long measuring time (300 sec) was adopted to
reduce statistical error.

Use of Correction Equation to Eliminate Background

In previous papers, it was shown that X-ray background intensity could be
predicted.[2] The equations are shown in Eqs. (1) and (2) below. The factors $I_{(\lambda)}$
and $C_{(\lambda)}$ in Eq. (2) represent, respectively, the distribution of wavelengths
striking the sample, and instrumental factors such as the reflectivity of an
analyzing crystal.

$$\frac{dW(\lambda, \phi)}{d\Omega} = \frac{\sum n_i d\,\sigma_i(\lambda, \phi)}{\rho \sum W_i \left(\frac{\mu}{\rho}\right)_i \left(\frac{1}{\sin \psi_1} + \frac{1}{\sin \psi_2}\right)} \tag{1}$$

$$I_{BG}(\lambda, \phi) = I_{(\lambda)} \cdot C_{(\lambda)} \frac{dW(\lambda, \phi)}{d\Omega} \tag{2}$$

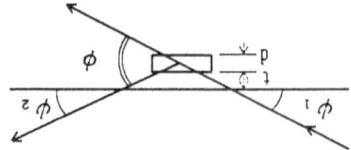

where

$$\phi = \psi_1 + \psi_2$$
n_i = number of atoms i in unit volume

$\dfrac{d\sigma_i}{d\Omega}$ = scattering cross-section of element i per unit solid angle $d\Omega$

$$d\sigma_i = d\sigma_{Thom.}^{(i)} + d\sigma_{Comp.}^{(i)}$$

$I_{(\lambda)}$ = irradiating X-ray intensity of λ
$C_{(\lambda)}$ = instrumental term

$\mu = \rho \sum_j W_j \left(\dfrac{\mu}{\rho}\right)_j$

ρ = sample density
W_i = weight fraction of element i
μ/ρ = mass absorption coefficient of element i

After measurement of a standard sample, the product of I and C is determined by comparing the measured intensity and that predicted by Eq.(1); this function can then be used as a proportional correction factor to predict background intensity for subsequent measurements.

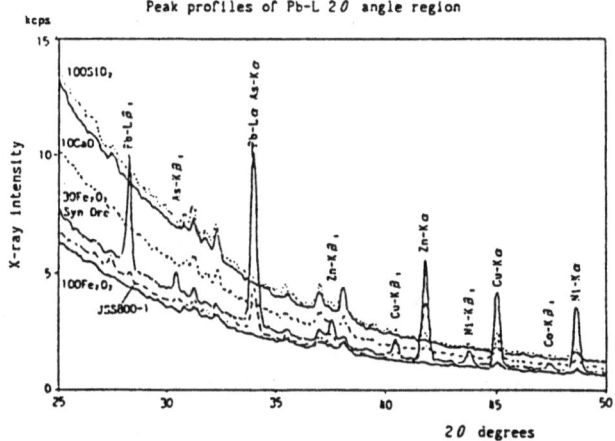

Fig. 2. Strip chart recording of various iron ores.

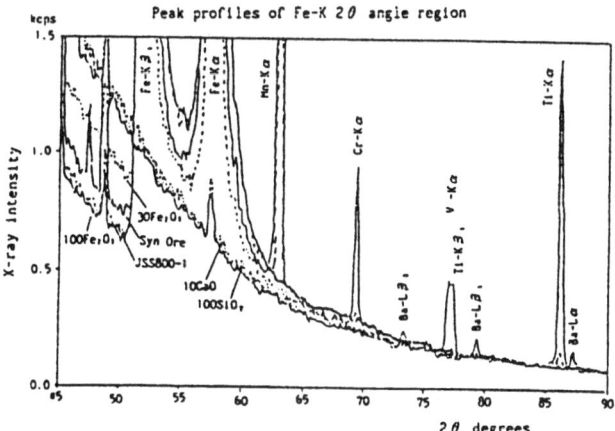

Fig. 3. Strip chart recording of various iron ores.

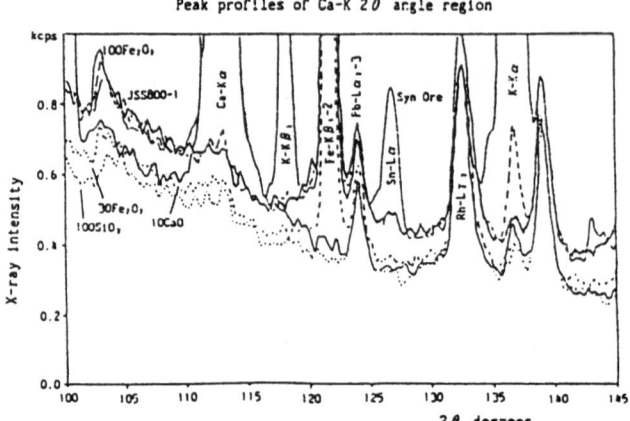

Fig. 4. Strip chart recording of various iron ores.

The intensities in the 25 - 50, 45 - 90 and 100 - 145 degree ranges of 2θ, obtained with an LiF(200) analyzing crystal on the RIX3000, are shown in Figs. (2), (3), and (4). Five different glass samples were made with the same flux matrials, using samples of pure Fe_2O_3, pure SiO_2, CaO, and a synthetic iron ore.

The data underscore the importance of understanding the background radiation. For example, in Fig.(2) the peak for Pb-Lβ_1, 0.2% concentration by weight in the ore, has a height that is roughly equalled by the background difference between pure Fe_2O_3 and pure SiO_2. The Zn-Kα peak is roughly four times the Fe_2O_3 and SiO_2 background difference at that 2θ position. High absorption by iron element can cause the sharp drop in background intensity, as seen in Fig. (3), where the background weakens considerably on the low angle side of the Fe-K absorption edge. Below the edge the background becomes more equal, becoming almost exactly equal above the V-Kα peak. At 136.7 degrees the samples have dramatically different K-Kα peak heights owing to greater absorption of K-Kα radiation by the SiO_2. These examples clearly illustrate the importance of taking background radiation into account in any sensitive quantitative analysis.

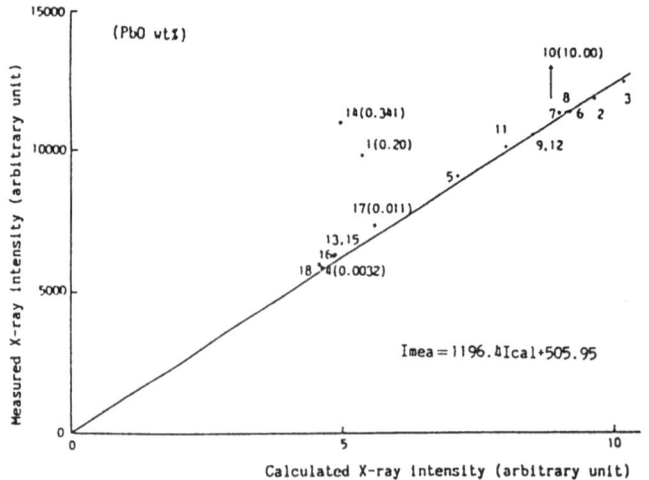

Fig. 5. Relationship between measured and calculated
background intensity of Pb Lβ_1 X-rays.

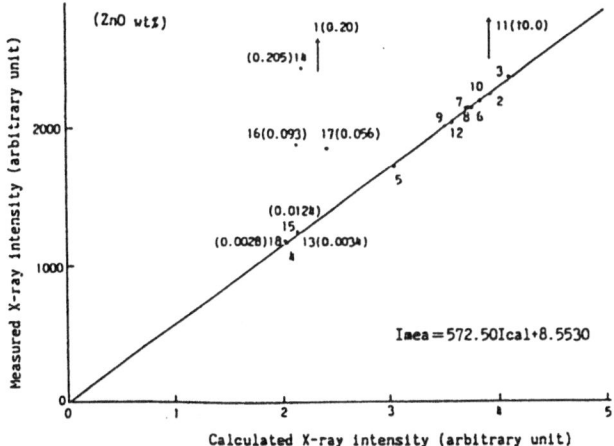

Fig. 6. Relationship between measured and calculated
background intensity of Zn Kα X-rays.

Evaluation of the Background Model

        The relationships between measured and calculated intensities for four
characteristic lines from a variety of samples are shown in Figs. (5), (6), (7) and
(8).   The measurements were performed with the multi-channel simultaneous
spectrometer (Simultix).   The calculated intensity is given as dW/dΩ from Eq.(1).
Since the measured signal was a combination of background and fluorescence from the
sample, lower concentrations of the fluorescing element provide an intensity closer
to the predicted background.   In the samples which had none of the fluorescing
element, a good linear relationship between measured and calculated intensities was
found.   In the samples which had the appropriate element, the discrepancy between
the measured intensity and the extrapolated background intensity can be assumed to
consist entirely of the appropriate fluorescence.   Its value, compared to the known
concentration of the fluorescing element, gauges the sensitivity of the measurement.

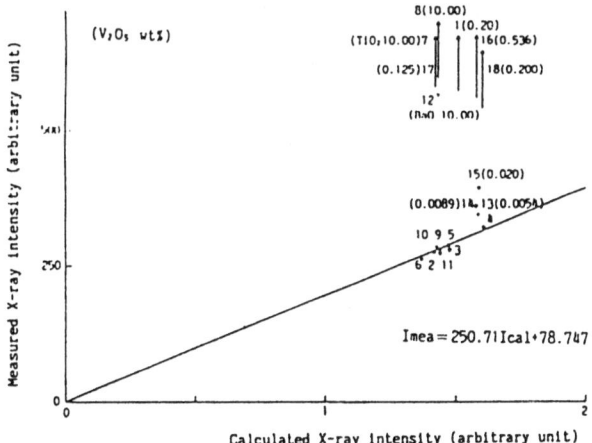

Fig. 7. Relationship between measured and calculated
background intensity of V Kα X-rays.

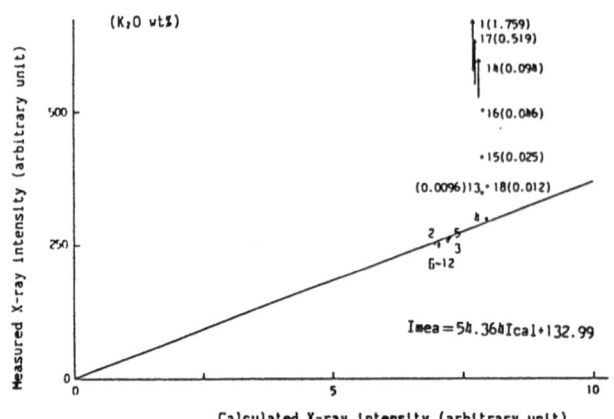

Fig. 8. Relationship between measured and calculated
background intensity of K Kα X-rays.

Higher Accuracy Achieved Using Background Correction

The working equations used in analysis of results are given in Eq.(3) and (4); they include background, matrix, and peak overlap corrections. The coefficients b and c were determined using a least-squares fit.

$$W = K(I_P - \sum_j \ell_j W_j)(1 + \sum_j d_j W_j) \tag{3}$$

$$I_P = I_{mea} - {}_{cal}I_{BG} = I_{mea} - (b\ I_{cal} + c) \tag{4}$$

K	: wt%/intensity
I_P	: net intensity of analyzing element
ℓ_j	: overlapping correction corefficient (intensity/wt%)
d_j	: matrix correction corefficent
W_j	: coexisting element concentration
I_{mea}	: measured X-ray intensity (including background X-ray intensity)
${}_{cal}I_{BG}$	: calculated background X-ray intensity using I_{CAL}
b,c	: constant of background intensity factor
I_{cal}	: calculated background X-ray intensity (dW/dΩ)

The degree of peak overlaps in 2θ for iron ore analysis was checked, with the results shown in Table (1).

Table 1. Overlapping X-Rays of Iron Ore Analysis

Mesuring Elements			
Spectrum	*¹ 2θ angle	Spectrum	*¹ 2θ angle
Mn-Kα	62.97	Fe-Kα background	
Ti-Kα	86.14	Ba-Lα	87.17
V -Kα	76.94	Ti-Kβ₁,Ba-Lβ₃	77.27, 77.36
Cr-Kα	69.36	V -Kβ₁	69.13
Co-Kα	52.80	Fe-Kβ₁	51.73
Ni-Kα	48.67	Co-Kβ₁	47.47
As-Kα	34.00	Pb-Lα	33.93
As-Kβ₁	30.45	Pb-Lβ Series background	
Ba-Kα	87.17	Ti-Kα	86.14

*1. Crystal : LiF(200), degrees

Table 2. Determination of PbO Contents from Measured Intensity
Using Calculated Background Intensity

Sample No.	Name	PbO contents (wt%)	I_{mea}	I_{cal}	$_{cal}I_{BG}$	I_P	Matrix correction coefficient (MC)	$I_P \times MC$	PbO analysis value (wt%)	Difference (Δ wt%)
1	Syn. Ore	0.20	9804	5.349	6906	2898.5	2.3935	6937.5	0.1963	-0.0037
2	100%SiO₂		11867	9.591	11981		1.2720			
3	50%SiO₂		12525	10.140	12638					
4	100%Fe₂O₃		5935	4.625	6039		2.881			
5	30%Fe₂O₃		9083	7.170	9084		1.755			
6	10%CaO		11434	9.143	11445		1.345			
7	10%TiO₂		11378	8.972	11240	137.9	1.348	185.8	0.0053	
8	10%V₂O₅		11408	9.095	11387	20.7	1.350	28.0	0.0008	
9	10%Co₃O₄		10628	8.455	10622	6.4	1.460	9.4	0.0003	
10	10%PbO	10.00	241830	8.795	11028	230801.6	1.531	353357.3	10.00	
11	10%ZnO		10113	7.976	10049	64.5	1.554	100.2	0.0028	0.00
12	10%BaO		10653	8.448	10613	39.8	1.505	59.9	0.0017	
13	76-6		6297	4.863	6324		2.709			
14	80-1	0.341	10978	4.940	6416	4561.8	2.652	12097.9	0.3424	0.001
15	83-4	0.0032	6322	4.873	6336		2.716			
16	83-9		6258	4.834	6289		2.734			
17	85-1		7352	5.579	7181		2.334			
18	85-4		5940	4.575	5980		2.919			
				$\times 10^{-3}$				PbO 35336. /1wt%		

In Tables (2), (3), (4) and (5) the measured levels of PbO, ZnO, V₂O₅ and K₂O respectively are shown compared to the chemically measured values. For each calculation, the background, overlapping, and matrix adjustments are shown.

For the samples used to obtain the results shown in Tables (2), (3), (4) and (5) the bulk of the sample consisted of SiO₂; the presence of this material influenced the background intensity. To investigate the effect of this background difference on the measured concentration of trace elements, the background intensities for several samples with no trace elements were compared. Pure Fe₂O₃ (sample No.4) was compared to a sample of pure SiO₂ (No.2) and a sample of 30% Fe₂O₃ (No.5). The equivalent concentration differences for these samples were calculated as the difference of intensity at the energy of the characteristic line divided by the intensity for a sample with a 1% by weight concentration of the measured element. These differences in equivalent measured concentrations are

Table 3. Determination of ZnO Contents from Measured Intensity
Using Calculated Background Intensity

Sample No.	Name	ZnO contents (wt%)	I_{mea}	I_{cal}	$_{cal}I_{BG}$	I_P	Matrix correction coefficient (MC)	$I_P \times MC$	ZnO analysis value (wt%)	Difference (Δ wt%)
1	Syn. Ore	0.20	2841	2.351	1338	1503.1	2.265	3404.5	0.196	-0.004
2	100%SiO₂		2236	3.929	2241		1.271			
3	50%SiO₂		2363	4.104	2342					
4	100%Fe₂O₃		1156	2.046	1163		2.677			
5	30%Fe₂O₃		1721	3.048	1737		1.6928			
6	10%CaO		2134	3.760	2145		1.3392			
7	10%TiO₂		2131	3.700	2110		1.3399			
8	10%V₂O₅		2124	3.741	2134		1.3425			
9	10%Co₃O₄		2000	3.515	2004		1.4381			
10	10%PbO		2185	3.838	2189		1.3588			
11	10%ZnO	10.00	133750	3.927	2240	131509.8	1.3183	173369.4	10.00	0.00
12	10%BaO		2036	3.575	2039		1.4763			
13	76-6	0.0039	1230	2.143	1219	11.2	2.5259	28.2	0.0016	-0.0023
14	80-1	0.205	2628	2.185	1243	1385.1	2.4847	3441.7	0.199	-0.006
15	83-4	0.0124	1307	2.148	1222	85.3	2.5332	216.1	0.0125	0.0001
16	83-9	0.093	1868	2.132	1213	655.5	2.5574	1676.3	0.0967	0.0037
17	85-1		1844	2.423	1379		2.2083	1026.6		
18	85-4		1170	2.027	1152		2.7132	47.7		
				$\times 10^{-3}$				ZnO 17337. /1wt%		

Table 4. Determination of V_2O_5 Contents from Measured Intensity Using Calculated Background Intensity

No.	Name	V_2O_5 contents (wt%)	I_{mea}	I_{cal}	$_{cal}I_{BG}$	I_p	Matrix correction coefficient (MC)	$I_p \times MC$	Ti TiO$_2$wt%	Ti $\Delta I \times M$	Ba BaO wt%	Ba $\Delta I \times M$	Corrected intensity	V_2O_5 analysis value (wt%)	Difference (Δwt%)
1	Syn. Ore	0.20	1065	1.506	299	766.4	1.2010	920.4	1.50	200.3	0.20	7.55	712.6	0.198	-0.002
2	100%SiO₂		277	1.417	277										
3	50%SiO₂		281	1.473	291										
4	100%Fe₂O₃		323	1.600	322										
5	30%Fe₂O₃		287	1.470	290										
6	10%CaO		264	1.366	264										
7	10%TiO₂		1342	1.421	278	1064.0	1.2550	1335.3	10.00	1335.3			0.0		
8	10%V₂O₅	10.00	28932	1.432	280	28651.7	1.2547	35949.3					35949.3	0.00	0.00
9	10%Co₃O₄		283	1.436	281										
10	10%PbO		205	1.422	278										
11	10%ZnO		276	1.437	282										
12	10%BaO		570	1.429	280	290.4	1.3007	377.7			10.00	377.7	0.0		
13	76-6	0.0054	316	1.584	318	28.0	1.1246	31.5	0.12	16.0			15.5	0.0043	-0.0011
14	80-1	0.0089	362	1.570	315	47.1	1.1440	67.9	0.075	10.0	0.391	14.77	43.1	0.0120	0.0031
15	83-4	0.020	394	1.584	318	75.8	1.1274	85.5	0.135	18.0			67.5	0.019	-0.001
16	83-9	0.536	2748	1.575	316	2431.2	1.1426	2778.8	6.34	846.6			1932.3	0.538	0.002
17	85-1	0.125	665	1.423	278	387.3	1.2762	494.2	0.317	42.3	0.0263	0.993	450.9	0.125	0.000
18	85-4	0.200	977	1.598	322	654.8	1.1222	734.8	0.25	33.4			701.5	0.195	-0.005
					$\times 10^{-3}$				TiO₂ 134. /1wt%		BaO 37.8 /1wt%		V₂O₅ 3595. /1wt%		

ΔI : Overlapping intensity   M : Matrix correction

Table 5. Determination of K_2O Contents from Measured Intensity Using Calculated Background Intensity

No.	Name	K_2O contents (wt%)	I_{mea}	I_{cal}	$_{cal}I_{BG}$	I_p	Matrix correction coefficient (MC)	$I_p \times MC$	K_2O analysis value (wt%)	Difference (Δwt%)
1	Syn. Ore	1.759	7486	7.693	285	7200.7	1.1550	8316.8	1.759	0.000
2	100%SiO₂		253	6.949	245					
3	50%SiO₂		258	7.209	259					
4	100%Fe₂O₃		299	7.940	299	0.04	1.1290	0.0401	0.00	
5	30%Fe₂O₃		261	7.239	261	0.64	1.1990	0.7730	0.0002	
6	10%CaO		250	7.035	249					
7	10%TiO₂		247	7.037	250					
8	10%V₂O₅		249	7.036	250					
9	10%Co₃O₄		251	7.046	250					
10	10%PbO		261	7.029	249					
11	10%ZnO		252	7.051	250					
12	10%BaO		246	7.097	252					
13	76-6	0.0096	348	7.853	294	53.7	1.1340	60.9	0.0129	0.0033
14	80-1	0.094	710	7.812	292	418.6	1.1440	478.9	0.101	0.007
15	83-4	0.025	413	7.854	294	119.2	1.1340	135.2	0.029	0.004
16	83-9	0.046	503	7.874	295	207.8	1.1370	236.3	0.050	0.004
17	85-1	0.519	2606	7.702	286	2312.0	1.1500	2668.0	0.564	0.045
18	85-4	0.012	358	7.940	299	59.6	1.1340	67.6	0.014	0.002
					$\times 10^{-3}$				K₂O 4728. /1wt%	

Table 6. Equivalent Concentration of Components Arising from Background X-Ray Intensity Difference

Unit : wt. percent

components	PbO	ZnO	V_2O_5	K_2O
Concentration between No.2 and No.4	0.17	0.062	-0.013	-0.0090
between No.5 and No.4	0.089	0.033	-0.010	-0.0080

divided intensity difference by intensity gradient of one wt. percent listed in Table (6).   The larger ratios for PbO and ZnO revael the importance of the background corrections for these elements.   In the case of V_2O_5 and K_2O the errors are smaller, but in the opposite direction due to the higher absorption of SiO_2. Clearly, background corrections are necessary to achieve accuracy equivalent to chemical techniques.

Conclusion

The variation of background intensities between fused glass sample materials can distort trace element fluorescence analysis if not accounted for.   The background intensities can be calculated with good accuracy using first-order models of coherent and Compton scattering from the atoms in the disk, with primary extinction taken into account.   This model can then be used to remove the background from the measured data if matrix and overlapping are considered.   Overlapping of characteristic peaks in iron ore specimens has been investigated.   Accurate determination of elemental concentrations has been demonstrated for a variety of elements.

Reference

1)   T. Arai ; X-Ray Spectrometry vol.20, (1991) 9 ~ 22.

A NEW INSTRUMENT FOR THE ENERGY DISPERSIVE X-RAY

FLUORESCENCE ANALYSIS OF OBJECTS OF ART AND ARCHAEOLOGY

Manfred Schreiner[1], Michael Mantler[2], Franz Weber[2],
Richard Ebner[2], Franz Mairinger[1]

1) Institute of Chemistry, Academy of Fine Arts
   Vienna/Austria

2) Institute of Applied and Technical Physics
   Technical University, Vienna/Austria

INTRODUCTION

Objects of art and archaeology are relics of the past, and
art historians, archaeologists and conservators are constantly
concerned with the questions of where, when or by whom such artifacts were made. Usually stylistic considerations can provide
answers to these questions, but as styles were sometimes copied at
locations and times quite different from those for which they were
most characteristic, material analysis is often essential when one
is attempting to infer how and of what materials an object was
made. The use of several compounds e.g. as pigments in paintings,
or the deliberate alloying of Cu with Sn, As, Sb and Pb, has
varied greatly from region to region and from time to time and can
be used to infer the geographic origin of an object or at least
the origin of the materials, out of which it was made.

The methods of X-ray fluorescence analysis play an important
role for the determination of the inorganic components within the
great number of analytical techniques applied to the study of
works of art or of archaeological artifacts[1,2]. The high advantage is their non-destructive applicability, which means that no
original sample material has to be taken from the object. The
energy dispersive detection of the characteristic X-rays enables
also a fast and nearly complete determination of the most important elements present as main components or as traces. Therefore,
energy dispersive XRF (EDXRF) with X-ray and also γ-ray excitation has often been used in the field of art and archaeology[3-5].

The new X-ray spectrometer designed for pixel by pixel
analysis along lines or across selected areas of artifacts mainly
consists of a spectroscopic X-ray tube and an energy dispersive
detection system. It enables the qualitative and semiquantitative
analysis based upon fundamental parameter models for multiple thin

Tab.1. Results of the qualitative analysis carried out by the
       new X-ray spectrometer at Indian miniature paintings of
       the Moghul period with European baroque additions of
       the "Millionenzimmer im Schloß Schönbrunn" in Vienna/
       Austria

Color	Element	Pigment

Indian miniature paintings

Color	Element	Pigment
White	Pb	Lead White: $2\ PbCO_3.Pb(OH)_2$
Red	Pb	Minium: Pb_3O_4
	Hg (S)	Vermillion: HgS
Blue	Cu	Azurite: $2\ CuCO_3.Cu(OH)_2$
	-	organic compound or Ultramarine: $3\ Na_2O.3\ Al_2O_3.6\ SiO_2.2\ NaS$
Green	Cu, (Pb)	Malachite: $CuCO_3.Cu(OH)_2$ - with Lead White
Gold	Au	Gold
Silver	Ag	Silver
Brown	Fe	Brown Ochre: Fe-oxide with Aluminasilicate

Baroque additions

Color	Element	Pigment
White	Ca	Chalk: $CaCO_3$
Red	Pb	Minium: Pb_3O_4
Blue	Cu, (Pb)	Azurite: $2\ CuCO_3.Cu(OH)_2$ with Lead white
Green	Cu	Malachite: $CuCO_3.Cu(OH)_2$

film layers, as well as the imaging of the elemental distributions
along lines or across selected areas. The technical details of
that instrument and the theoretical background of the quantification procedures are given in a separate paper of this volume [6].
This report emphasizes the results and interpretation of actual
data obtained at Indian miniature paintings of the 17th century.

ANALYTICAL PROCEDURES

    Measurements were carried out with an excitation energy of
20-30 kV and 10-20 mA using an X-ray tube with a tungsten target.
The primary beam was collimated to a diameter of 1 mm, and a
raster of 1 mm was chosen for the distance of each pixel. The
collecting time for each spectrum was usually 30 seconds and consequently the analyzing time varied from several minutes up to
several hours depending on the length of the lines and the size of
the area. The spectra were evaluated by defining regions of interest for the elements similar to commercially available programs.
Details of the software are described in a paper by Mantler[6].

RESULTS AND DISCUSSION

    The capabilities of the new X-ray spectrometer for the
material analysis could be demonstrated during the restoration of
the so-called "Millionenzimmer im Schloß Schönbrunn" in Vienna/
Austria. This room in the castle of Schönbrunn was decorated in
1762 using original Indian miniature paintings of the Mughal

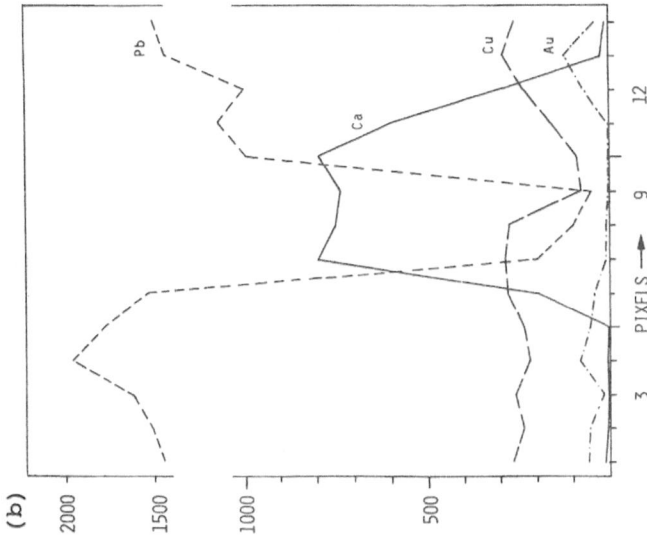

(b)

(a)

Fig.1. Detail of a baroque addition in an Indian miniature
        painting of the "Millionenzimmer im Schloß Schönbrunn"
        showing a white flower with golden ornaments on a blue
        background (a) and the distribution of Ca, Cu, Au and
        Pb (b) along the lines indicated in Fig.1a

period (17th century). The paintings on paper represent scenes of
the daily life and were accomplished and partly overpainted with
baroque additions. According the basic paint-structure a brush
drawing in black ink or in graphite is on the paper support[7].
This drawing is used to work out compositional problems and after
the design is established, it is covered with a translucent white
wash, which serves as the ground, on which the paint layers are
applied. Usually the paint layers consist of pure pigments or
mixtures of two pigments with an organic binding medium (e.g. gum
arabic).

It is not surprising that very little technical analyses have
been carried out on miniature paintings and very little knowledge
exists in the literature about the materials used. The scale of
this type of art object presents problems in examination and
usually it is impossible to take even small samples for analysis.
Therefore, some of the objects of the Millionenzimmer were investigated by the new X-ray spectrometer after demounting the paintings. The technical examinations included the determination of the
pigments used for the original Mughal artifacts and of the European baroque additions. Additionally, pixel by pixel analysis
along several lines and across selected parts of the representations were carried out for imaging the elemental distributions of
the characteristic elements.

Fig.2.   Representation of a shepard in the original Moghul
painting of the Millionenzimmer

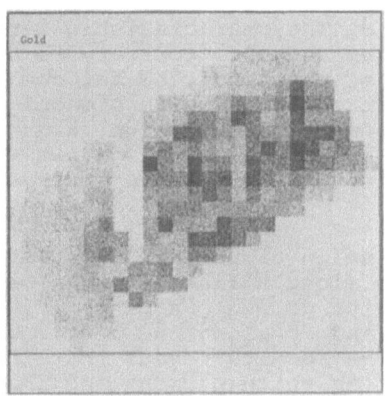

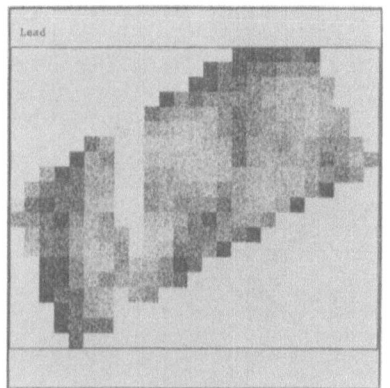

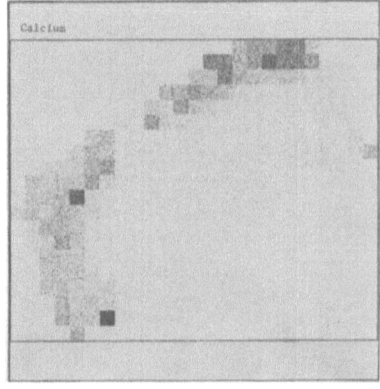

Fig.3. The distribution of Au, Pb and Ca across the area 1
(10 x 10 mm) indicated in Fig.2. The golden and white
(lead white) stripes are the original Moghul painting,
the chalk (CaCO$_3$) was used for the blue background
added in the baroque periode

The results obtained from the qualitative analysis are summarized in Tab.1. The palette of pigments identified by the new X-
ray spectrometer without taking any sample material consists of
lead white as white pigment, minium and vermillion as red pigments
and the copper containing pigment azurite as blue pigment. Cu was
also detected in the green parts of various representations, which
means that malachite was used as green pigment. Usually also the
characteristic lines of lead were present in the energy dispersive
X-ray spectra of the red, green and blue parts indicating a mixture of the pigments with lead white. For some of the blue as well
as the violet, pink and red parts no elements typical for pigments
or coloring agents could be detected by the new instrument. In
such cases the occurance of organic compounds or pigments containing mainly light elements such as Na, Al, Si or S (e.g. ultra-

marine) must be assumed. The characteristic X-rays for these
elements are absorbed by the air. Contrary to the results obtained
from the original Mughal paintings the baroque additions in the
representations contain chalk as white pigment. The red, blue and
green parts were painted by using minium, azurite and malachite
similar to the Indian miniature paintings but with different grain
size and grain size distributions as we observed under the optical
light microscope.

The lateral resolution of the new X-ray spectrometer can be
demonstrated by a scan along a line across a baroque white flower
on a blue background with golden ornaments (Fig.1a). The distributions of Ca, Au, Cu and Pb along the line indicated in Fig.1a
are shown in Fig.1b. The results reveal that the white flower was
painted by using chalk as white pigment and that the blue background was performed with a mixture of azurite and lead white. In
the ornaments only Au was determined.

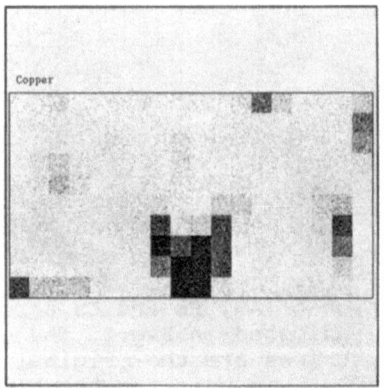

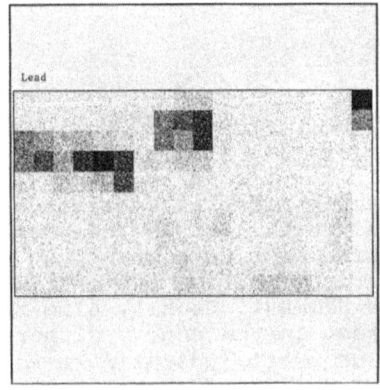
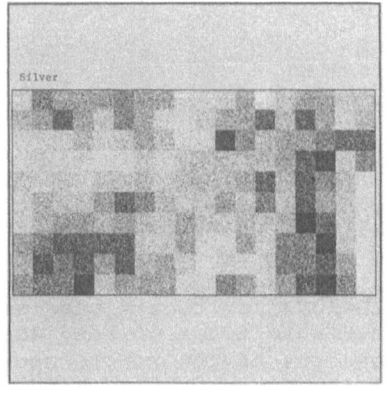

Fig.4.   Distribution of Cu, Pb and Ag across the area 2
          (15 x 10 mm) in Fig.2 showing a violett/white (Pb)
          flower with green (Cu) leaves on a gray (Ag) back-
          ground

The distribution of various elements present in the pigments in the areas indicated in Fig.2 are summarized in Fig.3 and 4. It should be mentioned that the size of scanned areas are 25 x 20 mm and 15 x 10 mm, and that a distance of 1 mm was chosen between the pixels. The collecting time for 1 spectrum was again 30 seconds so that a total analyzing time of 6 hours and 2 1/2 hours resulted for area 1 and area 2, respectively. Area 1 represents the head decoration with golden and white stripes whereas the scanned area 2 shows a violet flower with green leaves and a green stem surrounded by a gray background. The elemental distributions in Fig.3 reveal the presence of Au and Pb in the golden and white stripes and an enrichment of Ca from the baroque addition in the surrounding area, for the original green background was overpainted by azurite mixed with chalk. The leaves and the stem of the violet flower in Fig.2 was painted by using the Cu-containing pigment malachite, as it is evident from the results in Fig.4. The white spots in the blossom were performed with lead white and the gray coat of the shepard was carried out with powdered metallic silver.

CONCLUSION

The new X-ray spectrometer for the pixel by pixel analysis enables a qualitative analysis of the pigments used and an imaging of the typical elements at miniature painting without taking any original sample material. Further investigations will be carried out for the quantitative evaluation of the results using the fundamental parameter models and for imaging the elements of overpainted areas or underdrawings in easel paintings and graphic art objects.

REFERENCES

1) E. Sayre, P. Vandiver, J. Druzik, C. Stevenson (eds.): Materials issues in art and archaeology. Materials Research Society Vol. 123, Pittsburg, Pensylvenia (1988)
2) M. Schreiner, M. Grasserbauer: Fresenius J. Anal. Chem. 322 (1985) 181-193
3) C. Lahanier: PACT 1 (1977) 31-40
4) W.A. Oddy, F. Schweizer: Royal Numismatic Society London, special publications, Vol. 8 (1972) 171-183
5) F. Schweizer, A.M. Friedman: Archaeometry 14 (1972) 103-107
6) M. Mantler, R. Ebner, F. Weber, M. Schreiner, F.Mairinger: Advances in X-Ray Analysis, Vol.35 (1992)
7) B.B. Johnson: A preliminary study of the technique of Indian miniature painting. In: Pratapaditya Pal (ed.): Aspects of Indian Art, E.J.Brill, Leiden, 1972, p. 139-146

PROVENANCE OF KANJERA FOSSILS BY X-RAY FLUORESCENCE

AND ION MICROPROBE ANALYSES

A. M. Kinyua, T. Plummer, N. Shimizu, W. Melson, R. Potts

CNST, Faculty of Engineering
University of Nairobi, P.O. Box
30197, Nairobi, Kenya

Department of Anthropology
Yale University
New Haven, CT 06520 U.S.A.

Department of Geology and Geophysics
Woods Hole Oceanographic Institution
Woods Hole, MA 02543 U.S.A.

Department of Mineral Sciences
National Museum of Natural History
Washington, D.C. 20560

Department of Anthropology
National Museum of Natural History
Washington, D.C. 20560, U.S.A. and
National Museums of Kenya
P.O. Box 40658, Nairobi, Kenya

ABSTRACT

XRF and ion microprobe analyses of fossils of known and uncertain provenance from the Lower-Middle Pleistocene locality of Kanjera, Kenya, are reported. The goal of this study was to develop a nondestructive technique of provenancing fossils, which could be applied to the Kanjera sample. The fossils of known provenance were collected in the excavations of the 1987 Smithsonian Expedition. Three fossils of uncertain provenance, two specimens of *Theropithecus oswaldi* and a hominid fossil, were analyzed as test cases.

Both qualitative and quantitative XRF analyses of Kanjera fossils were carried out. In the qualitative analysis, the elemental peak areas from each fossil's XRF spectrum were calculated and normalized to the peak area of the incoherently scattered radiation. Results of the analysis showed that fossils from the Lower-Middle Pleistocene Kanjera Beds, for the most part, had higher levels of yttrium (Y) and zirconium (Zr) than those of the younger Apoko (Ap) Bed, black cotton soil (BCS) and modern bones (MD). The relative concentrations of uranium (U), strontium (Sr) and thorium (Th) were diagnostic of the Kanjera Bed of origin. These findings were confirmed by quantitative XRF and ion microprobe analyses of a subsample of Kanjera fossils. The *T. oswaldi* and hominid fossils had trace element concentrations suggestive of K2 and BCS provenances, respectively. These findings provide a framework for the qualitative XRF provenancing of other surface collected fossils from the locality.

Advances in X-Ray Analysis, Vol. 35
Edited by C.S. Barrett *et al.*, Plenum Press, New York, 1992

INTRODUCTION

Kanjera is situated on the foothills of Homa Mountain, Winam Gulf, Lake Victoria. Carbonatite volcanism in this region produced magma enriched with a variety of trace elements[1,2]. It is thus an ideal area to investigate trace element mobility and fixation in fossil bone.

The artifact and fossil-bearing exposures at Kanjera[3] came into prominence in 1911 when fossils, including the type specimen of *Theropithecus oswaldi*, were surface collected[4]. In 1932-33 and 1934-35, the Third and Fourth East African Archaeological Expeditions led by L. S. B. Leakey recovered additional fauna and the remains of at least four hominids[5]. Leakey[5-7] believed that the latter fossils were derived from the same beds and were thus the same age as the fossil fauna from the locality. His provenance of the hominids was soon questioned, however[8,9], throwing their age into doubt. More recent expeditions[10,11] also surface collected fossils, but it was not until the Smithsonian Expedition[3] that a large sample of *in situ* bone was recovered.

History of research at Kanjera has pointed out a persistent problem in paleontological analysis: the provenancing of surface collected fossils. Chemical provenancing studies in the past have relied largely on relative dating[12]. It is only reliable if the concentration of the element in question (often F and/or U) increases with increasing fossil age. Trace element uptake by fossils can be influenced by the supergene environment[13-15], confounding relative dating studies. In such circumstances, other provenancing methodologies must be employed.

With the improvements in available instrumentation[16], it has been possible to determine with great precision the concentrations of a variety of elements. This has led to a flourish of papers using the relative or absolute concentrations of trace and/or major and minor elements in the provenancing of pottery[17,18], lithic materials[19-21] and volcanic tuffs[22-23]. In these studies, specimens belonging to groups of known provenance are characterized chemically in order to find elemental values unique to each group. Often this characterization is based on a suite of elements, and is unrelated to the age of the specimen. The elemental profiles of specimens of uncertain history can then be compared to those of known provenance to determine their probable points of origin. Provenances are only conclusive if all relevant source groups are characterized. The same methodology can be employed in fossil provenancing. In this case the source groups are the fossil assemblages from the different stratigraphic units at a locality, and the specimens of unknown provenance are the surface collected fossils. The emphasis shifts from relative dating to establishing suites of elements diagnostic of fossils from a particular unit.

The use of scattered x-rays in material analysis dates back to the work of Anderman & Kemp[24]. Since then, research groups have utilized scattered radiation in different ways. Giauque *et al.*[25] have shown that the incoherent scattered radiation can serve as an internal standard to compensate for variations in sample mass and geometry. Nielson[26] normalized the x-ray fluorescent intensities to the scattered x-ray intensities to reduce errors due to instrumental and differences in absorption among samples and standards. Other groups[27-30] have also demonstrated quantitative analysis of samples using scattered radiation. Here we use the peak area of the incoherently scattered radiation as a proxy for mass in the analysis of bulk fossil samples, so that a qualitative evaluation can be performed without sample destruction. Quantitative XRF and ion microprobe analysis were performed on a subsample of fossils in order to check the results of the qualitative analysis.

EXPERIMENTAL

Sampling

Excavations of the 1987 Smithsonian Expedition recovered fossils from six stratigraphic units (from oldest to youngest Kanjera Beds K2-K5, Ap and BCS)[31]. An unconformity exists between the Kanjera Bed sequence and the overlying Ap. Two hundred and sixty-one fossils of known provenance (between 23 and 55 samples per stratigraphic unit) were analyzed qualitatively.

Quantitative XRF of 71 samples (between 10 and 13 specimens per stratigraphic unit) was carried out to evaluate the accuracy of the qualitative analysis. Two probable

Theropithecus femoral shaft fragments (KT62, KT64) collected by Leakey in 1958 were analyzed both qualitatively and quantitatively. These share the green color of much of the rest of the *T. oswaldi* sample. A small portion of a hominid tibia shaft surface collected in 1987 (National Museums of Kenya accession number KNM-KJ 7494A) was also analyzed. One sample from each stratigraphic unit, a modern bone from the Kanjera landsurface, KT62 and KNM-KJ 7494A were analyzed with the ion microprobe.

Sample Preparation

Bone samples were cleaned of adherent sediment and oven dried at 353 K for 24 hours. Sediment samples collected in excavation were dried at the same temperature and duration, and powdered with a FRITSCH pulverisette (W. Germany). Trace elements often have their highest concentrations in the periosteal (outer) rim of the compact bone[13,14,32]. Therefore, specimens analyzed qualitatively were placed on a mylar film and oriented with their periosteal surface perpendicular to the source.

Fossil samples analyzed by quantitative XRF were further cleaned in an ultrasonic bath of distilled water and sectioned in 1.5-2.5 mm increments from the periosteal surface inward with a 0.1 mm diamond impregnated saw. Manual pulverization of sediment and fossil samples was done to reduce the particle sizes to less than 50 microns[33]. Thin samples were prepared by pressing pellets of fossil or sediment diluted with starch powder for binding[33-36].

Ion microprobe analysis was carried out on transverse bone sections impregnated with an epoxy resin for 3 hours at 363 K.

Instrumentation

XRF. XRF analysis was carried out at the Centre for Nuclear Science Techniques, University of Nairobi, Kenya. A radioisotope source, ^{109}Cd (444MBq), was used for sample excitation. The x-rays were detected by a 30 mm^2 x 10 mm thick ORTEC Si(Li) detector located inside a cryostat with a 25 micron beryllium entrance window. The resolution of the detector was 184 eV at 5.9 keV Mn K_x line. The spectral data for analysis were collected by a Canberra S40 multichannel analyzer through a Canberra Model 2020 pile-up rejector system. Spectral analysis and data storage was done through a Digital Professional 350 microcomputer system. Analysis time was 200 seconds for qualitative XRF, and 2000 seconds for quantitative XRF.

Ion Microprobe. Ion microprobe analysis was carried out at the MIT-Brown-Harvard regional ion microprobe facility housed at MIT, using a Cameca IMS 3f ion microprobe. A beam of negatively charged oxygen ions with a net energy of 12.55-12.65 keV was focused on a spot 10-20 microns in diameter. Intensities of high-energy secondary ions were measured (energy filtering[37]) on ^{88}Sr, ^{139}La, ^{174}Yb and ^{238}U relative to ^{44}Ca. Concentrations were calculated using apatite from Durango, Mexico, as a standard. These elements have been analyzed in the Durango apatite by isotope dilution mass spectrometry by Grandjean[38] for Sr, La and Yb, and by Hauri[39] for U. Transects were made through the compact bone in three areas: the periosteal margin of the compact bone, the approximate middle of the compact bone, and the endosteal (inner) margin of the compact bone. Transects were oriented perpendicular to the medullary cavity. A minimum of ten readings were taken per transect. The average of each transect's elemental readings was used in the following discussion.

RESULTS

The elements which varied the most between samples and thus provide the greatest possibilities for provenancing are Y, Zr, U, Rb, Rn, Sr and Th (Figs. 1-5). The Rb and Rn values are highly correlated with and thus show the same trends as U, and so are not figured here. Quantitative XRF and ion microprobe values are compared in Table 1. To our knowledge, this was the first application of the ion microprobe to the analysis of fossil bone. There is general agreement in the results of the two techniques, with some discrepancy probably related to differences in the proportion of compact bone analyzed. Quantitative XRF concentrations are derived from compact bone sections several millimeters thick, while the ion microprobe results

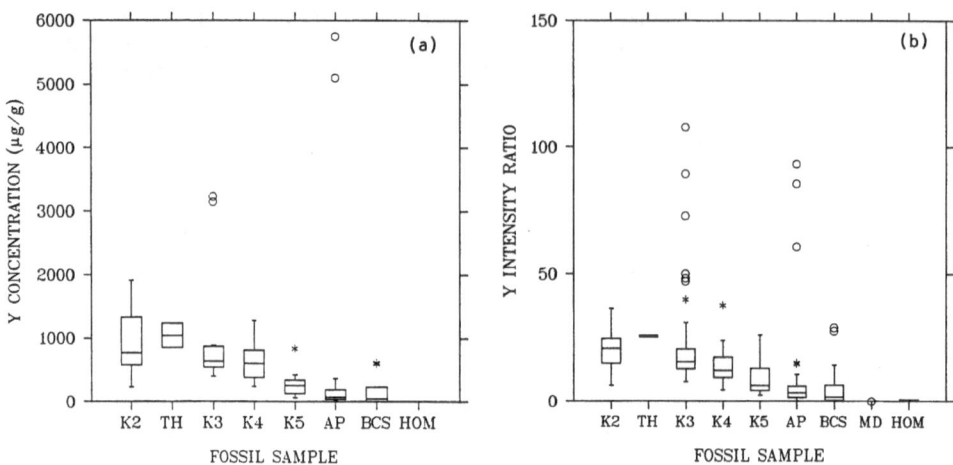

Fig. 1.   Box plots of periosteal Y values by fossil sample as determined by XRF.   Sample median marked by horizontal line in box. a)   Concentrations in ug/g.   b)   Y intensity ratio [(Y peak area/scatter peak area)x 100].   K2-K5, Ap, BCS refer to fossil samples derived from these stratigraphic units.   MD=modern bone collected from Kanjera landsurface, TH=*T. oswaldi* specimens, HOM=hominid specimen.

are from discrete points. Elemental values tended to be higher in the Kanjera Bed bone samples than the corresponding sediment samples (Table 2). For the powdered samples (n=71) the adjusted squared multiple R between pellet mass and scatter peak area was 0.92 (p<0.001). This high correlation is important, as the scatter peak area is used as a proxy for mass in the standardization of the qualitative data.

Sixty-six fossils were analyzed qualitatively and quantitatively. The two values determined for each element, one a normalized peak area and the other a µg/g concentration, were regressed against each other to determine whether the techniques produce comparable results. However, x-rays in the qualitative analysis almost certainly did not penetrate the full 1.5-2.5 mm depth into the compact bone analyzed as the periosteal transect in the quantitative analysis. Even with this reservation, the adjusted squared multiple R's from least squares regres-

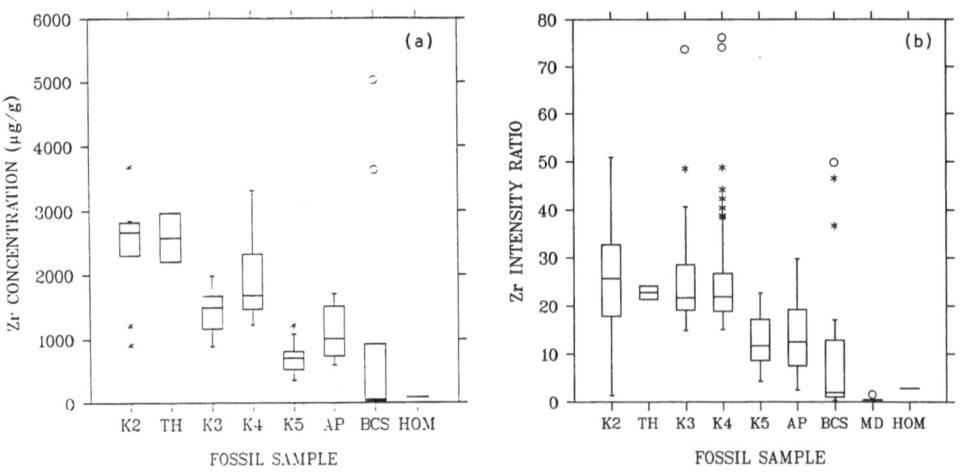

Fig. 2.   Box plots of periosteal Zr values by fossil sample as determined by XRF.   a)   Concentrations in ug/g.   b)   Zr intensity ratio. Samples defined in Fig. 1.

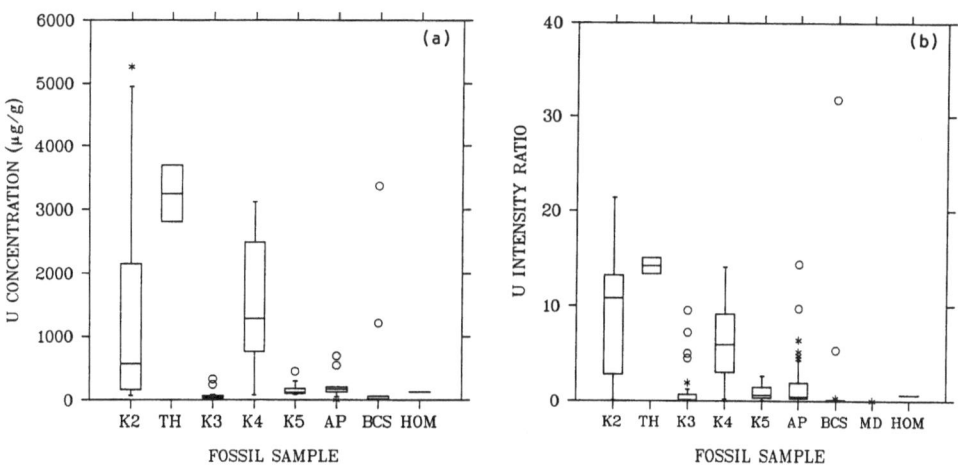

Fig. 3.   Box plots of periosteal U values by fossil sample as
determined by XRF.   a)   Concentrations in ug/g.   b)   U intensity
ratio. Samples defined in Fig. 1.

sions between quantitative and qualitative values of U, Rb, Sr, Zr, Th, Y and Rn were high
(0.92, 0.92, 0.93, 0.85, 0.84, 0.82 and 0.80, respectively). Moreover, the qualitative results
show the same general trends as those obtained from the quantitative analyses (Fig.s 1-5;
Table 1), though two or more times as many samples were analyzed qualitatively than quantitatively.

DISCUSSION

        The above results demonstrate that the qualitative values obtained by normalizing the
elemental peak areas with the scatter peak area accurately reflect fossil trace element concentrations, allowing the rapid, noninvasive analysis of bulk samples.

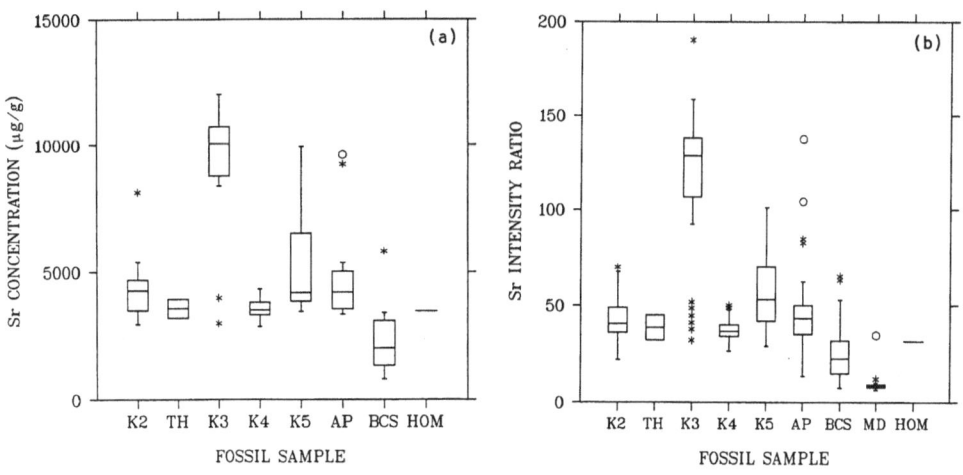

Fig.   4.   Box   plots   of   periosteal   Sr   values   by   fossil   sample   as
determined   by   XRF.   a)   Concentrations   in   ug/g.   b)   Sr   intensity
ratio. Samples defined in Fig. 1.

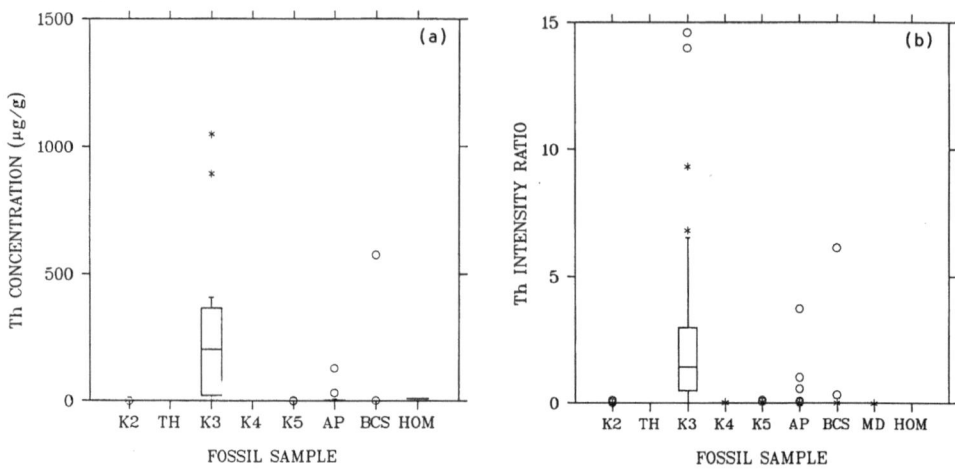

Fig. 5. Box plots of periosteal Th values by fossil sample as determined by XRF. a) Concentrations in ug/g. b) Th intensity ratio. Samples defined in Fig. 1.

The fact that values of Y, Zr, U, Sr and Th were higher in the Kanjera Bed fossils (Figs. 1-5; Table 1) than in the respective Kanjera Bed sediment samples (Table 2) demonstrates that high fossil elemental values could not have resulted from sediment contamination. The high U, La, Yb and Sr values obtained by ion microprobe analysis suggests that these elements, at least, have been incorporated into the apatite crystals. The quantitative analyses also suggest that concentration differences between high and low U fossils did not result from the leaching

Table 1. Comparison of samples analyzed by quantitative XRF and ion microprobe giving sample, bed of origin and portion of compact bone analyzed (values in µg/g dry weight). peri=periosteal (outer) rim of compact bone, mid=middle of compact bone, endo=endosteal (inner) rim of compact bone. ind=concentration below detection limit of apparatus.

Sample	Bed	Pos	XRF U	Sr	Y	Ion Microprobe U	Sr	La	Yb
KT62	Surf	peri	2810	3936	852	2755	3891	3441	283
"	"	mid	2198	3360	33	2158	3027	3	ind
"	"	endo	2176	3525	303	2741	3000	465	80
KJ 7494A	Surf	peri	132	3462	ind	141	4700	6	ind
"	"	mid	—	—	—	281	3172	1	ind
"	"	endo	—	—	—	320	3809	1	ind
SUR84 MD		peri	—	—	—	ind	1527	ind	ind
"	"	mid	—	—	—	ind	1182	ind	ind
"	"	endo	—	—	—	ind	1027	ind	ind
BCSI4	BCS	peri	9	1344	6	24	1364	6	ind
"	BCS	mid	—	—	—	5	1254	4	ind
"	BCS	endo	—	—	—	11	1254	22	ind
AP8	Ap	peri	696	3541	35	1074	4054	190	5
"	"	mid	—	—	—	869	3872	3	ind
K5GP5	K5	peri	—	—	—	66	2927	4613	ind
"	"	mid	—	—	—	104	3654	2	ind
"	"	endo	—	—	—	74	3427	5989	ind
K4AC20	K4	peri	2487	3795	569	2741	3591	2883	76
"	"	mid	—	—	—	3337	3200	ind	ind
"	"	endo	—	—	—	2613	3509	3701	81
K3SC1	K3	peri	43	12026	585	ind	10554	1570	119
"	"	mid	—	—	—	ind	8826	87	53
"	"	endo	—	—	—	ind	10726	1719	115
1987K2	K2	peri	5265	3670	1067	5240	3945	6119	330
"	"	mid	—	—	—	5978	2963	262	55
"	"	endo	—	—	—	5609	3009	1064	215

Table 2. Mean elemental concentration values of sediment samples collected from Kanjera Bed excavations (values in μg/g dry weight). ind=concentration below detection limit of apparatus.

Bed	Number of Samples	Y	Zr	Sr	Th	U
K5	2	32	412	498	2	ind
K4	2	38	424	697	7	ind
K3	4	31	561	1005	10	ind
K2	2	33	372	388	4	ind

of U from the latter group. This would have led to steeply decreasing U concentration gradients from the compact bone interior to the periosteal and endosteal surfaces. The quantitative analysis of 32 additional Kanjera fossils also failed to demonstrate significant U leaching[40].

Most Ap and BCS specimens had low elemental values relative to those of the Kanjera Bed sequence. Exceptions with higher values often had trace element chemistries similar to the fossils of the Kanjera Bed they overlay[40]. These may have been derived from the Kanjera Bed sequence and incorporated into Ap and BCS through channeling or pedogenesis, respectively. The sample of Kanjera Bed bones analyzed with the ion microprobe also had higher REE values than the fossils from above the unconformity. Within the Kanjera Bed sequence, the highest U, Rb and Rn values were found in K2 and K4 bones. As K2 and K4 outcrops are not contiguous, knowledge of where a fossil was surface collected plays a role in provenancing fossils to these beds. High Sr and Th values distinguish many K3 fossils from those of the rest of the section. These findings provide a framework for the provenancing of surface collected fossils.

Descriptions of past expeditions[4,41] suggest that most of the Kanjera *Theropithecus* was collected around a large outcrop of K2 sediment. The surface collections of the Smithsonian Expedition only recovered green fossils, such as KT62 and KT64, around this "K2 Island", suggesting that these specimens too were derived from this area. While the K2 Island is bordered by K3 deposits on one side, the high U (Fig. 3, Table 1) and moderate Sr values (Fig. 4) of KT62 and KT64 are clearly more similar to those of the K2 sample. This finding thus links Leakey's 1958 *Theropithecus* collection to K2, and provides a means of testing the provenance of the rest of the *Theropithecus* sample.

The hominid fossil (KNM-KJ 7494A) is distinct from the Kanjera Bed fossils and similar to the BCS and recent bones in having very low or indeterminate Y, Zr, La and Yb values (Figs. 1, 2; Table 1). Its Sr values, on the other hand, are comparable to those of some of the Kanjera Bed bones, and higher than the Sr values of the modern sample and many BCS bones (Fig. 4). Lastly, the hominid specimen's U values (Fig. 3, Table 1) are higher than the modern bone sample and some BCS bones, though much lower than the median K2 and K4 values. This overall trace element composition suggests that this specimen is from the BCS, and thus is younger than fossils from the Kanjera Beds. The provenancing of KNM-KJ 7494A to the BCS and the discovery of an anatomically modern human temporal in a nearby BCS outcrop in 1987 helps to tie the hominid material collected by Pilbeam[10], Pickford[11] and the Smithsonian Expedition[40] to the BCS, but does not indicate a provenance for Leakey's hominid fossils. The data presented here suggests that, like the elemental composition of pottery, lithics and volcanic tuffs, the trace element composition of fossils can be used to characterize excavated assemblages and link to them specimens of unknown provenance.

The high elemental values of the Kanjera Bed fossils almost certainly reflects the chemistry of the Homa Mountain carbonatites. Carbonatite lavas are known to be rich in U[42]. In addition, carbonatites contain the highest concentrations of REE and the highest ratios of light:heavy REE of any igneous rock[43]. The Homa carbonatites are no exception, exhibiting high REE and Sr values[1]. The high ratios of light:heavy REE in Kanjera Bed sediment samples[44] and suggested by the high La:Yb ratio of the fossils may be directly related to the chemistry of the local carbonatite deposits.

Some of the elemental concentration differences (e.g. U) noted between the Kanjera Bed samples may reflect fossilization under different ground water redox potentials[13-15] affecting elemental oxidation state and mobility. Alternatively, the differences may simply reflect differences in ground water trace element enrichment during fossilization. Similarly, the large elemental value ranges of the Kanjera Bed samples probably reflect microenvironmental differences within each bed during fossilization.

With the exception of Y and to some extent Zr, trace element concentration did increase with age through the Kanjera Bed sequence, suggesting that the environment of fossilization was the determining factor in trace element uptake. Prior to this analysis, it might have been tempting to postulate that Kanjera Bed bones would always have higher elemental concentrations than those from Ap and BCS. This was not found to be the case. Some poorly fossilized BCS fossils had U concentrations greater than K3 bones, fossils at least one million years old[40]. It is important to observe elemental concentration variation in fossils through the stratigraphic column before initiating a relative provenancing programme. A previous provenancing study of the Kanjera hominids collected by Leakey[45] was undertaken without the appropriate safeguards. Assays for F and U were used to infer the age of the hominid sample relative to that of a sample of undoubted Kanjera Bed fossils. While the F concentrations of the two groups were comparable, U concentration differences were thought to demonstrate that the human fossils were younger than those of the Kanjera Bed sequence. Lack of any detectable U in most K3 fossils and its low concentrations in K5 fossils (Fig. 3) indicates that U concentration cannot be used as a measure of age since deposition at Kanjera.

In summary, quantitative and qualitative XRF and ion microprobe analyses of fossils from Kanjera, Kenya were carried out. The correlations between the two types of XRF data were high, indicating that the relative elemental values of the qualitative analysis reflect the actual concentrations within the fossils. Differences in trace element concentration were useful in distinguishing Kanjera Bed bones from those of the more recent Ap and BCS and for provenancing fossils to particular Kanjera Beds. These results provide a framework for qualitative XRF provenancing of important fossils from the locality, including the complete collection of hominid and *T. oswaldi* fossils. These will be evaluated in a future paper.

ACKNOWLEDGEMENTS

We thank the Office of the President, Republic of Kenya, and R. E. Leakey and the National Museums of Kenya for permission and support in conducting the field and laboratory studies. We also thank the Kenya National Council for Science and Technology (NCST) and IAEA, for CNST funds and equipment support under project KEN/0/003. Erik Hauri (MIT/WHOI Joint Program) carried out U analysis by isotope dilution on short notice. Tom Plummer was funded by the Wenner Gren Foundation, Leakey Trust, L. S. B. Leakey Foundation, Boise Fund, Sigma Xi, Donner Foundation (through the Smithsonian Institution), the Smithsonian Institution and the J. F. Enders Fellowship Fund of Yale University. Funds from the International Science Programs (ISP), Sweden allowed A.M. Kinyua to present these research findings.

REFERENCES

1.   C. Barber, Lithos 7:53-63 (1974).
2.   M. J. Le Bas, "Carbonatite-nephelinite volcanism," John Wiley and Sons, N.Y. (1977).
3.   T. W. Plummer and R. Potts, J. Hum. Evol. 18:269-276 (1989).
4.   F. Oswald, Quart. J. Geol. Soc. Lond. 70:128-188 (1914).
5.   L. S. B. Leakey, "The Stone Age Races of Kenya," Oxford University Press, London (1935).
6.   L. S. B. Leakey, "Adams Ancestors," Methuen, London (1954).
7.   L. S. B. Leakey, Homo sapiens in the Middle Pleistocene and the evidence of Homo sapiens' evolution, in: "The Origin of Homo sapiens," F. Bordes, ed., UNESCO, Paris (1972).
8.   P. G. H. Boswell, Nature 135:371 (1935).
9.   P. E. Kent, Geol. Mag., 79:117-132 (1942).
10.   D. R. Pilbeam, Hominid-bearing deposits at Kanjera, Nyanza Province, Kenya, unpublished report (1974).
11.   M. Pickford, "Kenya Palaeontology Gazetteer," National Museums of Kenya, Nairobi (1984).
12.   K. P. Oakley, Bull. Brit. Mus. Nat. Hist. (Geol.) 34:1-63 (1980).
13.   P. Henderson, C. A. Marlow, T. I. Molleson and C. T. Williams Nature 306, 358-360 (1983).

14. C. T. Williams and C. A. Marlow, J. Archaeol. Sci. 14:297-309 (1987).
15. C. T. Williams and P. J. Potts, Archaeometry 30:237-247 (1988).
16. P. J. Potts, *A Handbook of Silicate Rock Analysis*, Chapman and Hall, N.Y. (1987).
17. J. D. Stewart, P. Fralick, R. G. V. Hancock, J. H. Kelley and E. M. Garrett, J. Archaeol. Sci. 17:601-625 (1990).
18. A. L. Wilson, J. Archaeol. Sci. 5:219-236 (1978).
19. P. Duerden, D. D. Cohen, E. Clayton, J. R. Bird, W. R. Ambrose, and B. F. Leach, Anal. Chem. 51:2350-2354 (1979).
20. H. V. Merrick and F. H. Brown, Afr. Archaeol. Rev. 2:129-152 (1984).
21. D. E. Nelson, J. M. D'Auria, and R. B. Bennett, Archaeometry 17:85-97 (1975).
22. T. E. Cerling and F. H. Brown, Nature 299:216-221 (1982).
23. F. H. Brown, I. McDougall, T. Davies and R. Maier, An integrated Plio-Pleistocene chronology for the Turkana Basin, in: "Ancestors: the Hard Evidence," E. Delson, ed., Alan R. Liss, N.Y. (1985).
24. G. Andermann and J. W. Kemp, Anal. Chem. 30:1306-1309 (1958).
25. R. D. Giauque, R. B. Giarett and L. Y. Goda, Anal. Chem. 49:511-516 (1979).
26. K. K. Nielson, Anal. Chem. 49:641-648 (1977).
27. M. M. Lavi, M.Sc. Thesis, University of Nairobi (1986).
28. D. S. Kendall, J. H. Lowry, E. L. Bour, and T. J. Meszaros, Adv. X-Ray Anal. 27:467-4723 (1984).
29. T. K. Smith, Adv. X-Ray Anal. 27:467-473 (1984).
30. G. E. Gigante, L. J. Pedraza and S. Sciuti, Nucl. Instr. Meth. Phys. Res. B12:229-234 (1985).
31. A. K. Behrensmeyer, R. Potts, T. W. Plummer and L. Tauxe (in prep.).
32. E. Badone and R. M. Farquhar, J. Rad. Anal. Chem. 69:291-311 (1982).
33. A. M. Kinyua, M.Sc. Thesis, University of Nairobi (1983).
34. B. Holynska, Spectrochim. Acta. 278:287 (1972).
35. P. Verbeke and F. Adams, Anal. Chim. Acta 109:85-95 (1979).
36. R. H. Leiser, Anal. Chem. 307:177-184 (1981).
37. N. Shimizu and S. R. Hart, Ann. Rev. Earth Planet. Sci. 10:483-526 (1982).
38. P. Grandjean, Ph.D. Thesis, University of Nancy.
39. Hauri, pers. comm. (1990).
40. Plummer, T., Ph.D. Thesis, Yale University (1991).
41. L. S. B. Leakey, J. E. Afr. Nat. Hist. Soc. 17:39-44 (1943).
42. E. M. Durrance, "Radioactivity in Geology", John Wiley and Sons, N.Y. (1986).
43. A. R. Woolley and D. R. C. Kempe, Carbonatites: nomenclature, average chemical compositions, and element distribution, in: "Carbonatites: Genesis and Evolution", K. Bell, ed., Unwin Hyman, Boston, (1989).
44. P. Henderson, M. Pickford and C. T. Williams, J. Afr. Earth. Sci. 6:221-227 (1987).
45. K. P. Oakley, J. Hum. Evol. 3:257-258 (1974).

FAST AND SIMPLE ROUTINE DETERMINATION OF BROMINE BY XRF IN WET BLOOD

SERUM MICROSAMPLES. EVALUATION OF ERRORS

C. Shenberg, J. Gilat and M. Mantel

Soreq Nuclear Research Center
Yavne 70600, Israel

ABSTRACT

A specific XRF method, developed at the Soreq Nuclear Research Centre, was applied to the determination of bromine in blood. The method is based on excitation with a Mo X-ray tube and detection of the fluorescent Br K X-rays by a Si(Li) detector. Serum microsamples (300 µL) are counted directly, without drying, for 100 sec. The detection limit obtained under these conditions is 0.6 ppm Br. The overall precision of the method was found to be ±3.1%. The different parameters which contribute to the total error of the method were studied. A survey of the bromine concentration in the blood serum of industrial workers exposed to bromine compounds was carried out.

INTRODUCTION

The rapid and accurate determination of bromine in blood serum is of importance for the diagnosis and follow-up of bromine poisoning. The latter may be a result of overdoses of bromine containing drugs or of overexposure to bromine compounds. Lately the follow-up of bromine in the blood of workers exposed to bromine compounds, has become compulsory. The method of X-ray fluorescence (XRF) was found to be most suitable for this purpose and an increasing number of investigators resorted to this technique[1-7].

The aim of the present study was to use a novel XRF system, developed recently in our laboratory[8-10] for the accurate, fast and simple routine determination of bromine in the blood serum of industrial workers exposed to bromine compounds. The method is based on excitation with a Mo X-ray tube and detection of the fluorescent X-rays by a Si(Li) detector, followed by identification and integration of the Br Kα peak. Serum microsamples are counted directly, without drying. A special filter and collimator assembly as well as the geometry of our system result in a detection limit of 0.6 ppm Br for 100 sec counting time, in spite of the use of wet microsamples.

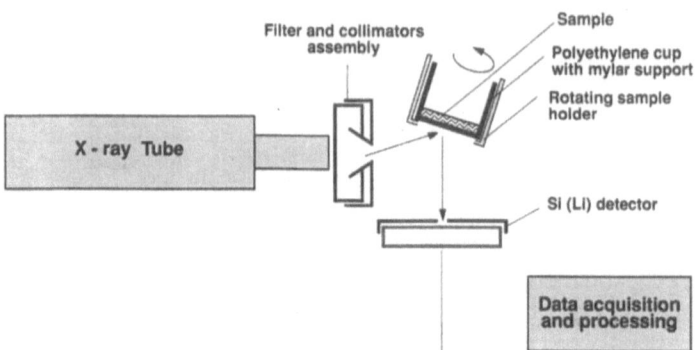

Fig.1. Experimental set-up

Spiked homogenized serum standards were used to obtain identical matrices for samples and standards and to avoid the need to correct for self-absorption In this way a linear calibration curve was obtained in the 0-165 ppm range.

An analysis of the various sources of error is presented. In particular, the use of the backscattered Mo K X-ray peak intensity to correct for geometrical irreproducibilities, is discussed. Finally, the precision and the accuracy of the method are established.

The new technique was applied to a survey of the bromine concentration in the blood serum of 460 industrial workers (males and females) exposed to bromine compounds.

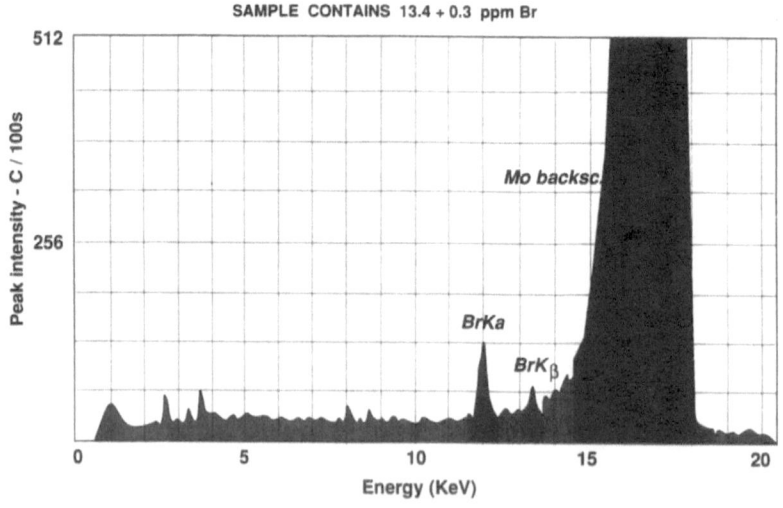

Fig. 2. Typical X-ray spectrum obtained from the analysis of a blood serum sample

Table 1. Variation of count rate (c/ppm Br) with Br
concentration in NH$_4$Br solutions in distilled
water

Concentration ppm Br	Number of samples n	Count rate[a] c/ppm Br
1	5	241 ± 7
2	5	235 ± 9
5	8	227 ± 7
10	6	214 ± 4
20	5	208 ± 7
50	5	199 ± 6
100	4	192 ± 7
200	6	194 ± 2

a - corrected according to MoKα bacscattered intensity.

EXPERIMENTAL

Apparatus

Excitation was carried out with a Mo X-ray tube with an optical focus of 0.4 x 0.8 mm. The tube was operated at 30 kV and 30 mA. The fluorescent Br K X-rays obtained from the sample are detected and counted by a Si(Li) detector of 25 mm^2 area coupled to a 4096 multichannel analyser and to a data processing system.

A schematic view of the experimental set-up is shown in Fig. 1.

Procedure

300 μL serum are pipetted onto a mylar support (the bottom of a small, 11 mm diameter, polyethylene cup) and counted directly, without drying, for 100 sec. The Br Kα X-ray peak is integrated and its intensity is normalized to the intensity of the backscattered Mo K X-rays in the 14-19 keV range.

A typical X-ray spectrum obtained from the analysis of a blood serum sample is shown in Fig. 2.

RESULTS AND DISCUSSION

Calibration

Table 1 shows the dependence of the counting rate on bromine concentration in aqueous solutions.

It may be seen that the count rate per ppm Br is constant above 20 ppm indicating a linear dependence. At lower concentrations the calibration curve is not linear due to concentration dependent self-absorption which saturates at the higher concentrations.

When serum standards (prepared by adding no more than 10% of bromine solution to homogenized serum) are used, the count rate is lower due to

Table 2. Errors due to counting statistics including peak integration and system stability

Sample[a]	Peak intensity - c/100s BrKα-uncorrected	
A 1	2604	
2	2502	
3	2405	
4	2630	
5	2566	Mean : 2542 ± 72
6	2516	(±2.8%)
7	2601	
8	2457	
9	2587	
10	2554	
B 1	2294	
2	2228	
3	2324	Mean : 2291 ± 38
4	2294	(±1.6%)
5	2315	

a - the same sample counted several times.

Table 3. Geometrical reproducibility

Sample[a]	Peak intensity - c/100s		
	BrKα-uncorrected	Backscatter	BrKα–corrected
C 1	2604	626453	4157
2	2451	600390	4082
3	2406	577896	4163
4	2294	598397	3834
5	2546	619297	4111
6	1740	423709	4107
7	2322	575777	4033
8	2444	583013	4192
9	2369	573676	4130
10	2380	570189	4174
Mean:	2356±236 (10%)		4098±104 (2.5%)
D 1	1008	378020	2667
2	1559	595040	2620
3	1561	619588	2519
4	1588	616972	2574
5	1597	615913	2593
6	1568	615293	2548
7	1613	625117	2580
8	1639	622129	2635
9	1636	621720	2631
10	1621	630226	2572
Mean :	1539±189 (12.3%)		2594±45 (1.7%)

a - different samplings of the same serum sample.

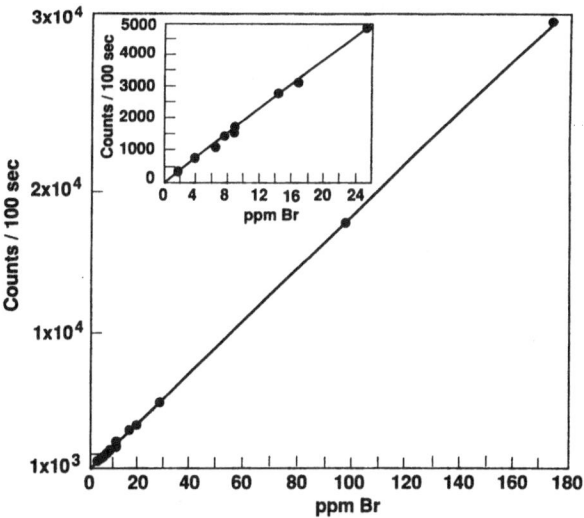

Fig. 3. Calibration curve obtained from Br standards in serum

higher self absorption (175c/ppm Br instead of 195c/ppm Br); however in this case the calibration curve is linear throughout the entire range, as shown in Fig. 3. Each result shown on the figure is the mean of 3 different samplings of the same bromine solution. The reproducibility of the results at each concentration is of the order of ±2 to ±3%.

The minimum detection limit (MDL) defined as $3\sqrt{B}$, (B = background) is 0.6 ppm Br for 100 sec counting time.

Evaluation of Errors

To test the effect of counting statistics, peak integration errors and overall system stability, the same sample was counted several times. Results are shown in Table 2. As can be seen the due to counting statistics is ±2 to ±3%.

To study errors introduced by geometrical effects (uniformity of polyethylene cups, positioning of cups in the system) a number of samplings of the same serum were measured. Results are shown in column 1 of Table 3. It can be seen that in both cases shown in the table (C and D), the measurements are reasonably close (within 13% or better), but two determinations (C6 and D1) deviate by as much as 30%. To correct for this random effect and to improve the reliability of a single determination, we use the intensity of the backscattered Mo K X-rays integrated over the 14-19 keV range. This value is shown in column 2 of Table 3. When the Br peak intensity is normalized to a constant backscattered intensity (in our case 10^6 counts), the discrepancy disappears and, as shown in column 3 of Table 3, the error is again ≤2.5%. Thus we find that the use of the backscattered X-ray intensity is an important means for improving the accuracy of the results in routine determinations.

Table 4 shows representative results for 30 samples out of 460 analyzed in our study. In most cases the backscattering correction greatly

Table 4. Representative results of bromine analyses in human serum

Sample	Peak intensity - c/100s			
	BrKα-uncorrected	δ (%)	BrKα -corrected	δ (%)
1	1441 1204	±13	2370 2315	±1.7
2	2779 2314	±13	4343 4228	±1.9
3*	2307 2043	±9	3443 3744	±5.0
4	1437 1224	±11	2277 2282	±0.2
5	1556 1160	±21	2416 2347	±2.0
6	1333 1665	±16	2737 2824	±2.2
7*	1540 1376	±8	2530 2199	±10.0
8*	1375 1172	±11	2185 1958	±7.7
9	1632 1366	±16	2499 2462	±1.1
10	1905 1638	±11	3157 3153	±0.1
11*	1805 1995	±7	3246 3462	±4.6
12	1574 1755	±8	2947 3018	±1.7
13*	1090 1333	±14	1868 2197	±10.6
14	1330 1112	±13	2219 2181	±1.2
15	1408 1123	±16	2302 2252	±1.6

improves the agreement of duplicate analyses. In some samples (indicated by asterisks) the discrepancies remain (±4.6% to ±10.6%) and we attribute them to serum inhomogeneity.

The precision of the method calculated from duplicate analyses of 460 blood serum samples was found to be ±3.1%.

To verify the accuracy of the above XRF technique, spiked serum samples were prepared and analyzed. The results obtained are shown in Table 5. The accuracy of the method was further tested by comparing our results with analyses of the same samples by high pressure liquid chromatography (HPLC)*. Results of the comparison are shown in Fig. 4.

As may be seen very good agreement was found in the determination of bromine in blood serum by the two methods. The HPLC technique is much more tedious and labor intensive than the XRF method. Also, the amounts of sample material required are larger by about a factor of 10.

---

*Ruth Ashkenazi. The Institute for Occupational Health, Tel-Aviv University, Sacler Faculty of Medicine.

Table 4.   (cont.)

| Sample | Peak intensity - c/100s | | | |
	BrKα-uncorrected	δ (%)	BrKα-corrected	δ (%)
16	1740 2249	±18	3492 3623	±2.6
17*	2576 2085	±15	4257 3920	±5.8
18	1031 1376	±20	2107 2119	±1.0
19*	1539 1377	±8	2677 2487	±5.2
20*	1574 1767	±8	2995 3222	±5.2
21	1205 1527	±17	3369 3416	±0.4
22	671 978	±26	1774 1841	±2.6
23	1153 1531	±20	2443 2513	±2.0
24	1687 1544	±6	2774 2766	±0.2
25*	1374 1032	±20	2151 1915	±8.2
26	1632 1856	±9	2889 2916	±0.7
27*	1869 1660	±8	2903 2709	±4.9
28	1585 966	±34	2757 2820	±1.6
29	1865 1465	±17	3084 3028	±1.3
30	1392 1510	±8	2503 2501	±0.1
Mean:		±14 [a]		±3.1 [b] ±1.3 [c] ±6.7 [d]

a - mean of 30 samples (uncorrected)
b - mean of 30 samples (corrected)
c - mean of 20 homogeneous samples (corrected)
d - mean of 10 inhomogeneous samples (corrected)
* - inhomogeneous serum

Table 5. Accuracy (A) of the XRF method

Br in serum[a] ppm	Br added ppm	Br found [a] ppm	A%
11.2±0.2 4.9±0.1	8.3 13.3	19.9±0.3 18.4±0.3	±2.1 ±1.1

a - mean value of 6 different samplings.

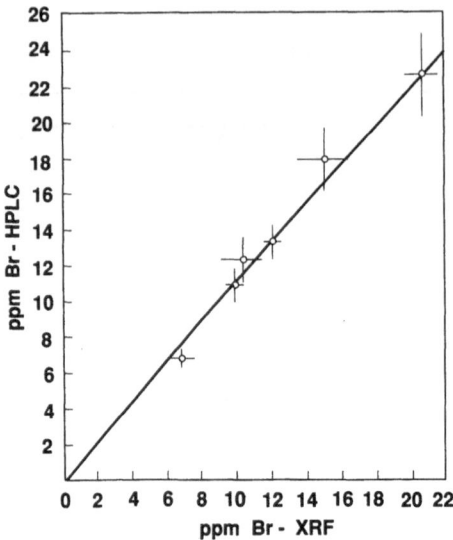

Fig. 4. Comparison between the results obtained for Br in serum by XRF and HPLC

CONCLUSION

The XRF method described permits the direct determination of bromine in wet serum microsamples without the need of any pretreatment. At the same time the method remains fast and accurate. Thus this technique is especially suited for routine surveys and other applications involving large sample populations.

REFERENCES

1. D. A. Applegarth and A.G.F. Davidson, Clin Biochem. 10:127 (1977).
2. H. H. Sky – Peck and B.J. Joseph, Clin. Biochem. 14:126 (1981).
3. M. S. Rapaport, M. Mantel and C. Shenberg, Med. Phys. 9:194 (1982).
4. G. E. Gigante and G. Varrassi, Boll. Soc. Ital. Biol. Sper. 23:1596 (1982).
5. F. Rastegar. E.A. Maier, R. Heimburger, C. Christophe, C. Ruch and M.J.F. Leroy, Clin. Chem. 30:1300 (1984).
6. L. Wielopolski, W.H. Adams and P.M. Heotis, Environ. Res. 41:91 (1986).
7. C. T. Yap, Appl. Spectroscopy 42:1250 (1988).
8. S. Chaitchik, C. Shenberg, Y. Nir-El and M. Mantel, Biol. T. Elem. Res. 15:205 (1988).
9. C. Shenberg, M. Mantel, T. Izak-Biran and B. Rachmiel, Biol. T. Elem. Res. 16:87 (1988).
10. C. Shenberg, T. Izak – Biran, M. Mantel, B. Rachmiel, J. Weininger and S. Chaitchik, J. Trace Elem. Electrolytes Health Dis. 3:71 (1989).

X-RAY MICROPROBE STUDIES OF HUNGARIAN BACKGROUND AND URBAN AEROSOLS

Sz. Török and Sz. Sándor

Central Research Institute for Physics
H-1525 Budapest, P.O. Box 49

C. Xhoffer and R. Van Grieken

Department of Chemistry, University of Antwerp (UIA)
B-2610 Antwerp-Wilrijk, Belgium

K. W. Jones

Brookhaven National Laboratory
Upton, NY 11973 USA

S. R. Sutton and R. L. Rivers

The University of Chicago
Chicago, Il 60637 USA

INTRODUCTION

In order to determine the polluting atmospheric sources in urban and background areas source apportionment of the air particulate matter is necessary. Hitherto these studies were mostly based on bulk composition measurements of the aerosol. Source profiles, i.e. the concentrations of several elements for air particulate matter originating from one source, can be deduced from the receptor data using a number of multivariate techniques among which the chemical mass balance. The application is limited by the large number of observations that must be made for each of the variables. Often an elaborated sample preparation is necessary for fractionating the sample into several sub samples, according to the density, particle diameter or other relevant properties . Often this may results in poorly resolved source profiles.

On the other hand, methods for single particle analysis provide direct information on chemical composition and morphology for each individual particle. In case of sufficient lateral resolution and chemical sensitivity this enables a ready fractionation on the basis of some measured or derived parameters, as well as an extraction of source profiles.

Automated electron probe x-ray micro analysis (EPMA) has sufficient spatial resolution (0.1-0.3 μm) to detect individual particles, and it has successfully been used to classify atmospheric aerosol particles from urban, remote, continental and marine areas, or suspension particles in estuaries and seas [1,2]. Such method was never before used to analyze aerosol samples from the area of Eastern Europe. In previous work only flyash particles originating from this geographical area were studied by single particle analysis [3]. The emitted stack flyash particles were classified into several chemically and morphologically different groups. The measurement data of summer and winter sampling showed that in the smaller(<2μm) particle size fraction two types of unexpected groups are present. One group has high barium content and the other group consists of calcium sulfate particles with high arsenic content. This arsenic is supposed to be on the surface of the particles due to the condensation in the cooled stack gas.

The aim of the present work is to find the relative abundance of the particle types originating from two different background monitoring stations in the middle of the Great Hungarian Plain. In urban areas most pollutants originate from traffic and municipal waste incineration. Since heavy metals play an important role in these samples the highly sensitive x-ray microscope (XRM) of the National Synchrotron Light Source (NSLS) of the Brookhaven National Laboratory was used. A feasibility study on individual aerosol particles sampled at the above background stations and in the urban area of Budapest is discussed.

EXPERIMENTAL

Sampling
Seven samples were taken in the spring of 1990 at the sampling site K-puszta, which is located nearly at the center of Hungary between the rivers Danube and Tisza, about 70 km south east of Budapest. The closest town of 100 000 inhabitants is about 10 km SE, and an iron work is 50 km from the station. The closest paved road (with very low traffic density) is at least 5 km away. The station is situated on a forest clearing in a pine forest but vineyards are in the neighborhood as well.

Sampling was performed at 2 m above ground level using Nuclepore filters of 0.4 μm pore size, with 1.5 cm^2 exposed area. The sampling time was 24 h and the sample volume varied between 3 and 8 m^3.

A similar sampling procedure was carried out in the Hortobágy National Park near Nagyiványi about 50 km west from Debrecen, 2 m above ground level. This campaign resulted in six samples taken in August 1990 .

Several samples were taken in the urban areas of Budapest at ground level, in traffic tunnel ventilation systems and on top of buildings. Only giant particles of these samples were studied by XRM.

EPMA

The aerosol loaded filters were measured by a JEOL 733 Superprobe equipped with a Tracor Northern particle recognition and characterization program (PRC) that facilitates a fully automated analysis on a preset number of particles. For x-ray micro analysis, 25 keV and 1 nA operating conditions were used. The PRC system operates in the following way. As the beam scans across the field of interest a particle is considered as detected when the digitized backscatter signal exceeds a preset threshold value. The coordinates of the contour points are determined and additional information such as particle diameter, perimeter and shape factor (perimeter^2/4πxarea) are calculated. Energy dispersive x-ray spectra are collected for 30 seconds with the electron beam positioned in the centroid of the particles. In each sample about 300 particles were measured. Measurement data were stored on floppy disc for "off-line" data processing on a VAX 11/780. The large data set consists of: net intensities for 18 elements, diameter and shape factor for each particle.

Classification of the particles was carried out by hierarchical cluster analysis based on the Ward's error sum strategy, that has previously been proven to be one of the most advantageous procedures for environmental applications [4]. A second hierarchical clustering was performed on the average composition data of the samples and resulted in a set of training vectors (centroids) that are relevant for the total sampling campaign. Finally a nearest centroid sorting is used to classify all particles from one campaign according to their distances from the centroids of the clusters. The method of Forgy [5] minimizes the sum of squares of the distances to the centroids for a fixed number of clusters. This procedure results in an average composition data set for each sampling site and the abundances of the particle groups in each sample.

XRM

Synchrotron radiation x-ray analysis has been performed on beam line X26 at the NSLS at Brookhaven National Laboratory. The NSLS source is characterized by the following parameters: electron energy 2.53 GeV, critical energy 5.0 keV, maximum ring current about 200 mA. The experimental methods adopted for this work are very similar to the ones described in [6]. The XRM utilizes the continuous x-ray spectrum produced by the bending magnet. The flux for photon energies about 4 keV is 2×10^6 photons/μm^{-2} s^{-1} mA^{-1} at 9 m from the x-ray ring. A set of four stepper-motor-driven tantalum slits can be used to produce collimated beams down to about 40 μm. A tantalum collimator placed 3.5 cm from the sample position was used in the present experiments to produce an x-ray beam of 4 μm x 7 μm.

RESULTS AND DISCUSSION

Size distribution curves of all samples were calculated. They show that 80 % of the detected particles is smaller than 3 μm. Under certain meteorological conditions as encountered this particle fraction might have long residence time in the atmosphere [7]. Hierarchical clustering of the chemical and morphological data showed that most particles (>0.3 μm) during this sampling period were of antropogenic origin, mainly from power stations and industrial emissions. Table I gives an overview of the 11 major particle types of the K-puszta sampling campaign.

The largest group seems to be industrial particles originating from power plants and metallurgy. Silicate particles of industrial origin are different in composition and shape from the soil silicates. They can originate from any type of coal burning. Their abundance did not depend

Table 1
Particle groups in background aerosols sampled in April-May
in K-puszta as detected by EPMA

Origin	Abundance (%)	Major components detected by EPMA	Diameter (μm)	Shape factor
Industrial				
Silicate	24	Si, Al, S, Fe	2.6	1.7
Gypsum	15	Ca, S, Si	2.3	2.2
Iron rich	12	Fe, Si	1.5	1.5
Pyrite	7	Fe, S, Si	1.2	1.6
Metal	1	Al or Ti or As		
Soil dust				
Silicate	22	Si, Al, K, Fe	2.7	2.6
Quartz	10	Si	1.5	2.3
Limestone	1	Ca	1.7	2.5
Sulphate	1.5	S	0.8	1.7
Traffic	1.2	Pb, Cl, Br	0.6	1.3
Biological	3	S, P, K	2.6	2.5

significantly on the wind direction. Very few particles were observed with a relatively high V and Ni content which would indicate oil fired power station as aerosol emission source.

The second group, gypsum, is very common in flyash since often lime is added to reduce gaseous sulphur emission. In some Hungarian power stations (Ajka) alkaline lignite with high Ca content is burned. However, gypsum particles might originate from various sources [8].

The iron rich particles might originate from various sources like steel smelters, corrosion and coal burning. The abundance of this group depends on the wind direction. The same holds for the iron sulfate group.

Soil dust silicates have rectangular or irregular shape giving shape factor >1.5. Their Na, Mg and K content is also higher than that of the flyash silicates. However if the measured data have higher statistical error the distinction of the two groups is very difficult. Quartz was present in all samples but the lime particle group was only occasionally observed.

Since the detection of the particles is under automatic control, very small aerosol particle groups about a few tenth of microns are not always observed. Moreover, in view of the energy deposited by the electron beam in the EPMA and the vacuum condition, unstable components will disappear. For this reasons the secondary $(NH_4)_2SO_4$ particles are not detected as abundant as they are present in the aerosol [9]. Since the sampling stage was far from paved road the presence of automobile exhaust particles was not significant.

Morphological visualization also confirmed that the samples contained numerous biological particles.

Relative abundances or particle number concentrations (expressed in percent) in each individual sample are given in Figure 1 for the K-puszta campaign.

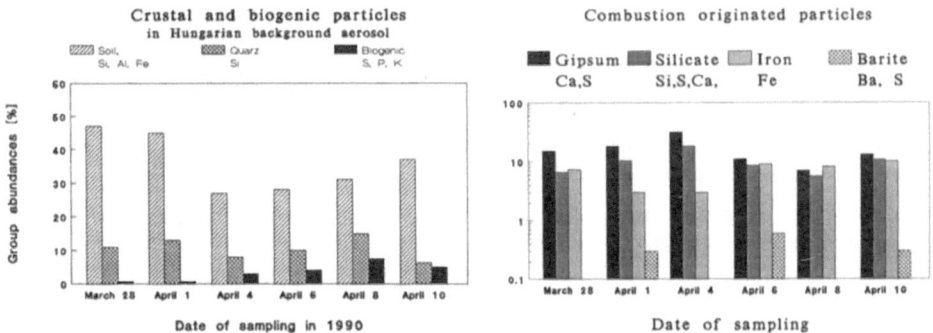

Figure 1.
Group abundances in % of crustal biogenic and combustion originated
particles in the K-puszta campaign.

The particles originating basically from coal burning have similar
tendencies during the sampling period.

The sampling campaign in Hortobágy showed very similar particle group
abundances except for biological particles that were more abundant there
due to the summer season. Silicates contained much less sulfur which
means that flyash was not as dominant as in the K-puszta sample, despite
that the site is much closer to the northern industrial area.

From the EPMA data it can be concluded that anthropogenic particles are
present at unexpectedly high abundance in the so called background
aerosol. In order to get better source profile a high performance micro
chemical technique , XRM was tested to study the detectability   of
various trace metals in atmospheric particles. Since the beamsize was
relatively large compared to the average particle diameter in the aerosol
at present only giant particles were measured. Table 2 shows
illustrative examples for some elements found in single aerosol
particles. The values correspond to 300 s counting time.

Table 2
Elements detected by XRM in individual aerosol particles

Particles	diameter [um]	Elements above determination limit
Background aerosol		
Soil	5	Si S K Ca Ti Mn Fe Zn
Flyash	3	S K Ca Ti Mn Fe Cu Zn Pb
Pyrite	6	S Ca Ti Cr Mn Fe Cu Pb
Urban		
Rust	4	S Ca Ti Cr Mn Fe Ni Cu Zn Pb Mo
Lead	6	S Ca Ti Cr Mn Fe Cu Zn Pb
Soot	5	S Ca Cr Mn Fe Ni Zn Pb
Agglomerate	20	S Ca Cr Mn Fe Ni Cu Zn Pb

XIV. XRS APPLICATIONS

Acknowledgement

This work was supported by Hungarian government (OTKAI/2 1030 and I/3 2984) by Belgium in the framework of a cooperation project between NFWO and the Hungarian Academy of Sciences.
The authors are indebted to the Atmospheric Physics Institute for kindly handing over the aerosol samples.

We acknowledge a grant from US-Hungarian Joint Board on Scientific and Technological Cooperation No. 111-91

References

1. Bernard,P., Van Grieken,R, Eisma,D, Environ. Sci.Technol. **20,** 457 (1986).
2. Artaxo, P.,Manhaut,W and Van Grieken, R, Tellus, in press. Török Sz, Sándor, S. and Rausch,H., Advances in X-Ray Analysis,33, 673 , Plenum Press , New York, 1990.
3. Török, Sz., Sándor, Sz., Xhoffer, C., Van Grieken, R., KFKI-1992-16-J Report, Central Research Institute for Physics, Budapest.
4. Van Espen, P., Anal. Chim. Acta, **165,** 31 (1984).
5. Forgy, E.W. , Biometrics **21,** 768 (1965).
6. Jones, K.W., Bockman, R.S; Gordon, B.M.,Rivers, M.L., Saubermann, A.J., Schildlovsky,G, and Spanne, P., in XRF and PIXE Applications if Life Science, Capri, Italy, eds. R.Moro and R. Ceareo (World Scientific Publishing Co., Singapore, 1990) p. 163.
7. Török,S., Sz. Sándor, Xhoffer, C, Van Grieken,R., Molnár, Á, Mészáros,E., Járai-Komlódi, M. Submitted to Meteorology.
8. Xhoffer,C, Brenard, P, Van Grieken, R and Van der Auwera, L, Envir. Technol. **25,**1470 (1991).
9. Bassette, M. and J.H.Seinfeld, Atmos. Environ. **17,**2237 (1984).

STASTITICAL COMPARISON OF ANALYTICAL RESULTS OBTAINED BY PRESSED POWDER
AND BORATE FUSION XRF SPECTROMETRY FOR PROCESS CONTROL SAMPLES OF
A LEAD SMELTER

Jorg G.H. Metz and David E. Davey

School of Chemical Technology
University of South Australia
The Levels, S.A. 5095, Australia

ABSTRACT

Two sample preparation procedures for XRF determination of
critical elements in process control samples of a lead smelter
have been compared. The two methods, pressed powder(PP) and
borate fusion(BF) were used in the analysis of (in order of
importance) lead, zinc, copper, sulphur, arsenic and the flux
elements (FeO, MnO, Al_2O_3, CaO, MgO and SiO_2). The techniques
were applied to three points in the overall lead-zinc-copper
production process.

The three sample types have been selected to allow comparison
between the BF and PP methods, where the PP method suffers from
difficulties with particle size, moisture absorption or
composition variation.

The BF technique was found to give better precision (typically
0.2 to 1% RSD) than the PP approach (typically 0.8 to 9% RSD),
and to give better accuracy as found using reference materials.
The stability of the BF bead over time is also of advantage in
quality control. The flexibility and reliability of the BF
technique has lead to its acceptance, and in part, to changes
in the overall smelter analysis structure.

INTRODUCTION

The borate fusion (BF) procedure for sample preparation in X-ray
spectrometry has been available for some time, and has gained
wide acceptance.[1-6] However, in metallurgical processing, some
problems are encountered which have hindered its full
acceptance, and kept the pressed powder (PP) technique popular.
PP-XRF[2,5] and BF-XRF[3,4,6] have both been used for a variety of
analyses for many years. The choice has been dependent on
material composition, accuracy and precision requirements and
the elements required for analyses, as well as their level,
i.e.trace, minor, or major. BF-XRF has been used in the iron
ore, cement and minerals sands industry for many years. But it
is only in recent times that it has become available to the
lead-zinc-copper industry. In the work done by Norrish and
Thompson[1], it is suggested that sulphur can be retained
quantitatively for a variety of mineral types. The potential
loss of volatiles in preparing the fused sample had until then
been a stumbling block in using the BF method with sulphide
minerals.

The PP method, however, was causing problems in process control
of lead-zinc-copper production at the Pasminco Metals/BHAS
smelter in Port Pirie, South Australia. (Further detail of the
process and selection of samples will be given later). The
problems observed, which may be of general interest to XRF
analysts were as follows. Firstly, moisture absorption on PP
surfaces seemed to affect SiO_2, Pb and S analyses. Secondly,
submicron particle requirements of some samples led to grinding
problems, and often gave gross errors. Finally, and crucially,
since the initial PP-XRF calibration some years ago, the
composition of input materials had changed considerably,
sometimes resulting in process materials being outside
calibration limits. It was thus an opportune time to carry out
a major recalibration. It was further decided to introduce the
BF procedure to the smelter operation on a trial basis.

In the tests initiated, the BF and PP sample preparation
procedures were compared over an extended period, in terms of
their influence on precision and accuracy for crucial elements.

EXPERIMENTAL

XRF Spectrometer and Fusion Apparatus

1. A Philips 1404 XRF Spectrometer incorporating a Sc/Mo X-ray
    tube was employed.

2. The Prometheus LXR6, DJC Fusion Apparatus, used to produce
    the lithium borate beads, was supplied by Automated Fusion
    Technology, Pty. Ltd., Melbourne, Australia, 3160. Heating
    in the apparatus is provided by oxygen-enriched liquid
    petroleum gas burners.

3. Melt temperatures were checked using a two-colour Hotspot
    pyrometer, the smallest aperture being selected to avoid
    reflectance errors.

Table 1. Briquette composition for control samples using the PP technique. Weights(g) for _dried_ material, time in seconds, pressure in tonnes per sq.inch.

Sample	Sample wt.	Boric acid	Grind time	Pressure
BLC	2	18*	60	3
BFS	10	10*	90	7
CDFM	2	18	60	3

* Note : Boric acid further dried at 80°C.

The Borate Fusion Method

The analytical sample was provided dry and less than 150 µm in particle size. Nominally, 0.66 g of sample was thoroughly mixed with 7.8 g of fusion flux (lithium tetraborate : lithium metaborate (12 : 22) with 12.8% sodium nitrate) in a Pt/Au crucible. This mixture was sintered in a muffle furnace at 700 $\pm$ 5°C for 10 to 20 minutes depending predominantly on the sulphur content.

The bead for XRF analysis was then produced by fusing the sintered material in the fusion apparatus, mentioned above, at 1060 $\pm$ 10°C. Ammonium iodide releasing agent was generally added to the melt before moulding to improve bead quality.

For samples above 5% but below 15% in Cu, the sample weight was reduced to 0.45 g, and for samples of even higher content, to 0.33 g. No ammonium iodide was added to the latter, since its addition led to low Cu recoveries, possibly due to the formation of copper iodide and its subsequent separation from the melt.

Table 2. BF calibration. **Each standard bead is based on a single elemental oxide or compound. Concentration ranges are given in wt % for the calibrant listed.**

Elemental Compound	Calibrant	Range (wt %)	Elemental Compound	Calibrant	Range (wt %)
As_2O_3	As_2O_3	0.01 - 60	$CaCO_3$	CaO	0.01 - 60
SeO_2	SeO_2	0.01 - 30	Sb_2O_3	Sb_2O_3	0.01 - 60
Bi_2O_3	Bi_2O_3	0.005 - 50	SnO_2	SnO_2	0.01 - 50
ZnO	ZnO	0.01 - 80	KCl	K_2O	0.01 - 50
CuO	CuO	0.01 - 60*	CdO	CdO	0.005 - 50
NiO	NiO	0.01 - 50	Ag_2O	Ag_2O	0.01 - 20*
CoO	CoO	0.01 - 20	KCl	Cl	0.01 - 20
Fe_2O_3	Fe_2O_3	0.01 - 60*	$CaSO_4.2H_2O$	SO_3	0.1 - 50
Mn_3O_4	Mn_3O_4	0.01 - 50	PbO	PbO	0.1 - 80
$K_2Cr_2O_7$	Cr_2O_3	0.01 - 20	SiO_2	SiO_2	0.1 - 80
$BaCO_3$	BaO	0.01 - 20	Al_2O_3	Al_2O_3	0.1 - 50
TiO_2	TiO_2	0.01 - 50	MgO	MgO	0.1 - 50

*Note: For Cu, Fe and Ag, the degree of absorption into the fusion crucible depends on the sample composition, completeness of oxidation, and the fusion conditions.

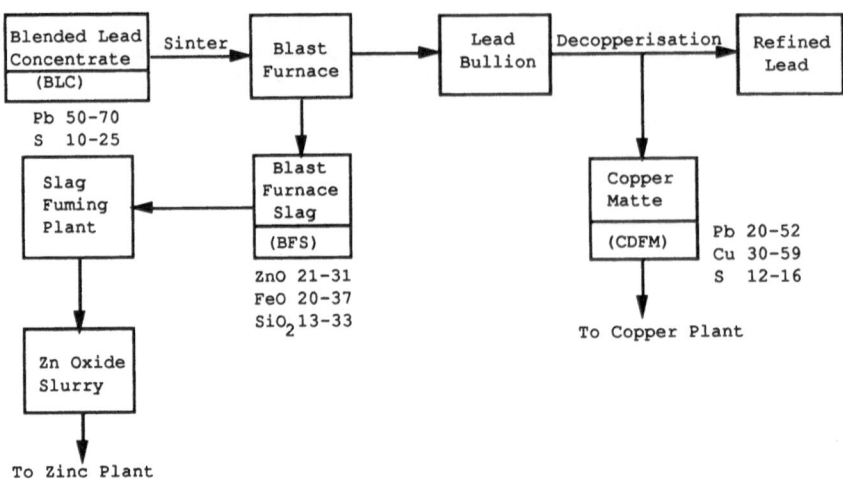

FIGURE 1.
The selection of control samples, BLC, BFS and CDFM, in
the smelter process. Critical concentrations (wt %) are
shown for the control samples.

The Pressed Powder Method

Sample, initially of particle size less than 150 μm, was ground
and mixed in a Rocklabs mill with boric acid diluent/binder. It
was then pressed into a "Somar" cup to form a briquette. The PP
technique requires individual procedures for each sample type, an
obvious disadvantage (Table 1).

XRF Calibrations

The initial BF-XRF Calibration Report[7] covers the global
analytical program (termed FUSION4); it includes the synthetic
primary calibration disc preparation, the count rate collection,
regression analyses, and accuracy and precision testing. The
calibration also contains information required to analyse a bead
for the following elements important in lead and secondary product
production: As, Bi, Zn, Cu, Fe, Mn, Ca, Sb, Sn, Cd, Ag, Cl, S, Pb,
Si, Al and Mg. Other elements included are: Ni, Co, Cr, Ba, Ti
and K. Se, lost on fusion, was excluded from the calibration.

RESULTS AND DISCUSSION

Sample Selection

In the smelter process illustrated in Figure 1, blended lead
concentrate (BLC) is sintered and the product fed into the
blast furnace. The blast furnace produces a lead bullion and a
silica slag (BFS). The lead bullion is then further refined;
the first stage being de-copperisation. Here copper matte (CDFM)
is obtained. The CDFM is transported to the copper plant where
electrowon copper is produced. The BFS, which is high in zinc,

Table 3. Statistical data on Norrish disc SU-25 for elements counted in the Drift Correction Monitor. Data collected at 5 a.m., from December 1990, over 34 days.

Element Channel	Mean (KCPS) (N=34)	SD	% RSD	Element Channel	Mean (KCPS) (N=34)	SD	% RSD
As	9.018	0.048	0.54	Ca	86.304	0.504	0.58
Se	3.223	0.023	0.73	Sb1	9.890	0.060	0.61
Bi	18.161	0.072	0.39	Sn1	5.724	0.041	0.72
Zn	64.642	0.158	0.24	K	14.750	0.150	1.02
Cu	80.605	0.189	0.24	Co1	7.422	0.022	0.30
Ni	5.291	0.012	0.22	Ag1	5.618	0.023	0.42
Co	2.020	0.017	0.83	Cl	10.947	0.072	0.66
Fe	25.497	0.073	0.28	S	23.122	0.093	0.40
Mn	2.647	0.010	0.39	Pb3	31.064	0.091	0.29
Cr	1.542	0.016	1.04	Si	13.923	0.066	0.48
Bal	0.900	0.010	1.15	Al	7.728	0.051	0.65
Ti	2.553	0.019	0.76	Mg	6.284	0.055	0.88

goes to the slag fuming plant, here a zinc oxide slurry is formed, and pumped to the zinc plant where, in turn, electrowon zinc is obtained.

Calibration Concentration Ranges for the BF Method

It needs to be again emphasised that for the PP method, a tailored calibration is required for each material type (Table 1). In the case of the BF method, calibration for each element is carried out using single element oxides (or compounds) in individual beads. Table 2 illustrates the practical concentration ranges employed within the Global program. Line overlap corrections and element-element alpha-correction alterations have been made for certain elements.

Stability of the Instrument

The Philips PW1404 XRF Spectrometer was found from internal calibration to be stable to better than 1.2% RSD for all elements, and to better than 0.5% RSD for major elements.

Table 4. Data stability test for a single selected BFS fused bead(BFSQC6).(a) Short term; 20 consecutive analyses; and, (b) long term; daily analysis over 26 days. Concentration in wt % for species listed.

(a)

	Pb	Zn	FeO	MnO	CaO	S	SiO_2	Al_2O_3	MgO
Reference	2.0	17.2	27.1	5.4	13.4	1.3	21.8	5.7	1.0
X	2.2	17.4	27.4	5.2	13.3	1.4	21.8	5.5	1.4
SD	0.02	0.3	0.7	0.2	0.03	0.009	0.08	0.04	0.03
% RSD	0.67	0.20	0.26	0.44	0.20	0.63	0.38	0.73	1.9

(b)

	Pb	Zn	FeO	MnO	CaO	S	SiO_2	Al_2O_3	MgO
Reference	2.0	17.2	27.1	5.4	13.4	1.3	21.8	5.7	1.0
X	2.2	17.6	27.7	5.3	13.5	1.4	21.3	5.6	1.3
SD	0.06	0.20	0.29	0.06	0.16	0.05	0.28	0.11	0.10
% RSD	1.8	0.43	0.45	0.65	0.33	1.6	0.52	1.2	5.5

Table 5. Data reproducability for the BF Method. Fresh daily analysis for each QC sample type over 30 days. Concentrations for listed species in wt %.

(a)   BLC samples - BLCQC1.

	Pb	As	Cu	Fe	SiO$_2$	S	Zn	Bi
Reference	68.1	0.13	1.0	4.1	1.6	16.3	5.3	0.009
X	69.1	0.14	1.1	4.3	1.6	16.7	5.4	0.009
SD	0.75	0.006	0.04	0.1	0.05	0.2	0.07	0.001
% RSD	1.1	4.0	3.8	2.5	2.8	1.0	1.4	11%

(b) BFS samples - BFSQC6.

	Pb	Zn	FeO	MnO	CaO	S	SiO$_2$	Al$_2$O$_3$	MgO
Reference	2.0	17.2	27.1	5.4	13.4	1.3	21.8	5.7	1.0
X	2.1	17.4	27.4	5.2	13.4	1.3	21.8	5.4	1.2
SD	0.03	0.12	0.18	0.05	0.1	0.02	0.2	0.06	0.03
% RSD	1.6	0.7	0.7	0.9	0.9	1.2	0.9	1.2	2.5

(c)   CDFM   samples - CDFMQC3.

%	Cu	Pb	S	As	Sb	Fe
Reference	32.0	39.5	10.7	4.37	1.35	0.94
X	32.1	40.5	10.7	4.78	1.46	1.0
SD	0.19	0.39	0.08	0.06	0.04	0.03
% RSD	0.6	0.9	0.7	1.2	2.5	3.1

Table 6. Data reproducability for the PP Sample Method. Daily analysis for: (a) 33 ; (b) 35 ; and, (c) 30 days.

(a) BLC QC sample - BLCQC1.

	Pb	As	Cu	Fe	SiO$_2$	S	Zn
Reference	68.1	0.13	1.0	4.1	1.6	16.3	5.3
X	66.4	0.13	1.1	4.0	1.6	14.6	5.1
S	1.5	0.02	0.05	0.1	0.1	1.3	0.07
% RSD	2.2	12.8	4.6	2.4	8.5	9.0	1.3

(b) BFS QC sample - BFSQC6.

	Pb	Zn	FeO	MnO	CaO	S	SiO$_2$	Al$_2$O$_3$	MgO
Reference	2.0	17.2	27.1	5.4	13.4	1.3	21.8	5.7	1.0
X	1.9	17.4	27.6	5.3	14.6	1.3	20.5	5.2	0.9
SD	0.1	0.1	0.1	0.02	0.1	0.05	0.5	0.2	0.06
% RSD	6.2	0.6	0.5	0.8	0.9	3.7	2.8	3.5	5.8

(c) CDFM QC sample - CDFMQC3.

	Cu	Pb	S	As	Sb	Fe
Reference	32.0	39.5	10.7	4.37	1.35	0.94
X	35.9	40.4	14.2	5.1	1.18	1.0
SD	0.7	0.3	0.08	0.3	0.03	0.04
% RSD	2.0	0.7	0.6	5.6	2.4	4.0

Quality Control Background

It was considered that available reference materials did not
closely match the lead concentrate, the blast furnace slag, and
the copper matte materials to be analysed. Hence, primary
reference standards were established on materials of appropriate
matrix by round-robin analyses with 4 laboratories. The
reference figures in Tables 4 to 6 were established in that
process.

Data Repeatability and Stability of the BF Sample Bead

The BF method provides an extremely stable matrix for each sample
as shown in Table 4. Repeatability is better than 1% RSD, and
long term stability better than 2% RSD, except for MgO. Precision
is comparable for the most part with that in Table 3, the drift
of the instrument itself.

Data Reproducability

Table 4 demonstrates the repeatability of the BF bead
determination, and the stability of the glass formed. Table 5
provides reference and statistical data (accuracy and precision)
for beads freshly prepared each day for the selected samples.
In Table 6, pressed powder sample data is presented for
comparison, again samples being prepared daily.

CONCLUSIONS

Some analytical observations are as follows :
*   Moisture absorption affects BFS results via the PP method,
     as evidenced by the low silica value and to a lesser extent
     by the high RSD's for Pb and S in Table 6 (b)
*   The BLC sulphur content is satisfactory by the BF technique
     but generally low by the PP method (Tables 5 and 6 (a))
*   High Cu and S levels were observed by the PP approach, with
     the CDFM sample difficult to grind (due to As levels). The
     fusion procedure was noticeably free of such particle size
     influences (Tables 5 and 6 (c))
*   No absorption of metals (such as Cu, Fe) into the
     crucibles is seen under the given BF conditions
*   Volatile elements, such as S and Cl, have been retained
     quantitatively
*   Although the BF process takes 20 minutes, the use of the
     automated fusion method (with reduced grinding time) was
     considered to provide, overall, a quicker technique than
     the alternative PP approach

Thus, the BF method has provided better accuracy and precision,
and solved a number of problems associated with the PP method.
Again, the introduction of new elements as single beads to the
analytical program, and associated ease of calibration for those

elements, are decided advantages. Finally, the Global Borate
Fusion method provides reliable results for 23 elements, over a
wide concentration range, from about 0.01 to 50% and beyond, and
for a variety of material types (Table 2).

REFERENCES

1. K.Norrish and G.M.Thompson, XRS Analysis of Sulphides by
   Fusion Methods, X-Ray Spectrometry, **19**(1990) 67-71.
2. A.R.Mauri, M.T.Domenech, M.de la Guardia, C.Mongay and
   M.C.Guillem, Neodymium Determination in Ceramic Pigments:
   A Comparative Study of Flame Emission and XRF Methods,
   Atomic Spectroscopy, **11**(1990) 90-95.
3. K.I.Mahan and D.E.Leyden, Simultaneous Determination of
   Sixteen Major and Minor Elements in River Sediments by
   Energy-Dispersive XRF Spectrometry after Fusion in Lithium
   Tetraborate Glass. Analytica Chimica Acta, **147**(1983)123-131.
4. R.A.Couture, An Improved Fusion Technique for Major-Element
   Rock Analysis by XRF. Advances in X-ray Analysis, **32**(1989)
   233-238.
5. R.H.Dow, A Statistical Comparison of Data Obtained from
   Pressed Disk and Fused Bead Preparation Techniques for
   Geological Samples. Advances in X-ray Analysis,**25** (1982)
   117-120.
6. G.S.Barger, A Fusion Method for the XRF Analysis of Portland
   Cements, Clinker and Raw Materials Utilizing Cerium(IV)
   Oxide in Lithium Borate Fluxes. Advances in X-ray Analysis,
   29(1986) 581-587.
7. J.G.H.Metz, Initial Borate Fusion (Philips PW1404) XRF
   Calibrations. Pasminco Metals/BHAS, 1991.

THE LOCAL STRUCTURE OF CHROMIUM(III) IN THE MIXED GLYCEROL AQUEOUS SOLUTION

T.Watanabe, K.Taniguchi, T.Ninomiya[*] and S.Ikeda[**]

Osaka Electro-Communication University, Osaka, Japan
* Hyogo Pref. Police H.Q., Hyogo, Japan
** Ryukoku University, Shiga, Japan

ABSTRACT

Chromium glycerate solutions are exposed to X-ray absorption fine structure analysis to clarify the local structure around the chromium atom by the use of a laboratory XAFS system. The chromium solution is stabilized by addition of water soluble polyvinyl alcohol so that after excess water, the solution is a kind of homogenized "sol" without any apparent change to the solution state. The new technique has been developed, because chromium is an element which has relatively small X-Ray absorption.

The comformations of two novel chromium glycerate compounds, which are prepared under both NH_4OH and $NaOH$ basic conditions, are examined with XANES and EXAFS techniques. From these experiments, it is suggested that these two chromium glycerates may have Cr-O-Cr bridge bonds consisting of chromium dimer complexes in common.

INTRODUCTION

During the course of the survey of new precursor for ceramics, glass and layered materials, it is found that a trialcohol, glycerol, coordinates with transition metal ions. There is a report[1] on the complexation reaction of Fe(III) ion with polyalcohol. Although the investigation of the structure about the Fe(III) complex has been studied, there is no report on the complexation reaction of Cr(III) with polyalcohol. When aqueous ammonia is added in excess to an aqueous solution of chromium(III) nitrate, usually precipitation of chromium hydroxide is observed. No precipitation, however, is observed in a basic solution of chromium(III) at pH near 12 when glycerol has been added previously to the chromium solution. Thus, the chromium solutions containing glycerol have been examined to clarify the local structure around the chromium atom.

Advances in X-Ray Analysis, Vol. 35
Edited by C.S. Barrett *et al.*, Plenum Press, New York, 1992

Recently XAFS (X-ray absorption fine structure) analysis has been applied to studies of the structure of metal complexes in solution. XAFS is an effective method for the studies of chemical species which may give neither electrical nor magnetic information.[2,3]   XAFS includes XANES (X-ray absorption near edge structure) and EXAFS (Extended X-ray absorption fine structure).   XANES is an absorption spectrum which occur in the approximately 50 eV range near the absorption edge.   XANES gives a characteristic spectrum reflecting specific chemical bonding of the compound, because it is very sensitive in the valence bond region.   EXAFS spectra generally refer to the region 40-1000 eV above the absorption edge.   The EXAFS phenomenon is due to a final state interferrence effect involving scattering from the neighboring atoms and the ejected photoelectrons.   Therefore EXAFS depends on the structural   information about the local environment around the absorbing atom.   Sincethe EXAFS is dominated by the structure within the mean freepath of the photoelectrons from the absorbers, the periodic structure for long range order is not necessary.   The technique is considered to be very useful toinvestigate the coordination structure of the metal complexes in the solutions.

In this paper, the local structure of chromium(III) solution containing glycerol is studied by the laboratory XAFS in order to make clear the interaction between chromium(III) ion and glycerol molecule.

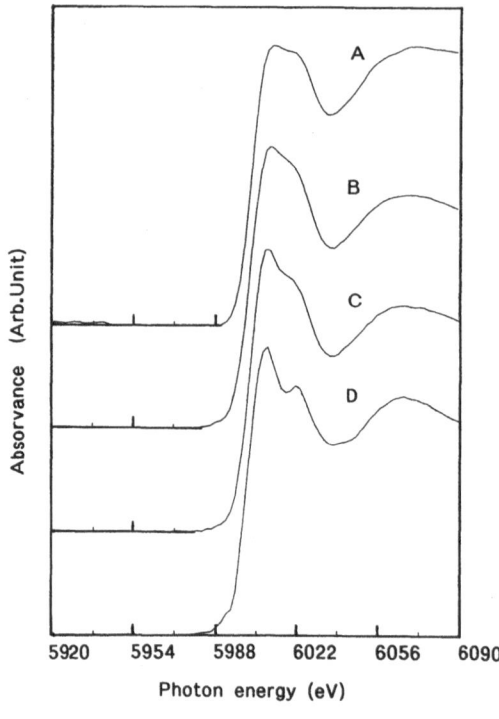

Figure 1. Cr K-edge XANES of (A) Cr(III)-Gly-NH4OH solution, (B) Cr(III)-Gly-NH4OH sol, (C) Cr(III)-Gly-NaOH solution and (D) Cr2O3 solids as reference. Gly: glycerol

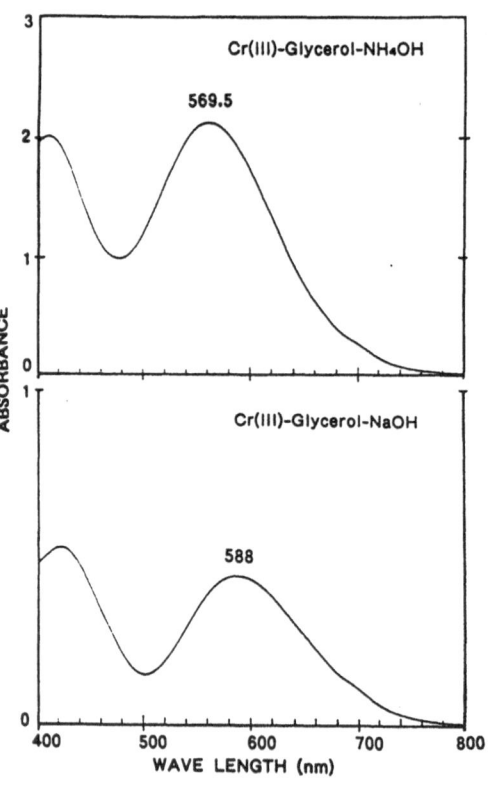

Figure 2. Visible light spectra of Cr(III)-Glycerol-NH4OH and Cr(III)-Glycerol-NaOH.

EXPERIMENTAL

The chromium(III) glycerate solutions are made as follows: four grams of Cr(NO3)3·9H O is dissolved in 20 ml H2O and 9.2g of glycerol is added. 1 mol. NaOH solution is added to keep the pH of the solution near 12, while for the case of using NH4OH, pH of the solution was near 10. The clear solution colored blue green when NaOH is used, while a deep-violet color is developed on addition of NH4OH.

A new techniqueidea is applied to mounting liquid samples for XAFS. The chromium solution is stabilized by addition of water soluble polyvinylalcohol (PVA) to keep a homogenized "sol" without any apparent change of the solution state. The sample thickness t is controlled to less than 1mm so that μt <2 (μ is X-ray absorption coefficient). Ten grams of PVA is dissolved in hot water (100ml) and 3ml of PVA solution is added into chromium-glycerol solutions. In the case of Cr(III)-Glycerol-NH4OH, the special holder is used in order to prevent the evaporation of NH3 and the solution is set between polystyrene films.

Samples are measured with a laboratory XAFS system (EXAC 800: Technos Inc.). A rotating anode (target is Mo) X-Ray tube is used, and emission current is controlled by a feed back system (5 to 100mA ,20kV) to keep the intensity at the constant level. Johann type Ge(400) is used as an analyzing crystal. Spectrometer is kept under vacuum conditions (10^{-3} Torr). The detector is a Si(Li) type SSD (solid state detector). The incident and transmitted X-ray intensities, I_0 and I, respectively, are measured by stepscanning around the Cr-K edge along the Rowland-circle (radius: 300 mm). Thus, $\mu t = \ln(I_0/I)$ are evaluated with I_0 and I.

RESULTS AND DISCUSSION

Near edge absorption spectrum

Figure 1 shows the near edge absorption spectra of Cr-K for Cr(III) materials (A: Cr(III)-Glycerol-NH4OH solution, B: Cr(III)-Glycerol-NH4OH "sol", C: Cr(III)-Glycerol-NaOH solution and Cr2O3 solid). It can clearly be seen that the spectrum about the NH4OH site is different from that about the sodium hydroxide site. It suggests that the spectrum should change along the due course (1) to (3), (1) elimination of NH. as a ligand (2) bonding OH ion to the ex-site of NH3 (3) condensation with dehydration. The XANES spectrum of Cr2O3 is used as a reference. The clearly be seen that the spectrum about the NH.OH site is different from difference of visible spectra between two samples as shown in Fig.2 also supports the above suggestions , that is , the maximal peak of visible spectrum of Cr(III) - Glycerol - NH4OH is 569 nm while that of Cr(III) - Glycerol - NaOH is 588 nm.

EXAFS data analysis

The EXAFS data are analysed with the curve fitting procedures utilizing the theoretical phase and amplitude functions of Teo and Lee. TM.EXE (made by T.Masuda: NOK Tsukuba Technical Laboratory Inc.) is used as software for data analysis.

The pre-edge absorption coefficient of each spectrum is estimated by Victreen's formula ($A\lambda^3 - B\lambda^4$), that is, fitting of linear least squares as

Table 1. The ranges used for Fourier transformation and the magnitude of RDF of each sample

Sample		kmin-kmax ($\overset{\circ}{A}^{-1}$)	F1max	F2max
Cr(III)-Gly-NH4OH	(solution)	3.074-10.348	12.449	4.398
Cr(III)-Gly-NH4OH	(sol)	3.694-11.097	13.881	5.065
Cr(III)-Gly-NaOH	(solution)	2.987-11.201	14.860	5.549
Cr2O3	(solid)	2.898-10.868	14.692	7.206

Table 2. The ranges used for Inverse Fourier transformation

Sample		The ranges(Å) of the first peak	The ranges(Å) of the second peak
Cr(III)-Gly-NH4OH	(solution)	1.080-2.111	2.307-4.398
Cr(III)-Gly-NH4OH	(sol)	1.080-2.062	2.307-5.065
Cr(III)-Gly-NaOH	(solution)	1.129-2.013	2.258-5.549
Cr2O3	(solid)	1.080-1.963	2.209-7.206

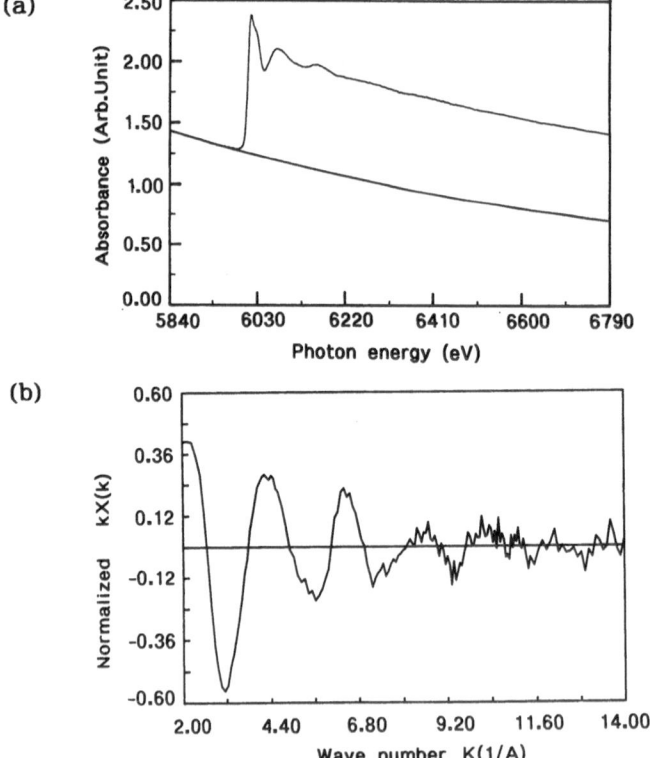

Figure 3. (a) is the method of background(dashed line) removal by the use of Victreen's formula about each sample(here is sample C) and (b) is EXAFS spectrum kX(k)vs k after normalization.

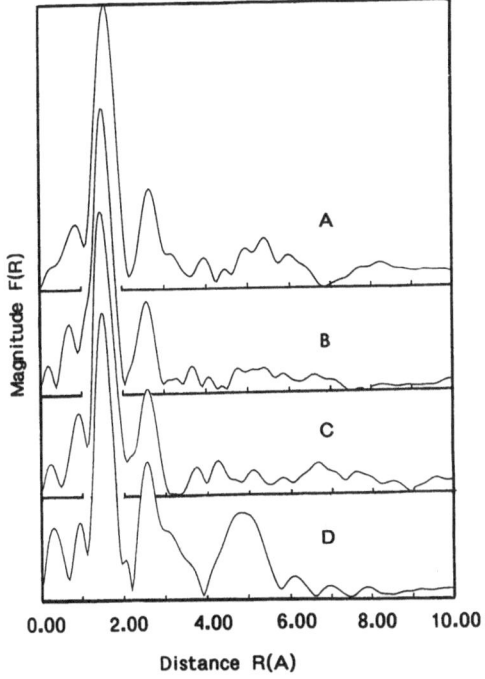

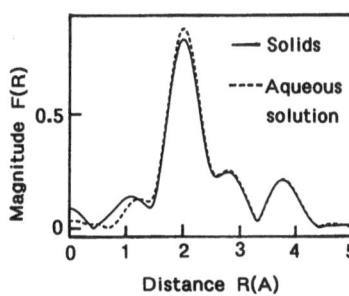

Figure 5.   The radial distribution function of $K[Cr(ox)_3]_3 \cdot 3H_2O$.

Figure 4.   The radial distribution functions of (A) Cr(III)−Gly−NH4OH solution, (B) Cr(III)−Gly−NH4OH sol, (C) Cr(III)−Gly−NaOH solution, and (D) Cr_2O_3 is reference.

shown in Figure 3 (a).   The net absorption is determined by substraction of background from measured X-ray absorption.   The energy threshold of the post-edge is estimated by the use of five-point removal smoothing method. $X(k)$ is exposed as follows, $X(k)=(\mu - \mu_o) / \{(A\lambda^3 - B\lambda^4)t\}$, where, μ is the net absorption coefficient , μ_o is the imaging free atomic absorption coefficient , A and B are Victreen's constant, k is wavelength, t is the sample thickness. μ_o is located ten points after point at whitch smoothing start and uses Victreen's parameter.

Figure 3 (b)   shows the EXAFS spectrum $kX(k)$ vs k.   The radial distribution function (RDF) of each samples is shown in Fig.4.   Hanning function is applied to filtering, and Cubic Lagrange method is applied to interpolating at $k^3X(k)$.   The k region which is used for Fourier transform of each sample is shown in table 1.   The magnitude of the firstpeak(F1max) and the second peak (F2max) of each sample are shown in table1 also. In figure 4, both distance of the first neighboring atom and of the second neighboring atom are neglected the phase shift effect.   To correct the errors on phase shift effect, mathematical data treatments areadopted as follows.   Inverse Fourier transformation(by the use of Hanning function) is done for the area as shown in table 2.   The paramaters obtained by curve-fitting analysis(see Fig.6) are shown in table 3.   Here, R, σ and N are the distance of the neighboring atom from absorption atom,Debye−Waller factors of the neighboring atom, and the coordination number, and also suffix 1 and 2 means 1st and 2nd neigboring atoms, respectively.   For Cr_2O_3,R ,Rand N are refered from literature.[4]   The coordination number N of eachsample is calculated by using the scale factor S as follows $N_1 = N_1$

Table 3. The paramaters obtained by curve-fitting

Sample	R1(Å)	σ1(Å)	N1	R2(Å)	σ2(Å)
Cr(III)-Gly-NH₄OH (solution)	2.02	0.001	5.51	3.02	0.071
Cr(III)-Gly-NH₄OH (sol)	1.99	0.009	6.08	2.99	0.069
Cr(III)-Gly-NaOH (solution)	2.00	0.012	6.27	3.00	0.071
Cr₂O₃ (solid)	1.99	0.015	6.00	2.96	0.075

Scale factor of the first neigboring atom of Cr_2O_3 is used as 2.79.

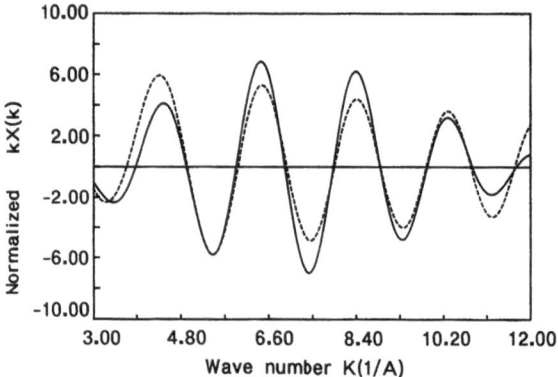

Figure 6. The minor peak in Fig.4 of each sample(here is the sample C) can be Fourier filtered and backtransformed into k space (solid line) and curvefitted with a different single-distance model (dashed line).

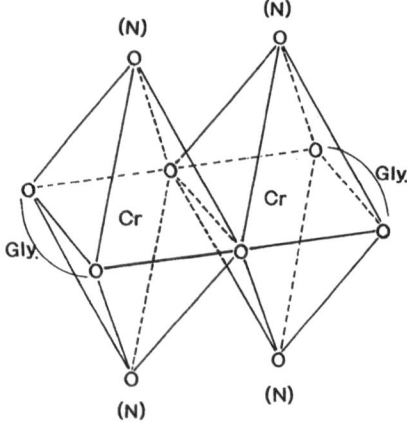

Figure 7. Skaleton-models of Cr(III)-Glycerol complexes.

(EXAFS analysis)·S where N₁ (EXAFS analysis) is the coordination number of 1st neighboring atom from the curve-fitting analysis. The scale factor S is obtained by S=N(literature)/N(EXAFS analysis) where N(literature) is refered from literature, N(EXAFS analysis) is obtained by EXAFS analysis for Cr_2O_3. The coordination number of second neighboring atom of each sample is not measured because it include so much multiple scattering effects.

The first peak of spectrum $D(Cr_2O_3)$ in Fig.4 conclude to be oxygen atom, and also the Cr-O distance is 1.99A. Thus, the first peak of spectrum C seems to be also oxygen atom. The spectrum A and B in Fig.4 may be supported that Ñ lone pair of NH_3 molecule may coordinate with chromium atom because basic NH_3 aqueous solution is used as a medium. It seems that the first peak in spectrum A and B corresponds to nitrogen atom. And also, this discussion is supported from the difference between the spectrum of Fig.2.

Each second peak in Fig.4 is observed sharply in comparison with the second peak of $K[Cr(ox)_3]_3 3H_2O$ which exists as-a shoulder of main peak asshown in Fig.5. The second peak of $K[Cr(ox)_3]_3·3H_2O$ in Fig.5 is correspond to curbon atom. The fact that the transparent solution is obtained even in the presence of 1 glycerol molecule per 1 atom of chromium suggests that each glycerol molecule coordinates with each Cr(III) atom and a bridge of Cr-O-Cr seems to be made by the dimerization condensation of two OH ligands. As a result, the scaleton-models of two novel complexes of chromium and glycerol are presented as shown in Fig.7.

REFERENCES

1. T.Mitani, Y.Mori, K.Yanagishita, H.Yokoi, The complexation reaction of polyalcohl and Fe(III) ion,Proceeding of Spring Annual Meeting of Chemical Society of Japan(1991).

2. N.Matsubayashi, Studies on the hydration and substitution reaction of Cu(II) and Cr(II) ions by EXAFS, Doctoral thesis of Osaka University (1986).

3. T.Miyanaga, N.Matsubayasi, I.Watanabe, S.Ikeda, EXAFS and XANES of Titanium(III),(IV) and the mixed valence complex in aqueous solution, Advances in X-ray Chemical Analysis Japan 19, 119-126, (1987).

4. Ralph W.G. Wyckoff, Crystal Structures, Second Edition, Volume 2, Interscience Publishers(1964).

5. H.Sakane, Studies on X-ray Absorption Fine Structure for Divalent and Trivalent First Transition Metal Complexes, Doctoral Thesis of Osaka University(1991).

USING *A PRIORI* INFORMATION IN ENERGY-DISPERSIVE X-RAY FLUORESCENCE
ANALYSIS OF COMPLEX SAMPLES

I. A. Kondurov, P. A. Sushkov, T. M. Tjukavina, G. I. Shulyak

Neutron Research Division
Leningrad Nuclear Physics Institute
Gatchina 188350, U.S.S.R.

INTRODUCTION

In multielement EDXRF analysis of very complex unknowns, some problems in data evaluation may be simplified if one can take into account *a priori* information on the properties of the incident and detected radiations, and also available data on the matrix of the sample. The number of variables can be drastically shortened in the LSM procedures in this case. One of the best examples of complex unknowns is the determination of the rare earth element content of ores, and most recently in samples of high temperature superconductors (HiTc).

In the case of HiTc samples consisting of a mixture of rare earth elements, the best way to excite characteristic x-rays is using radioactive sources, for example, ^{109}Cd, where one can use the 88 keV γ-line for the excitation of the whole spectrum of rare earths while light elements (Cu, Y, Ca) are excited by the Ag KX-rays. An example of such a complex spectrum is shown in Fig. 1. The spectrum consists of 41 lines from 11 rare earth elements.

SPECTRUM EVALUATION

A system of programs for the evaluation of complex spectra was developed based on the effective use of *a priori* information on the properties of the radiations used. The system was designed for using LSI-11 compatible computers.

The package includes both programs for maintenance and use of an x-ray data file, XRDF, and programs for evaluation of experimental data taking into account *a priori* data from the file. The x-ray data file consists of the following tables for all elements: energies and intensities of the x-ray series, mass absorption coefficients, energies and values of absorption edges, cross-sections of photoexcitation, and also energies and yields of γ-lines of radioactive sources to be used for XRF excitation. These data files can be shortened and reduced to the real experimental conditions (energy and intensity range, resolution, etc.) for elements under investigation.

Advances in X-Ray Analysis, Vol. 35
Edited by C.S. Barrett *et al.*, Plenum Press, New York, 1992

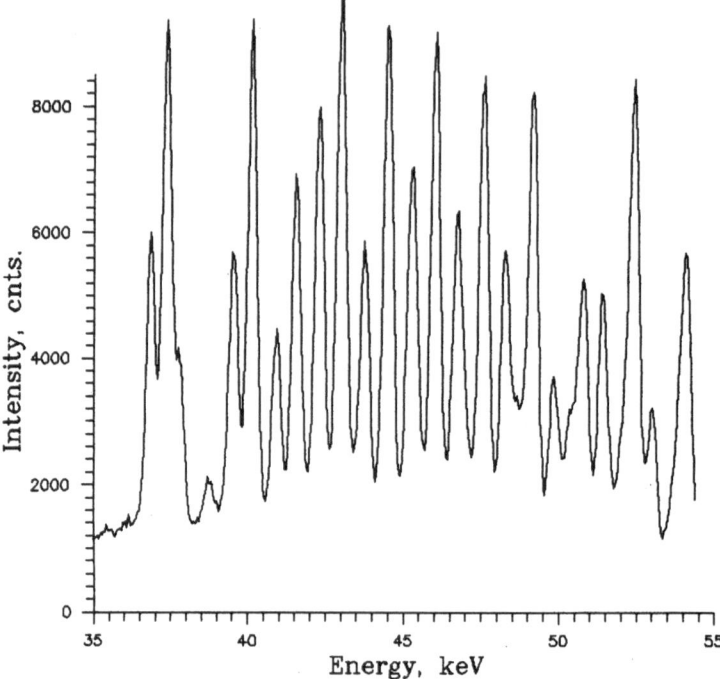

Fig.1. Part of a complex spectrum of the artificial mixture of 12
       rare earth elements.

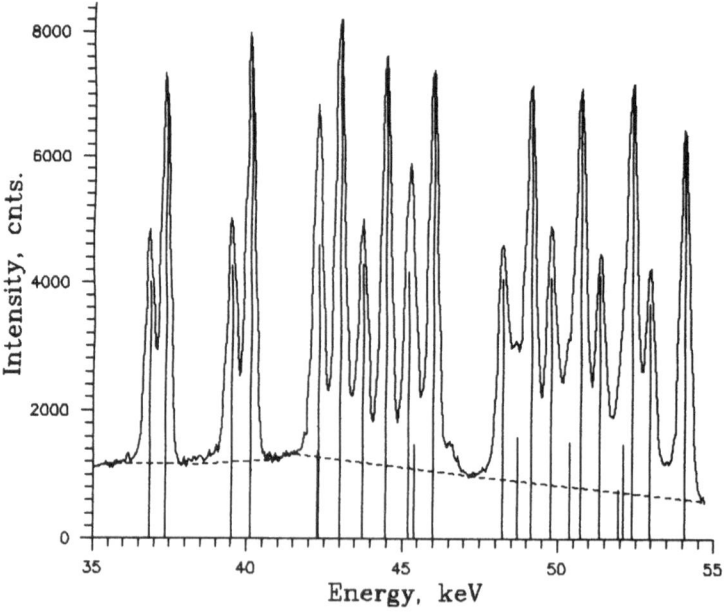

Fig.2. Decomposition of the complex part of spectrum. Vertical
       bars show positions and relative intensities of K-groups.

Programs for the analysis of experimental data include: approximation of informative parts of γ-spectra using several hypotheses of background and peak shapes, calibrations in energy, resolution, efficiency and the resultant analysis of experimental data yielding concentrations of the unknowns.

After a precision calibration of the spectrometer, it is possible to calculate positions of all the lines for each x-ray series using the data file XRDF. These positions are then fixed in the LSM procedure (vertical bars on Fig. 2). The ratio of the intensity of all the lines in the series also can be fixed. It is possible also to fix the intensity ratio for K_α and K_β groups independently.

As a result only 20 variables are required for an overestimated system of linear equations when fitting the spectrum of Fig. 1 by LSM. Note that the full number of components in the spectrum equals 41 as was mentioned above.

CALCULATIONS OF CONCENTRATIONS

The XRDF information is used automatically when calculating the elemental content from the intensities of the analytical lines. The method of calculation is based on the following formula written under the assumption that a sample in the form of a thick disk of area s is irradiated with a parallel beam of monochromatic (E_γ) photons of flux density Φ_γ at an angle ϕ. Secondary radiation is measured at an angle ψ at a distance R from the sample. The area of the i-th analytical line is:

$$A^z_{q,i} = \frac{\Phi_\gamma s}{4\pi R^2} \cdot C_z \cdot \frac{S^z_q - 1}{S^z_q} \cdot \tau^z_q \cdot \omega_{q,i} \cdot \varepsilon(E^z_{q,i}) \cdot \frac{1}{\dfrac{\mu(E_\gamma)}{sin\,\varphi} + \dfrac{\mu(E^z_{q,i})}{sin\,\psi}} \tag{1}$$

where: S^z_q - q-shell edge step ratio;

τ^z_q - total cross section for q-shell for the element Z;

$\omega_{q,i}$ - fluorescence yield for q-shell;

$\varepsilon(E)$ - energy dependence of the detector efficiency;

C_z - concentration of the element Z;

$\mu(E)$ - mass absorption coefficient at an energy E.

Let's rewrite (1) as:

$$A^z_{q,i} = K^z \cdot C \cdot M(E_\gamma, E^z_{q,i})$$

where $K^z = \dfrac{\Phi_\gamma \cdot s}{4\pi R^2} \cdot \varepsilon(E^z_{q,i})$, and M equals the last term of (1)

As was mentioned above, all the atomic constants are available from the XRDF file. The $\varepsilon(E)$ is measured using standard calibrated radionuclides. Effective values of ϕ and ψ can be calculated from the geometry. A dependence of K^z on Z is deducted from the relation

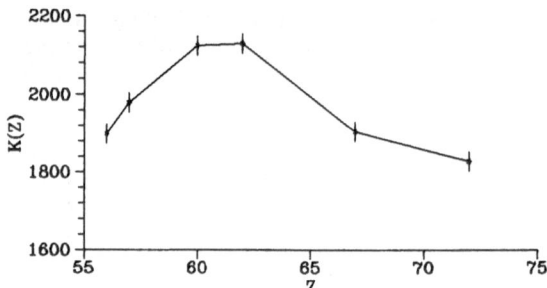

Fig.3. Dependence of K^Z from Z for K_α lines of rare earth elements.

$$K^Z = \frac{A^Z_{q,1}}{C^Z \cdot M(E_\gamma, E^Z_{q,1})} \quad ,$$

measured for each of the elements separately. This dependence for rare earth elements is presented in Fig. 3 for the experimental conditions under which the spectra of Fig. 1 and Fig. 2 were measured.

After measuring the unknown sample and deducting the area of the peaks the unknown concentrations can be calculated:

$$C^Z_{un} = \frac{A^Z_{un}}{K^Z \cdot M(E_\gamma, E^Z_{q,1})}$$

An example of such a calculation can be seen in Fig. 4. Presented are ratios of measured and original concentrations for an artificial sample prepared of a mixture of 9 rare earth elements. The error bars shown are only statistical ones. The total error after taking into account all systematic uncertainties is of order of 4 - 7 %. A visible trend in the ratio dependence can be connected with different experimental conditions: the standards were thin and diluted (90% of starch).

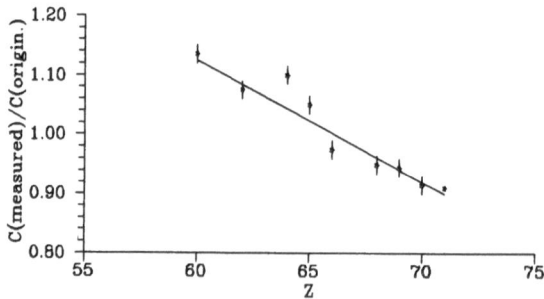

Fig.4. Comparison of measured and original concentrations.

REFERENCES

1. T. Browne, R. B. Firestone. Table of Radioactive Isotopes., G. Willey, 1986

2. M. A. Blokhin, I. G. Shweizer. Rentgenospektralnyj spravochnik., (in Russian), Moscow, Nauka publ., 1982

IN-PROCESS COATING LAYER ANALYSIS OF GALVANNEALED
STEEL SHEETS WITH MONOCHROMATIC INCIDENT X-RAYS

Hiroharu Kato, Kiyotaka Imai and Hideya Tanabe

Electronics Research Center
NKK Corporation
Kawasaki, Japan

ABSTRACT

A new method, for the on-line determination of the composition and the coating weight of galvannealed (Zn-Fe-alloy-coated) steel sheets using monochromatic incident x-rays with two optical systems, is described. In the case of galvannealed steel sheets, it is difficult to determine the composition and the coating weight precisely by simple XRF, because fluorescent x-rays of Iron are emitted not only from the coating layer but also from the underlying steel sheets. We have developed an on-line analyzer with two optical systems which are different in incident angles, take-off angles and wavelengths of monochromatic incident x-rays. We determine the composition and the coating weight by solving simultaneous equations of the data which we derive using two optical systems. Using monochromatic x-rays enabled us to obtain high precision with high speed. We considered error factors in on-line measurement such as statistical error or fluctuation of the distance between the sensor head and steel sheets. This on-line analyzer has been applied to the continuous galvanizing line in our Fukuyama works successfully. As a result, we have been able to significantly improve product quality.

INTRODUCTION

Recently galvannealed (Zn-Fe-alloy-coated) steel sheets, which are used for automobiles, have attracted a great deal of attention because of their high corrosion resistance, their good weldability and their good paintability. For the quality assurance and the high yield rate of high-quality galvannealed steel sheets, on-line determination of the composition and the coating weight of the coating layer has been required. In the case of galvannealed steel sheets, it is difficult to determine the composition and the coating weight precisely by simple XRF, because fluorescent x-rays of Iron are emitted not only from the coating layer but also from the underlying steel sheets.

PRINCIPLES

Principles of measurement of the composition and the coating weight of galvannealed steel sheets by XRF using monochromatic incident x-rays are shown in Fig. 1. Table 1 shows the measurement condition of this two optical systems. There are two optical systems (lower and higher angle systems) whose incident angles, take-off

Table 1   Measurement Conditions of
the X-ray optical systems

	Lower angle system	Higher angle system
X-ray tube	W	Mo
Incident X-ray	WI.β,(1.28A)	MoKα(0.71A)
Incident angle	15°	75°
Take-off angle	45°	60°
Detector	Si(Li) cooled by a Peltier element(Kevex)	

angles and wavelengths of monochromatic x-rays are different from each other. We can get the information about the relatively upper part of the steel sheets by using the lower angle system more than by the higher one, because the penetration depth of the former is shorter than the latter. We determine the composition and the coating weight by solving simultaneous equations of the data (intensity ratio of iron Kα to zinc Kα) which we obtain using these two optical systems (Fig. 2). These simultaneous equations are based on theoretical relationships given by Shiraiwa and Fujino[1]. We have also considered the secondary excitation of iron by K-line of zinc.

In order to determine the composition and the coating weight precisely through simultaneous equations, it is important to make the slope of the curve for the lower angle system as different from that of higher one as possible, or as flat as possible as well as to minimize the fluctuation of measured intensity caused by several error factors, because the same amount of fluctuation of measured intensity causes less analyzing error in the flat case than in the steep case as Fig. 2 shows. In order to make it as flat as possible, the penetration depth of the lower system's incident x-rays has to be as short as possible. Therefore we have made the incident angle and take-off angle of the lower system as low as the error caused by the fluctuation of the distance between the optical systems and steel sheets permits. We will discuss this error later. And we have made the wavelength of the lower system's incident x-rays WLβ1 (1.282Å), which has a little shorter wavelength than that of the absorption edge of Zn (1.283Å).

Using monochromatic incident x-rays offers the best measurement condition from the view point of incident x-rays' penetration depth or absorption in the coating layer. In the case of the measurement of galvannealed steel sheets using white x-rays, their penetration depths are much deeper than those of monochromatic x-rays of WLβ1, because the absorption coefficient of WLβ1 is nearly the largest of those x-rays which can excite

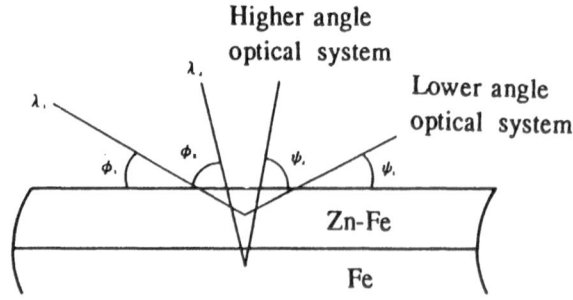

Fig. 1   Principles of the x-ray fluorescence method
using monochromatic incident x-rays for
Zn-Fe-alloy-coated steel sheets

Table 2 Calculated errors under the condition of
± 1 % intesity ratio fluctuation
W : White X-rays
M : Monochromatic X-rays

			Coating Weight ρt					
			30g/m²		60g/m²		90g/m²	
			△Fe%	△ρt	△Fe%	△ρt	△Fe%	△ρt
Fe Fraction Fe%	5 %	W	0.84	0.7	0.25	0.8	0.13	0.9
		M	0.32	0.4	0.11	0.6	0.08	0.8
	10%	W	0.86	0.7	0.32	0.9	0.19	1.3
		M	0.37	0.4	0.16	0.7	0.12	1.1
	15%	W	0.90	0.8	0.37	1.1	0.25	1.6
		M	0.41	0.5	0.19	0.8	0.16	1.3

△ denotes calculated error

iron K-line. Table 2 shows a comparison of calculated errors for nine samples (5, 10, 15Fe wt% × 30, 60, 90g/m²) under the condition of ±1% intensity ratio fluctuation for the cases of both monochromatic incident x-rays and white radiation. Errors of white x-rays were calculated using experimental data[2]. This result shows superiority in precision of the determination using monochromatic incident x-rays. The statistical error, which is one of the major intensity error factors, may be less in the case of white x-rays than monochromatic x-rays, but this cannot compensate the result under the measurement condition required in this application.

Using monochromatic incident x-rays has enabled us to make the calculation time of theoretical relationships suitable for on-line analysis. On the other hand, in the case of white x-rays we have to calculate the integration of intensity along wavelengths using a precise intensity distribution of incident x-rays. This is not easy and it takes too much time for on-line analysis. (We do not use effective wavelength, because it may vary by

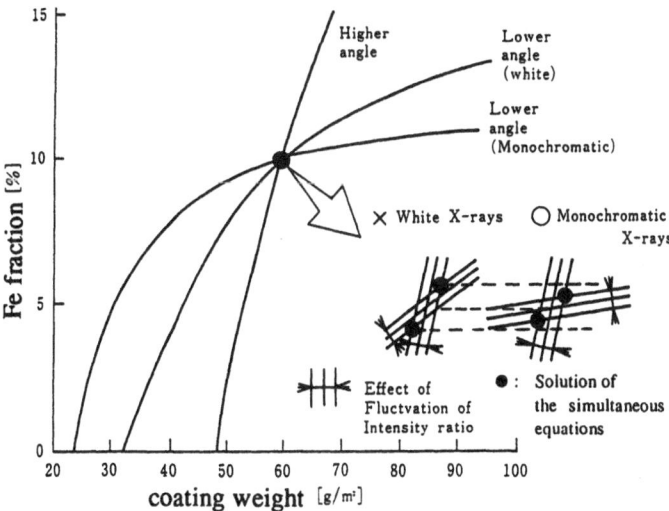

Fig. 2 Curves indicating composition-coating weight
combinations in which intensity ratio is constant

Table 3 Experimental conditions to evaluate the
precision of the system

Sample	Sample L : coating weight 84.9g/m², 9.0 Fe% Sample S : coating weight 40.7g/m², 12.1 Fe%
Measurement time	10seconds
Fluctuations in measurement distance	±1mm(3σ)
Fluctuations in measurement angles	±1°(3σ)
Fluctuations in ambient temperature	23±5°(3σ) humidity 50%
Fluctustions in ambient humidity	50±20%(3σ) temperature23°

ERROR FACTORS

We have estimated the effect of several error factors[3] and changed the measurement conditions to be suitable for on-line analysis. The error factors which we considered are as follows.

1. Statistical error
2. Fluctuation in measurement distance
3. Fluctuation in measurement angles around Y axis in Fig. 3
4. Fluctuation in measurement angles around X axis in Fig. 3
5. Fluctuation in ambient temperature
6. Fluctuation in ambient humidity

Under the conditions shown in Table 3 we estimated item 1 through 4 by carrying out experiments and item 5 and 6 by theoretical calculations. Table 4 shows the result of this estimation, for not only the fluctuation of the intensity of iron Kα but also the fluctuation of the intensity ratio of iron Kα to zinc Kα. In some error factors, we can reduce the fluctuation by using this intensity ratio. Therefore we adopted this intensity ratio instead of the intensity. We estimated total errors of intensity by calculating from these error factors. The precision in the composition and the coating weight corresponding to the total error are 0.3Fe wt% and 0.5 g/m² respectively for the sample S and 0.2Fe wt% and 1.0 g/m² respectively for the sample L.

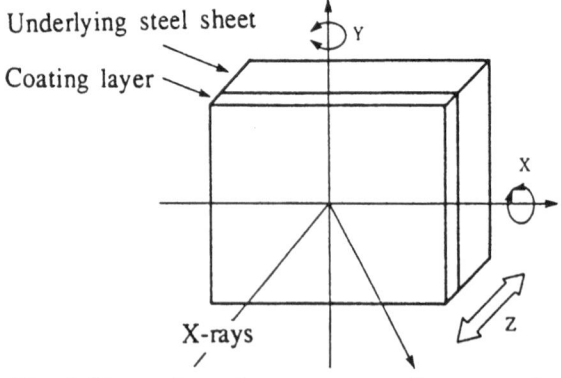

Fig. 3 Fluctuation of measurement distance and angles

Table 4 Fluctuations of X-ray intensities (%) caused by error factors

Error factors		Lower angle system		Higher angle system	
		Intensity of Fek α	$\dfrac{\text{Intensity of Fek } \alpha}{\text{Intensity of Znk } \alpha}$	Intensity of Fek α	$\dfrac{\text{Intensity of Fek } \alpha}{\text{Intensity of Znk } \alpha}$
Statistical error $3\sigma_1$	L	4.2	4.2	2.4	2.9
	S	3.3	3.3	2.0	2.6
Fluctuations in measurement distance $3\sigma_1$	L	4.5	1.4	1.8	0.31
	S	5.6	1.8	1.7	0.089
Fluctuations in measurement angles $3\sigma_1$ (around Y-axis)	L	1.2	0.89	1.5	0.99
	S	1.3	2.5	1.4	0.90
Fluctuations in measurement angles $3\sigma_1$ (around X-axis)	L	0.74	0.24	1.1	0.28
	S	0.35	0.0	0.46	0.48
Fluctuations in ambient temperature $3\sigma_1$		0.58	0.19	0.41	0.19
Fluctuations in ambient humidity $3\sigma_1$		0.028	0.0093	0.019	0.0093
Total $= (\sigma'_1 + \sigma'_2 + \sigma'_3 + \sigma'_4$ $\sigma'_5 + \sigma'_6)^{\frac{1}{2}}$	L	2.1	1.5	1.2	1.0
	S	2.2	1.5	1.0	0.95

L : 84.9g/m² 9.0 Fe% S : 40.7g/m² 12.1 Fe%

samples.) Thus, we can determine the composition and the coating weight with both high precision and high speed.

By using theoretical relationships, we can reduce the number of samples for calibration which we cannot easily get in actual practice.

ANALYZER CONFIGURATION

Table 5 shows the specifications of the analyzer and Fig. 4 shows the analyzer configuration. In order to measure under stable conditions, we arranged the two optical systems in an airtight and air-conditioned box called a sensor head. We can scan this sensor head along the transverse direction for directional data. In these optical systems we use solid state detectors which have higher resolution than other types of detectors have. Therefore, by using a single semiconductor detector we can detect iron Kα and zinc Kα separately without an analyzing crystal which reduces the intensity. Thus using solid state detectors enabled us to simplify the configuration and to obtain high x-ray intensity. The detectors are suitable for on-line use because they do not need liquid nitrogen or constant maintenance.

Table 5 Specifications of the coating layer analyzer

Standard measurement time	10 seconds
Measured side	Both (top & bottom)
Line speed	150m/min. (max)
Range of Fe content	5~18 Fe wt%
Range of coating weight	30~90g/m²
Manufacturer	Rigaku Industrial Corp.

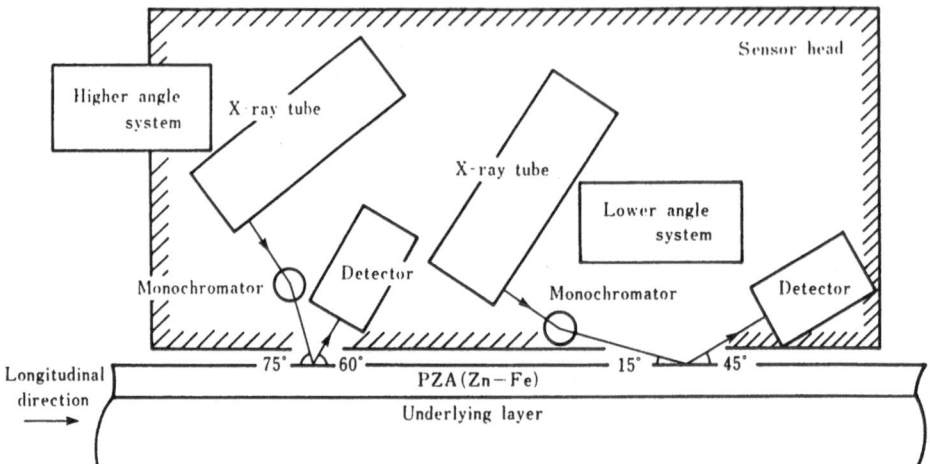

Fig. 4 Schematic drawing of the sensor head

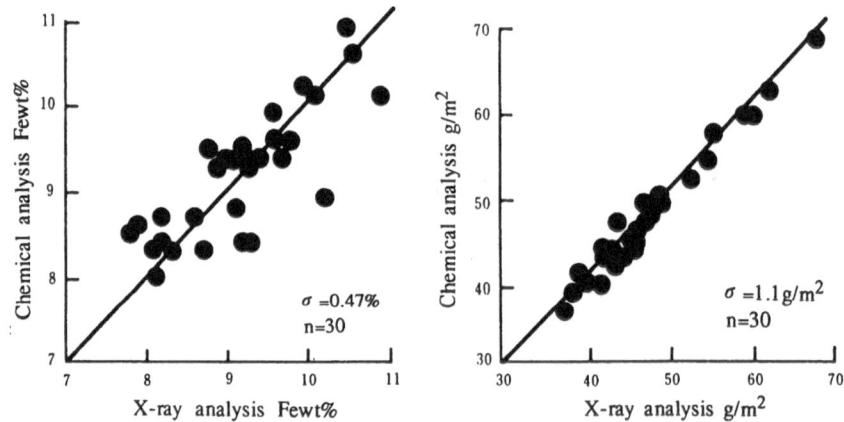

Fig. 5 Results of off-line analysis

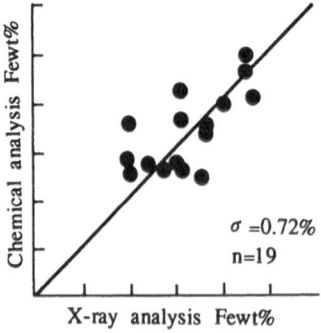

Fig. 6 Results of on-line analysis

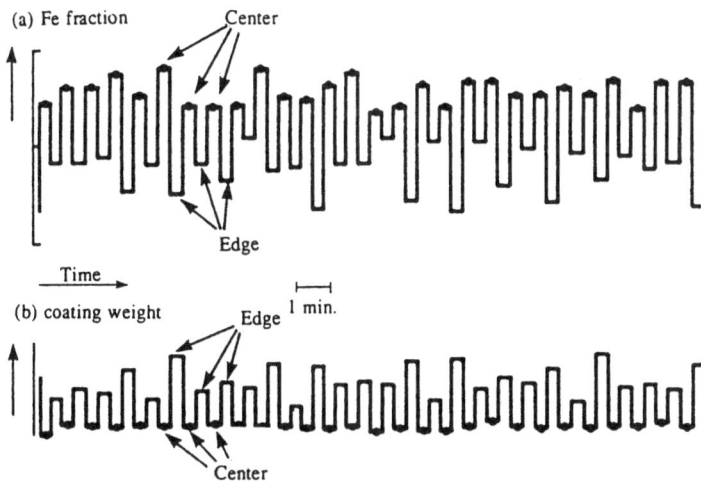

Fig. 7 An example of analysis with sensor head
scanning along the transverse direction
(3 point scanning mode)

RESULTS

Fig. 5 shows the result of the confirmation tests made by the analyzer off-line. Standard deviations of the difference between the data by chemical analysis and that by this method are 0.5Fe wt% in the composition and $1.1 g/m^2$ in the coating weight. Considering the analyzing error in chemical analysis, this result can be said to be precise. In order to obtain these data, we considered the surface characteristics of these steel sheets and modified this method[4] a little.

Fig. 6 shows the result of the on-line analysis of the composition. The standard deviation is 0.7Fe wt%, which means the analyzer can determine quite precisely; considering the on-line case, in which steel sheets are moving quickly, chemical analysis cannot be made for the same area as this x-ray analysis.

Fig. 7 is an example of the result of sensor head scanning. The sensor head was stopped for several seconds at three points along the transverse direction (one edge⇒ center⇒ the other edge⇒ center...) . The sample was made edge-over-coated on purpose. And the analyzer could detect the change along the transverse direction of the coating weight and the composition.

CONCLUSION

We have developed a new method for on-line determination of the composition and the coating weight of the coating layer of galvannealed steel sheets. We use two optical systems which are different in incident angles, take-off angles and wavelengths of monochromatic incident x-rays. The analyzer has been applied to the continuous galvanizing line in our Fukuyama works successfully. As a result, this analyzer has enabled us to improve and assure our product quality.

ACKNOWLEDGEMENTS

We would like to thank Naoki Matsuura of Rigaku Industrial Corporation for manufacturing this on-line analyzer and Sin-ichi Terada of Technos Co. Ltd. for arranging the electrical components of Kevex Instruments.

REFERENCE

1. T. Shiraiwa and N. Fujino : Theoretical Calculation of X-ray Intensities in Fluorescent X-ray Spectro Chemical analysis, J. Appl. Phys. Vol. 5, No. 10, 886-899 (1966)
2. T. Arai, T. Shoji and K. Omote : Measurement of the Spectral Distribution Emitted From X-ray Spectrographic Tubes, Advances in X-ray Analysis, Vol. 29, 413-422(1985)
3. K. Imai, H. Kato and K. Nishifuji : On-line Coating Layer Analysis of Galvannealed Steel Sheets by Monochromatic Excitation X-ray Fluorescence Method, Part 1 & 2, Proc. 32nd Japan Joint Automatic Control Conf., 417-418 (1989)
4. A. Honda, S. Harada, N. Taguchi, K. Yamauchi, H. Kato and K. Nishifuji : Development of On-line Alloy Coating Analyzer for CGL, CAMP-ISIJ, Vol. 3, 1284 (1990)

IMAGING THE THREE-DIMENSIONAL MICROSTRUCTURE OF MATERIALS

John C. Russ

Materials Science and Engineering Dept.
North Carolina State Univ., Raleigh, NC

Abstract

Three-dimensional tomographic reconstruction is capable of producing arrays of cubic voxels with resolution adequate to reveal important microstructural properties in materials. Visualization and measurement of these structures extends the types of information available from conventional 2D microscopy.

Practical Considerations for Industrial Tomography

Tomographic imaging as applied to materials and quality control inspection differs in several rather important respects from its medical uses. While the time required for analysis and the dosage used can be increased, the much greater variability in density and composition of the specimens poses problems for the reconstruction. Also, it is usually desirable to achieve much higher spatial resolution, and to produce 3D voxel images as opposed to a series of slices, in keeping with the dimensions and complexity of microstructural parameters in metals, ceramics and composites. Current industrial applications of tomography generally utilize X-rays and do little more than extend medical technology, but recent and ongoing developments in several research laboratories promise new methods for the near future.

Improvements in resolution, down to the range of 1 to 10 μm, utilize point sources of X-rays from synchrotron[1] and microfocus X-ray tube sources, as well as other particle beams such as ions or electrons. Electron beam tomography has demonstrated resolution of tens of nanometers[2,3]. The synchrotron is by far the brightest X-ray source[4], and can also be made monochromatic, but is inconvenient for routine use.

Microfocus sources can be filtered to produce quasi-monochromatic radiation, but with a significant drop in intensity and increase in exposure time. Using several different radiation energies can permit solving for the compositional as well as density distribution in the specimen[5]. Figure 1 shows a microfocus source used as an X-ray microscope able to collect projected images through a series of filters, which are then used to reconstruct the 3D structure of the sample from this "cone beam" geometry. The outstanding issues to be determined in these methods are the tradeoff between number of views and image noise, adequate detection hardware for the projection images, the best geometric manipulation of the sample to obtain optimal views, and the choice of algorithms to use the multiple types of information for direct 3D reconstruction.

Beam hardening - the change in the energy distribution of polychromatic radiation as it passes through the specimen - restricts the utility of white radiation sources. In principle, monochromatic radiation at two or more energies overcomes this problem and permits solving for both the density and composition of the individual voxels. Another approach, using a combination of attenuation and scatter tomography, also offers this capability and is fundamentally more sensitive. Practical implementation requires much study. The arrangement in Figure 1 uses the ratio of images acquired through pairs of filters with different absorption edges (balanced or Ross filters) to perform element-specific reconstructions[6].

Reconstruction in 3D

The traditional and relatively straightforward reconstruction methods - backprojection, Fourier and ART approaches - do not deal equally well with the problems of 3D geometries. These include missing directions due to mechanical limitations, variable statistical quality of projections, variable resolution in the voxel grid, and utilization of *a priori* information such as external sample dimensions or known compositions for phases present. Maximum likelihood techniques offer one, perhaps the best, solution to this dilemma. Several other new computer methods such as thermal annealing have been proposed as well.

A further complication in reconstructing 3D data is the very large numbers of voxels needed to adequately represent complex structures. Reconstruction of a conventional 2D slice image with 1000x1000 pixel resolution can be performed in a desktop computer, but a comparable 1000x1000x1000 voxel image is beyond the practical capability of the largest current machines available. The use of smaller array sizes permits development of methods, but limits the resolution needed to show small structures in the reconstruction. The figures shown in this paper are all 100x100x100; we are using a Cray YMP-24 to perform the reconstructions using standard ART and FBP techniques.

Reconstruction of a three-dimensional voxel array representing specimen microstructure raises several problems for display and interpretation not present for medical imaging of two-dimensional slices. Measurement of size, shape, position and density of features for the most part require only extensions of two-dimensional parameters, although these are not always straightforward to implement. However, characterization of the topological properties of networks and neighbor relationships among features is more difficult. Display of the data set for interactive human interpretation is making extensive use of small computers for volumetric and

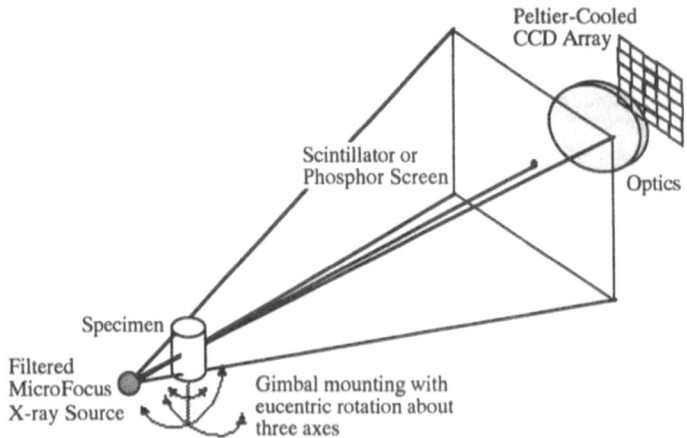

Figure 1. Diagram of a cone-beam imaging system. The projection image magnification is the ratio of source-screen and source-specimen distances. The attainable resolution is limited by the spot size of the microfocus X-ray source.

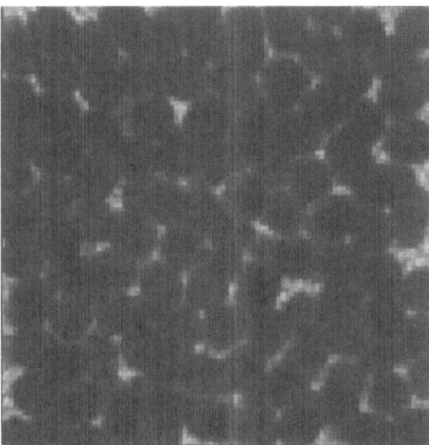

Figure 2. A single 2D projection view through the a sintered ceramic
consisting of 100 μm diameter alumina spheres.

surface shaded (rendered) displays, with considerable success. We are using a Macintosh to collect the raw data and generate the interactive displays shown in the figures.

Figure 2 shows a projected image through a microstructure of particles (approximately 100 μm diameter spheres) of alumina. This is a simulated test microstructure being used to develop methods for 3D tomography of microstructures of explosives and other similar, sintered materials. The projected image is shown here with 100x100 pixel resolution; with inexpensive cooled CCD cameras, higher resolutions from 256x256 and up are readily achievable. Each image can be integrated with the Peltier-cooled camera (designed for astronomical imaging) for up to 5 minutes without significant noise. A series of multiple views is then used to perform the reconstruction.

Figure 3 shows several planes of voxels from the reconstruction. The voxels are cubic, so the thickness of each plane is equal to the lateral dimensions. This is important for subsequent interpretation of the microstructure. Viewing the series of planes as an animation or movie enables the viewer to get some sense of the three-dimensional organization of the particles, by substituting time for one of the three spatial dimensions.

Volumetric reconstruction of the data allows viewing any internal plane section of the solid. Figure 4 shows an example in which several orthogonal plane surfaces have been interactively positioned in the reconstructed data, to show the particles. By positioning arbitrary planes, which can then be animated, it is possible to view the important points where these lightly sintered particles are in contact.

Human viewers are not familiar with cross-section images. Most of our real-world experience involves surface images. Selectively making the low density voxels transparent as shown in Figure 5 allows viewing the stack of planes in a way that helps us to "see" the 3D relationships between particles. In the figure, the planes of voxels have been spaced apart in the z direction to improve the visibility. Note that there is some uncertainty in the density of individual voxels, and that this results in a somewhat noisy definition of the particle boundaries which are actually rather smooth.

Figure 6 shows the data from the planes converted to a "rendered" or surface display showing the exterior of the particles. This view is most familiar to viewers, but actually hides much of the information. The interiors of particles cannot be observed (they could all be hollow!), and precedence causes particles and the important contact points between them to be hidden by other particles in front.

Structural measurements

A 3D array of voxels can be processed in ways analogous to the methods applied to conventional 2D images, to enhance edges, locate contact points, perform skeletonization, etc[7,8]. The skeleton in particular makes it possible to count the number of neighbors touching each particle (the number of branches which meet at each node). This kind of topological information[9] is not available from the 2D planar images obtained by conventional microscopy of sectioned surfaces, or from 2D tomographic images. It is very difficult to obtain from a series of 2D tomographic images, if the separation of the planes is greater than the lateral voxel resolution within the planes.

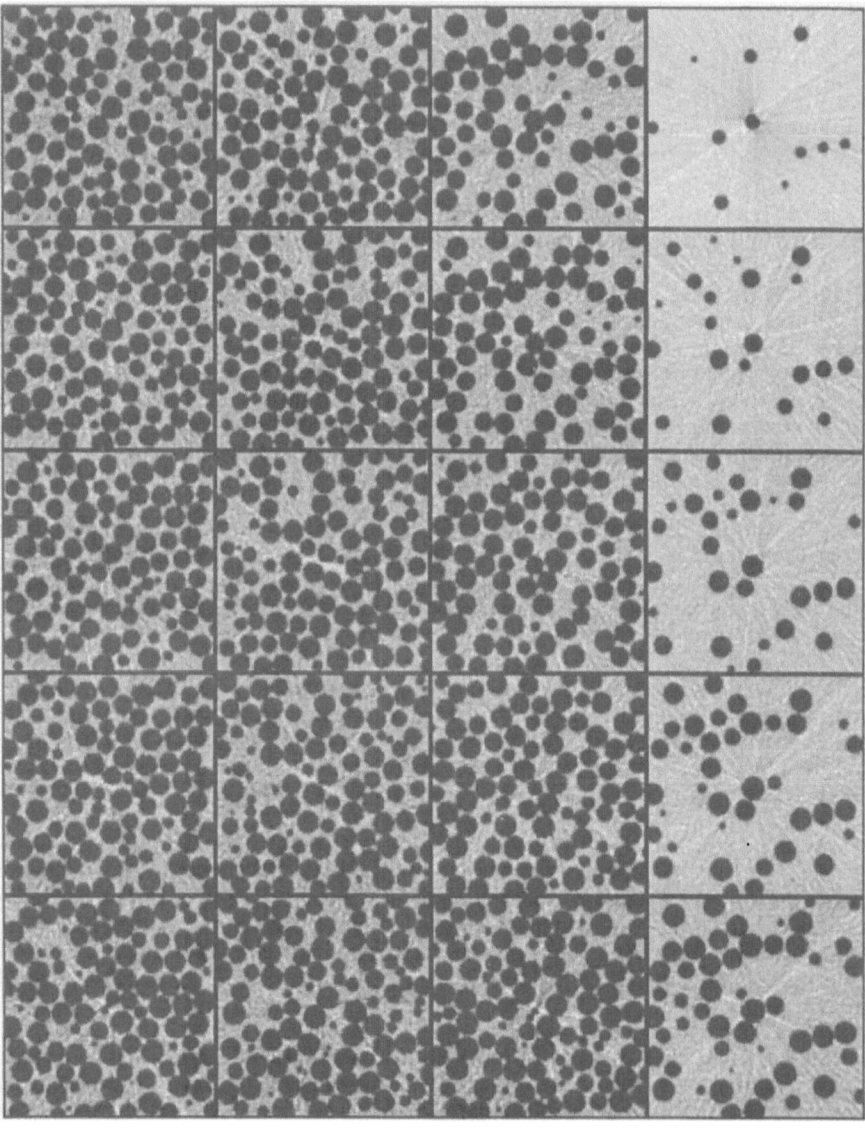

Figure 3. Twenty individual planes of reconstructed voxels showing sections through the spheres in the sample shown in projection in Figure 2.

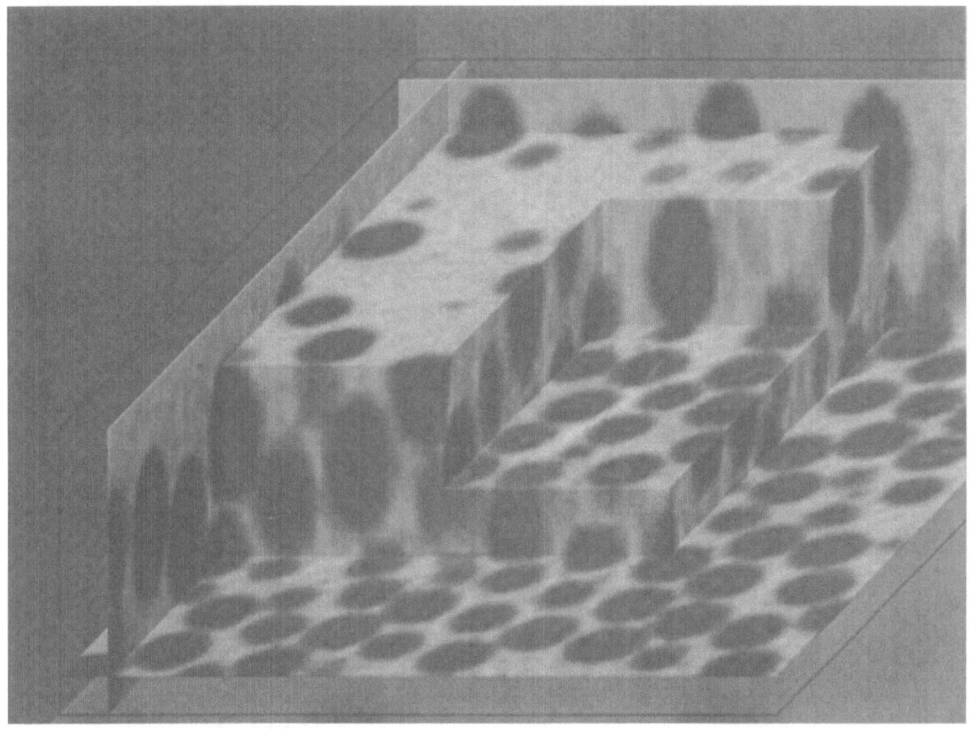

Figure 4. Three-dimensional presentation of the data set with section planes.

Figure 5. "Exploded" view of voxel layers, with low density voxels surrounding the particles shown transparent.

Figure 6. Surface-rendered view of particles from Figure 5.

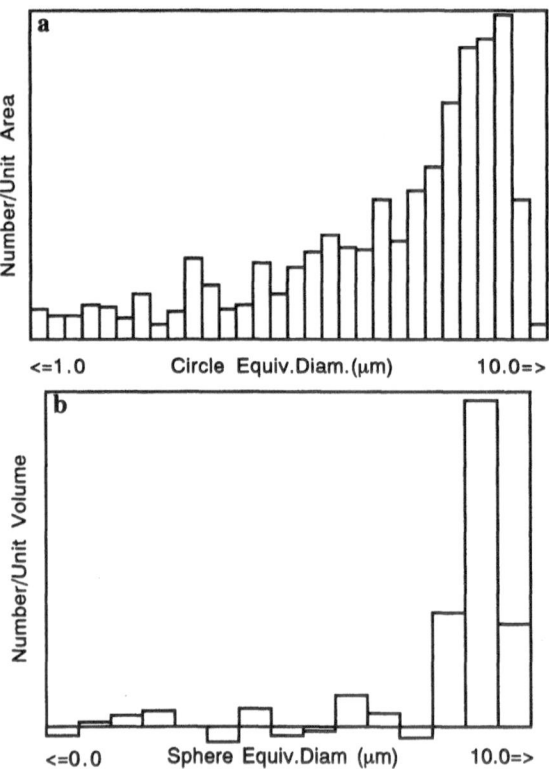

Figure 7. Comparison of 2D and 3D measurement of size of spherical particles in structure shown in Figure 5: a) size distribution of circles in 2D plane sections; b) estimated size distribution of spheres by unfolding the circle data in figure a (note negative values);

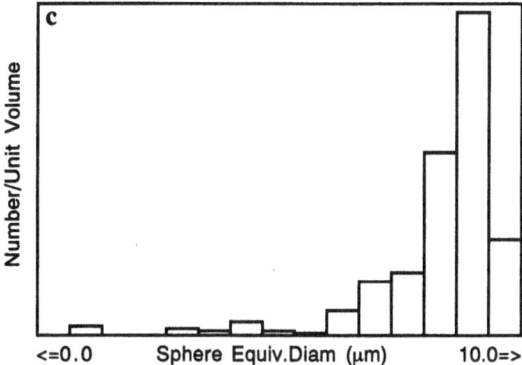

Figure 7 (cont'd). Comparison of 2D and 3D measurement of size of spherical particles in structure shown in Figure 5: c) directly measured size distribution of spheres from 3D voxel array.

There are some measurements, however, which need not be performed in 3D. One is the size distribution of the particles. In 2D section images, the spherical particles appear as circles. The size of the circle is generally smaller than that of the intersected sphere, because it is unlikely that the section plane will pass through the equator of the particle. However, by measuring the distribution of diameters of many circles, and using geometric probability to predict the size distribution of circles from random intersections of a sphere, the sphere size distribution can be determined. This is a standard stereological technique[10] which is well tested for simple shapes such as spheres, ellipsoids, cylinders, cubes, etc. It breaks down for mixtures of shapes, or more irregular shapes.

Figure 7 shows the results for the data from Figure 5. The size distribution of the circles shows many small intersections, which result from cutting a larger sphere at a high latitude. Unfolding this distribution produces a sphere size distribution which contains some negative values. These are not physically possible, of course, and arise from the statistical nature of the counting process. The magnitude of negative bins indicates the precision of the data. For comparison, a plot of the distribution of sphere sizes from the 3D voxel array is also shown. The overall agreement is good, and the measurement in 2D, which could be performed from section images using a light microscope or 2D tomographic reconstruction, is much quicker than cheaper than full 3D tomography. However, for more irregular shapes, or topological information such as the number of touching neighbors or the connectivity of the intervening pore space, three-dimensional imaging and measurement on the voxel array is essential.

Acknowledgements: Portions of the work described here have been performed under contracts funded by British Petroleum and the U. S. Dept. of Defense Mound Laboratories. Tom Prettyman and Young-Soo Ham have been involved in the design and implementation of hardware and software as part of their doctoral programs at NCSU.

References

1. H. W. Deckman, K. L. D'Amico, J. H. Dunsmuir, B. P.Flannery, S. M. Gruner (1989) Advances in X-ray Analysis 32:641
2. A. Engel, A. Massalski (1984)*3D reconstruction from electron micrographs: Its potential and practical limitations* Ultramicroscopy 13:71-84
3. R. Hegerl (1989) *Three-dimensional reconstruction from projections in electron microscopy* European Journal of Cell Biology 48 (Supplement 25):135-138
4. T. Prettyman, R. Gardner, J. Russ, K. Verghese (1991) "On the performance of a combined transmission and scattering approach to industrial computed tomography" in this volume

5. D. J. Schneberk, H. E. Martz, S. G. Azavedo (1991) "Multiple Energy Techniques in Industrial Computerized Tomography, in Review of Progress in Quantitative Nondestructive Evaluation (D. O. Thompson, D. E. Chimenti, ed.), Plenum Press, NY
6. J. C. Russ (1988) *Differential Absorption Three-Dimensional Microtomography* Trans. ANS 56 Spl. 3:14
7. J. C. Russ (1990) **Computer Assisted Microscopy**, Plenum Press, New York, NY
8. J. C. Russ (in press) **Image Processing Handbook**, CRC Press, Worcester MA
9. J. C. Russ, J. Ch. Russ (1989) *Topological Measurements on Skeletonized Three-Dimensional Networks* J. Comput. Assist. Microscopy 1:131-150
10.J. C. Russ (1986) **Practical Stereology**, Plenum Press, New York, NY

X-RAY OPTICS FOR SCANNING FLUORESCENCE MICROSCOPY
AND OTHER APPLICATIONS

Richard W. Ryon

Lawrence Livermore National Laboratory
Livermore, California USA

William K. Warburton

X-Ray Instrumentation Associates
Menlo Park, California USA

ABSTRACT

Scanning x-ray fluorescence microscopy is analogous to
scanning electron microscopy. Maps of the distribution of
chemical elements are produced by scanning the specimen with
a very small x-ray beam while collecting the XRF spectrum.
Our goal is to perform such scanning microscopy with
resolution in the range of <1 to 10 μm, using standard
laboratory x-ray tubes. In order to increase the radiation
flux on the specimen, we are investigating mirror optics in
the Kirkpatrick-Baez (K-B) configuration. K-B optics uses two
curved mirrors mounted orthogonally along the optical axis.
The first mirror provides vertical focus, the second mirror
provides horizontal focus. We have used two types of mirrors:
synthetic multilayers and crystals. Multilayer mirrors are
used with lower energy radiation such as Cu Kα. At higher
energies such as Ag Kα, silicon wafers are used in order to
increase the incidence angles and thereby the photon
collection efficiency. In order to increase the surface area
of multilayers which reflects x-rays at the Bragg angle, we
have designed mirrors with the spacing between layers graded
along the optic axis in order to compensate for the changing
angle of incidence. Likewise, to achieve a large reflecting
surface with silicon, the wafers are placed on a specially
designed lever arm which is bent into a log spiral by
applying force at one end. In this way, the same diffracting

angle is maintained over the entire surface of the wafer, providing a large solid angle for photon collection.

INTRODUCTION

Radiography is a widely used technique which readily shows flaws such as cracks, voids, inclusions, or density variations in materials and fabricated devices. In many of our applications, it is desirable to augment radiographic information with knowledge about the distribution of chemical elements such as that provided by x-ray fluorescence. Images of the distribution of elements can be obtained by scanning the specimen with a beam of electrons or other particles, or with x-rays. X-ray beams have advantages compared to particle beams such as: 1) low thermal loading of the specimen; 2) the specimen can be non-conducting; 3) the specimen can be examined in air to avoid loss of volatile components; 4) subsurface details can be imaged because x-rays penetrate some distance below the surface; and 5) minor and trace elements can be imaged because the signal to noise ratio is better. A simple technique is to collimate the x-rays to provide the resolution required. However, if the resolution needed is below 10's of micrometers, the time required becomes many hours for only moderately sized specimens when x-ray tubes rather than synchrotrons are used for excitation. Therefore, something more efficient than simple collimation is required for routine high resolution chemical imaging.

In order to increase resolution and imaging speed, we need to collect x-rays with the largest possible solid angles and focus them onto the smallest possible spots. Besides the primary aim of performing fluorescence imaging, related techniques such as transmission, backscatter, and even acoustic imaging (with a modulated beam) may be feasible. Additionally, fan beams (e.g., from a single mirror) can be used in radiography to reduce out of plane scatter. Fan beams and parallel beams can be used in tomography. We are also aware that the ability to magnify and demagnify objects through the use of x-ray optics can impact related fields such as projection imaging and lithography.

Scanning microscopies can be accomplished by a variety of means. Coded irradiation followed by image reconstruction can be used.[1] If the more direct method of raster scanning is employed, several choices can be made. A simple collimated beam can be used.[2] A more efficient means is to guide the x-rays to the specimen using hollow fibers[3,4] or fiber bundles.[5] Fresnel transmission optics have been used to advantage to focus x-rays,[6] even at energies up to 20 keV.[7] We have chosen to concentrate on Kirkpatrick-Baez (K-B) mirror optics[8,9] because this system is relatively easy to fabricate, it may have high efficiency, and it has desirable optical features.

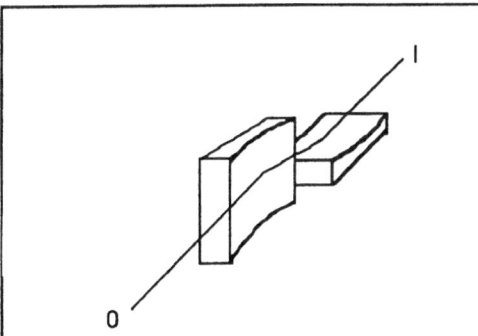

Figure 1. Schematic diamagram of Kirkpatrick-Baez optics, showing the central ray. Concave mirrors produce images of extended objects at small incidence angles[8].

DISCUSSION

A schematic diagram of K-B optics is shown in Figure 1, with an actual device shown in Figure 2. With the K-B arrangement, we can minimize astigmatism by separately adjusting the focus in the vertical and horizontal planes. (Some astigmatism remains when spherical rather than elliptical mirrors are used.) There is no theoretical limit to the size of the mirrors, so efficiency can be increased with only practical considerations in mind. The mirrors themselves can use either total specular reflection at longer wavelengths, Bragg reflection from multilayers at intermediate wavelengths, and Bragg reflection from crystals at shorter wavelengths. By careful design of multilayers on silicon wafers, all three ranges of wavelengths can be accomodated with a single mirror system. At longer wavelengths, the mirror behaves only as a smooth surface and total reflection results. At intermediate wavelengths, the multilayer is operative. At yet shorter wavelengths, the x-rays penetrate the multilayer and are diffracted by the silicon interatomic planes.

The focusing of rays is described by the Coddington equations[9]. The focus for tangential rays (in the plane of incidence), is

$$\frac{1}{f_t} = \frac{1}{u} + \frac{1}{v_t} = \frac{2}{R_t \sin\theta}$$

while the focus for sagittal rays (perpendicular to the plane of incidence) is

$$\frac{1}{f_s} = \frac{1}{u} + \frac{1}{v_s} = \frac{2\sin\theta}{R_s} \, ,$$

where θ is the glancing angle of incidence (given by the Bragg equation $n\lambda=2d\sin\theta$ for diffracting media), f is the focal length, u is the x-ray source to mirror distance, v is the mirror to focus distance, R is the radius of curvature of the mirror, and the subscripts t and s designate tangential and sagittal rays, respectively. The magnification factor for the focal spot M is equal to v/u. As an example, consider $CuK\alpha_1$ ($\lambda=1.5406\text{Å}$), a multilayer mirror with a period d of 12Å,

Figure 2. Kirkpatrick-Baez mirror system with benders for silicon wafers. The x-ray source is mounted on the end of the tube on the left.

R_t=1000mm, u=353mm, v_t=35.3. There is a 10-fold demagnification for the tangential rays. For the sagittal rays to focus at the same point, a radius of curvature of R_s = 4.1 mm would be required. For a spherical mirror where R_s and R_t are the same, the sagittal rays are unfocused, and a line image is formed by a single mirror. If a second mirror is placed orthogonally to the first and the reflected rays intercept it, the tangential rays from the first mirror become sagittal at the second mirror, and the sagittal rays from the first mirror become tangential. By appropriate selection of R (or θ via the d spacing) for the two mirrors, the beams collapse to a point focus.

In order to design multilayers for the mirrors, we use a Fortran program based upon the Fresnel equations which runs on a desktop computer.[10] Synthetic multilayers are produced by coating an atomically smooth substrate with alternating layers of high and low electron density materials with interlayer periods as low as 1 nm. We can graphically see the reflectivity as a function of angle or wavelength. We can study the effects of such parameters as composition, number of layers, and the ratio of heavy to light layer thicknesses on reflectivity and resolution. Judgements about design trade-offs can be made prior to fabrication. In Figure 3, we see the effect of multilayer pair compositon on resolution.

Another computational aid for mirror design is the ray tracing code "Shadow" which was developed at the University of Wisconsin Center for X-Ray Lithography.[11] A typical pair of output graphics is shown in Figure 5, where the improvement in focus using optimized geometric figure for the reflecting surface is shown. At present, we use spherical substrates for our multilayers, because they are readily available. The spherical aberration is apparent in Figure 5. Elliptical mirrors have been designed but not yet fabricated.

Given the narrow rocking curves shown in Figure 3 for multilayer mirrors, diffraction occurs over only a small band of the mirror surface when the source is at typical distances

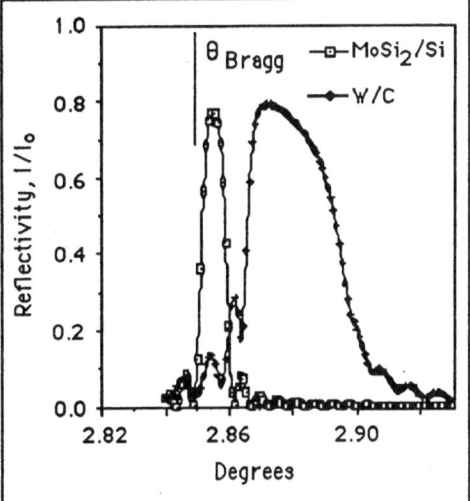

Figure 3. Rocking curves for two multilayer compositions ($MoSi_2$/Si pair on the left, W/C pair on the right)

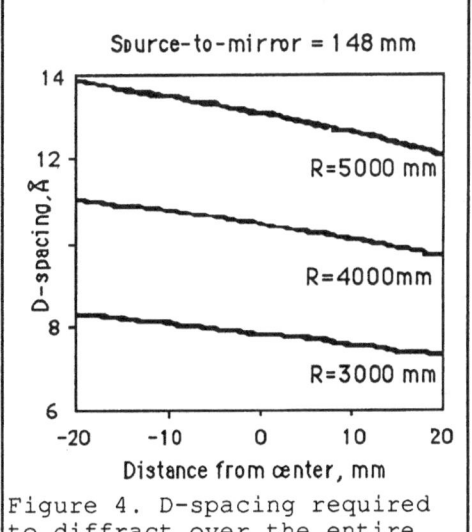

Figure 4. D-spacing required to diffract over the entire mirror surface. Designed to produce a parallel beam with Cu Kα radiation.

of 200 to 500 mm. This is because the curvature does not fully compensate for the change of incidence angle along the optic axis. One means to overcome this problem with multilayer mirrors is to vary the spacing between layers so that the Bragg condition for diffraction is satisfied at all points on the surface. A set of curves for spherical optics which satisfy both the Bragg condition and the tangential focusing condition are shown in Figure 4. Roughly speaking, the interlayer spacing needs to vary by a few tenths of a percent per mm along the optic axis. Such variations are within the capabilities of multilayer fabricators.

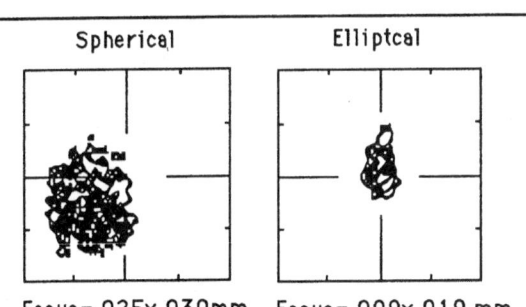

Figure 5. Output of "Shadow" ray tracing code, showing the improved focus obtained with an elliptical mirror. Geometric focus is .004x .010 mm.

For the even narrower rocking curves of crystalline reflectors, another technique must be used to compensate for the angular variation since the interatomic spacing

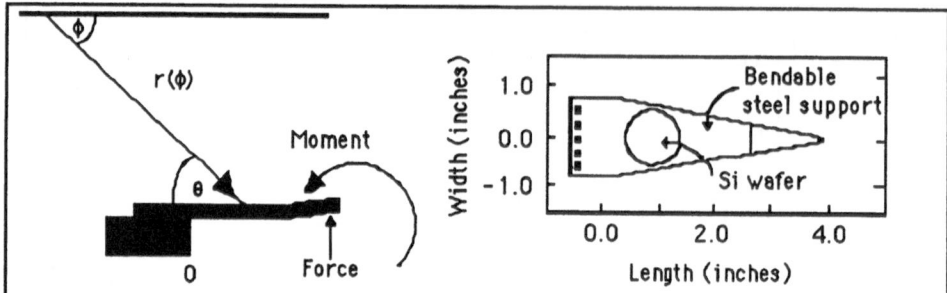

Figure 6. Side and top schematic views of the bendable wafer support which makes up one mirror of a Kirkpatrick-Baez focusing system.

is not a parameter that can be varied at will. The technique we have selected is to make all the incidence angles identical. The condition of invariant incidence angle is fulfilled for a log-spiral surface, as shown in Figure 6. The log-spiral figure provides 1-to-1 focusing, so we are no longer able to demagnify the anode focal spot as can be done with other figures used with multilayer mirrors. Without demagnification, a smaller focus (lower power) x-ray tube must be used. This drawback of a lower power tube is compensated by both the higher brilliance (i.e., higher heat load per unit area on the anode) and by the greater geometric efficiency due to a higher angle of incidence when using crystal mirrors.

CONCLUSION

We anticipate growing applications for efficient x-ray mirror optics. These applications include scanning x-ray fluorescence imaging (microscopy), analysis of very small objects, fan beam radiography and tomography, and parallel ("pencil") beam tomography. With a parallel beam, long distance alignment problems can also be approached using x-rays.

ACKNOWLEGEMENT

This work was performed under the auspices of the U.S. Department of Energy by the Lawrence Livermore National Laboratory under contract W-7405-Eng-48.

REFERENCES

1. N. Gurker, "Imaging Techniques for X-Ray Fluorescence and X-Ray Diffraction", Advances in X-Ray Analysis, Vol. 30, (1987), pp. 53-65.

2. Monte C. Nichols, Dale R. Boehme, Richard W. Ryon, David Wherry, Brian Cross, and Gary Aden, "Parameters Affecting X-Ray Microfluorescence (XRMP) Analysis", Advances in X-Ray Analysis, vol. 30 (1987), pp.45-51.

3. S. Larsson, P. Engstrom, and A. Rindby, "X-Ray Capillary Microbeam Spectrometer", Advances in X-Ray Analysis, Vol. 33, (1990), pp. 623-628.

4. D.A. Carpenter, M.A. Taylor, and C.E. Holcombe, "Applications of a Laboratory X-Ray Microprobe to Materials Analysis", Advances in X-Ray Analysis, vol. 32 (1989), pp. 115-120.

5. Muradin Kumakhov and Walter Gibson, as reported in "Piping X-Rays Through a Glass Brightly", Science, vol. 252 (12 April 1991), pp. 208-209.

6. B. Niemann, G. Schmahl, et al., "X-Ray Microscopy with Synchrotron Radiation at the Electron Storage Ring BESSY in Berlin", Nuclear Instruments and Methods in Physics Research A246 (1986), pp. 675-680.

7. Richard M. Bionta, Kenneth M. Skulina, et al., "Tabletop X-ray Microscope Using 8 keV Zone Plates", Optical Engineering, Vol. 29, No. 6 (1990), p576.

8. Paul Kirkpatrick and A.V. Baez, "Formation of Optical Images by X-Rays", Journal of the Optical Society of America, vol. 38, No. 9 (1948), pp. 766-774.

9. J.H. Underwood, A.C. Thompson, Y. Wu and R.D. Giauque, "X-Ray Microprope Using Multilayer Mirrors", Nuclear Instruments and Methods in Physics Research A266 (1988), pp. 296-302.

10. The multilayer code we use is an adaptation by Barry Jocoby of the Lawrence Livermore National Laboratory of a code by Underwood and Barbee. See James H. Underwood and Troy W. Barbee, Jr., "Synthetic Multilayers as Bragg Diffractors for X-Rays and Extreme Ultraviolet: Calculations and Performance", American Institute of Physics Conference Proceedings Number 75, Low Energy X-Ray Diagnostics—1981, pp. 170-178.

11. Franco Cerrina, Barry Lai, Karen Chapman, Chris Welnak, and Paul Runkle, "Shadow Primer", Center for X-Ray Lithography, University of Wisconsin, Madison, Wisconsin.

EVALUATION OF A COMBINED TRANSMISSION AND SCATTERING
APPROACH TO COMPOSITION IMAGING OF INDUSTRIAL SAMPLES

T. H. Prettyman, R. P. Gardner, J. C. Russ, and K. Verghese

Center for Engineering Applications of Radioisotopes
Box 7909, North Carolina State University
Raleigh, North Carolina 27695-7909

INTRODUCTION

Composition imaging of industrial samples has been reported using dual energy and multiple energy transmission computed tomography [1,2]. The simplest approach utilizes monoenergetic sources to obtain tomographs of a sample at two different energies. Each tomograph represents the linear attenuation coefficient distribution of the sample at the given source energy. The linear attenuation coefficient, μ, of each pixel can be expressed as a function of source energy, E, effective atom number density, n_{eff}, and effective atomic number, Z_{eff}:

$$\mu = n_{eff} g(Z_{eff}, E) \tag{1}$$

where, g can be given in terms of fitted expressions valid over a wide range of E and Z_{eff} [3]. Provided the source energies are above the absorption edges of the sample's constituents, two equations of the form of equation 1 are obtained for each pixel which can be solved — in this case, by forming the ratio of the two attenuation coefficients — to obtain n_{eff} and Z_{eff}.

An alternate way to measure distributions of n_{eff} and Z_{eff} is to combine information from a transmission scanner and a Compton scatterometer. As in the dual energy case, the transmission tomograph provides a relationship between n_{eff} and Z_{eff} in the form of equation 1 for every pixel in the image. The Compton scatterometer — which detects photons scattered from volumes or lines within a sample — provides additional information (typically electron density, $n_{eff}Z_{eff}$ [4]) needed to determine pixel composition. Potentially, this approach has some advantages over the dual energy method, including enhanced composition sensitivity. A comparison between the two techniques is presented along with a brief description of our work on the development of the combined approach.

SENSITIVITY COMPARISON

A convenient way to compare the sensitivity of both modes is to consider their ability to measure the composition of an inclusion of size Δ within a sample of otherwise known composition. Assuming that the position of the inclusion is known, two measurements are required to determine the atom number density, n_i, and atomic number, Z_i, of the inclusion. For the dual energy approach, two transmission projection measurements are required — one with source energy E_a, and one with source energy E_b. For the combined mode, one transmission projection is required with source energy E_t. A scattering measurement with source energy E_s yields the electron density, $n_i Z_i$, of the inclusion.

For both approaches, the variance in Z_i can be written in terms of the relative standard deviations in the measurements. For the dual energy approach, the variance in Z_i can be shown to be

$$\sigma_{Z_i}^2 = \left[\frac{1}{g(Z_i, E_b)\Delta n_i}\right]^2 \left[\epsilon_a^2 + \epsilon_b^2 R_{ab}^2\right] \left(\frac{\partial R_{ab}}{\partial Z_i}\right)^{-2} \tag{2}$$

where,

$$R_{ab} = \frac{g(Z_i, E_a)}{g(Z_i, E_b)} \tag{3}$$

and ϵ_a and ϵ_b are the relative standard deviations in the first and second transmission measurements, respectively. A similar expression for the combined mode can be shown to be

$$\sigma_{Z_i}^2 = \left[\epsilon_s^2 g^2(Z_i, E_t) + \epsilon_t^2 \left(\frac{1}{\Delta n_i}\right)^2\right] \left(\frac{\partial g}{\partial Z_i}\bigg|_{E_t}\right)^{-2} \tag{4}$$

where, ϵ_t and ϵ_s are the relative standard deviations in the transmission and scattering measurements, respectively. Note that equation 4 is not an explicit function of scatterometer source energy, E_s. Instead, E_s enters into the analysis implicitly along with the geometry of the device, source intensity and other parameters through the scattering measurement error, ϵ_s. For both modes, device dependent effects such as scanning geometry and source intensity can be included in the analysis by expanding the measurement error terms.

A good measure of device sensitivity is $\sqrt{2}\sigma_{Z_i}/Z_i$. This quantity, known as the minimum detectable change (MDC), is the minimum relative change in Z_i which can be detected with 68% confidence [5]. For illustration purposes, the MDC is plotted for several cases in figure 1 as a function of Z_i for both modes.

DISCUSSION

As shown in Figure 1, both methods exhibit a loss of sensitivity (increase in MDC) in the low Z_i range. In both cases, this loss of sensitivity is a result of the reduction of attenuation by the photoelectric effect at lower atomic numbers. Unlike the combined mode, however, the dual energy approach also exhibits a sharp loss of sensitivity at higher Z_i values. This loss of sensitivity occurs because R_{ab} is not a monotonically increasing function of Z_i. At a critical value, Z_c, R_{ab} exhibits a maximum (e. g., see Figure 1a in [2]). In the vicinity of Z_c, the slope of R_{ab} becomes small, which causes the sensitivity to be low. Also, in the region above Z_c, the sensitivity improves because, for fixed measurement errors, the uncertainty in R_{ab} diminishes with increasing Z_i. The value of Z_c depends on the source energy combination. For a given lower source energy, the minimum critical value occurs when E_b is closest to E_a.

The fact that R_{ab} is not a monotonic function points to a potential difficulty with the dual energy approach. Namely, for a given value of R_{ab} there exist two possible values of Z_i. In order for the method to be a reliable measure of composition, it must be known a priori on which side of Z_c each pixel lies. Despite the possibility of achieving higher sensitivity in the region above Z_c, practical implementation of the approach requires that source energies be selected so

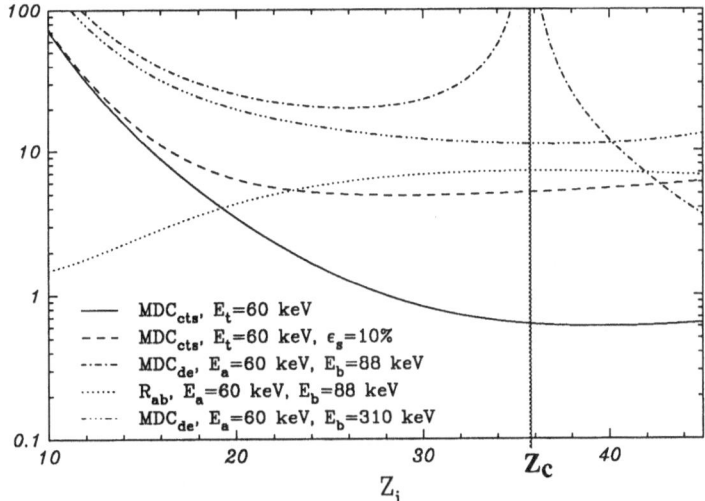

Figure 1. MDC values are plotted in units of percent for three source energy combinations as a function of Z_i. All curves are for $n_i = 4 \times 10^{22}$ $atoms/cm^3$, and $\Delta = 1$ mm. All measurement errors are fixed at 1% with the exception of one case where $\epsilon_s = 10\%$. R_{ab} is also displayed for one case. (de = dual energy, cts = combined transmission and scattering)

that the maximum Z_{eff} value of the sample lies below the minimum critical value. This is done in part because the transmission signal for the lower source energy is highly attenuated above Z_c, making it difficult to achieve the required measurement error, ϵ_a. Also, for lower source energies, oscillations in R_{ab} occur at higher atomic number due to the presence of absorption edges, resulting in the possiblity of multiple solutions.

To optimize the dual energy setup for a given application, the lower source energy should be selected so that contrast in the lower energy image is maximized. The upper energy E_b should be selected so that the maximum sample Z_{eff} is less than Z_c. Contrast in the upper energy image should also be high; however, E_b should be made as far above E_a as possible to ensure adequate composition sensitivity.

For the same application, an optimal arrangement for the combined mode would require that the transmission energy, E_t, be selected to maximize contrast in the transmission image. Hence, for the purpose of comparison, we shall assume that the lower energy for the dual energy approach will be selected to be the same as the transmission energy for the combined mode, $E_a = E_t$, for a given application. Furthermore, if the same scanning equipment is used to obtain the transmission images, it is reasonable to assume that $\epsilon_a = \epsilon_t$. The only remaining parameters to be specified for comparison are E_b, ϵ_b, and ϵ_s. We have found that when the measurement errors are equal ($\epsilon_b = \epsilon_s$), the combined mode is generally more sensitive than the dual energy approach, regardless of the selection of E_b (e. g., compare MDC values in Figure 1). In addition, relatively large errors in the scatterometer measurement can be tolerated without a substantial reduction in sensitivity.

DEVELOPMENT OF THE COMBINED MODE

The development of a scatterometer which can attain competitive values of ϵ_s presents significant design challenge. The primary limitation is that scatterometers are inherently low yield devices. Often, one or more X-ray or high specific activity isotopic sources are required to achieve an adequate signal. Unlike transmission scanners, scatterometers achieve maximum contrast when the source energy is high. The source energy, E_s is usually selected so that Compton scattering is dominant and so that attenuation in the sample is low. By selecting a high source energy, two problems are introduced. First, a significant portion of the measured signal can be caused by multiple scatters within the sample. Second, a higher degree of collimation is required to obtain the desired spatial resolution. Both higher spatial resolution and the reduction of the multiple scatter signal are usually at the expense of lower scatterometer yield. Fortunately, as suggested by the above analysis, the scatterometer need not be optimal in order to achieve higher composition sensitivity than the dual energy approach.

The purpose of our research is to evaluate the feasibility and performance of the combined mode. Emphasis has been placed on the study of systems that involve one or more projective scatterometers. These devices detect photons that scatter from lines within a sample. We have elected to study projective scatterometers for two reasons. First, unlike raster imaging systems, the scanning motion of a projective scatterometer is similar to that of the transmission mode, which means that the scatterometer and transmission scanner may share some common components. It may also be possible to obtain transmission and scattering data sets simultaneously. Second, each projection involves a large number of pixels, which means that fewer measurements are required to obtain information for the entire image. Reconstructions from limited data are hence possible. This is particularly important when the transmission measurements are of higher quality than the scattering measurements.

Our approach is to use Monte Carlo simulation to provide realistic simulations of projection data for device design and development. A Monte Carlo code, MCPT, which has been

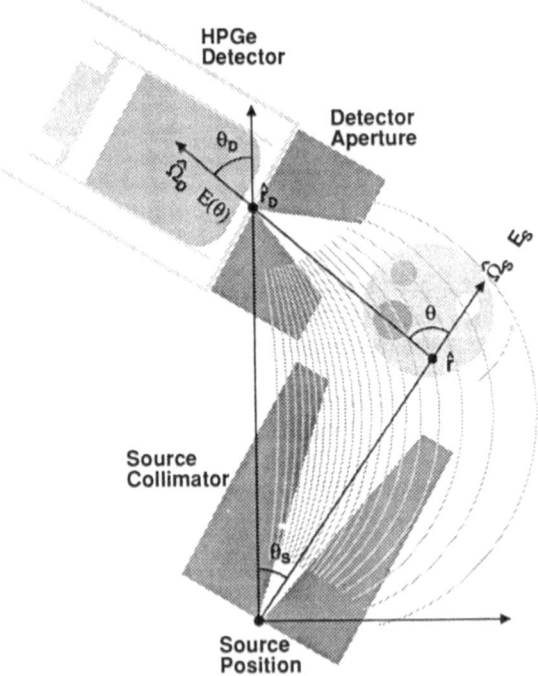

Figure 2. An energy dispersive, projective scatterometer. The device can measure projections of objects up to 5 cm in diameter, and can resolve features separated by 1.5 mm.

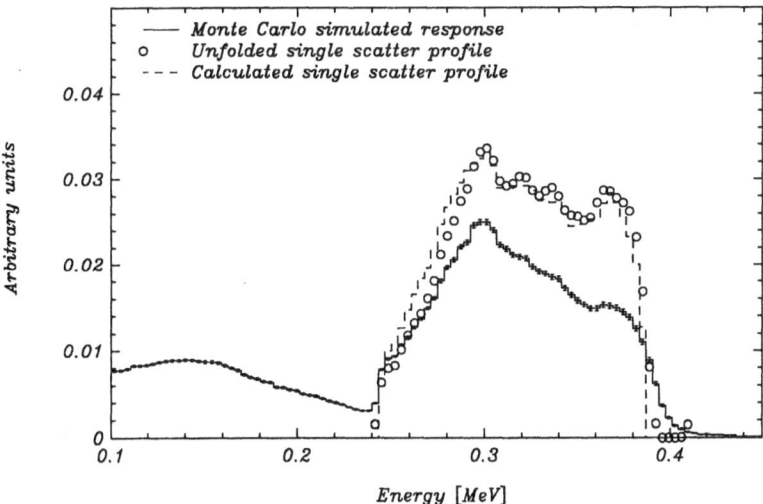

Figure 3. Monte Carlo simulated response of the scatterometer (Figure 2) for a concentric phantom (Table 1). Multiple scatters form a small portion of the response. The single scatter profile, or scattering projection, is unfolded from the scatterometer response.

benchmarked against experimental data, was developed specifically to model photon transport in tomographic scanners [6]. This code has been used to evaluate the feasibility systems with energy dispersive, projective scatterometers, similar to the device developed by Kondic for two phase flow mapping [7]. These devices make use of the unique relationship between the scattering angle and the energy of the scattered photon to measure photons that scatter from lines within a sample.

Primary components of an energy dispersive, projective scatterometer, including a high resolution detector with wide angle aperture, and a fan beam source collimator, are shown in Figure 2. The source is isotopic (Cs-137, 662 KeV). The geometry for the single scatter model is superimposed on the device. Under ideal conditions, only the response due to photons that scatter once in the sample is measured. In the model, all photons which arrive at the detector aperture with the same energy, E, scatter from points on an arc. The number of photons that pass through the aperture with energy, E, is represented by a line integral along the arc. The

Table 1. Concentric phantom components. The position of each component is specified by an inner and outer radius (r_1 and r_2).

Component	r_1 [cm]	r_2 [cm]	μ (at 60 KeV) [cm^{-1}]	n_{eff} [$\times 10^{24}$ $atoms/cm^3$]	Z_{eff}
SiO_2	0.0	0.6	0.598	0.0687	11.6
KI solution (15% KI by weight)	0.6	1.1	1.234	0.0202	22.6
Polyethylene	1.1	1.5	0.208	0.0864	4.37
Potassium	1.5	1.9	0.437	0.0133	19.0
Polyethylene	1.9	2.3	0.208	0.0864	4.37

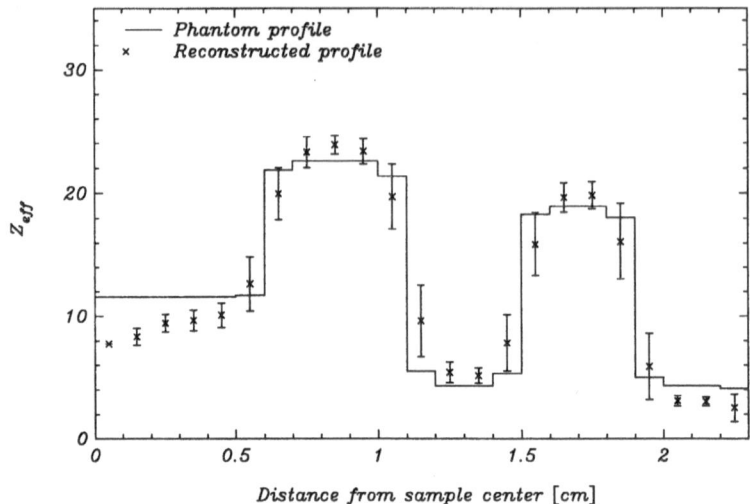

Figure 4. Comparison between reconstructed and phantom radial Z_{eff} profiles.

spectrum of incident photons, which can be unfolded from the measured response, is, hence, a projection of the sample. Different projections can be obtained by rotating the sample.

An algorithm, based on a single scatter model, was developed to reconstruct n_{eff} and Z_{eff} images from scattering projections and the transmission attenuation coefficient image. Both the nonlinear effect of attenuation of the scattered photons and the effect of the detector response function were accounted for by the algorithm. The algorithm was tested with data simulated by MCPT. As an example, a 4.6 cm diameter concentric phantom was reconstructed from a transmission image taken at 60 KeV and 25 scattering projections. The components of the phantom are listed in Table 1. A scattering projection of the phantom is shown in Figure 3. The radial distribution of Z_{eff} is shown in Figure 4 for both the phantom and reconstructed images. The distributions are in good agreement for the higher Z_{eff} components.

CONCLUSIONS

The combined transmission and scattering approach can be used to determine the composition of industrial samples with potentially higher composition sensitivity than the dual energy method. Using Monte Carlo simulation, we have found that quantitative measurements of sample composition can be made with devices which use energy dispersive, projective scatterometers. The optimum combination of transmission scanners and Compton scatterometers depends on the application of interest. Further work on the development of projective scatterometers with improved spatial resolution and scanning rates remains before the full potential of the method can be realized.

ACKNOWLEDGEMENTS

This work was supported by a research contract with BP America. We acknowledge the North Carolina Supercomputer Center for providing time on their Cray Y-MP.

REFERENCES

1. Engler, P., W. D. Friedman, and E. E. Armstrong (1989), "Determination of Material Composition using Dual Energy Computed Tomography on a Medical Scanner", *Proceedings of Industrial Computerized Tomography Topical*, ASNT, July 1989, 142-145.
2. Schneberk, D. J., H. Martz, and S. Azevedo (1991), "Multiple–Energy Techniques in Industrial Comuterized Tomography", *Review of Progress in Quantitative Nondestructive Evaluation*, Vol. 9.
3. Hubbel, J. H. (1969), "Photon Cross Sections, Attenuation Coefficients, and Energy Absorption Coefficients from 10 KeV to 100 GeV", NSRDS-NBS 29.
4. Harding G., "On the Sensitivity and Application Possibilities of a Novel Compton Scatter Imaging System", *IEEE Transactions on Nuclear Science*, NS-29, no. 3, (1982) 1260-1265.
5. Gardner, R. P., R. L. Ely, Jr. (1967), Radioisotope Measurement Applications in Engineering, Reinhold Publishing Corporation, New York, pp. 262–264.
6. Prettyman, T. H., R. P. Gardner, and K. Verghese (1990), "MCPT: A Monte Carlo Code for Simulation of Photon Transport in Tomographic Scanners", *Nucl. Instr. and Meth. in Phys. Res.*, A299, 516–523.
7. Kondic, N. (1983), "Three-Dimensional Density Field Determination by External Stationary Detectors and Gamma Sources using Selective Scattering", *Proc. of the 2nd International Topical Meeting on Nuclear Reactor Thermal–Hydraulics*, American Nuclear Society, Vol. II, pp. 1443–1455.

A NEW NONDESTRUCTIVE QUANTITATIVE COMPOSITION DEPTH

PROFILING TECHNIQUE BASED ON X-RAY EXCITED ELECTRON EMISSION

L.A.Backaleinickov, S.G.Konnikov, K.Ju.Pogrebitsky, Y.N.Yuriev, A.A.Vereninov

A.F.Ioffe Physical-Technical Institute
Academy Science USSR, St. Petersburg, Russia

R.Svagera

Institut für Angewandte und Technische Physik
Technische Universität Wien, Austria

R. Kaitna, G.Barnegg-Golwig

Rokappa G.m.b.H. Wien, Austria

Abstract

The physical basis for x-ray induced secondary electron emission in the range of the absorption edges is described. Applications of this principle for nondestructive composition determination and for nanometer resolved quantitative composition depth profiling are illustrated. A new commercial type of scientific instrument is presented which can be used for this new technique, and for almost all methods that require x-ray photon beam probing.

All the methods described in this paper are based on the same phenomenon, that is, the phenomenon of the jump-like increase of x-ray photoabsorption in the vicinity of the x-ray absorption edges of the chemical elements of the sample.

The total yield of photoelectrons æ (secondary electrons, photoelectrons, and Auger electrons) from a sample surface irradiated by an x-ray beam with a photon energy hν at an incidence angle φ is measured using channeltron. The dependence of æ on hν for a sample $Al_xGa_{1-x}As$ is presented in Fig. 1. A monotonic decrease of photoelectron emission with increasing photon energy in the regions where there are no absorption edges is observed. This response is due to the more rapid increase of the x-ray absorption path length in comparison to the increase in electron emission depth as the photon energy increases. This behavior of æ is interrupted in the regions of K-shell absorption edges of Al, Ga, and As. When a new electron shell is involved, the absorption rises sharply, the x-ray penetration depth decreases, new groups of electrons appear (K-shell photoelectrons Auger electrons) and consequently a jump-like increase of æ is observed.

The energies of absorption edges are characteristic of each chemical element. So, the presence of jumps at characteristic photon energies indicate the presence of certain chemical

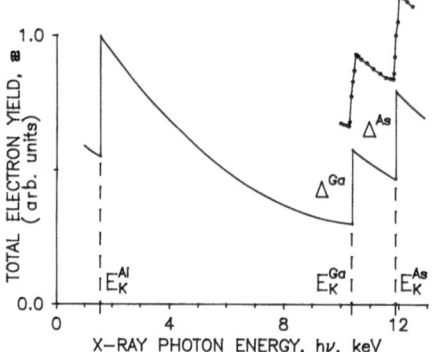

Fig.1 Total electron yield æ of
AlGaAs in dependence on the
x-ray photon energy

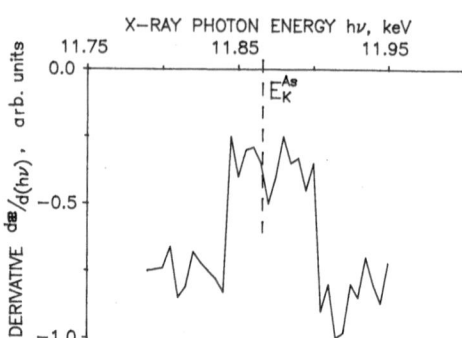

Fig.2 Derivative dæ/d(hv) of
blood sample in the
energy range of the As
K absorption edge

elements in the sample. This is the basis for a qualitative analysis of the sample. The
experimentally observed æ-jumps of Al, Ga, and As from the sample $Al_{0.31}Ga_{0.69}As$ are
presented in Fig.1. When the concentration of an element is much smaller, no distinct jump
may be observed and the derivative $\partial æ/\partial(hv)$ is used (Fig.2). The sample was a smear of
blood from a rat which had been poisoned by As. The concentration of As in the blood was
$2-5 \times 10^{18}$ atoms/cm³. The "rectangular peak" is due to the fact that an edge is not really
abrupt, but has a natural width of some dozens eV plus the resolution of the monochromator
of about 40eV.

The amplitude of a jump Δ (Fig.1) depends on the number of atoms of a given chemical
element in the sample and upon the yield depth. This is the basis for quantitative surface
analysis of the atomic composition. In this paper some results from $Al_xGa_{1-x}As$ ternary
specimens are given.

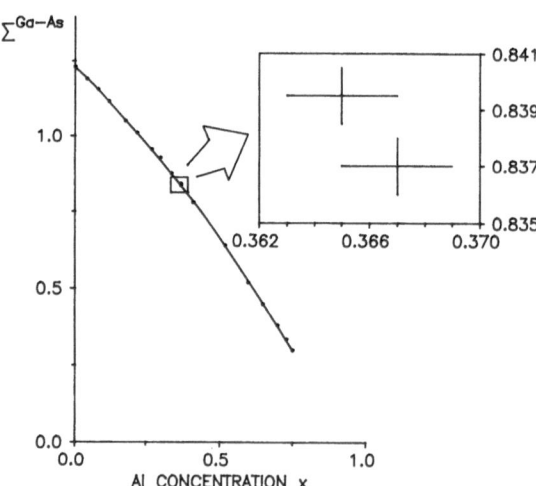

Fig.3 An example for the determination of Al
atomic concentrations from jump ratio
Σ^{Ga-As} data

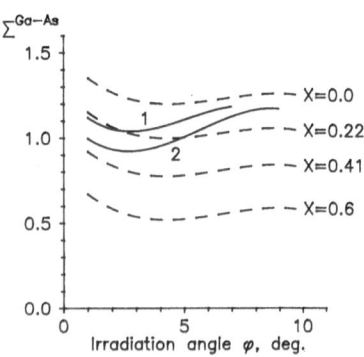

Fig.4 Σ^{Ga-As} responses of thinfilms (full
curves) and homogeneous standards
(dashed curves)

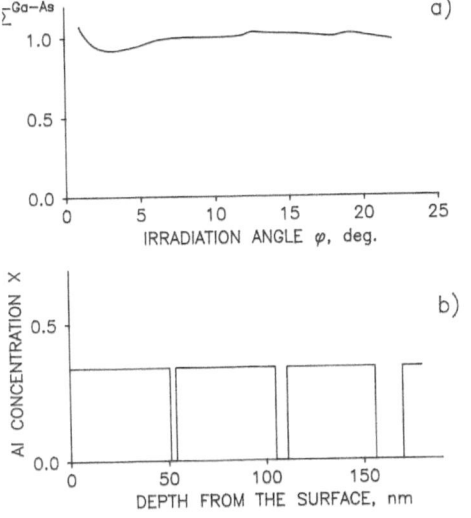

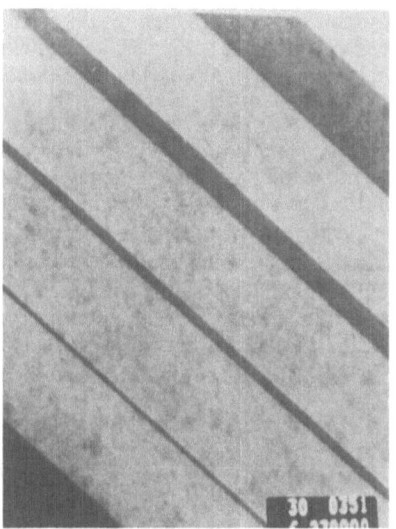

Fig.5 Measured angular Σ^{Ga-As} response of MOCVD AlGaAs sample (a) and concentration depth profile of Al evaluated from this response (b)

Fig.6 Transmission electron micrograph of the MOCVD AlGaAs sample

A characteristic quantity for an analysis is the jump ratio $\Sigma^{Ga-As} = \Delta^{Ga}/\Delta^{As}$. The dependence of Σ^{Ga-As} on atomic concentrations was experimentally determined on reference samples of defined composition. Results are depicted in Fig.3. An enlarged portion of the diagram demonstrates the resolution of the method in the composition range $x=0.365\pm0.002$ to $x=0.367\pm0.002$.

Quantitative nondestructive composition depth profiling is performed by determining the dependences of jump ratios on the incidence angles φ. When the sample under investigation has on its outermost surface one or more thin layers of different composition and the thicknesses are less than the average depth of electron emission, then the angular dependences of Σ^{Ga-As} of a thin film sample and a thick homogeneous standard are different. Fig.4

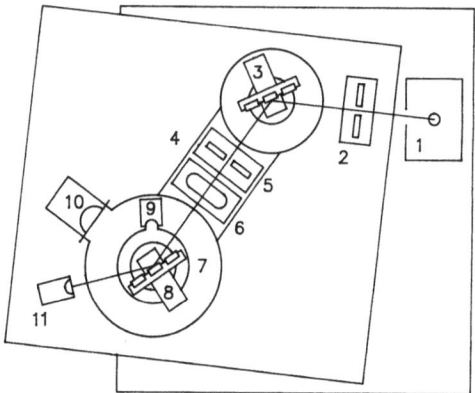

Fig.7 Schematic drawing of the experimental device

illustrates the experimental results from two single layer structures ($Al_{0.72}Ga_{0.28}As$ grown by MOCVD on GaAs substrate) with thicknesses of (26 ± 2)nm (curve 1) and (44 ± 2)nm (curve 2). Dashed curves represent the responses from thick standards of different composition.

Since a theoretical description of this effect is complicated and not yet solved, an empirical algorithm for quantitative composition depth profiles has been worked out. It can be used to determine composition depth profiles with a resolution of 0.3 nm in the region of 150 to 200 nm from the sample surface. This statement is confirmed by a comparison with the results obtained by transmission electron microscopy. From the response of the jump ratio Σ^{Ga-As} as a function of incidence angle φ shown in Fig. 5a, the layer structure of Fig. 5b has been evaluated and this result is compared to the transmission electron micrograph of Fig.6. TEM data for the thickness of the three first GaAs layers (dark-grey) are: (2.3 ± 0.2)nm, (5.1 ± 0.2)nm and (11.9 ± 0.2)nm. Electron emission depth profiling yields: (3 ± 1.5)nm, (6 ± 1.5)nm and (13 ± 2)nm. Other applications of this method were published previously[1-3].

As the principle described here is the basis of a new technique for concentration depth profiling and no commercially available instrument can be used, a new spectrometer has been developed for these investigations. Fig.7 shows a schematic drawing of the computer controlled equipment. The x-ray beam from the high power source RU-200 (1) is monochromatized by one of six crystal monochromators (3). This system allows scanning over a range of photon energies from 4 to 40 keV with resolution of 10 to 100 eV, depending on the choice of crystal and widths of the slits (2,5). The monochromatized beam is monitored by the ionization chamber (6). The electron emission system is installed on the 2θ drive of the goniometer (3). The sample is installed in the vacuum chamber (7) and can be replaced automatically by up to six samples or standards. The registration system includes an electron counter (9), detection of fluorescent radiation by an energy dispersive system (10), and an x-ray detector (11) for diffracted x-radiation. This instrumentation may be used for almost any method of x-ray analysis.

References

1. S.G.Konnikov and K.Ju.Pogrebitsky, Surf.Sci. 228,(1990) 532
2. S.G.Konnikov et.al.,Best of Sov.Semicond. Phys. & Tech.(1991) 263
3. Zh.I.Alferov et.al.,Best of Sov.Semicond. Phys. & Tech. (1991) 305

SOFTWARE DEVELOPMENT FOR X-RAY MICROBEAM SPECTROSCOPY

A. Rindby and P. Voglis

Chalmers University of Technology
Göteborg, Sweden

and

G. Nilsson and B. Stocklassa

National Laboratory for Forensic Science
Linköping, Sweden

Abstract

Software has been developed for performing image reconstruction of analytical results from an X-ray microbeam spectrometer. The software system has been developed on a conventional IBM-AT compatible computer and runs under WINDOW 3.0 system. The interactive part is developed as a complete "window-style" program where the operator controls the sample movement, the MCA and the micrograph image from the microscope.

A description of the spectrometer and the software system is given. Applications of the technique in analysing documents are presented.

The X-ray microbeam spectrometer

The spectrometer consist of six different parts; the X-ray source, the capillary focusing unit,the sample holder with it's positioning system, the X-ray detector, the monitoring system with the optical microscope etc and the software system controlling the whole spectrometer (see fig. 1).

The focusing unit, the sample holder system and the monitoring system are build on top of three different sledges which can be moved along a optical bench. The bench, the detector and the X-ray tube are attached to an optical table. A more comprehensive description of the spectrometer and especially of the capillary focusing technique is given in earlier papers[1,2,3,4].

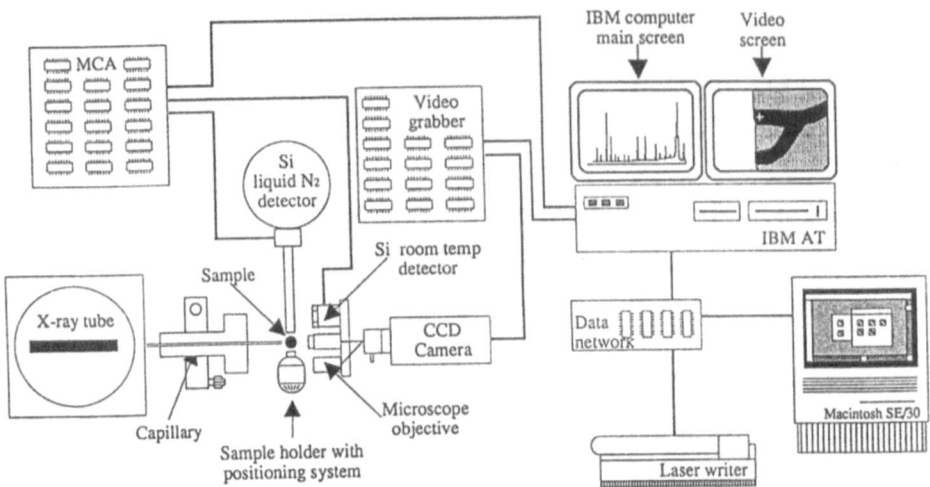

Fig 1   A schematic representation of the microbeam spectrometer

The focusing unit consists of a 5-axis gimbal which is used for aligning the capillary. The capillary is enclosed in a glass tube packed with highly absorbing materials. The glass tube is attached to the gimbal through a brass tube holder. The sample holder consist of 3 linear stages (X, Y and Z) and a rotational stage on top of that. All 4 stages are driven by DC motors which are controlled by the computer. The linear stages have a minimum increment of about 0.06 μm. The minimum rotational increment is about 10^{-6} rad. The interactive part of the software controlling the motorized stages in the sample

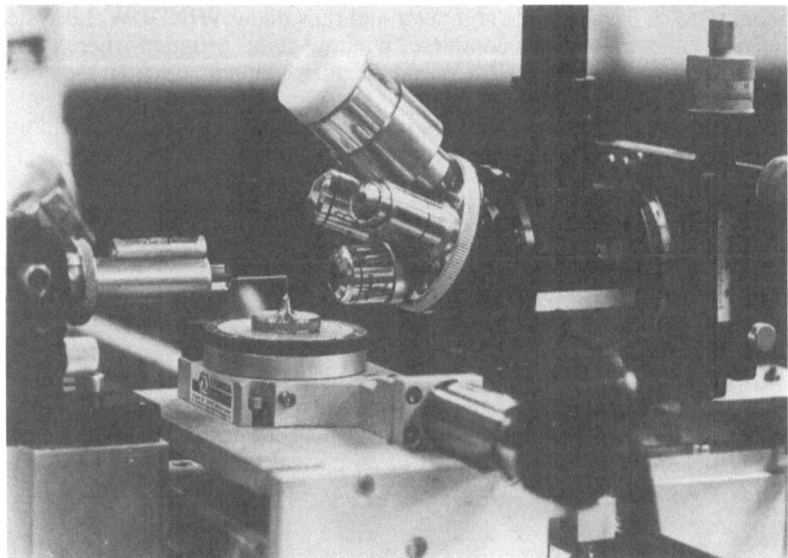

Fig 2   Photograph of the X-ray microbeam spectrometer. The capillary with the optical unit as well as the four stage sample holder are visible in the picture. Parts of the microscope can also be seen.

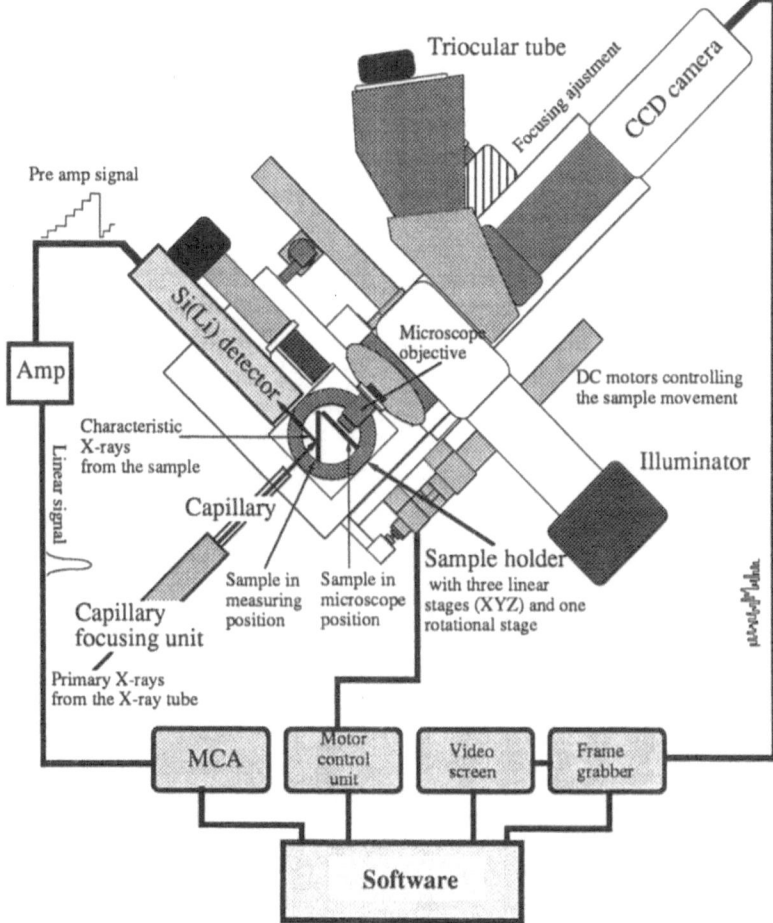

Fig 3 A schematic representation of the microbeam scanning technique is
given in this top view drawing of the spectrometer. Signals from the
Si(Li) detector and CCD camera are processed by the software system
which also at the same time controls the movement of the sample
holder. The result from the scans is presented as elemental mappings
on the video screen.

holder will enable the operator to focus and adjust the sample into proper position. The
area to be scanned is selected from the video screen and stored as a digitized image while
the front side of the sample is rotated back 180° (or less depending on the sample
geometry selected by the operator) to be exposed by the X-ray beam. Due to the high
accuracy of the movement the sample will be scanned in steps corresponding to each
pixel in the digitized image stored on the screen. This large degree of freedom will allow
the operator to rotate the sample around any axis in the focal plane (preferably the axi s
intercepting the beam direction).

The monitoring system consist of an optical microscope with conventional bright field
objectives and a room temperature Si detector (for monitoring the transmitted intensity)
mounted on the microscope revolver. The microscope is equipped with a CCD camera
connected to a frame grabber (see fig. 1 and 3). The photograph in fig. 2 shows the

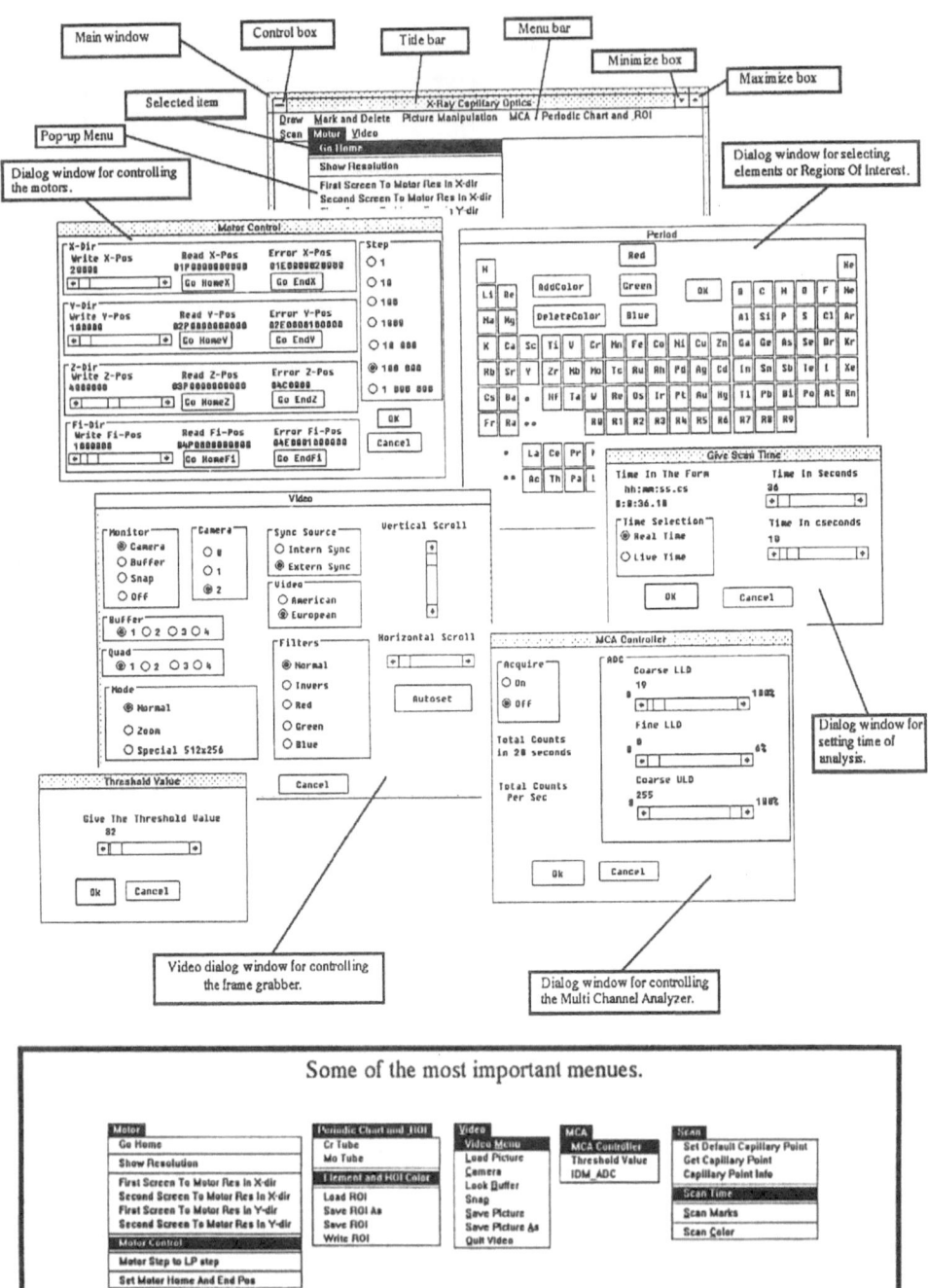

Fig 4   Figure 4 describes the operator interface, in terms of menus, windows
          and dialog boxes as it appear on the screen. Different windows or
          dialog boxes will appear on the screen depending on the menu item
          being selected. As the software is developed in a complete "window"
          style environment, every selection, data input etc is done by clicking in
          buttons, check boxes or control bars. Note, when the operator has to
          specify lines or areas in the micrograph image to be scanned, the mouse
          cursor automatically goes over to the video screen.

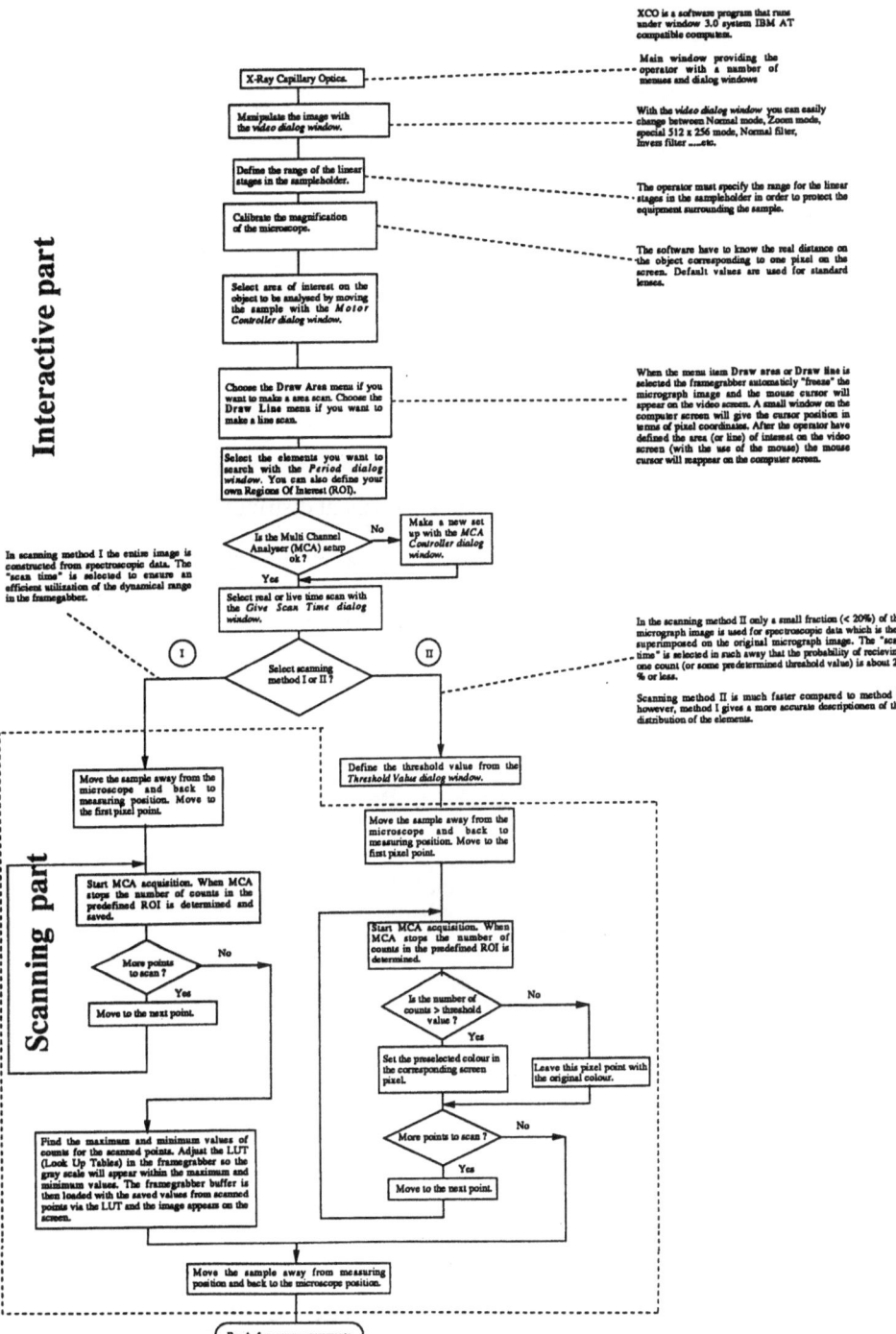

Fig 5   Flow chart diagram describing the main pathways in the software structure. The software can be divided into an "interactive" part, where the operator selects different analytical conditions, scanning area etc and a "scanning" part, where the software system is entirely occupied with controlling sample holder, reading and writing into the MCA and video screen etc.

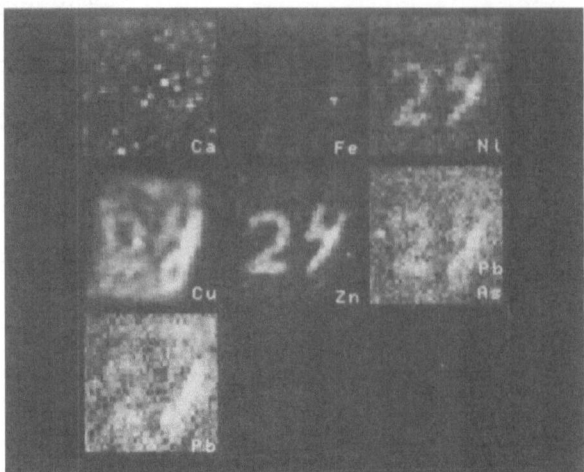

Fig 6   Image reconstructions from different element distributions. The figure
        shows the reconstructed text "24" which has been overwritten by
        another ink quality. The reconstructions were done at the microbeam
        spectrometer at the National Laboratory of Forensic Science in
        Linköping. Scanning method I (see figure 5) has been used.

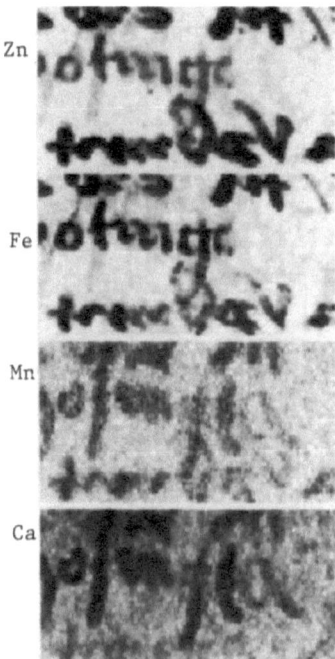

Fig 7   The reconstructed text from a parchment letter. On top the Zn and Fe
        reconstructions making up the word "Botinge" (the "B" in "Botinge" is
        not visible) then Mn and Ca at the bottom making up the falsified word
        "Gosmesta" (the "G" in "Gosmesta" is not visible). Note that the letter
        "s" in "Gosmesta" is written "f". The reconstructions were done at the
        microbeam spectrometer at the National Laboratory of Forensic Science
        in Linköping. Scanning method I (see fig. 5) has been used.

capillary focusing unit, the sample holder with it's three linear stages (X,Y and Z) and the rotation stage on top of that, and also the optical microscope.

Microbeam scanning technique

With the rise of Scanning Electron Microscope (SEM) a new microscopic principle was introduced; the micrograph image was constructed from spectroscopic data for each individual pixel in the image. By applying different types of spectroscopy different aspects of the sample were revealed. By using X-ray fluorescence spectroscopy the distribution of different elements could be studied. This scanning technique has now also been applied to microbeam PIXE (Proton Induced X-ray Emission) which has a much higher sensitivity for detecting different elements as compared to SEM[5,6]. With the use of X-ray induced X-ray emission very high sensitivity can be achieved but with much less energy deposited into the sample (as compared to particle induced X-ray emission). With X-rays the sample can be analysed at normal temperature and pressure. For charged particles the scanning can be performed by manipulating the beam by electrostatic means. When using X-rays, the sample has to be moved instead.

Several different X-ray microbeams have been constructed based on synchrotron radiation and with a resolution down to about 5 μm[7]. Figure 3 shows how such a scan is performed with the X-ray microbeam spectrometer as presented in figure 1 and 2.

Applications

Several applications of the scanning technique has been performed on the X-ray microbeam spectrometer at our lab as well as on the spectrometer at the National Laboratory of Forensic Science in Linköping, Sweden. Some of these applications concerns the analysis of biological, medical (like hair, leaves, tree rings etc) and other types of samples, and has been presented in earlier publications[8,9].

The microbeam spectrometer has also been used to analyse documents from an authenticity point of view. If someone has tried to falsify a document by writing a new text on top of the original or just add new text to an old document there is a great chance that the falsification will be made with a different ink quality. With the microbeam spectrometer differences in trace element content from different ink qualities can be determined. By scanning the beam over selected areas of the document old underlying text, or text which has been wiped out, can be reconstructed from it's trace element content. Fig. 6 shows an example of such a scan made on a piece of paper where the old text has been overwritten completely with another ink quality. As can be seen from the figure the number "24" appears very clear from the Zn mapping but is also visible from the Pb and Ni mappings. Fig. 7 shows the result from a scan made on a Swedish letter of possession made of parchment and which dates from the first of April 1499. This document was suspected to have been subjected to falsification during the 16'th century. However, the original text could not be revealed by any other technique. As can be seen from the figure below, the distribution of Ca and Mn corresponds to the text "Gosmesta" which are visible today, while the distribution of Fe and Zn makes up another word "Botinge" which was the original word.

Conclusions

It is quite clear that this type of X-ray microbeam spectrometry offers a very powerful technique for elemental mapping. As the sensitivity for conventional XRF (X-Ray Fluorescence) is very high, elemental mapping and image reconstruction can be performed even for trace elements. In combination with capillary optics an elemental map with a pixel size of about 100 μm can be generated in a short time due to the high intensity which is obtained by capillary optics. With conical capillaries microbeams (from conventional X-ray tubes) with the sizes of a few μm and with an intensity sufficient for trace element analysis have been achieved, thus elemental mapping on this level is possible but the time of analysis will be extensive. However, by using synchrotron radiation sub micron beams with very high intensity of X-rays have been achieved by capillary optics[10] and thus, fast mapping of trace elements on a microscopic level can easily be achieved with this kind of beam.

The ability to analyse samples in air at ambient pressure and without any sample preparation is of course of great advantages, especially for large- or odd-shaped samples (like old documents).

References

1   P. Engström, S. Larsson, A. Rindby and B. Stocklassa, Nucl. Instrum. and Meth. B26 (1989) 222

2·   P. Engström, S. Larsson, A. Rindby and B. Stocklassa, X-Ray Spectrometry 18 (1989) 109

3   P. Engström, S. Larsson, A. Rindby and B. Stocklassa, Conference Proceedings 25 from "2'nd European Conference on Progress in X-ray SR Research"(1990) 283.

4   P. Engström, S. Larsson, A. Rindby and B. Stocklassa, Advances in X-Ray Analysis 33 (1990) 623

5   A. Saint, G. S. Bench, M. Cholewa, S.Dooley, D. N. Jamieson and G. J. F. Legge. Nucl. Instrum. and Meth. B56/57 (1991) 717

6   J. L. Cambell, W. J. Teesdale and J. A. Maxwell Nucl. Instrum. and Meth. B56/57 (1991) 694

7   K. Jones et al Synchr. Rad. News 4 no:2

8   N. Shakir, S. Larsson, A. Rindby, P. Engström and A. Rindby Nucl. Instrum. and Meth. B52 (1990) 194.

9   B. Stocklassa, G. Nilsson, N. Paulsson, A. Rindby, P. Engström, S. Larsson, and P. Voglis "A new microbeam X-ray spectrometer for use in forensic applications" Oral presentation at the Interpol Conf. Lyon France (1990)

10   P. Engström, S. Larsson, A. Rindby, A. Buttkewitz, S. Garbe, G. Gaul, A. Knöchel and F. Lechtenberg. Accepted for publication in Nucl. Instrum. and Meth.

LARGE-AREA X-RAY MICRO-FLUORESCENCE IMAGING OF HETEROGENEOUS

MATERIALS

B.J.Cross, R.D.Lamb, S.Ma* and J.M.Paque*

Kevex Instruments, San Carlos, CA 94070

*Center for Materials Research, Stanford University
Stanford, CA 94305-4045

ABSTRACT

An X-ray Micro-Fluorescence (XRMF) spectrometer, with an analysis area of about 100 by 150 microns, has been used to collect 2-dimensional X-ray intensity maps over large-area (5 to 50 mm in X and Y) samples. These intensity maps were collected by scanning the sample on an XY stage, and converting X-ray Energy-Dispersive spectra to peak intensities for the elements of interest. The maps, when displayed using false-color or pseudo-gray scales, show the distribution of individual chemical elements over the analysis area. These maps can be collected at speeds from about 1 minute per frame (analyzing 25 elements simultaneously). Greater precision of chemical intensities, or larger area maps, may require several hours, particularly if extensive data processing is performed at each point. XRMF has advantages over more conventional SEM-EDS X-ray mapping, including sample preparation and presentation, as well as improved signal-to-noise ratios.

A technique is described which assists in analyzing the large amount of data which is collected in each map. Principal Component Analysis (PCA) is performed on all of the elemental maps simultaneously. This technique compresses the many elemental intensity maps into a few principal components, resulting in many fewer maps to evaluate. The intensity maps of these principal components display the most pertinent information. They can also be plotted as scatter plots which can help with the partitioning of the data into individual phases. This procedure can potentially be automated as a method for phase analysis. The selected pixels from the scatter plots can be averaged and converted into phase compositions, and the phase information re-displayed on the original elemental or principal component maps.

This technique has been applied to a thin section of rock, and to a synthetic multiphase alloy sample.

INTRODUCTION

There have been several papers, in recent years, describing the relatively new technique of X-Ray Micro-Fluorescence (XRMF). XRMF instruments are also known as laboratory X-ray microprobes. The technique of XRMF is based upon an XRF spectrometer, where the primary incident X-ray beam has been constrained to a very small size. The state-of-the-art of such spectrometers, currently produces a beam size (at the sample) with a diameter of about 100 microns or less. In order to position the sample accurately, these spectrometers usually employ an XY stage.

For example, Nichols et al.[1] introduced two spectrometers, and described techniques for measuring the important characteristics of these instruments. A more recent paper by Wherry et al.[2] reviewed much of the literature in this field, and described the characteristics and application of another XRMF instrument. The latter was a prototype of the instrument used in this study. A more recent paper describes the application of this particular instrument to trace analysis.[3] Other XRMF instruments, using the same technique of a short collimator to restrict the X-ray beam, have been described by Shiraiwa and Fujino,[4] Shimura and Kosaki,[5] and Watanabe and Kobayashi.[6] Other workers, such as Carpenter et al.,[7] Rindby et al.,[8] and Yamamoto et al.,[9] employ a narrow capillary to concentrate the X-ray beam to a small area.

A key technique, that many of these spectrometers use, steps the sample with an XY stage in a regular grid pattern, collecting information at each grid position. This information can be simple integrated peak intensities or processed data, such as background- or overlap-corrected intensities, or composition and thin-film thickness. The total number of points are often restricted for complete quantitative mapping, because of the long times required to acquiring and process the data. For example, quantitative results are often restricted to linescans or 2-dimensional topographic presentations where some interpolation is necessary to form a complete picture of the sample.[10] For good quantitative analysis, at least 10 seconds is required for the acquisition and subsequent data processing. For trace analysis, several minutes are usually required. For a 30x30 map, total collection times range from 3 to 50 hours.

By contrast, it is possible to acquire intensity maps, with little or no processing, at scan rates less than 10 ms per point. This means that a large area elemental "image" or map can be acquired at frame rates from about 10 seconds (30x30) up to 10 minutes (250x250), ignoring any time required to position the beam or stage. Of course, even for a qualitative image, better precision is usually required, and either several frames must be acquired, or the time per point increased. Therefore reasonable quality X-ray maps require typical collection times ranging from minutes to hours. This assumes that the sample has a high-contrast chemical distribution.

Now the problem becomes how to process this large amount of data, especially when there may be up to 25 elemental maps which have been acquired simultaneously. Conventional approaches to this problem often apply image processing techniques to make use of spatial information. That is, local clusters of data are identified which correspond to a particular phase or component of the heterogeneous material. In image processing terms, this is known as "Feature" or "Stereological" Analysis. However, this technique does not fully utilize the complete chemistry information (i.e. phase composition) that is available. Sometimes, simple logical operations are performed on individual maps, in order to obtain composite images, but this is usually just for visual presentation.[11]

PRINCIPAL COMPONENT ANALYSIS

A useful way to reduce the apparent size of the data set, is to use the technique of Principal Component Analysis (PCA). This technique was first proposed by Pearson,[12] in 1901, but really has only become popular in the last few decades with the advent of computers. It has frequently been used in geographical applications for multivariate analysis.[13] For an introductory review of the method, see the text by Manly.[14]

The objective of PCA is to take *n* variables (e.g. elemental maps) and produce *n* indices (components) that are as uncorrelated as possible. These indices are then ranked so that the first represents the largest amount of variation, and so on. First, this lack of correlation ensures that the indices are measuring different dimensions or vectors in the data set. Secondly, the aim of PCA is to reduce the large number of variables (i.e. maps) to just a few (e.g. 2 or 3) components.

The method only works if the original data set shows some correlation (i.e. two or more elements show some of the same spatial characteristics). This involves setting up a set of *n* simultaneous equations, and finding the eigenvalues of the covariance matrix. In order to avoid any one variable dominating the calculations, the variables (map intensities) are sometimes (but not in this work) normalized with a mean of zero, and a variance of one. After ordering the components by eigenvalues, they are summed, until the sum reaches a pre-determined threshold (e.g. 99%). Beyond this point, any further components are considered insignificant.

Principal Component (PC) maps are then calculated from the eigenvector coefficients, summing the products of these coefficients and the corresponding elemental intensities at the same X,Y coordinate. The eigenvalues must always be positive, but the individual eigenvector coefficients may be negative. Consequently certain PC maps will show an increasingly negative value, which corresponds to the lowest original intensity values. This results in these PC maps being inverted, in gray scale, compared with the original intensity maps. This inversion may be reversed.

A further step is necessary to make PCA more useful. This involves plotting two principal components as two-dimensional scatter plots.[15,16,17] In this process, for each component, a pixel is highlighted in the plot. As subsequent PC map pixel intensities are transferred to the scatter plot, any repeat "hits" on the same scatter plot pixel are shown by increasing the brightness of this pixel's color. This can also be done with two- or three-dimensional elemental scatter plots,[18] but using principal components makes it easier to identify phases. These are identified as clusters in the scatter plots. Because the PC plots represent more than just single elements, the clustering tends to be tighter. By using PCA to order the most important characteristics of the data set, it also offers the potential to automate the process of phase analysis. The next steps require automatic playback of the scatter plot clusters back onto the original elemental X-ray maps in order to locate the identified phases.

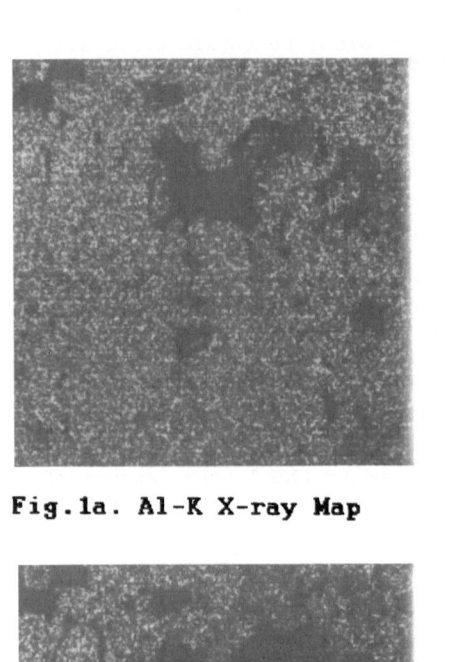

Fig.1a. Al-K X-ray Map

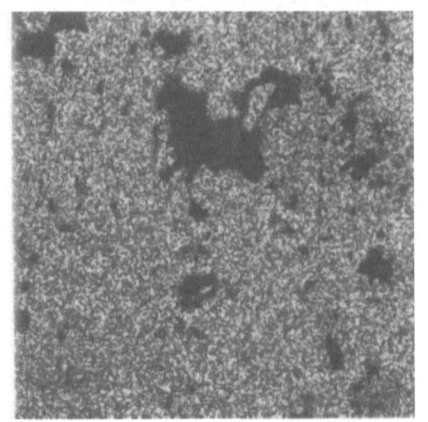

Fig.1b. Si-K X-ray Map

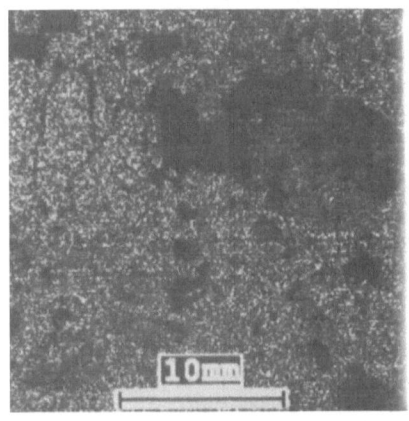

Fig.1c. K-K X-ray Map

Fig.1d. Ca-K X-ray Map

Fig.1e. Ti-K X-ray Map

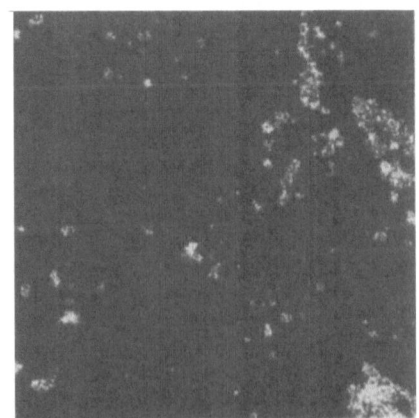

Fig.1f. Fe-K X-ray Map

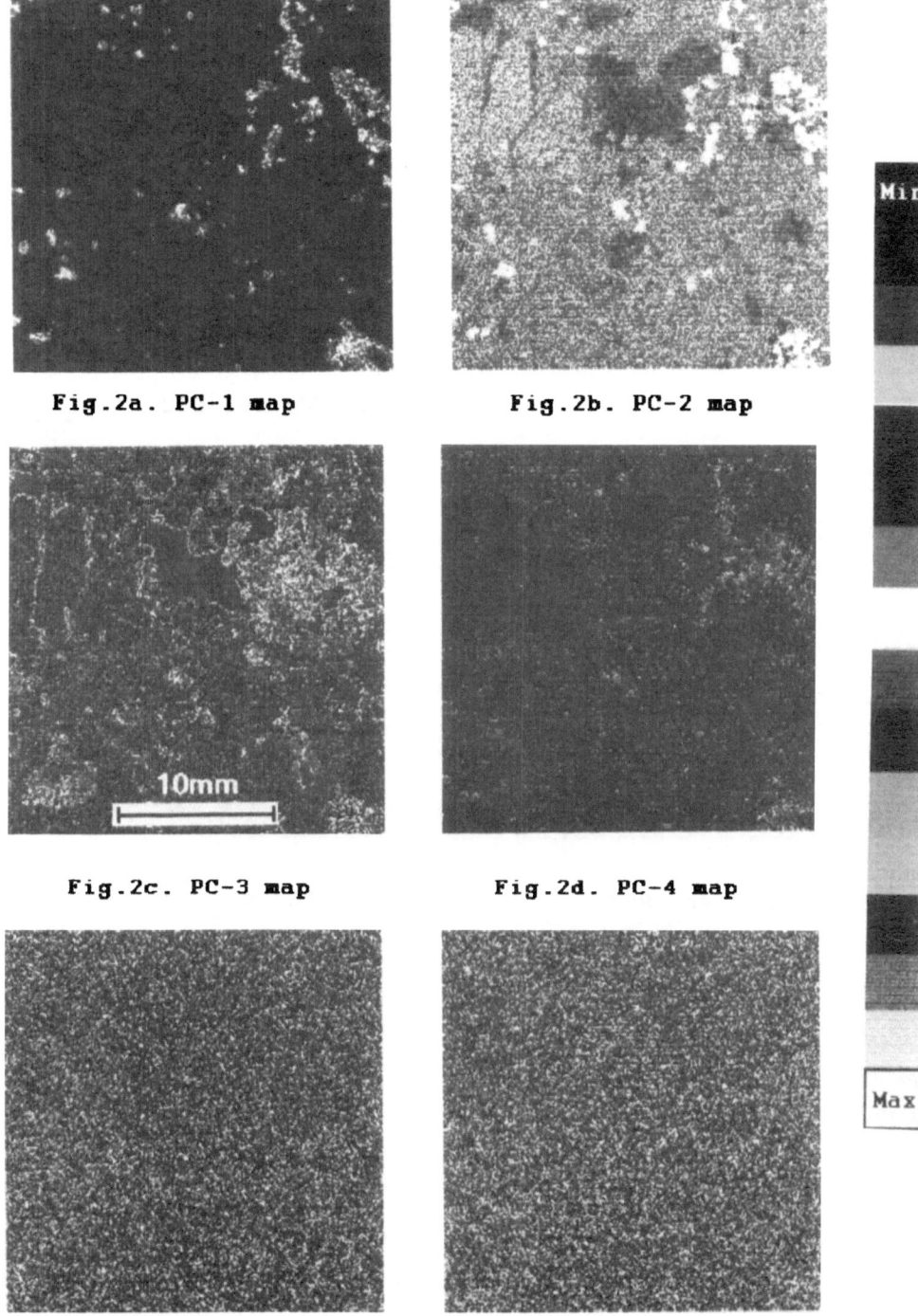

Fig.2a. PC-1 map

Fig.2b. PC-2 map

10mm

Fig.2c. PC-3 map

Fig.2d. PC-4 map

Fig.2e. PC-5 map

Fig.2f. PC-6 map

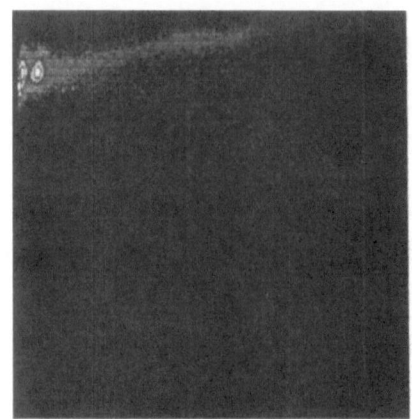

Fig.3a. PC1 vs. PC2   Fig.3b. PC1 vs. PC3

Fig.3c. PC2 vs. PC3   Fig.3d. PC1 vs. PC4

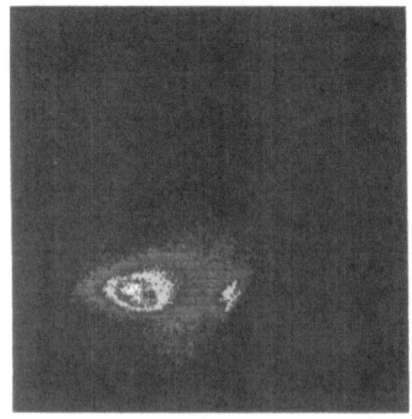

Fig.3e. PC3 vs. PC4   Fig.3f. PC2 vs. PC4

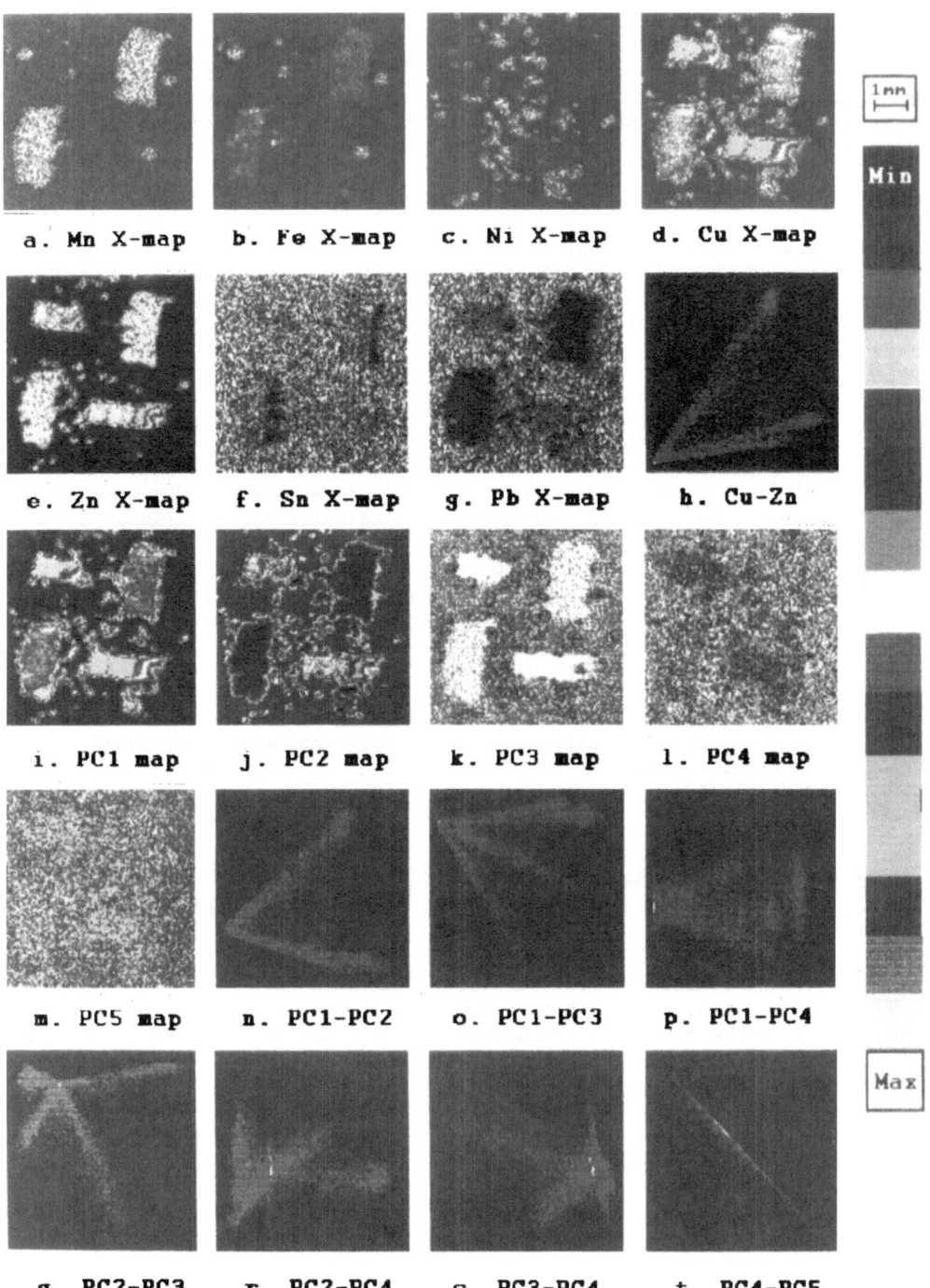

Fig.4. Elemental & Principal Component maps and scatter
       plots for M02 sample.
       Maps are 128*128 at 50 microns per pixel.

EXPERIMENTAL

The XRMF maps were obtained using a Kevex Omicron X-ray Microprobe spectrometer[2] equipped with a Mo anode small spot X-ray tube. The tube was operated at 50 kV and 1 mA, with no filtering. The final collimator size was 100 microns. It has been shown[2] that the spot-size (containing 90% of the flux) at the sample is about 75*105 microns. However, it is possible to resolve line pairs even closer than 75 microns.[1] For the results presented in this study, the data was acquired by scanning the sample stage at a constant speed (slew-scanning) in X, and chopping the acquisition into time slices, each of which correspond to both a certain distance in X and a pixel on each line of the X-ray maps. The stage employs microstepping stepper motors, where each microstep corresponds to 0.66 microns in X and Y. The scanning was performed bidirectionally in order to improve the data acquisition rate. Between each horizontal scan line, the Y motor is stepped to the next line position. In all cases shown in this study, the intensities are calculated from the gross peak integrals, without any background or overlap correction.

The first sample was a geological rock, prepared as a thin section on a glass slide. The feldspar-bearing rock contains various grains or phases of anorthoclase (a potassic feldspar), quartz, glass, plagioclase feldspar, clinopyroxene and some Fe-Ti oxide particles. The data were acquired in vacuo to improve the light element sensitivities. The complete map size was 250 by 250 points, with each point corresponding to 100 microns in X and Y. The X motor was scanned continuously at 1600 microsteps per second, and a total of 5 complete frames were acquired.

The second sample consisted of 4 separate copper-base alloy chips, imbedded in an epoxy resin matrix. Several chips of each alloy were included in the scan area. The map size was 128 by 128, with each point corresponding to 50 microns. The same X motor scanning speed was used, and a total of 10 frames acquired. The data were collected in air.

RESULTS

Figs. 1a-1f show some of the elemental X-ray maps for the thin section of rock. There is clearly some correlation between the Al, Si and K maps, and some anti-correlation with Ca. Also the Ti and Fe maps have some correlation with each other, and partly with Ca, but are anti-correlated with Al, Si and K. The PC maps, in Figs. 2a-2c, show very clear separation for the first three components, with mainly noise in the next three components (Figs. 2d-2f). Clearly this is an example where nine elemental maps (including As, Rb and Sr) have been condensed to three PC maps. The PC-1 map shows very good correlation with the Fe (and Ti) map, showing that Fe dominates this component. In these calculations the individual map variances were not normalized. The map of PC-2 is an example of an inverted color scale, because of negative eigenvector coefficients. Also the PC-3 map shows a strong anti-correlation with the Al, Si and K elemental maps.

Figs. 3a-3f show the scatter plots for the first four principal components. Each of the plots show very good clustering corresponding to 1, 2 or 3 phases. The outlying points probably correspond to interfaces or mixed compositions. Because of the finite size of the X-ray beam, it is often possible that the beam will be striking two different phases simultaneously. Also, it is possible that some phase mixing has occurred.

Fig. 4 shows the complete set of results for the copper alloy chips in epoxy resin. Figs. 4a through 4g show the elemental maps, 4h shows an example of an elemental scatter plot, 4i through 4m show the first five principal component maps, and 4n through 4t show the PC scatter plots. Cu is the element with the highest concentrations, with Zn the next highest. Therefore, the map of PC-1 correlates well with the Cu map. Similarly, the PC-2 map correlates with the Zn map, although not so strongly. Examining the scatter plots of Cu-Zn and PC1-PC2 one can see that the scatter patterns have merely been rotated about the origin. This is a natural consequence of the PCA approach, when two elements dominate the matrix.

It is noticeable in this sample data set that the scatter plots are more like vectors than clusters. Since these standards are assumed to be homogeneous, it is speculated that the vectors result from "edge effects" from the X-ray beam. Because the incident angle of the beam to the sample plane is 45 degrees, there can be substantial penetration of the beam through the epoxy resin matrix. This, in turn, causes fluorescence from the sides of the standard chips which exits relatively easily though the epoxy resin. This is evident on some of the small chips, but is more obvious on the larger ones.

CONCLUSIONS

A new technique of XRMF imaging, with X-ray elemental maps, has been described, which provides some new ways to analyze heterogeneous materials. This technique has certain advantages over SEM-EDS X-ray mapping (such as sample preparation, signal-to-noise ratios), although the spatial resolution is worse. In order to reduce the complexity of the data set, the technique of Principal Component Analysis has been applied to this data. This has the ability to simplify the interpretation of the data, and also the potential to yield automatic phase classification and analysis, after converting average phase intensities to composition.

ACKNOWLEDGEMENTS

The authors would like to thank Dr. G. R. Davies (Univ. Michigan) for the use of the rock thin section, and to Dr. L. Feng (Kevex Instruments) for the preparation of the NIST copper alloy standards sample.

REFERENCES

1. M. C. Nichols, D. R. Boehme, R. W. Ryon, D. C. Wherry, B. J. Cross and G. A. Aden, Parameters Affecting X-ray Micro-Fluorescence (XRMF) Analysis, Adv. X-Ray Analysis, 30:45 (1987).

2. D. C. Wherry, B. J. Cross and T. H. Briggs, An Automated X-Ray Micro-Fluorescence Materials Analysis System, Adv. X-Ray Analysis, 31:93 (1988).

3. B. J. Cross and J. E. Augenstine, Trace Analysis using EDS: Applications to Thin-Film and Heterogeneous Samples, accepted for publication in Adv. X-Ray Analysis, 34: (1991).

4. T. Shiraiwa and N. Fujino, Micro Fluorescent X-ray Analyzer, Adv. X-Ray Analysis, 11:95 (1968).

5. Y. Shimura and S. Kozaki, X-Ray Scanning Microanalyzer, "Proc. 6th Intl.
    Conf. X-Ray Optics and Microanalysis," eds. G. Shinoda, K. Kohra and
    T. Ichinokawa, Univ. Tokyo Press (1972).

6. H. Watanabe and Y. Kobayashi, X-Ray Induced Transport of Inorganic
    Elements in Living Rice Leaves Observed with X-Ray Fluorescence
    Element Mapping Spectrometry, Agric. Biol. Chem., 50:2077 (1986).

7. D. A. Carpenter, M. A. Taylor and C. E. Holcombe, Applications of a
    Laboratory X-ray Microprobe to Materials Analysis, Adv. X-Ray
    Analysis, 34:115 (1989).

8. A. Rindby, P. Engstrom, S.Larsson and B.Stocklassa, Microbeam Technique
    for Energy-Dispersive X-Ray Fluorescence, X-Ray Spectrometry, 18:109
    (1989).

9. N. Yamamoto and Y. Hosokawa, Development of an Innovative 5 micron
    diameter Focused X-Ray Beam Energy-Dispersive Spectrometer and its
    Applications, Jap. J. Applied Physics, 27:L2203 (1988).

10. D. C. Wherry and B. J. Cross, Applications of Small-Area X-ray
    Fluorescence Analysis Using a Microbeam X-ray Source, Spectroscopy,
    3:38 (1988).

11. D. R. Boehme, X-ray Microfluorescence Analysis of Thin- and Thick-
    Sectioned Geologic Materials, Sandia Report SAND87-8214 (1987).

12. K. Pearson, On lines and planes of closest fit to a system of points in
    space, Phil. Mag., 2:557 (1901).

13. S. Daultrey, Principal Component Analysis, in "Concepts and Techniques
    of Modern Geography," 8, Geo. Abstracts, Univ. East Anglia, UK (1976).

14. B. F. J. Manly, "Multivariate Statistical Methods: A Primer," Chapman
    and Hall, London (1986).

15. P. L. King, R. Browning, P. Pianetta, I. Lindau, M. Keenlyside and G.
    Knapp, Image processing of multispectral x-ray photoelectron
    spectroscopy images, J. Vac. Sci. Technol. A, 7:3301 (1989).

16. J. M. Paque, R. Browning, P. L. King, P. Pianetta, Quantitative
    Analysis of X-ray Images from Geological Materials, Microbeam
    Analysis, San Francisco Press, CA, 195 (1990).

17. R. Browning, P. L. King, J. M. Paque and P. Pianetta, EDS Image
    Classification by Recursive Partitioning, Microbeam Analysis, San
    Francisco Press, CA, 199 (1990).

18. D. S. Bright and D. E. Newbury, A Scatter Diagram Technique for Viewing
    Two or Three Related Images, Anal. Chem., 63:243A (1991).

COMPARISON OF SYNCHROTRON X-RAY MICROANALYSIS WITH ELECTRON AND PROTON

MICROSCOPY FOR INDIVIDUAL PARTICLE ANALYSIS

K.H. Janssens[1], F. van Langevelde[1], F.C. Adams[1], R.D. Vis[2],
S.R. Sutton[3], M.L. Rivers[3], K.W. Jones[3] and D.K. Bowen[4]

[1]Department of Chemistry, University of Antwerp (U.I.A.)
Universiteitsplein 1, B-2610 Wilrijk/Antwerp, Belgium

[2]Department of Physics and Astronomy, Free University of
Amsterdam, P.O. Box 7161, 1007 MC Amsterdam, The Netherlands

[3]Department of Applied Science, Brookhaven National
Laboratories, Upton, NY 11973, USA

[4]Department of Electrical Engineering, University of Warwick
Coventry CV4 7AL, Great Britain

INTRODUCTION

A considerable number of the elements and a profusion of organic compounds are emitted into the atmosphere in association with solid and liquid particles. More than 50 percent of all air pollutants are preferentially present in particulate matter rather than in the gas phase. The assessment of the potential environmental and toxic effects of particulate matter in the atmosphere requires a detailed physical and chemical characterization. Methods for analyzing aerosols have recently been extensively reviewed by Maenhaut[1]. Most of these techniques are trace-level bulk analytical methods such as ICP-MS (Inductively Coupled Plasma Mass Spectrometry)[2], AAS (Atomic Absorption Spectrometry) and its variations[3], INAA (Instrumental Neutron Activation Analysis)[4], (macro-)PIXE[5] (Particle Induced X-ray Emission) and conventional X-ray Fluorescence (XRF)[6].

An overview of techniques for the analysis of individual airborne particles is given in the book by Spurny[7]. Such techniques include EPMA[8] (Electron Probe Micro Analysis), micro-PIXE[9] and LAMMA[10] (Laser Mass Micro Analysis). By analyzing particles individually, after a classification step, the contributions and composition of each particle source at the sampling location (e.g., sea spray, soil dust, car exhaust, industrial fumes) can be determined directly. In this respect, charged particle-beam techniques such as EPMA have the obvious advantage that particle localization and data acquisition can be automated[8]. EPMA (and for the higher Z elements also μ-PIXE), however, are limited in sensitivity. Due to the high bremsstrahlung background in electron induced X-ray spectra, EPMA features minimum detection limits (MDL's) in the 0.1% range. A drawback of the nuclear microprobe is the considerable energy deposition in the sample, especially when non-thin target materials are employed[11].

Due to the high intensity of synchrotron radiation sources, yielding an increased sensitivity of analysis, and the high degree of polarisation of the radiation, causing a decrease in scattered background levels, sub-ppm detection limits for bulk analysis have been reported for synchrotron radiation induced X-ray fluorescence (SRXRF)[12-16]. In most of these studies, comparisons were made with other X-ray emission techniques for bulk analysis, such as PIXE[14,15], tube[13,14]- and radio-isotope excited EDXRF[16,17].

Relatively few X-ray microprobe facilities currently exist in the world. At Hasylab (Hamburg, FRG) and at the NSLS (Brookhaven, Upton, NY, USA), white light microprobes are in operation, using collimated pencil beams and attaining lateral resolutions in the order of 3 to 10 μm[18,19] with detection limits at the 10 ppm level. At SRS[20] (Daresbury, UK), SSRL[21] (Stanford, CA) and the Photon Factory[22] (Tsukuba, Japan), focused monochromatic microbeams are employed.

This paper is concerned with the evaluation of the use of μ-SRXRF as implemented at two existing X-ray microprobes for the analysis of individual particles. As representative environmental particulates, National Institutes of Science and Technology (NIST) K227, K309, K441 and K961 glass microspheres were analyzed using two types of X-ray micro probes: the white light microprobe at beamline X26A of the NSLS and the monochromatic (15 keV) X-ray microprobe at station 7.6 of the SRS. For reference, the particles were also analyzed with microanalytical techniques more commonly employed for individual particles analysis such as EPMA and micro-PIXE.

Evaluation of any microanalytical technique obviously involves assessing its sensitivity/limits of detection and its lateral resolution. Sensitive methods have the advantage that classification of particles can be done also on the basis of the trace element content of the particulates, making it possible to distinguish between two very similar aerosol sources. With respect to particle analysis, the lateral resolution is only important in the sense that particulates which are physically separated on a filter can also be analyzed separately. A much more important property of techniques suitable for individual particle analysis is the time required to obtain statistically meaningful information from each particle. This length of time obviously includes the data acquisition period (typically in the order of 20 to 100 sec/particle), but also entails the time required to locate and optimise the position of the next particle in the beam. As in atmospheric studies, in order to study, e.g., seasonal or regional variations in the aerosol composition, extensive numbers of samples are examined, while for each sample, several hundred of particles need to be analyzed, in practice, this property proves to be a very stringent one indeed.

For the four particle types and using the various microanalytical techniques mentioned above, X-ray spectra were collected from particles 20 to 30 μm in size in order to compare attainable detection limits and typical analysis times. Using the white light X-ray microprobe and the electron microprobe, also the size dependence of the fluorescent X-ray yield was experimentally determined.

EXPERIMENTAL

Sample material and sample preparation

As representative examples of coarse mode environmental samples with known composition, NIST K227, K309, K441 and K961 glass microspheres were

Table 1. Nominal Composition of NIST glass microspheres

Element	Concentration (%w)			
	K227	K309	K411	K961
O	16.4	38.82	42.88	47.00
Na	-	-	-	2.97
Mg	-	-	9.05	3.02
Al	-	7.94	-	5.82
Si	9.3	18.70	25.71	29.98
P	-	-	-	0.22
K	-	-	-	2.49
Ca	-	10.72	10.70	3.57
Ti	-	-	-	1.20
Mn	-	-	-	0.32
Fe	-	10.49	11.66	3.50
Ba	-	13.43	-	-
Pb	74.3	-	-	-

studied. Details on the preparation of these microspheres can be found in Ref. 23; the diameters of the spheres are in the range from 0.25 to 250 μm (corresponding to a weight range of 20 fg to 20 μg/particle) and were verified to be homogeneous and identical in composition to the bulk glass they were produced from using electron- and ion microscopy. The nominal composition of the various materials is listed in Table 1.

For the EPMA measurements, samples were prepared by dispersing the glass particles into an inert solvent (n-hexane) and filtering through a polycarbonate filter with pore holes of 0.4 μm. The particles were subsequently transferred onto marked electron microscopy grids coated with a formvar foil by pressing the grids against the filter. For the NSLS measurements, individual glass particles in the range 5 to 50 μm were mounted onto Kapton foil through micro-manipulation and held in place by means of droplets of silicon-oil. The kapton foil was attached to cardboard 5 mm slide frames which fit into the micro probe sample holder. For the SRS and PIXE experiments, glass spheres were dispersed into a 2 % formvar solution: by means of a rotating disk, droplets of the suspension were spread out and allowed to dry in the form of thin films. These films were subsequently mounted onto Al support rings fitting in the sample holder of the microprobes.

Instrumentation and Experimental Procedure

For the white light synchrotron excitation experiments, the X-ray microprobe at the X26A beamline of the NSLS (National Synchrotron Light Source) was employed. After emerging from the storage ring UHV, the beam is defined by four Ta slits and further collimated by a 5x8 μm^2 crossed slit system. The sample is positioned at 45° to the incoming beam; X-ray spectra are detected using a Si(Li) detector positioned at 90 degrees to the original beam.

The spectrum impinging on the sample is shown in Fig. 1, having a maximum in flux density of about 10^4 photons/sec/mA/μm^2 near 8 keV. Soft X-rays (0 < E < 5 keV) are heavily absorbed in the Be-end window of the beam pipe and in the air path between the collimator slits and the sample. Specimen can be viewed by a horizontally mounted stereozoom binocular microscope, equipped with a TV camera. X-ray spectra of the particles were collected by

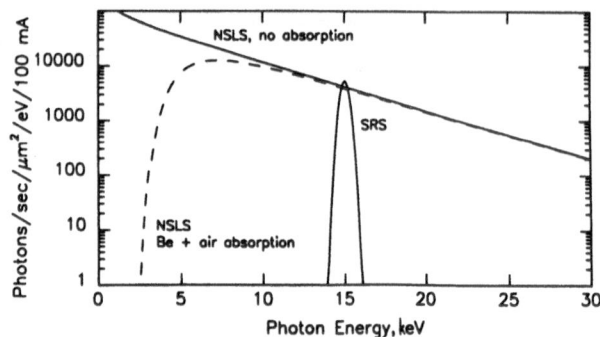

Fig. 1. Excitation Spectra at the NSLS and SRS XRM facilities.

localising the particles on the foil using the microscope, moving them into
the beam and maximizing the detectable count rate.

    Experiments using monochromatic synchrotron radiation excitation were
performed at station 7.6 of the Synchrotron Radiation Source (SRS),
Daresbury (UK). White radiation from the bending magnet beam line 7 was used
as primary source for the measurements. The basic component of the
microprobe is an ellipsoidally concave bent Si(111) crystal which
simultaneously monochromates and focuses 15 keV radiation. The crystal
passes a bandwidth of ca. 0.3 keV and produces a focused beam spot of 10x15
μm^2 FWHM; in the spot, flux densities of 3.4 10^4 ph/s/mA/μm^2 are achieved.
A 50 mm^2 area Si(Li) detector is located at 90° to the incoming beam at 35
mm from the sample. The sample is observed by means of a Zeiss Sv8 stereo
(zoom) microscope with 175mm working distance, equipped with a CCD camera.
Details on the optics, beam profiles and fluxes can be found elsewhere[20]. The
excitation spectrum of the SRS microprobe is also shown in Fig. 1. SR
induced X-ray spectra were obtained in a similar way as at the NSLS.

    Micro-PIXE measurements were performed at the Nuclear Microprobe setup
of the Free University, Amsterdam. 3 MeV protons are accelerated by a
Philips AVF cyclotron, yielding after focusing a microbeam of typically 2x5
μm^2 cross section at the focal spot[27], with a current density of 2 to 5
pA/μm^2. Similarly prepared samples as employed for the SRS measurements were
used.

    Using a Jeol 733B Superprobe microanalyser equipped with a Tracor
TN2000 computer system, electron induced X-ray spectra were acquired for
typically 120 sec per particles; data acquisition was started after manual
localisation of the particles and focusing of the beam at their centre. A
25 kV, 1 nA electron beam was used in all cases.

RESULTS AND DISCUSSION

Peak-to-Background ratios and Limits of Detection

    Fig. 2 shows X-ray spectra obtained from K961 particles of 20 to 30 μ
diameter using resp. the white and monochromatic X-ray microprobes and the
electron and protron microprobes. As can be seen from Fig. 2a, due to
scattering of the white spectrum of the NSLS-microprobe, a more or less
uniform background level of about 100 counts can be observed over the entire
energy range for a collection time of 300 sec. In contrast, in the SRS-

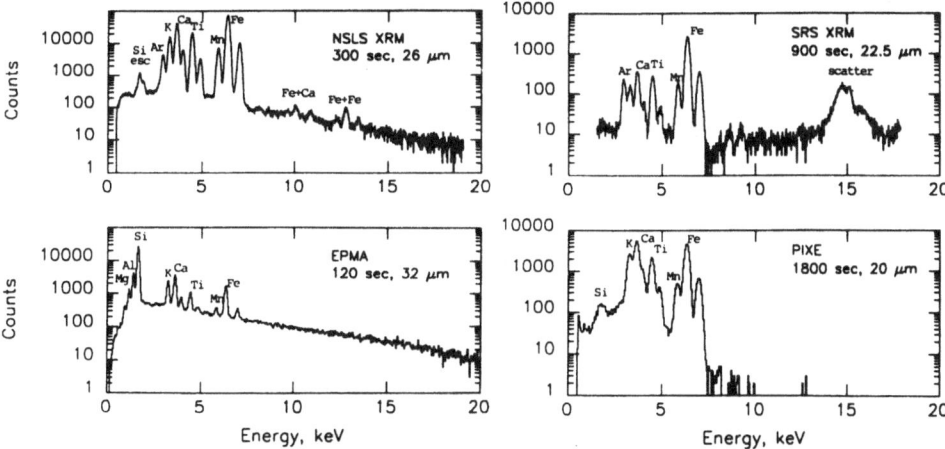

Fig. 2. X-ray spectra obtained from K961 glass microspheres
using the four different micro probes.

spectrum (Fig. 2b), almost no background in the region 2–14 keV is present.
Only near 15 keV, the background level rises due to the low-energy tail of
the incoherent scatter peak. As a results of the high count rates achievable
at the NSLS facility, in Fig. 2a also Ca+Fe and Fe+Fe sum peaks can be
readily observed near 8 and 10 keV. Because the measurements are performed
in air, also an appreciable Ar peak is present. The Si characteristic
radiation is heavily absorbed; no Al peak can be discerned. Fig. 2c
illustrates the (dis)advantages of electron induced X-ray emission. Overall,
a fairly high bremsstrahlungs background is observed, although on the other
hand, significant amounts of characteristic radiation of low-Z elements such
as Na, Mg, Al and Si can be detected in a relatively short counting time,
which is not the case for the XRM measurements. It should be noted however
that none of the XRM-instruments are optimized for light element detection
(detector windows, air operation) and that the comparison is only valid for
the specific conditions listed above. Finally, in Fig. 2d, the background
due to proton-induced secondary electron bremsstrahlung only gives rise to
an appreciable background in the 2–5 keV energy range.

     To allow for a more quantitative comparison, Table 2 lists for the four
microprobes the overall count rates and peak-to-background ratio's derived
from the spectra in Fig. 2. For Mn and Fe, the SRS microprobe features the
highest peak-to-background ratios, although for the lower elements, the NSLS

Table 2. Overall count rates and peak-to-background ratios derived from
the K961 particle spectra shown in Fig. 2

Micro probe	Count Rate		K_α Peak-to-background ratio							
	Total (cps)	% Net/ Backgr	Mg	Al	Si	K	Ca	Ti	Mn	Fe
NSLS XRM	7773	79/21	–	–	2.3	16	49	37	15	55
SRS XRM	70	64/36	–	–	–	7.6	22	23	21	330
EPMA	1646	61/39	1.2	4.5	28	2.7	4.8	1.5	0.3	3.4
μ-PIXE	59	89/11	–	–	1.2	6.5	15	14	9	39

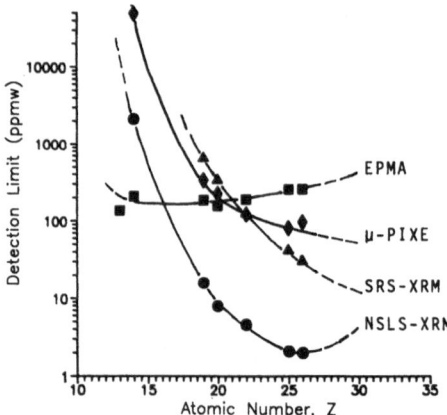

Fig. 3. Minimum Detection Limits obtainable using the four micro-probes at
100 sec counting times in K309 glass microspheres of 20-30 μ diameter.

micro probes offers better values. Despite the use of white light, the
overall ratio of characteristic to background count rate is better for the
NSLS than for the SRS case; the count rate is also a factor 100 higher at
the NSLS than at Daresbury due to the differences in available flux at both
facilities. For this type and size of particles, practical counting times
per particle are in the range of 100 sec for the NSLS and 1000 sec for the
SRS microprobe; for EPMA, typical counting times are shorter than 60
sec./particle. Of the four techniques used, the PIXE spectra offer the best
overall net-to-background ratio; however, due to the transparency of the
microspheres to the proton beam and the fact that the beam current had to
be limited to 25 pA in view of sample charging, an acquisition time of 30
min./particle was required in order to obtain statistically meaningful
spectra.

        Employing as figure-of-merit the minimum limit of detection (IUPAC
definition[25]) rather that the peak-to-background ratio, Fig. 3 shows MDL's
for the four techniques as a function of the atomic number Z. Whereas for
EPMA, a more or less uniform sensitivity in the 100 ppm range is obtained,
the MDL values of the other techniques vary considerable with atomic number.
Corresponding to the maximum in the excitation spectrum of the NSLS XRM near
8 keV, the lowest MDL values are obtained for elements such as Mn, Fe and
Co. Here for Fe, a relative MDL of 6 ppm at a counting time of 100 sec. is
achieved, corresponding in the present case to an absolute detectable amount
of 7.8 fg of Fe. In the case of the SRS microprobe, an MDL for Fe of around
60 ppm is obtained; for Zn, this value is expected to be around 5 ppm.

Size dependence of X-ray yields

        As a function of particle diameter, K961 particles were analyzed using
the NSLS XRM and the electron micro probe. The variation of the
characteristic X-ray intensities with size is shown in Fig. 4. As the
currently achievable synchrotron beam sizes are in the order of 5 to 10 μm,
the optical resolution of the sample viewing microcope at the NSLS facility
is also of this other. Consequently, only particles with radii down to 3.5
μm could be analyzed.

        For the EPMA data, a rise in sensitivity is observed up to a particle
diameter of around 5 μ. This behaviour can be explained by considering the

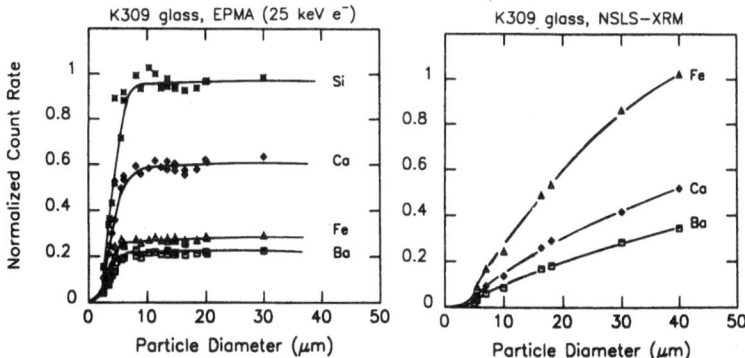

Fig. 4. Size-dependence of X-ray yields obtainable from K309 microspheres using EPMA and the NSLS XRM. Solid curves intended to guide the eye.

size of the interaction volume of the primary electrons with the particles which is of the order of 5 to 7 μm in the present case. As illustrated in Fig. 5, for particles with a radius larger than 2.5 μm, only a limited part of the particles is 'seen' by the electron beam, giving rise to a plateau in the size vs. intensity plots in Fig. 4.

In the case of the XRM data, in view of the much larger penetration depth of X-rays in comparison with electrons, a less outspoken dependence of the X-ray yield with size can be observed. From Fig. 4 it follows that for particles smaller than 10 μm radius, a much diminished X-ray yield can be expected, corresponding to MDL-values higher than those plotted in Fig. 3. This observation is important in view of the fact that the size distribution of e.g. ambient aerosols as collected on Nuclepore filters extends roughly from 0.1 to 10 μm. Whereas the NSLS XRM clearly features better relative MDL's in the case of particles which are larger than the beam dimensions (i.e., diameter > 10 μm), as shown in Table 3, for smaller particles, and when absolute detectable amounts are considered, not so large differences between EPMA and XRM are observed. Indeed, when the particles become smaller than the X-ray beam size, not all of the available photon flux is used effectively, while in the case of EPMA, the total electron flux will still impinge on the particle. Nevertheless, for the smallest particles which could be analyzed, the NSLS XRM still is 5 to 10 times more sensitive than EPMA for resp. Ca and Pb.

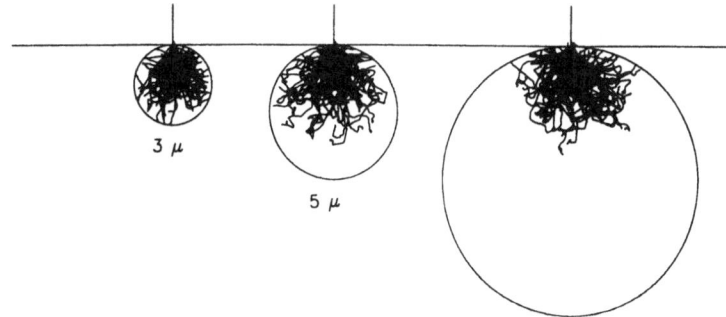

Fig. 5. Monte Carlo simulation of electron paths in K309 microspheres.

Table 3. Minimum detectable amounts of Ca, Fe, Ba and Pb
in 3.5 μm radius microscopic particles

Element	Minimum Detectable Amount (pg)	
	NSLS XRM	EPMA
Ca	0.10	0.5
Fe	0.04	0.8
Ba	0.2	1.6
Pb	0.4	4.1

CONCLUSIONS

In this paper, the possibilities of employing X-ray based micro beam instruments for performing individual particle analysis were evaluated by analyzing glass microspheres of known composition. In contrast to EPMA detection limits in the 0.1% range, for the elements Ca to Zn, the NSLS XRM offers MDL values at the ppm 1 to 10 ppm level for counting times of 1000 sec/particle and for particles of ca. 20 μm. In the case of the SRS microprobe, in view of limitations in flux, MDL values are a factor 10 higher in this range.

Considering the net count rate obtainable from microscopic particles at the NSLS, it can be concluded that performing individual particle analysis at the 10 to 100 ppm level on coarse fraction aerosols (diameter > 5 μm) using the NSLS XRM is feasible employing relatively short measuring times (typically 50 to 100 sec per particle) as required for individual analysis of large particle sets.

However, for (fine mode) particles whose diameter is smaller than the X-ray penetration depth and smaller than currently achievable beam sizes, a decrease in sensitivity with the third power of the particle diameter needs to be taken into consideration. In practise, analysis of the size fraction below 5 μm is hampered by limitations in the optical visualisation system used on the X-ray microprobes.

ACKNOWLEDGEMENTS

The authors are indebted to W. Dorrine for assisting with the EPMA measurements. K.J. wants to thank the Belgian National Science Fund for financial support. The research was supported in part by the US Department of Energy, Office of Basic Energy Sciences, Chemical Sciences Division, under Contract No. DE-AC02-76CH00016. M.R. wishes to thank the National Science Foundation (NSF) for support through grant No. EAR89-15699; S.S. acknowledges the support of the National Aeronautics and Space Administration (NASA) through grant No. NAG 9-106. The present research was also supported by the UK Science and Engineering Research Council (SERC) and the Netherlands organisation for advancement of Research (NWO) in connection with the agreement between SERC and NWO concerning the SRS.

REFERENCES

1.   W. Maenhaut, in "Controle and Fate of Atmospheric Trace Metals," J.M. Pacyna, B. Ottar, eds., Kluwer Academic Publishers, New York (1989).
2.   A.L. Gray, in "Inorganic Mass Spectrometry", F. Adams, R. Gijbels, R. Van Grieken, eds., Wiley, New York (1988).
3.   D.L. Fox, Anal. Chem., 59:280R (1987).
4.   A. Alian, B. Sansoni, J. Radioanal. Nucl. Chem., 89:191 (1985).

5.   W. Maenhaut, Anal. Chim. Acta, 195:123 (1987).

6.   R.E. Van Grieken, J.J. Labreque, in "Trace Analysis", Vol. 4, J.F. Lawrence, ed., Academic Press, New York (1985).

7.   K.R. Spurny, (ed.), "Physical and Chemical Characterisation of Individual Airborne Particles", Ellis Horwood, Chichester, UK (1985).

8.   K. Janssens, W. Van Borm and P. Van Espen, NBS J. Res., 93:260 (1988).

9.   S.H. Sie, C.G. Ryan, D.R. Cousens, W.L. Griffin, Nucl. Instr. Meth., B40/41:690 (1989).

10. R. Kaufmann, F. Hillenkamp, R. Wechsung, H.J. Heinen, M. Schürmann, in "Scanning Electron Microscopy", Vol. II, SEM Inc., A.M.F. O'Hare, IL 60666, USA, (1979).

11. M. Cholewa, G. Bench, B. Kriby, G.F.J. Legge, Nucl. Instr. Meth. B, 1990, in press.

12. A.L. Hanson, H.W. Kraner, K.W. Jones, B.M. Gordon, R.E. Mills, J.R. Chen, IEEE Trans. Nucl. Sci., NS-30:1339 (1983).

13. K.W. Jones, B.M. Gordon, A.L. Hanson, J.B. Hastings, M.R. Howells, H.W. Kraner, J.R. Chen, Nucl. Instr. Meth., B3:225 (1984).

14. A.J.J. Bos, R.D. Vis, H. Verheul, M. Prins, S.T. Davies, D.K. Bowen, J. Makjanic, V. Valkovic, Nucl Instr. Meth., B3:232 (1984).

15. F. van Langevelde, R.D. Vis, Anal. Chem., accepted (1991).

16. V.B. Baryshev, G.N. Kulipanov, E.I. Zavtsev, Y.V. Terekhov, V.I. Kalyuzny, Nucl. Instr. Meth., A261:279 (1987).

17. S. Török, Z. Szökefalvi-Nagy, S. Sándor, V.B. Baryshev, K.V. Zolotarev, G.N. Kulipanov, Nucl. Instr. Meth., A282:499 (1989).

18. K.W. Jones, B.M. Gordon, Anal. Chem., 61:341A (1989).

19. P. Ketelsen, A. Knöchel, W. Petersen, Z. Anal. Chem., 323:867 (1986).

20. F. van Langevelde, D.K. Bowen, G.H.J. Tros, R.D. Vis, A. Huizing, DKG de Boer, Nucl. Instr. Meth., A292:719 (1990).

21. J.H. Underwood, A.C. Thompson, Y. Wu, R.D. Giauque, Nucl. Instr. Meth., A266:296 (1988).

22. Y. Goshi, S. Aoki, A. Iida, S. Hayakawa, H. Yamaij, K. Sakurai, Jpn. J. Appl. Phys., 26:L1260 (1987).

23. J.A. Small, K.F.J. Heinrich, C.E. Fiori, R.L. Myklebust, D.E. Newbury, M.F. Dilmore, in "Scanning Electron Microscopy", Vol. 1, SEM Inc, AMF O'Hare, IL, USA (1987).

24. R.D. Vis, Fres. J. Anal. Chem., 337:622 (1990).

25. R. Jenkins, Spectrochimica Acta, 37:207 (1982).

A NOVEL SCANNING X-RAY DIFFRACTO-MICROSCOPE/X-RAY POWDER

DIFFRACTOMETER USING CONVERGED X-RAY BEAM

Ken Yukino, Fujio P.Okamura, Hiroshi Nozaki
Yuji Kobayashi* and Yoshiyuki Yamada*

National Institute for Research in Inorganic
Materials, Tsukuba, Ibaraki 305, Japan

Rigaku Corporation*, Akishima, Tokyo 196, Japan

ABSTRACT

A new type of scanning X-ray diffracto-microscope (SXDM) / X-ray powder diffractometer (XPD) which uses a converged incident beam, was designed, manufactured, and some of its basic characteristics were examined. The optical system consists of asymmetric reflection type curved crystal monochromators for both incident and reflection beams, a detector (PSPC, X-ray film, IP, nuclear plate), a translation mechanism for the specimen and also for the detector.

1.INTRODUCTION

The intensity of reflected X-rays measured by a powder diffractometer is significantly influenced by the distribution in size, orientation, shape and packing density of the constituent crystallites of specimen. Indirect methods to investigate the positional distribution of crystallites, grain-size distribution measurement, and observation by optical or electron microscope are known, but none of them directly observes the effect on the diffraction intensity by a diffractometer.

On the other hand, by ε-scan method proposed by one of the authors (K. Y.)[1], the angular dependency of crystallite distribution is directly examined by rotating the θ-axis off from its symmetrical reflection condition, while fixing the detector at a desired Bragg angle. From the observed angular dependency, the crystallite distribution is estimated, and in the case that there exists any preferred orientation, the necessary correction factor for it can be derived. Whereas by

τ-scan method[2], the specimen is translated to traverse the incident beam, in order to investigate the positional distribution of crystallites.

Thus the ε & τ-scan methods enable the direct observation of the effect of orientational and positional distribution of crystallites on the diffracted intensity by the diffractometer, but it is not possible to specify the position of individual contributing crystallite, because a divergent beam is usually employed. As a solution for this problem, the use of a thin parallel beam or an X-ray guide tube has been proposed. It is not applicable for general use because it requires a intense X-ray source.

Using a scanning X-ray diffracto-microscope with converging opitics that we developed recently, the incident beam is focused on the surface of a specimen which may be given ε and τ movements. This microscope can be used also as a diffractometer (XPD) by analysing the distribution of diffracted X-rays. Using this SXDM/XPD system, it is possible to identify, and observe the size, orientation, shape and packing density of crystallites as a function of position in the specimen. It is therefore useful for texture studies on materials such as changes in the texture of metal induced by fatique. In addition, the system allows the diffraction intensity measurement on small amount of specimen not only under normal conditions but also at low or high temprature, and under controlled atmosphere. For materials with higher symmetry, it can measure the integrated reflection intensities rapidly and with high precision.

2. PRINCIPLE

The princple of the new SXDM is illustrated in Fig. 1. It consists of an optical system, where the divergent X-rays from an X-ray source with line focus is converged to a line on the surface of the specimen by a curved crystal monochromator. The resultant divergent diffracted X-ray beam is reconverged at the detector position using another curved crystal monochromator. A PSPC (Position Sensitive Proportional Counter), X-ray film, IP

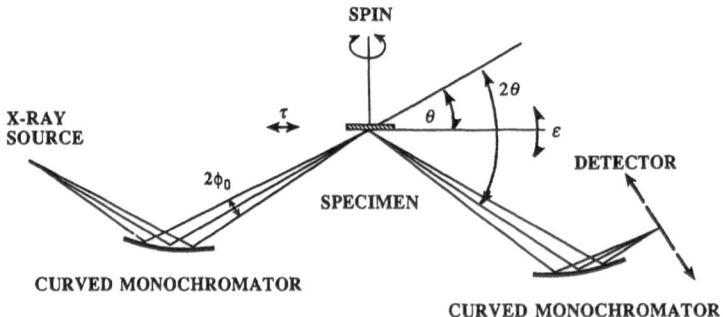

Fig. 1 Schematic diagram of scanning X-ray diffracto-microscope (SXDM)/X-ray powder diffractometer (XPD).

(Imaging Plate) or nuclear plate can be used as a detector. A direct image is obtained by translationally moving the specimen inclined by an angle ε off from symmetrical reflection condition, accompanied by a corresponding synchronous movement of the detector.

The reflection intensity, I, depends on the diffraction angle 2θ, the inclination angle of specimen ε, and the positional parameters x, h. Denoting the diffracted intensity from the specimen by I_r, then the intensity measured by the SXDM is given by,

$$I(2\theta,\varepsilon,x,h)=\int(\phi_0-|\phi|)\cdot I_r(2\theta+\phi,\varepsilon,x,h)\cdot A(\varepsilon,\phi)\cdot R(\phi,h)\cdot d\phi, \quad (1)$$

where h is the position along the focused line, $2\phi_0$ is the convergence (divergence) angle of the incident beam, ϕ is the angular deviation from the optical center line, $\phi_0-|\phi|$ is monochromator function, and the integral region is between $-\phi_0$ and ϕ_0. The factor A is twice the absorption factor assuming a large linear absorption coefficient sufficient for total absorption by the specimen, which is given as follows:

$$A(\theta)=1-\cot\theta\cdot\tan\varepsilon \quad (2)$$

$$-\theta+\phi_0\leqq\varepsilon\leqq\theta+\phi_0,$$

where R is the product of the X-ray source intensity, reflectivities of both the monochromators, the detector efficiency and the transmissivities of the windows within the beam path. The value R can be experimentally determined by rotating a single crystal about θ-axis and making appropriate intensity measurements.

When this SXDM is utilized as a powder diffractometer (XPD), the reflection intensity I and the integrated reflection intensity T are expressed as follows:

$$I(2\theta)=\int(\phi_0-|\phi|)\cdot I_r(2\theta+\phi)\cdot A(\phi)\cdot R(\phi)\cdot d\phi, \quad (3)$$

$$T(2\theta)=\int I(2\theta)d2\theta. \quad (4)$$

The factor $A\cdot R$ is also experimentally obtained through the intensity measurement by the θ-axis rotation of either a single crystal or a randomly packed fine powder specimen.

A typical diffraction profile from the new XPD is illustrated in Fig. 2 together with that by conventional Bragg-Brentano type diffractomer for comparison. In the latter case, assuming that the reflected X-rays converge to a line at the receiving slit position 2θ (Fig. 2a), the observed reflection gains the angular broadening equal to the width ω_0 of the receiving slit. It should be noted that the height of the reflection intensity stays unchanged. By the new XPD, on the other hand, the profile of the reflected X-rays in front of the second monochromator is given by a trapezoid indicated by the broken line in Fig. 2c, and its width is $2\phi_0+\omega_0$. The height of the intensity $I(2\theta)$ for a receiving slit wider than the convergence angle $2\phi_0$ is the same as that in the Bragg-Brentano case. If we take only the effect of $\phi_0 - |\phi|$ into account, the profile after being diffracted by the second monochromator is a triangle as shown in Fig. 2d. Namely, because the width of the reflected X-rays is not modified through the diffraction by the second monochromator, the length of the base of the triangle maintains a constant value of $4\phi_0$, twice the convergence angle, independent of diffraction angle. Provided that the reflectivity of the monochromator is equal to unity, the height of the triangle is also the same as for the above cases.

Therefore, the XPD diffractometry is suitable for the precise measurement of the integrated reflection intensity. The area of the triangle in Fig. 4d increases proportional to the convergence angle. In the strict sense, it is necessary to solve equation (3) to determine the diffraction angle, but

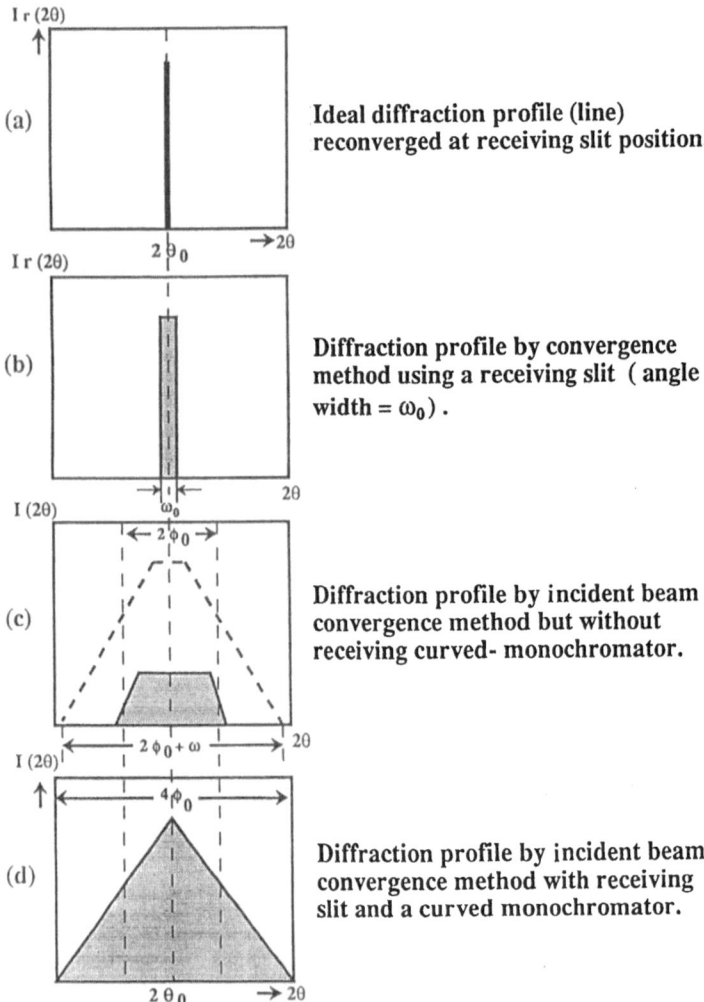

(a) Ideal diffraction profile (line) reconverged at receiving slit position.

(b) Diffraction profile by convergence method using a receiving slit (angle width = ω_0).

(c) Diffraction profile by incident beam convergence method but without receiving curved- monochromator.

(d) Diffraction profile by incident beam convergence method with receiving slit and a curved monochromator.

Fig. 2 Broadening of diffraction profile.

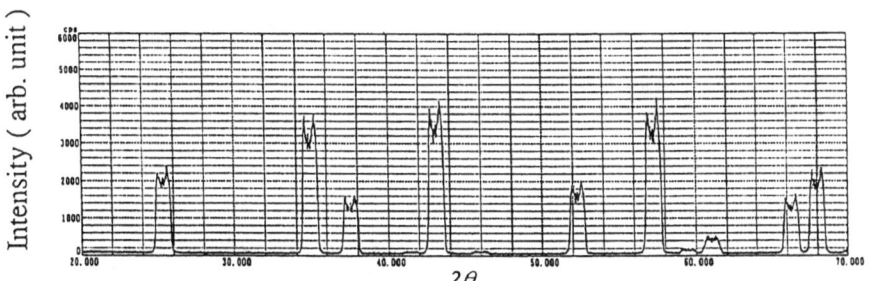

Fig. 3 $2\theta/\theta$ diffraction pattern of α-A$_2$O$_3$ by the convergence method (without receiving monochromator).

direct reading of the peak position is precise enough for the purpose. Thus this method is very simple and all the more practical.

3. EXPERIMENTAL

Some basic measurements were conducted at the stage when a prototype of the new SXDM/XPD system was manufactured by Rigaku., Ltd. The essential geometrical conditions of the system are as follows:

<u>curved crystal monochromators</u>:
   quartz (1011), asymmetric reflection and Johansson type
<u>specimen - monochromator distance</u>:
   first monochromator: 210mm,
   second monochromator: 225mm
<u>convergence angle</u>: 2°
<u>length of line focus</u>: 10mm.

3-1. $2\theta/\theta$ Scan of α-Al$_2$O$_3$

The incident beam was converged on the surface of α-Al$_2$O$_3$ specimen with convergence angle of 1°, and the diffracted X-rays were measured by the $2\theta/\theta$ method using a NaI(Tl) scintillation counter through a Soller slit and a receiving slit (width: 0.3mm). The scan pattern is shown in Fig. 3. The

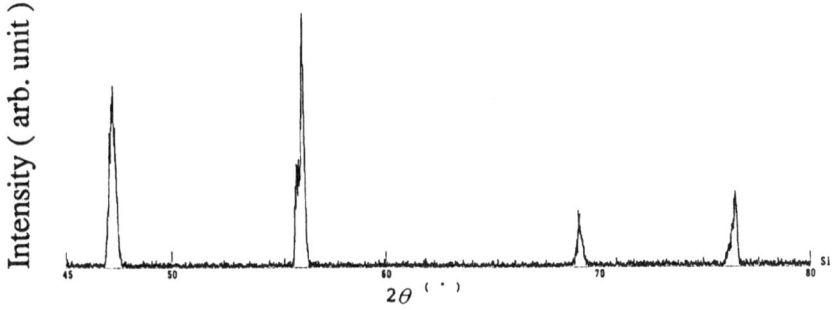

Fig. 4 $2\theta/\theta$ diffraction pattern of a Si pellet by the convergence method.

profile shows the peaks as were expected, except for some deviations presumably caused by the inhomogeneous reflectivity of the incident monochromator or insufficient adjustment of the optical system.

3-2. $2\theta/\theta$ Scan of Si Pellet

The diffraction pattern of a Si pellet was measured by $2\theta/\theta$ scan under the same irradiation conditions as 3-1, but replacing the Soller silt with a curved crystal monochromator of pyrolytic graphite (Fig. 4). The resultant profiles are similar to that expected from the equation (3). Some deviations in intensity are caused by coexisting coarser crystallites, and a shift of the peak towards the high-angle side, most possibly coming from the broadening of the reflection angle along the vertical direction, are observed.

3-3. τ-scan of Si Pellet

The detector was replaced with a PSPC and under the same experimental condition as 3-1, the 111 diffraction pattern of the Si pellet was measured by traversing it across the incident beam for 0.7mm of distance at intervals of $20\mu m$ (Fig. 5). The linear range of the specimen covered by the PSPC at each step of the measurement was 5mm. Although the effect of Soller slit is observed, the scan pattern shows a two dimensional distribution image of Si crystallites for which the (111) planes are within 0.5° of angular deviation from the specimen surface ($\varepsilon \leq 0.5°$). The dimension of the crystal that showed the maximum peak height was estimated ca. $100\mu m$.

3-4. Measurements on Bi-(Sr,Ca)-Cu-O Superconductor

The specimen used for this experiment is a multi-phase mixture of the Bi-(Sr,Ca)-Cu-O system, which was prepared on MgO(100) substrate by once melting its starting material followed by cooling.

The specimen was observed under the same condition as 3-1. Because the dimension of the specimen was 1.2mm, the width of the incident beam (along the focus direction) was limited to

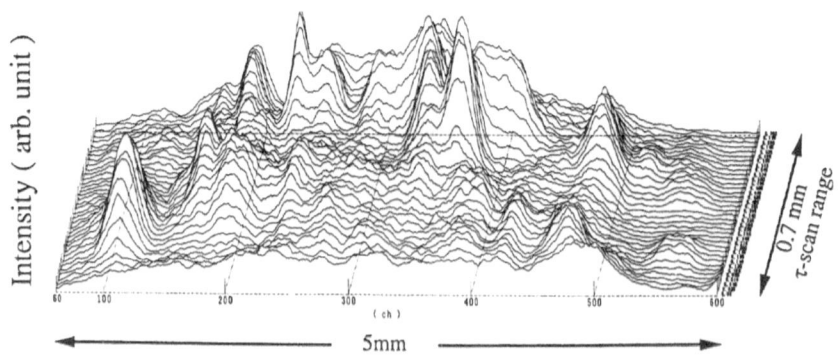

Fig. 5 111 τ-scan pattern of a Si pellet by the convergence method using PSPC (without receiving monochromator).

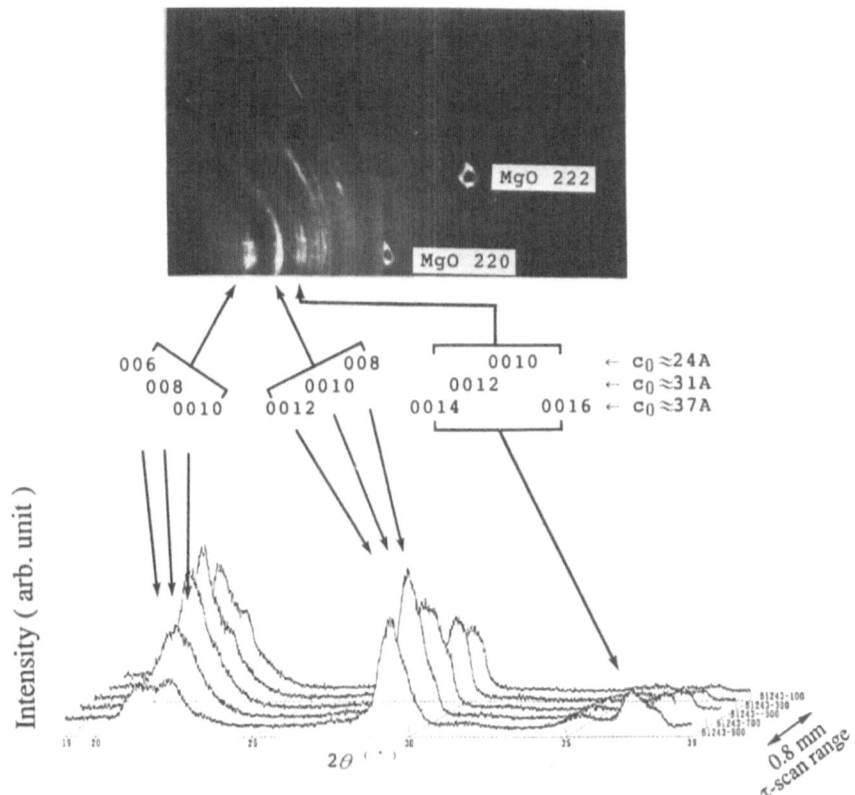

Photo 1 A diffraction photograph of Bi$_2$(Sr,Ca)$_{x+1}$Cu$_x$O$_y$ on
         MgO(100) when the incident beam is irradiating the
         central part (Bi243-700, θ=11.104°) and
Fig. 6   2θ/τ-scan patterns of Bi$_2$(Sr,Ca)$_{x+1}$Cu$_x$O$_y$
         θ=10.904°, 19°$\leqq$2$\theta$$\leqq$39°.

2mm. The diffraction patterns were recorded on Polaroid films
(#57) intercepting the diffracted beams from the specimen. A
typical example (for the glancing angle of 11.104°, Bi243-700)
of these photographs is shown in Photo 1. And 2θ/τ-scan
patterns corresponding to the photograph are shown in Fig. 6.
The interval of τ-scan (the specimen translation) was 200μm. In
the photograph, diffraction 'rings' from the specimen were
intermittently broken and mostly concentrated to limited areas
for stong orientational maldistribution of the crystallites.
Diffraction pattern of another non (00ℓ)-series was also
observed in a way similar to that of a single crystal by
rotation method. The diffraction patterns of (00ℓ)-series were
observed within the rotation plane of the goniometer. The
contributions from some coarser grains are also observed. And
two large diffraction spots from the substrate MgO with the
shadow of specimen were also observed in the photograph,
because of the significant absorption of X-rays by the
Bi-containing superconductor covering the central part. From
the variation of the sizes of the diffraction spots of MgO
against exposure time, it was found that regardless of the
rather small half width of the intensity distribution profile

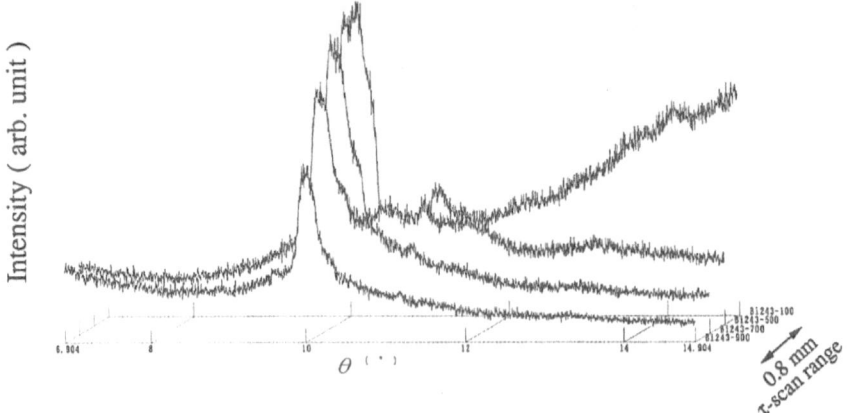

Fig. 7 ε/τ-scan pattern of $Bi_2(Sr,Ca)_{x+1}Cu_xO_y$
$2\theta_B=21.808°$, $6.904°\leq\theta+\varepsilon_B\leq14.904°$.

of the incident beam less than 150μm, the broadening of its
foot is fairly significant. The ε/τ-scan patterns with 2θ fixed
at 21.808° were shown in Fig. 7. From the 00ℓ ε/τ-scan pattern,
the peak position of the (001) planes of fine crystallites at
the central part specimen was found to be at $\varepsilon= 3.99°$. Both
$2\theta/\theta$- and $2\theta/(\theta+3.99)$-scan patterns are given in Fig. 8. The
latter reflection intensity is obviously stronger than the
former.

     The crystallites of Bi-(Sr,Ca)-Cu-O compounds generally
show strong preferred orientation, and some diffraction spots
from coarser grains are also recognized. From these photo and
figures, the orientational distribution of the (001) planes of
crystallites are largely effected presumably by epitaxial

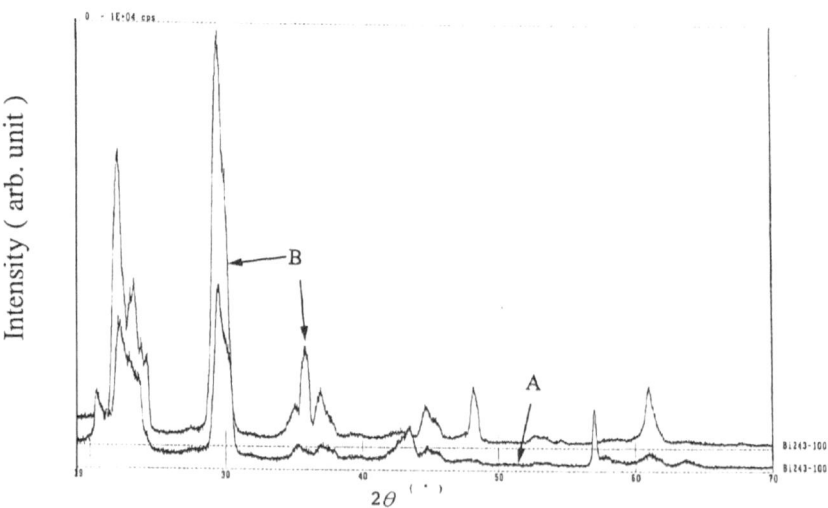

Fig. 8 Diffraction pattern of $Bi_2(Sr,Ca)_{x+1}Cu_xO_y$ (at center)
A:$2\theta/\theta$-scan,   B:$2\theta/(\theta+\varepsilon)$-scan, $\varepsilon=3.99°$.

mechanism with the (100) of the MgO substrate. This is in constrast with the result by Ono[3] that the (001) planes of Bi-(Sr,Ca)-Cu-O crystallites grown from melt without substrate were perpendicular to the surface.

4. CONCLUSIONS

The tests on the new SXDM/XPD system which uses a convergent X-ray beam on the sample has not been completed yet, but it was found that the system provides us with an effective means for the observation of the orientational and positional distribution of the crystallites in the specimen.

The SXDM as a scanning X-ray diffracto-microscope has the merit of allowing simple and rapid observations of the texture of a specimen, because of its use of a large convergence angle.

Moreover, because the system can be used also as a powder diffractometer (XPD), by combined use of $2\theta/\theta$-scan, 2θ-scan, ε-scan, τ-scan and photographic methods, it is possible to obtain the total image of distribution with more rapidity and certainty.

Because the convergence angle can be decreased, the system can be used as a scanning X-ray diffracto-microscope to get more detailed information on the orientational and positional distribution of the crystallites in the specimen. It should be noted that enough diffraction intensity is obtainable even for a small specimen of size $\sim$1mmϕ.

ACKNOWLEDGEMEMT

The authors are grateful to Professors S. Sueno of Tsukuba University and H. Komatsu of Tohoku University for the use of the specimen of Bi-(Sr,Ca)-Cu-O superconductor, with fruitful discussions and valuable comments. An extensive study on this material using the SXDM is now under way in cooperation with them. Present study was supported by the Superconductor Multi-core Project Fund by the Science and Technology Agency.

REFERENCE

1) K. Yukino and R. Uno: Jpn. J. Appl. Phys. 25: 661 (1986).

2) K. Yukino: FC Report 7: 401 (1989) [in Japanese].

3) A. Ono: Jpn. J.Appl. Phys. 27: L2276 (1988).

DEVELOPMENT OF A HIGH SPATIAL RESOLUTION X-RAY FLUORESCENCE ELEMENT MAPPING SPECTROMETER AND ITS APPLICATION TO QUANTITATIVE ANALYSIS OF BIOLOGICAL SYSTEMS

Natsuo FUKUMOTO, Yoshinori KOBAYASHl, Masayasu KURAHASHI
and Akira KAWASE

National Chemical Laboratory for Industry
Tsukuba, Ibaraki, 305, JAPAN

INTRODUCTION

We developed an X-ray fluorescence element mapping spectrometer (XEMS) based on commercially available energy dispersive XRF equipment several years ago. Using XEMS, we found that this technique is applicable to real-time observation of the elemental distributions in living biological samples. This kind of observation is almost impossible by conventional techniques such as EPMA, PIXE etc. But the spatial resolution of the previous system was about 200μm, inferior by almost two orders to that of EPMA for example. So we developed a new spectrometer with an improved resolution of better than 20μm and almost the same sensitivity.

EXPERIMENTAL

Fig.1 is the schematic diagram of the new system. An X-ray guide tube, which is a specially formed glass tube, was used to focus the X-rays effectively onto the sample using a commercially available Cu target X-ray tube for XRD (Philips PW2273/20). The usefulness of the guide tube technique has been demonstrated by several authors.[1] To detect fluorescence X-rays efficiently two Li doped Si solid state detectors (SSD) with a 30mm^2 sensitive area each were placed in close proximity to the sample. By measuring the polymer standard films with known amounts of various elements, the minimum detection limits of the new system for the elements from K to Ni were found to be comparable to the values achieved with conventional analytical methods such as atomic absorption and inductively coupled plasma emission spectroscopy.[2]

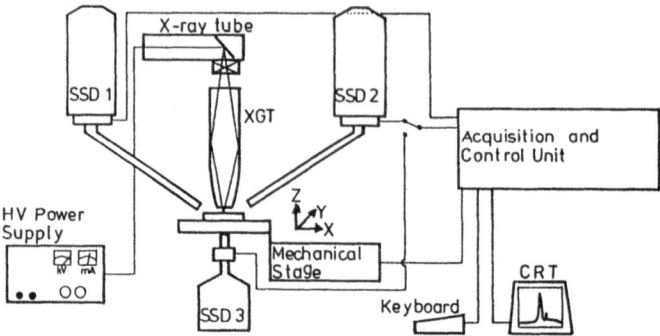

Fig.1   Schematic diagram of the XRF element mapping spectrometer.

RESULTS AND DISCUSSION

To determine the absolute amount or concentration of the element of
interest in a biological sample, it is necessary to measure the sample
thickness and relative sensitivity for each element. By measuring the
intensity of the X-rays from the light source with the SSD placed under the
sample, the thickness of living leaf could be obtained. The XRF count rates
of filter paper dipped into standard solutions were measured to determine
the relative sensitivity of this system for the element. From these studies
the Mn concentration in a living bird's foot trefoil(Lotus corniculatus L.,
BFT) leaf was determined to be about 1-5 mg/g, which agreed with the
previous atomic absorption result.

We also applied this mapping system to living BFT leaves damaged by the
treatment of artificial acid rain or exposure to severe X-ray irradiation.
The artificial acid rain was prepared by mixing sulfuric acid and nitric
acid in equal molar proportions and diluted to a pH value of 2. The leaf was
taped on polyethylene film and placed into the XRF spectrometer. A very
quick reduction of the potassium, followed by slower increase of calcium and
manganese concentrations around the X-ray damaged area was observed. The
distributions of K, Ca and Mn in the BFT leaf 6 days after the treatment are
shown in Fig. 2. In contrast to the X-ray irradiation the Ca concentration
was decreased in the damaged part.

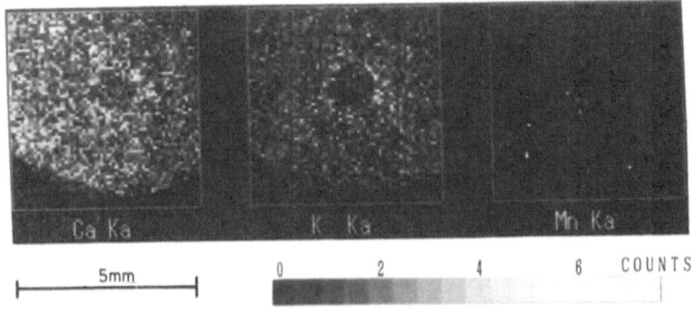

Fig. 2 Elemental distributions in the BFT leaf 6 days after treatment with
the artificial acid rain. The accumulation time was 4s for each point.

From these results, we found that the spatial resolution and the sensitivity of this system are suitable for studies to understand the physiological meaning of ion transport in a plant organ. We estimate that the spatial resolution of less than 10μm with some increase in the sensitivity may also be achieved by optimizing the X-ray optics.

REFERENCES

1. A. Rindby, P. Engström, S. Larsson and B. Stocklassa, Microbeam technique for energy-dispersive X-ray fluorescence, X-ray spectrometry 18:109(1989)
2. Y. Kobayashi, N. Fukumoto, and M. Kurahashi, X-ray fluorescence element-mapping spectrometer with improved spatial resolution, Meas. Sci. Technol. 2:183 (1991)

SCANNING X-RAY ANALYTICAL MICROSCOPE USING X-RAY GUIDE TUBE

Shuichi Shimomura and Hiromoto Nakazawa

National Institute for Research in Inorganic Materials
1-1 Namiki,Tsukuba-shi,Ibaraki-ken,305 Japan

ABSTRACT

A scanning X-ray analytical microscope was constructed using X-ray guide tube(XGT). XGT is a glass capillary which guides and focuses X-rays by total external reflection at the inner wall. The nature of X-ray beam passed through XGT is varied depending on the form of XGT. The cylindrical and conical types were used for the present setup. A small area (10μ m$\times 10\mu$ m) of the sample was irradiated by the X-ray microbeam formed by XGT. Fluorescent and diffracted X-rays from the small area were detected by SSD. By scanning the sample in the plane normal to the X-ray microbeam, the intensity distributions of such secondary X-rays were measured and used as picture elements for constructing X-ray mapping images. The sample was a thin-section of an old chinaware. The images suggest a wide application of this instrument.

INTRODUCTION

Various types of X-ray microscope have been reported using soft and hard X-rays. Because of shorter wave length, hard X-rays has advantage in mapping the intensity distribution of fluorescent and diffracted X-rays, but more difficult to focusing X-rays. We used X-ray guide tube (XGT)[1] for collecting hard X-rays, a simple X-ray optical component using total external reflection at the inner wall of the tube. We constructed a scanning X-ray analytical microscope (SXAM)[2] using XGT and pinhole (10μ m $\times 10\mu$ m). SSD was used as detector, so that fluorescent and diffracted X-rays could be detected simultaneously.

X-RAY GUIDE TUBE

Two types of XGT were made of pyrex glass tubes,(1) the cylindrical type, (inner diameter=0.4mmϕ ,length=220mm) and (2) the conical type[3], (inner diameter at the entrance=0.25mmϕ , that at the exit=0.45mmϕ and length=220mm). X-rays trace from source to the sample is schematically illustrated in Fig.1, assuming an ideal point source. X-rays passed

through the pinhole are in two traces. One is that X-rays come directly
to the pinhole and the other is that X-rays come to the pinhole after one
or more total external reflections at the inner wall of XGT. The
intensity of the reflected X-rays is added to that of the direct X-rays
which are only usable without XGT. The gain of X-ray intensity was
caluculated by the ratio of reflected X-rays solid angle to that of the
direct X-rays. For example, assuming that the reflectivity of the total
external reflection is equal to 1 and using following dimentions, the two
solid angles can be caluculated. The dimensions are: inner diameter of
XGT is 0.4mmϕ, pinhole size is 10μm, distance from X-ray generator to
pinhole is 280mm. Solid angle of direct X-rays which was shown as (a) in
FIg.1, is $\Phi_d = 6.28 \times 10^{-10}$rad. That of reflected X-rays which was shown as
(b) in Fig.1, is $\Phi_r = 3.20 \times 10^{-7}$rad. So the ideal gain of X-ray intensity at
the pinhole is $(\Phi_d + \Phi_r) / \Phi_d = 511$. XGT is, thus, an effective optical
component to condense radial X-rays.

It should, however, be kept in mind that divergence of X-rays passed
through a cylindrical XGT is larger than that without XGT. In the present
setup, divergence of the X-ray microbeam at pinhole is $\pm 1.45 \times 10^{-3}$rad. The
conical type XGT is a modification of XGT for the problem of larger X-ray
divergence. This has been described previously by Nozaki and Nakazawa
(1986). Ideally, an ellipsoidal XGT is the most effective form of XGT to
condense X-rays onto a small point. But because of small angle of total
external reflection of hard X-rays, it is difficult to fabricate an
ellipsoidal XGT with very small inner diameter. In a laboratory,
cylindrical and conical types of XGT are only presently available, and
they were used in the present experiment.

SETUP OF SCANNING X-RAY ANALYTICAL MICROSCOPE

SXAM was constructed by following components as shown in Fig.2.
X-ray generator was micro focusing type (Rigaku microflex CN4180E2). Its
focus size is 0.1mm$\times$1mm. X-ray take-off angle is $6°$. A cylindrical type
XGT was mainly used for making X-ray microbeam and a conical type one was
occationally used. After alignment of XGT, a pinhole was placed just
before the sample. The pinhole size is 10μm$\times$10μm. The sample was placed
on the x-z stage which traverse two-dimensionally in the plane normal to
X-ray beam using computer controlled stepping motor. SSD was used for
detecting secondary X-rays. (Li doped Si, 80mm^2 area, 3mm thickness) The
x-z stage and SSD were set on a goniometer to set the orientation of the
sample to a certain diffraction angle. In the case of mapping of
intensity distribution of fluorescent X-rays, SSD was placed as close as
possible to the sample. For mapping intensity distribution of diffracted
X-rays, SSD was set at a certain diffraction angle, 2θ, which corresponds

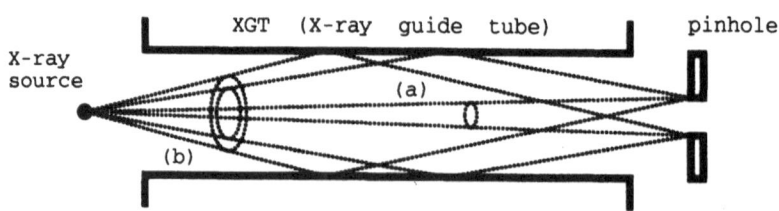

Fig. 1. X-ray traces of cylindrical type XGT
(a) direct X-rays   (b) reflected X-rays

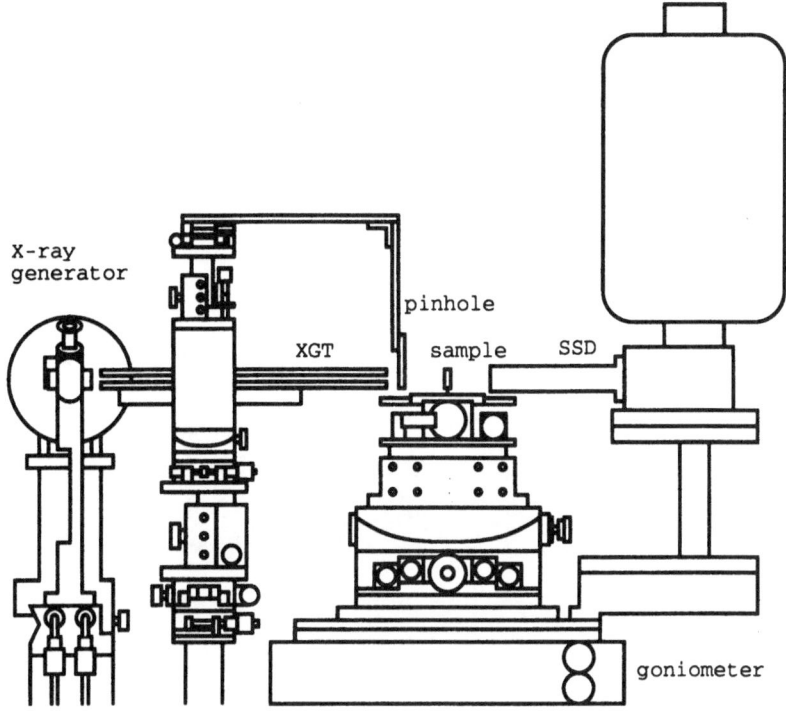

Fig. 2. Scanning X-ray analytical microscope

to the Bragg condition using the characteristic X-rays of X-ray source,
e.g. Cu-Kα . A small area of the sample was irradiated by the X-ray
microbeam. At that position, intensities of fluorescent and diffracted
X-rays were measured and stored in a computer memory. Then x-z stage
moved the sample for a small step, and the next position was measured
likewise. After measureing all position in a certain area on the sample,
mapping image could be obtained.

APPLICATION OF SXAM

     SXAM was applied to the old chinaware showing a special pattern on
its surface. It is a question which has long been standing how the
pattern was applied in the chinaware processing. To solve this,
fundamental characterizations are required on the elements consisting the
pattern and their crystal structure. The sample was prepared by cutting
and polishing the original chip in the form of a thin plate (10mm×10mm in
dimension and 100μ m in thickness)[4]. A preliminary experiment indicated
that the thin plate showed a small peak of X-ray diffraction at 2θ =27.8°
using Cu-Kα radiation as well as a halo pattern due to amorphous
structure. SSD was, thus, set at 2θ =27.8°. Measurements of intensity
distributions of the secondary X-rays, fluorescent X-rays of
Fe,Sr,Mn,Cr,Ca and diffracted X-rays were made on the sample under the
following condition: Target,Cu; accelerating voltage and current,30kV
3mA; measured area,2.56mm× 2.56mm; measured points,128× 128; x-z step
interval,20μ m; measurement time 20sec/point. A fluorescent X-ray images
of Fe-Kα is reproduced in Fig.3 (a) where the similar image using

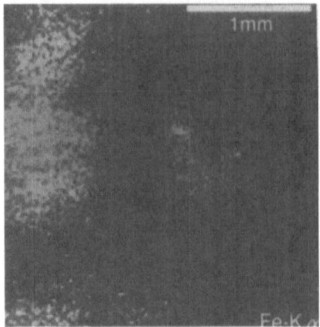

(a) Fluorescent X-ray image

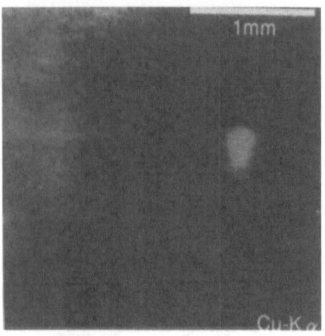

(b) Diffracted X-ray image

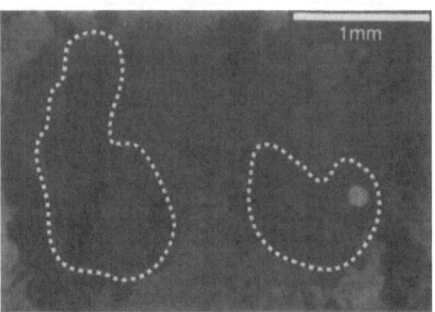

(c) Optical micrograph

Fig. 3. SXAM mapping images (a),(b) and optical micrograph (c).
A brown special pattern on the old chinaware was surrounded
with white broken line in (c) which corresponded to Fe
concentration in (a).

diffracted X-rays and optical micrograph of the same area are also
represented in Fig.3 (b) and (c), respectively. The special pattern,
brown part in the optical micrograph, corresponded well to that of Fe
concentration. Some iron oxide or silicate is, therefore, the material
drawing the pattern. Identification of the material has not been
successful.

SUMMARY

A SXAM was constructed using XGT which produced a X-ray microbeam of
$10\mu m \times 10\mu m$ in cross section. The SXAM is a useful instrument to obtain
images of diffracted and fluorescent X-rays from a small area of a sample
without any damage and pollution of the sample under atmospheric
conditions. The SXAM image of an old chinaware showed that a special
pattern of its surface corresponds to that of Fe concentration.

REFERENCES

1. H. Nakazawa, J. Appl. Cryst. (1983). 16, 239-241.
2. H. Nakazawa et. al. X-Ray Microscopy in Biology and Medicine, ed. by K. Shinohara et al., Japan Sci. Soc. Press. Tokyo/Verlin pp 81-86 (1990).
3. H. Nozaki and H. Nakazawa, J. Appl. Cryst. (1986). 19, 453-455.
4. Sample supply: Prof. Kazuo Yamazaki and Prof. Izumi Nakai.
5. A. Rindby, "Application of Fiber Technique in the X-Ray Region," Nuclear Instruments and Methods in Physics Research, A249 (1986) 536-540.
    (and many others from this group at Chalmers University, Sweden)
6. D. A. Carpenter and M. A. Taylor, "Fast, High Resolution X-Ray Microfluorescence Imaging," Adv. X-Ray Anal. 34 (1991), 217-221.

GAMMA RAY AND X-RAY IMAGING STUDIES OF THE LOCATION AND SHAPE OF THE MELT-SOLID INTERFACE DURING BRIDGMAN GROWTH OF GERMANIUM AND LEAD-TIN-TELLURIDE

R. T. Simchick[1], S. Sorokach[1], A. L. Fripp[*],
W. J. Debnam[*], R. F. Berry[*], and P. G. Barber[#]

[1]Lockheed Engineering and Sciences Corporation
Hampton, Va.23666

[*]NASA Langley Research Center
Hampton, VA 23665

[#]Longwood College
Farmville, VA 23901

INTRODUCTION

The success of new electronic materials has been due in part to the development of procedures that produce semiconductors of sufficient purity and perfection. These materials have been grown from the gas phase, solution, and melts. The Bridgman technique is one way semiconductor crystals are grown from the melt. In such furnaces the semiconductor material is usually sealed in an ampoule made of quartz or other suitable material, placed inside the tubular furnace, and heated to completely melt the sample. The ampoule with the molten material is slowly removed from the furnace by one of three ways. The furnace can be translated along the sample, the sample ampoule can be slowly extracted, or the temperature can be slowly lowered. The rate of crystal growth is typically only a few millimeters per hour, and all changes must be made smoothly and without any mechanical vibrations. A diagram of a typical Bridgman furnace can be found in the literature.[1] The furnaces used in such procedures are generally opaque to visible radiation.

This paper describes the development of procedures to visualize the melt-solid interface and to record its shape and movement during crystal growth. Although, the procedure has been developed for a particular application with Bridgman furnaces to grow germanium and lead-tin-telluride, the techniques are applicable to other crystal growth procedures. Other similar procedures have been reported in the literature for Czochralski growth.[2]

Advances in X-Ray Analysis, Vol. 35
Edited by C.S. Barrett *et al.*, Plenum Press, New York, 1992

HISTORICAL REVIEW OF INTERFACE OBSERVATIONS

The importance of the shape and location of the melt-solid interface during crystal growth stimulated the development of several techniques to observe it.[3-6] These techniques, however, are limited by the fact that the results are available only after the crystallization process has been finished and most measurement techniques are destructive to the sample. A quenching technique is one that has been used. In this procedure the sample is grown under well-defined experimental conditions and thermal gradients, quickly quenched in water or liquid nitrogen. The sample is then cut and carefully polished to observe the frozen interface shape.[3,4] A second technique is to grow the crystal and periodically impress current pulses or mechanical vibrations on the sample.[5,6] This has the effect of causing small variations in composition in the sample during growth and such variations can be observed after cutting, polishing, and etching the crystal.

ADVENT OF DIRECT OBSERVATION

Because of the importance of the melt-solid interface shape and location in determining the resulting crystal perfection and growth rate, observation techniques that provide results only after the growth has ceased do not enable the crystal growers to make furnace adjustments needed during the actual growth. Methods of directly observing the interface have been developed.[7] The early techniques used X-rays and gamma-rays to penetrate both the Bridgman furnace and the opaque semiconductor samples, and the images were recorded on photographic film. The use of the lowest energy radiation sufficient to penetrate the sample results in the greatest contrast between the melt and the solid. These early techniques allowed crystal growers to directly observe the interface, but they required exposures of eight minutes as well as processing times from several hours to days.

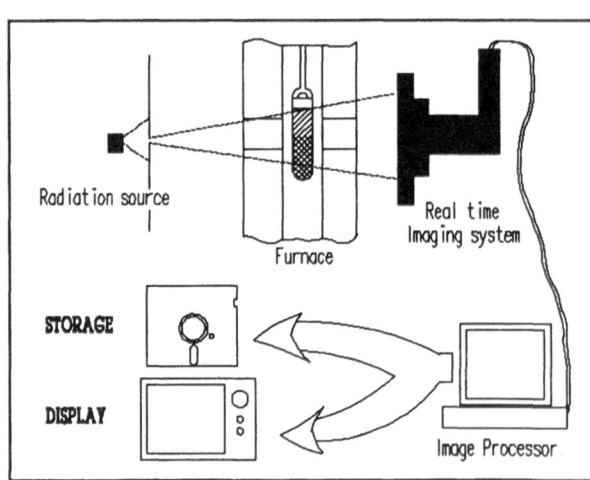

Figure 1. Radiation imaging system.

The real time imaging apparatus is shown schematically in figure 1. The crystal growth ampoule has been previously filled and sealed. The furnace is a standard three zone Bridgman design. The insulation zone is made from a low density, uniform silica fiber to facilitate the unhindered passage of the radiation. The interface images are generated from the passage of either x or gamma radiation through

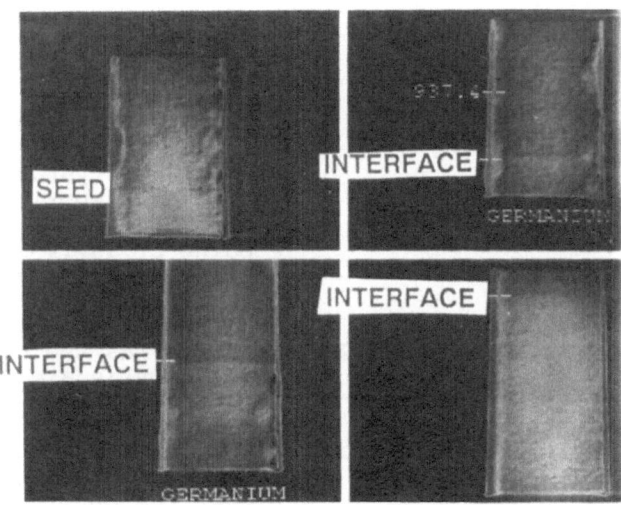

Figure 2. Real time images depicting interface movement during Bridgman growth of germanium.

the growing crystal onto a real time imaging system. (Gammascope GS220, RTS Technology, Inc., Sauerwein Group, North Andover, Massachusetts.)   This system is composed of a cesium iodide detector, an image intensifier and optics to shield the video camera from direct impingement from the radiation.   The image from the camera is then sent to an image processor which displays the radiation image in standard TV format.   The image can be stored for further processing.   During crystal growth the interface image can be captured.   While the interface is visible in real time, an integration of the image over eight seconds is used to increase the signal to noise ratio of the data. Additional image processing techniques used to improve the image visibility during crystal growth include contrast enhancement and edge filtering using a Sobel filter.[8] A column averaging technique was developed to provide a more complete view of the interface in the cylindrical shaped sample.[9]

The interface is recorded relative to both a fixed point in the furnace and fixed markers on the ampoule. This procedure produces independent measurements of both the ampoule pull rate and interface movement relative to the furnace. The interface shape measurements are made from the two dimensional image with the assumption of azimuthal symmetry. During actual growth rate measurements, the pull rate of the ampoule through the furnace remains constant, hence the real recorded data is the position of the interface relative to a fixed point in the furnace. This

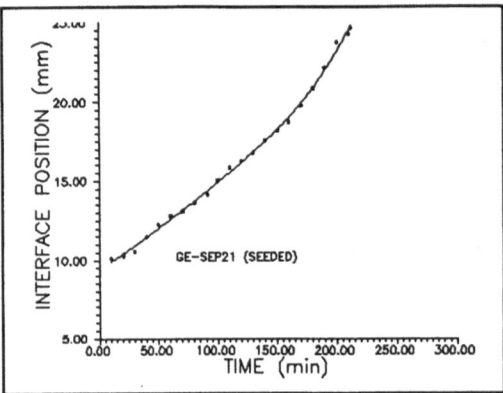

Figure 3. Interface position as a function of run time for a seeded germanium sample.

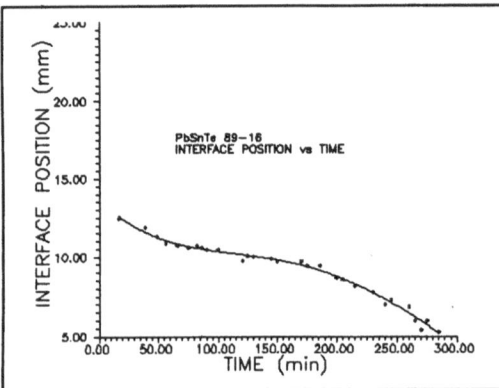

Figure 4. Interface position vs time for a PbSnTe crystal.

data is recorded from the inception of freezing throughout the run until the crystal is completely frozen. Data are collected every ten minutes during growth. Two types of crystals were grown in these experiments. Since germanium has a fixed melting point of 937.4°C, any interface movement is only a function of thermal loading due to ampoule position. PbSnTe was the other crystal grown and is a mixture of twenty per cent SnTe and eighty per cent PbTe with a pseudobinary substitutional phase diagram that extends over the entire constitutional range. The maximum solidus temperature of the mixture is 904°C but as growth proceeds the SnTe is preferentially rejected at the interface and into the liquid such that the solidus temperature continuously decreases during the run. Conversely the PbSnTe interface position is a function of both thermal loading and the continuously changing solidus temperature of the alloy.

The starting materials were loaded into a 16 mm ID fused silica ampoule, evacuated and sealed. The PbSnTe had been previously compounded from the elements in a 1000°C rocking furnace. The mass of the germanium boule was 65 grams and the PbSnTe boule was 90 grams.

EXPERIMENTAL RESULTS

Two experimental runs are described. A germanium crystal was grown from the seed and a PbSnTe crystal was grown unseeded. Figure 2 shows a series of pictures depicting the interface at different times during the growth of the germanium crystal. The interface position as a function of time is plotted in figure 3. As can be seen the interface is constantly moving away from the cold zone hence the growth rate is greater than the ampoule pull rate. Figure 4 shows a plot of the interface position as a function of time and a picture of PbSnTe during growth is shown in figure 5. The interface position moves toward the cold zone during the growth of PbSnTe hence the growth rate is less than the ampoule pull rate.

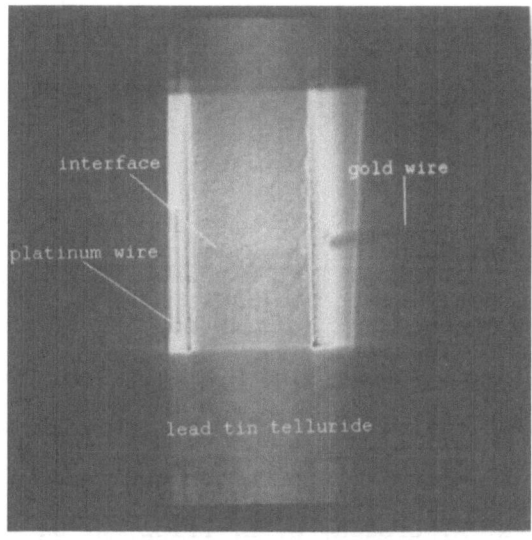

Figure 5. Real time image of PbSnTe during crystal growth.

    The interface shape is a function of position in the furnace
and this is shown for the extreme positions in figure 6.   The
convex interface was observed in germanium near the hot zone and
the concave interface was observed in PbSnTe near the cold zone.
The axial position has been normalized so that both curves can be
displayed on the same graph.

SUMMARY AND CONCLUSIONS

    The major conclusion
from this study is that real
time imaging using x-rays and
gamma rays can be an
effective tool in the crystal
growth process.   The
obtaining of real time images
has allowed the crystal
grower to observe the
interface,note it's shape and
position and using this data
along with the ampoule pull
rate calculate the true
growth rate.   The crystal
grower can also make
adjustments during growth
that can affect crystal
perfection.   Image

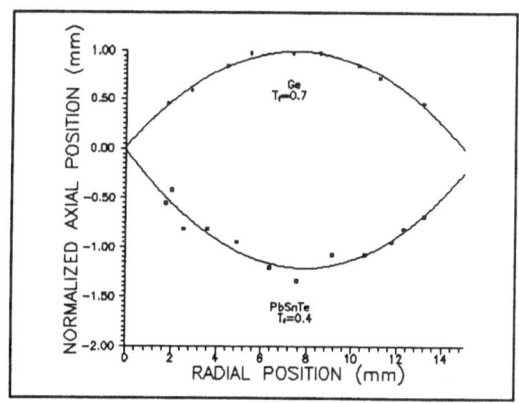

Figure 6.   Interface shape of a
germanium crystal and a PbSnTe
crystal at different location in
the insulation zone of the
furnace.

enhancement techniques have been developed to improve the
visibility of the images observed and give a truer picture of the
interface.

REFERENCES

1. P. G. Barber, R. K. Crouch, A. L. Fripp, I. O. Clark, W. J.
   Debnam, and R. T. Simchick, J. Mtls. Res. Soc. Europe, 1985.
2. K. Kakimoto, M. C. Eguchi, H. Watanabe, Taketoshi, and
   T.Hibiya, J. Cryst. Growth 91 (1988) 509.
3. Y. Huang, W. J. Debnam and A. L. Fripp, J. Cryst. Growth 104
   (1990) 315.
4. P. Capper, J. J. G. Gosney, and M. T. Quelch, J.Cryst. Growth
   63 (1983) 154.
5. A. F. Witt, H. C. Gatos, M. Lichtensteiger, M. C. Lavine, and
   C. J. Herman, J. Electrochem. Soc. 122 (1975) 276.
6. K. M. Kim, A. F. Witt, M. Lichtensteiger, and H. C. Gatos, J.
   Electrochem. Soc. 125 (1978) 475.
7. P. G. Barber, R. K. Crouch, A. L. Fripp, W. J. Debnam,R. F.
   Berry and R. T. Simchick, J. Cryst. Growth 74 (1986) 228-230.
8. E. R. Dougherty and C. R. Giardina, Image Processing
   Continuous to Discrete, Vol. 1 Prentiss-Hall, Inc. Newark
   (1987) 59.
9. P. J. Barber, R. F. Berry, W. J. Debnam, A. L. Fripp, Y.
   Huang, K. Stacy, and R. T. Simchick, J. Cryst. Growth 97
   (1989) 672.

MEASUREMENT OF MACROSEGREGATIONS OF STEELS BY X-RAY MICROFLUORESCENCE

J. Welfringer, P. Benoit, M. Guyon

IRSID 34, Rue de la Croix de Fer
78105 Saint-Germain-en-Laye, FRANCE

ABSTRACT

An apparatus based on X-ray microfluorescence has been developed at IRSID in order to routinely and quantitatively determine the state of segregation in steels. The equipment consists principally of an iron anticathode X-Ray tube, a Si(Li) energy - dispersive detector and a "on the fly" operation in a multi-scale analysis mode. The acquisition of a chromium or manganese mapping in a (150 x 150 mm^2) steel sample is possible over a total time of about one hour. The calibration curves for chromium or manganese are linear for the usual compositions of steels between 0,3% to 3%.

INTRODUCTION

Segregation in steel is the heterogeneity of composition produced by the concentration of elements into certain regions during the solidification process. This phenomenon is a major metallurgical problem.

Its detection and characterization are of importance in forecasting the characteristics of the product and adjusting continuous casting machines. There are currently two main methods of studying segregation of steels :

- Metallographic etching (Baumann Etching, Print Etching) is used for routine control of the state of segregation in steels. It has the major disadvantage of only providing qualitative information.

- The fine analysis of the varying degree of segregation in steels (micro - macro - mesosegregation) is performed by macroprobe, based on quantitative mapping. This requires very precise and highly sophisticated apparatus which is hardly suitable for routine control.

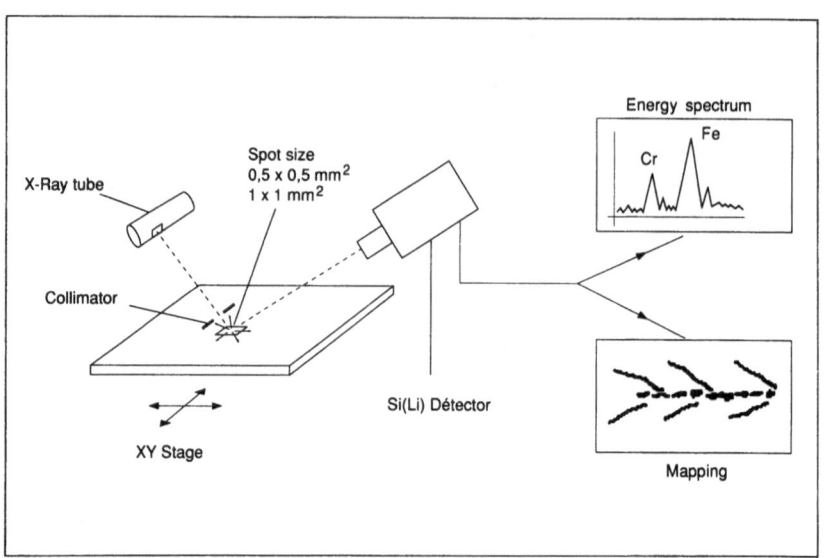

Figure 1 . Principle of X-Ray microfluorescence analysis apparatus.

X-Ray fluorescence analysis appears to be very attractive for developing an apparatus to routinely and quantitatively determine the state of segregation in steel.

The basic advantages of X-Ray probe compared to electron probe (principle of macroprobe) are the following :

- detection of most elements under atmospheric pressure,
- sample preparation is less important,
- greater depth can be analyzed.

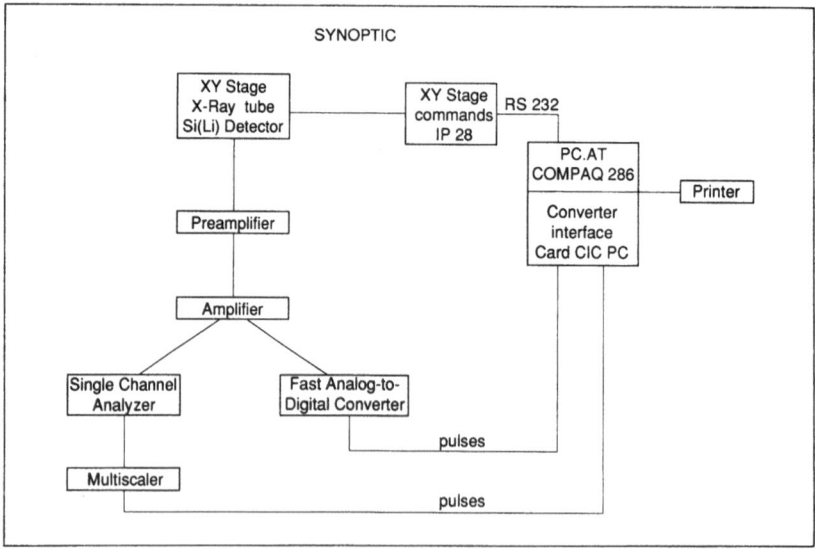

Figure 2 . Synoptic.

APPARATUS

Description of the experimental equipment

The equipement developed at IRSID is based on the energy dispersive X-Ray fluorescence analysis (figure 1).

It uses the following instrumentation.

X-Ray are generated by a KEVEX iron anticathode X-ray tube with a maximum power of 200 W (30 kV, 6,7 mA). The excitation beam is collimated to obtain an analysis spot of 500 x 500 μm^2 or 1 x 1 mm^2. The data are collected using a 30 mm^2 Si(Li) energy dispersive KEVEX detector with a resolution of 155 eV on the Kα peak of manganese. The sample, of up to a square of 150 mm side length in size, is moved by a microstepping motor-driven XY stage contained in a stainless steel chamber with vacuum possibility.

Analysis modes

Two analysis modes are possible with this equipment (figure 2).

In the amplitude analysis mode 1, a discrete movement of the sample is imposed and the energy spectrum is determined step by step.

In the multi-scale analysis mode 2, a window of energy corresponding to the Kα peak of the element to be analysed is selected and the acquisition in this window is made during sample movement. The moving speed is adjusted to the acquisition time and the analyser works in the multi-scale mode.

RESULTS

Chromium and manganese cartographies

The element that segregate strongly are sulfur and phosphorus but their content is kept as low as possible in steels. Consequently their detection is more difficult and requires analysis under vacuum.

However in most steels, other elements present in larger amounts, such as chromium or manganese, cosegregate with sulfur and phosphorus. So it is always possible to characterize the segregation with the analysis of elements which are more easily detectable.

The segregation of Mn and Cr is observed by X-Ray microfluorescence in a large sample of as cast 100C6 steel.

The conditions of the analysis are the following :

- multi-scale analysis mode,
- spot size 1 x 1 mm^2,
- sample size 150 x 150 mm^2,
- acquisition time 200 ms,
- mirror polishing surface finish,
- X-Rays produced under 10 kV-4 mA for Cr analysis and 20 kV-4 mA for Mn analysis.

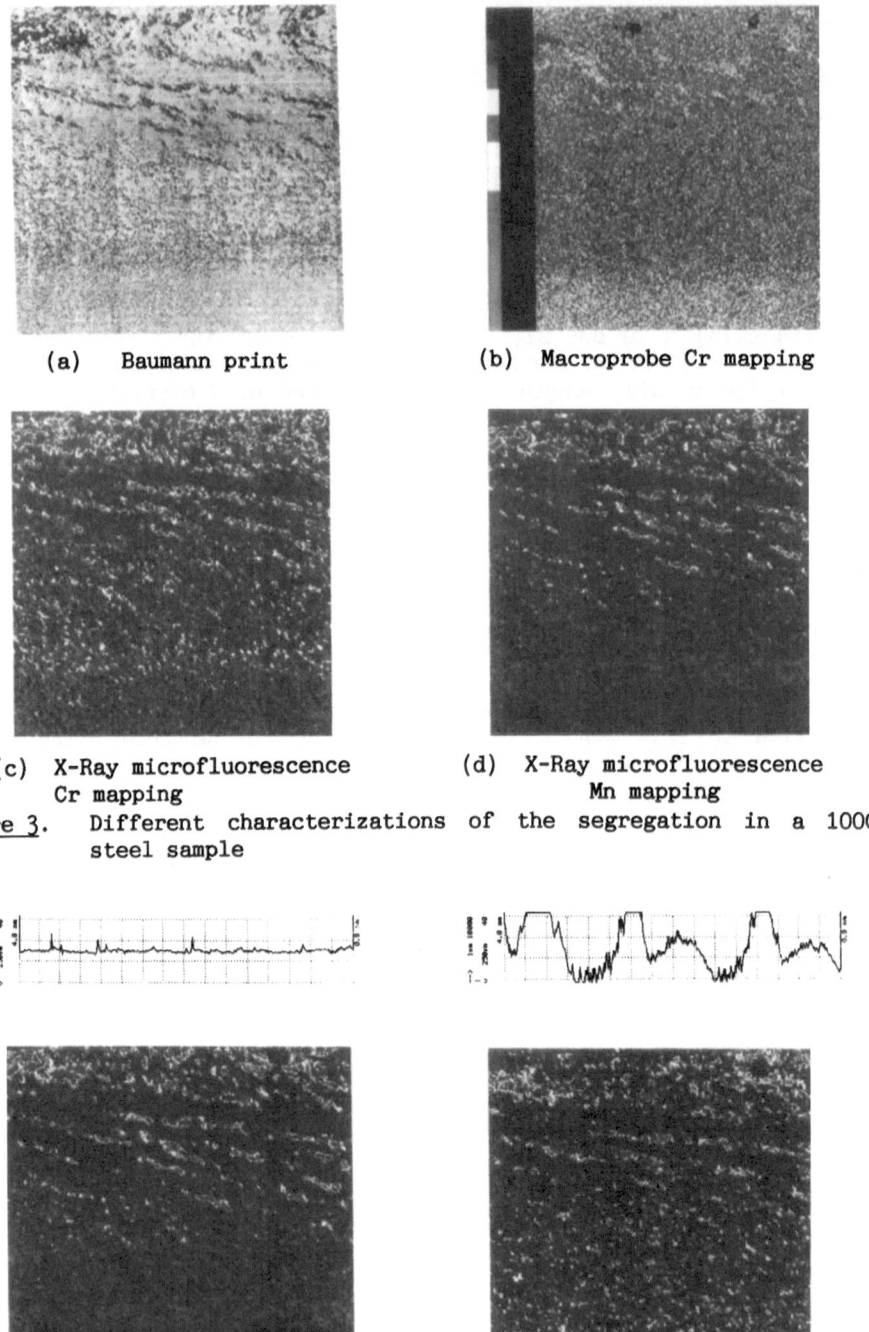

(a)   Baumann print   (b)   Macroprobe Cr mapping

(c)   X-Ray microfluorescence   (d)   X-Ray microfluorescence
       Cr mapping   Mn mapping

Figure 3.   Different characterizations of the segregation in a 100C6
             steel sample

(a) Mirror polishing surface finish   (b) Milling surface finish

Figure 4 .   X-Ray microfluorescence Cr mappings obtained with two
             different surface finishes - Corresponding roughness profiles
             (Hommel roughnessmeter).

The X-Ray mappings are compared with the Baumann print and the macroprobe chromium mapping from the same sample (figure 3). There is very good agreement between the different images.

Surface finish influence

The Cr segregation image was made on the 100C6 steel sample with a surface finish obtained by machine tool (milling) in order to define the minimum sample preparation required for X-Ray microfluorescence analysis.

This image offers the same visual quality as that obtained with a mirror polishing surface finish (figure 4).

A milled surface finish is largely sufficient for obtaining good quality X-Ray microfluorescence mapping.

Quantification

Calibration is necessary for converting peak intensity into element composition. Using standard samples with a similar matrix is a more convenient way to determine the calibration curves.

The Kα peak intensities have been plotted versus chromium or manganese compositions (figure 5).

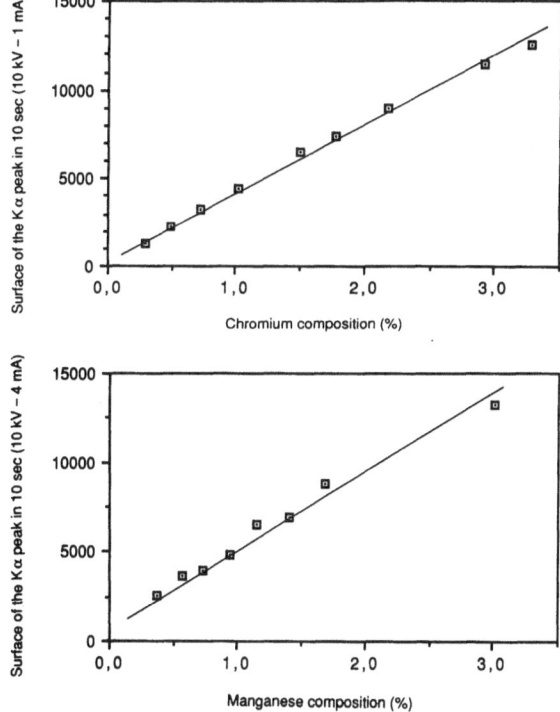

Figure 5 . Chromium and Manganese calibration curves

The calibration curves show a very good linearity in the region of interest to us (composition between 0,3% to 3%).

Mapping quantification is therefore a simple operation : linear transformation of the metering rate into the elementary composition.

Remember that a calibration curve is only valid for given measuring conditions (X-Ray tube excitation conditions, tube - sample - detector geometry).

CONCLUSION

An X-ray microfluorescence apparatus has been developed to routinely and quantitatively determine the state of segregation of steels which is a major metallurgical problem.

The analyser principally consists of an iron anticathode X-ray tube equipped with a collimator and a Si(Li) energy dispersive detector. It is suitable for the acquisition of a chromium or manganese mapping in a (150 x 150 mm^2) sample over a total time of about one hour with a (1 x 1 mm^2) spot size.

The analysis is performed under atmospheric pressure and on a surface finish obtained by machine tool (milling).

Mapping quantification is a simple operation - linear transformation of the Kα peak intensity into elementary composition from calibration curves obtained with standards.

REFERENCES

Wherry, D. C., B. J. Cross and T. H. Briggs, "An Automated X-Ray Micro-Fluorescence Materials Analysis System," Adv. X-Ray Anal., <u>31</u>, 93-98 (1988)

APPLICATION OF SR-XRF IMAGING AND MICRO-XANES TO METEORITES,

ARCHAEOLOGICAL OBJECTS AND ANIMAL TISSUES

Izumi Nakai and Atsuo Iida[*]

Department of Chemistry, University of Tsukuba, Ibaraki, 305 JAPAN

*Photon Factory, National Institute for High Energy Physics, Ooho, Tsukuba, Ibaraki 305 JAPAN

Abstract

Synchrotron Radiation Induced X-ray Fluorescence analysis was successfully applied to the analyses of meteorites, archaeological objects and animal tissues. Electronic states of Fe and Ti in chondrules of chondrites were clarified by micro-XANES to be Fe^{2+} and Ti^{4+}. Ancient iron implement (B.C. 3c-A.D. 3c) with zoning of rusts was characterized by micro-XANES. Two dimensional chemical state analyses of the sample were made by selective excitation of iron. Trace element analyses of Cu, Zn, Se, and Hg in the brain and kidney of rat, mouse, and guinea pig dosed with alkyl mercury and/or Se were made from a view point of mercury intoxication. Strong correlation between Hg and Cu, Zn, Se was observed by the correlation analysis of the elemental distributions. The XRF imaging also disclosed decrease of Cu and Zn level in cancerous tissues of human kidney compared with normal ones.

Advantages of Synchrotron Radiation (SR) as an X-ray source of X-ray Fluorescence (XRF) Analysis have been demonstrated by many pioneering researchers.[1-3] The advantages include wave-length tunability, high intensity and high resolution, which allow us to carry out nondestructive multi-elemental analysis of trace elements. However, practical application of the SR-XRF is still limited in numbers[4-8]. The authors expected that the application of SR-XRF is most promising in the fields of archaeological objects, geological samples, and biological samples and started the present research projects in 1987.[9-11] Now we have established the analytical procedures of these samples by SR-XRF that we report a part of our results, which so far we obtained from the analyses of geological samples, [12] archaeological objects[13] and biological samples.[14]

The techniques of chemical state analysis used in this study are based on the following two approaches: one is XANES (X-ray absorption Near Edge Structure) and the other is SIXES (Selectively Induced X-ray Emission Spectroscopy). It is well known that the former technique together with EXAFS has been developed rapidly with the development of intense SR beams.[15] SIXES was recently established by Sakurai et al.[7] It enables us to carry out chemical state mapping by X-ray fluorescence using absorption edge shifts. This method was first applied to practical problems in this study.

Advances in X-Ray Analysis, Vol. 35
Edited by C.S. Barrett *et al.*, Plenum Press, New York, 1992

INSTRUMENTATION

SR–XRF measurements have been made with monochromatized radiation with Si(111) double crystals from the bending magnet beam–line (BL–4A) at the Photon Factory, Tsukuba, using an XRF system with a Si(Li) detector. A fine parallel beam of desirable size was made by a set of vertical and horizontal slits and, if necessary, by pin hole. XRF imaging was carried out by scanning a sample on a computer controlled XZ stage. XANES spectra were measured by X–ray fluorescence detection using the Si(Li) detector. XANES spectrum was obtained by normalizing the intensity of the fluorescent X–ray with the incident X–ray intensity and by plotting the data against the X–ray energy.

APPLICATION TO GEOLOGICAL SAMPLES

Introduction

Knowledge of the chemical states of elements in a mineral is important to study its formation condition. However, it has been difficult to obtain two–dimensional information by conventional analytical techniques such as Mössbauer spectroscopy. It is known that XANES spectrum provides information on the electronic structure of an element.[16,17] A combination of the XANES technique with X–ray fluorescence detection using Si(Li) detector and SR X–ray source makes it possible to obtain XANES spectrum of trace element in a small sample area nondestructively. This method can be called as micro–XANES and is particularly useful in the analysis of geological samples and solid samples that have fine textures. Here, we report XANES analysis of Fe and Ti in some important meteorites, which belongs to stone meteorite and contain chondrules. These meteorites are the oldest objects so far known in our solar system.

Experimental

Roughly polished specimens of meteorites (chondrites: Allende, Murchison, Krymka and Semarkona) were used in the analysis. XANES spectra were measured with $100 \times 100 \mu m^2$ beam (5 – 20 sec/point, total 121 points for 1 spectrum).

Results and Discussion

Fe K–edge XANES spectra of some reference samples were given in Fig. 1. The XANES spectra of the meteorites were measured for the points of interest to examine electronic states of the

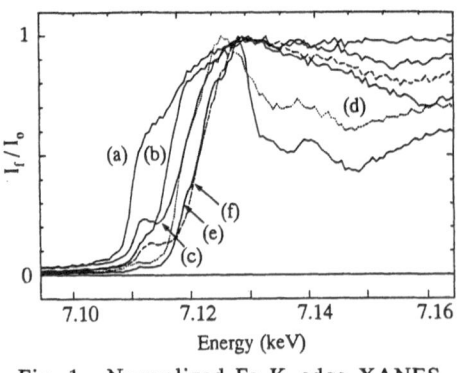

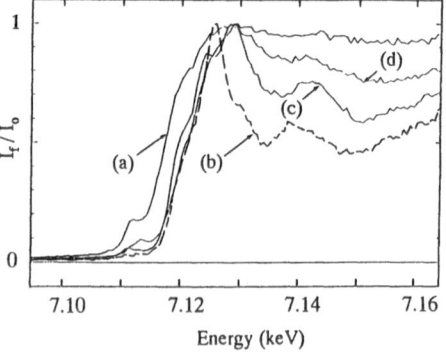

Fig. 1. Normalized Fe K–edge XANES spectra of (a) metallic iron, (b)phyrrhotite $Fe_{1-x}S$, (c)magnetite Fe_3O_4, (d) olivine $(Mg,Fe)_2SiO_4$, (e)antigorite $(Mg,Fe)_3Si_2O_5-(OH)_4$, (f) goethite FeOOH.

Fig. 2. Normalized Fe K–dge XANES spectra of the meteorite samples: (a) sulfide region, (b) chondrule, (c) fine grained CAI in the Allende meteorite and (d) unknown phase in the Murchison meteorite.

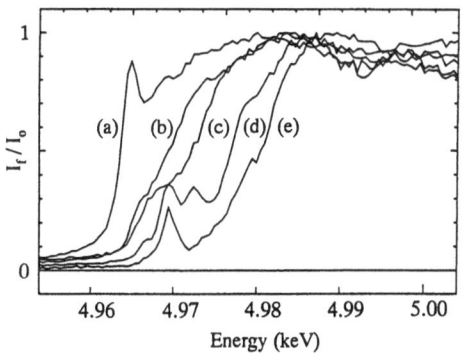

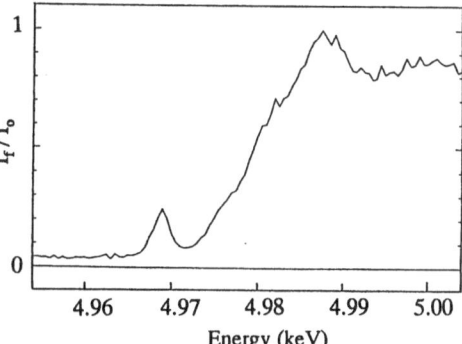

Fig. 3. Normalized Ti K–edge XANES spectra of (a) metallic titanium, (b) TiO, (c) Ti_2O_3, (d) TiO_2 (rutile), (e) $CaTiSiO_5$ (sphene).

Fig. 4. Normalized Ti K–edge XANES spectra of the chondrule in the Semarkona meteorite.

elements in chondrules and their matrix and are shown in Fig. 2. Since the chemical shift of Fe–K XANES spectrum of chondrule in the Allende meteorite (Fig. 2 (b)) is in the range of divalent iron. Since the spectrum is similar to that of olivine, iron is found to exist in the divalent state in the chondrule. The coordination structure of iron in the chondrule may be close to those in olivine, i.e., octahedral coordination with oxygen atoms. The XANES spectrum of CAI (calcium–aluminum–inclusion) region (Fig. 2(c)) in the Allende meteorite shows that it may be a mixture of olivine–like phase and antigorite–like phase. Antigorite (Fig. 1(e)) is a hydrated phyllosilicate $(Mg,Fe)_3Si_2O_5(OH)_4$.

Figure 3 shows Ti K–edge XANES spectra of reference samples showing shift of the absorption edge depending on the oxidation states of Ti. It is clearly seen that not only the chemical shift but also the shapes of the spectra are characteristic to each compound. For X–ray photon energies near the core-level binding energy, transitions occur to bound states of the metal atom, giving rise to characteristic spectral feature in the vicinity of the absorption edge. For example, the XANES spectrum of TiO_2 (rutile) has a triplet pre–edge absorption (Fig. 3(d)). Since the initial 1s state is a gerade state, the 1s → 3d transition is strictly dipole forbidden. However, the pre–edge absorption becomes dipole allowed due to a combination of stronger 3d–4p mixing and overlaps of the metal 3d orbitals with the 2p orbitals of the ligand.[16,17] It is found that this information can be used to characterize the Ti atom in a sample. Figure 4 shows Ti–K XANES spectrum of chondrule in the Semarkona meteorite. The concentration of Ti will be order of ppm levels. It is remarkable that such a trace amount of Ti gives a clear XANES spectrum. The spectrum closely resembles that of sphene with characteristic single pre–edge peak. From this information, it is found that Ti in the chondrule exists in the tetravalent state as a component of silicate but not in the form of simple titanium oxides.

APPLICATION TO ARCHAEOLOGICAL SAMPLES

Introduction

XRF technique has been widely used in the chemical analysis of ancient artifacts, because it enables us to carry out a nondestructive rapid multielemental analysis of the samples. It is expected that application of SR X–ray to the XRF analysis will greatly improve its analytical ability because of the high brightness, polarization, and wavelength tunability of the SR X–ray. Besides, there is a hidden advantage of the utilization of SR in the analyses of archaeological samples; i.e., the SR facility provides spatial freedom, which allows truly nondestructive analysis of large artifacts such as sculptures, paintings, etc. Taking these advantages, we started the following research programs in 1987 to establish the experimental procedures of archaeometric analysis and to apply the technique to the practical problems in archaeology.

1) Two dimensional chemical imaging of artistic pattern
2) Multielemental nondestructive bulk analysis for provenance investigation
3) Chemical speciation of component elements of archaeological objects

Here, we introduce the result of the chemical state analysis of iron implement.

Experimental

Figure 5 is a photograph of the ancient iron implements in Yayoi period (B.C. 3c – A.D. 3c) used in the present study. The sample is corroded showing zoning of rusts with different colors: the core part of the sample has metallic luster and is surrounded by zones of black and red rusts. Fe K–edge XANES spectra of each zone were measured with 200x200μm² beam (5 sec/point, total 121 points).

Results and Discussion

Fe K–Edge XANES spectra of each zone of the samples are given in Fig. 6. The spectrum (a) corresponds to the central metallic part and (b) to the black part and (c) to the reddish brown part. The observed chemical shifts in the spectra should correspond to the oxidation state or iron. As can be seen from Fig. 1, the reference spectra of Fe metal and goethite FeOOH agree well with those of part (a) and (c), respectively. Therefore, the metallic part is pure iron and reddish brown part is iron hydroxide having a chemical formula close to FeOOH. The XANES spectrum of the part (b) is close to that of magnetite Fe_3O_4. Therefore, it is found that the part (b) contains both divalent and trivalent irons.

Two dimensional analyses of iron in the sample were made at three different excitation energies, 7.110, 7.118, and 7.410 keV (see Fig. 6) and the results are given in Fig. 7 (a), (b), (c), where Fe concentration corresponds to 8 densities from black (highest conc.) to white (lowest conc.). An isolated image of each oxidation state of iron is obtained by a subtraction of two data measured at the different energy.[7] The results are shown in Fig. 7 (d), (e), (f), which show the distribution of iron at electronic state corresponding to those of (a), (b), (c) of Fig. 6, respectively. This type of two–dimensional chemical state analysis is only possible with using the tunable nature of the SR X–ray.

APPLICATION TO BIOLOGICAL SAMPLES

Introduction

Ingestion of toxic metallic elements causes intoxication after accumulation at various biological organs and tissues such as brain, liver, kidney. For example, intoxication of alkyl mercury is well

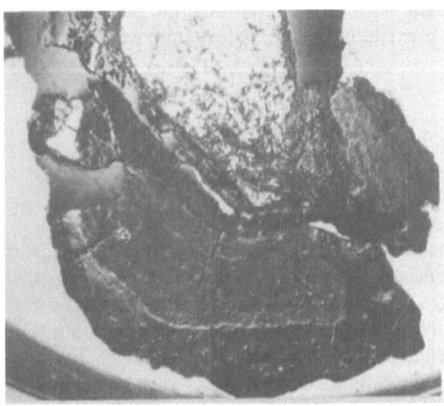

Fig. 5. Polished sample of Ancient corroded iron implement showing zoning of rusts.

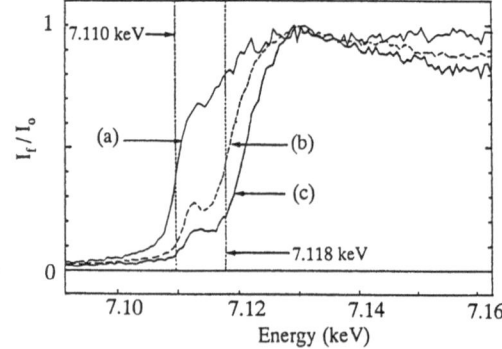

Fig. 6. Fe K–XANES spectra of (a) metallic, (b) black, and (c) reddish brown parts of the sample in Fig. 5.

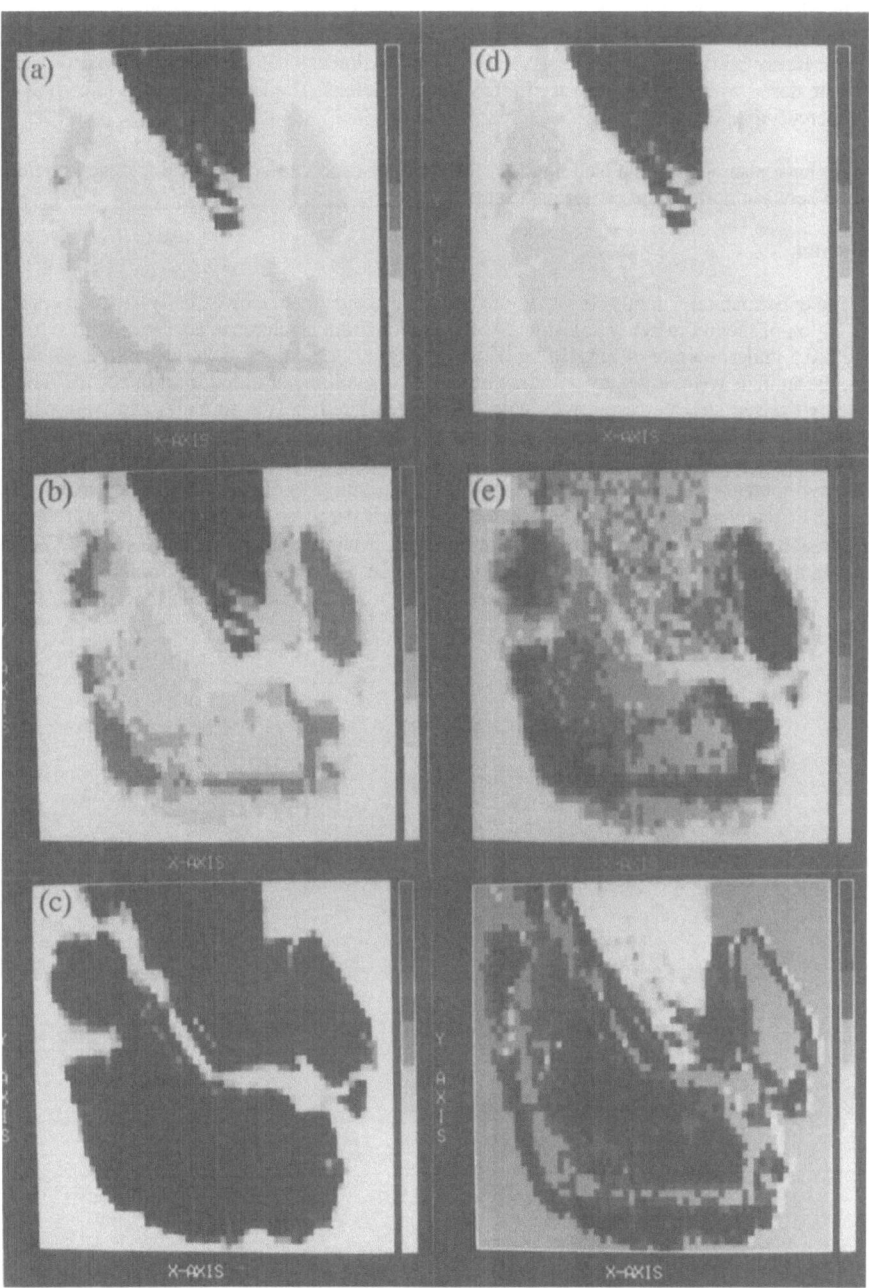

Fig. 7 Two dimensional chemical state analysis. The analyzed area corresponds to the picture in Fig. 5. These pictures show Fe K images excited at (a) 7.110keV, (b) 7.118keV, (c) 7.410keV (60x47 spots, 0.2mmx0.2mm step, 2sec/step) and their subtracted images: (d),(e), and (f), which give distributions of iron at the electronic states defined by the XANES spectra of (a), (b), and (c) in Fig. 6, respectively.

known as Minamata disease. In the study of metal intoxication, knowledge on the distribution of metal-lic elements after accumulation in each biological tissue is important. It is expected that the chemical imaging by SR–XRF is an ideal analytical technique for these samples, because it allows nondestructive two–dimensional multielemental analysis of trace elements. Thus, we have applied the technique to study the mercury intoxication. We report results of two–dimensional analysis of animal tissues. The animals were dosed with Hg and/or Se at different concentrations. The latter element is known as antag-onism in mercury intoxication.

We have also applied the technique to the study of cancer to reveal the difference in elemental distribution between normal and cancerous tissues.

Experimental

The experimental animals consist of rats and guinea pigs that received daily subcutaneous injec-tion of solution of methyl mercury chloride (MMC) and/or sodium selenious acid (Se) with ratio of 1:0, 1:0.5 and 0:0.5 under the rate of 3mgHg/kg of body weight. After the 7 days successive subcutaneous injection, the animals were killed by withdrawing an excess volume of blood from the heart. The brains were quickly excised after perfusion with cold saline, then fixed in 10% formalin and were paraffined. The brains were cut into slices of 1mm from center of the brains and subjected to the analysis. Another group of samples consists of kidneys of rat dosed with GSH (glutatione) plus MMC, EMC(ethyl–) or PMC (phenyl–mercury chloride) everyday for 7 days (s.c. 3mg as Hg/kg/day). Cancerous kidneys of men with different age, sex and stage of the disease were paraffined and cut into slices of 2mm thick. Two–dimensional analyses of the samples were made in air with monochromatic X–rays (14 or 16keV) of a few hundred micrometer size and distribution of Hg, Zn, Cu, and Se were examined.

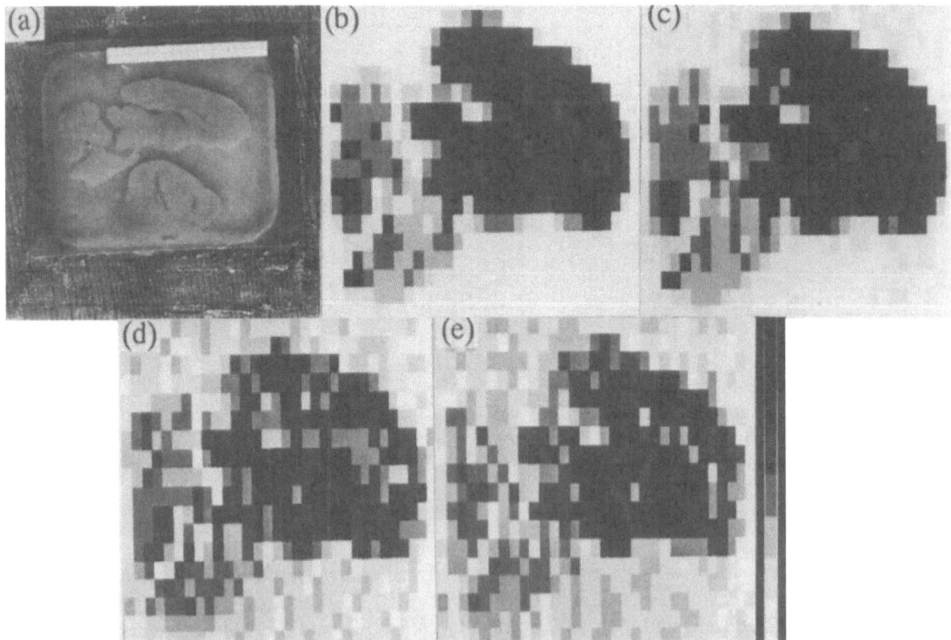

Fig. 8. Chemical imaging of brain of guinea pig dosed with MMC + Se. (a) the sample and distribution of (b) Hg, (c) Zn, (d) Cu and (e) Se with pixels of 32x18 spots (0.75mmx 0.75mm step, 10sec/step)

Table 1. Correlation coefficients between two trace elements for kidney sample

Sample		Hg-Zn	Hg-Cu	Hg-Se	Zn-Cu	Zn-Se	Cu-Se
Mouse	EMC	0.902	0.763	0.769	0.884	0.811	0.750
Rat	Cont.	0.105	0.112	0.091	0.925	0.552	0.589
	GSH	0.092	0.141	0.029	0.836	0.521	0.571
	MMC	0.920	0.911	0.913	0.860	0.840	0.867
	EMC	0.898	0.791	0.909	0.862	0.855	0.800
	PMC	0.656	0.895	0.924	0.763	0.497	0.848
Infant		/	/	/	0.460	0.448	0.332

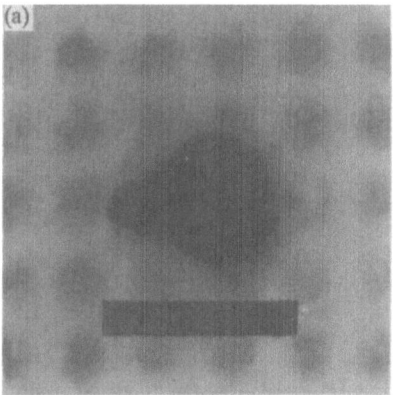
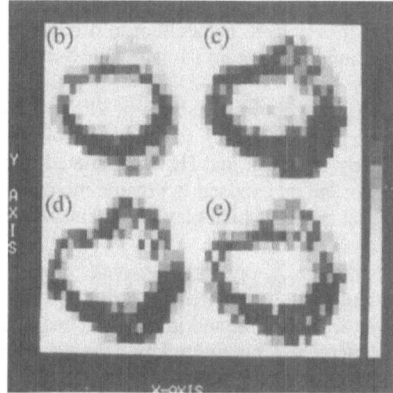

Fig. 9. Chemical imaging of kidney of rat dosed with EMC+GSH: (a) the sample and distribution of (b)Hg, (c) Zn, (d) Cu, and (e) Se with pixels of 24x16 spots (0.5mmx0.5mm step, 10sec/step).

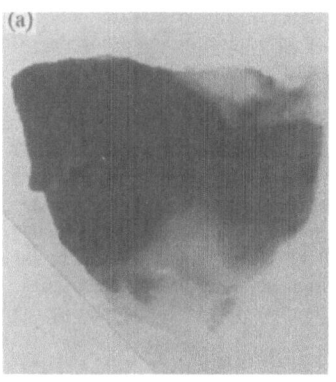
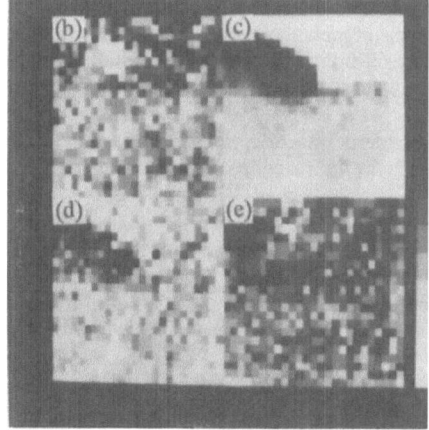

Fig. 10. Chemical imaging of cancerous kidney of human. (a) the sample and distributions of (b) Hg, (c) Zn, (d) Cu, and (e) Se (0.75mmx0.75mm step, 10sec/step).

Results and Discussions

Figure 8 (a) shows a photograph of a brain of guinea pig dosed with MMC and Se (The white bar is 2cm). The results of the chemical imaging are given in Fig. 8 (b) – (e). The XRF intensity was scaled with 8 densities of black to white in the image. The rats and guinea pigs dosed with Se only, Hg only and Hg + Se, and control samples were analyzed. The results are summarized as follows:

(1) The accumulation of both Hg and Se increased when Hg and Se were dosed together compared with the case when each element was dosed separately.
(2) There exists a correlation in concentration among Hg, Zn, Cu and Se.
(3) Concentrations of Hg in cerebellum and medulla of rat and guinea pig were lower than those in their cerebral cortex.

To examine the time–dependent distribution of the trace elements, the rat samples at 24hr after the final dose of alkyl mercury and the samples showing the signs of Hg poisoning were examined and the results were compared. It was observed that the level of Hg was high at cerebellum in the former case while it was high at cerebral cortex in the latter case. The present results suggest that with time passing Hg moves from cerebellum to cerebral cortex and increased accumulation of Hg in the cerebral cortex may induce the brain damage followed by the apparent signs of intoxication.

Fig. 9 (a) shows a sample of kidney of rat dosed with EMC+GSH and the results of chemical imaging are given in (b),(c), (d) and (e), which show distribution of Hg, Zn, Cu and Se, respectively. It can be clearly observed that Hg accumulated in renal cortex accompanying with other trace elements in the kidney of the rat exposed to organic mercury. This finding is interesting because there are glomeru-li in the renal cortex, which play important roles in cleaning waste materials from blood.

Using the X–ray intensity data of each analytical point, correlation coefficients between two trace elements were calculated for kidneys of mouse and rat dosed with MMC, EMC, MMC, GSH and for kidneys of control rat and infant. The result is listed in Table 1. In the kidneys of the rats exposed to organic mercury, high correlation between Hg and other elements was observed, whereas much lower correlation was observed for the sample without administration of organic mercury. Similar re-sults were obtained for the brain of the guinea pigs exposed to methylmercury. Therefore, we suggest that there is significant interaction between mercury and metalloenzyme system in the animals exposed to the organic mercury.

Figure 10(a) is the sample of human kidney. The lower part (2/3 of the net area) of the sample in the picture is cancerous tissue. Figures 10 (b),(c),(d), and (e) are results of the imaging showing the distribution of Hg, Zn, Cu, and Se, respectively. Concentrations of Hg and Se are negligible in the sample and no clear image was obtained for these elements. In contrast, the concentrations of Zn and Cu are significantly lower at the cancerous tissues than those at the normal tissues. Zinc and copper are central atoms of metalloenzyme. This indicates that cancer causes serious effects on the metalloenzyme system to reduce their concentration. Further analysis is now in progress to examine the effect of the age, sex and the stage of disease on the elemental distribution and to specify the enzyme.

In the analysis of biological samples, background due to scattering from the sample was serious compared with fluorescence X–ray of the analyzed elements and it determined the dead time of the detector system. This background should be minimized by setting the detector normal to the incident X–ray on the orbital plane of SR.[18] The present technique does not require such pre–treatment of sample as separation of each tissue or dissolution before the analysis. Thus, rapid analysis in ppm levels without destroying the form of tissues is possible with this technique. Consequently, it is expected that this technique will become a powerful tool in histochemical analysis of biological tissues. It is also beneficial point that the obtained data are suitable for correlation analysis of elemental distribution.

CONCLUSIONS AND FUTURE STUDIES

The present study clearly demonstrated the high potential of the SR–XRF techniques in the analyses of geological, archaeological and biological samples. Especially, we found micro–XANES is a powerful technique in the chemical speciation of solid materials. To study micro–region of the samples use of focusing beam is essential. Recently, micro–XAFS with spatial resolution of less than 20 μm has been attained in our beam line by Hayakawa et al.[19] For the analysis of geological sample spatial resolution of a few μm will be most desirable and we must await future progress of the microbeam technology. It should be also mentioned here that the intensity data used in the above study were raw intensity of X–ray fluorescence and no correction was made. Therefore, the result obtained above is semiquantitative from a view point of elemental concentration. Further study is necessary to establish quantitative analyses of these samples. This is another important point of future studies. So far, damage of samples due to X–ray irradiation was not observed in any samples used. Therefore, this method is truly nondestructive. From these experiences, we think feasibility of nondestructive chemical characterization of trace element in a micro–region is the most promising feature of the SR–XRF method.

ACKNOWLEDGMENTS

This study has been done as joint research projects with specialists of each field: i.e., Dr. A. Tsuchiyama, Dr. T. Noguchi (geological application), Dr. N. Shimojo and Miss S. Homma (biological applications), and Prof. I. Taguchi (archaeological applications). Their cooperation is greatly appreciated. We are also grateful to Drs. K. Sakurai and S. Hayakawa for their useful advice in the measurements and to Dr. S. Uehara for providing antigorite sample and to Mr. K. Imai for titanium oxides, and to Miss C. Numako for help in the experiment and the preparation of this manuscript. This work has been partially supported by a Grant–in Aid for Scientific Research from the Ministry of Education. The XRF experiments were performed under the approval of the PF Program Advisory Committee (# 87–056, 88–034, 89–070).

REFERENCES

1. P. Horowitz and J. Howell, Science 178:608 (1972)
2. C. J. Sparks, Jr., and J. B. Hastings, Oak Ridge National Laboratory Report ORNL–5089:8(1975)
3. C. J. Sparks, Jr., X–ray Fluorescence Microprobe for Chemical Analysis, in "Synchrotron Radiation Research", H. Winick, and S. Doniach ed., Plenum, New York (1980)
4. A. C. Thompson, J. H. Underwood, Y. Wu, R. D. Giauque, K. W. Jones, and M. L. Rivers, Nucl. Instr.and Meth. A266:318(1988)
5. A. Iida, K. Sakurai, T. Matsushita, and Y. Gohshi, Nucl. Instr. and Methods 228:556 (1985)
6. Y. Gohshi, S. Aoki, A. Iida, S. Hayakawa, H. Yamaji, and K. Sakurai, Jpn. J. Appl. Phys., 26:L1260(1987)
7. I. Sakurai, A. Iida, and Y. Gohshi, Anal. Sci. 4:37(1988)
8. V. B. Baryshev, N. G. Gavrilov, A. V. Daryin, K. V. Zortarev, G. N. Kulipanov, N. A. Mezentsev, and Ya. V. Terekhov, Nucl. Inst. Methods A282:570(1989)
9. I. Nakai, A. Mochizuki, T. Kawashima, S. Hayakawa, Y. Goshi, and A. Iida, PF Activity Report #5:135 (1987)
10. I. Nakai, Y. Suzuki, and A. Iida, PF. Activity Report:#6:141(1988)
11. I. Nakai, N. Shimojo, S. Homma, T. Kawashima, and A. Iida, PF Activity Report #6:94(1988)
12. I. Nakai, J. Mineral. Soc. Jpn. 20:147(1991)[in Japanese]
13. I. Nakai, I. Taguchi, and K. Yamasaki, Anal. Sci. in press (1991)
14. I. Nakai, S. Homma, N. Shimojo, and A. Iida, Adv. in X–ray Chem. Anal. Japan, 22:63(1991)
15. A. Bianconi, L. Incoccia, and S. Stipcich Ed., EXAFS and Near Edge Structure, Springer, Berlin(1983).
16. J. Wong, F. W. Lytle, R. P. Messmer, and D. H. Maylotte, Phys. Rev. B30:5596(1984)
17. R. B. Greegor, F. W. Lytle, D. R. Sandstrom, J. Wong, and P. Schultz, J. Non. Cryst. Solids 55:27(1983)
18. I. Sakurai, A. Iida, and Y. Gohshi, Anal. Sci. 4:3(1988)
19. S. Hayakawa, Y. Gohshi, A. iida, S. Aoki, and K. Sato, Rev. Sci. Instrum. in press (1991)

AUTHOR INDEX

SUBJECT INDEX